GROUTING IN ROCK AND CONCRETE
INJIZIEREN IN FELS UND BETON

PROCEEDINGS OF THE INTERNATIONAL CONFERENCE ON GROUTING IN
ROCK AND CONCRETE / SALZBURG / AUSTRIA / 11-12 OCTOBER 1993
BERICHTE DER INTERNATIONALEN KONFERENZ BETREFFEND INJEKTIONEN
IN FELS UND BETON / SALZBURG / ÖSTERREICH / 11-12 OKTOBER 1993

Grouting in Rock and Concrete
Injizieren in Fels und Beton

Editor / Herausgeber
RICHARD WIDMANN
Österreichische Gesellschaft für Geomechanik

A.A.BALKEMA / ROTTERDAM / BROOKFIELD / 1993

The texts of the various papers in this volume were set individually by typists under the supervision of each of the authors concerned
Die Beiträge dieser Ausgabe wurden unter Aufsicht der einzelnen Autoren geschrieben.

Published by / Veröffentlicht durch
A.A.Balkema, P.O.Box 1675, 3000 BR Rotterdam, Netherlands
A.A.Balkema Publishers, Old Post Road, Brookfield, VT 05036, USA

ISBN 90 5410 350 7
© 1993 A.A.Balkema, Rotterdam
Printed in the Netherlands / Gedruckt in den Niederlanden

Grouting in Rock and Concrete, Widmann (ed.) © 1993 Balkema, Rotterdam, ISBN 90 5410 350 7

Table of contents
Inhalt

2 Grouting procedures – General
Ausführung von Injektionen – Allgemein

3 Grouting procedures – Case histories
Ausführung von Injektionen – Beispiele

4 Research works – In situ tests
Forschungsarbeiten – In Situ Versuche

5 *Supplement*
Anhang

Grouting in Rock and Concrete, Widmann (ed.) © 1993 Balkema, Rotterdam, ISBN 90 5410 350 7

Preface

For almost two centuries grouting has been used for improvement of the underground and, more recently, also for the rehabilitation of cracked concrete. Various grout materials and procedures have been developed in order to obtain optimal results under a variety of conditions. However, all over the world, planning and execution of grouting works is still mostly done on an empirical basis while the theoretical fundamentals have remained at the stage of infancy and require a great deal of intensive research work.

Proper design presupposes the definition of the safest and most economic method for reaching the desired goal, in order to facilitate reliable cost estimates. In the case of grouting projects the first questions to be answered would be: which grout material, which grouting pressure, which spacing of boreholes and what grout take are needed to reach the desired aim in an optimal manner? In 1988, an interdisciplinary working group was founded in Austria for dealing with these very problems, comprising experts from the fields of design and construction as well as hydraulics, fracture mechanics and computing. In 1989, the International Commission for Rock Grouting was founded within the ISRM to provide a consistent continuation of the above concept. In 1992, the German 'Arbeitsgruppe für Feinstbindemittel' (working-party on microcement) also agreed to cooperate.

In order to find answers to the open theoretical questions at an international level an 'International Conference on Grouting in Rock and Concrete' was now held at which more than 60 papers from four continents were presented on the following subjects:

1. Properties of grouts: It is widely known that the properties of the mix of suspensions and solutions used for grouting depend not only on their composition but also on temperature, time and flow velocity unfortunately also seem to depend on the chosen test methods.

2. Grouting methods in general: Reports provided an overview on a variety of available grouting methods for specific applications such as sealing, consolidation and lifting of structures.

3. Grouting procedures – Case histories: Papers related to the successful use of different grouting methods under a variety of conditions.

4. Laboratory and in situ tests, mathematical models: Papers report on the flow processes in joints or cracks which depend on flow cross-sections, flow properties, effective pressures and deformation properties of the medium to be grouted. The possibility of in-process assessment of the success of grouting was also considered.

The organizers would like to thank all those who have contributed to the conference for their work and commitment, and hope that this synopsis of research and practical experience from all over the world will constitute a valuable aid for many specialists in the field.

Grouting in Rock and Concrete, Widmann (ed.) © 1993 Balkema, Rotterdam, ISBN 90 5410 350 7

Vorwort

Injektionen werden seit fast zwei Jahrhunderten zur Verbesserung des Baugrundes und in neuerer Zeit auch zur Sanierung von Rissen in Beton eingesetzt. Verschiedenste Injektionsmaterialien und -Verfahren wurden entwickelt, um unter den unterschiedlichsten Bedingungen optimale Ergebnisse zu erzielen. Nach wie vor erfolgen Planung und Ausführung weltweit überwiegend auf empirischer Basis, während die theoretischen Grundlagen in ihren akademischen Kinderschuhen stecken und daher noch intensiver Forschungsarbeit bedürfen.

Gegenstand jeder Planung muß die Festlegung des sichersten und wirtschaftlichsten Weges zur Erreichung des Planungszieles sein, um die erforderlichen Kosten verläßlich abzuschätzen zu können. Für Injektionsarbeiten bedeutet dies zunächst die fundierte Beantwortung der Frage: Mit welchem Injektionsgut, welchem Injektionsdruck, welchem Bohrlochabstand und welcher Injektionsgut-Aufnahme kann das angestrebte Ziel auf optimalem Weg erreicht werden?

Zur Bearbeitung dieser Probleme wurde 1988 in Österreich eine interdisciplinäre Arbeitsgruppe gebildet, der Fachleute für Planung und Ausführung ebenso wie für Hydraulik, Bruchmechanik und Computer-Berechnungen angehören. In konsequenter Weiterführung dieser Grundgedanken wurde 1989 die Commission on Rock Grouting im Rahmen der ISRM gebildet. 1992 konnte schließlich auch die deutsche 'Arbeitsgruppe für Feinstbindemittel' zur Mitarbeit gewonnen werden.

Um die Beantwortung der offenen theoretischen Fragen auf eine möglichst breite internationale Basis zu stellen, wurde nun die 'Internationale Konferenz betreffend Injektionen in Fels und Beton' organisiert. In über 60 Beiträgen aus vier Kontinenten werden folgende Themen behandelt:

1. Eigenschaften der Injektionsmaterialien: Sowohl die Eigenschaften der Suspensionen wie auch jene der Lösungen, die bei Injektionen Verwendung finden, hängen bekanntlich außer von ihrer Zusammensetzung auch von der Temperatur, der Zeit und der Fließgeschwindigkeit, aber – leider – auch zumindest scheinbar vom Prüfverfahren ab. Die Beiträge befassen sich mit diesem Problemkreis.

2. Injektionsverfahren – Allgemein: In diesem Abschnitt geben die Beiträge einen Überblick über die Vielfalt der verfügbaren Injektions-Verfahren zur Erreichung der verschiedenen Injektionsziele, wie Abdichtung, Verfestigung und Hebung von Gebäuden.

3. Injektionsverfahren – Ausführungsbeispiele: Über die erfolgreiche Anwendung verschiedenster Injektionsverfahren unter den unterschiedlichsten Verhältnissen wird in den Beiträgen berichtet.

4. Versuche im Labor und in situ, mathematische Modelle: Hier werden in den Beiträgen die Fließvorgänge in Klüften oder Rissen untersucht, die von den Durchflußquerschnitten, den jeweils repräsentativen Fließeigenschaften und wirksamen Drücken ebenso wie von den Verformungseigenschaften des zu injizierenden Mediums abhängen. Auch die Möglichkeit einer Beurteilung des Injektionserfolges bereits während der Injektion wird behandelt.

Der Veranstalter möchte allen, die mit ihren Arbeiten zur Konferenz beigetragen haben, insbesondere den Autoren und Generalberichterstattern, für ihre Mühewaltung danken und hofft, daß diese Sammlung von Erfahrungen aus Forschung und Praxis vielen Fachleuten eine Hilfe für ihre Arbeiten sein möge.

1 Properties of grout material
Eigenschaften von Injektionsmaterial

Rheological properties of cement and bentonite grouts with special reference to the use of dynamic injection

Die rheologischen Eigenschaften von Zement- und Bentonit-Injektionsgut insbesondere in bezug auf die Verwendung bei dynamischer Injektion

Lennart Börgesson
Clay Technology AB, Lund, Sweden

ABSTRACT: Grouting of very fine fractures in rock with a hydraulic conductivity of 10^{-10} to 10^{-7} m/s has been made in the Stripa project with the purpose to minimize the water flow in disturbed and naturally fractured zones in nuclear waste repositories. This purpose has put special demands on the injected material to ensure good sealing properties and longevity. These demands mean that cement with very low w/c ratio or bentonite slurry with a low water ratio is required. In order to be able to inject such materials, a new dynamic injection technique has been invented and tested.

The rheological properties of cement-based and bentonite-based grouts have been investigated by laboratory tests. The influence of different factors on the rheological properties of these materials are shown with special reference to the influence of vibrations. Mathematical models of the flow properties and the influence of vibrations are presented as well as examples of laboratory tests which show that the shear resistance of a grout can be reduced by up to 100 times with the dynamic technique.

ZUSAMMENFASSUNG: Das Vergiessen feiner Bruchflächen mit einer Wasserleitfähigkeit von 10^{-10} bis 10^{-7} m/s wurde im Stripa-Projekt durchgeführt, um den Wasserfluss in gestörten Bereichen und solchen mit natürlichen Bruchflächen in Deponieren für radioaktive Abfälle zu minimieren. Das eingespritzte Material musste daher ganz bestimmte Forderungen erfüllen, um gute Dichtegenschaften und Langleibigkeit zu gewährleisten. Diese Forderungen bedeuten, dass Zement mit einem niedrigen w/c-Verhältnis oder Bentonit-Schlamm mit einem niedrigen Wassergehalt benötigt wird. Um in der Lage zu sein, solche Materialien einzuspritzen, wurde eine dynamische Einspritztechnik erfunden und getestet.

Die rheologischen Eigenschaften der Vergiessmörtel auf Zement- und Bentonit-Basis wurden im Labortests untersucht. Der Einfluss verschiedener Faktoren auf die rheologischen Eigenschaften dieser Materialien wird unter besonderem Hinweis auf die Schwingungen dargestellt. Mathematische Modelle der Fliesseigenschaften und der Einfluss von Schwingungen werden ebenso aufgeführt wie Beispiele der Labortests, die zeigen, dass die Scherfestigkeit des Vergiessmörtels mit der dynamischen Technik um bis zu 100mal reduziert werden kann.

1 INTRODUCTION

The need for grouting very small fractures with a material that can survive for thousands of years has arisen in connection with the development of methods for disposal of radioactive and other toxic wastes. Granitic rock surrounding waste packages at a depth of ≈ 500 m often contains fractures with apertures smaller than 100 µm that can lead water. If these small fractures can be efficiently sealed, the function of the respository will be considerably improved.

The reological properties of viscous materials are improved by the dynamic technique. Dynamic pressure pulses on the grout with a frequency of 40-200 Hz are superimposed on a static pressure. When the pulses reach a fracture they are transformed to shear strain pulses of a magnitude that are determined by the pressure amplitude, the frequency and the fracture aperture. The oscillating shear strain reduces the shear resistance in the grout (decreases the viscosity). The shear strain amplitude and the frequency control the decrease in shear resistance.

Laboratory as well as field tests of two possible grout materials have been performed: cement and bentonite slurries. The rheological properties of fresh cement and bentonite and the influence of vibrations on these properties are similar, but the change in properties with time differs. The test technique and the common properties will be examplified in some cement tests.

2 RHEOLOGICAL TESTING

The rheological properties were tested in a Brookfield viscometer with a rotating vertical bob inside a cylindrical cup. The cup was vibrated vertically by an electromagnetic vibrator with frequencies f that can be varied between 20 and 10 000 Hz and displacement amplitudes δ_a that can be varied between 0 and 1.5 mm. The amplitude was measured by a small accelerometer fixed to the cup. Fig 1 shows the arrangement.

The width b of the slot between the cup and the bob was varied in a number of tests. These tests showed neither b nor the vibrating amplitude δ_a but the ratio δ_a/b, which is equal to the amplitude of the vibrating shear strain γ_a, is the parameter that controls the influence of the vibrations on the rheological behavior. The width b was then set at 3 mm and the maximum vibrating shear strain amplitude used in the tests was $\gamma_A=0.5$.

3 CEMENT GROUT MATERIALS

3.1 *Grout composition*

The following 4 different types of cement have been investigated:

1. Canadian finely ground Portland cement
2. Swedish finely ground Portland cement
3. Microdur
4. Alofix

These cements were mixed to different proportions with the following components:

1. Superplasticizer, i.e. sodium napthalene formaldehyde condensate (SP=0-5%)
2. Silica fume (SF=0 and 10%)
3. Water (w/c=0.2-1.0)

3.2 *Hardening of cement*

The hardening process of cement starts directly after mixing. The rate of this process is a function of many parameters, the most important ones being:

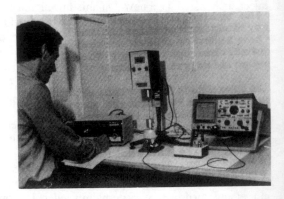

Fig. 1 Arrangement for viscometer tests with vibrated grout. A sine wave generator is seen to the left while a Brookfield viscometer and the vibrator attached to the cup are seen in the center of the picture. A small Piezotron accelerometer is mounted on the cup. The right hand side shows the Piezotron coupler and the oscilloscope, used for measuring the vibrations.

Anordnung für Viskosimeter-Tests mit eingerüttelten Vergiessmörtel. Ein Sinuswellengenerator ist links zu sehen, während ein Brookfield-Viskosimeter und der Rüttler - angeschlossen an den Becher - in der Mitte des Bildes zu sehen ist. Ein kleiner Piezotron-Beschleunigungsmesser ist am Becher montiert. Auf der rechten seite sind eine Piezotron-Kuppler und das Oszilloskop dargestellt, das zur Schwinungsmessung eingesetzt wird.

4

1. Temperature
2. Superplasticizer content
3. Cement type
4. w/c ratio
5. Stirring

The hardening process is important for the decisions respecting:

- the time from mixing that can be allowed to pass until the cement grout is no longer useful for injection
- the required time until the cement is hard enough to allow for release of the packer in the injection hole
- the grout composition for achieving proper viscous properties

Very strong stirring of a fresh cement mixture delays hardening. However, after 1-2 hours the viscosity starts to increase, even when stirring the mixture in a colloid mixer, and maximum time for using the cement was therefore set at 1 hour after mixing. Fig. 2 illustrates the time dependence (Swedish cement, w/c=0.35, SF=10%, and SP=0.75%). The figure shows the shear resistance at different time periods after mixing for unvibrated grout.

3.3 *Rheological properties of unvibrated cement grout*

General

The viscometer tests made on the different cement types and different mixtures have yielded the following very important findings for fine-grained injection cements:

1. The viscous properties of a particular cement type may vary from shipment to shipment
2. The age of the cement before mixing with water affects the viscous properties. The viscosity seems to decrease with increasing age due to aggregation by hydration
3. The storing technique affects the visous properties due to hydration. Hence, cement in open paper bags change its properties quicker than cement in air-tight plastic bags
4. The mixing technique affects the properties.

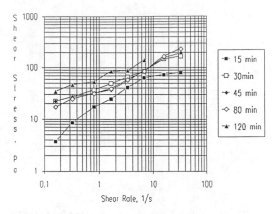

Fig. 2 Results from viscometer tests on cement at different time periods of rest after mixing.

Ergebnisse der Viskosimeter-Tests an Zement bei unterschiedlichen Ruhezeiträumen nach der Mischung

Proper colloid mixers must be used both in the field and for the cement tested in laboratory
5. The superplasticizer undergoes aging and will loose its viscosity-reducing properties with time
6. The viscometer used for determining the properties of the mixture *must be equipped with a ribbed bob* to prevent slippage between the bob and the grout material. A smooth bob with which all standard viscometers are equipped can give inaccurate flow data for these materials

Viscous behavior

Fig. 3 shows typical examples of the rheological behavior of unvibrated cement. The figure shows the influence of the shear rate $\dot{\gamma}$ on the shear resistance τ in a double logarithmic diagram for Swedish finely ground Portland cement.

The figure shows that the $\dot{\gamma}$-τ relation is almost linear and that it can be suitably expressed according to Eqn. 1.

$$\tau = m\left(\frac{\dot{\gamma}}{\dot{\gamma}_0}\right)^n \tag{1}$$

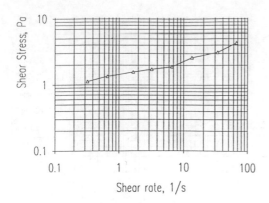

Fig.3 Example of the shear resistance of Swedish finely ground Portland cement plotted as a function of the shear rate (w/c=0.4, SF=10% and SP=1.5%)

Beispiel für die Scherfestigkeit von schwedischem, fein gemahlenem Portland-Zement, dargestellt als Funktion des Schergefälles (w/c=0.4, SF=10% und SP=1,5%)

where τ = shear stress (Pa)

$\dot{\gamma}$ = shear rate (1/s)

$\dot{\gamma}_0$ = normalized shear rate (=1.0 l/s)

m = parameter (Pa)

n = parameter

The parameter n corresponds to the inclination of the curve in the double logarithmic diagram and is thus a measure of the rate dependence of the shear resistance, while m is a measure of the gel stiffness and gel strength. If n=1.0 the material is Newtonian and the ratio $m = \tau/\dot{\gamma}$ will be equal to the true viscosity.

None of the investigated cement mixtures is Newtonian and their n-values vary between approximately 0.2 to 0.9 depending primarily on the w/c ratio and the SP-content. In principle, an increase in w/c ratio and SP-content will increase the n-value.

3.4 Rheological properties of vibrated cement grout

General

Different cement compositions with varying

water/cement ratio, superplasticizer content and silica fume content have been tested at different frequencies and amplitudes, and after different times after mixture. However, most tests were performed 30 minutes after mixture.

Results

An example of different series of viscometer tests performed under vibrations with the frequency f=40 Hz and amplitudes $0 \leq \gamma_A \leq 0.5$ is shown in Fig. 4.

The figure shows that the non-newtonian material turned almost newtonian when vibrations were applied ($n \approx 0.9$).

Several tests have shown that n varies between 0.9 and 1.0 when the vibrations are large enough to break down the structure of the material. The vibrations thus make the material almost newtonian. Fig. 4 also shows that vibrations reduce the shear resistance: the higher the shear strain amplitude γ_A, the lower the shear resistance.

Fig. 5 shows the parameter m in Eqn. 1 plotted as a function of the applied oscillating shear strain amplitude γ_A at different applied frequencies f. The m-γ_A relation is obviously also a straight line in a double logarithmic scale. The relation can be written according to Eqn. 2:

$$m = a \cdot \left(\gamma_A / \gamma_{A_0} \right)^b \qquad (2)$$

where

γ_A = oscillating shear strain amplitude

γ_{A_0} = normalized γ_A (=1.0)

a = parameter (Pa) (m at γ_A= 1.0)

b = parameter

Fig. 5 also shows that a decrease in γ_A can be well compensated by an increase in f up to 300 Hz. However, the corresponding effect is not valid at higher frequencies as can be seen for f=1000 Hz which would require an amplitude γ_A as high as for f=300 Hz. An increased frequency up to f=300 Hz as well as an increased oscillating shear strain amplitude thus reduces the shear resistance.

The effect of an increased amplitude versus an increased frequency can be compared if f is plotted as a function of γ_A for the same rheological properties, i.e. for the same m-value. Fig. 6

6

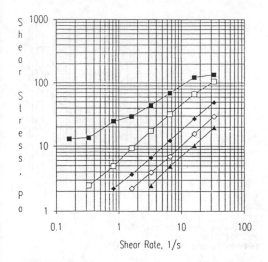

Fig. 4 Shear stress measured at viscometer tests as a function of shear rate for different applied vibrating shear strain amplitudes γ_A at the frequency f=40 Hz. (Swedish cement; w/c=0.35; SP=0.75%; SF=10%). ■ γ_A=0.0, □ γ_A=0.067, ◆ γ_A=0.2, ◊ γ_A=0.33, ▲ γ_A= 0.5

Scherbeanspruchung gemessen mit Viskosimeter-Tests in Abhängigkeit vom Schergefäll für verschiedene angewandte Rüttel-Scherbeanspruchungsamplituden γ_A bei einer Frequenz von f=40 Hz (schwedischer Zement; w/c=0.35, SP=0.75%, SF=10%)

shows the γ_A-f relation at m=1.0 and one finds that a 10-fold change in amplitude corresponds to a 5-fold change in frequency at f<300-400 Hz. This means that the effect of doubling the frequency is stronger than the effect of doubling the amplitude. However, the required power is direct proportional to f^3 and γ_A^2, which means that more power is required if the frequency is increased than if the amplitude is increased in order to reach the same drop in shear resistance. Thus, a low frequency is favourable and it should not exceed ≈300 Hz in order to get a good effect on the shear resistance.

4 BENTONITE GROUT MATERIALS

4.1 *Grout composition*

The reference clay grout material used in the

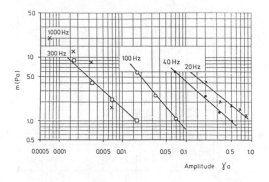

Fig. 5 The parameter m plotted as a function of the oscillating shear strain amplitude γ_A at different frequencies. The same cement mixtures as in fig. 4.

Der Parameter m wird als Funktion der Rüttel-Scherbeanspruchungsamplitude γ_A bei verschiedenen Frequenzen gezeichnet. Gleiche Zementmischung wie in Abb.4

laboratory tests and in the field tests in Stripa, was Tixoton. It is a bentonite with sodium as dominating exchangeable cation converted from Ca-bentonite through Na treatment. Tixoton is a bentonite of very high quality with a high liquid limit and a favourable particle sizes distribution.

The properties of Tixoton vary somewhat with delivery, due to the natural variation in the raw material. Two deliveries have been achieved. One before autumn 1988 (called old Tixoton) and one after (called new Tixoton).

The following mixtures that may be used for grouting have been tested:

- Pure tixoton
- Tixoton mixed with 25-75% finely ground quartz powder
- Tixoton mixed with quartz powder and 0.1-2.0% salt

The reason for mixing the tixoton with quartz powder is to decrease the erodability while the reason for adding salt is to temporarily lower the liquid limit and thus to be able to inject a grout with a lower water ratio. When the salt has diffused into the rock, the bentonite slurry has regained its original liquid limit and the injected slurry will be very stiff.

The cone liquid limit w_L of different mixtures

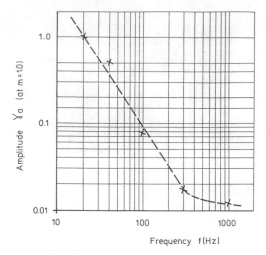

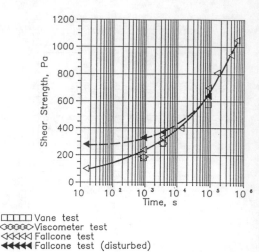

Fig. 6 The vibrating shear strain amplitude as a function of the corresponding frequency at equal viscous behavior (m=1.0).

Rüttel-Scherbeanspruchungsamplitüde als Funktion der entsprechenden Frequenz bei gleichem Viskositätsverhalten (m=1.0)

Fig. 8 Shear strength of tixoton measured with different methods at different times after mixing. The filled triangles denote a tixoton that has been strongly disturbed after about 3 days ($2.4 \cdot 10^5$ s)

Scherbeanspruchung des Thixoton gemessen mit vershiedenen Methoden zu unterschiedlichen Zeitpunkten nach der Mischung. Die ausgefüllten Dreiecke markieren eine starke Störung des Thixotons nach ca. 3 Tagen.

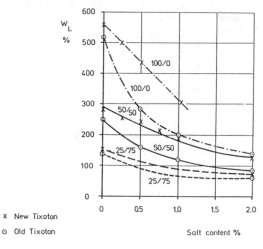

x New Tixoton

o Old Tixoton

Fig. 7 Liquid limit w_L as a function of the salt content in the pore water for different mixtures tixoton (%)/quartz (%). The difference between the two deliveries of tixoton is clear.

Fliessgrenze w_L in Abhängigkeit vom Saltzgehalt im Porenwasser für verschiedene Mischungen von Thixoton (%) und Quartz (%). Der Unterschied zwischen den beiden Thixoton-Austrägen ist klar

has been determined. The liquid limit w_L is the water ratio at which transition of the material from plastic to liquid consistency takes place according to the soil mechanical definition. Fig. 7 shows the liquid limit as a function of the salt content for different mixtures of tixoton and quartz powder. The added salt is in the form of NaCl.

4.2 Rheological properties

Thixotropy

A bentonite gel is thixotropic with a primary, very quick regain in structure immediately after disturbance and a secondary slow regain that goes on for several days or weeks. The properties of a bentonite gel is dependent on its state of deformation and the time after mixing or stirring. Four different states can be identified:

- Before the structure is broken down

- At very small strains after a <u>long</u> time of rest
- At very small strains after a <u>short</u> time of rest

- After the structure is broken down

- At large strain when the gel is continuously sheared
- At vibrations

A number of tests were performed in which the shear strength was measured at different times after mixing. Laboratory vane tests and viscometer tests as well as cone tests were made. Fig. 8 shows the results. The strength increases very much with time or a factor of 10 in a week. The measured strength at the different tests agree very well. The higher value for the cone test can be explained by a higher rate of shear. The regain in strength after a disturbance is also shown.

Viscous properties

Many series of viscometer tests with vibrated and unvibrated bentonite slurries suitable for grouting have been made. They have all been carried out on Tixoton with the following variables:

- The water ratio (usually expressed in terms of the relation between water ratio and the liquid limit w/w_L)
- The salt content of the final pore water
- The content of added quartz powder
- The frequency of the vibrations
- The amplitude of the vibrations

Admixtures of quartz and salt to the slurries are primarily changing the liquid limit. If different mixtures are compared in relation to the liquid limit, the rheological behaviour does not change very much. Typical flow curves are shown in Fig. 9 which refers to a mixture of 50% tixoton and 50% quartz (T/Q=50/50) tested at the frequency f=40 Hz with different amplitudes. The figure shows that the effect of vibrations on the viscous properties of a bentonite-based grout can be modeled in the same way as for a cement-based grout with the shear resistance expressed according to Eqn. 1 and the influence of vibrations expressed according to Eqn. 2.

    Example of the relation between the m-values

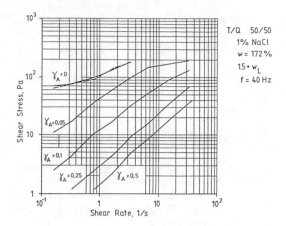

Fig. 9 Flow curves from viscometer tests with varying vibrating shear strain amplitudes

Fliesskurve der Viskosimeter-Tests mit verschiedenen Rüttel-Scherbeanspruchungsamplituden

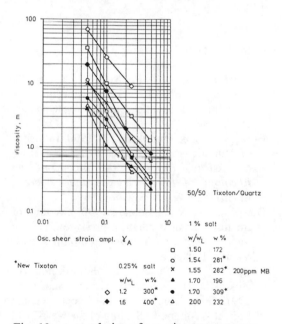

Fig. 10 γ_A-m relations from viscometer tests on a tixoton/quartz mixture at different salt contents and water ratios. 200 ppm of methylene blue is added to one slurry

γ_A-m-Verhältnis der Viskosimeter-Tests an einer Thixoton/Quartz-Mischung bei unterschiedlichem Saltzhalt und Wasserverhältnis.

9

and the oscillating shear strain amplitude γ_A for different w/w_L ratios at the salt contents 0.5 and 1.0% are shown in Fig. 10. The primary factor for the position of the curve is the w/w_L ratio, but the salt content and the amount of added quartz also have an influence.

5 CONCLUSIONS

The laboratory tests show that the rheological properties are strongly influenced by vibrations and that the influence can be modeled according to Eqns 1 and 2 both for cement-based and bentonite-based grouts.

The laboratory tests also demonstrate the potential of a bentonite slurry as a grout material. The tixotrophy of the material will increase the shear strength a factor ≈ 10 after a week compared to the strength just after mixing. The vibrations achieved at dynamic injection can further reduce the shear resistance by a factor $\approx$ 100.

6 REFERENCES

Börgesson, L. & Fredrikson, A. 1990. Rheological properties of Na-smectite gels used for rock sealing. *Engineering Geology*, 28:Elsevier

Börgesson, L. & Fredrikson, A. 1990. Influence of Vibrations on the Rheological Properties of Cement. *Proc. of the Int. Conf. on Rheology of Fresh Cement and Concrete*, Liverpool: E.&F.N. SPON

Börgesson, L. & Jönsson, L. 1990. Grouting of fractures using oscillating pressure. *Proc. of the Int. Conf. on Mechanics of Faulted Rock*, Vienna: Balkema

Börgesson, L., Pusch, R., Fredrikson, A., Hökmark, H., Karnland, O. & Sandén, T. 1991. Fina Report of the Rock Sealing Project - Volume I - Sealing of the Near-Field Rock Around Deposition Holes by Use of Bentonite Grouts. *Stripa Project Technical Report 91-34*, SKB, Stockholm

Pusch, R., Erlström, M., & Börgesson, L. 1985. Sealing of Rock Fractures, A Survey of Potentially Useful Methods and Substances, *SKB Technical Report 85-17*, Stockholm

Pusch, R. Börgesson, L., Fredrikson, A., Markström, I., Erlström, M., Ramqvist, G., Gray, M & Coons, W. 1988. Rock Sealing - Interim Report on the Rock Sealing Project (Stage 1). *Stripa Project Technical Report 88-11*, Stockholm

Grouting in Rock and Concrete, Widmann (ed.) © 1993 Balkema, Rotterdam, ISBN 90 5410 350 7

Untersuchungen zum Eindringvermögen von Injektionsmitteln

Investigations on the penetration capacity of injecting agents

Martin Bolesta & Karl-Heinrich Zysk
DMT-Gesellschaft für Forschung und Entwicklung mbH, Essen, Deutschland

In den letzten Jahren haben Maßnahmen zur Gebirgsverfestigung im deutschen Steinkohlenbergbau stark an Bedeutung gewonnen. Hierfür steht eine große Produktpalette an Injektionsmitteln, insbesondere Zementsuspensionen und Kunstharze, zur Verfügung, für die es lange Zeit keine Beurteilungskriterien gab. Der Einsatz und die Auswahl des Injektionsmittels erfolgte dabei nach den empirischen Erfahrungen des Anwenders. In Forschungsarbeiten sollten bei der DMT die Zusammenhänge einer erfolgreichen Injektionsmaßnahme und den technologischen Eigenschaften des Injektionsmittels untersucht werden. Das Eindringvermögen der Injektionsmittel spielt hierbei eine wichtige Rolle, da die wirksame Gebirgsverfestigung nur an verfüllten Rissen beobachtet werden konnte. Ein für Untersuchungen des Eindringvermögens entwickelter Versuchsstand zeigt, daß die im deutschen Steinkohlenbergbau eingesetzten Injektionsmittel sich hinsichtlich ihres Eindringvermögens erheblich unterscheiden.

In the course of the past years, strata consolidation measures gained strongly increasing importance in German coalmining. For these measures, a large range of injection agents, in particular cement slurries and synthetic resins for which no assessment criteria were on hand for a long time, are available. The selection and use of the injection agent was decided upon by empirical criteria of the user. The DMT investigations were to find out about the correlations between a successful injection measure and the technical properties of the agent injected. The penetration capacity of such an agent plays an important part, since consolidation measures become effective only - as observed - in filled-up cleat and fissure systems. A test rig particularly developed for penetration capacity tests gave evidence that the injection agents used in German coalmining exhibit definitely different features as far as their penetration capacity is concerned.

1 EINLEITUNG

Injektionsmittel werden im deutschen Steinkohlenbergbau hauptsächlich im Übergangsbereich Streb-Strecke eingesetzt [1].

Hier wird der schützende Ausbau der Strecke teilweise gelöst, um den reibungslosen Ablauf der mechanischen Kohlegewinnung zu ermöglichen (siehe Abbildung 1). Dadurch können durch hohen Überlagerungsdruck zerstörte Dachschichten hereinbrechen, was neben vermehrten Stillstandzeiten zu Förderausfall führt und auch im Besonderen die Unfallgefahr für die Belegschaft erhöht. Um diesen Bereich erfolgreich zu stabilisieren, müssen die Injektionsmittel verschiedene technologische Eigenschaften besitzen.

Hierzu zählen die Klebfestigkeit, die Verformbarkeit und das Eindringvermögen.

Die Klebfestigkeit ist eine in Laborversuchen ermittelte Größe und setzt sich aus Anteilen von

Abb.1: Fotografie eines Streb-
Streckenübergangs. Die Aufnahme
wurde von der Strecke aus in Richtung Streb gemacht. Man erkennt
den geöffneten Ausbau und das auf-
gelockerte Gebirge.

Fig.1: Photograph of a face-end
zone. The photo was taken from the
roadway in direction of the coal-
face. The open support and the
loosened strata can be recognized.

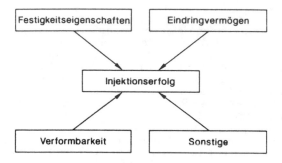

Abb.2: Technologische Eigenschaften von Injektionsmitteln, die den
Injektionserfolg beeinflussen.

Fig.2: Technical properties of injection agents which control the
injection success.

Haftzug-, Biegezug-, Zug- und
Druckfestigkeit zusammen. Sie gibt
an, welche Spannung notwendig ist,
um gebrochene und mit dem Injektionsmittel wiederverklebte
Prismen erneut zu brechen. Daß die
Wirksamkeit einer in diesen Versuchen ermittelten hohen Klebfestigkeit zu besseren Injektionserfol-

gen führt als eine niedrige, wurde
in Versuchen bestätigt. In diesen
Versuchen dienten die gemessenen
Ausbruchsvolumina als Index für
den Injektionserfolg [2]. (Siehe
Abbildung 3)

Obwohl Injektionsmaßnahmen im
Streb-Streckenübergang durchgeführt werden, lassen sich durch
den Abbau der Kohle und den dadurch entstehenden hohen Zusatz-
drücken hervorgerufene Gebirgbewe-
gungen nicht vermeiden. Um den Zusammenhalt des Verbundes Gebirge-
Injektionsmittel-Gebirge aufrecht-
zuerhalten und damit ein Heraus-
fallen von Steinen zu verhindern,
müssen die in diesem Bereich eingesetzten Injektionsmittel in der
Lage sein, den Gebirgsbewegungen
zu folgen, ohne dabei ihre Klebfestigkeit einzubüßen. Als technologische Größe kann hierfür die Verformung des Injektionsmittels dienen, die es bei den Klebfestig-
keitsversuchen erfährt. Injektionsmittel, bei denen keine Verformungen festgestellt werden
konnten, wurden als spröde bezeichnet. Im Gegensatz dazu wurden
Injektionsmittel mit meßbaren, unterschiedlich hohen Anteilen an
elastischen Verformungen als
elasto-plastisch bezeichnet. In
untertägigen Versuchsreihen im Bereich Streb-Strecke konnte bei
spröden Injektionsmitteln deutlich
mehr Ausbruch festgestellt werden,
als bei elasto-plastischen Injektionsmitteln, auch bei gleichen
Klebfestigkeiten. (Siehe Abbildung
3)

Die dritte für einen Injektionserfolg wichtige technologische
Eigenschaft, das Eindringvermögen,
ist von besonderer Bedeutung. Der
für die Bestimmung dieser Eigenschaft entwickelte Versuchsstand
wird im folgenden ausführlich behandelt.

2 EINDRINGVERMÖGEN

Für eine erfolgreiche Gebirgver-
festigung ist die Anzahl derjenigen Risse, die mit einem
Injektionsmittel verfüllt werden
konnten die bestimmende Größe des
Injektionsmittelerfolges. Erst
bei idealer Rißinjektion können
die bereits genannten Klebfestigkeiten und Verformbarkeiten

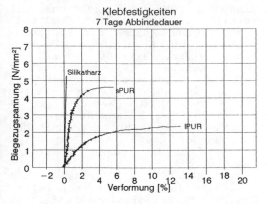

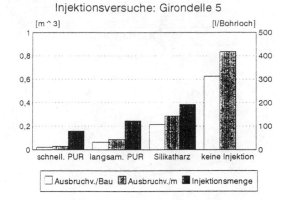

Abb.3: Gegenüberstellung ausgewählter Spannungs-Verformungs-Kurven (links) und Ausbruchsvolumina unter Tage in einer mit diesen Injektionsmitteln injizierten Strecke (rechts).

Fig. 3: Comparison of selectes stress-deformation curves (left side) and mineral fall volumes underground for a roadway in which these agents were injected (right side).

von Injektionsmitteln auf eine verklebte Fäche wirken. Wichtig sind hierbei die verfahrenstechnischen Einflüsse und technologische Eindringeigenschaften des Injektionsmittels.

Bei den verfahrenstechnischen Einflüssen spielen insbesondere der Injektionsdruck und der Volumenstrom eine Rolle. Sie hängen von der Injektionsausrüstung und der Pumpenkennlinie ab.

Mit den für Injektionsmaßnahmen zur Verfügung stehenden Pumpen kann ein Injektionsmittel, unabhängig davon, ob es eine Suspension oder Flüssigkeit ist, mit nahezu jeder Druck Volumenstrom-Kombination in das zu verfestigende Gebirge eingebracht werden. Für Vergleiche der Eindringvermögen verschiedener Injektionsmittel ist es daher erforderlich, die Injektionsdrücke und Volumenströme konstant zu wählen, so daß unter Berücksichtigung aller weiteren Einflußfaktoren, wie z.B. Durchmesser der Injektionssonde, Länge aller Pumpleitungen, etc., das Eindringen eines Injektionsmittels in einen Spalt als technologische Eigenschaft des Injektionsmittels angesehen werden kann.

Diese Eigenschaft des Injektionsmittels wird im Weiteren als Eindringvermögen bezeichnet [3].

3 INJEKTIONSVERSUCHE

Für die Entwicklung eines Versuchsstandes, der möglichst praxisnahe Bedingungen des Eindringvermögens von Injektionsmitteln anstreben sollte, war es erforderlich, den genauen Aufbau des zu injizierenden Gebirges zu untersuchen. Um Aufschluß über die Öffnungsweiten und Verläufe von Rissen in den mit Injektionsmitteln zu verfestigenden Bereichen zu erfahren, wurden Unter Tage eine Reihe von Endoskopuntersuchungen durchgeführt. Diese Untersuchungen zeigten, daß der überwiegende Teil der Risse (ca. 75%) Öffnungsweiten von unter 1mm hatten. Von diesen Rissen entfiel der größte Teil auf Risse mit Öffnungsweiten kleiner 0,1mm.

Weiter fiel bei Untersuchungen mit einem Endoskop, wie es auch zur Bauschadenserkundung verwendet wird, auf, daß die Mehrzahl der Risse nahezu rechtwinklig zur Bohrlochachse, also konzentrisch um die Streckenachse ausgerichtet waren [2].

3.1 Aufbau des Versuchsstandes und Versuchsdurchführung

Der entwickelte Versuchsstand

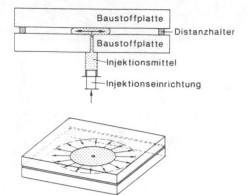

Abb.4: Schematische Darstellung des Injektionsversuchs.

Fig.4: Scheme of an injection test.

Abb.5: Injektionsmittelprüfstand: Stahlrahmen (a), Kraftmeßdose (b), Prüfplatten (c), Spannzylinder (d)

Fig.5: Injection agent test rig: steel frame (a), force transducer (b), test plates (c), tensioning ram (d)

wurde einer untertägigen Rißinjektion nachgebildet mit dem Ziel, auch kleinste Risse (Zehntel-Millimeter-Bereich) definiert nachstellen zu können. Zwei Platten mit einer definierten Oberflächenrauhigkeit, die die Flanken eines natürlichen Risses simulieren sollten, wurden hierzu parallel in einen Stahlrahmen eingespannt. Durch die Vorspannkraft wurde ein Aufweiten des zu injizierenden

Risses durch den Injektionsdruck verhindert. Die lichten Spaltweiten konnten hierbei mit Abstandshaltern aus Aluminium- und Messingstreifen variiert werden. Um realitätsnahe Klimaverhältnisse bei den Versuchen zu gewährleisten, wurden die Versuchsplatten vor Einbau in die Versuchsapparatur in einem Klimaschrank unter abbaubedingten wettertechnischen Bedingungen, wie einer Temperatur von 30°C und einer relativen Feuchte von 80%, gelagert. Hierdurch konnten auch klimatische Einflüsse auf die sehr temperatur- und feuchtigkeitsempfindlichen Kunstharze minimiert werden. Durch eine in der unteren Platte eingegossenen Sonde wurden dann die Injektionsmittel in den Spalt zwischen den Platten injiziert.

Bei der Durchführung der Versuche wurden die Injektionsmittelmengen für alle untersuchten Spaltweiten so berechnet, daß sich bei einer vollständigen Injektion jeweils der gleiche Durchmesser des Injektionskörpers ergab.

Die Injektionen mit den verschiedenen Gebirgsverfestigungsmitteln wurden jeweils bei den größten untersuchten Spaltweiten (1mm) begonnen. Ließ sich in diese Spalte die gesamte Menge mit der pneumatischen Injektionseinrichtung injizieren, so wurde für den nächsten Versuch die Spaltbreite um 0,1mm verringert.

Der genaue Aufbau des Versuchsstandes ist der Abbildung 3 zu entnehmen. Man erkennt den Stahlrahmen (a), die Kraftmeßdose (b) zur Messung der Vorspannkraft, die Prüfplatten (c) und die Spannzylinder (d), mit denen die Prüfplatten eingespannt werden. Die Injektionseinrichtung befindet sich unterhalb der Prüfplatten.

3.2 Meßwertaufnahme

Während des Injektionsvorgangs wurden verschiedene Kenndaten, hierzu gehören unter anderem die Injektionsdauer, der Injektionsdruck und der Injektiosvolumenstrom, aufgenommen. Aus diesen Daten ließ sich dann unter Berücksichtigung der Spaltbreite das Eindringvermögen verschiedener Injektionsmittel feststellen und somit die Kriterien für die Beurteilung des Eindringvermögens ableiten. Die beiden wichtigsten Kriterien sind dabei die Grenzspaltbreite und die spezifische Injektionsarbeit.

Die Grenzspaltbreite ist die Spaltbreite, in die sich ein Injektionsmittel in der geforderten Menge injizieren läßt.

Die spezifische Injektionsarbeit beschreibt die von der Injektionspumpe bei dieser Injektion zu verrichtende mechanische Arbeit, bezogen auf das bei einer Spaltbreite verfüllte Rißvolumen. Der Effekt, daß einige Injektionsmittel durch ein Aufschäumen selbst Injektionsarbeit verrichten, wird dabei berücksichtigt. Die Berechnung der spezifischen Injektionsarbeit erfolgt rechnergestützt unter Berücksichtigung der bei der Versuchsdurchführung aufgezeichneten Daten.

Das Wechselspiel zwischen Injektionsdruck und Volumenstrom hat auf das Ergebnis keinen Einfluß, da beide Größen zu jedem Zeitpunkt der Injektion in die Injektionsarbeit eingehen.

4 ERGEBNISSE DER EINDRINGVERSUCHE

Die untersuchten Injektionsmittel unterschieden sich, wie bei einer derartig großen Palette an Injektionsstoffen nicht anders zu erwarten, hinsichtlich ihres Eindringvermögens zum Teil erheblich. Die Abbildung 6 zeigt die Ergebnisse ausgewählter Injektionsmittel. Man kann deutlich erkennen, daß die geforderte Injektionsmittelmenge einer Portlandzement-Suspension in einen Spalt mit der lichten Breite von 0,3mm noch vollständig injiziert werden konnte. Wurde die Spaltbreite verringert, so stieg die Injektionsarbeit sprunghaft an, in eine Spaltbreite von 0,2 mm drang die geforderte Menge nicht mehr vollständig ein. Die Grenzspaltbreite lag demnach zwischen 0,2 und 0,3mm.

Die geprüfte Feinstzement-Suspension wies ein deutlich besseres Eindringvermögen als die Portlandzement-Suspension auf. Mit ihr war eine vollständige Injektion in den 0,2mm Spalt möglich. Bei kleineren

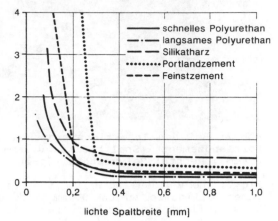

spez. Injektionsarbeit [J/ml]

lichte Spaltbreite [mm]

legend:
— schnelles Polyurethan
—·— langsames Polyurethan
— — Silikatharz
········· Portlandzement
---- Feinstzement

Abb.6: Spezifische Injektionsarbeit über die Spaltbreite bei der Injektion verschiedener Gebirgsverfestigungsmittel.

Fig.6: Specific injection work over the fissure width for various consolidation agents injected.

Spaltbreiten stieg die Injektionsarbeit, ähnlich des Verhaltens der Portlandzement-Suspension, sprunghaft an. Eine Spaltbreite von 0,1mm war soeben nicht mehr vollständig injizierbar, die Grenzspaltbreite lag daher knapp oberhalb von 0,1mm.

Neben dem besseren Eindringvermögen wies die Feinstzement-Suspension gegenüber der Portlandzement-Suspension noch einen weiteren Vorteil auf. Während es bei Injektionsvorgängen in Spalten nahe der Grenzspaltbreite bei der Portlandzement-Suspension zu Entmischungen kam - das Wasser wurde durch Fließkanäle ausgepreßt - konnte bei der Feinstzement-Suspension ein solches Verhalten nicht beobachtet werden.

Ein vergleichbares Ergebnis mit der Feinstzement-Suspension brachte die Injektion mit Silikatharz. Die Grenzspaltbreite des Silikatharzes lag bei 0,1mm und stimmte in etwa mit der der Feinstzement-Suspension überein. In der Nähe der Grenzspaltbreite war die spezifische Injektionsarbeit des Silikatharzes kleiner als die der Feinstzement-Suspension, bei größeren Spaltbreiten als

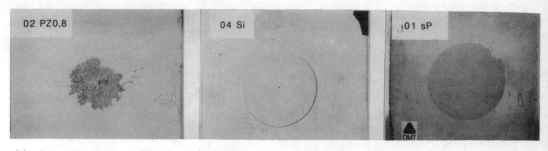

Abb.7: Bei den Injektionsversuchen entstandene Injektionskörper. Links: Portlandzement (Spaltbreite 0,2mm), Mitte: Silikatharz (Spaltbreite 0,4mm), Rechts: Polyurethan (Spaltbreite 0,1mm)

Fig.7: Body resuling from the injection tests. Left side: Portland cement (fissure width 0.2mm), center: silicate resin (fissure width 0.4mm), right side: polyurethane (fissure width 0.1mm).

0,2mm war sie jedoch deutlich größer.

Das beste Eindringvermögen zeigten die Polyurethane. Sie waren in die 0,1mm Spalten mit deutlich geringerer spezifischer Injektionsarbeit vollständig injizierbar. Selbst bei einem Aufeinanderlegen der Versuchsplatten ohne Abstandshalter war eine Injektion mit einem langsamen Polyurethan noch fast vollständig möglich, die Oberflächenrauhigkeit der Platten reichte für das Eindringen der Lösung aus.

Die Abbildung 7 zeigt die nach einem Injektionsversuch auseinandergeklappten Platten. Das linke Bild zeigt den Injektionsversuch einer Portlandzement-Suspension in einen 0,2mm-Spalt. Die Fließkanäle des ausgepreßten Wassers sind deutlich zu erkennen, die Suspension konnte nicht vollständig injiziert werden.

Die mittlere Abbildung zeigt den Injektionskörper eines Silikatharzes, 0,4mm Spaltbreite, die rechte Abbildung den Injektionskörper eines schnellen Polyurethans bei einer Spaltbreite von 0,1mm.

5 AUSBLICK

Mit dem Injektionsprüfstand sollten die im Bergbau am häufigsten eingesetzten Injektionsmittel unter gleichen Bedingungen geprüft werden, so daß sich Aussagen über deren Eindringvermögen gewinnen lassen.

Der Injektionsmittelprüfstand der DMT, er wurde so konzipiert, daß die Strömungsverhältnisse in der idealisierten Kluft weitgehend denen realer Injektionen entsprechen, ermöglicht dieses.

Dieser Prüfstand wurde auf die untertägigen Verhältnisse im deutschen Steinkohlenbergabu zugeschnitten, doch kann er auch zur Prüfung von Injektionsmitteln, die für einen anderen Einsatzbereich gedacht sind, angewandt werden. So sind Grundsatzprüfungen und ein Produktvergleich unter gleichen Bedingungen von Injektionsmitteln für den Einsatz im Tunnelbau und zur Bauwerkssanierung möglich. Die hierfür nötigen Änderungen, wie die Verwendung anderer Versuchsplatten (z.B. mit keilförmig zulaufenden Rissen, Füllung der Risse mit körnigen oder wäßrigen Medien oder die Verwendung speziell für einen Anwendungsfall ausgewählter Gesteine für die Prüfplatten) und Einstellung anderer Injektionsparameter (z.B. Injektionsdruck, Injektionsvolumenstrom), sind ohne weiteres durchführbar.

Literaturverzeichnis:

[1] J.Werfling
    Injektionstechnik: Kosten

16

der Gebirgsverfestigung
Diplomarbeit, Bergakademie
Freiberg, 1991,
unveröffentlicht

[2] D.Gemmel
Anforderungen an die technologischen Eigenschaften von Injektionsmitteln zum Verfestigen der angeschnittenen Hangendschichten in Abbau- und
Basisstrecken
Dissertation, RWTH Aachen,
1992, Verlag Dr.Shaker, Aachen

[3] D.Kölbl
Entwicklung eines Prüfverfah-
rens zum Eindringvermögen von
Gebirgsverfestigungsmitteln
Diplomarbeit, RWTH Aachen,
1991, unveröffentlicht

Erprobung und Anwendung der Feinstzementinjektion zur kraftschlüssigen Rißschließung in Beton

Testing and application of concrete crack grouting with very fine cement suspensions

H. Budelmann & A. Brandau
Universität Kassel, Deutschland

K.-H. Fromm
BS Betonschutz, Hofgeismar, Deutschland

ABSTRACT: Compared with chemical injection media, suspensions of very fine cements show important advantages for injection into concrete crack spaces. Therefore they are increasingly used for the force-locking filling of cracks in concrete members. Very fine cements used for the crack injection must have a high specific surface of about 15.000 cm^2/g; at least 95 % by weight should have a grain size below 16 µm. A series of experiments has been performed at the University Kassel, to show the suitability of fine cement suspensions for the force-locking filling of small cracks in concrete members. Cement suspensions were injected into flex cracks of 0.3 mm crack width of nine small sized reinforced concrete beams. It could be shown that a complete filling of the cracks down to 0.05 mm crack width can be achieved and that the flexural tensile strength of the injected beams is in the same magnitude as that of uncracked beams. In the second part of the contribution two examples of application, recently realized, are reported to give an impression of the capability of this new injection technology.

ZUSAMMENFASSUNG: Die Eignung von Feinstzementsuspensionen für die kraftschlüssige Rißfüllung in Beton wurde in einer Versuchsreihe an Stahlbeton-Biegebalken untersucht. Zwei ausgeführte Praxisbeispiele vermitteln einen Eindruck von der Leistungsfähigkeit und Vielseitigkeit dieser neuen Injektionstechnik.

1 EINLEITUNG

Die kraftschlüssige Verfüllung von Rissen in Beton wird heute üblicherweise mit Epoxidharz vorgenommen. Die Injektion von Feinstzement-suspensionen kann eine geeignete Alternative sein. Denn diese weisen gegenüber Epoxidharzen einige materialtypische Vorzüge für den Einsatz in Betonbauteilen auf wie z. B.: Weitgehende Unempfindlichkeit gegen Feuchtigkeit im Riß, korrosionsschützende Wirkung für die Bewehrung im Riß, ökologische Unbedenklichkeit wie Beton, Anwendbarkeit auch bei niedriger Bauteiltemperatur, annähernd gleichbleibende Viskosität während einer ausreichend langen Verarbeitbarkeitszeit.

Die Injizierbarkeit in Risse mit Breiten im Zehntelmillimeterbereich setzt äußerst fein gemahlene Zemente voraus. Nach /1/ kann für die Mindestabmessung der noch füllbaren Rißbreite etwa die Hälfte des Größtkorndurchmessers, der zu einem Siebdurchgang von 95 % gehört, angesetzt werden. Feinstzemente für die Rißinjektion weisen Korngrößenverteilungen

auf, bei denen zumindest 95 Gew.-% einen Korndurchmesser kleiner als 16 µm haben. Die Grenze des Eindringvermögens liegt damit bei etwa einem hunderstel Millimeter. Die Mindestrißbreite an der Bauteiloberfläche muß allerdings deutlich größer sein, um eine vollständige Rißfüllung zu erreichen. Hier ist etwa vom zehnfachen Wert des 95 %-Größtkorns auszugehen, also etwa 0,1 bis 0,2 mm.

Im heutigen Regelwerk für die Rißinjektion in Beton /2, 3/ ist die Injektion von Zement-suspensionen noch nicht berücksichtigt. In einer Versuchsreihe, die – veranlaßt von der bs betonschutz gmbh, Hofgeismar – an der Amtlichen Baustoff- und Betonprüfstelle der Universität – Gh Kassel durchgeführt wurde, konnte die Eignung von Feinstzementsuspensionen zur kraftschlüssigen Rißinjektion im Konstruktionsbeton gezeigt werden. Zur Anwendung kam eine zweikomponentige Suspension der Heidelberger Baustofftechnik GmbH, bestehend aus einer den Feinstzement enthaltenden Trockenkomponente und einer Flüssigkomponente, die neben demineralisiertem

Längsschnitt

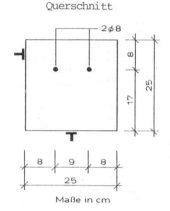

Maße in cm

Bild 1: "Kleiner Biegebalken" für die Rißinjektion

Figure 1: "Small-sized" Concrete Beam for Crack Injection

Wasser verschiedene Zusatzmittel enthält.

Nachfolgend wird über die Versuche, deren Ergebnisse und über praktische Anwendungsbeispiele berichtet.

2 VERSUCHSKÖRPER UND VERSUCHSPARAMETER

Es wurden insgesamt 9 sog. kleine Biegebalken entsprechend Bild 1 hergestellt, an deren Unterseite im Dreipunktbiegeversuch jeweils ein Riß mit einer Rißbreite zwischen 0,3 und 0,5 mm an der Betonoberfläche erzeugt wurde. Durch eine Verspannung wurde die Rißbreite auch nach Lastentfernung bei im Mittel 0,3 mm konstant gehalten. Die Rißlänge (von der Balken-Unterseite bis zur Rißwurzel) betrug im Mittel etwa 200 mm.

Für den Beton der Biegebalken wurde ein üblicher B 35 (360 kg PZ 35 F, A/B 16 Quarzkies, w/z = 0,50, β_{w20} (28d) = 41 N/mm^2) verwendet. Die Rißerzeugung erfolgte im Alter von 7 d.

Drei unterschiedliche Feuchtezustände im Riß (je drei Biegebalken) wurden vor den Injektionen eingestellt: trocken (Feuchte des Prüfkörpers im Alter von ca. 10 Tagen bei 20°C/ 65% r.F.); feucht (verdämmter Riß, 1 Tag wassergefüllt, vor Injektion abgelassen); wasserführend (verdämmter Riß, 1 Tag wassergefüllt; Injektion gegen Druckwasser). Die Risse wurden vollständig verdämmt. Für die Injektion wurden Sho-Bond Bics "BL-Injektor" Klebepacker verwendet, je ein Packer mittig auf dem Riß auf der Balkenunterseite und ein Packer seitlich etwa in Höhe der Rißwurzel. Die Mischung der Suspensionskomponenten erfolgte mit einem Kolloidalmischer, injiziert wurde über die unteren Packer. In allen Fällen blieb die Packerschwellung bis zum Erstarren etwa konstant, so daß ein Nachverpressen nicht erforderlich war.

3 EIGENSCHAFTEN DER SUSPENSION

Von entscheidender Bedeutung für den Injektionserfolg sind u.a. die Viskosität, die Mischungsstabilität (Gefahr des Sedimentierens) und eine ausreichende Verarbeitbarkeit. Die

Tabelle 1: Eigenschaften der Suspension

Table 1 : Properties of the Suspension

Kenngröße	Dimension	Mittl. Ergebnis
Wasser-Bindemittel-Wert	-,-	0,85
Auslaufzeit Marsh-Trichter Anfangswert	sec.	42
Wert nach 60 Min.	sec.	43
Absetzen (1.000 ml-Zylinder nach 120 Min.	mm	kein Absetzen
Rohdichte ρ der Festsuspension	kg/dm^3	1,52
Biegezugfestigkeit b_{bz} im alter 7 d, 4/4/16-Prismen	N/mm^2	2,8
Druckfestigkeit β_D im Alter 7 d 4/4/16-Prismen	N/mm^2	25

20

erhärtete Suspension sollte eine Druckfestigkeit von mindestens 20 N/mm^2 und eine Biegezugfestigkeit von mindestens 2 N/mm^2 im Alter von 7 d erreichen. Die für die verwendete Suspension ermittelten Kenngrössen enthält Tabelle 1.

4 PRÜFUNG DES INJEKTIONSERFOLGES

Sieben Tage nach der Injektion wurden der Füllgrad der Risse und das Verhalten des Füllgutes im Riß (Haftzugfestigkeit, Biegezugfestigkeit) geprüft. Der Füllgrad wurde an Bohrkernen (d = 50 mm) geurteilt, die vertikal in der Balkenmitte, etwa mittig über dem injizierten Riß und durch die gesamte Balkenhöhe im Alter von sieben Tagen entnommen wurden.

Bei 100facher Vergrößerung unter dem Mikroskop konnte festgestellt werden, daß alle Risse unabhängig vom Feuchtezustand im Riß bis zu Rißbreiten von 5/100 mm homogen und vollständig verfüllt wurden. Bild 2 zeigt beispielhaft den verfüllten Riß auf einem Bohrkern, der vor der Injektion "feucht" konditioniert wurde.

senkrecht zur Rißebene gezogen. Das Prinzip ist in Bild 3 skizziert.

An den Biegebalken wurden im Alter von sieben Tagen nach der Rißinjektion erneute Dreipunktbiegeversuche vorgenommen, um die Biegezugfestigkeit zu bestimmen. Der Fehlquerschnitt infolge der Bohrkernentnahme wurde bei der Ermittlung der Widerstandsmomente berücksichtigt.

Die Ergebnisse der Prüfung der zentrischen Schalenzugfestigkeit und der Biegezugfestigkeit sind in Tabelle 2 zusammengefaßt. Die ausführlichen Ergebnisse enthält /4/.

Der Bruch erfolgte entweder ausschließlich durch Kohäsionsversagen im Riß, überwiegend aber mit etwa gleichen Flächenanteilen im Riß und im Beton. Die Ergebnisse zeigen, daß eine kraftschlüssige Rißverfüllung bis in feinste Riß-

Bild 3: Prüfung der Zentrischen Schalenzugfestigkeit (Prinzipskizze)

Figure 3: Test-setup for the Tensile "Cup" Strength (Principal Sketch)

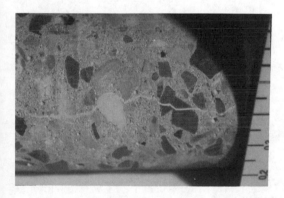

Bild 2: Verfüllter Riß auf einem Bohrkern; "feucht" konditioniert

Figure 2: Concrete Drill Core with a Filled Crack

An den Bohrkernen, in denen längs und etwa mittig der verfüllte Riß verläuft, wurde die Qualität des Haftverbundes im Riß mittels der sog. zentrischen Schalenzugfestigkeit geprüft. Die Bohrkerne wurden dazu in etwa 40 mm hohe Kernabschnitte zerteilt. Anschließend wurden auf die Abschnittsmantelflächen zylindrische Halbschalen aus Stahl geklebt. Über gelenkig angeschlossene Zugstangen wurde

breiten gelang. Die Biegezugfestigkeit über dem verfüllten Riß ist praktisch ebenso hoch wie die Biegezugfestigkeit des ungerissenen Betons. Die zentrische Schalenzugfestigkeit über dem injizierten Riß liegt im Mittel bei nur etwa 2/3 der zentrischen Zugfestigkeit des ungerissenen Betons. Allerdings ist dabei zu beachten, daß die Aussagefähigkeit dieses Versuches wegen der nicht konstanten Spannungsverteilung auf der Rißfläche (Spannungsspitzen außen) nur begrenzt ist.

5 PRAKTISCHE ANWENDUNG

Neben den an der Universität-Gh Kassel durchgeführten Versuchsreihen wurden auch bereits mehrere Objekte in der Praxis ausgeführt. Zwei Beispiele sollen angesprochen werden:

Kunsthalle Wolfsburg

Bei diesem Bauvorhaben handelt es sich um Verbundstützen Stahl/Stahlbeton mit einer

Höhe von ca. 19 m, die, bedingt durch Ausführungsfehler bei der Betonage, sowohl Fehlstellen als auch Risse im Betonquerschnitt aufwiesen.

Nach einer eingehenden Schadensanalyse durch das Fachgebiet Stahlbau der Universität Gh Kassel wurde ein Instandsetzungskonzept erarbeitet, in dem auch brandschutztechnische Aspekte zu berücksichtigen waren.

Es wurden Injektionsbohrungen d = 6 mm durch den Stahlmantel (15 mm) bis ca. 5 cm in den Betonquerschnitt geführt. Die Anzahl der Bohrungen über den Umfang betrug 6 (a $\cong$ 32 cm), der Abstand in der Höhe 30 cm.

Tabelle 2: Zentrische Schalenzugfestigkeit und Biegezugfestigkeit

Table 2 : Tensile "Cup" Strength and Flexural Tensile Strength

Kenngröße		Rißkonditionierung			Unge-rissener Beton
	Dimens.	trocken	feucht	wass. gefüllt	
Zentr. Schalen-zugfestig-keit β_{HZ} im Alter 7d im Mittel	N/mm²	2,1	1,8	2,5	3,3
Biegezug-festigkeit β_{bz} bei Riß-erzeugung im Mittel	N/mm²	2,6	2,8	2,7	-,-
Biegezug-festigkeit β_{bz} des in-jizierten Balkens im Alter 7 d im Mittel	N/mm²	2,5	2,7	2,7	-,-

Wie in den geschilderten Versuchen an der Gh Kassel wurden Sho-Bond Bics "BL-Injektor" Klebepacker gesetzt und abgedichtet.

Für die Aufbereitung des Injektionsgutes auf Feinstzementbasis wurde hier ein von der Firma Heidelberger Baustofftechnik GmbH entwickelter hochtourig arbeitender Mischer mit Druckluftantrieb eingesetzt. Für die Förderung des aufbereiteten Injektionsgutes diente eine Doppelplunger-Pumpe der Firma K. Obermann GmbH.

Die Injektion des aufbereiteten Injektionsgutes (Mischdauer 5 Min. nach Zugabe der Feststoffkomponente in die vorgegebene Flüssigkomponente) erfolgte von unten beginnend über die auf den Einfüllstutzen aufgeschraub-

ten Injektoren. Zur augenscheinlichen Prüfung eines eventuellen Materialdurchganges wurden die über der Injektionsebene liegenden Einfüllstutzen zunächst nicht mit den Injektoren versehen. Diese wurden nur dann aufgeschraubt wenn, bedingt durch die Größe der Fehlstellen, ein Materialdurchgang stattfand. In diesem Ablauf wurde kontinuierlich über die gesamte zu injizierende Stützenhöhe vorgegangen.

Die Einfüllstutzen und das Abdichtungsmaterial wurden am darauffolgenden Tag von der Stahloberfläche entfernt. Eine durch die Universität - Gh Kassel durchgeführte Überprüfung der injizierten Bereiche bestätigte die Erfüllung der gestellten Anforderungen.

Kraftwerksgruppe Waldeck

Bei der Kraftwerksgruppe Waldeck der Preussen Elektra AG handelt es sich um Kraftwerke im Bereich der Edertalsperre. Dort zeigten sich im Krafthaus Waldeck 1 im Bereich der Stahlbetonaußenwände wasserführende Risse. Die Wände grenzen unterhalb der Geländeoberkante liegend direkt an die Eder an. Die Wandhöhe beträgt ca. 5,00 m und die Wanddicke i. M. 2,00 m. Die Risse sollten gegen den anstehenden Wasserdruck kraftschlüssig verschlossen werden. Dazu wurde in Abstimmung mit dem Auftraggeber folgende Verfahrensweise gewählt:

Im Anschluß an die Rißaufnahme wurde die Betonoberfläche von Verschmutzungen und Ausblühungen gereinigt. Im Abstand von ca. 25 - 30 cm wurden Sho-Bond-Bics "BL-Injektoren" auf die Risse geklebt und diese verdämmt.

Für die Feinstzementsuspension wurden folgende Kenngrößen gewählt: W/B = 0,85; Auslaufzeit mit dem Marsh Trichter bei Beginn 42 sec.; nach 60 Min. 44 sec.. Es kam die Feinstzementsuspension Microcem B der Heidelberger Baustofftechnik GmbH zum Einsatz. Der Flüssigkeitskomponente aus demineralisiertem Wasser wurden 2 Gew.-% Injektionshilfe IH 3 zugegeben.

Die Komponenten der Suspension wurden mittels eines hochtourigen Mischgerätes (Ultra Turrax mit Dispergierwerkzeug) bei max. 8.000 U/min. aufgeschlossen. Die Mischzeit betrug 8 Minuten einschließlich der Zugabezeiten der Einzelkomponenten. Die Suspension wurde mit einer Doppelplunger-Pumpe DP 36 gefördert und wiederum über die BL Injektoren von unten nach oben mit einem maximalen Druck von 3 bar eingebracht.

An entnommenen Bohrkernen wurde der Verfüllgrad geprüft und eine vollständige Verfüllung festgestellt. Die langzeitige Dichtigkeit der injizierten Risse ist zu erwarten.

22

6 LITERATUR

/1/ Ivanyi, G.; Rosa, W.: Füllen von Rissen und Hohlräumen im Konstruktionsbeton mit Zement-suspension. Beton- u. Stahlbetonbau 87 (1992), H.9, S.224-229.

/2/ ZTV-Riss 88: Zusätzl. techn. Vorschriften und Richtlinien für das Füllen von Rissen in Betonbauteilen. BMV, Verkehrsblattverlag, 1988.

/3/ Rili-SIB: Richtlinie für Schutz und Instand-setzung v. Betonbauteilen. DAfStb, 4 Teile, 1990, 1991. 1992.

/4/ Die Injektion von Microrissen in Stahlbe-ton. Prüfbericht der Amtlichen Baustoff- u. Betonprüfstelle, Universität - Gh Kassel, 1993.

Evaluation von Harzen zur Immobilisierung von sorbierten Markierstoffen

Evaluation of resins to immobilize sorbed tracers

Ch. Bühler
Solexperts AG, Schwerzenbach, Schweiz

J. Dollinger
Geotechnisches Institut AG, Bern, Schweiz

ZUSAMMENFASSUNG: In einer wasserführenden Kluft werden Migrationsexperimente mit radioaktiven Tracern durchgeführt. In einem Folgeprojekt ist geplant, die Kluft mit stärker sorbierenden Tracern zu markieren und anschliessend den gesamten Porenraum der Kluft und wenn möglich einen Teil der umgebenden Gesteinsmatrix mit Harz zu verfüllen, zu exkavieren und im Labor zu untersuchen. Zu diesem Zweck wurden verschiedene Harze untersucht. Injektionsversuche in Bohrkernen mit einer künstlichen Kluft erlaubten, Harze mit unterschiedlichen Eigenschaften zu vergleichen. Speziell entwickelte Epoxidharze mit niederer Viskosität bei tiefen Umgebungstemperaturen von 12 - 13°C liessen sich mit relativ geringem Injektionsdruck (4-10 bar) einpressen. Dabei wurde ein hoher Verfüllungsgrad erreicht. Die Epoxidharze mit Viskositäten von anfänglich weniger als 100 mPa · s erreichen bei Umgebungstemperaturen von 13°C erst nach 3-4 Wochen Zugfestigkeiten, die ein Weiterverarbeiten erlauben (Herstellen von Dünnschliffen, Anschliffen u.a.m.).

ABSTRACT: Migration experiments are presently being conducted in a water bearing shear zone in an underground rock laboratory in Switzerland (GTS). The aim of these experiments is to study the transport of sorbing radionuclides. In the future it is planned to immobilize sorbed tracer by resin injection and to excavate the shear zone to "ground-truth" the observations. Persuant to this, a series of tests were conducted to identify the most suitable resin. A special technique was developed to compare different epoxy resins under identical conditions. Granodioritic drillcores were sawn into 10 cm long sections. A plate of approx. 12 mm thickness was cut out along the axis and was replaced with an artificial fault gouge. Both core halves enclosing the synthetic fracture were shrouded with heat-shrink plastic tubing. These drill cores were then placed in an impregnation device. The test procedure included establishing a continuous stationary flow of water through the synthetic fracture prior to resin injection. A series of experiments were performed using resins of different viscosity. Guided by the results, new resins of lower viscosity were developed. These low viscosity resins (initially less than 100 mPa · s) showed good penetration of the narrow pore space, a high degree of filling and retarded hardening. The only unfavorable characteristic is that of a reduced strength.

1 EINLEITUNG

Im Felslabor Grimsel der Nagra (Nationale Genossenschaft zur Lagerung radioaktiver Abfälle) in den Schweizer Zentralalpen findet gegenwärtig ein Radionuklid-Migrationsexperiment statt. Eine wasserführende Scherzone in granitischem Gestein wurde vom Laborstollen aus mit mehreren Bohrungen durchörtert. Die Intervalle der Scherzone wurden mit hydraulischen Packern isoliert. Die vor den eigentlichen Migrationsversuchen durchgeführten Hydrotests ergaben Transmis-

sivitätswerte für die Scherzone von 2×10^{-6} m^2s^{-1}(Hoehn et al., 1990). Sie besteht aus einem feinen Netz wasserführender Klüfte in relativ undurchlässigem, mylonitisiertem Granit. Zum Teil sind die Klüfte mit einer hochporösen, glimmerreichen Kluftbreccie ("fault gouge") gefüllt. Als ganzes ist dieses Kluftnetzwerk das Resultat jüngster tektonischer Spröddeformationen von älteren, duktilen Scherzonen (Bossart und Mazurek, 1991). Für die Migrationsexperimente wird zwischen jeweils zwei Bohrungen (in wechselnder Konstellation, in Abhängigkeit vom eingesetzten Tracer) ein Dipolfliessfeld erzeugt. In dieses werden verschiedene Tracer über die Injektionsbohrung injiziert und bei der Entnahmebohrung extrahiert und analysiert. Ziel der Experimente ist es, die Wechselwirkung zwischen Tracer und Fels zu beobachten. Die gewonnenen Resultate werden als Inputparameter für Modellierungen verwendet und dienen der Fragestellung, inwieweit Labordaten auf Feldbeobachtungen übertragen werden können. Eine Uebersicht über die bisher durchgeführten Untersuchungen im Migrationsexperiment geben Frick et al. (1992).

Ein Folgeprojekt sieht vor, nach Abschluss des Migrationsexperiments die Kluft mit stärker sorbierenden Tracern zu markieren (z.B. Cs) und anschliessend mit Hilfe von Harzen zu konservieren, um die Tracer unter Bewahrung der Fliesswegstrukturen zu fixieren. Dies ermöglicht es, die Scherzone zu exkavieren, so dass sie im Labor im Mikro- und Makrobereich untersucht werden kann.

Zum Konservieren bieten sich grundsätzlich zwei verschiedene Verfahren an. Die eine besteht im Abkühlen des ganzen Fels- kompartiments und dem Gefrieren des Kluft- wassers, die andere im Verfüllen des Porenraums durch Harzinjektionen.

2 ANFORDERUNGEN AN INJIZIERBARE HARZE

Das Verfüllen der Scherzone, mit dem Ziel ihre Struktur zu konservieren sowie die zuvor applizierten Tracer zu fixieren, stellt spezielle Anforderungen sowohl an ein zu verwendendes Harz als auch an die Injektionstechnik.

- Bei der Injektion soll ein hoher Verfüllungs- grad bei geringen Injektionsdrücken erreicht werden (im Gegensatz zu den üblichen Anwendungen im Tiefbau).

- Die Viskosität des Harzes soll bei tiefen Umgebungstemperaturen möglichst niedrig sein. (Temperatur des zu injizierenden Felskompartiments: 12 - 13°C)

- Volumenänderungen z.B. durch Wasserauf- nahme/-abgabe oder durch Gasentwicklung sind unerwünscht.

- Die Polymerisation und das Aushärten muss ohne Volumenänderung erfolgen, so dass keine strukturelle Veränderung der Kluft eintritt.

In einer ersten Studie (Hufschmied et al. 1991)

wurden verschiedene Harze zusammengestellt, die im Tiefbau für Injektionen verwendet werden. Dabei zeigte sich, dass Polyurethan- harze wegen der bei der Reaktion mit Wasser entstehenden CO_2-Entwicklung für den vorge- sehenen Zweck nicht in Frage kommen. Poly- esterharze wurden nicht in die Untersuchungen einbezogen, da sie eine relativ hohe Viskosität aufweisen, rasch polymerisieren und einen erheblichen Volumenschwund aufweisen. Beide Harzarten lassen zudem keine kraft- schlüssige Verbindung erwarten. Grosse Vorteile von Acrylharzen sind die tiefe Viskosität, die ähnlich derjenigen von Wasser ist, und die Fähigkeit, Porenwasser in ihrer Struktur einzubauen (Hydrophilie). Allerdings bleibt das Wasser nicht im Harz eingebaut, sondern muss zum Beispiel durch Verdunstung wieder weggeführt werden, damit das Harz aushärten kann. Auch bei kleinen Proben dauert es mehrere Tage bis Monate, bis das Wasser aus der Harzstruktur verdunstet ist. Das polymerisierte Acrylharz kann bei Kontakt mit Wasser zusätzlich noch bis 15% (Minimum 5%) in seine Struktur einbauen, was eine Volumenzunahme bewirkt. Ebenfalls wurde eine starke pH-Erniedrigung (pH 2.2) und eine Abgabe von organischen Verbindungen im Wasser beobachtet (Berry und Cowper, 1992). Gegenüber Acrylharzen zeichnen sich Epoxid- harze generell als höher viskos aus und errei- chen ausserdem eine deutlich höhere Festig- keit. Bei den ersten Laboruntersuchungen liessen sich Harze ausscheiden, die das Volumen beim Aushärten änderten oder bei Anwesenheit von Wasser nicht richtig aushärteten (Dollinger et al., 1992). Mit Harzen, die keine Schwundeffekte aufwiesen, wurden letztendlich die Injektionsexperimente in Bohrkernen durchgeführt.

3 INJEKTIONEN IN BOHRKERNE MIT KÜNSTLICHER KLUFT

3.1 Experimenteller Aufbau

Epoxidharze der Firma Sika AG zeigten bei Laboruntersuchungen günstige Eigenschaften. In einer ersten Phase wurden vier Harze mit unterschiedlichen physikalischen Eigenschaf- ten getestet (Harze 1 bis 4, Tabelle 1). Für vergleichbare Resultate ist es unerlässlich, nach einem Verfahren zu suchen, welches die verschiedenen Harze unter identischen Bedin- gungen testen kann. Dazu wählten wir das im folgenden beschriebene Vorgehen.

Da gleichartige Bohrkerne, die eine Scher- zone schneiden, schwierig zu gewinnen sind, werden Granit-Bohrkerne (Ø 72 mm) aus Bohrungen vom Felslabor Grimsel in 10 cm lange Kerne zersägt. Anschliessend ist längs des Kerns eine planparallele Platte von 10 mm Dicke herausgesägt worden. Mit dem Schnittverlust beträgt die entfernte Schichtdicke ca. 12 mm. Eine Scherzone, welche vom Laborstollen des Felslabors angeschnitten wird, lieferte Mylonit und Kluftmaterial (sogenannte "fault gouge") zur Herstellung der künstlichen Kluftbreccie. Das Kluftgestein ist im Labor getrocknet und in mehreren Schritten immer wieder mechanisch zerkleinert und gesiebt worden. Durch Zusammenmischen der Fraktionen lassen sich identische Teilproben mit der typischen Kornverteilung einer realen "fault gouge" herstellen. Das Kluftmaterial wird mit wenig Wasser benetzt und gut vermischt. Mit einem Spatel wird es auf eine liegende Kernhälfte aufgetragen und glattgestrichen, so dass eine Schicht von gleichmässiger Dicke (ca. 12 mm) entsteht. Danach wird die zweite Kernhälfte daraufgelegt und zur Fixierung ein Stück Schrumpfschlauch (ø 76 mm) über den Kern geschoben. Dieses umschliesst nach dem Erwärmen mit einer Gaslötlampe beide Kernhälften satt.

Zur Harzimprägnation werden die Bohrkerne mit der künstlichen Kluft in ein speziell für dieses Experiment entwickeltes Imprägnationsrohr eingespannt (Figur 1). Dieses besteht aus einem Stahlrohr (Länge 160 mm, Innendurchmesser 115 mm), welches mit zwei Deckeln mit Flansch unten und oben verschlossen wird. Im Innern des Rohres wird der Kern beidseits durch Silikongummi-Manschetten fixiert. Vor dem Einbau des Bohrkerns in das Imprägnationsrohr wird der Schrumpfschlauch auf beiden Seiten auf der Höhe der Kernenden abgeschnitten. Die Silikongummi-Manschetten werden durch das Anziehen der Gewindestangen über die beiden Deckel hart an die glatten Endflächen des Kerns angepresst und dichteten diese dabei ab. Um ein Auspressen des Kluftmaterials zu verhindern, wird bei der Ein- und Auslassöffnung in der Gummimanschette je ein Filter montiert (Stahllochplatten und Geotextil).

Die Imprägnation erfolgt mit Hilfe eines Injektionszylinders, da die Epoxidharze nicht direkt mit der verwendeten HPLC-Pumpe injiziert werden können. Im Injektionszylinder unterteilt ein Kolben das beidseitig mit einem Deckel versehene Rohr in zwei Kammern. In die vordere Kammer wird bei zurückgeschobenem Kolben das gemischte Harz eingefüllt. Durch ein vorsichtiges Nachschieben des Kolbens wird die Luft aus der Kammer ausgetrieben, bis sie nur noch mit

Harz gefüllt ist. Danach wird die Kammer hinter dem Kolben mit Wasser gefüllt.

Der Injektionszylinder, das Imprägnationsrohr und die HPLC-Pumpe werden in einem Flowboard eingebaut. Nach dem vollständigen Sättigen des Kerns mit Wasser und dem Einbau des mit Harz gefüllten Injektionszylinders werden die Ventile des Flowboards umgestellt, so dass das von der Pumpe geförderte Wasser in die hintere Kammer des Injektionszylinders gepresst wird. Anstelle von Wasser wird nun Harz anfänglich mit derselben Fliessrate in die künstliche Kluft injiziert (Fliessrate 10 ml/min). Ein angeschlossenes Datenerfassungssystem zeichnet den Druckverlauf während der Wasser- und Harzinjektion kontinuierlich auf. Bei der Aufsättigung der Kluft mit Wasser steigt der Druck in der Injektionsleitung auf ca. 2-4 bar, je nachdem wie dicht die künstliche Kluft gepackt ist. Da Epoxidharze gegenüber Wasser eine wesentlich höhere Viskosität aufweisen, wird bei der Harzinjektion der Maximaldruck jeweils auf 4 - max. 10 bar begrenzt, um Deformationen in der künstlichen Kluft und/ oder ein Ausschwemmen des Kluftmaterials zu verhindern. Um den vorgegebenen Maximaldruck nicht zu überschreiten, muss die Fliessrate periodisch erniedrigt werden (minimal 0.1 ml/min). Bei den meisten Versuchen tritt nach ca. 1 Stunde Harz im Ueberlauf aus.

Nach dem Aushärten des Harzes (Harz 1-4 nach 1 Woche, Harz 5 nach 3 Wochen, Harz 6 nach 4 Wochen bei Umgebungstemperaturen von 13°C) werden die Kerne aufgesägt und sowohl makroskopisch wie auch mikroskopisch untersucht.

3.2 Untersuchte Epoxidharze

Für die ersten Injektionsversuche wurden von der Firma Sika AG, Zürich, drei neue Epoxidharze formuliert. Als Referenzharz diente das Standardharz Sikadur 52, mit welchem bei den ersten Laborversuchen die besten Resultate erzielt worden waren. Sikadur 52 besteht aus dem Epoxidharz Sikadur 281 und dem "long pot life" Härter Sikadur 341. Die Ergebnisse der Versuche mit diesen vier Harzen führte sodann zur Entwicklung der niederviskosen Harze 5 und 6. Ausgewählte physikalische Eigenschaften der Harze 1 - 6 sind in Tabelle 1 zusammengestellt.

TABELLE 1: Physikalische Eigenschaften der untersuchten Harze und ihre Beurteilung aufgrund von Injektionsversuchen

TABLE 1: Physical properties of tested resins and performance evaluation based on injection experiments

Physikalische Eigenschaften

	Referenzharz (Sikadur 52)	Harz 2	Harz 3	Harz 4	Harz 5	Harz 6
Dichte: Harz [g/cm²] Härter	1.125 / 0.987	keine Bestimmung (Mischung ~1.1)	keine Bestimmung (Mischung ~1.1)	keine Bestimmung (Mischung ~1.1)	1.142 / 0.987	1.124 / 0.987
Viskosität [mPa·s] 10 Min	495 (13°C) ~200 (20°C)	~1700 (20°C)	~200 (20°C)	~1000 (20°C)	83 (13°C)	86 (13°C)
30 Min	713 (13°C)				173 (13°C)	236 (13°C)
40 Min	913 (13°C)					
60 Min					452 (13°C)	243 (13°C)
70 Min					781 (13°C)	
90 Min						463 (13°C)
Stress [MPa] 3 Tage	1.4 (13°C) 7.9 (23°C)	keine Bestimmung	keine Bestimmung	keine Bestimmung	2.0 (23°C)	1.5 (23°C)
7 Tage	4.2 (13°C) 18.0 (23°C)				7.5 (23°C)	5.2 (23°C)
14 Tage	8.8 (13°C) 22.3 (23°C)				2.0 (13°C) 12.9 (23°C)	1.4 (13°C) 8.8 (23°C)
21 Tage					4.9 (13°C)	3.4 (13°C)
28 Tage					14.6 (13°C)	7.9 (13°C)

Beurteilung der Harzinjektion aufgrund von aufgesägten Bohrkernen (makroskopisch) und von Dünnschliffen (mikroskopische)

	Referenzharz	Harz 2	Harz 3	Harz 4	Harz 5	Harz 6
Imprägnation der künstl. Kluft bei 13°C	sehr gut	nicht getestet	gut	genügend bis gut	gut bis sehr gut (enthält nicht imprägnierte Stellen in Form von Blasen)	gut (enthält relativ viele n cht imprägnierte Stellen n Form von Blasen)
Imprägnation der kluftnahen Korn-grenz- und trans-granularen Poren	häufig	selten	mittelmässig	mittelmässig	häufig	mittelmässig bis selten
Eindringtiefe des Harzes in das Neben-gestein der künstl. Kluft	maximal 1.4 mm	maximal 0.9 mm	maximal 1.6 mm	maximal 0.8 mm	maximal 6.3 mm durchschnittlich 1.4 - 1.8 mm	maximal 3.0 mm durchschnittlich 1.4 mm
Bemerkungen		auch bei 27°C kein vollständiges Verfüllen der künstl. Kluft	reagiert mit Wasser, was vollständiges Aushärten verzögert	bei 13°C kein vollständiges Verfüllen der künstl. Kluft		

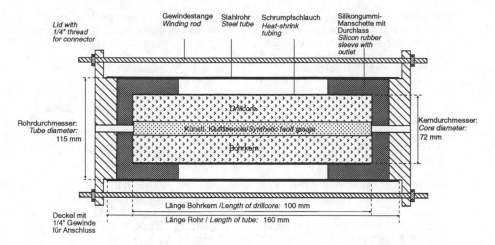

Fig. 1:   Imprägnationsrohr   -   *Impregnation device*

4 ERGEBNISSE

Aus der Fliessrate und der Schichtdicke kann für die künstliche Kluft ein Durchlässigkeitsbeiwert nach dem Gesetz von Darcy berechnet werden. Falls die ganze Querschnittsfläche von 12x72 mm durchströmt wird, entspricht dies bei einer Fliessrate von 10 ml/min einer theoretischen Filtergeschwindigkeit von 1.9×10^{-4} m/s. Mit Hilfe des Druckabfalls von 3 bar über die Länge des Bohrkerns lässt sich ein Durchlässigkeitsbeiwert innerhalb der künstlichen Kluft von mindestens 6.3×10^{-7} m/s errechnen. Da die Injektionsquelle nicht über die ganze Breite der Kluft verteilt war, und somit die durchströmte Querschnittsfläche kleiner ist, kann angenommen werden, dass der effektive Durchlässigkeitsbeiwert höher ist. Die Transmissivität von 7.6×10^{-9} m²/s ist wesentlich tiefer als jene der Migrationskluft (2×10^{-6} m²/s). Dabei gilt es allerdings zu bedenken, dass sich diese beiden Kluftsysteme bezüglich Skala und Homogenität stark unterscheiden und deshalb ein direkter Vergleich nur bedingt gemacht werden darf.
Die Ergebnisse aller Harzinjektionen sind in Tabelle 1 zusammengefasst. In den Experimenten mit den ersten vier Harzen wurde die künstliche Kluft mit dem Referenzharz (Harz 1) am besten verfüllt. Dieses Harz drang auch am tiefsten in die kluftnahen Korngrenz- und transgranularen Poren der umgebenden Gesteinsmatrix des Bohrkerns (Granodiorit) ein. Generell wurde jedoch beobachtet, dass die Viskosität der getesteten Harze 1 - 4 für die Umgebungstemperaturen im Felslabor von

13°C noch zu hoch waren. Bei den höher viskosen Harzen 2 und 4 war ein vollständiges Verfüllen der künstlichen Kluft bei Drucken unter 10 bar nicht möglich. Aus diesem Grund wurden die niederviskosen Harze 5 und 6 entwickelt, welche in einer zweiten Versuchsserie getestet wurden. Harz 6 liess sich sogar bei einem Druck von ca. 4 bar injizieren. Da die physikalischen Eigenschaften wie Viskosität und Dichte von Harz 6 denjenigen von Wasser angeglichen wurden, vermag das Harz das Wasser nicht mehr überall vollständig zu verdrängen. Die künstliche Kluft enthält relativ viele nicht imprägnierte Stellen in Form von Blasen. Am besten scheint sich zur Zeit das Harz 5 zu eignen, mit dem eine relativ gute Imprägnation des Kluftmaterials erreicht wurde und das tief in die kluftnahen Korngrenz- und transgranularen Poren der umgebenden Gesteinsmatrix eindrang. Diese beiden niederviskosen Harze zeigen jedoch eine langsamere und nur beschränkte Festigkeitsentwicklung.

5 SCHLUSSFOLGERUNGEN UND AUSBLICK

Das gewählte Vorgehen, die Harze in Bohrkerne mit künstlicher Kluft zu injizieren, erwies sich als geeignetes Instrument, verschiedene Harze unter identischen Bedingungen zu testen und miteinander zu vergleichen. Wie die vorliegenden Untersuchungen zeigen, ist es möglich, den wassergefüllten Porenraum von geklüftetem Fels zu einem hohen Grad zu verfüllen, auch wenn das Harz mit vergleichsweise niedrigen Drücken injiziert wird. Gewisse Epoxidharze erfüllen die Anforderungen, ohne Volumenveränderung unter Wasser auszuhärten und zugfeste Verbindungen herzustellen. Die Ergebnisse deuten darauf hin, dass mit solchen Harzen das Konservieren einer Kluft machbar ist.

Zur Zeit der Niederschrift dieses Artikels sind weitere Untersuchungen in Vorbereitung oder geplant. Mit Hilfe einer an der Stollenwand über einer Kluft angebrachten Adapterkappe soll nach dem Aufsättigen mit Wasser das niederviskose Harz 5 injiziert werden. Die nach dem Aushärten erbohrten Kerne werden Aufschluss über den Verfüllungsgrad in einer natürlichen Kluft geben, wo im Vergleich zu den beschriebenen Experimenten weit weniger homogene Verhältnisse vorherrschen. In Planung befindet sich zudem ein Experiment, bei welchem ein Bohrkern mit künstlicher Kluft vor der Harzinjektion mit Radionukliden markiert wird. Diese Untersuchung soll Aufschluss darüber geben, ob zwischen sorbiertem Tracer und injiziertem Harz Wechselwirkungen bestehen, die nicht erwünscht sind.

VERDANKUNGEN

Für die Finanzierung dieser Arbeiten danken wir der NAGRA, Wettingen und der PNC, Japan. Im weiteren möchten wir Herr Dr. M. Meyer, SIKA AG, unter dessen Leitung die neuen Harze formuliert wurden sowie den Herren Dr. B. Frieg und Dr. U. Frick, NAGRA, und Dr. P. Bossart, Geotechnisches Institut AG, für die kritische Durchsicht einer ersten Fassung dieses Artikels danken.

LITERATUR

Berry, J.A. and Cowper M.M. 1992. Grimsel test site excavation project: Evaluation of sorption properties of possible resins for the impregnation of the migration shear zone. Nagra Internal Report, Nagra,Wettingen, Switzerland.

Bossart, P. and Mazurek, M. 1991. Grimsel test site: Structural geology and water flowpaths in the migration shear-zone. Nagra Technical Report NTB 91-12, Nagra, Wettingen, Switzerland.

Dollinger, H., Bossart, P., Bühler, Ch. and Mazurek, M. 1992. Grimsel test site excavation project: Resin impregnation experiments in granodiorite borecores. Nagra Internal Report, Nagra, Wettingen,Switzerland.

Frick, U., Alexander W.A., Bayens, B., Bossart P., Bradbury, M.H., Bühler, Ch., Eikenberg, J., Fierz, Th., Herr, W., Hoehn, E., McKinley, I.G., and Smith, P.A. 1992. Grimsel test site: The radionuclide migration experiment - Overview of investigations 1985 - 1990. Nagra Technical Report NTB 91-04, Nagra, Wettingen, Switzerland.

Hoehn, E., Fierz, Th. and Thorne, P. 1990. Grimsel test site: Hydrological characterization of the migration experimental area. Nagra Technical Report NTB 89-15, Nagra, Wettingen, Switzerland.

Hufschmied, P., Benkert, J.-P., Straumann, U., Kottmann, H., Bossart, P. und von Zeerleder, M.E. 1991. Migrationsexperiment im Felslabor Grimsel: Exkavation der Migrationskluft, Machbarkeitsstudie. Bericht z.H. Nagra von Emch und Berger Bern AG.

Grouting in Rock and Concrete, Widmann (ed.) © 1993 Balkema, Rotterdam, ISBN 90 5410 350 7

Prüfprogramm zur Abschätzung der ökologischen Verträglichkeit von Injektionsmitteln zur Schadensbehebung in Kanalisationen

Test program to evaluate the environmental compatibility of grouts to the rehabilitation of sewer channels

Frank von Gersum
Ruhr-Universität Bochum, Deutschland

Bedingt durch zahlreiche umweltbeeinträchtigende Einflüsse auf unseren unmittelbaren Lebensraum findet das Thema Umweltschutz in zunehmendem Maße Beachtung. Bei dieser Thematik darf natürlich der Bau und die Instandhaltung von Kanalisationen nicht ausgeklammert werden. Um das von einer schadhaften Kanalisation ausgehende Umweltgefährdungspotential zu reduzieren, ist in erster Linie deren dauerhafte Dichtheit zu fordern. Im Falle einer Schadensbehebung an undichten Kanälen reicht eine solche alleinige Forderung jedoch keinesfalls aus. Vielmehr muß darüber hinaus sichergestellt sein, daß zugunsten einer dichten Kanalisation auch von der Schadensbehebungsmaßnahme und den dabei zum Einsatz kommenden Materialien keine Umweltgefährdung ausgeht und sich kein erneuter "ungesetzlicher" Zustand einstellt. Nachfolgend wird über ein diesbezügliches Prüfverfahren für Injektionsmittel, die zur Schadensbehebung in Kanälen eingesetzt werden, berichtet und bereits vorhandene Ergebnisse aus einem Forschungsvorhaben vorgestellt.

Die Überlegungen bezüglich des Boden- und Grundwasserschutzes und des damit verbundenen Prüfprogramms sind jedoch keineswegs auf Kanalsanierungsmaßnahmen beschränkt. Sie lassen sich ebenso auf Injektionen zur Bodenverfestigung und -abdichtung übertragen.

Because of many environmental impairments the subject of environmental protection becomes more and more significant. In this subject the construction and renovation of sewer systems has to be included, too. In order to reduce the potential of environmental pollution which is based on damaged sewer systems the durable tightness of sewers is to demand. With regard to sewer renovation this demand has still to be extended. It has to be garanteed that in favour of tight sewers, sewer renovation methods and emploied materials as well do not produce environmental pollutions. In this essay is now introduced a test program for injection materials which are emploied for the purpose to eleminate damages in sewer systems. With this test program it can be estimated if these materials are ecologically friendly. Existing results of a research project are as well introduced.

But this test program is not at all restricted to sewer renovation. A transfer to injection materials emploied for soil conditioning and soil sealing is also possible.

1. ANLASS DER UNTERSUCHUNGEN

1.1 Gesetzliche Anforderungen

Zur Behebung örtlich begrenzter Schäden an Kanalisationen und insbesondere von Undichtigkeiten in Rohrverbindungen haben sich seit langem die Injektionsverfahren etabliert. Jede Injektion, bei welcher der Baugrund und/oder das Grundwasser beeinträchtigt werden kann, unterliegt im Prinzip dem Wasserhaushaltsgesetz (WHG). Es handelt sich hierbei um eine Maßnahme, mit der unter Umständen eine Einwirkung auf das Grundwasser (Schutzgut) verbunden sein kann. Somit ist eine nach den Umständen erforderliche Sorgfalt anzuwenden, um nachteilige Veränderungen der Grundwassereigenschaften zu vermeiden (§ 1a WHG). Da eine solche Injektion gemäß § 3, Abs.1, Punkt 5 WHG ein Einleiten von Stoffen in das Grundwasser ist, wird eine behördliche Erlaubnis oder Bewilligung für den Einsatz dieses Injektionsmittels erforderlich (§ 2, Abs. 1 WHG).

1.2 Genehmigungsmöglichkeiten

In Ermangelung vereinheitlichter Richtlinien erfolgt die Entscheidungsfindung bezüglich einer Bewilligung für den Einsatz eines Injektionsmittels im konkreten Fall in der Regel durch die untere Wasserbehörde auf der Grundlage vorhandener Einzelgutachten. Eine solche Regelung kann jedoch unter Umständen zu regional verschiedenen Erlaubniserteilungen führen, was wiederum zur Verunsicherung der Auftraggeber führt. Viel sinnvoller im Hinblick auf vereinheitlichte Richtlinien und Prüfgrundsätze wäre es somit, eine Bewilligung aufgrund einer vorhandenen ökologischen Verträglichkeit des Injektionsmittels ebenfalls im Rahmen einer bauaufsichtlichen Zulassung zu regeln. Eine Einbettung in das Zulassungsverfahren könnte nach dem in Bild 1 [1] dargestellten Ablaufschema stattfinden. Die Referate III 2 und II 6 des Instituts für Bautechnik (IfBt), Berlin, haben sich diesbezüglich bereits mit der Erstellung von Beurteilungskriterien für die allgemeine bauaufsichtliche Zulassung von Bauteilen, -stoffen und -arten für die Sanierung und Instandsetzung von Abwasserkanälen und -leitungen der Grundstücksentwässerung befaßt, in denen auch die Dichtstoffe aufgenommen worden sind.

Bis jetzt gab es jedoch bezüglich der Abdichtung von undichten Rohrverbindungen mit Injektionsmitteln kein behördlich verbindliches Prüfprogramm. Lediglich innerhalb des Lebensmittel- und Bedarfsgegenständegesetz [2] ist ein genormtes Prüfverfahren für sich in Trinkwasserbereichen befindliche Bedarfsgegenstände wie z.B. Rohre oder Dichtungen angegeben. Da es sich bei einer Abdichtung von Rohrverbindungen unter anderem um eine vor Ort vorgenommene Injektion handelt, bei der das Injektionsmittel unter Umständen sofort mit dem das Rohr umgebenden Boden und Grundwasser in Kontakt kommt, vor Ort aushärtet und auch die Verfahrenstechnik einen Einfluß auf die Prüfergebnisse haben kann, ist dieses Prüfverfahren nicht auf Injektionen zur Rohrverbindungsabdichtung übertragbar.

Aus diesem Grund wurde im Rahmen eines vom Umweltbundesamt, Berlin geförderten Forschungsvorhabens von der Ruhr-Universität Bochum und dem Institut für wassergefährdende Stoffe an der Technischen Universität Berlin unter Beteiligung des Referates II 6 des Instituts für Bautechnik, Berlin, dem Institut für Wasser-, Boden- und Lufthygiene des Bundesgesundheitsamtes, Außenstelle Langen, und dem Hygiene Institut Gelsenkirchen ein Prüfprogramm zur Abschätzung einer Umweltbeeinträchtigung infolge von zur Abdichtung von Rohrverbindungen eingesetzten Injektionsmitteln erarbeitet.

1.3 Bisherige Untersuchungen

Bei den zur Muffenverpressung verwendeten Injektionsmitteln handelt es sich um Kunststofflösungen. Entsprechend den in der Praxis zum Einsatz kommenden zweikomponentigen Injektionsmitteln wurden im Rahmen des Forschungsvorhabens Polyurethane und Acrylate in das Prüfprogramm aufgenommen. Neben einer Stofferfassung müssen jedoch auch mikrobiologische Untersuchungen, Toxizitätstests und Abbauversuche durchgeführt werden. Die diese Materialien betreffenden bisherigen Untersuchungen bezüglich einer Einschätzung der

von ihnen möglicherweise ausgehenden Umweltbeeinträchtigungen bestanden größtenteils aus Eluatanalysen bzw. aus Analysen des Kontaktwassers [2] eines labormäßig hergestellten, ausreagierten Probekörpers. Da jedoch davon auszugehen ist, daß eine mögliche Umweltbeeinträchtigung mehr vom noch nicht ausgehärteten bzw. ausreagierten Injektionsmittel hervorgerufen wird, bestand der Anlaß, erstmalig eine auf diese Phase abgestimmte Methode zur Beurteilung der ökologischen Verträglichkeit und eine dazugehörige Prüfeinrichtung zu konzipieren. Aufgrund der Beeinflussung dieser Phase durch die Verpreßtechnik bzw. den Mischvorgang der beiden Komponenten muß die Prüfeinrichtung auch den Einsatz des zum jeweiligen Injektionsmittel zugehörigen Verpreßgerätes ermöglichen.

2. UNTERSUCHTE SYSTEME

2.1 Injektionstechnik

Für die Abdichtung von Rohrverbindungen im nichtbegehbaren Querschnittsbereich kommt es in der Regel zum Einsatz von Packersystemen (Bild 2) deren Grundprinzip das Penetryn-Verfahren ist. Mit dieser Verpreßtechnik läßt sich neben der Injektionsmöglichkeit eine sofortige Kontrolle der Muffenabdichtung in Form einer Dichtheitsprüfung durchführen. Diesen Tatsachen entsprechend wurden im Rahmen des Forschungsvorhabens Injektionsmittel ausgewählt, die mittels einer solchen Packertechnik verpreßt werden (Tab. 1). Die Prüfeinrichtung wurde aus den bereits erwähnten Gründen in Art und Größe für die Aufnahme eines Packers ausgelegt.

2.2 Injektionssysteme

Im Rahmen dieses Forschungsvorhabens wurden nur Injektionssysteme zur Muffenverpressung auf der Basis von Kunststofflösungen untersucht. Von den möglichen wasserlöslichen Systemen (Acryl- und Phenol-Harze, Aminoplaste) wurden 3 verschiedene Acrylharzprodukte und von den nicht wasserlöslichen Systemen (Epoxid- und Organomineral-Harze, Polyurethane) 4 verschiedene Polyurethane in das Untersuchungsprogramm aufgenommen. Es handelt sich bei allen untersuchten Systemen um 2-Komponenten-Gemische (Komponente 1 und 2 in Tabelle 2) in flüssiger Form.

3. PRÜFUNG

3.1 Versuchsaufbau

Die Integration des gesamten Reaktionszeitraumes eines Injektionsmittels in das Prüfprogramm sowie der Einfluß der Mischtechnik auf diesen Zeitraum und somit auf die Prüfergebnisse schließen kleinformatige Laborversuche aus. Im Rahmen des Vorhabens kam daher erstmalig eine Apparatur (Bild 3) zum Einsatz, die eine Muffenverpressung im Maßstab 1:1 unter insitu ähnlichen Bedingungen im Labor ermöglicht. Die Anlage ist ausgelegt für Kreisquerschnitte der dominierenden Nennweiten ≤ DN 300. Ferner verfügt sie über Möglichkeiten, den das Rohr umgebenden Sand mit Wasser zu durchströmen, wodurch Grundwasserströmungen simuliert werden können.

Um die große Bandbreite der unterschiedlichsten Randbedingungen bzw. Situation der Praxis mit einem Versuchsauf- bzw. -einbau abzudecken, war die Schaffung eines "worst case" erforderlich. Dieser wird durch Faktoren, die den Stoffeintrag ins Grundwasser fördern sowie einen möglichst großen Injektionsmittelverbrauch bewirken, geprägt. Im einzelnen sind das:

- Rohrwerkstoff
- Sieblinie des eingebauten Sandes
- Schadensart und -ausmaß
- Fließgeschwindigkeit des den Sand durchströmenden Wassers.

Aufgrund der glatten Außenwandbeschaffenheit fiel die Wahl auf ein Steinzeugrohr. Aus praktischen Überlegungen beträgt die Nennweite DN 300. Die Körnungslinie (0,2 ÷ 2,0 mm) (Bild 4) des eingebauten gewaschenen Sandes mit einer Porenzahl n=0,35 entspricht einer Prüfkörnungslinie nach DIN 4093 [3]. Die Betrachtung möglicher Schadensbilder im Muffenbereich [4],[5] sowie die Analyse verschiedener Schadens- bzw. Inspektionsberichte zeigten, daß durch einen fehlenden Dichtring sowie durch eine gleichzeitige Muffenspaltvergrößerung auf 2 cm (planmäßiges Mindestmaß nach [6]: 5mm) ein ausreichend gravierender Schaden in Hinblick auf einen möglichst hohen Injektionsmittelverbrauch dargestellt wird. In Anpassung an reale Fließgeschwindigkeiten des Grundwassers wurde die Strömungsgeschwindigkeit im Rahmen des Prüfprogrammes über den gesamten Prüfzeitraum auf 0,75 m/Tag festgelegt. Als Durchströmungswasser diente das Bochumer Leitungswasser.

Während des gesamten Prüfzeitraumes von 12 Tagen erfolgte die permanente Durchströmung des das Rohr umgebenden Sandes (kein Strömungskreislauf). Die Prüfungen auf Umweltbeeinträchtigungen erfolgten am Durchströmungswasser, wobei die Beprobungsintervalle mit zunehmendem zeitlichen Abstand zum Injektionszeitpunkt größer wurden.

3.2 Tests und Analysen

In Zusammenarbeit mit dem Institut für Bautechnik, Berlin, dem Institut für Wasser-, Boden- und Lufthygiene des Bundesgesundheitsamtes, Außenstelle Langen, und dem Hygieneinstitut, Gelsenkirchen wurde das folgende Prüfprogramm zur Erfassung der Umweltbeeinträchtigung durch Injektionsmittel entwickelt.

3.2.1 Zusammenstellung der erforderlichen Tests

A) Stofferfassung

1) Analytische und mengenmäßige Erfassung der Ausgangsstoffe

2) Analytische und mengenmäßige Erfassung der Reaktionsprodukte sowie auch der Nebenprodukte, sofern Hinweise auf deren Entstehung vorliegen.

3) Summenparameter TOC (gesamter organisch gebundener Kohlenstoff)

B) Wirkung und Abbau

1) Mikrobiologische- bzw. Toxitätstests
   - Koloniezahl
   - Bakterientoxizität bezüglich Escherichia coli und Pseudomonas aeruginosa
   - Leuchtbakterien-Test
   - TTC-Test

2) Biologischer Abbau

3) Mutagenitätstest (Ames-Test)

C) Physikalisch-chemische Parameter
   ph-Wert, Temperatur, Redoxopotential, elektrische Leitfähigkeit (Die beiden letzten Parameter sind nur zu messen, falls die Gelbildung bei einem stark veränderten pH-Wert stattfindet)

3.2.2 Kurzbeschreibung der Tests

TOC (engl. Total Organic Carbon)

Hierbei handelt es sich um eine Maßzahl zur Angabe des Anteils organischen Kohlenstoffs von Wasserinhaltsstoffen, der sowohl in gelösten Verbindungen als auch in festen, kohlenstoffhaltigen Schwebeteilchen gebunden ist. Seine Messung erfolgt durch katalytische Verbrennung zu CO_2 bei 900°-1000°C, das dann quantitativ durch Infrarotmessung bestimmt wird. Er ist ein Summenparameter, und es können daraus keine Schlüsse auf einzelne kohlenstoffhaltige organische Verbindungen und ihre Konzentrationen gezogen werden. Er wird in mg/l angegeben.

Koloniezahlbestimmung, Bakterientoxizität gegenüber Escherichia coli und Pseudomonas aeruginosa

Diese Versuche sollen die Wirkung des Durchströmungswassers auf die Mikroorganismen erfassen, die neutral, fördernd, hemmend oder toxisch sein kann. Diese Tests lehnen sich an die Trinkwasserverordnung (TrinkwV) an. Für die Koloniezahlbestimmung wurde Oberflächenwasser und für die Bakterienansätze Reinkulturen von E. coli ATCC 11229 und P. aeruginosa ATCC 15442. verwendet. Eine toxische Wirkung soll durch das Ausbleiben der Koloniebildung angezeigt werden und eine positive oder negative Wirkung des Eluats durch Abweichung des Testergebnisses des Versuchsansatzes um mindestens eine Zehnerpotenz im Vergleich zum Kontrollansatz.

Leuchtbakterien-Test

Durch toxische Stoffe wird die Biolumineszenz der Leuchtbakterien gehemmt (das Leuchten beruht auf der Oxidation spezifischer Leuchtstoffe unter der katalytischen Wirkung des Enzyms Luziferase). Die Hemmung (in %) der Biolumineszenz einer Konzentrationsreihe des zu prüfenden Stoffes wird gemessen.

TTC-Test

Im Zusammenhang mit dem biologischen Abbau werden Bakterien aus Belebtschlamm mit dem Farbstoff TTC (Triphenyltetrazoliumchlorid) inkubiert. TTC wirkt dabei als Reduktionsindikator, da er durch Reduktion vom farblosen Zustand in das tiefrote, gegen Luftsauerstoff beständige Formazan übergeht. Diese Reaktion weist auf eine verminderte Atmungsaktivität (toxische Wirkung) hin.

Biologischer Abbau

Mikroorganismen von Belebtschlamm werden mit dem Durchströmungswasser zusammengebracht und der Abbau des Stoffes über das Verhältnis von CSB zu BSB (chemischer Sauerstoffbedarf zu biologischem Sauerstoffbedarf) bestimmt. Ein 100%iger Abbau entspricht einem CSB (in mg O_2/l) zu BSB (in mg O_2/l) von 1:1.

Ames-Test

Da bei einigen Bausteinen von Polyurethanen mutagene Potentiale (insbesondere nach einer metabolischen Aktivierung) nachgewiesen wurden, scheint es angezeigt, eine In-vitro-Mutagenitätsprüfung mit Hilfe des sog. Ames-Tests durchzuführen.

Die erbgutverändernden Wirkungen, d.h. die Mutagenität eines Stoffes, kann durch den Nachweis von Wechselwirkungen des zu prüfenden Stoffes mit der DNS (Desoxyribonucleinsäure, englisch Desoxyribonucleinacid DNA) getestet werden. Die Nucleinsäuren treten in allen Zellen des Tier- und Pflanzenreiches auf. Sie regeln den Aufbau der Proteine, die für das Leben und die Funktion jeder Zelle notwendig sind. Die Baupläne für die Proteinsynthese sind in bestimmten Abschnitten der Chromosomen, den Genen, enthalten, die aus langen Ketten von Desoxyribonucleinsäuren bestehen. Für die Bestimmung der Punktmutationen, d.h. kleinsten mikroskopisch nicht sichtbaren Veränderungen im molekularen Aufbau der DNS, haben sich Bakterien als nützlich erwiesen. Beim sogenannten Ames-Test sind es Bakterien, die von der Aminosäure Histidin abhängig sind. Untersucht wird nun, ob auf histidinarmen Nährböden die Bakterien unter Einwirkung des zu prüfenden Stoffes so mutieren, daß sie auf dieser Nährlösung wachsen.

Da der Nachweis einer Wechselwirkung eines Stoffes mit der DNS als ursprünglich für die Krebsentstehung angesehen wird, wird ein positiver Ames-Test als Hinweis auf eine mögliche krebserzeugende Wirkung gewertet.

3.3 Ergebnisse

Im Rahmen des Forschungsvorhabens wurden je Injektionsmittel drei Muffenverpressungen durchgeführt. Das komplette Prüfprogramm, das die TOC- und pH-Wert-Ermittlung, mikrobiologische Untersuchungen, Toxizitäts-, Abbau- und Ames-Tests sowie Analysen auf Ausgangsstoffe umfaßt, wurde an einer Injektionsmaßnahme je Injektionsmittel durchgeführt. Bei den beiden restlichen Injektionsmaßnahmen (Kontrollinjektionen) wurden nur TOC- und pH-Werte bestimmt. Im Hinblick auf die einzusetzende Analytik zur Bestimmung der im Eluat (Durchströmungswasser) befindlichen Ausgangsstoffe des Injektionsmittels, konnten diese Analysen auch vom Hersteller durchgeführt werden.

Der Prüfzeitraum von 12 Tagen erwies sich als sinnvoll. Innerhalb dieses Zeitraums war in allen Fällen eine Anstieg-, Maximum- und Abklingphase der Kohlenstoffkonzentrationen im Prüfwasser zu beobachten (Bild 5). Nach Abschluß des Prüfzeitraums wurden annähernd die Kohlenstoffkonzentrationen im Prüfwasser erreicht, die vor der Injektion vorherrschten. Das TOC-Maximum ist in jedem Fall innerhalb des Zeitraumes eines kompletten Wasserumsatzes zu erwarten, wodurch die Notwendigkeit, den Injektionsvorgang, d. h. die flüssige Injektionsmittelphase, in das Prüfprogramm einzubeziehen, bestätigt wird.

Die maximalen absoluten TOC-Mittelwerte (Tab. 3) aus den drei Meßreihen je Injektionsmittel reichten von 21 ppm bis 277 ppm. Die Bandbreite der normierten, d. h. auf den Liter Injektionsmittel bezogenen, maximalen TOC-Mittelwerte erstreckten sich von 3 ppm/l bis 63 ppm/l. Mit jeder Injektion war ein Absinken des pH-Wertes in den sauren Bereich verbunden (Bild 6). Die pH-Mittelwerte aus den drei Meßreihen je Injektionsmittel sanken unmittelbar nach der Injektionsdurchführung auf Werte zwischen 4,8 und 6. Bei allen Rezepturen war jedoch bereits innerhalb des Prüfzeitraumes ein Anstieg auf einen pH-Wert um 7, dem Wert vor der Muffenverpressung, festzustellen.

Die mikrobiologischen Untersuchungen, die die Koloniezahlbestimmung sowie die Wirkung auf die

Bakterien Escherichia coli und Pseudomonas aeruginosa beinhalten, hatten sowohl zum Zeitpunkt des TOC-Wert-Maximums als auch in der Abklingphase des TOC-Wertes für alle untersuchten Injektionsmittel eine neutrale Wirkung. Die Ergebnisse der In-vitro-Mutagenitätsprüfung (Ames-Test) zeigten bei allen analysierten Rezepturen kein mutagenes Potential.

Im Rahmen des TTC-Tests wurde für das Produkt PUR A der Bayer AG sowohl zum Zeitpunkt des TOC-Maximums als auch in der TOC-Abklingphase keine Toxizität festgestellt. Im Falle des Produktes PUR B zeigte sich beim TTC-Test in der TOC-Maximumphase eine Restaktivität der Bakterien von 85,0 % gegenüber dem Kontrollansatz. In der TOC-Abklingphase wurde wie beim Produkt PUR A keine Toxizität bezüglich des TTC-Tests festgestellt. Bezüglich der Bayer - Rezeptur A zeigte sich eine biologische Abbaubarkeit von 97 % (TOC-Maximumphase) bzw. 97,8 % (TOC-Abklingphase) im Verhältnis zum CSB. Der maximale Abbau wurde nach 20 (TOC-Maximumphase) bzw. nach 8 Tagen (TOC-Abklingphase) erreicht. Bei der Rezeptur B der Firma Bayer AG betrug die biologische Abbaubarkeit in der TOC-Maximumphase nach 7 Tagen 97,5 % und in der TOC-Abklingphase nach 7 Tagen 95 %. Für die unverdünnten Wasserproben beim Produkt PUR A ließ sich eine Hemmung der Biolumineszens im Leuchtbakterientest von weniger als 40 % berechnen. Beim Produkt PUR B betrug die Hemmung der Biolumineszens ca. 30 % (TOC - Maximum) bzw. 10 % (TOC-Abklingphase).

In dem von der Firma Bayer AG auf Ausgangsstoffe analysierten Durchströmungswasser konnten keine aromatischen Amine nachgewiesen werden. Bei den in geringen Mengen im Eluat gefundenen Rezepturbestandteilen handelt es sich nach Aussage der Firma Bayer AG um Verbindungen, die untoxisch und nach wenigen Tagen biologisch abbaubar sind.

Für das Produkt Scotch 5610 der Firma 3 M betrug die Abbaubarkeit 96,6 % nach 15 Tagen (TOC-Maximum) bzw. 95 % nach 9 Tagen (TOC-Abklingphase). Hierbei zeigte sich im TTC-Test zum Zeitpunkt des TOC-Maximums eine Restaktivität der Bakterien von 81 % gegenüber dem Kontrollansatz. In der TOC-Abklingphase betrug die Restaktivität 83,7 %. Während die Wasserprobe aus der TOC-Abklingphase nur geringen Einfluß auf die Leuchtintensität ausübte, wies die unverdünnte Probe aus der TOC Maximumphase eine Hemmung der Biolumineszens von 40 % auf.

Im Hinblick auf die Analyse bezüglich der Ausgangsstoffe im Durchströmungswasser konnte Aceton sowohl in der TOC-Maximumphase als auch in der TOC-Abklingphase festgestellt werden.

Die Abbaubarkeit beim Produkt Scotch 5614 der Firma 3 M betrug 30,8 % nach 16 Tagen (TOC-Maximum) bzw. 98 % nach 16 Tagen (TOC-Abklingphase). Im TTC Test wurde in der TOC-Abklingphase keine Toxizität festgestellt. Im TOC-Maximum betrug die Restaktivität der Bakterien im TTC-Test 75,8%. Aufgrund einer fehlenden Analytik konnte in diesem Fall keine Analyse auf Ausgangsstoffe durchgeführt werden. Für die unverdünnte Wasserprobe der TOC-Maximumphase ließ sich eine 70 %ige Hemmung der Biolumineszens berechnen. Der entsprechende Wert für die TOC-Abklingphase betrug 20 %.

Bezüglich des Produktes AC 400 der Firma Geochemical Corporation wurde nach 10 Tagen eine 97,1 %ige biologische Abbaubarkeit für die Inhaltsstoffe des Durchströmungswassers in der TOC-Maximumphase gemessen. In der TOC-Abklingphase betrug die biologische Abbaubarkeit nach 20 Tagen 83 %. Zum Zeitpunkt des TOC-Maximums zeigte sich im TTC-Test eine Restaktivität der Bakterien von 8,8 % gegenüber dem Kontrollansatz. In der TOC-Abklingphase war die Restaktivität mit 76,9 % wesentlich höher. Es konnten keine organischen Ausgangsstoffe nachgewiesen werden. Die an der Reaktion nicht teilnehmenden anorganischen Persulfate konnten erfaßt werden. Die Hemmung der Biolumineszenz für die unverdünnte Probe der TOC-Maximumphase betrug annähernd 100 %. In der TOC-Abklingphase war die Leuchtintensität deutlich erhöht (ca.30 %).

Im Falle des Produktes ECO 1 der Firma Rhone-Poulenc betrug die biologische Abbaubarkeit nach 15 Tagen 21,3 % (TOC-Maximumphase) bzw. nach 12 Tagen 40 % (TOC-Abklingphase). Die im TTC-Test ermittelte Restaktivität der Bakterien betrug 21,5 % (TOC-Maximum) bzw. 42,9 % (TOC-Abklingphase) gegenüber dem Kontrollansatz. Es konnten keine organischen Ausgangsstoffe nachgewiesen werden. Die an der Reaktion nicht teilnehmenden anorganischen Persulfate konnten erfaßt werden. Im Leuchtbakterientest zeigte sich bei der Probe aus der TOC-Maximumsphase eine Hemmung von über 90 %. In der TOC-Abklingphase wurde eine Hemmung von 40 % errechnet.

Bezüglich des Produktes PLEX 6803-O der Firma Röhm betrug die Abbaubarkeit 33,3 % nach 28 Tagen (TOC-Maximum) bzw. 13 % nach 8 Tagen (TOC-Abklingphase). Im TTC-Test ergaben sich Restaktivitäten der Bakterien von 46,4 % (TOC-Maximum) und 61,6 % (TOC-Abklingphase). Die Analysen auf Ausgangsstoffe ergaben, daß der TOC-Wert des Durchströmungswassers im wesentlichen auf die eingesetzten Monomere zurückzuführen ist. Die berechneten Hemmungen der unverdünnten Proben im Leuchtbakterientest betrugen ca. 70 % (TOC-Maximum) bzw. ca. 50 % (TOC-Abklingphase).

4. ERGEBNISINTERPRETATION UND AUSBLICKE

Es muß betont werden, daß es sich bei der folgenden Diskussion der Ergebnisse nicht um eine gutachterliche Stellungnahme handelt und daß für die bauaufsichtliche Zulassung und die dazu erforderliche Beurteilung der verschiedenen Injektionsmittel allein das Institut für Bautechnik in Berlin zuständig ist.

Der TOC-Wert allein erlaubt keine Aussage über die eventuelle Umweltbeeinträchtigung der angewendeten Injektionsmittel. Es ist also z.B. nicht sinnvoll, eine obere noch duldbare Grenze allein für den TOC-Wert festzulegen. Entscheidend dabei ist, wie sich diese zweifelsfreie Belastung des Bodens mit organischen Stoffen, dessen Maß der TOC-Wert ist, im Boden auswirkt.

Bei allen untersuchten Injektionsmitteln ist keine Mutagenität im Rahmen des Ames Tests festgestellt worden. Die mikrobiologischen Untersuchungen bzw. Toxizitätstests durch Koloniezahlbestimmung zeigten in allen Fällen eine neutrale Wirkung bzw. keine Toxizität. Bei PUR B, Scotch 5610 und Scotch 5614 wurde eine geringe und bei Rocagil ECO 1 und PLEX 6803-O eine stärkere Aktivitätshemmung im TTC-Test festgestellt. Die biologische Abbaubarkeit war im Falle von PUR A, PUR B, Scotch 5610 sowie von AC 400 in der TOC-Maximumphase wie in der TOC-Abklingphase als leicht einzustufen. Im Falle von Scotch 5614 wiesen die Inhaltsstoffe des Umströmungswassers eine schwerere Abbaubarkeit im TOC-Maximum, aber eine leichte Abbaubarkeit in der TOC - Abklingphase auf. Im Falle von Rocagil ECO 1 und PLEX 6803-O waren die Inhaltsstoffe des Durchströmungswassers schwerer abbaubar.

In Anbetracht der durch die schadhafte Kanalisation andauernden Belastung des Bodens und Grundwassers mit einer Vielzahl von Stoffen mit zum Teil hohem Schadenspotential muß nach der ökologisch relativ besten Lösung für die Instandsetzung von Kanalisationen gesucht werden. Dabei sollte jedoch auch darüber diskutiert werden, ob die offensichtlich unvermeidliche, aber geringe Belastung des Bodens - ausgelöst durch die Injektionsmittel - mit nichttoxischen, im allgemeinen leicht abbaubaren Stoffen in Kauf genommen werden kann.

Aufgrund der erarbeiteten Ergebnisse bescheinigte das Institut für Bautechnik, Berlin bereits einigen Injektionsmitteln die ökologische Verträglichkeit. Näheres hierzu und zu weiteren Untersuchungen, die nicht im Rahmen des erwähnten Forschungsvorhabens vorgenommen wurden, kann beim Institut für Bautechnik, Referat II 6, in Berlin erfragt werden.

Das dargestellte Untersuchungsprogramm ist jedoch keineswegs auf den Bereich Kanalsanierung beschränkt. Vielmehr ist eine Übertragung auf andere Injektionsmaßnahmen zur Bodenverfestigung und Abdichtung möglich. Ein hierzu erforderlicher Versuchsaufbau (Bild 7), in den ebenfalls die Verpreßtechnik integriert werden kann, steht bereits zur Verfügung

LITERATUR

[1] Grunder, H. Th.: Anforderungen an Dichtungsmaterialien aus der Sicht des Boden- und Grundwasserschutzes. Boden-/Grund-wasser - Forum Berlin 1988. IWS-Schriftenreihe des Instituts für wassergefährdende Stoffe an der TU Berlin - Band 5.

[2] Gesundheitliche Beurteilung von Kunststoffen und anderen nichtmetallischen Werkstoffen im Rahmen des Lebensmittel- und Bedarfsgegenständegesetz für den Trinkwasserbereich. Bundesgesundheitsblatt, 20. Jahrgang, 1977.

[3] DIN 4093 (09.87): Einpressen in den Untergrund - Planung, Ausführung, Prüfung

[4] ATV-Merkblatt M 143, Teil 1 (12.89): Inspektion, Instandsetzung, Sanierung und Erneuerung von Abwasserkanälen und -leitungen Grundlagen

[5] ATV-Merkblatt M 143, Teil 2 (06.91): Inspektion, Instandsetzung, Sanierung und Erneuerung von Abwasserkanälen und -leitungen - Optische Inspektion.

[6] ATV-Arbeitsblatt A 139 (11.88): Richtlinie für die Herstellung von Entwässerungskanälen und -leitungen.

Anschrift des Referenten:

Dipl.-Ing. Frank von Gersum
Ruhr-Universität Bochum
D-44780 Bochum
Tel.: 0234-700 6102

oder:

Höchstener Straße 26
D-44267 Dortmund
Tel.: 0231-48 47 06

1) Eine ausführlichere Darstellung der Ergebnisse kann dem Forschungsbericht 92-102 04 504 des Umweltbundesamtes, Berlin - **Gersum, F. von; Grunder, H. Th.; Lühr, H.-P.; Stein, D.: ENTWICKLUNG UND ERPROBUNG UMWELTFREUNDLICHER INJEKTIONSMITTEL UND -VERFAHREN ZUR BEHEBUNG ÖRTLICH BEGRENZTER SCHÄDEN UND UNDICHTIGKEITEN IN KANALISATIONEN UNTER BERÜCKSICHTIGUNG DES GEWÄSSERSCHUTZES** - entnommen werden.

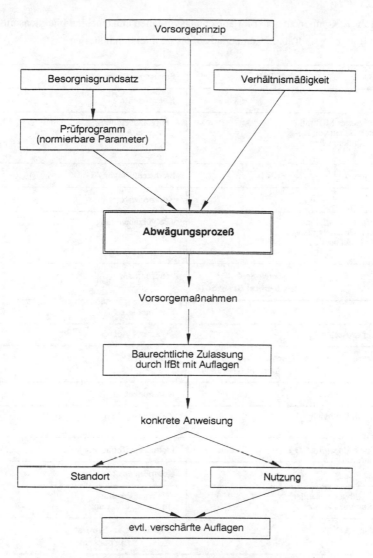

Bild 1: Flußdiagramm eines möglichen Prüfverfahrens

Fig. 1: Flow chart of a possible test program

Tab. 2: Komponenten der im Forschungsvorhaben untersuchten Injektionsmittel

Tab. 2: Compounds of tested injektion materials

Name des Produktes:	PUR A	Produzent: **Bayer AG**
Komponente 1		**Komponente 2**
- Flüssiges, modifiziertes MDI mit einem bestimmten Gehalt von Isomeren und höher funktionellen Homologen		- Polyolzubereitung Variante A

Name des Produktes:	PUR B	Produzent: **Bayer AG**
Komponente 1		**Komponente 2**
- Flüssiges, modifiziertes MDI mit einem bestimmten Gehalt von Isomeren und höher funktionellen Homologen		- Polyolzubereitung Variante B

Name des Produktes:	**Scotch-Seal Chemical Grout 5610**	Produzent: **3M**
Komponente 1		**Komponente 2**
- PUR-Prepolymer in Aceton		- Wasser + Additiv

Name des Produktes:	**Scotch-Seal Chemical Grout 5614**	Produzent: **3M**
Komponente 1		**Komponente 2**
- PUR-Prepolymer in Diethylenglykolethyletheracetat		- Wasser + Additiv

Name des Produktes: **Rocagil ECO 1**		Produzent: **Rhône-Poulenc**
Komponente 1		**Komponente 2**
- Methoxytriethoxyethylmethacrylat - Polyethylenglykoldimethacrylat		- Natriumpersulfat in wässriger Lösung

Name des Produktes: **AC 400**		Produzent: **Geochemical Corporation**
Komponente 1		**Komponente 2**
- Magnesiumacrylat in wässriger Lösung - Triethanolamin - Aethylenglykol		- Ammoniumpersulfat in wässriger Lösung

Name des Produktes: **PLEX 6803-O**		Produzent: **Röhm GmbH**
Komponente 1		**Komponente 2**
- 2 Hydroxyethylmethacrylat und basisches Methacrylsäurederivat		- Natriumpersulfat in wässriger Lösung

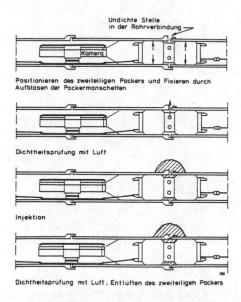

Undichte Stelle
in der Rohrverbindung

Kamera

Positionieren des zweiteiligen Packers und Fixieren durch
Aufblasen der Packermanschetten

Dichtheitsprüfung mit Luft

Injektion

Dichtheitsprüfung mit Luft ; Entlüften des zweiteiligen Packers

Bild 2: Arbeitsablauf einer Injektion mittels Packer

Fig. 2: Prinzip of an injection by packer

Tab. 1: Im Bochumer Forschungsvorhaben eingesetzte Injektionsmittel und teilnehmende Verpreßfirmen

Tab. 1: Tested injection materials in the research project of Bochum

INJEKTIONSMITTEL	INJEKTION DURCH
PUR A	Bayer AG
PUR B	Bayer AG
Scotch 5610	Buchen GmbH
Scotch 5614	Kanal Müller Gruppe
AC 400	Kanal Müller Gruppe
Rocagil ECO 1	Kanal Müller Gruppe
PLEX 6803-O	MS-Hohlraum TV

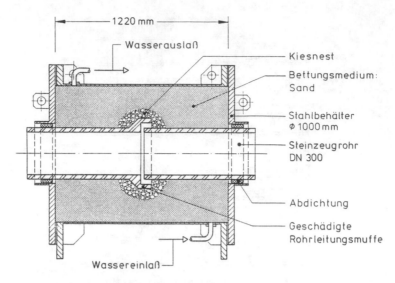

Bild 3: Prüfeinrichtung für Injektionen zur Abdichtung von Rohrverbindungen

Fig. 3: Test stand for materials which are emploied for pipe joint sealing

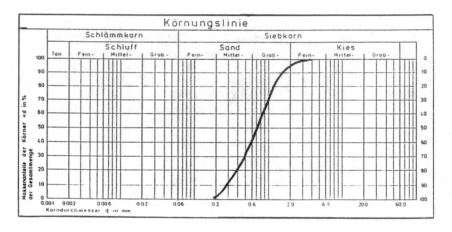

Bild 4: Körnungslinie des verwendeten Sandes

Fig. 4: Grain size distribution in the test program

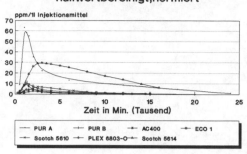

TOC-MITTELWERTE
nullwertbereinigt,normiert

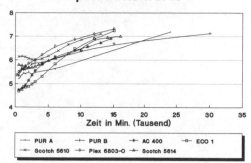

pH-Mittelwerte

Bild 5: Mittlere, normierte TOC-Verläufe

Fig. 5: Diagram of averaged TOC-values

Bild 6: Mittlere pH-Wertverläufe

Fig. 6: Diagram of averaged pH-values

Tab. 3: Maximale TOC-Werte

Tab. 3: Maximum TOC-values

TOC - Mittelwerte / gemittelt aus 3 Injektionen		
Produkt	Max. TOC - Mittelwert absolut, nullwertbereinigt [ppm]	Max. TOC - Mittelwert nullwertbereinigt, normiert [ppm /l]
PUR A	216	63,0
PUR B	21	4,6
Scotch 5610	71	7,0
Scotch 5614	277	30,0
AC 400	70	3,0
Reoagil ECO 1	277	11,5
PLEX 6803-O	150	10,8

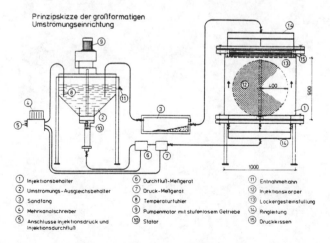

Bild 7: Prüfeinrichtung für Injektion zur Bodenverfestigung und -abdichtung

Fig. 7: Test stand for injections which are emploied for soil conditioning/soil sealing

Research in cement grout and the mechanical properties of the set grout

Forschungen betreffend Zementsuspensionen und deren mechanische Eigenschaften in erhärtetem Zustand

Guo Yuhua
Information Research Institute of Water Resources & Electric Power, People's Republic of China

ABSTRACT: Cement grouting is a very important measure in foundation treatment for large dems and other buildings. The mechanical properties of the cement grout and the grout set play an important role in the efficiency of the cement grouting. Therefore the research in this topic is very important for practice. Laboratory research in the influence of the factors of water: cement (w/c) ratio, mixing time and the action of pressure filtration of the cement grout on the physical-mechanical properties of the grout and its set is summarized and analyzed in this paper. The conclusions are drawn as follows.
A. The apparent viscosity of the cement grout remarkably decreases with the increase of the w/c ratio only if the ratio is less than 1.5. The apparent viscosity slightly decreases with the increase of the w/c ratio, but mechanical properties of the set grout decreases very much, when the ratio is more than 1.5. It is further proved that the trend of using thick cement grout in recent years in the world is correct.
B. Grout mixing time has very important impact on the mechanical properties of the set grout. It is very important to control the mixing time properly for the cement grouting.
C. The compressive strength and impermeability of the set grout are remarkably improved under the action of pressure filtration in cement grouting, e.g. the compressive strength of the set grout after 28-day cure will reach 10,000 N/cm² if it is formed under the pressure of 70 N/cm² with filtration. It is proved that high pressure cement grouting should be adopted if possible and we must block out the major passage in the foundation in advance to prevent the grout leakage, providing a pressure filtration condition for the formation of the set grout. Then the cement grouting will start. And merely in this way can we ensure a good quality of grouting.

ZUSAMMENFASSUNG: Zementinjektionen sind eine sehr wichtige Maßnahme zur Behandlung des Untergrundes großer Talsperren und anderer Bauwerke. Die mechanischen Eigenschaften der erhärteten Zement-Mischung spielen eine bedeutende Rolle bei der Wirksamkeit von Zement-Injektionen. Daher ist die Forschung auf diesem Gebiet für die Praxis sehr wertvoll. In diesem Beitrag werden Forschungsarbeiten im Labor über den Einfluß des Wasser-Zementfaktors, der Mischzeit, der Wirkung des Druckes bei der Filtration der Zement-Mischung auf die physikalischen und mechanischen Eigenschaften der flüssigen und erhärteten Mischung zusammengefaßt und ausgewertet. Daraus können folgende Schlußfolgerungen gezogen werden:
   A. Die scheinbare Viskosität der Zement-Mischung sinkt wesentlich mit steigendem Wasser-Zementwert, wenn dieser kleiner als 1.5 ist. Bei Wasser-Zementwerten über 1.5 steigt die scheinbare Viskosität nur mehr wenig, aber die mechanischen Eigenschaften verringern sich sehr stark. Damit wurde die Richtigkeit der neueren, weltweiten Tendenz zur Verwendung dickerer Zementmischungen bestätigt.
   B. Die Mischzeit hat einen wesentlichen Einfluß auf die mechanischen Eigenschaften des erhärteten Injektionsmaterials. Daher ist die Kontrolle der Mischzeit während der Ausführung sehr wesentlich.
   C. Die Druckfestigkeit und Dichtigkeit des Injektionsmaterials wird durch die Filtration unter Druck während der Injektion wesentlich verbessert; so erreicht die die Druckfestigkeit nach 28 Tagen 10 kN/cm², wenn die Bildung unter einem Druck während der Filtration von 70 N/cm² erfolgt. Daher sollten möglichst Hochdruckinjektionen eingesetzt werden. Dabei muß man vor der eigentlichen Injektion größere Durchtritte im Untergrund abblocken, um höhereDrücke während der eigentlichen Injektion zu ermöglichen.

1 INTRODUCTION

The cement grouting has a very long history. Though various new materials and new technologies have been adopted for chemical grouting, the cement grouting is mainly used in foundation treatment till now for the economical and environmental reasons. The properties of cement grout and its set have a very important impact on the quality and efficiency of cement grouting. So it is very important to make researches on these properties to improve the technology and efficiency of cement grouting.

The hydration and set of cement is a very complicated physical and chemical process. In the period of grouting the cement hydration conditions are even more complicated than that of the concrete, because in grouing process the following problems exist: the w/c ratio is large and variable; the mixing time is not fixed and grout is subject to the disturbances from the grouting pressure and the hydrodynamical pressure in the grout settling. In practice we find these factors affect properties of the cement grout and its set. So we have carried out tests and researches on the influence of w/c ratio, mixing times and the effection of pressure filtration on the properties of them (cement grout and set grout).

2 UNDERSTANDINGS ABOUT THE FORMED MECHANISMS OF THE CEMENT GROUT SET

Basically there are two theories about the formation of the set cement grout for a very long time. the one is the theory of sedimentation and the other is theory of pressure filtration. I consider that, the process of the grout setting is more complicated, because the grout will meet obstructions from the viscous drag of itself and from the fillings in the cracks in its way when it flows radially from the grouting holes, resulting in the following velocities being reduced rapidly. The cement particles will be deposited gradually when the velocity is reduced to some extent. The deposited particles will exert drags against the coming grout, which will enhance the sedimentation process of the cement particles in the coming grout and form the fresh grout set. Tests show that the fresh grout set is permeable and the coeficient of permeability is more than 10^{-5} cm/s. Therefore the excess water and the water in the coming grout can be filtered out under pressures and then the grout set is formed. The process of sedimentation is not independent and seperative process.

In fact, the grouting process consists of two stages. Grouting begins with sedimentation stage and then continuously and gradually changes into pressure filtration stage. So the grout set is formed by the comprehension actions of these processes. If we consider the grout sets being formed in the sedimentation process and in the pressure filtration process as two extreme cases, the grout set being formed in practice is in a intermediate situation between the two extreme conditions. Therefore for the convenience of the research work, we use two methods of the free sedimentation and the pressure filtration forming the grout set to perform the research on the influences of the cement grout properties on the grout set qualities.

3 THE CEMENT GROUT SET FORMATION METHOD

3.1 Free sedimentation method

We mount the top box frame on the 7x7x7cm³ specimen mold, and pour the ready cement grout as calculated volume into the mould, and then take off the top box frame, and smooth the specimen when the sedimentation and bleeding approach to a stable state, and then cure the specimen in 20°C water until taking off the mould.

3.2 Pressure filtration method

We have taken a pressure filtration formation apparatus developed by ourselves in our researches. The apparatus is shown in fig 1.
Put the text specimen mould (without the bottom plate) on the bottom of the formation tube and pump the ready cement grout as caculated volume into the formation tube. Then the grout is sedimented and filtered under the pressure. When the outlet water approaches the bleeding water, then close the outlet water valve and maintain pressure for several minutes, then slowly releasing the pressure, removing the cover of the formation tube and the tube itself, taking out the specimen carefully, smoothing the top of the specimen and taking off the mould and then putting the formed specimen into 20°C water for curing.

4 ANALYSIS OF THE EXPERIMENT RESULTS

4.1 Water/cement ratio

The influence of w/c on the viscosity of grout and on the mechanical properties of

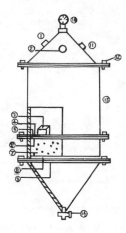

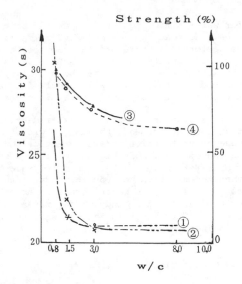

Fig.1 Sketch of the pressure filtration apparatus
    (1)   air inlet valve
    (2)   adjusting pressure valve
    (3)   test speciment frame
    (4)   filter
    (5)   geotextile
    (6)   copper wire mesh
    (7)   fine sand
    (8)   copper wire and filter
    (9)   punched plate
    (10) pressure meter
    (11) pressure raising valve (air)
    (12) bolt
    (13) formation tuble
    (14) water outlet valve

Bild.1   Der Druckfilter
    (1)   das EinlaBventil
    (2)   das Druckregulierventil
    (3)   der Proberahmen
    (4)   das Filterpapier
    (5)   das organische Gewebe
    (6)   Gewebe mit Kupferdrahtsnetz
    (7)   feiner sand
    (8)   Kupferdrahtshetz mit Filterpapier
    (9)   Gestanzte Platte (mit ⌀0.5mm Löcher)
    (10) der Druckmesser
    (11) das Drucksteigerungsventil (Luftdr-
          uck)
    (12) die Schraube
    (13) das Formrohr
    (14) das AuslaBventil

Fig.2 Influences of w/c on the viscosity of the grout and on the strength of the set grout.
    (1)   viscosity curve of 525# Portland cement
    (2)   vicosity curve of 425# Portland cement
    (3)   set grout strength curve of 525# Portland cement
    (4)   set grout strength curve of 425# Portland cement

Bild.2 w/c's Wirkung auf die Mörtelvikosität und die Versteinerungsstärke
    (1)   die Viskositätskurve von Silikatzem- entbrei Nr. 525
    (2)   die Viskositätskurve von Silikatzem- entbrei Nr. 425
    (3)   die Versteinerungsstärkeskurve von Silikatzementbrei Nr. 525
    (4)   die Versteinerungsstärkeskurve von Silikatzementbrei Nr. 425

the set grout (under free sedimentation conditions) are shown by four curves in fig 2.
From fig.2 we can see that if the w/c of the grout is more than 1.5, the grout vis- cosity will slightly decrease with increase of its w/c and the strength of the set grout decrease remarkably. The antiseepage of the set grout also decreases with the incr- ease of the cement grout w/c (See Page 6

for Table 1). It is proved by the analysis and experiments that the main reasons of the mechanical properties of the set grout decreasing with the increase of the cement grouts w/c are: the w/c of the set grout will increase with the increase of the cem- ent grouts w/c (See Page 6 for Table 2), and the water of no chemical reaction will increase, resulting in more the remain ca- pillary water and pores formed in the set while the formation of the cement grout co- nstruction and the strength of the set de- pend mainly on the change of the capillary pore and water. In the theory, the cement may completely be hydrated if the w/c of the cement grout is equal to 0.38, but in fact there must be some capillary pores in the cement gel to play the role of passage for water diffusing into the cement particles. So that the complete hydration can be performed. Therefore in practice, the w/c for complete hydration is 0.437. So the more the w/c, the more the capillary pores and water, which will result in the decrease of the strength and antiseepage of the set.

It is also proved by many experimenties that the set formed with various w/c will have no contraction in its curing period, on the contrary, most of them will have some cubical expensions. This is mainly due to that the reaction heat during hydration and set process of the cement grout is absorbed by the curing water and the gel water and the capillary water in the set will not be vaporized in the wet condition.

4.2 Mixing time

The grout is mixed with a blade-mixer. The blade-mixer runs at 100 rpm. The influence of the mixing time on the viscosity of the cement grout and the strength of the set (formed in free sedimentation condition) is shown in fig.3
From fig.3 we can see that the influence of the mixing time is very little on the apparent viscosity (by funnel viscosimeter) of the grout, but is very strong on the strength of the set. Analyzing the reasons, we find that the mixing destroys the gel crystalline stracture formed in the grout during the hydration of the cement. According to the recent cement – hydration – hardening hypotheis, two cohesive structures, the coherent structure and the crystalline structure, are simultaneously formed in the grout during the cement hydration process. The grout is not only a hydration production but also a solid rigid structure in the cement hydration process. A series of macro behavious of the set grout are determined not only by the hydration productions, but also, more directly by the grout stru-ctures, including the distribution of the pores and the relations among the solid productions. In the four stages indicating the courses of the cement hydration, the second stage (the 40th – the 120th minutes interval) is the important period for the formation of the cement grout structures. Therefore if grout is mixed during that period or after that period, the cement grout structure formed in the cement hydration process will be destroyed, which will result in the decrease in the mechanical properties of the set grout. Experiments have proved that, the strength of a set grout mixed for 3 hours is 30% less than that of mixed only one hour and the streng-

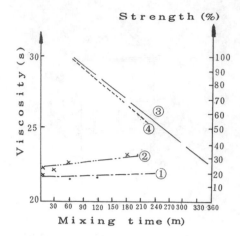

Fig.3 Influence of the mixing time on the viscosity of the grout (w/c = 1.5) and on the strength of the set (grout w/c = 0.8)
(1) 525# Portland cement grout viscosity curve
(2) 425# Portland cement grout viscosity curve
(3) 525# Portland cement set strength curve
(4) 425# Portland cement set strength curve

Bild.3 die Wirkung der Mischzeit auf die Mörtelviskosität (w/c = 1.5) und die Versteinerungsstärke (Mörtel's w/c = 0.8)
(1) die Viskositätskurve von Silikatzementbrei Nr.525
(2) die Viskositätskurve von Silikatzementbrei Nr.425
(3) die Stärkekurve von Silikatzementbrei Nr.525
(4) die Stärkekurve von Silikatzementbrei Nr.425

th of a set grout mixed for 4 hours is 50% less than that of mixed only one hour. So it is necessary to strictly control the mixing time of the grout in the grouting process (especially for the high speed mixer).

4.3 Effects of the pressure filtration

Grouting is a process of injecting the grout under pressure, into the faults, such as cracks, voids etc. in the injected body, and forming the set, to consolidate injected body and raise its impermeability. To understand the influence of pressure on the grout and the quality of the set, we take four groups of cement grout, the w/c ratios of 0.8, 1.5, 3.0 and 8.0 respectively, and

Table 1 Antiseepage of set grout formed from cement grouts with various w/c

w/c k(cm/s) Cement sort	0.8			1.5		
	3d	7d	14d	3d	7d	14d
525# Portland cement	3.7×10^{-7}	7.1×10^{-9}	S_g	5.3×10^{-7}	1.9×10^{-8}	S_g
425# Portland cement	3.1×10^{-7}	1.3×10^{-8}	S_g	1.3×10^{-6}	5.9×10^{-8}	1.6×10^{-8}

Table 2 w/c of set grout formed from cement grout with various w/c

w/c of set grout / cement sort	0.8	1.5	3.0	8.0
525# Portland cement	0.50	0.56		0.90
425# Portland cement	0.64	0.57	0.60	0.80

Table 3 Set time of set under various pressures

Cement sort	w/c set time pressure (N/cm²)	0.8		1.5		3.0		8.0	
		initial set time	final set time	initial set time	final set time	initial set time	final set time	initial set time	final set time
525# slag Porland cement	0*	10:43'	13:18'	13:10'	17:10'	18:00'	23:20'	20:15'	27:15'
	10	6:50'	10:50'	8:05'	13:55'	10:15'	14:45'	9:10'	15:10'
	30	4:45'	9:50'		11:35'	6:35'	12:15'	3:00'	13:20'
	50	2:10'	8:25'	2:40'	10:15'	3:40'	11:10'		12:45'
525# Portland cement	0*	7:25'	9:45'	6:55'	10:30'		11:10'		
	10	4:30'	7:20'	4:51'	7:05'		6:55'	4:45'	7:00'
	30	3:10'	6:15'	2:20'	7:00'		6:10'	3:15'	4:45'
	50			2:45'	5:45'		6:05'		

* Free sedimentation formation

put them in the pressure filtration apparatus to from the specimens under various pressure-filtration conditions, and then measure the set time, strength of the set, and their impermeabilities.

1. Set time

The set time for the set formation under the action of pressure filtration is much less than that under free sedimentation and is inverse proportional to the pressure. The experiment results are shown in table 3. From table 3 we can see that when 525# portlant cement grout forms the set under the action of pressure (30 N/cm²) fitleration, its initial set time is 4 hours earlier than that in free sedimentation condition and its final set time is 3 hours earlier than that of free sedimentation set.

2 Strength

It is proved by experments that the compressive strength of the set is greatly increased with increase of the formation pressure (see table 4).

The compressive strength of 525# Portland cement set formed under the action of 70 N/cm² pressure filtration if three times higher than that of the set formed in free sedimentation conditions. Sets formed from cement grout with various w/c taken in experiment groups and under the formation pressure of 70 N/cm² have a compressive strength higher than 7,000 N/cm² after 28 days cure. Especially for the two sets,

47

Table 4 Compressive strength of set grout formed under various pressure filtration actions (N/cm²)

Cement sort	w/c cure age (d) strength (N/cm²) pressure (N/cm²)	0.8		1.5		3.0		8.0	
		7	28	7	28	7	28	7	28
525[#] Port- land cement	0*	2,200	4,040		3,200		2,960	970	2,600
	10	4,920	6,400	4,640	5,760		7,760	3,060	3,680
	30	5,180	5,820	5,370	7,280		7,960	4,700	5,680
	50	5,640	6,840	5,500	7,640		8,310	5,290	7,800
	70	5,860	7,080	7,020	10,840		9,220	5,760	8,400
525[#] slag Port- land cement	0*	1,720	3,420	1,200	3,080	540			
	10	2,770	5,460	1,660	3,640	1,300	2,840		
	30	3,300	5,680	2,560	4,640	1,460	2,760		
	50	3,200	5,680	2,680	4,840	1,760	3,350		

Table 5 Relations between the porosity of the 525[π] slag Portland cement set and the formation pressures

w/c porosity (%) pressure (N/cm²)	0.8	1.5	3.0	8.0
0*	0.2497	0.2923	0.3778	
10	0.1647	0.1785	0.2647	0.2635
30	0.1414		0.1864	0.1929
50	0.0707	0.1353	0.0771	0.1394

*Free sedimentation formation

w/c 1.5 and 3.0 respectively, their compre-sive strength will reach approximately 10,000 N/cm² after 28 days curing.

3 Impermability

Set formed under action of pressure (10 N/cm² to 50 N/cm²) filtration will be cured 7 days, and then is performed a permeability test on SS15 sand grout permeability apparatus. The results show that the set is impermenable under a pressure of 80 N/cm².

4 Contractibility

It is proved by many experiments that for specimens formed in pressure-filtration condition with various pressure there are no contraction during the curing period in the water, on the contray, there are some little expansion instead. In short, cement set grout formed in pressure-filtration condition with various pressures will have excellent mechanical properties. Analysis of the reason and experiments show that the pressure-filtration process will raise the density of the set. The porosities of set formed by pressure-filtration method with various w/c vatios are decreasing with the increase of the formation pressure, as being shown in table 5 (See Page 7 for Table 5).

5 SUMMARY

On the basis of experiments and reasearch, we primarily understand the laws of influ-ences of w/c ratio, mixing time and forma-tion pressure, the three factors on the properties of the cement grout and the mechanical properties of the set which provides a scientific basis for improving the efficiency of the cement grouting, expecially the material datas of the mech-anical properties of the set formed by pressure-filtration method, which is very interest for engineering practice. The main and basis laws are shown as following:

5.1 The apparent viscosity of the cement grout decreases with the increase of the w/c of the grout, if the w/c is less than 1.5, if w/c increases beyond 1.5, it decreases very slowly with the increase of w/c, but the mechanical properties of the set greatly decrease, no matter by free sedimentation method or pressure filtration method. Therefore the thick grout with small w/c ratio, not only can guarantee a good groutability of the cement grout, but also play an important role in guaranteeing the quality and durability of the cement grouting. It is proved that the tendency of adopting thick grout recenty in many countries in the world is very correct.

5.2 It is proved by experiment that too long mixing time will result decrease of the mechanical properties of the set. Therefore, to guarantee the cement grouting quality we must strictly control the mixing time in the grouting process, or strictly control the expire time of the grout, expecially when the high speed mixing.

5.3 For the fresh set formed by the pressure filtration method the set time will be reduced. All the sets are formed under the closed grouting pressures in engineering practice. So all sets formed during grouting process are under the action of pressure-filtration and the closer to the grouting hole, the higher is the pressure. Therefore, it is suggested to reduce the set waiting time for the efficiency of the grouting process.

5.4 The compressive strength and the impermeability of the set formed under the action of pressure - filtration are greatly improved eg. the compressive strength of a set formed under the formed pressure of 70 N/cm^2 reachs 10,000 N/cm^2 (for 28 days), which is even much higher than the label of the same cement conctete. Therefore the pressure filtration is very important to increase the mechanical properties and durability of the set. The high pressure grouting will be implemented if possible, which can greatly improve the quality of grouting in the foundation. On the other hand, in the region where grout intake is very large, we had better adopt proper

procedure to seal the leakage, cutting off the larger grout leakage pass first, to provide the pressure - filtration condition for the set forming, then performing grouting. It is very good for the quality of the grouting.

It is proved by experiment that the cement set will not contract in the water or wet envirement, on the contrary, it will expand a little. So there is nothing to be worried about the contraction of set grout in the wet ground.

REFERENCES

Houlsby A.C.
  Cement Grouting for Dams, Proceeding of the Conference on Grouting in Geotechnical Engineering, New Orleans, 1982.
Lombard G.   Grout Slurries Thick or Thin? Proceeding of the Session Sponsored by the Geotechnical Engineering Division of American Society of Civil Engineering in Conjunction with the ASCE. April, 1985.
Houlsby A.C., The Durability of Cement Grouting, ANCOLD Bulletin, April, 1986.
Aleksandar Bozovie, General Report Q_{58} Fifteenth International Congress on Large Dams, Lausame-swit, June, 1985.
Cheng Zuowei and Chen Zhenghong, Technique of Concrete Vacuum Dehydration, China Construction Industry Publishing House, October, 1986.

Grouting in Rock and Concrete, Widmann (ed.) © 1993 Balkema, Rotterdam, ISBN 90 5410 350 7

Studies on high permeable epoxy resin grouting materials YDS

Untersuchungen an Kunstharzen mit hoher Eindringfähigkeit

He Yongsheng, Xian Anru, Zhang Guangzhao, Yie Linhong & Zhao Rujian
Guangzhou Institute of Chemistry, Academia Sinica, People's Republic of China

ABSTRACT The compound reinforcer agent YDS can activate swiftly carbonyl groups of furfural and acetone molecules to produce what can solidify with epoxy resin. The grouting materials YDS can keep high mechanical properties with a large amount of diluent added and has some characteristic properties including their low initial viscosity, strong affinity to mediums, excellent permeability and good ageing property.
KEY WORDS: Chemical grout, Epoxy Resin, Activation, Affinity, Permeation

INTRODUCTION The epoxy resin grouting materials with furfural/acetone mixed solvent as their diluent have been widely used to repair rifts in concrete and weak bases of dams and other buildings since 1967[1]. In order to improve the permeability of these grouting materials by reducing their viscosity, a large amount of diluent is often added. But this leads to a sharp drop of the mechanical properties of the solidified grouting materials. To satisfy both sides, we have done much research and prepared several kinds of grouting materials such as Zhonghua-798 and AA, they have been successfuly applied to treatment of weak foundations of hydroelectric stations.

This article deals with the grouting materials YDS (YDS), which has the most excellent properties of all the grouting materials we have studied.

1 Results and discussion

1.1 Principle

YDS were prepared from epoxy resin, furfural, acetone, diethylenetriamine and a little amount of compound reinforcer YDS(CRY), which plays a very important role in the system. CRY can activate the carbonyl groups of furfural and acetone molecules and results in a series of reaction, including self- resinification of furfural or acetone and resinification of furfural with acetone, which produce linear, cyclic or network oligomers. Having hydroxy groups, double bonds and α - hydrogens in their molecules, the oligomers continue to react with one another forming larger molecules or react with diethylenetriamine which will also react or have reacted with epoxy groups and solidify with epoxy resin.

We have reported that Additive A could activate molecules of furfural and acetone, but it did slowly[2]. The mixture of furfural with CRY, or additive A had no sooner been determined with IR and GC than they were mixed in comparsion with the determination of pure furfural (Fig.1 and Fig.2). Aldehyde group has

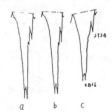

Fig.1 Infrared spectrum

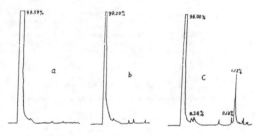

Fig.2 Gas chromatogram

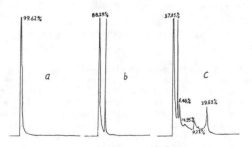

Fig.3 Gas chromatogram

51

Fig. 4 Oligomers from self-condensation of acetone

properties		YDS-1	YDS-1'	YDS-4	ZH-798
density	(g/cm³)	1.0490	1.0460	1.0338	1.0697
surface tension	(10⁻⁸N/cm)	32.99	32.81	32.28	34.42
contact angle to quartz sand		10.8		1.0	56.2
initial viscosity(mPa·s)		2.63	4.62	2.05	5.35
gelation time (day)		4	2	4.5	4
affinity (10⁻⁸N/cm)		32.5		32.3	19.2

Table 1.a The exterior changes of furfural (35°C in air)

Activated time (day)	pure furfural	mixed with Additive A	mixed with CRY
0	light yellow and transparent	light yellow and transparent	black and opaque, gave out heat instantly
3	yellow and transparent	brown and transparent	black and opaque, viscosity increased
8	light brown and transparent	ditto	black and opaque, viscosity increased more
31	ditto	brown and opaque	black and opaque, viscosity increased even more

Table 4. Compression strength of the soaked cement block

property		soaked in YDS-5	soaked in ZH-798	unsoaked
compression strength (MPa)		40.6	36.4	30.3
increament	(%)	34.0	20.1	

Table 1.b The exterior changes of acetone (35°C in air)

Activated time (day)	pure acetone	mixed with additive A	mixed with CRY
0	colourless and transparent	colourless and transparent	brown and transparent, gave out heat instantly
3	ditto	light brown and transparent	brown and opaque, viscosity increased
8	ditto	brown and opaque	dark brown and opaque, viscosity increased more
31	ditto	dark brown and opaque	dark brown and opaque, viscosity increased even more

Table 5. Mechanical preperties of YDS

mechanical properties	YDS-1	YDS-1'
sandgel compression strength (MPa)	81.9	101.1
sandgel shearing strength (MPa)	49.6	56.4
sangdgel tensile strength (MPa)	2.4	14.9

Table 2   The Effect of CRY and Additive A on the sandgel compression strength of grouting materials

formala	1	2	3	4	5	6	7	8
with additive A	44.0	27.7	19.0	20.1	16.2	5.4	2.8	0
with CRY (MPa)	81.9	82.0	73.1	72.0	70.8	43.8	30.3	8.6
increament of compression strength (%)	86.1	81.0	306	258	337	707	982	

* percentage of epoxy resin is cut down in order to 0

The exterior changes of furfural and acetone after they were mixed with CRY also suggest the swift activation of CRY(Table 1).

Additive A activates molecules of furfural and acetone slower than diethylenetriamine reacts with epoxy resin.So only little of them can join the chain of epoxy resin. Unlike Additive A, CRY is a swift activator and can make most of the furfural and acetone molecules bond with epoxy resin bridged by diethylenetriamine. That's why YDS have excellent mechanical properties.

1.2 Properties of YDS

.1) The reinforcement of CRY

CRY reinforces the grouting materials strongly (Table 2). Though the percentages of epoxy resin in the grouting material are cut down progressively, the increments of compression strength rise. Even more, when no epoxy resin is added, the grouting material can also solidify and its sandgel compression strength can arrive at 8.58 MPa. It is very CRY that makes furfural, acetone and diethylenetriamine react into large molecules and solidify.

2) Physical properties of YDS

The permeability of grouting material depends on its initial viscosity and affinity which is the product of cosine of the contact angle of the

its absorption peaks at 2816 and 2724 cm⁻¹ . Fig. 1 shows the amount of aldehyde group in C is least. From Fig.2 we can also know the percentage of furfural in C is least.No doubt, CRY can activate carbonyl groups of furfural molecules in a short time and promote the resinification of furfural, So it can do acetone (Fig.3).Fig.4 shows oligomers from the self-condensation of acetone.

grouting material to the grouted medium and the surface tension($\gamma \cdot \cos\theta$)[3]. The less the initial viscosity and the more the affinity the grouting material has, the better its permeability is From this point, YDS are more permeable than Zhonghua-798, the most permeable grouting materials in the past which can be grouted into muddy sandwich with the permeable coefficient 10^{-8}cm/s.Table 5 shows the mechanical properties of two kinds of YDS. Obviously, they are satisfying.

compression
strength(MPa)

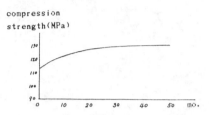

Fig.5 Ageing test in oven at 35℃

compression
strength(MPa)

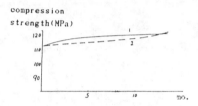

Fig.6 Ageing test
1) soaked in water   2) in exposure

By soaking cement block in YDS-5 and Zhonghua-798, measuring the compression strength of the block after the grouting material solidified, we confirmed this(Table 4).

3) Ageing properties
Two kinds of YDS have been layed in oven at 35℃, in exposure and soaked in water at 35℃ respectively. No drop of compression strength has been found on the above condition strength during the past 70 months (Fig.5 and Fin.6).

CONCLUSION

YDS not only have excellent permeability, mechanical properties and ageing properties but are relatively cheap grouting materials. They have the promise to be grouted into weak bases with the permeable coefficient less than 10^{-8} cm/s.

REFERENCES

Press for irrigation and Electricity 1980.   Chemical Grouting Technology, P128.
He Yongsheng ect. Property and application of epoxy resin grouting meterial No. 400 system.   Chemical Materials for Construction, 1991 (6),17.
Ren Kechang ect. Penetration mechanism of furfural -acetone-epoxy resin grouting, Yangtae River 1988 (12),28.

Bleed and rheology of cement grouts

Das Bluten und die Rheologie von Zementmörtel

Raymond J. Krizek, Lois G. Schwarz & Stanley F. Pepper
Department of Civil Engineering, Northwestern University, Evanston, Ill., USA

ABSTRACT: Described in this paper are the bleed characteristics and rheological behavior of ordinary and several microfine cement grouts. The major variables investigated are the water:cement ratio, the effect of adding a small amount of bentonite, the role of mixer type, and the influence of temperature. In general, the microfine cement grouts manifested less bleed than the corresponding grouts composed of ordinary Type I cement and those containing portland cement bled less than the ones containing only microfine slag cement. With a few exceptions, the addition of 1% bentonite to the mixes caused less bleed in the grouts composed only of portland cement and little change or increased bleed in the grouts containing slag cements; with larger percentages of bentonite, both the total bleed and rate of bleed decreased considerably. High energy mixing reduced bleed and the temperature of the mixing water had little effect. Decreasing the water:cement ratio of a given grout caused significant increases in the apparent viscosity, somewhat lesser increases in the yield stress, and little change in the plastic viscosity; as the percentage of slag in the grout increased, both the apparent viscosity and the yield stress dramatically decreased. Additions of bentonite increased both the yield stress and the apparent viscosity, with larger increases noted for the grouts composed of the finer cements. Mixer type produced only small changes in the rheological parameters, and varying the temperature of the mixing water changed the yield stress and apparent viscosity by as much as five-fold.

ZUSAMMENFASSUNG: In dieser Abhandlung werden die Charakteristiken der Klärung und des rheologischen Verhaltens von ordinärem und einigem ultrafeinem Zementmörtel beschrieben. In diesem Versuchsprogramm wurden Variablen wie das Wasser/Zement Verhältniss, der Einfluss einer kleinen Menge von Bentonit im Mörtel, die Rolle des Typs der Mischmaschinen, und der Einfluss der Temperatur erforscht. Im allgemeinen zeigten ultrafeine Zementmörtel weniger Klärung als der entsprechende Mörtel, der mit Typ I Zement hergestellt wurden. Mörtel, die mit Portland Zement gemischt wurden, stellten weniger Klärung aus, als Mörtel die nur ultrafein Schlacke-Zement erhalten hatten. Mit wenigen Ausnahmen die Beimischung von 1% Bentonit veruhrsachte weniger Klärung, als Mörtel, die mit Portland-Zement hergestellt wurden. Wenig Änderung oder wachsende Klärung wurde in Mörtel mit Schlacke-Zement bemerkt. Mit grösserem Prozentgehalt von Bentonit nahm sowohl die gesamte Klärung als auch die Klärungsziffer bedeutend ab. Höhere Energiemischung hat die Klärung herabgesetzt, aber die Temperatur des Mischwassers hat wenig Einfluss gezeigt. Der Annahme des Wasser/Zement Verhältnisses eines bestimmten Mörtel liess die offenbare Viskosität bedeutend, die Fliessgrenze etwas weniger zunehmen und die plastische Viskosität wurde nur wenig geändert; mit zunehmendem Prozentgehalt der Schlacke nahmen sowohl die offenbare Viskosität als die Fliessgrenze dramatisch ab. Der Zusatz von Bentonit vermehrte sowohl die Fliessgrenze als die offenbare Viskosität; eine deutliche Vermehrung wurden bei Mörtel bemerkt, die mit feinem Zement gemischt wurden. Der Typ der Mischmaschine hat nur kleine Änderungen im rheologischen Parameter erzeugt, jedoch änderte sich die Fliessgrenze und die offenbare Viskosität bei wechselnden Wassertemperatur um das Fünffache.

1 INTRODUCTION

Cement grouts must be sufficiently "fluid" to allow efficient injection into rock and soil formations and sufficiently "stable" to resist sedimentation prior to setting. During the mixing process, cement particles are dispersed in an aqueous suspension which is often unstable and results in the accumulation of bleed water at the tops of void spaces (Helal and Krizek, 1992). The amount of bleed is dependent in large part on the water:cement ratio of the grout, the reactivity of the cement particles, and the particle size. Rheologically, cement suspensions are frequently modelled as Bingham fluids, which exhibit a yield stress and a viscosity, both of which are strongly dependent on the concentration of particles and the type of grout. The pressure required to maintain flow is controlled by the plastic viscosity (Deere and Lombardi, 1985).

Notwithstanding the generally acknowledged positive effects of high energy shear mixing on cement grouts (Houlsby, 1982, 1990), some argue that low energy paddle mixers can furnish similar performance if dispersants are used to decrease particle sedimentation. Schwarz and Krizek (1992) have shown phenomenologically that variations in the rheology and bleed characteristics of particulate grouts are influenced by the type of mixer used; this suggests the use of caution when attempting to predict the performance of field-mixed grouts prepared in one type of mixer with data obtained from laboratory-mixed grouts prepared in a totally different table-top mixer.

2 OBJECTIVES

The primary objectives of this study are (a) to quantify the bleed characteristics and early age rheological behavior (viscosity and yield stress) of ordinary and microfine cement grouts as functions of particle size, material constituents, and a small amount of bentonite; and (b) to investigate the effects of mixer type (including lab versus field size) and mixing water temperature (from 5°C to 35°C).

3 SCOPE OF STUDY

The experimental program includes sedimentation and viscosity tests on four cement grouts (three microfine cements and one ordinary portland cement) mixed in three different mixers (a high energy laboratory-size mixer, a high energy field-size mixer, and a low energy laboratory-size mixer) at three water:cement ratios (1:1, 2:1, and 3:1). In addition, a series of tests was performed on a microfine cement grout obtained by blending two of the basic microfine cement grouts. A small amount of bentonite was added to approximately a quarter of the tests, and the effect of mixing water temperature was investigated for one microfine cement.

3.1 Materials Used

The cements tested were ordinary Type I portland cement (manufactured by Continental in Hannibal, Missouri), MC-100 (a microfine slag cement marketed by Geochemical Corporation of Ridgewood, New Jersey), MC-300 (a microfine portland cement obtained from Blue Circle Atlantic of Sparrows Point, Maryland), MC-500 (a proprietary blend of microfine slag and portland cements manufactured by Onoda Cement Company of Japan), and a blended microfine cement (consisting of 50% MC-100 and 50% MC-300). It is noteworthy to mention that, based on experience over recent years, the variations in measured properties caused by different supplies of cement (often from the same producer) are sometimes of the same order of magnitude as the variations caused by the parameters being studied; therefore, the results of this study, as well as any other study) should be accepted largely as relative relationships and not absolute.

A 1% slurry of Wyoming bentonite (obtained from the Jenson Drilling Company of Eugene, Oregon) was high-energy mixed for 2 minutes, allowed to hydrate for 2 hours, and then remixed for 30 seconds prior to adding it to the grout mixes based on a 1% proportion of dry bentonite to dry weight of cement. A complementary study was conducted to quantify the effect of adding varying percentages of bentonite to a microfine cement grout. The bentonite mixing procedures were patterned after studies reported by Jefferis (1982) and Weaver (1991). All grout mixes contained 1% (by dry weight of solids) of the superplasticizer naphthalene sulphonate formaldehyde (NS-200) to decrease interparticle forces and enhance thorough wetting of the cement particles.

Figure 1 shows a typical particle size

Table 1. Particle size distributions of materials used

Material	Cummulative Volume (%) at Particle Diameter (μm) of:											Peak		Mean (μm)	Median (μm)
	0.1	0.2	0.5	1	2	5	10	20	50	100	>100	D (μm)	V (%)		
MC-100	<1	1	9	23	49	82	98	100	100	100	100	2.3	3.5	3.2	2.3
MC-300	<1	4	16	24	38	74	99	100	100	100	100	4.6	4.2	3.7	3.2
MC-500	<1	3	12	21	36	69	97	100	100	100	100	5.5	4.0	4.2	3.5
Blend	<1	3	13	21	37	74	99	100	100	100	100	4.2	4.4	3.8	3.2
Type 1	<1	2	6	9	13	23	42	62	94	100	100	28.3	3.4	20.2	14.6
Unhydrated Bentonite	0	<1	1	3	7	18	33	52	88	100	100	44.6	4.0	25.4	20.2
Hydrated Bentonite	0	0	2	13	38	74	92	99	100	100	100	2.6	3.9	4.5	2.9

distribution curve obtained by use of a laser particle size analyzer. The data given in Table 1 for each of the materials tested include, in addition to certain statistical parameters, the cummulative percentage of a given material with a particle diameter less than that indicated in the column heading.

Microfine cements generally have average particle sizes approximately one-half to an order of magnitude smaller than ordinary portland cement. The smaller size particles react faster due to their significantly larger surface area which exposes more material to water. In general, blast furnace slag cement is less reactive than ordinary portland cement, but its chemical composition can vary widely from source to source; the specific gravity of slag generally ranges from 2.85 to 2.94 as compared to 3.15 for portland cement (Weaver, 1991).

Bentonite is often added to grout mixes to stabilize the suspension and reduce sedimentation; however, it simultaneously increases the yield stress significantly and, to a somewhat lesser extent, the viscosity of the grout. The use of bentonite in cement grouts is met with divided opinion; the major negative arguments are (a) the increased stiffness (yield stress and viscosity) that it induces and (b) the inconsistency in grout performance that often accompanies its use, and many believe that these negatives offset the benefit of lesser bleed that may be realized. Mineralogically, bentonite is an impure montmorillonite clay that is capable of absorbing water into its crystalline lattice structure and delaminating into more numerous, but smaller clay particles. Evidence of this mechanism is apparent by comparing data in Table 1 for unhydrated and hydrated bentonite. It is highly advantageous, therefore, to prehydrate bentonite before adding it to a cement grout.

3.2 Mixers and Mixing Procedures

Three different mixers with widely different characteristics were used for various tests in this study. The first was a laboratory-size standard blender (Oster Model 460-01), which had an estimated rotational speed of 5,000 to 10,000 rpm under load; 1 minute mixing times were used for a batch volume of 0.65 liter. The second was a commercially available field-size Colcrete mixer (Model 5D4 manufactured by Colcrete in England), which had a rotational speed of 1500 rpm; batch volumes of 80 liters were mixed for 1 minute. The third was a laboratory-size paddle mixer (manufactured by Mixing Equipment Company of Rochester, New York), which caused a stirring action within the suspension by means of a propeller with a diameter about one-third the diameter of the drum. This mixer was

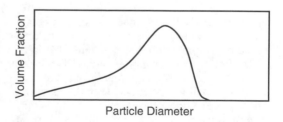

Figure 1. Typical particle size distribution curve

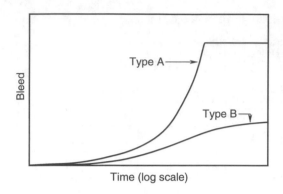

Figure 2. Typical bleed response curves

Table 2. Bleed data for various blender-mixed grouts

Material	Water: Cement Ratio	No Bentonite		1% Bentonite	
		Time (minutes) to Maximum Bleed	Maximum Bleed (%)	Time (minutes) to Maximum Bleed	Maximum Bleed (%)
MC-100	1:1	1440	27	1440	5
	2:1	480	45	1440	53
	3:1	420	57	1440	60
MC-300	1:1	45	0.8	45	0.4
	2:1	240	2.9	60	1
	3:1	360	22	420	9
MC-500	1:1	150	2	1440	2.5
	2:1	360	20	360	31
	3:1	300	36	300	45
Blend	1:1	15	0.4	30	0.4
	2:1	30	0.8	210	2
	3:1	210	12	360	13
Type I	1:1	1440	38	1440	26
	2:1	1440	70	1440	58
	3:1	1440	80	1440	74

operated for 2 minutes (time necessary to obtain a relatively homogeneous mixture) at a rotational speed of 50 rpm with a batch volume of 1 liter.

The first two mixers may be termed high speed mixers because of the high rotational speeds and generally small tolerances between the rotor and housing. The shearing action imparted by any given mixer will depend on the rotational speed of the mixer, the mixing duration, the batch volume, and the concentration of the mix; the specific energy imparted to the mix will be influenced by the physical size of the mixing drums, rotors, and tolerances between the respective rotor and housing (these parameters were not addressed in this study). Regardless of the water:cement ratio, the total volume of grout per batch was nominally constant for a given mixer. The NS-200 dispersant was first mixed thoroughly with the mixing water; then, the cement was added and the suspension was mixed for the designated time. For the cases where bentonite was added to the grout, the prehydrated bentonite slurry was first mixed with the dispersant and mixing water (about 10 seconds), after which the cement was added and mixed for one minute.

4 SEDIMENTATION

According to Stoke's law, the fall velocity of a particle in a viscous fluid is directly proportional to its size and specific gravity and inversely proportional to the viscosity of the fluid. Therefore, larger particles will settle faster than smaller ones, and more viscous fluids will hold particles in suspension longer than thinner fluids. Immediately after mixing the cement with water, the particles begin to settle; however, due to the increase in grout viscosity as a result of attractive surface forces and chemical reactions, the settling velocity of suspended particles will decrease with time. The addition of a dispersant temporarily suppresses the interparticle attractive forces, thereby allowing the particles to remain in a relatively dispersed state (particles are separated by repulsive forces between negatively charged surfaces) without the presence of supernatant water for a longer period of time prior to set.

Sedimentation tests were performed to evaluate the stability of the grout mixes as a function of time. Stability is defined here as the volume of bleed water above the grout suspension, ΔV, expressed as a percentage of the total initial slurry volume, V_o. Immediately after mixing, slurries were poured into transparent tubes having an inside diameter of 2.5 cm and a length of 25 cm. The top of each tube was gently covered with a plastic cap to prevent evaporation (but the cap was not sufficiently tight to create a vacuum), and the tube was stored vertically in an area of negligible vibration at a temperature of 20°C ± 1°C for a period of 24 hours, during which time periodic readings of ΔV were taken. Figure

2 shows two typical relationships that were obtained for sedimentation (or bleed) as a function of time over a period of 24 hours (1440 minutes). Curve A characterizes the data from all of the 2:1 and 3:1 grouts and many of the 1:1 grouts (which manifest maximum bleed within a few hours), whereas Curve B better describes the 1:1 grouts composed of MC-100 and Type I cements, both with and without the addition of bentonite, and MC-500 cement with bentonite (for which maximum bleed is usually not attained in 24 hours). In the case of Curve A response, the bleed rate is quantified by the time to reach maximum bleed (i.e. the data from a given test are represented by the "break point"), and in the case of Curve B response the maximum bleed is designated as the bleed at 24 hours (which is within a percent or two of the longer term value) and the time to reach maximum bleed is simply recorded as "greater than 1440 minutes".

4.1 Water:Cement Ratio

The data in Table 2 for sedimentation tests on blender-mixed grouts without bentonite verify the expected trend, wherein grouts with higher water:cement ratios manifest higher maximum bleed; however, the magnitude of the difference in bleed with changes in the water:cement ratio is drastically different for various cements. In contrast, the time to attain maximum bleed for the various cements shows no consistent trend as a function of the water:cement ratio of the grout.

4.2 Type of Cement

Reference to the maximum bleed in Table 2 for blender-mixed grouts without bentonite indicates that the type of cement strongly influences the bleed capacity of grouts. Comparing the results for microfine slag (MC-100) and microfine portland (MC-300) cements shows that MC-300 has much less bleed than MC-100 (3%, 6%, and 39% for the 1:1, 2:1, 3:1 mixes, respectively). On the other hand, a 50-50 blend of the two cements produces considerably less bleed than either one individually (about one-half the bleed measured for the MC-300 alone). The MC-500 grouts had a bleed capacity somewhat higher than their MC-300 counterparts, and the Type I portland grouts, as expected, manifested the highest bleed of all. The sedimentation rates are strongly dependent on the cement type, both in magnitude and trend;

Table 3. Bleed data for 2:1 MC-500 blender-mixed grouts with various percentages of bentonite

Bentonite (% dry weight)	Time (minutes) to Maximum Bleed	Maximum Bleed (%)
0	360	20
1	360	31
2	240	10
4	180	1.3
6	120	0.4
8	120	0.4

Table 4. Effect of mixer type on bleed characteristics of MC-500 grouts

Type of Mixer	Water: Cement Ratio	Time (minutes) to Maximum Bleed	Maximum Bleed (%)
Blender	1:1	150	2
	2:1	360	20
	3:1	300	36
Colcrete	1:1	360	1.5
	2:1	360	15
	3:1	360	20
Paddle	1:1	300	3.5
	2:1	300	27
	3:1	360	47

the MC-100 exhibits an increasing sedimentation rate with increasing water:cement ratio, whereas MC-300 and the 50-50 blend show the opposite trend and the MC-500 and Type I manifest no well defined trend. Since all the microfine cements have essentially the same mean particle size (about 3 to 4 microns), the sedimentation rate, as well as maximum bleed, is attributed to the slower reactivity of slag and its different affinity for dispersant adsorption. The Type I cement is considerably coarser (mean particle size of about 20 microns) than the microfine cements, and this is undoubtedly a major reason for its higher sedimentation rate and overall bleed.

59

Table 5. Effect of mixing water temperature on bleed characteristics of blender-mixed MC-500 grouts

Water:Cement Ratio	Mixing Water Temperature (°C)	Time (minutes) to Maximum Bleed	Maximum Bleed (%)
1:1	35	360	1.3
	20	150	2.1
	5	300	2.9
3:1	35	240	44
	20	300	36
	5	240	48

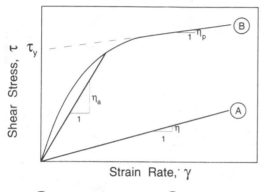

A) Newtonian
$$\tau = \eta \dot{\gamma}$$
η = dynamic viscosity
η_a = apparent viscosity

B) Bingham
$$\tau = \tau_y + \eta_p \dot{\gamma}$$
η_p = plastic viscosity
τ_y = yield stress

Figure 3. Parameters for different rheologic materials

4.3 Addition of Bentonite

The addition of a small amount (1% by dry weight) of bentonite to cement-based grouts generally had little effect on the sedimentation rate and decreased the bleed of portland cement grouts, whereas it tended to decrease the sedimentation rate and increase the bleed of grouts containing slag; this can be seen by comparing the corresponding data in Table 2 for grouts with and without bentonite. As illustrated in Table 3, increasing the percentage of bentonite (up to 8% by dry weight of cement) increased the

sedimentation rate as much as three-fold and decreased substantially (as much as 98%) the bleed (except for the grout with only 1% bentonite) relative to the corresponding values for grouts without bentonite.

4.4 Type of Mixer

Table 4 gives comparative data for the settling behavior of MC-500 grouts mixed by the various mixers. High energy mixing is clearly more effective than low energy paddle mixing (e.g. paddle-mixed grouts exhibited about twice the bleed of blender-mixed grouts) and the Colcrete mixer is 25% to 50% more effective than the blender for reducing the bleed. However, the type of mixer appears to have little influence on the sedimentation rate.

4.5 Mixing Water Temperature

As shown in Table 5, decreasing the temperature of the mixing water from 35°C to 5°C appears to have little effect on the sedimentation rate and to increase somewhat the bleed of MC-500 grouts.

5 RHEOLOGY

Grout suspensions are often modelled as Bingham fluids, whereby the viscosity (represented by the coefficient of proportionality between the velocity gradient and the applied shear force) is dependent on the shear rate and no flow occurs until the shear force exceeds a minimum yield value. Macroscopic flow generally alters the microstructure of a suspension; dispersed particles may align along flow lines or flocculated particles may be separated (although such effects may be offset by Brownian motion and reflocculation). Which effects predominate depends on the particle size, strength of interparticle forces, and applied strain rate. Viscosity tests (ASTM D2196-86) were performed by using a Brookfield rotational viscometer, and the rheological parameters describing the behavior of these grouts were plotted in formats depicted by Figures 3 and 4. Figure 3 shows the basic type of graph obtained from a viscosity test, where shear stress is plotted versus strain rate; specifically illustrated are (a) the relationships for a Newtonian fluid (primarily for reference) and a pseudo-Bingham fluid and (b)

60

Table 6. Rheology data for various blender-mixed grouts

Material	Water: Cement Ratio	No Bentonite			1% Bentonite		
		Yield Stress (mPa)	Plastic Viscosity (mPa·s)	Apparent Viscosity (mPa·s)	Yield Stress (mPa)	Plastic Viscosity (mPa·s)	Apparent Viscosity (mPa·s)
MC-100	1:1	20	85	3.7	130	86	6
	2:1	18	68	2.1	20	68	2.2
	3:1	17	76	1.8	19	76	1.8
MC-300	1:1	4700	76	552	14200	84	1560
	2:1	550	74	62	1550	81	153
	3:1	250	81	34	750	76	67
MC-500	1:1	310	81	35	430	72	38
	2:1	85	72	3	150	72	5
	3:1	25	79	2	55	81	3
Blend	1:1	4400	72	410	8700	80	974
	2:1	600	68	60	1600	81	158
	3:1	215	72	25	625	83	66
Type I	1:1	42	76	2.7	72	76	3.6
	2:1	24	84	1.6	22	82	1.7
	3:1	20	82	1.4	21	83	1.4

the definitions for yield stress, plastic viscosity, and apparent viscosity. Figure 4 shows typically the change in apparent viscosity as a function of strain rate; measurements were made over approximately a ten-fold change in the strain rate. The values reported in the following tables are those obtained after 1 minute at the highest spindle rotation rate of 60 rpm, which corresponds to strain rates between 15 sec^{-1} and 75 sec^{-1}, depending on which spindle is used in the viscometer. The higher strain rates were used for the thinner grouts.

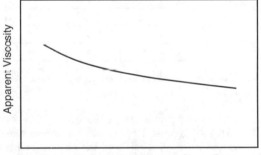

Figure 4. Typical relationship for apparent viscosity as a function of strain rate

5.1 Water:Cement Ratio

Reference to Table 6 for each individual cement grout without bentonite at water:cement ratios of 1:1, 2:1, and 3:1 shows that (a) the yield stress decreases (sometimes substantially, with factors as high as 20) as the water:cement ratio increases, with most of the change occurring between the 1:1 and 2:1 mixes, (b) there is relatively little change in the plastic viscosity for a given grout at different water:cement ratios, and (c) the apparent viscosity decreases with increasing water:cement ratio in a proportion

roughly equivalent to that for the yield stress. Values of the apparent viscosity given in Table 6 were measured at relatively high strain rates (see Figure 4); as the strain rate decreased about ten-fold, the apparent viscosity increased for all grouts except the one with the Type I cement (for which it remained essentially the same). These rate-dependent increases for corresponding grout mixes ranged from 50% to 75% for the MC-100 grouts, 500% to 600% for the MC-300 and

61

Table 7. Rheology data for 2:1 blender-mixed grouts with various percentages of bentonite

Bentonite (% dry weight)	Yield Stress (mPa)	Plastic Viscosity (mPa · s)	Apparent Viscosity (mPa · s)
0	85	72	3
1	150	72	5
2	450	72	41
4	2500	72	205
6	7800	72	778
8	13000	84	1600

Table 8. Effect of mixer type on rheology data for MC-500 grouts

Type of Mixer	Water:Cement Ratio	Yield Stress (mPa)	Plastic Viscosity (mPa · s)	Apparent Viscosity (mPa · s)
Blender	1:1	310	81	35
	2:1	85	72	3
	3:1	25	79	2
Colcrete	1:1	340	81	39
	2:1	70	81	4
	3:1	40	79	3
Paddle	1:1	380	84	43
	2:1	80	84	4
	3:1	30	82	2

blended grouts, and between 100% and 700% for the MC-500 grouts (the smaller increase was measured for the 3:1 mix and the largest for the 1:1 mix).

5.2 Type of Cement

As seen from the data summarized in Table 6 for grouts composed of different cements, but including no bentonite, the type of cement has a strong influence on the yield stress and the apparent viscosity but little effect on the plastic viscosity (values of the latter are about 76 mPa•s ±10% for all grouts of all cements). The MC-100 grouts exhibit relatively low values for both the yield stress and the apparent viscosity. In contrast, the yield stresses and apparent viscosities of the MC-300 grouts are higher than those of the MC-

100 grouts by approximately 2.5, 1.5, and 1.0 orders of magnitude for the 1:1, 2:1, and 3:1 mixes, respectively, thus providing strong quantitative evidence of the difference in particle surface reactivity for portland and slag cements. Grouts prepared by blending MC-100 and MC-300 microfine cements behaved similar to the MC-300 grouts; yield stresses nominally decreased about 10% and apparent viscosities decreased about 35%. A comparison of the behavior of the MC-500 grouts with that of the MC-300 or blended grouts shows that the corresponding yield stress of the former is about 5% to 15% that of the latter and the apparent viscosity of the former is about 5% to 10% that of the latter. The effect of particle size can be most readily distinguished by comparing the behavior of MC-300 and Type I grouts; the MC-300 grouts manifested yield stresses and apparent viscosities between one and two orders of magnitude higher than their counterpart grouts of Type I cement.

5.3 Addition of Bentonite

As illustrated in Table 6, the effect of adding 1% bentonite to the grout mixes resulted in increasing the yield stress and apparent viscosity about two- to three-fold, with virtually no change in a few cases (2:1 and 3:1 mixes of MC-100 and Type I cements) and little change in the apparent viscosity of MC-500 grouts at all water:cement ratios. When the amount of bentonite added to the 2:1 mix of MC-500 grout was increased incrementally to 8%, both the yield stress and the apparent viscosity increased dramatically, as shown in Table 7; on the other hand, the plastic viscosity exhibited little change. When the cement comes into contact with the prehydrated bentonite in the mixer, the slurry undergoes a rapid stiffening caused by mutual flocculation of the negatively charged bentonite particles and the positively charged cement particles; in addition, the cement releases calcium and other ions into solution, and this will also promote flocculation of the bentonite. A combination of these mechanisms, together with particle size and concentration, determines the time-dependent fluidification of the grout mix.

5.4 Type of Mixer

The effect of mixer type on the rheological behavior (yield stress, plastic viscosity, and

Table 9. Effect of mixing water temperature on rheology data for blender-mixed MC-500 grouts

Water: Cement Ratio	Mixing Water Temperature (°C)	Yield Stress (mPa)	Plastic Viscosity (mPa · s)	Apparent Viscosity (mPa · s)
1:1	35	1100	72	105
	20	310	72	35
	5	210	72	31
3:1	35	50	72	2.2
	20	25	81	2.2
	5	10	70	2.5

apparent viscosity) of MC-500 microfine cement grouts is summarized in Table 8. In general, for a given grout mix there is relatively little mixer related variation (generally less than 10% using the Colcrete values as the basis) in the corresponding values of a particular material property.

5.5 Mixing Water Temperature

As shown in Table 9, increasing the temperature of the mixing water from 5°C to 35°C increased the yield stress five-fold, with the majority of the change occurring between 20°C and 35°C. The plastic viscosity was relatively insensitive to temperature changes, as was the apparent viscosity for the 3:1 mix; however, the apparent viscosity of the 1:1 mix increased three-fold as the temperature of the mixing water increased from 5°C to 35°C, with more than 90% of the increase occurring between 20°C and 35°C.

6 CONCLUSIONS

The bleed of cement grouts is strongly influenced by the concentration, size, and reactivity of the particles in suspension. Low water:cement ratio grouts bled much less than high water:cement ratio grouts, and they generally reached their maximum bleed in a noticeably shorter time. Particle size exerted a stronger effect on the rate of bleeding than the total bleed capacity. Among portland cement grouts, MC-300 with more sub-micron size particles settled over a shorter time period with less bleed than the coarser Type I.

Of the microfine cement grouts with similar particle size distributions, the slag cements bled considerably more than the portland cements due to their slower reactivity; accordingly, the larger the percentage of slag relative to portland in blended microfine cement grouts, the more bleeding will occur. The addition of 1% bentonite (by dry weight) decreased slightly the bleed of portland cement grouts, but did not affect or actually increased somewhat the bleed of slag and blended cement grouts. With larger percentages of bentonite to one particular blended cement (MC-500), the maximum bleed decreased substantially and the time to reach maximum bleed was reduced three-fold. The effect of mixer type on bleed is most evident for higher water:cement ratio grouts; Colcrete-mixed grouts settled the least and paddle-mixed grouts settled the most, but sedimentation rates were generally unaffected. Increasing the temperature of the mixing water from 20°C to 35°C decreased slightly the bleed of 1:1 mixes and increased the bleed of 3:1 mixes, whereas decreasing the mixing water temperature to 5°C produced increased bleed for both water:cement ratio mixes; sedimentation rates were not significantly affected.

The rheological properties of cement grouts are governed largely by the same factors as bleed. Decreasing the water:cement ratio from 3:1 to 1:1 increased significantly the apparent viscosity and, to a somewhat lesser extent, the yield stress, but it had little, if any, effect on the plastic viscosity. Coarser-sized portland cement (Type I) grouts had significantly lower apparent viscosities and yield stresses than microfine portland (MC-300) grouts, but particle size had rather little effect on the plastic viscosity. Increasing the slag in a grout dramatically reduced both the apparent viscosity and the yield stress. Grouts with additions of bentonite showed increased apparent viscosity and yield stress, with larger increases being noted for mixes with greater fineness. Grouts prepared by high energy mixing generally had lower apparent viscosities and yield stresses, with the greater differences being manifested by the lower water:cement ratio mixes. In general, the yield stress was most sensitive to mixer type and the plastic viscosity was relatively insensitive. Increasing the temperature of mixing water from 5°C to 35°C tripled the apparent viscosity of the 1:1 mix (little change was measured for the 3:1 mix) and increased the yield stress five-fold for both 1:1 and 3:1 mixes.

REFERENCES

Deere, D.U. and Lombardi, G. (1985), *Grout slurries - thick or thin?,* Issues in Dam Grouting, W.H. Baker, ed. American Society of Civil Engineers, New York, 156-164.

Helal, M. and Krizek, R.J. (1992), *Preferred orientation of pore structure in cement-grouted sand,* Proceedings of the American Society of Civil Engineers, Conference on Grouting, Soil Improvement, and Geosynthetics, New Orleans, 526-540.

Houlsby, A.C. (1982), *Cement grouting for dams,* Proceedings of the American Society of Civil Engineers Specialty Conference, Grouting in Geotechnical Engineering, New Orleans, 1-34.

Houlsby, A.C. (1990), Construction and Design of Cement Grouting, Wiley and Sons, New York, 10-25.

Jefferis, S.A. (1982), *Effects of mixing on bentonite slurries and grouts,* Proceedings of the American Society of Civil Engineers Specialty Conference, Grouting in Geotechnical Engineering, New Orleans, 62-76.

Schwarz, L.G. and Krizek, R.J. (1992), *Effects of mixing on rheological properties of microfine cement grout,* Proceedings of the American Society of Civil Engineers, Conference on Grouting, Soil Improvement, and Geosynthetics, New Orleans, 512-525.

Weaver, K.D. (1991), Dam Foundation Grouting, American Society of Civil Engineers, New York, 19-38.

Grouting in Rock and Concrete, Widmann (ed.) © 1993 Balkema, Rotterdam, ISBN 90 5410 350 7

Injection of rock with microcements

Injektion von Fels mit Mikrozementen

T.A. Melbye
MBT - Europe, Switzerland

ABSTRACT: Grouting and injection are, in many cases, looked upon as a kind of theology; one must have a certain amount of faith in order to be able to trust the materials and technology. In the last few decades, very little development has taken place, either on the product, or the equipment side; attempts have been made to solve the problems using traditional methods with the help of cement and water-glass (sodium silicate).
Of-late, cements have come onto the market which have a much finer grain than the high, early strength Portland cement (HPC), called microcement.
 Microcements are cements produced with special mills and special mill techniques. They can be based on ordinary Portland cement clinker or mixed with slag. The largest grain, in some microcements, is less than 20 microns and with a fineness, of 1500 Blaine (1500 m^2/kg), while an HP-cement, has a grain size of 70-90 microns or even bigger (more than 20% bigger) and a Blaine of 400-480.
This means that the infiltration capacity in very fine fissures is considerably better than ordinary HP-cement and, through practical experience and tests, shows that they have more or less the same infiltration capacity to infiltrate very fine fissures, just like chemical injection fluids (solutions), such as silicate and acrylic resins.
This new cement, microcement, therefore opens up new possibilities in the injection world. We now have an efficient tool based on cement materials and a technique to tighten very fine fissures. In certain circumstances, it can replace the chemical injection fluids (solutions) completely.
The paper will also show examples of microcement in use with pre-injection in tunnels with very though water tightness requirements, in leakage less than 2l/min per 100m tunnel.

Felsinjektion mit Mikrozement: Die Sicherheit von Felsinjektionen wird in vielen Fällen als Glaubenssache betrachtet; Materialien und die damit verbundene Technologien fordern daher ein nicht unerhebliches Mass an Vertrauen.
Während der vergangenen Jahrzehnte hat auf diesem Gebiet sowohl bei den Produkten wie auch bei den Maschinen nur wenig Entwicklung stattgefunden. Aufkommende Probleme wurden mit herkömmlichen Methoden wie der Kombination von Zement mit Wasserglas (Natriumsilikat) zu lösen versucht.
 Seit kurzem sind nun unter dem Namen Mikrozement neue Zementsorten auf den Markt gekommen, die eine viel kleinere Korngrösse aufweisen als beispielsweise hochfrühfester Portlandzement. Mikrozemente werden durch spezielle Mahlverfahren hergestellt. Sie basieren entweder auf gewöhnlichem Portlandzementklinker oder werden mit Schlacke vermischt. Einige Mikrozementsorten weisen eine maximale Korngrösse von weniger als 20 Mikron und eine spezifische Oberfläche von 1500 m2/kg auf. Im Vergleich dazu hat hochfrühfester Portlandzement eine maximale Korngrösse von 70-90 Mikron und eine spezifische Oberfläche von 400-480 m2/kg. Die

Struktur der Mikrozemente bewirkt daher ein viel grösseres Penetrationsvermögen selbst in feinste Risse. Praktische Erfahrungen und Tests haben sogar gezeigt, dass das Penetrationsvermögen mit jenem von chemischen Injektionsflüssigkeiten wie zum Beispiel Silikat- und Akrylharzen gleichgesetzt werden kann.

Dieser neuartige Zement, genannt Mikrozement, eröffnet somit neue Möglichkeiten im Bereich Felsinjektion. Wir verfügen damit über ein wirksames Verfahren, auf Zementbasis bis in feinste Risse injizieren zu können, so dass in gewissen Fällen chemische Injektionsflüssigkeiten (Lösungen) vollkommen ersetzt werden können.

Im folgenden Bericht werden auch Beispiele erläutert, in denen Mikrozement zur Vorinjizierung in Tunnels mit sehr hohen Dichtigkeitsanforderungen (weniger als 2 Liter pro 100 Meter) zur Anwendung kommt.

General Description

Grouting and injection are, in many cases, looked upon as a kind of theology; one must have a certain amount of faith in order to be able to trust the materials and technology. In the last few decades, very little development has taken place, either on the product, or the equipment side; attempts have been made to solve the problems using traditional methods with the help of cement and water-glass (sodium silicate).

In the meantime, extensive research and development work has been underway, necessitated by the large amount of activity, in connection with oil, hydro power installations, railways, road tunnels and sub-sea tunnels. These fields are concerned with the sealing of leakages in rock, and the requirements for new products/methods/equipment had to be met.

Leakage Problem Definition

As a rule, leakage in rock must always be sealed from the wrong side, i.e. the air-face side.

Arguments to the effect that the price of the water lost due to leakage is small in relation to the price of the necessary sealing, do not always hold good.

As a structural engineer, one should not perform sealing on one's hope and faith in chemical facts. Here it is not always necessary to make stabs in the dark in the world of knowledge. As a structural engineer, one must have respect for the facts of chemistry.

Situation Description

The sealing method and choice of materials will depend on the cause of leakage and tightness conditions.

The market displays a bewildering array of different materials, some of which are advertised as if they were washing powders containing high-powder granules. We have to get a grip on ourselves and look reality straight in the eye: concrete cannot be sealed with a "cure-all potion administered by the laying on of hands".

Structural engineering consultants and building owners are being forced into an "orienteering race" in a jungle of materials proliferating with chemical engineering aids. Dangerous plants are found here, sprouting from sales enthusiasm, good faith and anxiety about water

Injection Fluids

The choice of injection fluids depends on water quantity, pressure, fissuring structure, tightness stipulations, hardening time, etc.

Water-loss measurement is a good aid in choosing the agent. We wish here to briefly differentiate between two main types: cement-based suspensions and chemical solutions. The difference between the two groups is that the suspensions contain particles which are held in limbo in an aqueous phase, whereas the chemical solutions are free of particles. Consequently, the chemically soluble ones have the best infiltration capacity in fine cracks, and hardening is effected by the solution forming a gel that sets after a certain period of time and wedges fast in the crack.

With the cement-based suspensions, what happens is that the cracks are filled with solid particles. In the case of systematic pre-injection where there is a stringent tightness requirement, interaction between these two types of injection fluids is necessary to achieve entirely tight injection.

Cement-based

The most common practice is to use high, early strength cement, owing to its fineness, faster setting and hardening, as well as the belief that it penetrates finer cracks better than coarser standard cement. Pure cement slurry is the simplest and cheapest injection material.

Cement injection is hardly relevant in cracks of less than 1-2 mm.

Microcement

General

Of late, cements have come onto the market which have a much finer grain than the high, early strength Portland cement (HPC), called microcement.

Microcements are cements produced with special mills and special mill techniques. They can be based on poor Portland cement clinker or mixed with slag. The largest grain, in some microcements, is less than 20 microns and with a fineness, of 1500 Blaine (1500 m^2/kg), while an HP-cement, has a grain size of 70-90 microns or even bigger (more than 20% bigger) and a Blaine of 400-480.

This means that the infiltration capacity in very fine fissures is considerably better than ordinary HP-cement and, through practical experience and tests, shows that they have more or less the same infiltration capacity to infiltrate very fine fissures, just like chemical injection fluids

(solutions), such as silicate and acrylic resins.

This new cement, microcement, therefore opens up new possibilities in the injection world. We now have an efficient tool based on cement materials and a technique to tighten very fine fissures. In certain consistences, it can replace the chemical injection fluids (solutions) completely. Practice shows that microcement can enter into fissure smaller than 0.25-0.3mm. In connection with injecting cement suspension into fissures, it will easily build up a front in the drilling holes/fissures. Only water from the suspension is able to enter through this barrier. From this barrier through to the front of the drilling/injection holes, there will be a shut down of suspension, as it gradually sediments. How good the tightness will be depends mostly on the grain size, w/c, types of admixtures that have been used and of course the form and size of the fissures. Microcement will always give a better final tightness result because it can penetrate the fissures better.

Criteria cement particle size

When using Microcement, it is of great importance to relate the particle size to the fissures/ voids in the soil or rock to be grouted.

Usually, as mentioned before, one criterion is that a fissure with smaller aperhire than two - three times the maximum grain size of the cement can not be injected. This "rule of thumb" is a very rough one and is based on experience with ordinary HP-cement The criteria should, therefore, be revised to account for the "new" microcements. In a test done by Fell 1989, it was found that the "real" particle size, in the suspension, may differ significantly from the dry cement powder. This is due to the fact that you use an efficient superplasticizer as a dispenser with the microcement technique. As a powder, the fine microcement grains bind together into electrically charged lumps. When the superplasticizer is added to the mix, it will neutralize and disperse the lumps into single grains.

When using microcement for injection, also valid for ordinary cement injection, there are some important factors which influence the final result of injection.

* Bleeding stability
* Max. particle size of the suspension
* Rheological properties of the suspension.

Bleeding

The particles in a cement suspension will, under certain conditions, separate and sedimentation will occur due to gravity. A criterion to measure this is to measure the amount of free water left on top of the cement mix after 2.5 hours. If there is less than 5% of free water of the total volume, then the mix can be considered as "stable".

The mixing equipment has a big influence on the stability, therefore it is important to use mixing equipment with the proper performance. Another important factor is the use/adding of superplasticizers like Rheobuild 1000 to the mix, as it will affect the sedimentation. This is due to what happens during the injection process - the very fine grains become well dispersed, forming a stable mass in the liquid, causing blockage/sediment; this will not take place under pressure.

Rheological properties

Because cement suspension contains particles, the suspension will achieve a shear strength, which must be achieved in order to indicate the flow. When the shear strength is obtained, then the flow will be in accordance with the characteristics of the fluid phase.

The thixotropy influence is also an important parameter for the suspension. The thixotropy effect is reduced by attaching a steaming devise to the mixing drum. This means that the suspension can be agitated, for a certain time, as required. The agitation gives the material a "fresh" performance.

Fineness of cement (specific surface)

Both the shear strength and the plastic viscosity is influenced by the fineness of the cement. If the fineness is increased, it will result in an increase of shear strength and viscosity.

Water/Cement - ratio

When the w/c ratio is decreased, then the shear strength and plastic viscosity will increase. Normally speaking, we have to use w/c ratio of 1-4 with microcement, in order to achieve the fineness of the cement. The w/c ratio depends on the microcement type, fineness, amount, of superplasticizers and the fissures form and size.

Superplasticizers

The main effect of the superplasticizers is to disperse the cement, i.e. particles which have joined together, forming agglomerates, become separated. A radical increase of water in specific surface available will take place, leading to an increase of activity during the hydration process. Superplasticizers also have the effect of negatively charging the particles, causing the particles to repel each other; that means that the particles can not form agglomerates. Superplasticizers will also decrease the shear strengths and the viscosity of the suspension. The viscosity can, with the right amount and the right type of superplasticizers, become close to that of the properties of water. When adding superplasticizers, you can also get a stable mass without bleeding (it is also stable when under pressure). Injection into a fissure is a very important procedure. Microcement, without

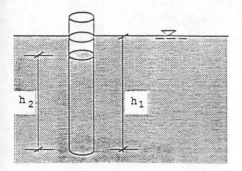

Figure 1.

superplasticizers, will get blocked very quickly. Because of the pressure used, the water is squeezed out of the suspension. Normally speaking, you will always have to use high amounts of superplasticizers when using microcement.

The best effect, with good stabilization, dispersion and a decrease of viscosity is found by using naphthalene (BNS) like Rheobuild 1000. Practical experience and tests have shown that 3 % of c/w is the most suitable dosage rate for superplasticizers.

Field equipment for testing the suspension rheological properties, Ulf Häkansson, Lars Hässler, Häkan Stille, Royal Institute of Technology, Dept. of Soil and Rock Mechanics, Stockholm, Sweden (during a R&D and testing) found an appropriate method for the field testing of microcement.

The method, which can easily be described theoretically, is to immerse a pipe, with known radius and rough inner wall, into the fluid to a certain level (h_1). By observing how high the fluid reaches inside the pipe (h_2), the yield value can be estimated.

The yield value can be estimated by observing how high the fluid reaches inside the pipe.

When the propagation inside the pipe stops the "plug" reaches the wall of the pipe where the shear stress at this moment equals the yield value.

Mathematically the method can be described by performing a vertical

force balance, whereby the yield value can be evaluated from

$$\tau_0 = \frac{\yen R}{2} \left(\frac{h_1}{h_2} - 1 \right)$$

where $\yen$ is the specific weight of the fluid and R is the pipe radius. Naturally, it can be noted that if the fluid reaches all the way up inside the pipe ($h_2 = h_1$), the yield value is zero. Also, of course, the shorter the distance the fluid reaches, the higher the yield value. The above equation can be written:

$$\tau_0 = \frac{\yen R}{2} \left(\frac{\Delta h}{h_1 - \Delta h} \right)$$

where Δh is $h_1 - h_2$. The graphical representation is shown below.

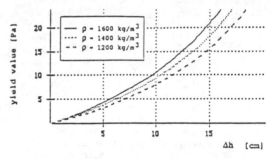

Figure 2.

Yield value as a function of Δh for different densities, pipe radius R = 3 mm

A long pipe with a small radius will give better precision than a short pipe with a large radius.

The density of the fluid, which must be known, can be found by using a simple device for field use from the mud drilling industry, called a "mud balance" (Deere, 1982).

Some preliminary results from a comparison between yield values evaluated with the viscometer and pipe, respectively, are shown in fig 3.

Yield value from viscometer and pipe, respectively. Also shown is the "perfect fit" line.

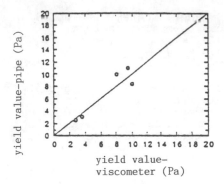

yield value-
viscometer (Pa)

Figure 3.

Mixing procedure

Always start with the thinnest mix
design, w/c (4:1/or 3:1) and if the
suspension goes into the fissure
easily, without building up a
certain pressure, you must slowly
thicken the suspension (decrease
the water) lowest c/w is 1. Always
add the amount of water into the
mixer first. The second step is to
add the superplasticizers (normally
speaking, 3% of c/w) and finally,
add the microcement. Mix it well at
a high speed with a colloidal mixer
for 4-5 minutes. Do not use too
large a mix at any one time, in
order to be sure that you get a
proper mixing. Pour the finished
mix into an agitator, which will
keep the mix turning at a slow
agitation at all times during
injection.
  It is of great importance that you
use a proper injection pressure, so
that you can be sure that you can
place the suspension into the small
fissures. Normally, you have to use
a higher injection pressure than
usual, especially in the final part
of the injection to insure complete
tightness. Often, the final
injection pressure is 40-50 bars or
even higher, depending on the water
pressure and rock conditions. At
the very least, the final pressure
should be twice the water pressure.
Of course, you must take into
consideration the rock conditions
and adjust the injection pressure
to this.

Mixing equipment

Should consist of the following

units and properties

* high speed colloidal mixer
* agitator
* grouting pump (piston)

Same conditions as for the normal
cement injection. It is of great
importance that you have a very
efficient colloidal mixer, which
should operate at least 1500 r.p.m.
or another type of mixer which can
give similar mixing results. At the
entrance of the agitator there
should be a strainer mounted to
prevent large solid particles from
reaching the pump/mix.

Advantages of using microcement

* Ordinary cement injection,
  technology and equipment
  can be used
* Better tightness performance
  than with ordinary cement
* Can replace chemical fluids/gels
* Better working environment/no
  toxic problems
* Higher durability
* Higher strengths than with
  chemical gels
* More economical, compared to a
  chemical injection

Where to use it?

- Rock injection

* in all underground
  construction work,
  like tunnels, caverns, etc.
* pre-injection
* post-injection
* stabilization of rock

- Soil injection
* stabilization
* water tightness
* prevent ground water lowering
* Injection of cracks in concrete
* Injection into injection hoses
  like Fuko
* Jet grouting
* Mud Jack grouting
* Grouting of anchors in rock and
  soil
* Prepacked grouting - fine
prepacked
  materials
* Dam injection
* Contact injection

MBT - products

MBT - Rheocem system is a special microcement system for injection of rock, soil and concrete. The system consists of two components one part is Rheocem-microcement and the other is superplasticizer.

Rheocem 650 SR

A super-fine sulphate which resists Portland cement with a low C_3A and a low alkaline content and complies to BS 4027 "Sulphate resisting Portland cement".

Typical properties

* Surface Area m^2/kg   650
* Max. grain size   30
* Initial set   80-160 min.
Final 140-210 min.

Compressive strengths
(BS 4550 concrete test)
1 day	15-20 Mpa
3 days	28-34 Mpa
7 days	36-42 Mpa
28 days	45-52 Mpa

Rheocem 900

Rheocem 900 is an ultrafine Portland cement which is ground and classified from a Portland cement clinker to a low particle size graded cement

Typical properties
Surface Area (m^2/kg)	875-950
Setting Time	
Initial (mins)	60-160
Final (mins)	120-240

Particle Size Distribution
finer than
40 microns (%)	100
15 microns (%)	93
10 microns (%)	80
5 microns (%)	49
2 microns (%)	22

Properties

Main particle size 5 microns

Health and safety for Rheocem

Contact with Microcem, mixed with body fluids (eg. sweat or eye fluid) or with concrete or mortar should be avoided as it may cause irritation, dermatitis or burns.
If such contact occurs, then the affected area should be washed with plenty of clean water. In case of eye contact, seek immediate medical advice.

Rheobuild 1000

Rheobuild 1000 is a water-soluble, sulphonated polymer-based admixture, which allows mixing water to be reduced considerably and injection grout strength to be accelerated significantly, particularly at early stages.

Advantages

Rheobuild 1000 provides high quality, rheoplastic - flowable and non-segregating-grout. Dosage: 3% c/w.

Compatibility

Rheobuild 1000 is compatible with all comments and mixtures meeting the ASTM and UNI standards.

Dispensing

Rheobuild 1000 is introduced into the mixer together with water.

Closing remarks

With this information, it is our intention to give an overview of injection techniques and to give ideas on the procedures, in a practical way, without going too much into detail.
We hope that this information has given you some basic input and ideas, and that your next grouting project will be successful.

References: Tom Melbye Grouting and Injection of Rock 87 Örjan Sjöström Rock Team AB, Sweden Grouting, Sealing, Strengthening and Stabilizing of Rock and Soil Ulf Häkansson 1) Rheological properties

Lars Hässler 2) of Microfine
cement
Håkan Stille 3) with Additives

1) Royal Institute of Technology,
Dept. of Soil and Rock
Mechnanics, Stockholm, Sweden.
2) Golder Associates Inc., Uppsala,
Sweden.
3) SKANSKA AB, Stockholm, Sweden

The Snake in Stockholm

In cycles of 5 years heavy rain
falls over the city of Stockholm
are causing problems as the
capacity of the network of the
pipelines for rain and waste water
is insufficient. To reduce
overflows into the surrounding
rivers and lakes, a tunnel has been
excavated to serve as a temporary
storage of surplus water which is
kept in this reservoir until the
pressure on the pipelines and the
waste water treatment plants is
reduced. Eight raised bared shafts
lead the rain water from the
streets down to this tunnel. Due to
its winding form the tunnel is
called "Ormen" (ormen = snake).

Sensitive excavation area

"The Snake" was constructed at a
depth of 40 to 60 metres beneath
the central part of Stockholm city
in an extremely sensitive area of
the old town where many of the
houses are supported on wooden
piles. Any lowering of the ground
water level in the vicinity of
theses buildings would have
resulted in serious subsidence,
causing settling and cracking of
the buildings.
  For this reason the level of
permitted leakage of water into the
tunnel was set at a maximum of 2
litres per minute per 100 metres of
tunnel length. This is a very
severe requirement which
practically means 100%
watertightness.
  Two procedures were applied in
order to reach this requirement:

- Continuous preinjection of the
tunnel alignment was necessary in
order to close any cavities and
keep any water out of the

excavation area ahead of the tunnel
boring machine.

- The use of the full face boring
method eliminated the risk of
vibration damages to overlying
structures and was the least likel
to cause the type of damage to the
rock resulting in a collapse or
water ingress through fissures.

Waterproofing and injection by
preinjection

All waterproofing and injection
operations have been executed by
preinjection.
  The tunnel boring machine was
furnished with drilling and
injection equipment. While the
boring operation advanced with a
speed of 2.3 metres per hour eight
injection holes were driven. All
injection holes were subjected to
a water pressure test to determine
the adequate mix design and
injection material for every
single ring of holes. The choice
of mix design and w/c ratio also
depended upon the porosity of the
strata.

Rheocem: the solution for
preinjection

Based on former experience and
some previous tests, Rheocem
microcements were chosen as
injection materials. Both Rheocem
650 and Rheocem 900, in some
extreme cases in combination with
Delvocrete, were used.

The combination of Rheocem and
Delvocrete: a revolution in the
injection world

With Rheocem expensive and toxic
resins can be replaced. Rheocem
microcements have the same
penetration into fine materials as
resins, but make a much more
durable, more economical and more
environment friendly solution.
  Rheocem in combination with
Delvocrete gives a cementitious
compound that allows adjusting of
the setting time exactly as with
resins. The use of this
combination proved to be very
efficient in the event of open

voids in the rock and of surface
leakages: a barrier is easily
built up into the rock which
prevents the uncontrolled spread
of the injection material.
  We believe this technology is the
future for all tunnel injections.
It even solves the difficult
problem of waterproofing road
tunnels: instead of placing
membranes, waterproofing can be
done by preinjection with Rheocem
microcement. Preinjection with
Rheocem is a cheaper and
technically better solution.

Result

  The measuring of the final water
leakage into the finished tunnel
was impressive. The measured values
of less than one litre per minute
per 100 metres of tunnel were well
below the required two litres per
minute per 100 metres of tunnel.
  Even the estimate of the quantity
of injection material as stipulated
in the tender proved to be
accurate: The stipulated
consumption of 62 kg per metre of
tunnel came impressively close to
the 60 kg per metre which were
actually consumed in the project.

Conclusion

"The Snake" project is a good
reference for our new microcement
technology consisting of Rheocem
microcement, our cementitious
injection equipment and Delvocrete,
which is not only revolutionary for
shotcreting, but also an excellent
tool for injection.
  Bo Carlsson, Site Manager of the
Ormen project concluded: "The
Snake" project is not a tunnel job,
but really an injection job. The
result of the injection decided on
the tunnelling advances und the
final result.

Injektion mit Feinstbindemittel in der Geotechnik

Grouting with microfine binders in geotechnical applications

Peter Noske & Georg Kühling
Heidelberger Baustofftechnik GmbH, Leimen, Deutschland

Zusammenfassung

Feinstbindemittel – mikrofeine, hydraulische Bindemittel – werden als mineralische Suspensionen für Injektionszwecke in der Geotechnik angewendet (Verfestigungen, Abdichtungen). Weiterentwicklungen der vergangenen Jahre führten zur Erweiterung der Anwendungsgebiete. Mittlerweile gelten Sie als Alternative zu den sogenannten chemischen Injektionsgütern und besitzen Vorteile hinsichtlich der Umweltverträglichkeit.

Durch Modifikation der Suspensionen können unterschiedliche Eigenschaften beeinflußt werden. Dies ermöglicht den Einsatz in verschiedenen Anwendungsgebieten.

Summary

Microfine hydraulic binders are being applied as mineral suspension for grouting purposes in geotechnics (consolidation, impermeabilization). Fields of application were extended by proceedings in the last few years. Meanwhile they are an alternative to the socalled chemical grouts and are provided with clear advantages concerning environmental compatability.

By modification of the suspensions, different properties can be influenced. Various applications can be operated in that way.

Was sind Feinstzemente und wie unterscheiden sie sich von Standardzementen (Normzementen)?

Der Blaine-Wert, als Maß für die spezifische Oberfläche, ist die zunächst auffälligste Kenngröße, die Feinstzemente von Standardzementen unterscheidet.

Standardzemente weisen je nach Sorte einen Blaine-Wert von 2.700 cm^2/g – 5.700 cm^2/g auf. Je nach Typ des Feinstzements liegt der Blaine-Wert zwischen 12.000 cm^2/g und 16.000 cm^2/g.

Resultierend aus dieser hohen spezifischen Oberfläche ist der Wasseranspruch nach DIN 1164 auf ca. 60 % gestiegen; für Injektionen mit Feinstzementen sollten Wasserbindemittelwerte (W/B-Werte) > 0,8 gewählt werden.

Es ist von sehr großer Wichtigkeit, daß die hohe Feinheit ausschließlich über den Kornaufbau des Feinstzementes erzielt wird. Das bedeutet, daß Kornaufbau und Größtkornanteil die entscheidenden Kriterien sind, welche einen Feinstzement charakterisieren.

Es wäre z.B. möglich, einen herkömmlichen Portlandzement mit Silikatstäuben (z.B. Silica-Fume) zu versetzen, um so einen drastischen Anstieg des Blaine-Wertes zu erzielen. Bei diesen "abgemischten Bindemitteln" wäre aber noch der Grobanteil des Standardzementes enthalten. Dies führt jedoch zu keiner Verbesserung der Penetrationsfähigkeit der Suspension in den Poren und Kapillaren des Bodens. Je höher der Anteil grober Partikel des abgemischten Bindemittels ist, um so größer ist der die Penetration behindernde

Kenngrößen verschiedener Zemente

Kenngrößen	Zemente	PZ 35 F	PZ 55	ADDIMENT Microcem *) A	B
Spez. Oberfläche	cm²/g	2.700 – 3.300	5.400 – 5.700	11.000 – 12.000	15.000 – 16.000
Schüttdichte	kg/dm³	ca. 1,30	ca. 1,30	ca. 0,70	ca. 0,70
Dichte	kg/dm³	ca. 3,10	ca. 3,10	ca. 3,00	ca. 3,16
Wasseranspruch für Normsteife	Gew.-%	25 – 31	31 – 34	ca. 55	ca. 60

*) Werte ermittelt in Anlehnung an DIN 1164

Abb. 1 : Unterschiede zwischen Standardzementen und mineralischen Feinstbindemitteln

Fig. 1 : Differences between standard cements and microfine mineral hydraulic binders

Abb. 3 : REM-Aufnahmen von Zementpartikeln

Fig. 3 : REM-photographs of cement particles

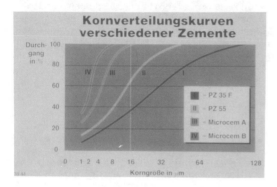

Kornverteilungskurven verschiedener Zemente

Durchgang in %

- I – PZ 35 F
- II – PZ 55
- III – Microcem A
- IV – Microcem B

Korngröße in µm

Abb. 2 : Unterschiede der Kornverteilungskurven von Standardzementen und Feinstzementen

Fig. 2 : Differences of particle size distribution of standard cements and microfine mineral hydraulic binders

Effekt. Dies ungeachtet des willkürlich erhöhten Blaine-Wertes infolge zugemischter Feinstanteile.

Feinstzemente sollten daher möglichst nur Grobanteile von < 16 µm enthalten.

Als Richtwerte können hierbei die Durchgangswerte bei 16 µm betrachtet werden. Bei Feinstbindemittel sollte hier der Durchgangswert > 95 Gew.-% betragen, d.h. nur maximal 5 % der Zementpartikel dürfen größer als 16 µm sein.

Die unterschiedliche Partikelgröße von Normzement (HOZ 35 L) und Feinstbindemittel ist in Abb. 3 zu sehen. Der HOZ 35 L, mit einem w/z-Wert von 0,6 (links) enthält Partikel, die deutlich

größer sind als 16 µm (2.000-fache Vergrößerung, 24 h Hydratation, 5U = 5 µm). Das Feinstbindemittel mit einem W/B-Wert (Wasserbindemittelwert) von 2,0 weist keine groben Partikel, also kein Sperrkorn auf (Mitte). Rechts ist das sehr dichte Gefüge des Feinstbindemittels mit einem W/B-Wert von 1,0 zu erkennen (2.500-fache Vergrößerung, 24 h Hydratation, 5U = 5 µm).

Feinstzemente wurden u.a. für geotechnische Anwendungen konzipiert, um aus Gründen des Umweltschutzes eine mineralische Alternative zu den chemischen Injektionsmitteln zu haben. Mit Feinstzementen gelingt es, in nahezu alle Poren- und Kapillarräume zu penetrieren, die bislang nur durch chemische Injektionsmittel penetrierbar waren.

Bei der Injektion mit Feinstzementen sind einige grundlegende Gedankengänge notwendig.
Die bodenphysikalischen Parameter müssen berücksichtigt, können aber nicht beeinflußt werden. Die Beeinflussung des Penetrationsverhaltens muß sich daher auf die Steuerbarkeit der Fließeigenschaften der Suspension beziehen.

Im Gegensatz zu den "einphasigen" chemischen Lösungen handelt es sich bei Feinstzementsuspensionen um ein "2-Phasen-System" aus feinsten Zementpartikeln und Wasser.
Bei der Injektion einer Suspension treten in den Poren- und Kapillarräumen grundlegende bodenphysikalische Vorgänge in Erscheinung.

Während der Injektion werden Durchfluß-

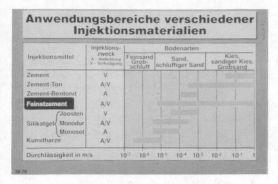

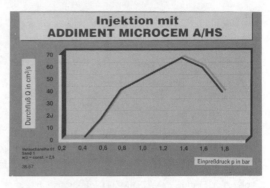

Abb. 4 : Einfluß der Bodenart auf die Wahl des Injektionsmittels

Fig. 4 : *Influence of type of soil concerning the selection of the grouting material*

Abb. 5 : Einfluß von Durchflußmenge in Abhängigkeit vom Einpreßdruck

Fig. 5 : *Influence of percolation depending from grouting pressure*

mengen und Fließgeschwindigkeiten zunächst proportional zum Einpressdruck ansteigen. Diese Phase ist durch ruhiges, laminares Fließen mit geringen Fließwiderständen charakterisiert.

Bei Überschreitung eines bestimmten Einpressdruckes wird aus dem ruhigen, laminaren Fließvorgang fast schlagartig ein turbulentes Fließen. Der Bereich des Darcy'schen Gesetzes wird verlassen. Die Fließwiderstände steigen drastisch. Infolge von Entmischungsvorgängen der Suspension werden Durchflußmengen und Fließgeschwindigkeiten, trotz gesteigerter Einpressdrücke, nicht weiter ansteigen.

Die Höhe des Einpressdruckes ist abhängig vom Bodenaufbau und dem Aufbau der Suspension. Beide stehen in gegenseitiger Wechselwirkung.

Durch Überschreiten des "Druckoptimums" können weder große Fließweiten der Suspension noch gesteigerte Füllgrade des Bodens erzielt werden. Dies hat Nachteile auf die Festigkeit und Dichtigkeit der injizierten Böden.

Als Nebeneffekt kann es möglicherweise zu Zerstörungen der Bodenstruktur infolge innerer Umlagerungen von Feinmaterial des Bodens durch übersteigerte Strömungsvorgänge kommen (Suffosion).

Durch Zugabe besonderer Additive können u.a. die Fließeigenschaften der Suspension entscheidend beeinflußt werden, so daß bei höheren Einpressdrücken immer noch laminare Fließvorgänge möglich sind.

Durch die Anwendung von Injektionshilfen (IH) kann die Neigung zur Agglomeratbildung, die zu einer "Vergrößerung" der Teilchen führt, in erheblichem Umfang re-

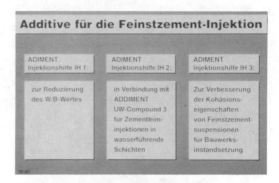

Abb. 6 : Unterschiedliche Injektionshilfen für Feinstzementsuspensionen

Fig. 6 : *Different types of ADDIMENT Pressure Grouting Agents on suspensions of microfine hydraulic binders*

duziert werden. Die hier verwendeten Additive werden nämlich an der Oberfläche der Zementpartikel bzw. an den Reaktionsprodukten des Zementes absorbiert und laden die Teilchen gleichsinnig auf (Neutralisation). Da sich gleiche Ladungen abstoßen, kommt es nur in sehr viel geringerem Umfang zu Agglomerationen. Die abstoßenden Kräfte bleiben auch dann erhalten, wenn die Teilchen von Wasser umhüllt sind, womit sich die innere Reibung des Zementleimes gleichzeitig reduziert.

Für die Eigenschaften der Injektionssuspension bedeutet dies, daß:

1. die Fließgrenze γ_0 der Frischsuspension, die u.a. ein Maß für die Verarbeitbarkeit der Mischung ist, kann durch den Zusatz dieser Addi-

77

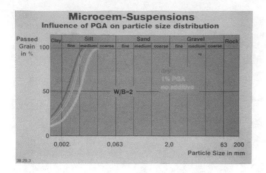

Abb. 7 : Einfluß der ADDIMENT Injektionshilfe auf die Partikelgrößenverteilung in der Suspension

Fig. 7 : *Influence of ADDIMENT Pressure Grouting Agents on particle size distribution in suspension*

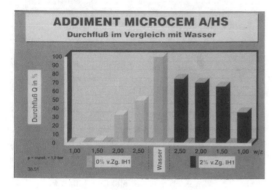

Abb. 8 : Feinstzement-Suspension mit/ohne Injektionshilfe

Fig. 8 : *Suspensions of microfine hydraulic binders with/without Pressure Grouting Agent*

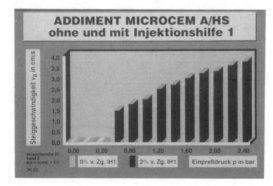

Abb. 9 : Abhängigkeit zwischen Fließgeschindigkeit, Einpressdruck und IH1

Fig. 9 : *Interaction between velocity of flow, grouting pressure and PGA1 (IH1)*

tive erheblich reduziert werden,

2. die Partikelgrößenverteilung des Feinstzementes in der Suspension bei Zugabe dieser Additive nahe an die feinere Korngrößenverteilung des trockenen Feinstzementes herankommt,

3. feststoffreiche Suspensionen gezielt injiziert werden können.

Der Zusatz von ADDIMENT Injektionshilfe bewirkt, daß die Partikelgrößenverteilung in der Suspension weitgehend identisch mit der des Trockenmaterials bleibt. Die guten Penetrationseigenschaften bleiben dadurch erhalten.

Der entscheidende Einfluß der Injektionshilfen zeigt sich bei den rheologischen Eigenschaften der Frischsuspension. Besonders markante Ergebnisse sind nachstehend dargestellt. Die Dichte, der pH-Wert, die elektrische Leitfähigkeit, die Sedimentationsstabilität und die Erhärtung des Injektionsleimes werden durch die Injektionshilfe nicht nachteilig beeinflußt.

Markante Suspensionseigenschaften
(W/B = 1) [4]

Parameter	ohne Zusatz	+ 1 % IH (1)	+ 2 % IH
Fließgrenze	45 Pa	36 Pa	14 Pa (2)
Marshzeit	n.m.	900 s/1	61 s/1

(1): Zugabe von Injektionshilfe 1, bezogen auf das Feststoffgewicht

(2): gemessen mit einem Rotationsviscosimeter der Fa. Haake

Bei W/B-Werten ≥ 2,0 hat die Injektionshilfe einen größeren Einfluß auf das Dispergieren der Feinstzementpartikel als auf die "Fließzeiten" der Suspension. Der Einfluß auf die Fließ- und Penetrationsfähigkeit wird auch bei höheren W/B-Werten deutlich (Abb. 8).

Wenn Wasser eine 100 %ige Penetrationsfähigkeit hat, ist diese Eigenschaft bei Suspensionen vom W/B-Wert und von der Zugabe von IH1 abhängig. Suspensionen mit

Injektionshilfe penetrieren bei geringen W/B-Werten, während Suspensionen ohne Injektionshilfe nicht penetrieren. Die Erhöhung des Wassergehaltes bringt nur geringe Verbesserungen der Penetrationsfähigkeit.

Im Gegensatz zur bisherigen Meinung ist es möglich, auch feststoffreiche Suspensionen einzubringen, um Festigkeiten und Dichtigkeiten der injizierten Böden zu steigern.
Die einst üblichen W/B-Werte 5,0 - 8,0 können bis zu 1,0 reduziert werden.

Die Anwendung spezieller Zusatzmittel erlaubt höhere Einpressdrücke, bevor turbulente Fließvorgänge beginnen; d.h. der Bereich laminaren Fließens mit geringen Fließwiderständen erweitert sich.

Labor- und Feldversuche zeigen, daß analog zu den Fließgeschwindigkeiten (Steiggeschwindigkeit) die Durchflußmengen gesteigert werden können. Durch Erweiterung des laminaren Fließbereiches kommt es zur vollständigeren Ausnutzung der Kapillarräume als Fließwege. Dies könnte zu größeren Bohrlochabständen führen und die anfallenden Bohrkosten senken.

Die Dichte, der pH-Wert, die elektrische Leitfähigkeit, die Sedimentationsstabilität und die Erhärtung des Injektionsleimes werden durch die Injektionshilfen nicht nachteilig beeinflußt.

Suspensionen mit speziellen Injektionshilfen können eine geringere Fließgrenze als Suspensionen mit höherem W/B-Wert und ohne Zusätze haben.
Damit die rheologischen Eigenschaften der Suspensionen gezielt beeinflußt werden, ist es notwendig, Kenntnis über die anstehenden geologischen Parameter des Bodens zu haben. Geologische Parameter und Aufbau der Suspensionen stehen in enger Wechselwirkung.
Der Kornaufbau des Bodens wird im Diagramm als Sieblinie dargestellt. Ihre Lage, Verlauf und Neigung mit den daraus resultierenden Parametern, sowie Form und Oberfläche der Partikel bestimmen primär das Porengesamtvolumen, Dimensionierung und Form der Kapillaren sowie Kapillarwirkung.
Porenvolumen, Porenraumverteilung und Verteilung der kapillaren Engstellen des Bodens sind Hauptfaktoren zur Beurteilung der Injizierbarkeit.
Ebenso wichtig sind Kornaufbau der Feinstzemente, insbesondere deren Grobanteil.

Alle diese Parameter stehen untereinander in Wechselwirkung und in ihrer Summe in Wechselwirkung zum Aufbau der Suspension.
Signifikante Parameter der Suspension sind dabei W/B-Wert sowie die rheologischen Eigenschaften.

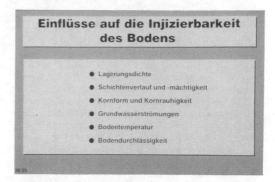

Abb. 10 : Verschiedene Einflüsse des Bodens
Fig. 10 : Several influences of soil

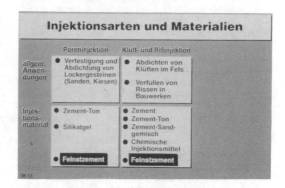

Abb. 11 : Verschiedene Einsatzgebiete von Feintzementen
Fig. 11 : Several fields of application of micofine hydraulic binders

Für die Beurteilung, ob ein Boden injizierbar ist oder nicht, reicht die übliche Kornverteilungskurve in vielen Fällen nicht aus.

Die wesentlichen Anwendungsgebiete bei der Feinstzementinjektion sind die Abdichtungs- und Verfestigungsinjektion.
Unterschiedliche Arten von Poren, Klüften und Hohlräumen können abgedichtet werden, andererseits lassen sich poröses Mauerwerk, geklüfteter Fels oder Lockergestein verfestigen.
Mögliche Einsatzgebiete bestehen z.B. im Erd- und Grundbau; in der Verfestigung von Lockergestein (z.B. Hangsicherung), bei Gebäudeunterfangungen sowie beim Herstellen von Injektionssohlen von Baugruben. Weiterhin ist es möglich, Dichtungsschleier im Dammbau herzustellen, zur Verhinderung von Wasserum- und -unterläufigkeiten. Im Felsbau wird Feinstzement

zum Verfüllen von engen Felsklüften bzw. Rissen eingesetzt. Die Rißsanierung im Beton- und Mauerwerksbau war bislang eine Domäne für chemische Injektionsmaterialien, hier liegt aber ein zukunftsträchtiges Einsatzgebiet für Feinstzement liegt. Dazu zählt u.a. auch die Sanierung von Staumauern. Einsatzgebiete in der Umwelttechnik sind z.B. die Sicherung von Altlasten in feinklüftigem Tongestein sowie die Sanierung von Abwasserkanälen. Die Abdichtung von Abwasserrohren mit Rissen ist mit speziellen Feinstzement-Suspensionen möglich.

Hierbei muß ein wesentliches Augenmerk auf die Durchlässigkeit und chemische Widerstandsfähigkeit der Feinstbindemittel-Suspensionen bzw. des ausgehärteten Materials gelegt werden.
Die Merkmale werden im Langzeitversuch untersucht, wobei verschiedene Prüfkörper aggressiven Flüssigkeiten ausgesetzt werden. Es konnte deutlich gemacht werden, daß erhärtete Prüfkörper an feststoffreichen Suspensionen (geringe w/b-Werte) dem chemischen Angriff standhielten, während dies bei Prüfkörpern aus feststoffarmen Suspensionen nicht der Fall war. Sie versagten dem chemischen Angriff bereits nach wenigen Tagen.

Durch eine Erhöhung des Feststoffanteils wird das Gefüge des Feinstbindemittel-steins erheblich dichter und der verfügbare Porenraum nimmt in nennenswertem Maße ab. Dies bewirkt eine deutliche Reduzierung des Durchlässigkeitsbeiwertes sowohl gegen Wasser als auch gegen aggressive Flüssigkeiten.

In Langzeit-Durchlässigkeitsversuchen (350 d) wurden, selbst bei massivem chemischem Angriff durch synthetische Sickerwässer, Durchlässigkeitsbeiwerte für den feststoffreichen Feinstbindemittel-stein von $k_{10} = 1 \cdot 10^{-10}$ bis 10^{-11} m/s mit nahezu gleichbleibender Tendenz ermittelt (bei W/B = 1,0).
Chemisch-mineralogische Untersuchungen bestätigen, daß bei den feststoffreichen Suspensionen im allgemeinen lediglich in der obersten, direkt mit der Prüfflüssigkeit beaufschlagten Schicht geringe Umsetzungsprozesse stattgefunden haben. Ansonsten ist der Feinstbindemittelstein nahezu völlig intakt.
Lediglich bei Beaufschlagung mit einem künstlichen Sickerwasser, bestehend aus starken Mineralsäuren, treten, infolge von Lösungs-, Transport und Fällungsvorgängen neu entstandener Produkte, deutlichere Veränderungen in der chemisch-mine-

ralogischen Zusammensetzung auf, doch auch in diesem Fall verhält sich die feststoffreiche Suspension günstiger.

Literatur

[1] Huth, R.; Kühling, G.:
Injektion mit Feinstbindemittel; 8. Nat. Ing.-Geol. Tagung, Tagungsband S. 162-166, Berlin 1991

[2] Kühling, G.:
Feinstzemente - mikrofeine hydraulische Bindemittel; TIS 11/90, S. 782-784

[3] Noske, P.:
Injection with Microfine Hydraulic Binders; La Meccanica del Rocce a Piccola Profondità; Turin 10/1991

[4] Schulze, B., Kühling G., Tax, M.:
Neue Zusatzmittel für feststoffreiche Feinstzement-Suspensionen; Bauingenieur 11/92

[5] Schulze, B.:
Resistenz von Feinstzementsuspensionen gegen chemischen Angriff; Bauingenieur 66 (1991)

[6] Schulze, B.:
Injektionssohlen: Theoretische und experimentelle Untersuchungen zur Erhöhung der Zuverlässigkeit; Veröffentlichungen des Instituts für Bodenmechanik und Felsmechanik - Universität Karlsruhe, Hest 126

Longevity of bentonite and cement grouts

Dauerhaftigkeit von Bentonit- und Zement-Injektionen

Roland Pusch
Clay Technology AB & Lund University of Technology, Sweden

ABSTRACT: Physical and chemical stability of grouts are required for use as seals in repositories. Physical stability is needed for avoiding piping and erosion and the thixotropic behavior smectite gels offers quick initial hardening to resist considerable hydraulic gradients. Cement grouts harden considerably after a few hours except at low temperatures and can sustain high gradients thereafter.

Chemical stability is needed for long term performance and smectite and cement grouts appear to behave similarly although the degradation processes are quite different. Smectite gels degrade by dissolution and transformation to non-expanding hydrous mica (illite) at a rate that is controlled by temperature and access to potassium. Cement degrades by dissolution of the portlandite, ettringite and calcite components, a process that initially creates self-sealing but later leads to an increase in porosity and hydraulic conductivity.

A major difference between smectite and cement grouts is that the former may sustain some fracture displacement while the brittle cement does not.

ZUSAMMENFASSUNG: Physikalische and chemische Stabilität von Injektierungsmitteln sind Voraussetzungen für deren Anwendung in Anlagen für Aufbewahrung von radioaktiven Abfälle. Physikalische Stabilität is notwendig um "Piping"-Erscheinungen und Erosion zu vermeiden, und das thixotropische Verhalten smektitischer Gele bieten eine schnelle initielle Härtung an, die erhebliche hydraulische Gradienten zu wiederstehen vermag. Zementbasierte Injektierungsmittel härten erheblich nach einige Stunden wenn die Temperatur nicht zu niedrig ist und können danach hohe hydraulische Gradienten wiederstehen.

Chemische Stabilität ist für die langfristige Funktion erforderlich, und Smektit- und Zement-basierte Injektierungsmittel weisen diesbezüglich gleichartige Verhalten auf, auch wenn die Degradierungsprozesse völlig underschiedlich sind. Smektitische Gele degradieren durch Auflösung und Umwandlung in nich-quellfähige Hydroglimmerminerale (Illit). Zementbasierte Injektierungsmittel degradieren durch Auflösung der Portlandit-, Ettringit- und Calzitkomponenten, eine Prozesse die anfangs eine Selbstabdichtung bewirkt, aber später zu erhöhten Porosität und Wasserdurchlässligkeit führt.

Ein bedeutender Underschied zwischen Smektit- und Zementbasierte Injektierungsmittel ist dass die frühgenannte Verschiebung entlang Gefügetrennflächen wiederstehen können, was für den spröde Zement nicht der fall ist.

1 INTRODUCTION

Grouts will be used for several purposes and their properties have been investigated in detail in the Stripa project (Pusch et al., 1991). Thus, sealing of major water-conducting fracture zones that are intersected by tunnels and shafts must be made in the construction phase and additional sealing of such zones and certain minor zones and discrete fractures is required for making the waste application rational. A second application is for sealing at strategic sites for retarding and shunting off groundwater flow. There are quite different demands with respect to long-term performance of these two applications. Thus, while sealing for constructing repositories and applying the waste and its embedment does not have to be intact for more than a few decades, seals for minimizing groundwater flow and radionuclide transport have to be intact for many thousands of years.

2 SHORT-TERM STABILITY OF INJECTED GROUTS

2.1 *General*

Piping is the term for local, fast penetration of water creating continuous passages through a soil exposed to a hydraulic gradient. In soil engineering practice particular emphasis has been put to the development of techniques for measuring this soil property, and comprehensive theoretical and experimental work on the subject have been conducted and reported in literature. Still, the exact mechanisms involved are not known with certainty but for smectite gels it is indicated from microscopy studies (Pusch, et al., 1987 and Pusch, 1983) that piping can take place in the two ways indicated in Fig. 1 in homogeneous soft gels.

Applying basic fracture theory of elastoplastic media and estimated strength and stress/strain data, it has been concluded that piping should occur at a fairly high pressure gradient, and experiments in the form of observation of the development of piping using light microscopy have confirmed this.

2.2 *Smecite clay*

Smectite clay grouts undergo very quick thixotropic strength regain after the completion of the injection and further increase in strength takes place for days and weeks thereafter. Thus, the risk of piping and erosion, which both depend on the shear strength of the clay, is highest immediately after the injection. It appears that gradients higher than about 30 may produce piping and formation of continuous channels where flow of water may produce erosion if the rate of flow exceeds about 10^{-4} m/s (Pusch et al., 1991). Self-sealing of such channels tends to occur especially in salt

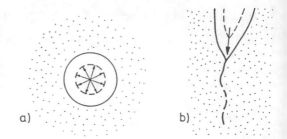

Fig. 1 Piping in soft smectite gels by a) radial expansion of void, or b) "hydraulic fracturing".

Wasserdurchbruch in weichem Ton durch a) radielle Expansion von einem Porenraum, oder b) "hydraulischer Bruch"

groundwater since the clay torn off by erosion has the form of large aggregates which may be stuck and cause clogging in constrictions.

2.3 *Portland cements*

While smectite clay grouts undergo considerable thixotropic stiffening almost instantaneously when the mechanical agitation caused by the injection is stopped, freshly injected cement slurry with 1-1.5% superplasticizer undergoes some minor strengthening after about 10 minutes but does not harden significantly until after 10 hours at room temperature and 20 hours at 10°C rock temperature. Pinhole-type tests under the microscope have shown that piping takes place at much lower critical hydraulic gradients than in smectite grouts and strong erosion is caused by water flow rates as low as 10^{-5} m/s in the first 5-10 minutes at room temperature (Pusch et al., 1987). This means that freshly injected cement grout films are very sensitive to water flow along or through them.

3 LONG-TERM STABILITY

3.1 *General*

The chemical stability of smectite and cement grouts under repository conditions was one of the major subjects in the Stripa Rock Sealing Project, and it was considered both by applying current and new degradation models, hydrothermal experiments, and reference to geological evidence. The matter turns out to be very complex both for smectites and cements and rather crude estimates have to be made for prediction of the long-term performance.

3.2 Smectite clay

The major threat to long-term performance of smectite clay grouts in repository environment is conversion to 10 Å non-expanding hydrous mica (illite) by collapse of smectite stacks through uptake and fixation of potassium.

The Stripa Project comprised hydrothermal experiments that were conducted using clay grouts prepared with a water content of 1.3 times the liquid limit, i.e. on the same order of magnitude that is required to make them injectable. Both pure clay samples and mixtures with very fine-ground quartz powder were investigated (Pusch et al., 1991).

The testing involved exposure of the samples to hot pressurized distilled water or strongly brackish solutions with sodium or calcium as major cations and chlorine as dominant anion, for periods of up to 270 days, with subsequent testing of the hydraulic conductivity and shear strength as well as of changes in mineral content and composition.

The amount of montmorillonite appeared to have been almost unchanged in the Na bentonite clay samples saturated with distilled water and exposed to temperatures up to 200°C, as concluded from X-ray diffraction and chemical analyses. The hydraulic conductivity of these samples was also similar to that of untreated clay with the same density (1.08 g/cm^3), i.e. around $5 \cdot 10^{-9}$ m/s. 75/25 and 50/50 mixtures of bentonite/quartz saturated with distilled water and somewhat higher density (1.1-1.15 g/cm^3) showed even lower values. The shear strength of the materials prepared with distilled water increased by several hundred per cent at 90°C and even more at 130°C, while it dropped at higher temperatues.

The samples exposed to strongly brackish solutions showed no clear change in montmorillonite content except for the clay exposed to the most potassium-bearing solution, which gave indication of some very slight formation of 10 Å minerals in 270 days at 200°C. Separate experiments with very potassium-rich solutions and heating at 200°C for 1 month showed clear but small conversion of montmorillonite to hydrous mica, possibly through beidellitization, but more probably by neoformation of hydrous mica.

The hydraulic conductivity of the clay samples that had been hydrothermally treated with strongly brackish solution changed early and consistently with the temperature. The treatment led to conductivity values of 10^{-7}-10^{-6} m/s and occasionally up to 10^{-5} m/s at 90°C and higher temperatures, which is ascribed to microstructural alteration in the form of coagulation and aggregation caused by the increased electrolyte content, particularly when Ca was dominant in the porewater. Addition of quartz gave almost the same conductivity as the pure clay samples. The strength increase of the samples exposed to brackish solutions was very significant and the fact that it was highest for the mixtures with 50% quartz indicates that precipitation of silica released by the heat treatment and precipitated at cooling contributed to the strengthening.

A general conclusion of the tests was that there were only very slight changes in the smectite contents and only insignificant changes in sealing properties. Very slow and slight conversion to hydrous mica, and cementation by released silicon and precipitated silica were concluded to have been the major alteration processes. Dissolution and cementation were concluded to be different in the case of Na montmorillonite below and above about 130°C, which was therefore taken as a critical temperature for short term heating (Pusch et al., 1991).

3.3 Portland cement

Relationships between the hydraulic conductivity, leaching and porosity characteristics of intact hardened cement grouts were derived for materials with w/c in the range 0.4 to 1.2. The conductivity tests showed that at 20°C with w/c < 0.7 the water saturated grouts were virtually impermeable when the hydraulic gradient was less than about 15 000 (Onofrei et al., 1991).

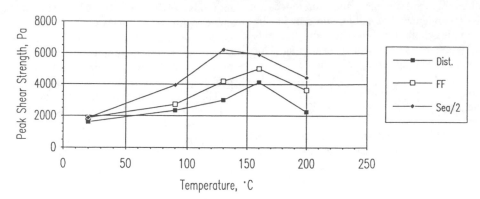

Fig. 2 Shear strength of 100% Na clay prepared with 736% water content and 1.08 g/cm³ density, hydrothermally treated with distilled water (D), Ca-rich (FF) and Na-rich (Sea/2) solutions for 270 days

Scherfestigkeit von 100% Na Ton mit einem Wassergehalt von 736% und 1.08 g/cm³ Dichte, hydrotermal behandelt mit destilliertem Wasser (D), und Ca-reich (FF) and Na-reich (Sea/2) Lösungen

Leaching tests were carried out under both chemically open and closed conditions at temperatures from 10 to 150°C using a number of solutions representing possible groundwater compositions. These tests showed that calcium and magnesium hydroxides and calcium carbonate were formed on particle surfaces and tended to minimize dissolution of the primary C-S-H and C-A-H phases. Also, the mean pore size and total pore volume tended to decrease with leaching time. Examination of the microstructure showed that unhydrated material existed with the hardened grout.

A thermodynamic approach was made in order to test and demonstrate the feasibility of applying numerical modelling techniques to the assessment of the longevity of performance of repository seal elements composed of Portland cement grout and to develop a methodology which can be used to calculate a seal's performance with time (Alcorn et al., 1991). Chemical reactions between grout and groundwater were calculated initially using the geomechanical computer code PHREEQE and primarily using EQ3NR/EQ6. Changes in porosity of the grout were calculated by summing the volume of dissolved grout with the volume of secondary phases formed and incorporating that net volume change into the overall grout porosity. A

relationship was derived from experimental data in the literature to convert porosity to hydraulic conductivity. The new conductivity was used to calculate groundwater travel time through the seal under a realistic repository hydraulic gradient. The calculations were iterated to generate a quasi-continuous change in hydraulic conductivity with time, from which the time period of acceptable performance of the seal could be estimated.

4 PREDICTION OF LONGEVITY

4.1 Smectite clay

The minimum time for conversion of montmorillonite to non-expanding hydrous mica can be estimated by use of a simple conversion model that implies that the conversion is entirely controlled by the rate of uptake of potassium from the groundwater at temperatures exceeding 50-60°C. One finds that for very low potassium contents and completely stagnant ground-water conditions the montmorillonite content may be preserved for many hundred thousand to several million years at 90°C or lower temperatures, while percolation of clay grout by groundwater may yield shorter life-times. Thus, at hydraulic gradients of 10^{-2} to 10^{-1} and a potassium content of 5 to 100 ppm, complete conversion would theoretically take between 10 millions and 50 000 years (Pusch et al., 1991).

However, these considerations imply that grouted fracture channels are completely filled with clay, and this is not correct as concluded from the observations at the field testing. Thus, groundwater will flow along the clay grout in the channels and potassium uptake will take place by diffusion from this water into the thin clay film, which has a very small width. Applying reasonable geometrical conditions, it was concluded that complete conversion to hydrous mica at around 90°C will be a matter of a few hundred years if the potassium concentration of the groundwater is high, like in ocean water, and a few thousand years when the concentration is very low. Reduction of the groundwater flow rate, leading to more or less stagnant conditions, may increase the time for complete conversion to tens to hundreds of thousands of years.

4.2 *Portland cement*

Based on calculations using current state-of-the-art thermodynamic data, it is reasonable to conclude that cement grout seals exposed to percolating groundwater will maintain an acceptable level of performance for many tens of thousands to millions of years, providing that the repository is sited in an area of low hydraulic gradient. Calculations incorporating host rock as well as grout and groundwater suggest that rock does not adversely affect the longevity of grout and in fact may enhance it (Onofrei et al., 1991 and Alcorn et al., 1991).

While the laboratory studies and theoretical modelling of the degradation of cement grouts susggest that they will serve as effective seals for very long periods of time, a number of factors may reduce the operational lifetime considerably like in the case of clay seals. Thus, the retardation of the hardening caused by the superplasticizer may lead to a low shear strength with low resistance to piping and erosion of the cement grout early after injection. Also, expansion of fractures caused by stress changes and temperature effects may separate hardened grout films from the fracture surfaces and expose

them to water flow over the entire basal surfaces, which speeds up the disintegration. Hence, it is concluded that, depending on the prevailing hydraulic gradients and magnitude of rock movements, the sealing function of cement grouts may last for millions of years under very favourable conditions and for hundreds or thousands of years or even less than that under more severe conditions.

5 REFERENCES

Alcorn, S.R., Coons, W.E., Christian-Frear, T.L. & Wallace, M.G. 1991. Theoretical Investigations of Grout Seal Longevity. *Stripa Project Technical Report 91-24*, SKB, Stockholm

Onofrei, M., Gray, M. & Roe, L. 1991. Cement Based Grouts - Longevity Laboratory Studies: Leaching Behaviour. *Stripa Project Technical Report 91-33*, SKB, Stockholm

Pusch, R. 1983. Stability of Bentonite Gels in Crystalline Rock - Physical Aspects. *SKBF/KBS Technical Report 83-04*, SKB, Stockholm

Pusch, R., Erlström, M. & Börgesson, L. 1987. Piping and Erosion Phenomena in Soft Clay Gels. *SKB Technical Report 87-09*, SKB, Stockholm

Pusch, R., Karnland, O., Hökmark, H., Sandén, T. & Börgesson, L. 1991. Final Report of the Rock Sealing Project. *Stripa Project Technical Report 91-30*, SKB, Stockholm

Grundlagen zu Anwendung und Nutzen von Zement- und Polyurethan-Injektionen im Berg- und Tunnelbau

Application and advantages of cement and polyurethane injections in mining and tunnelling

Archibald Richter & Wolfgang Cornely
CarboTech – Berg- und Tunnelbausysteme GmbH, Essen, Deutschland

ZUSAMMENFASSUNG: Polyurethanharze und Feinstbindemittel auf der Basis von ultrafein aufgemahlenem Zement gehören zu den Injektionsstoffen, die in letzter Zeit wachsende Bedeutung erlangt haben. In diesem Beitrag werden sie hinsichtlich einiger für Felsinjektionen entscheidenden Eigenschaften = Viskosität, Injizierbarkeit, mechanische Festigkeit, Klebfestigkeit und Verformungsarbeit - miteinander verglichen und die Vorteile beider Stoffklassen in der Anwendung herausgestellt.

ABSTRACT: During the recent years, some progress has been made in the field of organic an cementious grouts, in particular two component polyurethanes and microcements. In spite of the wide use in tunnelling and mining, comprehensive information on their characteristics relative to injection is still missing. This contribution tries to draw a comparision for these two more advanced grout types. The focus is on penetration capacity, mechanical strength, adhesive properties and deformation work. For the penetration capacity of cementious grouts, water solid ratio as well as the maximum grain size of the cement are decisive. Microcem, the grout which is presented here, can enter cracks of down to 0.2 mm. In cracks, the final adhesive strength of 2 N/mm² if formed almost within 24 hours. Previous exposure to compressive stress, however, causes immediate failure in adhesion tests.

Two component polyurethane grouts normally form a cellular material with approx. double the original volumne. Thus, they are able to yield slightly. Adhesive strength of 2 - 5 N/mm² is formed in 3 - 12 h. Even after exposure to 40 N/mm² compressive stress, their favourable adhesive strength is maintained . After exceeding its elastic deformation, it will not break but yield without losing cohesion ("deformation work"). PU is capable of penetrating minute cracks of less than 0.1 mm width.

In mining applications at longwall face ends with high tear and shear stresses, polyurethane injections have proven to be superior in technical and in economical respect.

Microcem should preferably be used in applications where requirements as high compressive strength, low viscosity at low temperatures are prevailing. The use of polurethane resins is recommended in situations where chemical resistance, high setting speed, high tensis and adhesive strength are needed, especially when the strata is under pressure or is subject to shock waves.

1.EINLEITUNG

Injektionen in Gebirge und Bauwerke zählen seit vielen Jahren zu den etablierten Arbeitsmethoden in Tunnel- und Grundbau sowie im Bergbau. Der eigentliche Vorgang entzieht sich der unmittelbaren Anschauung; über dem Aufbau eines zu injizierenden Lockergesteins mag man noch ein vergleichsweise zuverlässiges Bild besitzen; über Rißweiten und -verlauf im Festgestein geben Morphologie des aufgeschlossenen Gesteins, Endoskopie und Lugeon-Versuch zwar Anhaltspunkte, aber keine Gewißheit. Die quantitative Erfassung der physikalischen und chemischen Vorgänge während der Injektion ist selbst bei Kenntnis des Untergrundes nur mit vereinfa-

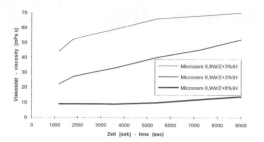

Bild 1 Viskositäts-Zeit-Kurve für verschiedene
Feinstbindemittelsuspensionen mit Zusatz von
"Injektionshilfe" bei 25 ° C
Fig. 1 Viscosity vs. time plot for several
microcement suspensions with addition of
injection booster at 25 °C

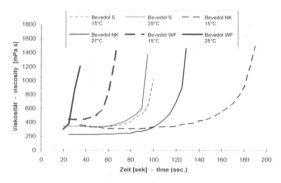

Bild 2 Viskositäts-Zeit-Kurve für verschiedene
Zweikomponenten-Polyurethanharze bei 15° und
25 °C.
Fig. 2 Viscosity vs. time plot for several two
component polyurethane grouts at 15° and 25° C

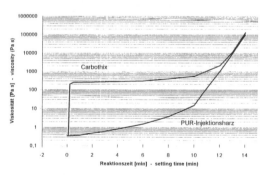

Bild 3 Viskositäts-Zeit-Kurve für ein Thixo-
Polyurethanharz
Fig. 3 Viscosity and time plot for a Thixo-
polyurethane grout

chenden Ausnahmen möglich.

Gegenstand dieses Beitrags ist eine vergleichende Betrachtung zwischen Feinstbindemitteln und Zweikomponentenpolyurethanen als Injektionsstoffen hinsichtlich einiger zentraler Parameter.

2. FEINSTBINDEMITTEL

In früheren Jahren wurden für Injektionen handelsübliche Zemente eingesetzt. Je nach Anwendung wurden Mörtel, Pasten (W/Z < 0,5) und Suspensionen eingesetzt. Auch mit Suspensionen stieß man bald an die Grenzen der Injizierbarkeit. Nachdem neue Mahl- und Klassierverfahren zur Verfügung standen, konnte man die Feinheit der Zemente wesentlich steigern. Materialien mit Blaine-Werten zwischen 10.000 und 16.000 cm^2/g stehen von verschiedenen Herstellern zur Verfügung. Entscheidender für die Injizierbarkeit ist jedoch die sog. maßgebende Korngröße von 85 % der Baustoffmasse D_{85}. Sie sollte unter 6-8 μ liegen.

2.1 Injizierbarkeit

Als Grenze für die Injizierbarkeit von Rissen gilt, daß das Größtkorn einen Durchmesser von höchstens einem Drittel und der D_{85} einen von höchstens einem Fünftel der Spaltbreite haben darf. Bei größeren Durchmessern kommt es zu einer Brückenbildung zwischen den Rißflanken, es bildet sich ein Filterkuchen, der Riß verstopft.

Ein weiterer wichtiger Parameter für die Injizierbarkeit ist die Viskosität. Maßgebend für die Viskosität ist der W/Z-Wert. Bei Suspensionen ist die Viskosität nur wenig von der Temperatur abhängig. Bekanntlich handelt es sich bei Zementsuspensionen um Bingham'sche Flüssigkeiten. Ein Injektionszement sollte zwar eine niedrige Viskosität, aber eine hohe Fließgrenze haben, damit die Suspension in größeren Klüften möglichst nicht schon durch Gravitation wegfließt. Die Fließgrenze kann durch entsprechende Additive beeinflußt werden (Bild 1). In der Regel liegt sie unter 10 Pa, lediglich Microcem weist einen Wert von 25 Pa auf.

Allgemein verfolgt die Anwendung von Additiven bei Feinstbindemitteln drei Ziele:

a)   Verringerung von Wasserbedarf und /
     oder Viskosität
b)   Erhöhung der Kohäsion der Suspension
c)   Stärkung der Resistenz gegen
     chemischen oder mikrobiellen Angriff.

Für die praktische Durchführung ist es natürlich wichtig zu wissen, welche Suspensionen mit welchem Energieaufwand in Risse welcher Weite injiziert werden können. Dazu gibt es ein Verfahren der DMT (DeutscheMontanTechnologie, Essen), über das hier an anderer Stelle berichtet wird.

Aus Messungen nach diesem Verfahren läßt sich folgendes ableiten:

Während Portlandzement PZ 55 mit einer geringen Injektionsarbeit bis in Risse von 0,3 mm Weite eindringen kann, erreicht man mit einem Mittel wie Microcem auch Spaltbreiten von 0,2 mm (s. Bild 4). Das Verfahren bestätigt im Ergebnis also den bekannten Zusammenhang zwischen D_{85} und Spaltbreite.

2.2 Festigkeit

Die Festigkeitsentwicklung der Feinstbindemittel entsprecht im wesentlichen der anderer Zemente. Durch die große spezifische Oberfläche werden relativ hohe 28-d-Festigkeiten erreicht, bei Microcem liegen sie um 65 N/mm².

Ein Injektionsmittel soll aufgeklüftetem Gebirge seine Verbandsfestigkeit wiedergeben. Dazu ist die Klebfestigkeit eine entscheidende Größe. Auf folgendem Wege läßt sie sich vergleichsweise einfach bestimmen:

Ein ausgehärtetes Zementprisma von 40 x 40 x 160 mm wird in seiner Mitte gebrochen, mit dem zu prüfenden Injektionsmittel in einer einzustellenden Spaltbreite wieder verklebt und zwar möglichst unter definierten Temperatur- und Luftfeuchtigkeitsbedingungen. Nach unterschiedlichen Aushärtezeiten wird ein solches Prisma im Dreipunktbiegeversuch (DIN 52112 Biegefestigkeit) wieder gebrochen. Die Durchbiegung des Prismas an der Klebestelle kann zusätzlich gemessen werden (siehe 3.2. c). Die in die Klebenaht eingeleitete Spannung setzt sich in Druck-, Zug-, Haftzug- und Scherspannungen um, wie sie auch in der Realität auftreten.

Mit diesem Versuchsaufbau lassen sich für Zemente nach 24 Stunden bereits Klebfestigkeiten angeben, die sich mit fortschreitender Zeit aber kaum weiter erhöhen. Deutlich ist die Abhängigkeit vom W/Z-Wert: Mit W/Z = 0,5 erreicht man bei PZ 55 4 N/mm², mit W/Z = 0,8 sinkt der Wert unter 2 N/mm². Injizierbarkeit und Festigkeit verlaufen also gegenläufig.

Mit einem Feinstbindemittel wie Microcem lassen sich, bezogen auf den W/Z-Wert, etwas höhere Festigkeiten erreichen (s. Bild 5)

3. POLYURETHANE

Wie bei hydraulisch abbindenden Baustoffen spielt auch für Polyurethane (PU) Wasser zwar

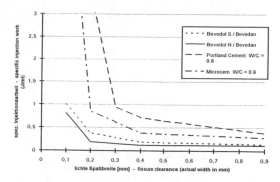

Bild 4 Spezifische Injektionsarbeit verschiedener Injektionsstoffe in Abh. der Spaltbreite d. h. Angabe der Arbeit, die pro ml Injektionsgut in einer bestimmten Spaltbreite erforderlich ist.
Fig. 4 Specific injection work vs. crack width for several grouts i.e. indication of the work required for injecting 1 ml grout in certain crack widths.

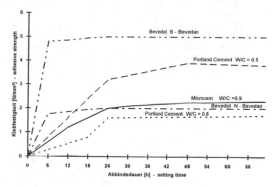

Bild 5 Entwicklung der Klebefestigkeit verschiedener Injektionsstoffe bei 30° C und 80 % rel. Luftfeuchte.
Fig. 5 Development of adhesive strength of several grouts at 30° C and 80 % rel. air humidity

eine wichtige Rolle im Abbindeprozeß, aber in einer ganz anderen Funktion, und zwar nach folgenden Reaktionsgleichungen:

Isocyanat + Wasser → Polyharnstoff + CO_2
Isocyanat + Polyol → Polyurethan.

An dieser Stelle soll ausschließlich von sog. Zweikomponenten-Polyurethanen wie den verschiedenen Bevedol-Bevedan-Systemen die Rede sein, bei denen die Isocyanat-Komponente mit einem gegebenenfalls wasserhaltigen Polyol zu einem festen und meist durch CO_2 geschäumten Feststoff aushärten. Standardeinstellungen schäumen etwa auf das doppelte Volumen auf, Schaumfaktoren von 15 -

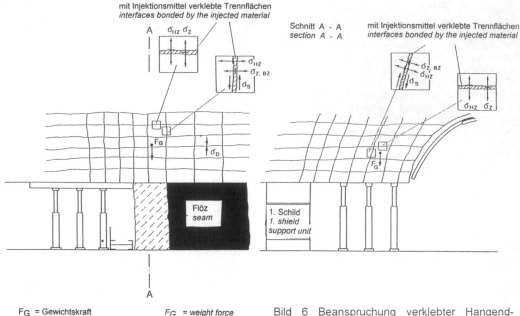

mit Injektionsmittel verklebte Trennflächen
interfaces bonded by the injected material

Schnitt A - A
section A - A

mit Injektionsmittel verklebte Trennflächen
interfaces bonded by the injected material

Flöz
seam

1. Schild
1. shield
support unit

F_G = Gewichtskraft	F_G = weight force	Bild 6 Beanspruchung verklebter Hangend-schichten am Streb/Streckenübergang im Streb und in der Strecke
σ_{HZ} = Haftzugbeanspruchung	σ_{HZ} = adhesion load	
σ_Z = Zugbeanspruchung	σ_Z = tensile stress	Fig. 6 Stresses in glued roof layers at the face and in longwall mining
σ_{BZ} = Biegezugbeanspruchung	σ_{BZ} = flexural stress	
σ_S = Schubbeanspruchung	σ_S = shear stress	
σ_D = Druckbeanspruchung	σ_D = compressive stress	

20 stellen die obere Grenze für die Praxis dar.

Einkomponentig verpreßte, rein wasseraktive PU-Harze dienen bei der Anwendung in Rißsystemen lediglich zur temporären Abdichtung.

3.1 Injizierbarkeit

Voraussetzung für die Wirkung von Injektionsmitteln ist deren Eindringen in die im Gebirge entstandenen Risse und Klüfte. Die Viskosität der Ausgangskomponenten, in der Regel Newton'sche Flüssigkeiten, kann dabei zur ersten Abschätzung dienen. Die Viskositäten sind stark temperaturabhängig; als Faustregel gilt, daß sich mit 10 K Temperaturunterschied die Viskosität etwa verdoppelt bzw. halbiert.

Diese Daten reichen jedoch nicht, um das Eindringverhalten zu beschreiben. Die hohe Reaktivität der Kunstharzmischungen bringt in der Regel schon während der Injektion eine Erhöhung der Viskosität mit sich (s. Bild 2). Bei chemisch schäumenden Harzen wie den Polyurethanen (Wasserreaktion) kommt noch der in der Kluft entstehende Schäumdruck hinzu, der die sog. Selbstinjektion bewirkt. In der (s. 2.1) Versuchsanordnung der DMT wird auch dieser Effekt wirksam. PU-Harze dringen selbst in feinste Risse vor (s. Bild 2).

Einen anderen Viskositätsverlauf erhält man bei der Reaktion der sog. Thixopolyurethane wie Carbothix. Hier steigt die Viskosität sofort nach dem Mischen zu einer pastenartigen Konsistenz, der ein verzögertes Aushärten folgt (s. Bild 3). Dadurch kann es z. B. als Ankerharz über Kopf eingesetzt werden.

3.2 Festigkeit

a) Druckfestigkeit

Die mechanischen Festigkeiten vernetzter Polyurethane sind sehr hoch, die Druckfestigkeiten können 100 N/mm² überschreiten. Durch das Aufschäumen kommt es jedoch zu einem überproportionalen Absinken der Festigkeits-Werte, so daß wir im normalen Einsatzbereich mit Festigkeiten zwischen 5 und 30 N/mm² rechnen können. Diese Werte werden, je nach Reaktivität des Harzes, innerhalb von Zeiten zwischen 1 und 24 h erreicht. Selbstverständlich spielt auch hier die Temperatur eine wesentliche Rolle; als Faustregel gilt in der organischen Chemie wieder eine Verdopplung der Reaktionsgeschwindigkeit bei 10 K Temperaturerhöhung.

Die Temperatur beeinflußt auch die mechanischen Eigenschaften: Die

Glastemperatur von Bevedol-Bevedan - Systeme sowie Carbothix H liegt bei etwa 60° C: oberhalb dieser Temperatur sinken die Verformungsmodule auf ein wesentlich niedrigeres Niveau. Bei höheren Gebirgstemperaturen als 60° C sollte PU deshalb nicht als Injektionsmaterial verwendet werden. Durch aggressive Medien wie Schwefelsäure von pH = 3 oder Natronlauge von pH= 10 kommt es zu keinen Schwächungen des Schaumkörpers, wie einjährige Lagerversuche gezeigt haben.

Der E-Modul der PUR-Systeme liegt, je nach Rohdichte, zwischen 300 und 3.000 N/mm². Beim Ankerverpressen mit Bevedol S - Bevedan können wir von einem Modul in Höhe von 800 N/mm² ausgehen, bei Carbothix H von 2500 N/mm². Dadurch wird z. B. bei Böschungsankerungen, die aufgrund der Trennflächen in der Böschung enstandene Belastung nicht nur gleichmäßiger sondern auch zu einem weit höheren Anteil als bei zementösen Bindemitteln auf den Ankerstahl in seiner gesamten Länge übertragen, ohne daß das Bindemittel selbst bricht (Korrosion!). Dies konnte durch Ankerzugversuche im Anhydrit-Rohr bei der DMT und in FEM-Berechnungen durch das Büro Wittke gezeigt werden.

b) Zugfestigkeit

Die Druckfestigkeit des Injektionsmittels gilt im Bauingenieurwesen gemeinhin als die wesentliche Kerngröße. Wie Bild 4 zeigt, gibt es jedoch auch Situationen, vor allem während der Bauphase, in denen Zug- und Scherspannungen des Gebirges eine entscheidende Rolle spielen. Anders als hydraulisch abbindende Baustoffe, in denen Diskontinuitäten, vor allem kleine und kleinste Schwindrisse, relativ schnell zum Versagen bei Zugbelastung führen können, erreichen Polyurethane hier Werte, die in derselben Größenordnung wie die Druckfestigkeiten liegen.

Die Klebfestigkeiten der Polyurethane, wie sie nach der in 2.2 wiedergegebenen Anordnung gemessen werden, reichen nicht an die Zugfestigkeiten heran, erreichen aber fast die Biegezugfestigkeit des Vergußmaterials der Zementprismen (s. Bild 5). Die Meßergebnisse zeigen, daß innerhalb von Stunden Werte erreicht werden, die etwa doppelt so hoch liegen wie die mit hydraulischen Bindemitteln erreichbaren. Die Rauhigkeit der Oberfläche hat nur geringen Einfluß; Feuchtigkeit oder Nässe vermindert die Klebewirkung etwa auf die Hälfte, weil das PU-Harz an der Haftfläche etwas stärker schäumt.

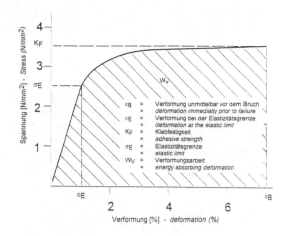

Bild 7 Spannungs-Verformungsdiagramm für einen mit Bevedol S - Bevedan verklebten Probekörper
Fig. 7 Stress-deformation plot for a specimen glued with Bevedol S - Bevedan

c) Verformungsarbeit

Aus dem Bergbau ist bekannt, daß Gebirgsbewegungen nicht gänzlich aufgehalten werden können. Diese führen zu Verformungen, die ihrerseits aber nicht zur Zerstörung der Verklebung führen dürfen. Für diese Fälle hat DMT die Forderung aufgestellt, daß Injektionsmittel für den Bergbau folgendes Verhalten bei der Prüfung auf Klebfestigkeit zeigen sollten: Nach einem elastischen, d. h. auch reversiblen Bereich, der bis zu einer möglichst hohen Spannung gelten sollte, folgt ein plastischer Bereich, in dem weitere Bewegungen möglich sind, ohne daß es zum Versagen kommt. Das Integral unter der Spannungs-Verformungskurve hat die Dimension einer Arbeit, d. h. hier ist die Arbeit angegeben, die der Injektionsstoff gegen die Gebirgsverformung bis zum Versagen leisten kann. Wie sich gezeigt hat, sind bislang nur zellige Injektionsstoffe in der Lage, diesem Bild näherungsweise zu entsprechen (Beispiel siehe Bild 7). Ungeschäumte spröde Stoffe sind praktisch inkompressibel und nur unter Bruch verformbar, während andererseits weichelastische Stoffe, wie sie zum Abdichten nach ZtV-Riß dienen, zu niedrige Ansprechfestigkeiten haben.

Aufgrund dieser Eigenschaft sind PU-Verfestigungen - anders als mit sprödharten Stoffen verfüllte Fugen - auch in der Lage, nach vorheriger Druckbelastung (gemessen bis 40 N/mm²) unveränderte Klebefestigkeit zu zeigen.

Aufgrunddessen ist es auch möglich, daß Polyurethanverklebungen größeren Erschütterungen widerstehen; es erlaubt

Anwendungen wie Ausbesserungsarbeiten an Brücken, die unter Schwingungsbelastung stehen, in Zonen, in denen gesprengt wird u. a.

3.3 Einfluß von Wasser und Druck

Wie aus den Reaktionsgleichungen in 3. hervorgeht, ist Wasser ein potentieller Partner der Reaktion, und zwar nicht nur das Wasser, das in der Polyol-Komponente enthalten ist, sondern auch solches, das erst im Gebirge mit dem Flüssigharz in Kontakt kommt. Hierdurch wird ein stärkeres Aufschäumen bewirkt. Dieser Effekt kann durch Additive gedämpft werden. Bei Carbothix verhindert die pastenartige Konsistenz ein nachträgliches Einmischen von Wasser.

Ein ungehindertes Aufschäumen ist jedoch nur dann möglich, wenn freies Volumen zur Verfügung steht. Unter Druck bleibt Kohlendioxid in der Harzphase gelöst, der Ablauf der Wasserreaktion wird verzögert. In der Praxis wirkt sich das beispielsweise so aus, daß flüssiges Bevedol S - Bevedan beim Austreten aus Rissen, d. h. beim Entspannen, spontan aufschäumt und aushärtet, aber flüssig bleibt, solange es unter Injektionsdruck steht.

3.4 Umweltverträglichkeit

Polyurethanharzen wird eine schlechte Umweltverträglichkeit nachgesagt. Die dunkle Isocyanatkomponente wird als Gesundheitsrisiko angesehen.

Tatsache ist, daß als Isocyanatkomponente bei Anwendungen im Bauwesen grundsätzlich MDI (4.4'-Diphenylmethan-Diisocyanat) eingesetzt wird (Ausnahme PU-Hydrogele bei der Kanalsanierung). Bei normalen Verarbeitungstemperaturen kann schon wegen des niedrigen Dampfdrucks des MDI der MAK-Wert nicht erreicht werden.

Auch im Baugrund konnten bislang kein MDI oder irgendwelche Umwandlungsprodukte gefunden werden. Dies ist durch Untersuchungen der Ruhr-Universität Bochum und das Hygieneinstitut Gelsenkirchen belegt. Das Isocyanat setzt sich mit Wasser zum inerten Polyharnstoff um. Wenn diese Reaktion der Polyol-Isocyanat-Reaktion zuvorkommt, also je langsamer die Harzeinstellung ist, desto eher bleibt ein Quantum Polyol ungebunden und gelangt so ins Grundwasser. Dessen biologische Abbaubarkeit (BSB) wird aber als gut bezeichnet und läuft unter Testbedingungen in 6-10 Tagen vollständig ab. Aufgrund einschlägiger Untersuchungen werden Bevedol - Bevedan - Formulierungen inzwischen z. B. im Einzugsbereich von Thermalquellen akzeptiert. Zulassungen durch das Umweltbundesamt bzw.

das Institut für Bautechnik stehen bevor.

Andererseits sind ausgehärtete Polyurethane wie Bevedol - Bevedan gegenüber mikrobiellem Angriff beständig.

4. ANWENDUNGEN DER VERFESTIGUNG UND ANKERUNG (ANHAND VON BEISPIELEN)

4.1 Durchörterung druckhaften Gebirges

Zur ersten größeren Anwendung von Polyurethan-Zweikomponenten-Harzen im Tunnelbau kam es beim Bau des Furka-Basis-Tunnels.

Ein wesentliches Problem waren Gebirgsdeformationen und Abplatzungen, die in Zonen mit steil gelagertem Granit von bis zu 1.500 m Überlagerung auftraten. Der Gebirgsdruck führte zum sukzessiven Bruch der jeweils inneren intakten Schale. Schwerer Anker-Spritzbeton-Ausbau konnte dem nicht erfolgreich engegenwirken. Erst Injektionsanker in Verbindung mit PU-Verpressungen, konnten durch Abbau der Lastspitzen und Vertiefung des Gebirgsdrucks, ein stabiles Gewölbe aufbauen, das seine Dauerhaftigkeit inzwischen bewiesen hat.

Andere wichtige Anwendungen im Furka-Tunnel betrafen Abdichtungen von Wasserzuflüssen und Stabilisierung gestörter, wasserdurchflossener Zonen.

Seit etwa 4 Jahren sind Injektionsanker mit integrierten Blähpackern im Einsatz. Diese zeichnen sich aufgrund der Blähpackern dadurch aus, daß eine weitreichende Injektion und eine solide Ankervermörtelung sowohl mit Microcem als auch mit beschleunigtem PU-Harz möglich ist.

4.2 Durchörterung von geologischen Störungen

Störungzonen stellen insbesondere bei Vollschnittmaschinen (TBM) ein großes Problem dar. Injektionen mit Polyurethan haben sich hier sowohl als Präventiv- als auch als Sanierungsmaßnahme bewährt. Vorausinjektionen werden entweder von Pilotstollen oder, sofern möglich, aus der Maschine heraus vorgenommen. Die kurzen Aushärtezeiten und die hohen Festigkeiten, verbunden mit einer Abdichtung, ermöglichen ein schnelles und sicheres Durchörtern der Störzone.

Als Beispiel sei der Tunnel Trin im Tal des Vorderrheins angeführt. Sowohl beim Auffahren des Pilottunnels mit 3 m Durchmesser als auch beim nachfolgenden konventionellen Vortrieb wurden die vorhergesagten Kluftzonen von bis zu mehreren Metern Weite mit

Hermann Gustav

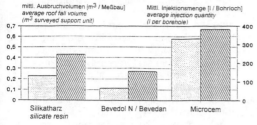

Girondelle 5

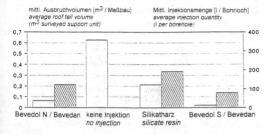

Groß Mühlenbach

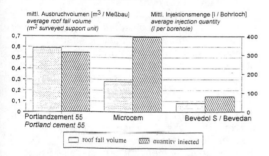

Bild 8 Vergleichende Darstellung des Injektionserfolges von Streckensaumstabilisierungen mit
verschiedenen Injektionsstoffen.
Fig. 8 Comparative presentation of grouting effect in roadside stabilisations (coal mining) with
several grouts

schnellreagierenden Polyurethanen verfestigt.
Der Schaumfaktor des Polyurethans wurde dafür
den Erforderungen angepaßt, d. h. beim
Kalottenvortrieb reichte ein Material mit dem
Schaumfaktor von rund 15 aus, um mit wenig
Mitteln einen sehr großen Effekt zu erzielen.

4.3 Kalottenfußstabilisierungen

Dieselbe Einstellung wurde im Portalbereich
dieses Tunnels benutzt. Vor Erreichen des festen
Gebirges mußte - typisch für Alpentunnel - eine
ca. 180 m lange Zone mit nicht kohäsivem
Hangsturzmaterial durchfahren werden. Unter
diesen Umständen fehlte dem Kalottenausbau

ein standfester Fußbereich, was zu
Konvergenzen der Firste und Divergenzen der
Stöße führt. Zementinjektionen liefen aufgrund
der hohen Porösität in untere Bereiche weg. Mit
schellen Zweikomponenten-Polyurethan konnten
auch die oberen Bereiche stabilisiert werden,
wodurch das Gebirge im Kämpferbereich wieder
eine ausreichende Verbandsfestigkeit erhielt.
Ähnliche Anwendungen gab es z. B. im
Clasaurer-Tunnel und im Pradella -Druckstollen
(Graubünden).

4.4 Tunnel- und Streckensanierung

In einem Bergwerk der Ruhrkohle wurde eine zu
ca. 65 % konvergierte Strecke durch Injektionen
mittels Injektionsankern (4.1) und Microcem im
First- und Stoßbereich durch 2 m bis 4 m tiefe
Bohrungen saniert. Eine weitere Konvergenzzu-
nahme konnte erfolgreich gestoppt werden.

So vielfältig die Verfahrensweisen bei der Tun-
nelsanierung sind, so zahlreich sind auch die
Anwendungsmöglichkeiten von Polyurethanen,
von denen einige genannt seien:

⇒   Sicherung gegen Herauslaufen
      von Lockergestein hinter der Tunnel-
      ausmauerung
⇒   Abdichtung gegen Wasserzuflüsse
⇒   Sicherungen mit Injektionsankern
⇒   Sanieren von Rissen

5. VERGLEICHENDE BETRACHTUNG

5.1 Injizierbarkeit

Während es sich bei den Harzkomponenten um
Newton`sche Flüssigkeiten mit einer ausgepräg-
ten Abhängigkeit der Viskosität von der
Temperatur handelt, ist die Viskosität der
Bingham`schen Suspensionen im wesentlichen
vom Wasserzementwert abhängig. Während
Harze auch in feinste Risse z. B. unter 0,04 mm
Breite eindringen können, vorausgesetzt, der
Druck ist ausreichend, gibt es bei Suspensionen
eine Grenzspaltweite, unterhalb derer die
Festbestandteile ausfiltern. Bis dahin liegt die
Injektionsarbeit in etwa derselben
Größenordnung. Bei den Harzen muß eine
weitere Größe in Betracht gezogen werden: die
Reaktion während es der Injektion . Während es
beim Zementverpressen immer wieder
vorkommt, daß Injektionsgut in Bereiche
außerhalb der vorgesehenen Zone abfließt, wird
dies bei schnell eingestellten PU-Harzen durch
den steilen Viskositätsanstieg am Ende der
Schäumzeit vermieden.

93

5.2 Festigkeiten

Polyurethane erreichen nur im ungeschäumten Zustand oder bei niedrigen Schaumfaktoren die Druckfestigkeit von Injektionszementen, dies jedoch bereits in kürzester Zeit, je nach Einstellung bereits nach einer Stunde. Die Biegefestigkeiten liegen bei Zementen bei einem Fünftel bis Achtel der Druckfestigkeiten, bei Kunstharzen wird sie jedoch ähnlich hoch wie die Durckfestigkeiten. Die Klebfestigkeiten liegen nach praxisorientierten Messungen bei Polyurethanen etwa doppelt so hoch wie bei Zementen. Dies ist vor allem für die Fälle von Bedeutung, in denen Druckspannungen des Gebirges in Zug- und Scherspannungen transformiert werden.

Ein entscheidender Vorteil der geschäumten PU-Harze wurde von der DMT als Verformungsarbeit definiert, d. h. die Prüfkörper brechen nicht bei Überschreiten der reversiblen elastischen Verformung, sondern haben auf hohem Spannungsniveau ein erhebliches Verformungspotential. Spröde, nichtgeschäumte Bindemittel, so auch Zemente, versagen bereits bei geringen Verformungen, so daß sich ein Integral unter der Spannungs-/Verformungskurve nicht sinnvoll angeben läßt. In ähnlicher Weise findet man, daß bei zementverklebten Prismen nach vorheriger Druckbelastung keine Klebfestigkeit mehr meßbar ist, während Verklebungen mit PU auch nach Druckbelastung von einer 40 N/mm² weiterhin unveränderte Klebfestigkeiten aufweisen.

5.3 Anwendungskriterien

Zementinjektionen gehören seit mehr als hundert Jahren zum vertrauten Stand der Technik. Die Verwendung von Feinstbindemitteln wie Microcem erfordert allerdings ein vorheriges kolloidales Aufmischen. Demgegenüber werden den Arbeiten mit Polyurethanen gegenüber - zumindest anfängliche - Vorbehalte entgegengebracht. Der Umgang mit Polyurethanen, erfordert ein anderes spezielles Know-How, vor allem , wenn man die Vielfalt der verfügbaren Typen und Anwendungmöglichkeiten nutzen will. Die erforderlichen Pumpaggregate sind sehr handlich, kleinbauend und robust.
Zemente sind hydraulisch abbindende Baustoffe. Das Wasser kann zum Quellen von tonigen Gesteinen und zum Wasserentzug aus der Suspension führen. Polyurethane hingegen sind nichtwäßrige Flüssigkeiten und können deshalb auch gut in trockene tonige Sande injiziert werden. Sie reagieren jedoch mit dem Gebirgswasser unter Aufschäumen, was die erreichbaren Festigkeiten halbieren kann, so daß sie dann in etwa in der Größenordnung der Zemente liegen.
Durch ihr schnelles Abbinden lassen sich die Injektionsreichweiten der Polyurethane eng begrenzen, während Zemente, auch thixotropierte, oft in entfernte Bereiche abfließen. Zementinjektionen gelten allgemein - trotz deutlicher Erhöhung der pH-Werte - als ökologisch verträglich. Für schnellabbindende Polyurethane ist die ökologische Unbedenklichkeit nachgewiesen. Auf der anderen Seite werden ausgehärtete Polyurethane weder von Mikroben noch von den üblichen Säuren, Laugen und Salzlösungen angegriffen, was bei Zementen Probleme aufwerfen kann. Ausschlaggebendes Kriterium sind in aller Regel die Kosten. Man spricht bei Zementen vom Preis pro Tonne , bei PU vom Kilopreis. Die für eine Injektion erforderliche Menge und damit auch der Zeitaufwand unterscheiden sich jedoch erheblich, bedingt durch niedrigere Dichte, Aufschäumen und begrenzbare Reichweite, wodurch die Kosten stark relativiert werden.

Im deutschen Steinkohlenbergbau wurden vor einigen Jahren Vergleichsversuche in verschiedenen Abbaufeldern gleichen Zuschnitts bei der Stabilisierung des Streckensaums durchgeführt (Bild 6). Folgende Ergebnisse wurden erzielt (Bild 8).
Ohne Injektionsmittel ereignen sich, sobald die Abbaufront diesen Streckenbereich durchfährt, durch den hohen lokalen Gebirgsdruck große Ausbrüche. Durch Injektionsmaßnahmen können diese Ausbrüche verringert werden. In der Reihenfolge Portlandzement, Microcem, Silikatharz, Bevedol N-Bevedan, Bevedol S-Bevedan sinkt die Ausbruchsmenge. Interessanterweise sinkt auch die Menge des injizierten Materials in derselben Reihenfolge.

6. SCHLUSSBETRACHTUNG

Bei sprödharten Injektionsmitteln stellen die Feinstbindemittel wie Microcem das bisher erreichte Optimum dar, gekennzeichnet durch hohe Fließgrenze, niedrige Viskosität, gute Injizierbarkeit und hohe Endfestigkeit.
Die sog. plastoelastischen Injektionsstoffe wie die Zweikomponenten-Polyurethane Bevedol-Bevedan zeichnen sich aus durch Chemikalienbeständigkeit, hohe Reaktivität, d. h. schnellen Viskositäts- und Festigkeitsanstieg, gute Injizierbarkeit, Selbstinjektion, hohe Zug- und Klebfestigkeit und die Fähigkeit, in druckhaftem Gebirge Verformungsarbeit zu leisten.

LITERATUR

Dietrich-Wilhelm Gemmel, Anforderungen an die technologischen Eigenschaften von Injektionsmitteln zum Verfestigen der angeschnittenen Hangendschichten in Abbau- und Basisstrecken, Dissertation, Aachen, 1992

Christian Burgstaller, Numerische Simulation von Schauminjektionen auf der Basis von experimentellen Untersuchungen, Diplomarbeit, Leoben,1989

Selection criteria for Portland and microfine cement-based injection grouts

Auswahlkriterien von Injektionsmischungen auf der Grundlage von portland-
und Mikrozementen

K. Saleh & J. Mirza
Hydro-Québec, Varennes, Que., Canada

G. Ballivy & T. Mnif
University of Sherbrooke, Que., Canada

ABSTRACT: This paper presents the results of a study conducted in the framework of research into the injection of microcracks in concrete dams to determine the rheological, physical and mechanical characteristics of a number of Portland and microfine cement-based grouts as a result of variations in the water/cement (W/C) ratio, the amount of superplasticizer added and the temperature ($20°$ C and $4°$ C). The aim of this study is to establish the selection criteria for injection grouts based on characterization tests on fresh grouts (viscosity, setting time, stability) and hardened grouts (compressive strength, modulus of elasticity). The results were verified by injecting fractures in concrete slabs with type I Portland cement grouts and type I Portland cement containing a superplasticizer, using different W/C ratios.

ZUSAMMENFASSUNG: Diese arbeit ist das resultat einer studie, internommen im rahmen einer forschung von Einspritzung in mikrorisse in betongdämme, um in einer anzahl von Portland und Mikrfeinen Zementlagigen Mörtel die rheologischen, physischen und mechanischen Eigensphaften festzustellen, die durch die schwänkungen im Wasser/Zement verhältnis, von zugefügten mengen von "Superplastizer" und in der Temperatur ($4°$ C und $20°$ C) entstanden sind. Der Zweck dieser studie ist die Auswahl Criteria für Einspritz-Mörtel festzustellen, die auf Tests von Eigenschaften in frischen Mörtel (Grad des Zähflüssigkeit, Festsetzung der Zeit, Stabilitat) und gehärtiten Mörtel (Compressive Stärke, Modulus von Elastizität) begründet sind.
Die resultât waren geprüft durch einspritzung der Mörtel in die risse von Zementplatten mit Type I Portland Zement Mörtel und Type I Portland Zement mit "Superplastizer" von verschiedenen Wasser/Zement verhältnissen.

1 INTRODUCTION

Questions abound when the time comes to doing maintenance work or repairs on concrete structures. What is the best method to apply? Which is the best product for the job? Which is the best equipment to use? Considering the economic and social importance of such structures, a poor repair job can have serious consequences, which is why the method must be selected in terms of its effectiveness.

Cracking is esteemed to be one of the most serious forms of degradation that threatens the stability of concrete structures, especially dams. The appearance of cracks (macro or micro), whatever their origin, affects the mechanical,

hydraulic and hydrogeological behavior of the medium (ACI Committee 224, 1984). Injecting a liquid (grout) that will harden with time is the most popular approach to deal with this kind of degradation in cracks and, if well done, will have the required effect. Over the last 30 years, injection techniques have evolved to such an extent that they are now suitable for any type of hydraulic concrete structure that develops cracks. The choice of grout and technique depends largely on the configuration and state of the crack to be repaired. It is therefore essential to trace the origin of the cracks before selecting the injection technique, procedure and, in particular, the equipment and product for any new repair job.

The characteristics of injection products have changed over the last 200 years from simple clay suspensions to cement suspensions, including Portland cement-based grouts and countless organic products (polyurethane, epoxies, etc.) and new products based on microfine cements. Because Portland cement-based grouts are easy to apply and a great deal of experience has been acquired in this field, it has always been recommended to employ this type of grout, but for micro cracks, it is not suitable and resins or microfine cements must be used instead. The use of resins is nevertheless limited to the injection of stable cracks where the ambient temperature is higher than 8° C and shows little variation. For active microcracks or injection work at low temperatures and, also, to better preserve the integrity of the structure, it is recommended to use microfine cements. However, little is known as yet about these cements and the use of these new products is not necessarily preceded by a study of the rheological, mechanical and physical characteristics in comparison to commonly employed cements (Portland). Such a study would not be complete without an analysis of the behavior of these grouts during and after injection into micro cracks, i.e. a study of the penetrability, pressure distribution in the cracks, and the injection efficiency.

2 TEST PROGRAM

The main parameters governing the choice of grout are the gradation, viscosity, stability, setting time, and mechanical and physical characteristics. These parameters are influenced by various factors related to the products used and their preparation method; for example:
• chemical and mineralogical composition of the cement
• particle size and fineness of the cement
• W/C ratio
• presence of additives or admixtures
• ambient temperature and water temperature
• mixing method and time
• presence of inert particles, etc.

Five types of cement were selected for this test program in order to study the influence of the particle size, W/C ratio, presence of additives, and temperature on the grout selection parameters. The program begins with grout identification tests (gradation) then a study of the characteristics of fresh and

hardened cement at 20° C and at 4° C with different W/C ratios. The latter temperature was selected because it represents the year-round temperature of water at a depth of 3 m or over in some Canadian reservoirs.

The five cement types selected for testing comprised:
• types I and III Portland cement (ASTM)
• microfine cements M1, M2 and M3.

In order to study the behavior of these grouts during and after injection, we prepared large slabs of concrete, induced fissures in them by a hydraulic fracturing technique and injected the fissures with different cement grouts.

3 GRADATION

The particle size is one of the limiting factors with regard to the use of grouts based on cement suspensions. Several studies, in fact, have shown the relationship between the minimum opening of fissures and the diameter of the cement grains (Houlsby, 1990). The cement must be sufficiently fine to ensure the best possible penetrability but the finer the grains, the greater the specific surface area and the faster the hydration reaction, which increases the viscosity and reduces the setting time.

On the other hand, with a fine cement, it is essential to make sure that the mechanical characteristics are acceptable. A good grain distribution, with optimum fineness, obviously

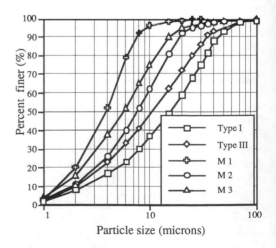

Figure 1. Gradation of types I and III Portland cement and microfine cements.

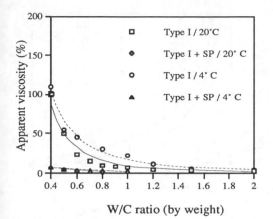

Figure 2. Variation in the viscosity as a function of the W/C ratio at 4° C and 20° C (type I Portland cement and type I + superplasticizer)

will give the best results (Tsivilis, 1990). We observed three different particle sizes in our study (see Figure 1):
• types I and III Portland cements, which are comparatively coarse, the average particle size being about 11-18 μm with grains exceeding 100 μm. Therefore microcracks cannot be injected with this type of cement, which is a major drawback of these products;
• microfine cement M1, which is very fine: the average particle size (D50) is approximately 3 μm while the maximum does not exceed 30 μm;
• M2 and M3 microfine cements of intermediate size. The average particle size (D50) is approximately 6-8 μm while the maximum does not exceed 70 μm.

4 EXPERIMENTAL TESTS

4.1 Tests Performed

A wide range of W/C ratios from 2/1 to 0.4/1 (by weight) was used, depending on the cement type. For some mixtures, a dry extract of naphthalene-based superplasticizer was used in a proportion of 1% of the weight of the cement to increase the grout fluidity.

Immediately after the different ingredients had been mixed for 3 min, viscosity, stability and setting-time tests were performed. Cylindrical (Nx; 52 mm in diameter) and cube-shaped (50 mm x 50 mm x 50 mm) samples were then taken and stored for 28 days in a

humid chamber for testing the modulus of elasticity and compressive strength. Low-temperature tests (4° C) on fresh grouts were performed in a temperature-controlled chamber. Grout samples for tests in the hardened state were stored in the same environment for 28 days' curing.

4.2 Characterization of Fresh Grout

4.2.1 Viscosity

The cohesion and viscosity are the two main parameters to be considered when studying the rheology of stable grouts (Lombardi, 1985).

The shearing limit (cohesion) is the minimum force needed to set in motion a stable grout subjected to a shear stress. Once this threshold has been exceeded, the grout could be supposed to behave like a Newtonian fluid with a plastic viscosity (Shaughnessy and Clark, 1988).

According to the definition by the ASCE injection committee, viscosity is the internal strength of a fluid which allows it to resist flow. In the case of stable grouts, which have a Bing-hamian behavior, two types of viscosity can be defined: plastic and apparent and the latter was adopted for our viscosity tests.

A Brookfield rotating viscosity meter was used to measure the viscosity. However, since calibration of this instrument with a Newtonian fluid of known viscosity followed by application of the results to a Binghamian fluid (cement grout) is not the most rigorous approach, all results are expressed with respect to a type 10 reference cement grout with a W/C ratio=0.4 at 20° C, whose relative viscosity was considered equal to 100% (Ballivy et al., 1991).

Figures 2 to 4 present variations in the viscosity as a function of the W/C ratio for all cements at the temperatures studied. As expected, the viscosity of the different grouts was inversely proportional to the W/C ratio. However, after a certain ratio specific to each cement, a change in slope may be observed. On the contrary, the viscosity value becomes almost invariable for fairly high W/C ratios. On the basis of this observation, two limits were proposed for each cement (Table 1): an upper limit representing the W/C ratio marking the point at which the variation in the viscosity becomes almost negligible, and a lower limit representing W/C ratios below which the

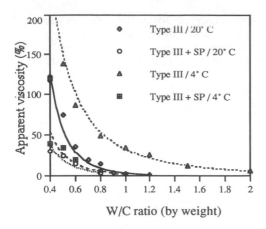

Figure 3. Variation in the viscosity as a function of the W/C ratio at 4° C and 20°C (type III Portland cement and type III + superplasticizer)

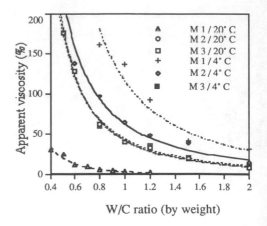

Figure 4. Variation in the viscosity as a function of the W/C ratio at 4°C and 20°C (microfine cements M1, M2, M3)

Table 1. Range of W/C ratios proposed on the basis of viscosity tests at 4° C and 20° C

Cements	Tests at 20° C		Tests at 4° C	
	lower limit	upper limit	lower limit	upper limit
T I	0.6	1.4	0.8	1.6
T III	0.6	1.2	1.0	1.2
T I+SP	0.4	0.8	0.5	1.0
T III+SP	0.5	0.8	0.6	1.2
M 1	0.4	1.2	1.5	3
M 2	1	2	1.2	2.5
M 3	1	2	1.2	2.5

viscosity exceeds 40%.

The main causes for the difference in viscosity between the different cements are the particle size and the chemical composition (Figures 2-4). For a same W/C ratio, the Portland cements (with a larger grain size) have a lower viscosity than microfine cements.

The addition of a superplasticizer greatly reduces the viscosity (Figure 2), which means that it is possible to work with low W/C ratios yet still have an acceptable fluidity.

The temperature drop increased the viscosity possibly owing in part to the increased viscosity of the water and the superplasticizer and partly due to the slower hydration reaction at lower temperatures.

The significance of these two phenomena emphasizes or reduces the difference between the viscosity at 4° C and that at 20° C, a difference which is felt mainly at the level of the chemical composition and particle size. In the case of type III cement, which has a faster hydration rate than type I, the difference in viscosity is greater at 4° C than at 20° C.

4.2.2 Setting time

The setting-time test was performed as per ASTM C-953-87 on the limit W/C ratios proposed on the basis of the viscosity test (see Table 1). The initial setting times at 20° C varied between 7 h and 16 h for lower limits and between 12 h and over 24 h for upper limits. The inconvenience of using superplasticizers is that they increase the cement setting time, as confirmed by the results obtained from grouts with and without superplasticizer. The initial setting time increased from 7 h for type I cement without superplasticizer to 12 h for the same cement with superplasticizer and a lower W/C ratio. This phenomenon may cause problems when these products are used in cold regions where the risk of the water freezing must be taken into consideration, or in the case of injection with a water flow, which may result in significant leaching of the grout if it takes a long time to set.

Furthermore, as seen in Table 2, at low temperatures (4° C) the setting time of all cements is practically twice as long, especially for a high W/C ratio, which raises serious doubts about the advantages of injections with high W/C ratios in colder parts of the world.

On the basis of these results, a range of W/C ratios was proposed for each cement, as for the viscosity test (Table 3). The two arbitrary limits considered are:
• upper initial setting time of 3 h to ensure adequate fluidity of the grout during injection;
• a lower final setting time of 24 h to avoid problems related to leaching and freezing water, as mentioned above.

4.2.3 Stability

The stability of a grout is represented by the quantity of cement particles remaining in suspension in the mix a certain time after mixing. A grout is considered stable if bleeding does not exceed 5% of the total volume of the mix 2 h after the ingredients are mixed.

Tests on cements (Figure 5) have revealed the close relationship between their fineness and their stability: the finer a cement, the more it is stable. Since the specific surface area of fine cement is larger, there are more water molecules surrounding the cement particle and contributing to the hydration reaction. Hence less bleeding occurs.

Any drop in ambient temperature brings about a slight decrease in the stability as a result of the slow hydration of cement grains (Figure 6).

Taking account of the 5% bleeding limit (Figures 5 and 6) and the viscosity limits and initial and final setting times, a range of W/C ratios is proposed in Table 3 for each cement, based on the viscosity, setting-time and stability tests.

4.3 Tests on hardened grout

Tests on hardened cement provide a means of characterizing grouts during injection. The results obtained in this study allowed us to determine the range of W/C ratios for each cement. However, the quality of each grout after injection and its behavior in its injected medium cannot be known without performing characterization tests on the hardened product. Taking field conditions into account, therefore, the following tests can be envisaged: compressive strength, modulus of elasticity, Poisson coefficient, tensile strength, shear strength, permeability, etc. In the present research

Table 2. Initial setting times for the proposed limit W/C ratios at 4° C and 20° C

| Cements | Measurements at 20° C | | | |
| | Lower limit | | Upper limit | |
	T_{init}*	T_{fin}*	T_{init}*	T_{fin}*
Type I	7.0	10	17	> 24
	(0.5)**	(0.5)	(1.2)	(1.2)
Type I + SP	12,0	14	>18	--***
	(0.4)	(0.4)	(0,6)	
Type III	7,0	10	12	14
	(0.6)	(0.6)	(1.2)	(1.2)
Type III + SP	9.5	-	12	13.5
	(0.5)		(0.8)	(0.8)
M 1	16	20	>27	--
	(0.5)	(0.5)	(1.2)	
M 2	10.5	14.5	>24	>24
	(1.0)	(1.0)	(2.0)	(2.0)
M 3	7.5	--	>24	>24
	(1.0)	(1.0)	(2.0)	(2.0)
Cements	Measurements at 4° C			
	Lower limit		Upper limit	
	T_{init}*	T_{fin}*	T_{init}*	T_{fin}*
Type I	9	13	>24	>24
	(0.8)	(0.8)	(1.5)	(1.5)
Type I + SP	15	17	>24	>24
	(0.5)	(0.5)	(1.0)	(1.0)
Type III	9	11.5	>24	>24
	(1.0)	(1.0)	(1.2)	(1.2)
Type III + SP	11	15	14	--
	(0.6)	(0.6)	(1.2)	(1.2)
M 1	--	--	--	--
	(1.5)	(1.5)	(3.0)	(3.0)
M 2	13	16.5	>24	>24
	(1.2)	(1.2)	(2.5)	(2.5)
M 3	15	18	>24	>24
	(1.2)	(1.2)	(2.5)	(2.5)

* T_{init}/T_{fin}: initial and final setting times (h)
** (0.5):W/C ratio
***measurements not performed

program, most of these tests were indeed performed but the present paper is limited to the results of the first two, i.e. the compressive strength f'c and the modulus of elasticity E.

4.3.1 Compressive strength

The test was performed on cubes (50 mm x 50 mm x 50 mm) in accordance with ASTM C942-86 after 28 days' curing at ambient temperature

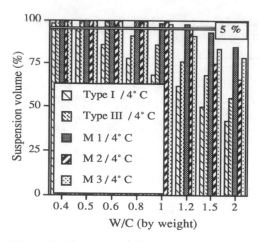

Figure 5: Cement stability at 4° C as a function of the W/C ratio.

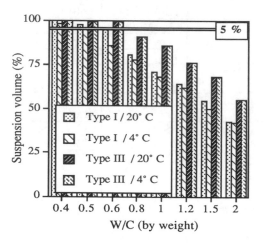

Figure 6: Cement stability as a function of the W/C ratio and temperature.

Table 3. Range of W/C ratios proposed on the basis of viscosity, setting-time and stability tests at 4° C and 20° C.

Cements	Tests at 20° C		Tests at 4° C	
	lower limit	upper limit	lower limit	upper limit
T I	0.6	1.1	0.8	1.0
T III	0.8	1.2	0.8	1.0
T I+SP	0.4	0.6	0.5	0.8
T III+SP	0.6	0.8	0.6	1.2
M 1	0.4	1.2	1.5	2.0
M 2	1.0	1.5	1.2	1.5
M 3	1.0	1.5	1.2	1.5

(20° C) and at 4° C. Curves of the variations in f'c as a function of the W/C ratio were plotted using a log regression model (Figures 7-9). These variations show similar trends in all the cements tested, although microfine cements show a higher compressive strength.

The addition of a superplasticizer had a slight influence on the compressive strength. Even in the case of type III cement, only a super-plasticizer barely changed its compressive strength, yet it allowed grouts with a lower W/C ratio to be prepared (Figure 8). On the other hand, the drop in temperature causes a decrease in the compressive strength of cement grouts.

4.3.2 Modulus of elasticity

This test was performed on cylindrical samples (52 mm in diameter) after 28 days' curing at ambient temperature (20° C) and at 4° C. Curves of the variations in E as a function of the W/C ratio were plotted using a log regression model (Figures, 10-12) and, as in the previous tests, show the same trends. The superplasticizer again seems to have little effect.

5 INJECTION TEST

Once the grouts have been defined, their effectiveness in real conditions has to be assessed. We therefore prepared large slabs of concrete (242 cm x 140 cm x 40 cm) in order to study the injection pressure and penetrability of the grout in a fissure. A horizontal crack was made in the centre of each slab using the hydraulic fracturing technique and the slabs were then injected with different types of grout (cement, epoxy resin, etc.) (Ballivy et al., 1992).

So far, ten slabs have been injected, including two with a type I cement grout, one with a superplasticizer, the other without. Injection begins on the downstream side of the slab and continues until a refusal pressure is reached. An outlet on the upstream side of the slab allows first the water to escape, then the grout, in order to simulate the injection of a crack in contact with the reservoir. A total of 24 pressure sensors and eight displacement sensors are installed on the slab to monitor the pressure variations and the crack opening during injection.

To simulate the conventional injection

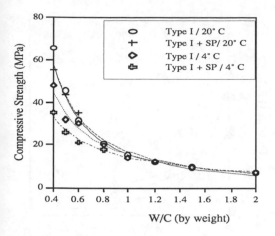

Figure 7: Variations in f'c as a function of the W/C ratio at 20° C and 4° C (type I cement with and without a superplasticizer)

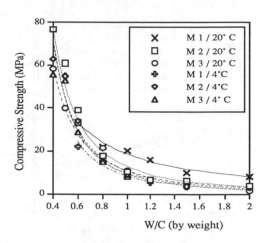

Figure 9: Variations in f'c as a function of the W/C ratio at 20° C and 4° C (microfine cements M1, M2, M3)

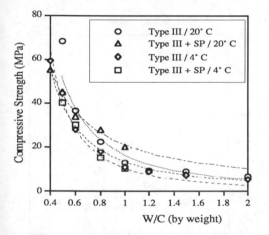

Figure 8: Variations in f'c as a function of the W/C ratio at 20° C and 4° C (type III cement with and without a superplasticizer)

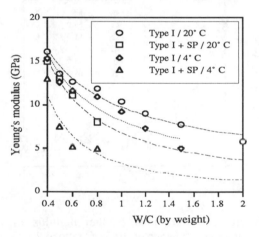

Figure 10 Variations in the modulus of elasticity as a function of the W/C (type I cement)

process, grout with a high W/C ratio was used and each time the grout penetrated the entire crack, the W/C ratio was decreased. Injection of slab #5 with a type I cement grout thus began with a W/C ratio of 3.30 (by weight) and continued with lower ratios. The ratios decreased successively from approximately 3.30 > 2.0 >1.3 >0.85 >0.70 (Figure 13).

Similarly, slab #27 was injected with a type I cement grout with a superplasticizer added. The W/C ratios used are 3.0, 2.0 and 1.2. It should be pointed out that changing from one

grout to another increased the pressure.

In order to analyze the behavior of cement grouts in a crack, we studied the effects of both the variations in the W/C ratio and the grout characteristics on the pressure distribution in the crack.

A lower W/C ratio results in a higher pressure at the injection point (Figure 13) and a greater difference in pressure between this point and other points on the slab (Figure 14). In other words, the higher viscosity (stable grout) of the grout raises the pressure around the injection

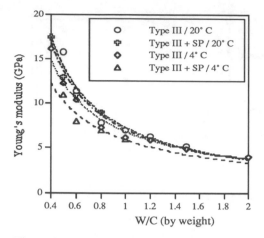

Figure 11: Variations in the modulus of elasticity as a function of the W/C (type III cement)

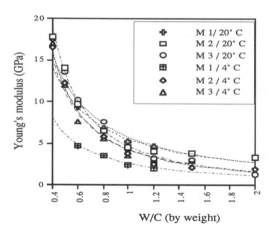

Figure 12: Variations in the modulus of elasticity as a function of the W/C ratio at 20° C and 4° C (microfine cements M1, M2, M3)

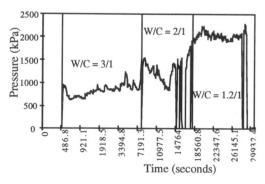

Figure 13 Injection pressure vs. time (type I cement grout with superplasticizer added)

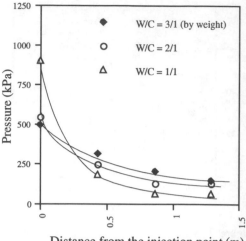

Distance from the injection point (m)

Figure 14: Pressure variations in the slab as a function of the W/C ratio.

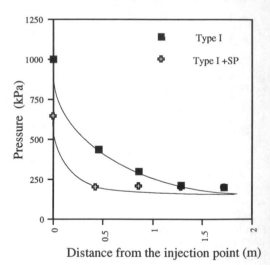

Distance from the injection point (m)

Figure 15: Pressure variations in the slab as a function of the grout characteristics.

point and generally lowers the pressure at points further away, causing the grout to flow in the case of an open crack. The use of a superplasticizer enhances the grout fluidity. The pressure at the injection point is higher for type I grout than for type I with superplasticizer and, in the case of the latter, the pressure attenuation is also greater with increasing distance from the injection point (Figure 15).

The next stage of this project will involve injecting the slabs with the selected products using acceptable limit W/C ratios. Comparison of the results will furnish data on the effectiveness of the selected products with lower W/C ratios than those used in conventional injection techniques.

6 CONCLUSION

Tests were performed at 4° C and 20° C on various cements that differed with respect to fineness, chemical composition and type (Portland cement, microfine cement) in an aim to determine their rheological, physical and mechanical characteristics. The results obtained allowed us to determine the behavior of these grouts after variations in the W/C ratio, the amount of additive or admixture, and temperature, and in some cases to establish W/C ratio ranges beyond which grouts are considered inappropriate to inject. In order for injection to succeed, it is important to know the field conditions, i.e. the fissure configuration and condition, and the temperature variations. The equipment and injection products will be selected on the basis of this information. The study allowed us to select a number of ranges of the W/C ratio for some injection grouts under very specific conditions and with arbitrarily selected limits. Obviously, under other circumstances, these ranges would have to be modified by carrying out other tests to better simulate local conditions.

In conclusion, the most important points that will influence the final choice of grout injection for a particular job are the crack configuration and condition as well as the local field conditions.

Injection of slabs with cement grouts emphasizes the links between laboratory studies on cement grouts and injection work in the field. The relationship between the viscosity and the pressure distribution in the crack can be established by means of additional injection tests.

ACKNOWLEDGMENTS

This study would have been impossible without the financial support of Hydro-Québec (Manicouagan Region) and the cooperation of its staff J. Maniez, L.M. Landry and M. Stirbu. M. Nadeau, in charge of the instrumentation, and J.-P. Lacasse must be thanked in particular. We also thank M.A. Langevin, student, and A. Watier, technician as well as G. Lalonde and M. Lizotte, technicians at the Université de Sherbrooke, for their active participation throughout the project. A final word of thanks goes to L. Kelley-Régnier for the English translation of the text.

REFERENCES

ACI Committee 224, 1984. Causes, evaluation and repair of cracks in concrete structures. Committee report n°2241.R-84, ACI Journal: 211-230.

Ballivy, G., Saleh, K., Mnif, T., Rivest, M., Popiel, M., Nadeau, M. 1992. Restauration of concrete dams: laboratory simulation of cracking and grouting, Soil Improvement and Geosynthetics Conference, ASCE, New-Orleans: 12 p.

Ballivy, G., Saleh, K., Mnif, T., Mirza, J., Rivest, M. 1991. Coulis de ciment pour injection de microfissures, 2ème Colloque canadien sur le ciment et le béton, Vancouver, 13p.

Houlsby, A.C. 1990. Construction and Design of Cement Grouting, a Guide to Grouting in Rock Fondations, Ed. John Wiley and sons,

Lombardi, G. 1985. The role of cohesion in cement grouting of rock, Q. 58, R.13, Lausanne, Vol. III, 235-262.

Tsivilis, S., Tsimasm, S., Hamiotakis, E., Benetatou, A. 1990. Study on the contribution of the fineness on cement strength, Zement - Kalk - Gips, vol. 43, N° 1, p 26-29.

Neuere Untersuchungen über die Injizierbarkeit von Feinstbinde-mittel-Suspensionen

Recent investigations on the groutability of microcement-suspensions

Bertram Schulze

Bilfinger + Berger Bauaktiengesellschaft, Mannheim, Deutschland

ZUSAMMENFASSUNG: Die Injizierbarkeit von Feststoff-Einpreßmassen in vorhandene Hohlräume (Poren, Klüfte, Risse) hängt ab vom Verhältnis der Größen der Feststoffpartikel und der Größe der vorhandenen Hohlräume. Insbesondere beim Einsatz neuartiger Feinstbindemittel-Suspensionen lassen Erfahrungswerte oder grobe Abschätzungen nur in günstigen Fällen zuverlässige Prognosen für das Eindringverhalten zu. Da bei den zu erwartenden Einsatzbereichen für solche Suspensionen (feinkörnige Böden mit einem Feinsandanteil bis zu 30 %, Klüfte und Risse mit Öffnungsweiten von bis zu 0,1 mm) Erfolg und Mißerfolg sehr nahe zusammen liegen, werden Feinstbindemittel hinsichtlich ihrer Eindringfähigkeit gelegentlich nicht ausgenutzt und damit unwirtschaftlich eingesetzt. Der vorliegende Beitrag zeigt, daß die Rezeptur der Suspension (Wasser-Bindemittel-Wert, Zugabe von Additiven etc.) stärker als bisher in "Injizierbarkeits-Prognosen" eingehen muß, da die Größe der Feststoffpartikel (man spricht auch von "wirksamer Partikelgrößenverteilung") maßgeblich durch diese Parameter beeinflußt wird. Darüberhinaus kann die Vorhersage des Injektionserfolges verbessert werden, wenn zusätzlich die Größe und Verteilung der maßgeblichen Hohlräume berücksichtigt werden. Am Beispiel einer Injektion in Lockergestein wird gezeigt, wie sich diese Kennwerte, die zur sog. Porenengstellenverteilung führen, ermitteln lassen und welche Auswirkungen sich daraus für den Injektionserfolg ergeben.

ABSTRACT: Recent investigations of the groutability of microfine cement suspensions The penetrability of suspension grouts in existing cavities in the subsoil (such as pores, joints, fissures) depends on the relation of the particle size of the solids to the dimension of these cavities. Especially if grouts with the new developped microfine cements are used, a reliable prediction of the penetration behaviour based on empirical data or rough estimation is possible only under favourable conditions. Since success and failure are very close to each other in the interesting fields of application (these are fine-grained soils with up to 30 % fine sand, joints and fissures with a width of up to 0,1 mm), microfine cement grouts are often used inefficiently. In much cases it is hardly possible to take full advance of the good penetrability of such grouts. The following contribution shows that the composition of the grout (water : cement ratio, addition of injection helpers a.s.o.) has to be taken into stronger consideration in "penetrability predictions" than it is the case now, because the particle size of the solids (the so-called effective particle size distribution) is influenced mainly by these parameters. Moreover the prediction of grouting success can be improved, if the size and the distribution of the cavities concerned are taken into account additionally. With the example of an injection in loose soil it is shown how these parametes which lead to the so-called distribution of the pore constrictions can be determined and which effects on the grouting success result from this proceeding.

1 BISHERIGE VORGEHENSWEISE BEI DER BEURTEILUNG DER INJIZIERBARKEIT VON ZEMENT-SUSPENSIONEN

1.1 Grundsätzliche Überlegungen

Viele Bauaufgaben sind ohne eine vorauseilende Baugrundverbesserung durch Injektionen nicht durchführbar. Bei diesem Verfahren werden vorhandene Hohlräume (Poren im Lockergestein, Klüfte im Festgestein) mit einer geeigneten Technik zugänglich gemacht und unter Druck mit geeigneten Materialien verfüllt, wodurch eine Verfestiqung und/oder eine Abdichtung des behandelten Bereiches erfolgt.

Bekanntlich werden zwei prinzipiell unterschiedliche Gruppen von Materialien für Injektionszwecke eingesetzt. Diese sind

1. chemische Lösungen (i. d. Regel auf Wasserglasbasis), die hier außer acht bleiben sollen und
2. Suspensionen, also Wasser-Feststoff-Gemische, wobei als Feststoffe Tonmehle, Bentonite, Zemente und neuerdings Feinstbindemittel eingesetzt werden.

Charakteristisch für Suspensionen und für die Beurteilung ihres Einsatzbereiches von enormer Bedeutung sind die mehr oder weniger ausgeprägte Neigung zum Entmischen (Sedimentation) und das Vorhandensein von Partikeln bestimmter Größe im Injektionsgemisch (im Gegensatz zu Lösungen, bei denen man von Partikeln mit für baupraktische Belange relevanter Größe nicht sprechen kann).

Die Größe dieser Partikel bestimmt den Anwendungsbereich bzw. die untere Anwendungsgrenze für das betrachtete Einpreßmittel, da die vorhandenen Fließwege im Untergrund so beschaffen sein müssen, daß ein Eindringen des Einpreßgutes in das Hohlraumsystem geometrisch möglich ist (s. u. Bild 6).

Die demzufolge erforderlichen geometrischen Betrachtungen werden dadurch erschwert, daß die Partikel des Einpreßgutes naturgemäß nicht alle gleich groß sind, sondern daß hier - wie auch bei der Beschreibung von Böden - von einer Partikelgrößenverteilung gesprochen werden muß. Noch schwieriger ist die Frage nach der Größe der vorhandenen Hohlräume des Untergrundes zu beantworten. Der Porenraum im Lockergestein ist ein Gebilde aus unterschiedlich großen Hohlräumen, die durch unterschiedlich weite Kanäle, die die sog. Porenengstellen bilden, miteinander verbunden sind. Auf die gleichen Probleme stößt man - allerdings in noch größerem Maße - bei Kluftsystemen im Festgestein. Dort haben die Klüfte unterschiedliche Öffnungsweiten und Ausdehnungen, sind u. U. vielfach verästelt, kreuzen einander oder laufen im Gesteinskörper aus, so daß eine für die Beurteilung der Injizierbarkeit mit einem bestimmten Einpreßgut hinreichend genaue Beschreibung der Fließwege bislang kaum möglich ist.

Die genannten Probleme haben in der Vergangenheit und in der Gegenwart dazu geführt, daß verschiedene Wege beschritten wurden, um die Injizierbarkeit von Locker- und Festgestein einigermaßen zutreffend beurteilen zu können. Auf die wichtigsten

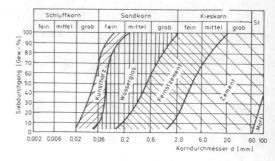

Bild 1: Anwendungsgrenzen von Injektionsmitteln (Tausch/Teichert (1990))
Fig. 1: Range of application of injection materials (Tausch/Teichert (1990))
(Siebdurchgang: percent finer by weight, Korndurchmesser: grain size, Kunstharz: resins, Wasserglas: sodium silicate, Feinstzement: microfine cement, Zement: cement, Mörtel: mortar)

der sog. Injektionskriterien und die Probleme, die mit der Anwendung dieser Kriterien verbunden sind, wird im folgenden kurz eingegangen.

1. 2 Lockergestein (Porensysteme)

Entscheidungshilfen für die Beurteilung der Eindringfähigkeit von Injektionssuspensionen werden vielfach in Form von Grenz-Körnungslinien angegeben (vgl. Bild 1).

Gelegentlich wird zur Abschätzung der Wasserdurchlässigkeitsbeiwert des zu injizierenden Bodens herangezogen (vgl. Bild 2). Hier ist anzumerken, daß solche Abschätzungen strenggenommen nur für Injektionsmassen zulässig sind, die ein dem Wasser vergleichbares Fließverhalten zeigen (NEWTON'sche Fluide) und deren Eigenschaften durch die Strömung durch das Korngerüst nicht beeinflußt werden; solche Mittel sind etwa die chemischen Injektionslösungen (Silikatgele).

Injektionssuspensionen (Fluide vom BINGHAM-Typ) erfüllen diese Voraussetzungen nicht; Aussagen über einen Injektionserfolg mit Suspensionen aufgrund solcher Bodenkennwerte sind daher kritisch zu treffen.

Durchgesetzt haben sich auch Kriterien, bei deren charakteristische Kennwerte des zu injizierenden Bodens und des Injektionsgutes miteinander verglichen werden, bei denen also zumindest angedeutet wird, daß es sich bei der Beurteilung der

Eindringfähigkeit um ein geometrisches Problem handelt (z. B. Nonveiller (1989)). Betrachtet wird hier bekanntlich die Verhältnisgröße

$$N = \frac{d_{15, \text{ Boden}}}{d_{85, \text{ Verpreßgut}}} \qquad (1)$$

Wenn N > 25, dann gilt eine Penetration des Verpreßgutes durch das Korngerüst als möglich und wenn N < 9, dann wird eine Penetration für ausgeschlossen gehalten.

Über die Probleme und die Ungenauigkeiten, die mit der Anwendung dieser drei unterschiedlichen Injektionskriterien verbunden sind, wurde bereits ausführlich berichtet (Schulze (1993)). Gemeinsam sind diesen Kriterien zwei prinzipielle Fehler, die hier nochmals kurz dargestellt seien.

1. Zur Beurteilung der Injizierbarkeit eines Bodens wird dessen Korngrößenverteilung herangezogen. Erkenntnisse über die zur Verfügung stehenden Hohlräume (Poren bzw. Porenengstellen), die das zu injizierende Medium darstellen, lassen sich damit nur indirekt und in qualitativer Art gewinnen. Für die Größe der Hohlräume wichtige Bodenkennwerte wie die Lagerungsdichte und der Ungleichförmigkeitsgrad des Bodens werden nicht berücksichtigt. Die damit verbundenen Probleme werden noch vergrößert, wenn lediglich ein Wert (z. B. d_{15}) die Korngrößenverteilungskurve charakterisiert.

2. Das Injektionsmittel geht lediglich

in allgemeiner Form in die Injektionskriterien ein. Diese Vorgehensweise beinhaltet die Vorstellung, daß die zu injizierenden Feststoffe immer in der gleichen Form (Partikelgröße) vorliegen. Tatsächlich aber hat die Rezeptur der Injektionssuspension entscheidenden Einfluß auf die Partikelgrößenverteilung der Feststoffe in der Suspension (s. Abschnitt 2.1) und muß daher bei allen Injektionskriterien berücksichtigt werden. Bei der - zumindest denkbaren - Betrachtung des reinen Pulvers (also ohne Zugabe von Wasser) wird die Eindringfähigkeit des betrachteten Stoffes erheblich überschätzt.

1.3 Festgestein (Kluftsysteme)

Die Erkundung und die (rechnerische) Beschreibung der Fließwege im Festgestein über ganze Systeme von glatten oder rauhen Schichtflächen, teilverfüllten oder unverfüllten Klüften unterschiedlicher Ausdehnung ist noch wesentlich schwieriger als im Lockergestein. Aus diesem Grunde wird auf Injektionskriterien in der oben für Lockergestein beschriebenen Art und Weise bei der Planung von Injektionsarbeiten im Festgestein bislang verzichtet. Da Injektionen im Festgestein in aller Regel der Abdichtung dienen, wird vielmehr vorab geprüft, ob eine Abdichtung (d. h. eine Injektion) überhaupt erforderlich ist. Hierzu werden bekanntlich Wasserabpreßversuche nach Lugeon durchgeführt. Aus den Charakteristiken der Wasseraufnahme und der Druckentwicklung beim Verpressen kann man zumindest qualitativ auf die vorhandene Durchlässigkeit des Gebirges und damit auf die Notwendigkeit einer (Zement-) Injektion schließen. Aufgrund umfangreicher Untersuchungen und langjähriger Erfahrung kommt Kutzner (1991) zu folgenden Aussagen:

- "Bei LU-Werten unter 5 (oder 10) ist das Gebirge als hinreichend dicht anzusehen. Eine Injektion von Zement ist zwar auch ohne Aufreißen möglich, aber die Verminderung der Sickerwassermenge ist unbefriedigend (...)".
(Anm.: 1 LU entspricht einer Wasseraufnahmerate von 1 l/min je m Bohrloch bei einem Druck von 1,0 MPa).
- Bei LU-Werten über 25 (oder 30) ist das Gebirge gut injizierfähig und verspricht brauchbaren Erfolg im Versperren der wesentlichen Sickerwege und damit Herabsetzen der Sickerwassermenge (...)"

In günstigen Fällen (LU-Werte über 25

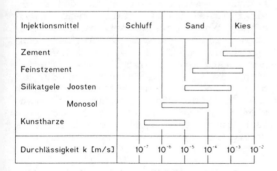

Bild 2: Anwendungsbereiche von Injektionsmitteln in Abhängigkeit vom k-Wert (Tausch (1981))
Fig. 2: Range of application of injection materials dependent on the coefficient of permeability k (Tausch (1981))
(Injektionsmittel: injection material, Schluff: silt, Sand: sand, Kies: gravel, Durchlässigkeit: permeability)

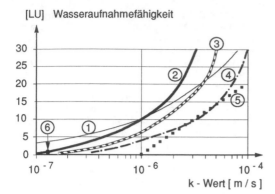

[LU] Wasseraufnahmefähigkeit

k - Wert [m / s]

Bild 3: Beziehungen zwischen LU-Wert und Gebirgsdurchlässigkeit (Kutzner (1991)) 1. Heitfeld (1965), 2. USBR, 3. Rissler (1980) anisotrop, 4. Rissler (1980) isotrop, 5. Krauß-Kalweit (1987), 6. Nonveiller (1968)
Fig. 3: Relation between Lugeon-value (LU) and permeability of rock masses (Kutzner (1991)) (Wasseraufnahmefähigkeit: water taking capacity, k-Wert: k-value)

oder unter 5) läßt sich also beurteilen, ob eine Injektion erforderlich (sinnvoll) ist oder nicht. Hinweise auf die Art des Injektionsmittels oder gar auf eine erfolgversprechende Rezeptur werden nicht erhalten. So kann z. B. eine hohe Wasseraufnahmefähigkeit des Gebirges (aus der man auf die Notwendigkeit einer Injektion schließen würde) aus wenigen Klüften mit großer Öffnungsweite oder aus einer Vielzahl von Klüften mit geringer Öffnungsweite resultieren. Im ersten Falle kann eine Injektion mit herkömmlichen Zementen Erfolg bringen, im zweiten Fall kann eine Injektion nur mit Feinstzementen möglich oder gänzlich ausgeschlossen sein.

Aufgrund dieser Problematik gibt es - in Ergänzung zu der bisher weitgehend empirischen Vorgehensweise bei der Planung von Injektionsarbeiten im Fels Überlegungen, wie aus den Ergebnissen von Wasserabpreßversuchen auf die Größe und Ausbildung der vorhandenen Fließwege geschlossen und geeignete Injektionsmischungen ausgewählt werden können. Hierbei wird versucht, zunächst eine Abhängigkeit zwischen LU-Wert (Transmissivität in Lugeon) und der Gebirgsdurchlässigkeit herzustellen. Diese Abhängigkeit ist allerdings nicht eindeutig und wird stark von der Tropie des Gebirges beeinflußt (vgl. Bild 3).

In einem zweiten Schritt werden diesen Durchlässigkeiten Spaltweiten zugeordnet, wobei die Anzahl der Klüfte (pro m) und die Ausbildung der Kluftwandungen (rauh

oder glatt) von enormer Bedeutung sind. Aufgrund von Erfahrungen und Laborversuchen kann man nun in einem dritten Schritt die in Frage kommenden Injektionsmaterialien diesen Spaltweiten zu ordnen, so daß eine Penetration gerade noch möglich ist (vgl. Bild 4). Für konventionelle Zemente gelten Klüfte mit Öffnungsweiten von 0,3 bis 0,5 mm als noch injizierbar, unter Verwendung von Feinstzementen darf erwartet werden, daß Klüfte mit Öffnungsweiten bis etwa 0,1 mm injizierbar sind (Kutzner (1991)).

Wird beispielsweise in einem anisotropen Gebirge ein Wert für die Transmissivität von 10 LU ermittelt, bedeutet das nach Bild 5 eine Gebirgsdurchlässigkeit von rd. $2 \cdot 10^{-4}$ cm/s. Wenn gleichzeitig im interessierenden Bereich bei einer Kluftkörperaufnahme durchschnittlich 3 Klüfte pro Meter mit nicht nennenswerter Rauhigkeit festgestellt wurden, dann läßt sich diesen Klüften eine mittlere Spaltweite von 0,1 mm zuordnen. Damit wird die Notwendigkeit einer Injektion mit Feinstzement deutlich, die Rezeptur einer eindringfähigen Feinstzement-Suspension bleibt allerdings außer acht.

In den Einsatzgebieten von konventionellen Zementen (Kiesböden mit geringem Sandanteil, Klüfte mit Öffnungsseiten im [mm]-Bereich) genügten die vorgenannten Injektionskriterien (Abschnitte 1.2 und 1.3) durchaus für die Vorhersage eines Injektionserfolges. Durch die Entwicklung der Feinstzemente werden allerdings Untergrundbereiche für die Feststoff-Injektion erschlossen, in denen Erfolg (Penetration) und Mißerfolg (Ausfiltern) sehr nahe beisammen liegen. So kann es für den Erfolg einer Injektion mit Feinstzement-Suspensionen beispielsweise entscheidend sein, ob der zu injizierende Boden einen Feinsandanteil von 5 oder von 10 % aufweist. Aus diesem Grunde erscheint es erforderlich, die entscheidenden Parameter für die Beurteilung der Injizierbarkeit, nämlich die Partikelgrößenverteilung der Feststoffe im Injektionsgut und die Größe und Verteilung der maßgeblichen Hohlräume im Untergrund, etwas genauer zu betrachten.

2   BISLANG NICHT BERÜCKSICHTIGTE EINFLÜSSE AUF DIE INJIZIERBARKEIT VON SUSPENSIONEN

2.1 Wirksame Partikelgrößenverteilung der Feststoffe

Die Feststoffe, die für Injektionszwecke in Frage kommen (Bentonite, Zemente, Feinstzemente), liegen zunächst als Pulver mit sehr geringen Partikelgrößen vor. Beim

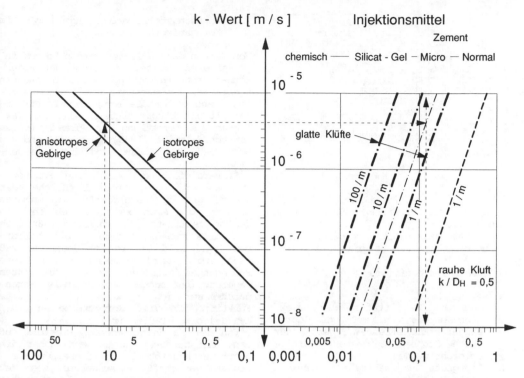

Bild 4: Zusammenhang zwischen Transmissivität (LU-Wert), Gebirgsdurchlässigkeit, Spaltweite und geeignetem Injektionsgut (leicht modifiziert nach Widmann (1993))
Fig. 4: Relation between transmissivity (LU-value), permeability of rock mass, effective width of the joint and suitable injection material (slightly modified after Widmann (1993))
(Spaltweite: width of joint, anisotropes Gebirge: anisotropic rock mass, glatte Klüfte: even joints, rauhe Kluft: rough joint)

Kontakt dieser Stoffe mit Wasser finden sofort Quell-, Agglomerations- bzw. Hydratationsprozesse statt, die zwar ein Zeichen ihrer Eignung als Einpreßmittel sind (Abdichtungs- und /oder Verfestigungswirkung), die sich aber zunächst negativ auf die Partikelgrößenverteilung der Feststoffe in der Suspension und damit negativ auf die Eindringfähigkeit auswirken.

Aus diesem Grunde genügen Partikelgrößenverteilungskurven der trockenen Stoffe für die Beurteilung ihrer Eindringfähigkeit nicht. Vielmehr sind Aussagen über die "wirksame Partikelgrößenverteilung" der Feststoffe, also diejenige, die unmittelbar vor dem Eindringen der Suspension in die Hohlräume herrscht, erforderlich. Wie an anderer Stelle ausführlich erläutert (SCHULZE (1993)), hängt diese ab (vgl. Bild 5):

1. von der Art der Ausgangsstoffe,

2. vom Wasser-Zement-Wert,
3. von einer eventuellen Bentonitzugabe,
4. von der Zugabe von Additiven,
5. von der Zeitdauer zwischen dem Herstellen der Suspension und dem Verpressen,
6. von der Art der Aufbereitung der Suspension

zu 1. Feinstbindemittel weisen erheblich feinere Partikel auf als konventionelle Zemente.

zu 2. Je größer der Wasser-Zement-Wert ist, desto mehr nähert sich die Partikelgrößenverteilungskurve der des trockenen Stoffes an.

zu 3. Die Bentonitzugabe wirkt sich wegen der starken Agglomerationsneigung dieser Stoffe äußerst nachteilig aus. Bei der Verwendung von Feinstbindemitteln muß daher auf den Einsatz von Bentoniten verzichtet wer-

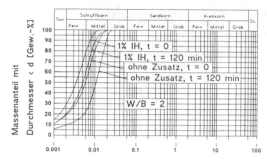

Bild 5: Wirksame Partikelgrößenverteilungskurven von Feinstzement-Suspensionen (beispielhaft)
Fig. 5: Effective grain size distribution curve of microfine cement grouts (exemplary)
(ohne Zusatz: no additive)

den. Die Stabilität solcher Injektionssuspensionen muß ggf. durch einen reduzierten W/B-Wert oder durch Zugabe geeigneter Additive erreicht werden.

zu 4. Geeignete ausgewählte und sinnvoll dosierte Additive (sog. Injektionshilfen) wirken sich positiv auf die Partikelgrößenverteilung des Einpreßmittels aus (vgl. Schulze et al (1992)).

zu 5. Die Hydratations- und Agglomerationsprozesse führen mit der Zeit zu immer größeren Partikeln bzw. Partikelansammlungen. Deswegen müssen Verpreßvorgänge zügig ausgeführt werden.

zu 6. Insbesondere Feinstzemente bedürfen einer ausreichenden Mischzeit und einer ausreichenden Mischintensität, damit die Stoffe ordnungsgemäß aufgeschlossen werden und die Feinheit des Produktes genutzt werden kann.

Abhängig von den Randbedingungen und insbesondere der Rezeptur gibt es also verschiedene "Feinheitsgrade" von Feinstzement-Suspensionen, so daß Pauschalurteile über das Eindringvermögen dieser Stoffe ("Feinstzemente sind geeignet für die Injektion von Mittelsand") nicht angebracht sind. In allen Injektionskriterien, die charakteristische Untergrundkennwerte mit charakteristischen Suspensionskennwerten vergleichen, sind die wirksamen Partikelgrößenverteilungskurven heranzuziehen.

2.2 Porenengstellenverteilung (PEV) im Kornhaufwerk

Von Seiten des zu injizierenden Bodens interessieren weniger die Größe der Bodenkörner, sondern vorwiegend die vorhandenen Hohlräume, insbesondere die Porenengstellen, wenn es um die Frage der Penetrierbarkeit mit Einpreßmitteln geht. Ein Verfahren zur Ermittlung der Porenengstellen ist von Silveira (1965) entwickelt und von Schulze (1992) ergänzt und erweitert worden.

Silvera folgend, wird eine Porenengstelle aus drei unterschiedlich oder gleich großen, kugelförmig angenommenen Körnern, die sich berühren, gebildet. In den so gebildeten Zwischenraum läßt sich ein Kreis mit dem Radius r_p einbeschreiben (vgl. Bild 6). Es ist deutlich, daß sich ein Erdstoff, dessen Körner diese Bedingungen erfüllen, in der dichtesten Lagerung befindet. Der minimale Porendurchmesser ($d_p = 2 \cdot r_p$) ist eine Funktion der drei Korndurchmesser (d_1, d_2, d_3) bzw. -radien (r_1, r_2, r_3).

Von dem zu untersuchenden Erdstoff muß man sich zunächst Anhaltswerte über die Auftretenshäufigkeit der einzelnen Korngrößen verschaffen. Ausgehend von der Korngrößenverteilung (Siebung) und unter den Annahmen gleicher Wichten sowie der Kugelform der Körner läßt sich in einfacher Weise (vgl. Ziems (1968)) die sog. Anzahlverteilungskurve (AV-Kurve) des Erdstoffes erstellen. Diese Kurve wird

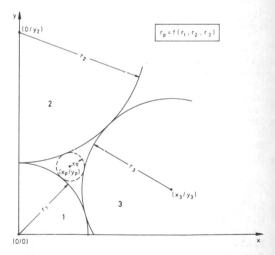

Bild 6: Ebener Schnitt durch eine Porenengstelle (Modell von Silveira (1965))
Fig. 6: Plane section through a pore constriction (after Silveira (1965))

dann in prinzipiell beliebig viele Klassen
mit einem maßgebenden Korndurchmesser d_i
unterteilt (d_i kann z. B. das geometri-
sche Mittel aus Klassenober- und -unter-
grenze (d_o, d_u) sein. Weniger als 5
Klassen sind nicht sinnvoll, mehr als 10
nicht erforderlich. Nun werden alle möglichen Dreier-Kombinationen von Korngrößen, die sich aus der gewählten Klassenstärke bilden lassen, angegeben und das
dazugehörige d_p ermittelt (Schulze
(1992)). Bei einer Klassenstärke von 5
gibt es 35 Kombinationsmöglichkeiten, bei
einer Klassenstärke von 10 bereits 225.
Jeder Dreier-Gruppe (und damit jedem berechneten d_p) wird eine bestimmte Auf-
tretenswahrscheinlichkeit P zugeordnet
(SILVEIRA (1965)), wobei die Auftretens-
wahrscheinlichkeit einer Korngröße aus der
zuvor ermittelten AV-Kurve bestimmt wird.
Die Wertepaare (P, d_p) werden nun nach
steigendem d_p geordnet und die zugehörigen Wahrscheinlichkeiten sukzessive aufaddiert. Das Ergebnis läßt sich (am aussa-
gekräftigsten in üblichen Korngrößenver-
teilungs-Diagrammen) in einer Kurve (P
über d_p) darstellen, an der man unter
Vorgabe eines beliebigen Durchmessers
(d_p) ablesen kann, wieviel Prozent des
Porenengstellen einen kleineren (oder
größeren) Durchmesser haben (s. u.).
Wie vor allem experimentelle Untersuchungen (Witt (1986)) gezeigt haben, hat
die Form der Bodenkörner keinen nennenswerten Einfluß auf die Porenengstellenverteilung. Deshalb und aus Platzgründen soll
diesem Umstand hier nicht weiter nachge-
gangen werden.
Beispielrechnungen - zweckmäßigerweise
mit Hilfe eines Rechenprogrammes - zeigen
folgende Ergebnisse:

- Der Boden, der durch die Kornverteilungskurve 2 charakterisiert ist,
  weist die kleinsten Porenengstellen
  auf, ist also am schwersten injizierbar. Boden 1 hat (bei gleichem d_{10})
  erheblich gröbere Poren (Bild 7).
  Der kleinste Porenengstellendurch-
  messer hängt also - wie erwartet -
  vom kleinsten vorhandenen Korndurchmesser ab. Der Feinanteil des Bodens
  wird durch die Angabe von d_{10} offensichtlich nicht ausreichend berücksichtigt. Der Boden nach Kurve 3
  hat eine PEV-Kurve, die von der des
  Bodens 1 erst im oberen Bereich (ab
  etwa 80 % Auftretenshäufigkeit) mar-
  kant abweicht; erst dort macht sich
  das Vorhandensein grober Bodenkörner
  (hier: Kies) bemerkbar.
- Ermittelt man aus den Kornverteilungskurven (1, 2 und 3) die Wasser-

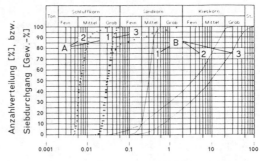

Bild 7: Verlauf der PEV-Kurve (A) in Abhängigkeit vom Verlauf der Kornverteilungskurve (B)
Fig. 7: Shape of the PCD-curve (A) dependent on the shape of the grain size
distribution curve (B)
(Anzahlverteilung: grain size distribution
by number (ND), Porendurchmesser: pore
size, Porenengstellenverteilung (PEV):
pore constriction distribution (PCD))

durchlässigkeitsbeiwerte, so ergibt
sich die gleiche Reihenfolge wie bei
den PEV-Kurven (k_2 am geringsten,
k_3 am größten). Der Unterschied ist
aber bei weitem nicht so gravierend,
was seine Ursache darin hat, daß der
k-Wert aus d_{10} berechnet wird und
vor allem, daß die Lagerungsdichte
bei dieser k-Wert-Bestimmung keine
Rolle spielt. Gerade für weitgestufte
Erdstoffe (Kurven 2 und 3) ist es
aber für die Größe der Porenengstellen und damit für die Injizierbarkeit
von entscheidender Bedeutung, ob der
Boden locker oder dicht gelagert ist.
- Am Beispiel des Boden 3 wird der Unterschied zwischen Wasserdurchlässig-
  keit und Injizierbarkeit mit Feststoffen deutlich. Der Durchlässigkeitsbeiwert von Boden 3 ist fast
  eine Zehnerpotenz größer als der von
  Boden 1. Dies hat seine Ursachen im
  Vorhandensein größerer Hohlräume ab
  etwa 80 Anzahl % (siehe Bild 7). Zum
  größten Teil (und vor allem im unteren Bereich) stimmen die PEV-Kurven
  beider Böden - dichte Lagerung vorausgesetzt - jedoch überein, deswegen
  ist unter diesen Bedingungen auch die
  Injizierbarkeit beider Böden vergleichbar.
- Zwischen der vergleichsweise einfach
  zu ermittelnden AV-Kurve und der zugehörigen PEV-Kurven besteht folgende
  vereinfachende (empirische) Beziehung:

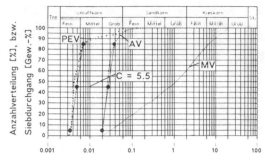

Bild 8: Näherungsweise Ermittlung der PEV-Kurve aus der AV-Kurve
Fig. 8: Approximate determination of the PCD-curve from the ND-curve
(Kornmassenverteilung (NV): grain size distribution by weight (WD)

$$d_p \sim \frac{1}{c} \cdot d_{AV} \qquad (2)$$

Der Proportionalitätsfaktor C liegt dabei zwischen 5,5 (bei ungleichförmigen Böden, vgl. Bild 8) und 6.5 (bei gleichförmigen Böden). Die Abweichung vom Proportionalitätsfaktor C im oberen Bereich der Kurven ist für die Beurteilung der Injizierbarkeit ohne Belang.
- PEV-Kurven für den Fall lockerer Lagerung lassen sich nur näherungsweise und vergleichsweise grob bestimmen (Schulze (1991)). Darauf soll hier nicht eingegangen werden.

3   VORSCHLAG FÜR EIN NEUARTIGES INJEKTI-ONSKRITERIUM FÜR LOCKERGESTEIN

Gemäß Bild 6 kann eine Penetration stattfinden, wenn die größten Partikel des Einpreßgutes kleiner sind als die kleinsten zur Verfügung stehenden Hohlräume. In Anlehnung an Gl. 1 gilt dann:

$$N_i = \frac{d_{0,P}}{d_{100,sus}} \qquad (3)$$

Hier ist:

N_i eine Injektionskennzahl
$d_{0,P}$ der kleinste Durchmesser der zur Verfügung stehenden Hohlräume (s. u. Bild 9)
$d_{100,sus}$ der Durchmesser der größten Partikel der Suspensionsfeststoffe (s. u. Bild 9)

Wenn $N_i \geq 1$, dann findet eine Penetration statt (vgl. Bild 9)

Bedingt durch die rechnerische Annahmen, die zur Ermittlung der Porenengstellenverteilung führen, stellt dies ein sehr strenges Kriterium dar. Es ist physikalisch kaum vorstellbar, daß alle Körner eines Erdstoffes so dicht gepackt sind wie in Bild 6 dargestellt ist. Wie Labor- und Feldversuche gezeigt haben, ist ein Injektionserfolg auch möglich, wenn $N_i < 1$ ist. In einem solchen Falle müssen aber die Verläufe der wirksamen Partikelgrößenverteilungskurve und der PEV-Kurve betrachtet werden. Ein Injektionserfolg ist umso wahrscheinlicher, je näher N_i an 1 liegt und je kleiner die Fläche ist, die von der wirksamen Partikelgrößenverteilungskurve und der PEV-Kurve gemeinsam eingenommen wird (s. u., Bild 10).
Besonders vorteilhaft wird das Injektionskriterium nach Gl. 3 (bzw. Bild 9) bei ungleichförmigem Boden eingesetzt, da die wahrscheinliche Lage der PEV-Kurve in Abhängigkeit von der Lagerungsdicke abgeschätzt werden kann. Ein Anwendungsbeispiel ist in Bild 10 dargestellt.
Hier ergibt sich ein (von der Lagerungsdichte unabhängiger) Wert von N = 17. Gemäß dem Injektionskriterium nach Gl. 1 ist die Frage nach einem Injektionserfolg nicht eindeutig zu beantworten. Wie ein Feldversuch in diesem Boden zeigte, fand

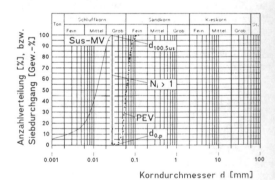

Bild 9: Graphische Darstellung des Injektionskriteriums nach Gl. 3 (Injektionserfolg)
Fig. 9: Graphical presentation of the injection criterion after eq. 3 (grouting success)
(Sus-MV: grain size distribution of the solids in the suspension, N_i: a characteristic injection value, $d_{0,P}$: the smallest pore size, $d_{100,sus}$: the diameter of the largest particles of the solids in the suspension)

bei lockerer Lagerung eine Penetration statt, bei dichter Lagerung (in einem Laborversuch) konnte der Boden nicht injiziert werden. Dieses Ergebnis deckt sich mit den Schlußfolgerungen, die man aus der Betrachtung der Porenengstellenverteilung unter Berücksichtigung der Lagerungsdichte ziehen kann. Bei dichter Lagerung sind nur etwa 30 % der vorliegenden Hohlräume größer als die gröbsten Feststoffpartikel => keine Penetration; bei lockerer Lagerung sind etwa 85 % der vorliegenden Hohlräume größer als die gröbsten Feststoffpartikel => Penetration.

4 VERSUCH EINER ÜBERTRAGUNG DER ERGEBNISSE AUF DIE INJEKTION VON FESTGESTEIN

Hinsichtlich der wirksamen Partikelgrößenverteilung der Suspensionsfeststoffe können die Ergebnisse natürlich direkt übertragen werden, da diese unabhängig vom Untergrund ist. Eine Übertragung der Ergebnisse im Hinblick auf eine "Kluftengstellenverteilung" ist dagegen nicht ohne weiteres möglich. Während im Lockergestein die hydraulische Leitfähigkeit einer Bodenschicht recht zutreffend durch Korngrößenverteilung und Lagerungsdichte beschrieben werden kann und diese Kenngröße wegen der im Vergleich zu Festgestein recht ausgeprägten Durchlässigkeitsisotropie - auch für einen vergleichsweise großen Bereich gilt, werden für die Beschreibung der hydraulischen Leitfähigkeit von Festgestein erheblich mehr Informationen notwendig.

Sorgfältig durchgeführte Wasserabpreßversuche liefern dabei erste Hinweise über das hydraulisch wirksame Kluftsystem, wobei nochmals erwähnt werden muß, daß ein markanter Unterschied zwischen der Durchlässigkeit für Wasser und der für Suspensionen besteht. Um diese Ergebnisse im Hinblick auf die Injizierbarkeit des Gebirges auswerten zu können, muß ein besonderes Augenmerk auf die Aufnahme und Ansprache der Bohrkerne aus dem betreffenden Bereich gelegt werden.

Eine solche Ansprache muß enthalten:
- die Anzahl der Klüfte
- die Raumstellung der Klüfte bzw. der Kluftscharen
- die Öffnungsweiten der Klüfte (wobei Entspannungsvorgänge während der Bohrkernentnahme nicht vernachlässigt werden dürfen)
- Angaben über Kluftrauhigkeit und -besatz (Genese) und
- sonstige Hinweise auf die hydraulische Leitfähigkeit des Gebirges

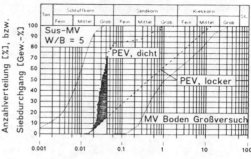

Poren-, Korndurchmesser d [mm]

Bild 10: Injektionsversuch in ungleichförmigem Boden, Penetration nur bei lockerer Lagerung
Fig. 10: Test injection in poorly sorted soil, penetration only if soil is poorly compacted
(dicht: well compacted soil, locker: poorly compacted soil, MV Boden Großversuch: grain size distribution of the soil in the test field)

Aus der Vielzahl der Daten muß ein hydraulisches Modell des Untergrundes erstellt werden. Die statistische Auswertung in der oben beschriebenen Art und Weise kann dann eine Häufigkeitsverteilung der wirksamen Kluftweiten liefern, wobei die Frage nach dem Gültigkeitsbereich dieser Verteilung durch die allgemeine geologische Ansprache beantwortet werden muß.

Abschließend sei bemerkt, daß die Bestimmung der Porenengstellenverteilung für ein (zu injizierendes) Kornhaufwerk auf statistischer Basis zu zuverlässigen und direkt für die Praxis verwendbaren Ergebnissen führt. Ob dies auch für die Beschreibung der hydraulischen Eigenschaften und damit der Injizierbarkeit von Festgestein zutrifft, muß sich erst zeigen. Insofern sind die vorstehenden Äußerungen in erster Linie als Denkanstoß gedacht.

LITERATURVERZEICHNIS

Kutzner, C. (1991): Injektionen im Baugrund. Ferdinand Enke Verlag Stuttgart

Nonveiller, E. (1989): Grouting - Theory and Practice. Developments in Geotechnical Engineering, 57. Elsevier Amsterdam - Oxford - New York - Tokyo

Schulze, B. (1992): Injektionssohlen: Theoretische und experimentelle Untersuchungen zur Erhöhung der Zuverlässigkeit. Veröff. d. Inst. f. Boden- und Felsmechanik, Universität Karlsruhe, Heft 126

Schulze, B. (1993): Der Einfluß der Rezeptur auf das Eindringverhalten von Injektionssuspensionen. Seminar der Technischen Akademie Esslingen, Feb. 1993

Schulze, B./Kühling, G./Tax, M. (1992): Neue Zusatzmittel für feststoffreiche Feinstzement-Suspensionen, Bauingenieur 67, S. 499-504

Silveira, A. (1965): An analysis of the problem of washing through in protective filters. Proc. 6th ICSMFE, Montreal, Vol. 3

Tausch, N. (1981): Schäden infolge falscher Injektionen - Sanierungsmöglichkeiten durch Injektionen. Seminar der Technischen Akademie Wuppertal, Juni 1981

Tausch, N./Teichert, H. (1990): Injektionen mit Feinstbindemittel. 5. Christian Veder Kolloquium "Neue Entwicklungen in der Baugrundverbesserung" in Graz, 1990

Widmann, R. (1993): Injektionen: Aus der Erfahrung zur Theorie, Seminar der Technischen Akademie Esslingen, Februar 1993

Witt, K.-J. (1986): Filtrationsverhalten und Bemessung von Erdstoff-Filtern. Veröff. d. Inst. f. für Boden- und Felsmechanik, Universität Karlsruhe, Heft 104

Ziems, J. (1968): Beitrag zur Kontakterosion nichtbindiger Erdstoffe. Dissertation Technische Universität Dresden

Die rheologische Charakterisierung von Feinstzement-Suspensionen Untersuchungen mit einem schubspannungsgesteuerten Rheometer

Rheological characterisation of microfine cement suspensions – Examinations with a shear stress-controlled rheometer

R. Umlauf
Wilhelm Dyckerhoff Institut für Baustofftechnologie, Wiesbaden, Deutschland

ZUSAMMENFASSUNG: Zementsuspensionen haben die Eigenschaft, Fließgrenzen auszubilden, die für die Injektionsreichweite bestimmend sind. Die Fließgrenzen und sehr kleinen Schergeschwindigkeiten in ihrer Umgebung wurden mit einem schubspannungsgesteuerten Rheometer unter Verwendung eines aufgerauhten Kegel-Platte-Meßsystems gemessen. Untersucht wurden Feinstzement-Suspensionen (Mikrodur R-U) mit Wasser/Bindemittelwert: 0.8 mit und ohne Verflüssiger. Bestimmt wurden 1.) die statische Fließgrenze durch kontinuierliche Steigerung der Schubspannung von 0.0615 auf 60 Pa in 120 sec und 2.) die dynamische Fließgrenze durch kontinuierliche Verringerung der Schubspannung von 60 auf 0.0615 Pa in 120 sec. Die Fließgrenzen wurden sofort, 30 und 60 min nach dem Anmachen je dreimal, die Fließkurve nur einmal gemessen.
Die statische Fließgrenze der Suspension ohne Verflüssiger steigt über den Versuchszeitraum nur wenig an. Die momentanen Schergeschwindigkeiten zeigen Kriechen an. Der Verflüssigerzusatz verringert die statische Fließgrenze nur wenig, senkt aber die momentane Viskosität drastisch. Die dynamischen Fließgrenzen der Suspension ohne Verflüssiger sind über den Versuchszeitraum etwa halb so groß wie die entsprechenden statischen Fließgrenzen. Bis zur Erreichung der Fließgrenze wird ein Kriechbereich durchlaufen. Die Verflüssigerzugabe halbiert die dynamische Fließgrenze und senkt die momentanen Viskositäten deutlich. Im Gegensatz zu der statischen Fließgrenze steigt die dynamische Fließgrenze der verflüssigerhaltigen Suspension über den Versuchszeitraum stark an. Fließgrenzen, die nach den rheologischen Modellen von Herschel-Bulkley und Casson aus den Fließkurven errechnet wurden, zeigen Übereinstimmungen mit den gemessenen dynamischen Fließgrenzen.

ABSTRACT: The yield value is a characteristic feature of cement suspensions and determins the radius of permeation. The yield values and very small shear rates, which can occur, when yield values are exceeded, were measured with a controlled stress rheometer, using a roughened cone-and-plate measuring system. Suspensions of a microfine cement (Mikrodur R-U) with W/C: 0.8 were prepared with and without a dispersant. Measurements included 1.) determination of the static yield value by increasing the shear stress from 0.0615 to 60 Pa within 120 sec. and 2.) the determination of the dynamic yield value by decreasing the shear stress from 60 to 0.0615 Pa within 120 sec. Immediately, 30 and 60 min after mixing, static and dynamic yield values were measured three times each, the flow curve only once. The static yield value of the suspension without dispersant increases during the measuring period only to a small amount. The instantaneous shear rates show the occurence of creep. Addition of a dispersant has only little influence on the static yield value but lowers the instantaneous viscosity drastically. The dynamic yield values of the suspension without dispersant are only half the corresponding static yield values. In the dynamic yield tests, creep is observed, before the yield point is reached. Addition of a dispersant halves the dynamic yield value and decreases the instantaneous viscosities. Unlike the static yield value, the dynamic yield value of the dispersant containing suspension increases strongly during the measuring period. Yield values calculated from flow curves using model equations of Herschel-Bulkley and Casson show coincidences with the measured dynamic yield values.

1. EINLEITUNG

Ein wichtiges Kriterium für den Erfolg einer Injektion ist das Fließverhalten des verwendeten Injektionsmaterials.

Das Fließverhalten von Zementinjektionen wird neben der Korngröße des verwendeten Materials von komplexen Reaktionen bestimmt, die einsetzen, sobald der Zement mit Wasser in Berührung kommt.

In wässriger Suspension bilden die Zementteilchen innerhalb des ganzen Ansatzes über elektrostatische Anziehung eine gerüstartige Struktur aus. Unterhalb einer bestimmten Schubspannung, die als statische Fließgrenze bezeichnet wird, verhält sich die Suspension festkörperartig. Nach Überschreitung dieser Fließgrenze wird die Struktur zerschert und die Zementsuspension fließt. Bei sinkenden Schubspannungen orientieren sich die Teilchen erneut, bis bei einer bestimmten Schubspannung, der dynamischen Fließgrenze, das festkörperartige Gerüst wieder aufgebaut ist. Mit Hilfe von Verflüssigern kann man die elektrostatische Anziehung verringern und so das Fließverhalten beeinflussen.

Dieser eigentlich reversible Vorgang wird überlagert durch den Einfluß der fortschreitenden Hydratation des Zements. In der Praxis charakterisiert man Fließgrenzen z.B. mit der Kugelharfe, dem Pendelgerät, dem Kasumeter oder dem Tauchrohr. Im Labor nimmt man mit Rotationsviskosimetern Fließkurven auf, aus denen man durch Extrapolation die Fließgrenze ermittelt.

2. AUFGABENSTELLUNG

Ziel der Untersuchung war, das Fließverhalten von Feinstzementsuspensionen in Rissen oder Klüften in einem Rheometer zu

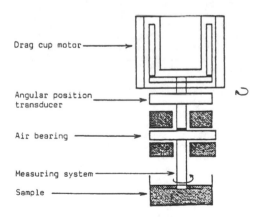

Abb.1 Hauptbauteile des Rheometers
       Principal components of the
       rheometer

simulieren. Besonderes Augenmerk galt der Beobachtung der Strömung bei niedrigen Schergeschwindigkeiten, also in der Umgebung der Fließgrenzen.

Bei einer Injektion gibt man über den Injektionsdruck eine Schubspannung auf das Injektionsmaterial, die eine Scherdeformation und schließlich eine Fließbewegung bewirkt. Will man diese Verhältnisse in einem Rotationsviskosimeter nachvollziehen, muß dieses Gerät nach dem Prinzip der Schubspannungsvorgabe arbeiten. Dabei werden über das Meßsystem Schubspannungen auf die zu untersuchende Substanz vorgegeben und die resultierende Deformation über eine Winkelauslenkung ermittelt. Der gebräuchliche Viskosimetertyp arbeitet umgekehrt nach dem Prinzip der Deformationsvorgabe. Zum Beispiel wird beim Fann-Viskosimeter über die Rotation des Außenzylinders eine Schergeschwindigkeit auf die Flüssigkeit gegeben und die resultierende Schubspannung am Innenzylinder gemessen.

Ein schubspannungsgesteuertes Rheometer wurde bereits erfolgreich für Untersuchungen an Suspensionen von Tiefbohrzementen eingesetzt (Banfill & Kitching, 1991). Banfill und Kitching untersuchten den Anstieg der Fließgrenzen von Zementsuspensionen, der bei Pumpstillständen während der Bohrloch-zementierung eintritt.

3. BESCHREIBUNG DES RHEOMETERS

Für die Untersuchungen wurde ein Bohlin CS 50 Rheometer eingesetzt, das nach dem Prinzip der Schubspannungsvorgabe arbeitet.

Der Aufbau des Geräts ist in Abb. 1 skizziert. Von einem Glockenläufermotor wird über ein Meßsystem eine definierte Schubspannung auf eine Probe aufgegeben und die resultierende Bewegung mit einem Meßwertaufnehmer registriert. Um geringe Bewegungen auch bei niedrigen Schubspannungen messen zu können, ist das Gerät mit einem reibungsarmen Luftlager ausgestattet.

Technische Spezifikation des Rheometers (lt. Datenblatt):
Drehmomentbereich: 0.001 - 50 mNm
Drehmomentauflösung: < 0.0002 mNm

Winkelauflösung: $1 \cdot 10^{-6}$ rad
Schergeschwindigkeit:
        10^{-6} - 100.000 sec^{-1}

Als Meßsystem wurde ein Kegel-Platte-Sy-
stem gewählt:
Kegeldurchmesser: 40 mm
Kegelwinkel : 4°
Kürzung der Kegelspitze: 0.15 mm
Rauhigkeit der Kegelmantelfläche: ca.
0.020 mm

Die Verwendung eines Kegel-Platte-Systems
hat den Vorteil,daß die Scherdeformation
im Meßspalt, unter der Voraussetzung eines
kleinen Kegelwinkels, unabhängig vom
Radius und konstant ist. So wird zum Beispiel bei der Schubspannungserhöhung aus
der Ruhe die Zementsuspension gleichmäßig
bis zur Überschreitung der Fließgrenze beansprucht.
Aus der homogenen Scherdeformation im gesamten Spalt folgt die Konstanz der
Schergeschwindigkeit und Schubspannung.
Die Fließkurven nicht-newtonscher Flüssigkeiten können deshalb ohne Korrekturen
ermittelt werden.

Es ist bekannt, daß Zementsuspensionen
besonders bei niedrigen Schergeschwindigkeiten zu Wandgleiten neigen. Das heißt,
daß sich zwischen der Suspension und der
Wand des Meßsystems ein Wasserfilm bildet,
auf dem die Suspension gleitet. Eine
Maßnahme zur Verbesserung der Wandhaftung
ist die Aufrauhung der Meßflächen. Für
diese Versuche wurde die Mantelfläche des
verwendeten Meßkegels mit SiC-Pulver ge-
strahlt.

4.   VERWENDETE MATERIALIEN

Als Suspensions-Feststoff wurde   Mikrodur R-U, ein hüttensandreicher Feinstzement mit folgender Spezifikation verwendet.

Als Verflüssiger wurde der Betonver-flüs-
siger BV 80 der Fa. Tricosal verwendet.

5.   VERSUCHSDURCHFÜHRUNG

Mikrodur R-U und deionisiertes Wasser wurden mit einem Ultra-Turrax-Dispergierer
bei 3000 Upm 5 min aufgeschlossen. Die

Tab. 1 Technische Daten Mikrodur R-U
Specification of Mikrodur R-U

Dichte	ca. 3.0 g/cm^3
Spezif. Oberfläche	ca. 16 000 cm^2/g
Kornverteilung	
< 2 µm	ca. 25 %
< 4 µm	ca. 55 %
< 8 µm	ca. 90 %
< 9.5 µm	ca. 95 %
d_{50}	< 3.5 µm

Suspensionsmenge betrug 2 Liter bei einem
Wasser/Bindemittelwert von 0.8. Die
Versuche erfolgten bei 20°C. Zur Untersuchung der Verflüs-sigerwirkung wurde die
Gesamtmenge an Verflüssiger (1% auf Zement
bezogen) dem Anmachwasser vor dem Zement
zugegeben.

Meßtermine:
1. sofort nach Beendigung des Ansatzes
2. 30 min nach Beginn des Ansatzes
3. 60 min nach Beginn des Ansatzes
Zwischen der zweiten und dritten Messung
wurden die Suspensionen mit einem Propel-
lerrührer (10 cm Durchmesser) bei 400 Upm
ständig gerührt. Zu jedem Meßtermin wurde
das Meßsystem neu befüllt. Innerhalb einer
Meßfolge verblieb die Probe im Rheometer.

6.   RHEOMETERMESSUNG

Jede Meßfolge gliederte sich in eine Aufnahme der Fließkurve und je drei abwech-
selnde Messungen der statischen und dynamischen Fließgrenzen. Die Steuerung des
Rheometers und die Auswertung der Versuche
erfolgte über einen PC.

Meßfolge:

1.   Fließkurve :
von 10 bis 511 sec^{-1} in 7 Schritten. Das
Rheometer regelt programmgesteuert die
Schubspannung so nach, daß konstante
Schergeschwindigkeiten verwirklicht werden. Bei jeder Geschwindigkeit wurde 20
sec geschert wobei nur die Messungen der
letzten 5 sec erfasst und gemittelt wurden.

2.   1.Statische Fließgrenze:
unmittelbar im Anschluß an die Aufnahme

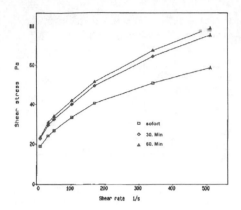

Abb. 2 Fließkurven-fallende Scherge-
schwindigkeit(ohne Verflüssiger)
Flow curves-descending shear
rates(without dispersant)

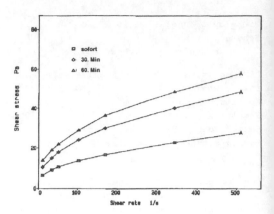

Abb. 3 Fließkurven-fallende Scherge-
schwindigkeit(mit Verflüssiger)
Flow curves-descending shear
rates(with dispersant)

der Fließkurve erfolgte 15 sec lang eine
Vorscherung der Probe mit 60 Pa, dann 15
sec Ruhe-danach kontinuierliche Steigerung
der Schubspannung von 0.0615 auf 60 Pa in
120 sec. Nach Überschreitung der Fließgrenze erfolgte manueller Abbruch der
Messung.

3. 1.Dynamische Fließgrenze:
unmittelbar im Anschluß an die Messung der
statischen Fließgrenze wurde 15 sec lang
die Probe mit 60 Pa vorgeschert-danach
erfolgte die Verringerung der Schubspannung von 60 auf 0.0615 Pa in 120 sec.
Nach Erreichen der Fließgrenze (Blockieren
des Meßsystems) wurde die Messung manuell
abgebrochen

4. 2.Statische Fließgrenze:
wie 1.Statische Fließgrenze

5. 2.Dynamische Fließgrenze:
wie 1. Dynamische Fließgrenze

6. 3.Statische Fließgrenze:
wie 1.Statische Fließgrenze

7. 3.Dynamische Fließgrenze:
wie 1. Dynamische Fließgrenze

Bei den Versuch zur Bestimmung der Fließgrenzen wird im Meßspalt keine stationäre
Strömung ausgebildet wird. Die rechnerisch
ermittelten Werte für die Schergeschwindigkeit und Viskosität werden mit
dem Zusatz -momentan- bezeichnet.

7. ERGEBNISSE

7.1. Fließkurven

Abb. 2 zeigt die Fließkurvenzweige mit
fallenden Schergeschwindigkeiten sofort,
30 und 60 min nach dem Anmachen für die
Mikrodur R-U Suspension (ohne Verflüssigerzusatz). Die Zunahme der Viskosität mit

Tab. 2 Berechnete Fließgrenzen
Calculated yield values

Ohne Verflüssiger   Without dispersant

	sofort	30 min	60 min
Bingham	22.8	26.9	28.1
Herschel -Bulkley	10.2	15.3	14.4
Casson	15.4	17.9	18.5

Mit Verflüssiger   With dispersant

	sofort	30 min	60 min
Bingham	8.2	13.9	17.5
Herschel -Bulkley	4.1	4.2	6.7
Casson	4.9	7.9	10.4

Rheologische Modelle:
Bingham: $\tau = \tau_0 + k \cdot D$
Herschel-Bulkley: $\tau = \tau_0 + k \cdot D^n$
Casson: $\sqrt{\tau} = \sqrt{\tau_0} + k \cdot \sqrt{D}$
τ = Schubspannung
τ_0 = Fließgrenze
D = Schergeschwindigkeit

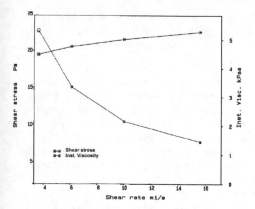

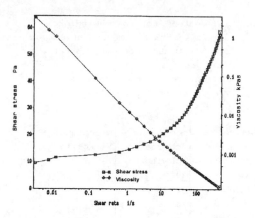

Abb. 4 Diagramm: 1.statische Fließgrenze (ohne Verflüssiger)
Diagram: 1. static yield stress
test (without dispersant)

Abb. 5 Diagramm:1.dynamische Fließgrenze (ohne Verflüssiger)
Diagram: 1.dynamic yield stress
test (without dispersant)

der Versuchszeit ist an der zunehmenden
Steigung der Fließkurven erkennbar.

Der viskositätssenkende Einfluß des
1 %igen Verflüssigerzusatzes und die
zeitabhängige Viskositätsentwicklung sind
aus der Abb. 3 zu ersehen. Alle
Fließkurven zeigen strukturviskoses
Fließverhalten.

Die Parameter der rheologischen Modelle
nach Bingham, Herschel-Bulkley und Casson
wurden aus dem Fließkurvenzweig mit fallenden Schergeschwindigkeiten nach der Methode der kleinsten Fehlerquadrate errechnet. In der Tabelle sind nur die Werte
für die Fließgrenzen wiedergegeben.

7.2. Statische Fließgrenzen

Als Beispiel ist in Abb. 4 das Diagramm
der 1. Fließgrenzenmessung sofort nach dem
Ansatz (ohne Verflüssiger) dargestellt.
Unter den gewählten Versuchsbedingungen
ist keine elastische Deformation unterhalb
der Fließgrenze zu beobachten. Das
Zementgel hält der aufgebrachten Schubspannung stand, bis bei 19.43 Pa die erste
Schergeschwindigkeit mit $3.65 \cdot 10^{-3}$ sec^{-1}
registriert wird. Die resultierende Mo-
mentanviskosität beträgt 5327 Pas. Die
Schubspannung der ersten Fließbewegung ist
besser als Kriechgrenze statt als Fließgrenze zu bezeichnen.

Tab. 3 Statische Fließgrenzen mit
momentanen Schergeschwindigkeiten und Viskositäten
Static yield values with
instantaneous shear rates and
viscosities

Ohne Verflüssiger	Fließgr. (Pa)	Scherge. (sec^{-1})	Viskos. (Pas)
sofort	19.43	$3.7 \cdot 10^{-3}$	$5.3 \cdot 10^3$
	18.45	$3.8 \cdot 10^{-3}$	$4.9 \cdot 10^3$
	18.44	$2.9 \cdot 10^{-3}$	$6.3 \cdot 10^3$
30 min	20.45	$1.8 \cdot 10^{-3}$	$1.2 \cdot 10^4$
	20.44	$1.1 \cdot 10^{-3}$	$1.8 \cdot 10^4$
	21.43	$1.3 \cdot 10^{-3}$	$1.7 \cdot 10^4$
60 min	20.44	$4.9 \cdot 10^{-3}$	$4.1 \cdot 10^3$
	21.45	$4.5 \cdot 10^{-3}$	$4.7 \cdot 10^3$
	-	-	-

Mit Verflüssiger	Fließgr. (Pa)	Scherge. (sec^{-1})	Viskos. Pas)
sofort	17.45	17.92	0.97
	16.44	0.24	67.7
	15.44	0.25	61.8
30 min	19.43	2.76	7.1
	17.45	$8.5 \cdot 10^{-4}$	$2.0 \cdot 10^4$
	17.45	$7.3 \cdot 10^{-3}$	$2.3 \cdot 10^3$
60 min	18.44	$1.1 \cdot 10^{-3}$	$1.6 \cdot 10^4$
	19.43	$9.5 \cdot 10^{-3}$	$2.0 \cdot 10^3$
	19.43	$1.7 \cdot 10^{-3}$	$1.1 \cdot 10^4$

Tab. 5 Momentane Schergeschwindigkeiten
und Viskositäten im ersten Versuch
zur Bestimmung der dynamischen
Fließgrenze
Instantaneous shear rates and
viscosities of the first yield
stress test

Schub-spannung	momentane Scherge-schwindig-keit	momentane Viskosität
40.57 Pa	134.8 sec^{-1}	0.30 Pas
30.52	59.1	0.52
15.49	2.0	7.8
10.44	0.007	1489

Tab. 6 Dynamische Fließgrenzen mit
momentanen Schergeschwindigkeiten und Viskositäten
Dynamic yield values with
instantaneous shear rares and
viscosities

Ohne Verflüssiger   Without dispersant

	Fließgr. (Pa)	Scherge. (sec^{-1})	Viskos. (Pas)
sofort	9.45	$3.0 \cdot 10^{-3}$	$3.1 \cdot 10^3$
	9.45	$4.0 \cdot 10^{-4}$	$2.4 \cdot 10^4$
	10.44	$2.5 \cdot 10^{-3}$	$4.3 \cdot 10^3$
30 min	10.45	$3.4 \cdot 10^{-3}$	$3.0 \cdot 10^3$
	10.45	$2.0 \cdot 10^{-3}$	$5.3 \cdot 10^3$
	11.44	$3.0 \cdot 10^{-3}$	$3.8 \cdot 10^3$
60 min	11.44	$4.2 \cdot 10^{-3}$	$2.7 \cdot 10^3$
	12.46	$3.6 \cdot 10^{-3}$	$3.5 \cdot 10^3$
	13.45	$4.9 \cdot 10^{-4}$	$2.7 \cdot 10^4$

Mit Verflüssiger   With dispersant

	Fließgr. (Pa)	Scherge. (sec^{-1})	Viskos. (Pas)
sofort	4.45	$2.7 \cdot 10^{-2}$	164
	5.50	0.12	4.5
	6.50	0.29	22.6
30 min	7.50	0.41	18.2
	7.50	$7.5 \cdot 10^{-2}$	100.2
	8.51	0.39	22.1
60 min	8.49	$1.8 \cdot 10^{-2}$	476
	9.50	$6.4 \cdot 10^{-2}$	149
	9.50	$1.3 \cdot 10^{-4}$	$7.2 \cdot 10^4$

In Tab. 3 sind die statischen Fließgrenzen mit den zugehörigen momentanen Schergeschwindigkeiten und Viskositäten aufgeführt.

Es überrascht, daß die statischen Fließgrenzen über die Versuchszeit nur gering-
fügig ansteigen. Die Zugabe des Verflüssigers bewirkt im Vergleich nur eine Verringerung der Fließgrenzen um ca. 2 Pa,
aber innerhalb der ersten 30 min nach dem
Beginn des Ansatzes eine deutliche Senkung
der momentanen Viskosität.
In wieweit die Versuche zur Ermittlung der
statischen Fließgrenzen das Fließverhalten
von Zementsuspensionen aus dem Ruhezustand
(z.B. bei dem Start einer Injektion) wiedergeben, wird Gegenstand zukünftiger
Untersuchungen sein.

7.3. Dynamische Fließgrenzen

Als Beispiel für eine Bestimmung der dynamischen Fließgrenze ist das Diagramm der
1. Messung (ohne Verflüssiger) sofort nach
dem Anmachen dargestellt.

Wird die Schergeschwindigkeit von
1 sec $^{-1}$ unterschritten, stagniert die
Fließbewegung endgültig erst nach Senkung
der Schubspannung um weitere 4 Pa. Offensichtlich dominiert der Einfluß von
Kriechvorgängen. In Tab. 5 ist für diesen
Versuch eine Auswahl von Schubspannungen
mit den resultierenden momentanen Schergeschwindigkeiten und Viskositäten aufgeführt.

Abb. 6 zeigt zusätzlich die Diagramme der
Proben , die nach 30 und 60 min der Suspension (ohne Verflüssiger) entnommen
wurden. Sie zeigen das gleiche, von
Kriecheffekten dominierte Fließverhalten.
Infolge fortschreitender Hydratation sind
die Kurvenverläufe zu höheren Viskositäter
verschoben.

Die dynamischen Fließgrenzen sind in Tab.
6 aufgelistet.

Der Verflüssigerzusatz beeinflußt die dynamische Fließgrenzen deutlicher als die
statischen. Die dynamische Fließgrenze
sinkt bei Verwendung des Verflüssigers auf
die Hälfte des Vergleichswertes ohne

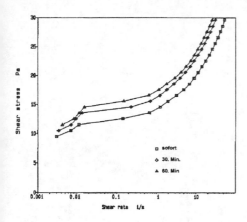 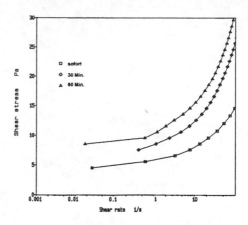

Abb. 6 Diagramme der jeweils ersten
Versuche(ohne Verflüssiger) zur
Bestimmung der dynamischen
Fließgrenze sofort, 30 und 60
min nach dem Anmachen
Diagrams of the first dynamic
yield stress tests (without
dispersant)immediatelly, 30 and
60 min after mixing

Abb. 7 Diagramme der jeweils ersten
Versuche(mit Verflüssiger)zur
Bestimmung der dynamischen
Fließgrenzen sofort, 30 und 60
min nach dem Anmachen
Diagrams of the first dynamic
yield stress tests (with
dispersant)immediatelly, 30 and
60 min after mixing

Verflüssiger, steigt aber im Lauf einer Stunde wieder auf den Ausgangswert an. Die Meßdiagramme (sofort, 30 und 60 min nach dem Anmachen) des Versuchs unter Verwendung des Verflüssigers sind in Abb. 7 dargestellt. Bei niedrigen Schubspannungen treten die Kriecheffekte nicht so deutlich in Erscheinung wie in dem Versuch ohne Verflüssiger.

Im Vergleich gemessener und errechneter Fließgrenzen ergeben sich Übereinstimmungen zwischen den gemessenen dynamischen Fließgrenzen mit den nach rheologischen Modellen von Casson und Herschel-Bulkley errechneten.

Aus den Untersuchungen zur Ermittlung der dynamischen Fließgrenze deutet sich die Möglichkeit an, Informationen über das Stagnationsverhalten von Injektionsmaterialien zu erhalten. Damit könnte zum Beispiel eine Hilfe zur besseren Abschätzung von Injektionsreichweiten gegeben werden.

8.   AUSBLICK

Die Untersuchungen zeigen, daß es sinnvoll ist, nicht nur Zahlenwerte für Fließgrenzen zu erhalten, sondern auch die Strömungsverhältnisse in ihrer Umgebung zu berücksichtigen. Es bietet sich deshalb aus der Sicht des Bindemittelherstellers an, dieses Gerät beispielsweise für die Optimierung von Bindemittel-Systemen einzusetzen. So ist vorgesehen, praxisorientierte Injektionsversuche an Bauteilen durchzuführen und die eingesetzten Feinstzementsuspensionen in beschriebener Weise rheologisch zu charakterisieren. Ziel ist, Beziehungen zwischen den Eigenschaften zementgebundener Injektionssysteme und rheologischen Kennwerten zu erfassen und für die Weiterentwicklung zu nutzen.

Literatur

Banfill,P.F.G., & D.R. Kitching. Use of
a controlled stress rheometer to
study the yield stress of oilwell
cement slurries. Proceedings of the
International Conference of the
British Society on Rheology of Fresh
Cement and Concrete, Liverpool 16-29
March 1990 London: E.&F.N. Spon

Grouting in Rock and Concrete, Widmann (ed.) © 1993 Balkema, Rotterdam, ISBN 90 5410 350 7

The effect of surface treatment agents on epoxy resin grouts and their rules of application

Wirkung von Oberflächen-aktiven Zusätzen auf Kunstharze und Regeln für die Anwendung

Xian Anru, He Yongsheng, Yie Linhong, Qiu Xiaopei, Zhang Yirong & Zhang Guangzhao
Guangzhou Institute of Chemistry, Academia Sinica, People's Republic of China

ABSTRACT The influence of purity and amount of silicane surface treatment on mechanical properties of epoxy resin grouts, the agents' hydrolysis and the reactions among epoxy resin, acetone and the agents have been studied. The effects of some coupling agents such as titanate and silicane upon several kinds of mediums have also been investigated and their selected rules have been determined.
KEY WORDS Surface treatment agent, Chemical grout, Epoxy resin, Acetone

INTRODUCTION Though surface treatment can play important parts in improving mechanical properties of epoxy resin grouts, so far little has been reported about this. One of the main kinds of them is the group of coupling agents including silicane, titanate, zirconium, organic complex of Chrome, organic magnesium, organic aluminium, ect. This article deals with the two formers.

A silicane coupling agent (SCA) molecule is generally considered to have two active end groups. One end can hydrolyze into silanol which will react hydroxyl forming the structure-Si-0- medium(usually inorganic compound), the other end can bond with organic compounds, i.e. "coupling". SCA have not been thought to react with epoxy resin[1], but we confirmed they could do both epoxy and acetone. There have not been rules for selecting surface treatment agents for defferent mediums. From the experiments to determine the surface treatment agents for SiO_2 and $CaCO_3$ widely distribution in the earth. We give the general rules.

1 EXPERIMENT

Materials: Bisphenol-A epoxy resin, Acetone, Furfural, Diethylenetriamine, Silicon dioxide and calcium carbonate were all commercial. Commercial titanate coupling agents including T_2, T_4, T_6, and, SCA including γ-aminopropyltriethoxysilane(γ-APTES), anilinomethyltriethoxysilane(AETES), γ-glycidoxyproltrimethoxysilane(γ-GPTES) were purified before use.

Main equipments: Fourier transform infrared spectrophotometer(FTIR), Nuclear magnetic resonance spectrometer(NMR), Gas chromatograph(GC), X-ray diffractometer(XRD), Abbe refractometer and Material testing machine.

2 RESULTS AND DISCUSION

1) Relationship between compression strength of grouts and the amount or purity of SCA

The more the SCA is, the higher the compression strength of grout will be(Fig. 1). The compression strength of every grout reaches maximum at almost same amount of SCA, and, they will keep constant xeceeding the amount. Also, the compression strength increases with silicane's purity(Fig. 1), because the condensation of Silanol from hydrolysis of silicane in preparation and reservation can inactivate silanol.

2) Hydrolysis of SCA

SCA can absorb moisture from air and hydrolyze into silanol and methanol or ethanal and other(Eq. 1. 2) which has big vapor pressure, their hydrolysis process

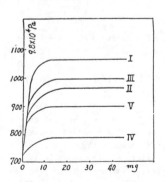

Fig.1 Relationship between compression strength of grouts and the amount or purity of SCA.
I. γ-APTES(purified)
II. γ-APTES(commercial)
III. AETES(purified)
IV. AETES(commercial)
V. γ-GPTES(commercial)

nan chareeterized with the curve of silicane weight to time (Fig.2). OAB in curve I shows that purified a sample absorbs more water, the purified and commerical sample reserver in air at experiment, but the latter has already had absorbed some water before experiment.

Hydrolysis
$$R-Si(OC_2H_5)_3 + H_2O \longrightarrow R-Si(OH)_3 + C_2H_5OH \quad (1)$$
Polycondensation

$$R-Si(OH)_3 \longrightarrow R-(-\overset{|}{\underset{|}{Si}}-O-)_n + H_2O \quad (2)$$

Results from GC (Fig.3) are in consistent with that in Fig.2. Besides, hydrolysis rate would change with the coupling agents. For instance, with regard to hydrolysis rate, γ-APTES > γ-GPTES> AETES. Since materials for preparing grouting are commercial and have some water in them, coupling agents with slow hydrolysis should be selected.

We had discovered a periodic lattice arrangment in hydrolysis product of SCA with X-ray diffraction. For example, γ-APTES reflection angle θ =11° , then lattice distance between two adjacent planes (d)

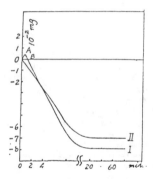

Fig.2 The weight changes of SCA in the air
I purified sample II commercial sample

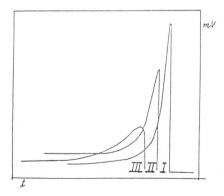

Fig.3 GC of hydrolysis products
I, γ- APTES II, γ-GPTES III, AETES

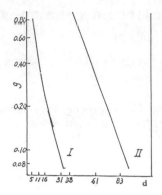

Fig.4 Relationship between gelation time of grouts and the weights of coupling agents.
I AETES II γ-APTES

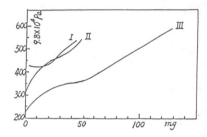

Fig.5 Dependence of grouts' compression strength upon the weights of coupling agents.
I T2 II γ-APTES(purified) III γ-APTES(commercial)

was calcutated to be 4 A according to Bragg equation $n\lambda = 2ds\sin\theta$. Obviously, the distance is not that is silicon dioxide or covalent radius between Si and O [2], we think the ordered arrangement is as follows.

$$-\overset{R}{\underset{|}{Si}}-O-\overset{R}{\underset{|}{Si}}-O-\overset{R}{\underset{|}{Si}}-O-$$

3) Role of SCA on cross-linking of grouts

SCA can play a role as a cross-linking agent without hardener added to a grout and the gelation time at the decreases with its weight increases (Fig.4). With insufficient hardener they can play a role as a hardener instead of the absent part. The more SCA is, the compression strength of the grout is.

With IR and NMR, we discovered SCA with primary amine groups (PSCA)or secondary amine groups (SSCA) could react with epoxy at room temperature or 70 ℃ (Fig.6,7), this made the yield strength of some gels reach ten MPa (Tab.1). Also, SCA can react with acetone(Fig.8).Anilino-methyl-triethoxy-silicane and can even produce a gel.

Table 1 Influence of coupling agents on
Yield strength of solidified epoxy resin

Epoxy resin model	E44		E51	
$H_2NR'Si(OR)_3$/epoxy (wt%)	14.98	30.12	—	—
γ-APTES/Epoxy (wt%)	—	—	14.97	30.12
Yield strength (MPa)	76.21	43.50	47.99	42.59

Table 2   Effect of coupling agents upon
grout's sandgel compression strength(basic medium)(MPa)

coupling agent	γ-APTES	γ-GPTES	AETES	T2	T4	T6
without coupling agent	62.0	87.6	66.9	65.4	64.1	65.7
with coupling agent	62.1	66.2	69.5	73.7	76.4	65.8
increament(%)	0	0	3.9	12.7	19.2	20.7

* calcium carbonate as medium
* Acidic property of coupling agent from weak to
strong in order of γ-APTES to T6

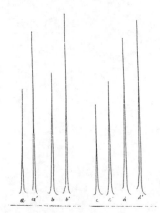

Fig.6 Hydroxyl group changes in NMR of mixtures of
epoxy resin and coupling agents   (20℃).
a,a')mixture of SSCA and epoxy resin(0.91%)
b,b')mixture of SSCA and epoxy resin(11.46%)
c,c')mixture of PSCA and epoxy resin(1.30%)
d,d')mixture of PSCA and epoxy resin(9.56%)

Table 3 Effect of coupling agents upon silicon dioxide
grout's sandgel compression strength(acidic medium)(MPa)

coupling agent	T4	T2	AETES comme-rical	AETES purif-ied	γ-GPTES	γ-APTES comme-rical	γ-APTES puri-fied
without coupling agent	75.7	70.7	74.4	73.6	74.8	74.5	72.2
with coupling agent	91.8	88.1	81.9	92.4	93.0	99.5	106.6
increament(%)	21.3	24.8	10.1	25.5	24.4	33.6	47.6

* silion dioxide as medium
* Basic property of coupling agent from weak to
strong in order of T4 to γ-APTES

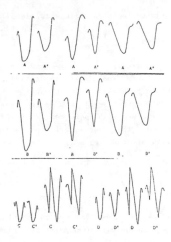

Fig.7 Epoxy group changes in FTIR of mixture of
coupling agents and epoxy resin (Heated 30min at 70℃)
a,a')mixture of PSCA and epoxy resin (10.4%)
b,b')mixture of PSCA and epoxy resin (10.0%)
c,c')mixture of SSCA and epoxy resin (2.40%)
d,d')mixture of SSCA and epoxy resin (10.7%)

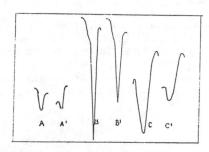

Fig.8 Carbonyl group changes in FTIR of mixtures of
coupling agents and Acetone (Heated 30 min. at 35℃)
a,a')mixture of PSCA and acetone (2.8%)
b,b',c,c')mixture of SSCA and acetone (3.4%)

Non-interfacial failure of the solid-ified grouts in
material tests as well as suggests SCA can cross with
compounds in the system.

4) The effect of coupling agents upon different
mediums

SCA were found to be good coupling agents for silicon
dioxide but poor ones for calcium carbonate. On the
contrary, titanates were good for calcium carbonate
and poor for silicon dioxide. This can be explained
with the "Acid-basic match rule". That is basic
coupling agents including γ-APTES, γ-GPTES and
AETES match acidic medium such as silicon dioxide.
but don't do basic medium like calcium carbonate well.
Being acidic, Titanated coupling agents match calcium
carbonate which is basic. Certainly, this rule is a

guide for selecting coupling agents.(tab.2,3)

5) Application

Basing on the studies above we have prepared the compound reiforcer YDS[83]. It can activate acetone and furfural of which mixture is usually used as a diliuent of epoxy resin and make them network forming structure with epoxy resin. By this way, it can improve the mechanical properties and permeability of epoxy resin as well as lower their cost.

CONCLUSION

1 Surface treatment agents can play an important role in improving mechanical properties of epoxy resin grouts.
2 Rules for selecting surface treatment agents are of their favourable rate in hydrolysis, excellent purity and lewis acid-basic match between the agent and the medium.

REFERENCES

Li Tzedong, Thermosetting Resin, 1, 1989.
М. В. ВОЛЬКЕНШТЕЙН, СТРОЕНИЕ И ФИЗИЧЕСКИЕ СВОЙСТВА МОЛЕ-КУЛ 107P
He Yongsheng, Xian Anru, ect. International conference on Grouting in Rock and Concrete, Austria, 1993.

2 Grouting procedures – General
Ausführung von Injektionen – Allgemein

Aktive Eindringung, eine neue Dimension in der Injektionstechnik

Praktische Beispiele für die Nachinjektion mit Feinstbindemittel und Polyurethan

Active penetration, a new dimension in grouting technique – Case histories in postgrouting with microcement and polyurethane

Peter Borchardt
Göteborg, Schweden

ZUSAMMENFASSUNG: Unter einer erfolgreichen Injektion, besonders bei einer Vorinjektion im Felsen, versteht man die vollständige Füllung der offenen Klüfte und Gesteinsrisse in einem abgegrenzten Bereich. Wenn auch die Zementsuspension immer noch das Basmaterial für die Felseninjektion ist, gibt es Situationen, wo der Zement nicht mehr als Injektionsmateial ausreicht. Eine Situation ist, wenn bei einem starkem Wasserandrang in den Klüften und den Zertrümmerungszonen die Suspension vor ihrer Härtung weggespült wird. Die Fähigkeit der Suspension in feine Gesteinsrisse einzudringen bedeutet ein anderes Problem.

Für diese beiden Situationen wird die Lösung der Probleme beschrieben, wobei Polyurethaninjektionen angewendet wurden. Die Polyurethane reagieren, wenn diese in Kontakt mit Wasser kommen, wobei bei der Reaktion sich Gas bildet. Der Gasdruck paßt sich an die Größe der Klüfte an, und das Injektionsmaterial dringt aktiv in die feinen Risse ein. Dieses Phänomen ist die Ursache, warum die Polyurethane die Lösung der Probleme sind, wenn nichts anderes mehr hilft. In diesem Text wird beschrieben, wie die Polyurethane in drei verschiedenen Fällen für die Nachinjektion im Felsen angewendet wurden.

ABSTRACT: A successful rock grouting corresponds to the grouting of a specific, pre-determined volume of rock, with complete filling of open fractures and voids. Although cement suspension is still the base material in rock grouting there are situations when grouting with cement is not adequate. One situation is when there is a strong water flow in fractures and fault zones and the suspension simply is flushed away before hardening. Another such situation is when the fractures are too fine for the cement particles to enter.

In both situations described the solution of the problem should be polyurethane grouting. The polyurethane reacts when in contact with water and during the reaction a gas is emitted. The pressure from this gas adapts to the size of the fractures and the injection material penetrates actively into small fissures. This phenomenon is the reason to why polyurethane solves the problem when nothing else helps. The text describes the use of one such polyurethane in three different cases of post grouting.

EINLEITUNG

In einem schwedischen Kompendium kann man folgenes lesen:

"Systematische Vorinjektion ist der einzige mögliche Weg, eine hohe Dichtigkeit im Felsen zu erreichen, um die Grundwassersenkung in der Umgebung zu verhin-' dern. Eine Nachinjektion ergibt selten oder nie ein Resultat"

Ist dies eine Resignation oder ist es wirklich unmöglich eine Feinabdichtung des Felsen auszuführen?

Leckende Tunnel, die unter den Jahren 1970-85 in den schwedischen Städten gebaut wurden, verursachten Setzungen in den Tonschichten aufgrund von Grundwassersenkungen mit Schäden an den Gebäuden zur Folge, wenn die Abdichtungsarbeiten kein zufriedenstellendes Resultat zu angemessenem Kosten ergaben. Hohe Beträge wurden an die Eigentümer bezahlt, um die Gebäude zu reparieren und zu verstärken. Um die weite-ren Grundwassersenkungen zuverhindern, wird heute Trinkwasser in die permeablen Schichten infiltriert.

In den Tunneln werden die Installationen vom Sickerwasser mit "Plastikdächern" geschützt, und das Wasser wird abgeleitet.

Die Verkehrstunnel mit tropfendem Wasser in der Firste oder rinnendem Wasser entlang den Wänden verursachen in der kalten Periode Eisprobleme. Zum Beispiel, wenn es bei stromführenden Hochspannungsleitungen zu Überschlägen kommt, wird der Zugverkehr unterbrochen. Wieviele Arbeitsstunden müssen aufgebracht werden, um Eis zu knacken oder wegzuräumen ? Auch hier wird die einfache Methode gewählt, nämlich das Wasser unter isolierten Schalen abzuleiten. Eine Montage von elektischen Wärmekabeln ist in gewissen Fällen erforderlich. Aber was geschieht, wenn die Sickerwassermengen durch Ausspülen der Klüfte ansteigen oder der Felsen nachbricht ?

Mit der Herstellung der Feinstbindemittel und den dazugehörigen Zusatzmitteln erhielt die Injektionstechnik eine neue Möglichkeit.

Viele chemische Injektionsmittel haben nämlich eine nicht ausreichende Langzeitbeständigkeit und bedeuten oft eine Gefahr aufgrund der Giftigkeit für die Umwelt oder auch für den Anwender.

Aus diesem Grunde verbleibt die Suspension mit Partikeln d.h. Zement und Feinstbindemittel die Basis der Injektion.

Aber es besteht immer eine Unsicherheit bei den Injektionsarbeiten im Felsen ein zufriedenstellendes Abdichtungsresultat zu verträglichen Kosten zu erhalten, obwohl mit erfahrendem Personal, bester Qualität der Injektionsausrüstung und des Injektionsmateriales gearbeitet wird.

- Ist es nun wirklich unmöglich eine Felsenoberfläche mit tropfendem Wasser zu verträglichen Kosten abzudichten?

Unter den letzten Jahren wurde in Skandinavien ein in

Anisotropisches Gebirge - Direkter Kontakt:

Ein dichtes Netz von Gesteinsrissen mit vielen Schnittpunkten reduziert stark den Druck hinter der Bohrlochwand. Viele Gesteinsrisse werden um das Bohrloch mit der Suspension gefüllt. Ein gutes Abdichtungsresultat kann erwartet werden.

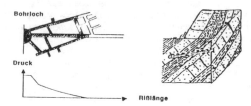

Isotropes Gebirge - Indirekter Kontakt:

Bei den einzelnen Gesteinsrissen oder einer kleinen Anzahl Risse ohne Schnittpunkte wird der Druck erst nach einer langen Strecke aufgrund der Reibung reduziert. Die Füllung des Gesteinsrisses kann unglücklicherweise außerhalb des Sickerbereiches geschehen.

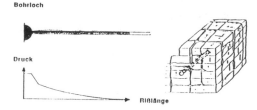

Figur 1   Schematische Darstellung des Kontaktes der Gesteinsrisse mit dem Bohrloch

Figure 1   Schematic illustration of contacts between rock farctures and bore hole

Japan entwickeltes Injektionsmaterial auf Polyurethanbasis mit speziellen Eigenschaften zum Feinabdichten des Felsens mit Erfolg angewendet. Diese Material reagiert mit dem Kluftwasser in dem Felsen unabhängig von dem Wasserzugang in einem kontrollierten Prozeß, wobei das Material unter der Zeit durch die Variation des auftretendem Gasdruckes das noch flüssige Material aktiv in die Gesteinsrisse einpreßt. Aufgrund der Expansion des Materiales werden die Klüfte vollständig verfüllt. Mit diesen Eigenschaften dringt das Material kontrollierbar in die Klüfte und verhindert eine Wegspülung aufgrund von strömmendem Sickerwasser, so daß das Verfüllen der Klüfte in einem abgegrenzten Bereich geschieht. Um die Materialkosten minimieren zu können, wurde dieses Material den Suspensionen als aktives Zusatzmittel zugegeben.

Mit einigen Beispielen von Nachinjektionsarbeiten im Felsen wird diese neue Dimension der Injektionstechnik, nämlich die aktive Eindringung, beschrieben.

1.   DIE ABDICHTUNG VON FELSENKLÜFTEN MIT DER NACHINJEKTION

1.1   Bohrung

Die Bohrungen werden in der Regel so angesetzt, daß möglist viele Klüfte getroffen werden. Die Länge der Bohrlöcher bei den Nachinjektionen ist abhängig von den örtlichen Verhältnissen und besonders von der Intensität der Leckagen. In der Regel wird 2.4-4.0 m tief gebohrt. Doch können für ergänzende Arbeiten die Längen der Bohrlöcher zwischen 0.4 bis 0.6 m liegen.

Schwierig wird es, wenn die Felsenoberfläche mit Spritzbeton versehen ist. Dadurch sind die Kluft- und Rißstrukturen verdeckt. Oft dringt das Wasser durch den Spritzbeton weit von den wasserführenden Klüften. Eine gute Hilfe, um die Lage der wasserführenden Klüfte feststellen zu können, ist der Zugang zu den geologischen Kartierungen.

1.2   Kennzeichen der wasserführenden Klüfte oder Hohlräume

Das Ziel mit einer Nachinjektion ist, die Klüfte, die das Wasser an die Felsenoberfläche leiten, durch Füllung dicht zu bekommen. Diese wasserführenden Klüfte, die im direkten oder indirektem Kontakt mit dem Bohrloch stehen, haben verschiedene Breiten und Formen und damit verschiedene hydraulische Eigenschaften.

In einem anisotropen Gebirge, einem Gebirge mit einer Schichtstruktur und mit günstigen Sickerverhältnissen, gibt es entlang dem gebohrtem Loch für die Injektion viele charakteristische Öffnungen. Das Bohrloch hat einen "direkten" Kontakt mit den durchlässigen Schichten (Figur 1).

Dies bedeutet, daß der Druck bei der Injektion, der die Strömung der Suspension unter dem Injektionsprozeß aufrechterhält, sich stark aufgrund der dichten Struktur der Gesteinsrisse an den Öffnungen und hinter der Bohrlochwand reduziert. Dieser Druckfall ist hauptsächlich ausschlaggebend für die Ausbreitung der Suspension.

Bei einem isotropen Gebirge sind die Sickerverhältnisse ungünstig, da die Kluftstruktur oft aus großen Blöcken besteht, die oft mit feinen Gesteinsrissen durchsetzt sind. Dieses Kluftsystem kann als Flächenstruktur bezeichnet werden. Die Gesteinsrisse entlang des Bohrloches werden "flächenmäßig" gefüllt und haben nur einen indirekten Kontakt zu den naheliegenden eventuell wasserführenden Gesteinsrissen.

Die Ausbreitung der Suspension in diesen Gesteinsrissen wird nicht nur stark beeinflußt durch den Druckfall an den Öffnungen, sondern auch durch die Reibung entlang der Oberfläche des Gesteinsrisses. Mit dem erforderlichen Druck für die Eindringung der Suspension in dem größten Gesteinsriß erhält man mit Sicherheit eine größere Ausbreitung der Suspension vom Bohrloch als bei einem Kluftsystem mit "direktem" Kontakt. Der Abstand der Bohrlöcher kann somit größer gewählt werden.

Gemäss dieser Überlegung sollte man bei der Injektion die hydraulischen Eigenschaften für die einzelnen Gesteinsrisse oder dem Kluftsystem kennen. Um nämlich eine Strömung der Suspension in einem Gesteinsriß nach dem Druckfall am Bohrloch in Bewegung zuhalten, ist eigentlich für jeden Gesteinsriß ein bestimmter Druck erforderlich. Ist der Injektionsdruck zu hoch, so fließt die Suspension zu schnell vom Bohrloch weg. Die Ausbreitung wird sehr groß oder das Material fließt an der Oberfläche ab.

Bei einem zu niedrigen Injektionsdruck kann der Druckfall an der Öffnung nicht überwunden werden, so daß keine Eindringung in dem Gesteinsriß eintrifft. Um ein gutes Dichtungsresultat zu erhalten, sind die Öffnungen entlang des Bohrloches von großer Bedeutung. Feine Gesteinsrisse können unverfüllt bleiben, und der Felsen verbleibt permeabel.

Dieses Problem wird in der Figur 2 dargestellt. Man sieht deutlich, wie das Injektionsmaterial mit Hilfe des Injektionsdruckes zuerst in den größten Gesteinsriß eindringt.

Ein Eindringen des Injektionsmateriales in die feineren Gesteinsrisse entlang des Bohrloches ist nur dann möglich, wenn

- die Drucksteigerung ein Niveau erreicht hat, das erforderlich ist, um feinere Gesteinsrisse zu penetrieren
- die feinen Kluftöffnungen sind offen und werden nicht durch die Eigenschaften der Suspension abgedichtet (Korngröße,Filtration, Sedimentation, beschleunigtes Ansteigen der Viskosität)
- die Klüfte in den schwachen Zonen des Felsens durch ein hydraulisches Aufpressen erweitert werden
- die Klüfte enthalten keine Tonpartikel

Es gibt noch andere Faktoren, die den Injektionsprozeß erschweren. Wird bei den Injektionsarbeiten ein Injektionsdruck aufgrund schlechtem Eindringen in der größten Kluft überschritten, so daß ein hydraulisches Aufreißen eintrifft, so ist es fast unmöglich noch feinere Risse abzudichten. Beim hydraulischem Aufpressen besteht auch die Gefahr, daß Schalenbrüche auftreten können.

Bei stark rinnendem Gebirgswasser in den Klüften oder rinnendem Sickerwasser zur Felsenoberfläche trifft immer ein Wegspülen der Suspensionen oder der chemischen Injektionsmittel ein. Um dieses rinnende Wasser bei der Injektion zum Stillstand zu bringen, wird in der Regel der Suspension einen Beschleuniger zugegeben. Bei den chemischen Injektionsmitteln kann die Gelzeit drastisch verkürzt werden.

In den letzten Jahren wurden auch aufschäumende Polyurethane angewendet. Mit dieser Verfahrensweise erhält man gewiß einen Stopp des Wasserstromes in den größten Klüften, aber auch eine Blockierung des Injektionsmateriales, um weiter in die Klüfte einzudringen. Gleichzeitig werden die Öffnungen der feineren Risse verstopft. Diese Gesteinsrisse verbleiben permeabel.

Es müssen nun neue Bohrlöcher gebohrt werden, um die nicht gefüllten Gesteinsrisse abzudichten.
Unkontrolliertes Verstopfen im Kluftsystem bedeutet aber eine Erschwernis der weiteren Injektionsarbeiten.

Diese oben beschriebene Probleme, die bei der konventionellen Injektionstechnik eintreffen, erfordern eine radikale Veränderung, eine Entwicklung, eine neue Dimen-

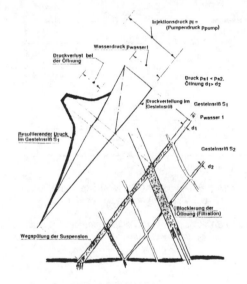

Der Gesteinsriß mit der Größten Öffnung und rinnendem Wasser wird zuerst mit der Suspension gefüllt.

Gesteinsriß S_1 mit der größeren Öffnung d_1 als der Gesteinsriß S_2 mit kleinerer Öffnung d_2 erfordert nur den Druck p_{S_1}. Druck $p_{S_1} < p_{S_2}$.

Figur 2   Die Eindringung des Injektionsmateriales erfordert für die verschiedenen Rißöffnungen einen veränderlichen Injektionsdruck.

Figure 2   Penetration of the grouting material requires variable injection pressure for different fracture openings.

oion in der Injektionstechnik wenn ein Abdichtungserfolg zu angemessenen Kosten verlangt wird.

Gibt es ein Injekterinsmaterial mit Fließeigenschaften, die sich unter dem Injektionsprozeß an die hydraulischen Eigenschaften der Kluftstruktur anpassen?

2. KENNZEICHEN DER SUSPENSIONEN UND CHEMISCHEN INJEKTIONSMITTELN

2.1 Suspensionen

Obwohl ein Injektionsexpert mit vielen praktischen Erfahrungen über die grundlegenden theoretischen Betrachtungen, über die Fließeigenschaften der Suspension in den Felsenstrukturen mit "direktem oder indirektem Kontakt", über theoretische Kenntnisse von Viskosität, Scherspannungen und Thixotropie der Suspensionen, Partikelgröße, u.s.w. informiert wird, hat er keine Zeit sich mit diesen Problemen eingehend zu beschäftigen. Diese Kenntnisse sind für ihn von unterordneter Bedeutung.

Für die Durchführung der praktischen Arbeit mit

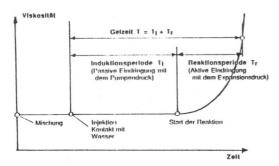

Figur 3A Prinzip der TACSS-Methode. Veränderung der Viskosität abhängig von der Zeit.

Figure 3A Principle of the TACSS method. Viscosity increase with time.

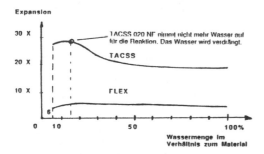

Figur 3B Expansion von TACSS und FLEX im Verhältnis zum Wasserzugang.

Figure 3B Expansion of TACSS and FLEX in relation to water access.

Suspensionen ist folgenes interessant:

- die wasserführenden Gesteinsrisse mit dem Bohrloch zu treffen
- eine Eindringung der Suspension zu erhalten
- ein Wegspülen der Suspension in dem Felsen zu vermeiden und den Ausfluß an der Oberfläche zu stopfen
- den Injektionsdruck bei dem Start richtig zu wählen (kein hydraulisches Aufpressen)
- einen Druckanstieg mit verträglichen Materialverbrauch unter dem Injektionsprozeß zu erhalten

Kann dieses nicht erfüllt werden, so wird mit chemischen Injektionsmitteln gearbeitet.

2.2 Chemische Injektionsmittel

Die chemische Injektionsmittel sind reine Lösungen (enthalten keine Partikel), wo die Eindringung und Ausbreitung vom Bohrloch abhängig von der Viskosität und dem Injektionsdruck ist. Mit der Steuerung der Gelzeit kann die Ausbreitung im Gebirge begrenzt, der Ausfluß zur Felsenoberfläche und ein Wasserstrom gestopft werden.

Generell wird zwischen wasserlöslichen und nicht wasserlöslichen Injektionsmitteln unterschieden. Die Eigenschaften der Injektionsmittel sind detailiert in den Datenblättern der Lieferanten beschrieben.

Für einen Praktiker ist es immer schwierig, die richtige Gelzeit für den Injektionsprozess zu wählen. Auch sollten die hydraulischen Eigenschaften des Kluftsystemes bekannt sein, da jeder Gesteinsriß "seinen eigenen" Druck verlangt. Im Gegensatz zu den Suspensionen gibt keine Filtration an den Öffnungen der feinen Risse, und die Ausbreitung wird durch die Steuerung der Gelzeit begrenzt. Wird eine zu kurze Gelzeit gewählt, so wird zum Beispiel der Ausfluß zur Oberfläche gestopft, aber die Eindringung in die feineren Gesteinsrisse wird gleichzeitig blockiert.

Die unmögliche Situation kann bei der Wahl von einfachen Polyurethanen eintreffen, wo eine Reaktion trotz gewählter Gelzeit unkontrolliert durch das Aufschäumen beim Kontakt mit dem Wasser eintrifft. Außerdem oben genannten Problem, daß die Ausbreitung des Materiales sehr begrenzt wird, schrumpft das aufgeschäumte Material nach einer gewissen Zeit.

Die vorhandenen Wassermengen beeinflussen die Reaktionsgeschwindigkeit des Aufschäumens und den Schaumfaktor, welches zu einer labilen Struktur des ausgehärteten Materiales führt, und somit ein Schwinden verursacht. Dieses Schwinden geschieht doch nicht direkt, so daß die Leckagen erst nach Tagen oder Wochen eintreffen. Diese neuen auftretenden Leckagen sind dann sehr schwer abzudichten, da viele Kanäle durch den Schaum blockiert sind.

2.3 Eigenschaften des Injektionsmateriales mit aktiver Eindringung

Aus praktischen Gründen können keine verschiedene

Drücke für jeden einzelnen Gesteinsriß entlang des Bohrloches geschaffen werden. Man kann doch unter dem Injektionsprozeß den Druck im Bohrloch mit Hilfe eines wasserreaktiven Injektionsmateriales ändern, da eine Gasentwicklung bei der Reaktion entsteht.

Um bei dem Injektionsprozeß die Fließeigenschaften des Materiales an die hydraulischen Eigenschaften des Kluftsystemes anzupassen, muß ein wasserreaktives Injektionsmaterial folgene Forderungen erfüllen:

(i) Der Reaktionsprozeß soll erst beim Kontakt mit dem anstehendem Wasser nach einer im voraus bestimmten Induktionszeit starten (Figur 3A).

-> *Dies bedeutet, eine Sicherstellung einer guten Ausbreitung des Injektionsmateriales vom Bohrloch in den Gesteinsriß, sowie die Verhinderung von ungewünschten Blockierungen der anderen feineren Klüfte.*

(ii) Unter der Reaktion soll ein Gasdruck geschaffen werden, der mit dem Druckunterschied unreagiertes Material aktiv in die Klüfte einpreßt.

-> *Bei der Reaktion ensteht ein veränderlicher Druckgradient angepasst an die hydraulischen Eigenschaften des Kluftsystemes.*

(iii) Der Reaktionsprozeß soll kontrolliert unabhängig vom Wasserzugang geschehen (Figur 3B).

-> *Das fließende Wasser im Kluftsystem und zur Felsenoberfläche wird gestopft, ohne daß die Öffnungen zu den feineren Gesteinsrissen aufgrund eines zu schnellen Reaktionsverlautes blockiert werden.*

(iv) Der Gesteinsriß soll vollständig unabhängig von der eingepreßten Menge, die sich im Gesteinsriß befindet, gefüllt werden.

-> *Das Material expandiert bei der Reaktion.*

Das Prinzip mit dem veränderlichen Druck unter der Zeit, der sich an die hydraulischen Eigenschaften des Kluftsystemes anpaßt, wird in der Figur 4 erläutert.

An der Chalmers Technischen Hochschule in Göteborg wurden diese oben gestellten Eigenschaften bei dem Material TACSS nachgewiesen.

Der Injektionprozess mit dem Material TACSS besteht somit aus zwei Phasen. Unter der passiven Phase wird das Material mit Hilfe des Pumpendruckes zuerst in die gröbsten Klüfte gepreßt, wobei das Wasser verdrängt wird.
An der Injektionsfront kommt das Material in den Kontakt mit dem Wasser, oder es vermischt sich mit dem rinnendem Wasser im Felsen.

Nach Ablauf der Induktionszeit beginnt die aktive Phase nämlich die Reaktion des Materiales mit einer Gasbildung. Diese Gasbildung erzeugt einen Gasdruck,

der das Material weiter aktiv in das Kluftsystem hinein-treibt.
Nach Ablauf der Reaktionszeit kann das Material aufgrund des Ansteigens der Viskosität nicht mehr bewegt werden. Das noch flüssige Material hinter dieser Front wird mit dem Gasdruck, der sich dem Pumpendruck überlagert, in die Öffnungen hineingepresst, die einen höheren Druck für die Eindringung verlangen.

Dabei kommt dieses Material in Kontakt mit Wasser und der Vorgang wiederholt sich, wobei der Gasdruck weiter ansteigt, und es wird wieder flüssiges Material in noch feinere Öffnungen aktiv hineineingepresst.

Der "Eindringungsprozeß" wird durch die Veränderung des Gasdruckes an die an die hydraulischen Eigenschaften des Kluftsystemes entlang des Bohrloches angepaßt. Unter der Injektionszeit überlagert sich der entstehende Gasdruck (Expansionsdruck) mit dem Pumpendruck, um dem Grundwasserdruck und den Druckfall an den Öffnungen der Gesteinsrisse zuüberwinden.

Wenn alle zugänglichen Kanäle und Hohlräume in dem abgegrenzten Bereich gefüllt sind und der resultierende Druck bestehend aus Pump- und Gasdruck nicht mehr einen Gegendruck überwinden kann, d.h. der Druckgradient ist höher als 0.18 MPa, so kommt die Reaktion zum Stillstand. Dadurch besteht kein Risiko, daß es im Felsen aufgrund eines unkontrollierten Gasdruckes zu gefährlichen Schalenbrüchen kommt.

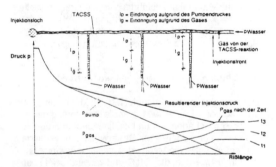

Die Gasbildung bei der TACSS-reaktion kann den Injektionsdruck in den Rissen hinter dem Gaskissen vergrößern. Der Gasdruck p_{gas} vergrößert sich mit der Zeit (t_1 - t_2) und wird zu dem Pumpendruck p_{pump} addiert. In den Rissen, wo die Gasbildung hat, wird die Eindringung durch das Gas entgegenwirkt. Die aktive Eindringung des Injektionsmateriales mit den veränderlichen Drücken, die sich unter der Reaktion bilden, sind angepaßt an die hydraulisvhen Eigenschaften des Kluftsystemes.

Figur 4   Modell für aktive Eindringung gemäß Rapport B 90:1 "Fintätning av berg", Ingvar Bogdanoff, Chalmers Technische Hochschule, Göteborg

Figure 4   Model for active penetration according to Rapport B 90:1 "Fintätning av berg", Ingvar Bogdanoff, Chalmers University of Technology, Göteborg

135

Dieser Prozess der zielgerichteten Injektion mit der aktiven Eindringung ist eine neue Dimension in der Injektionstechnik, nämlich alle Gesteinsrisse entlang des Bohrloches innerhalb eines abgegrenzten Bereiches mit einem veränderlichen Druck unter der Injektionszeit vollständig abzudichten.

3. AUSGEFÜHRTE INJEKTIONSARBEITEN MIT AKTIVER E INDRINGUNG

In den letzen 14 Jahren wurden viele interessante Injektionsarbeiten in Skandinavien im Lockergestein, Felsen und Beton ausgeführt, wo diese neue Dimensión in der Injektionstechnik ausgenutzt wurde. Als Injektionsmaterial wurden die wasserreaktiven Produkte TACSS und FLEX angewendet, die die oben beschriebenen Forderungen unter anderem, die aktive Eindringung, erfüllen.

Die ersten Injektionsarbeiten hatten als Ziel, eine Abdichtung unter schwierigen Bedingungen zu erhalten, wo alle konventionellen Injektionsmethoden versagten. Trotz hoher Materialkosten im Vergleich zum Zement wurde diese Methode oft vom Auftrageber gewählt aufgrund der einfachen Hantierung mit leichter Injektionsausrüstung und dem erwartenden Abdichtungserfolg.

Bei den Nachinjektionsarbeiten im Felsen ist es immer schwierig, die im voraus veranschlagten Kosten nicht zu überschreiten.

Anhand von drei ausgeführten Arbeiten, die spezielle Schwierigkeiten für die Durchführung der Arbeiten hatten, wird die Möglichkeit gezeigt, wie effektiv die Nachinjektionen im Felsen durchgeführt wurden.

3.1 Versorgungstunnel in Göteborg, Schweden

Das Ziel dieser komplizierten Nachinjektion, ausgeführt am Ende des Jahres 1991, war die Abdichtung der Firste, um Schäden an den Installationen aufgrund von rinnendem und tropfendem Sickerwasser zu verhindern. Außerdem sollte die Menge des Infiltrationswassers vermindert werden, die den Grundwasserdruck in den permeablen Schichten unter den mächtigen Tonschichten aufrechterhält.

In diesem Tunnelabschnitt von ca 20 m Länge führten früher ausgeführte konventionelle Injektionsarbeiten zu keinem Erfolg. Der aktuelle Tunnelabschnitt liegt ca 80 m unter der Oberfläche, und es wurde gegen einem Wasserdruckes von ca 0.7 MPa gearbeitet.

Die Figur 5 zeigt übersichtlich den Bohrplan in der Firste. Die Lage der Bohrlöcher wurde aufgrund von Sprizbeton auf der Felsenoberfläche willkürlich, aber immer mit Hinsicht auf die Leckagestellen, gewählt.
In der ersten Injektionsetappe war die Länge der Bohrlöcher 2.4 m.

Die auftretenden neuen Leckagen aufgrund der Drucksteigerung nach der Injektion und die in der ersten Etappe nicht abgedichteten "Tropfen" wurden in der zweiten und dritten Etappe mit kürzeren Bohrungen

angebohrt und abgedichtet.

Der Abdichtungserfolg war nicht nur in der Firste zusehen, sonder auch eine drastische Verminderung der Menge des Infiltrationswassers wurde festgstellt.

3.2 Transporttunnel Juktan Kraftwerk, Nordschweden

Schon 1990 wurde bei einer Besichtigung des Tunnels

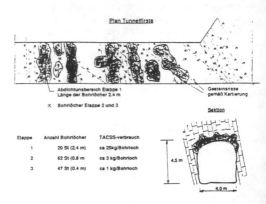

Figur 5　Versorgungstunnel in Göteborg, Schweden. Abdichtung der Firste.

Figure 5　Supply tunnel in Göteborg, Sweden. Sealing of tunnel roof.

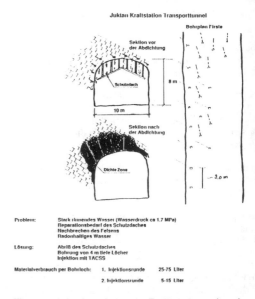

Figur 6　Injektionarbeiten im Transporttunnel nach dem Abriß des Schutzdaches

Figure 6　Grouting work in transporttunnel after removal of cover overhead.

136

die Möglichkeit diskutiert, das einleckende radonhaltige Sickerwasser im Felsen in der Firste und den Wänden abzudichten. Die Reparation der Anordnungen für die Ableitung des Sickerwassers und auch der Schutzdächer zum Schutz der Fahrbahn waren für den Auftrageber ausschlagebend, eine Nachinjektion des Felsens durchzuführen. Die Schutzdächer müssen nach einem gewissen Zeitintervall aus Sicherheitsgründen demontiert werden, um die Felsenoberfläche zu besichtigen. In gewissen Fällen muß der Felsen dann nachgebrochen werden.

Stark einströmmendes Wasser bedeuten auch sehr schlechte Arbeitsbedingungen für die Durchführung dieser Arbeiten..

Bei einer Probeinjektion, ausgeführt im Herbst 1990, wurden mit Bohrlöchern von ca 0.6 m Länge das einströmmende Sickerwasser in den Klüften bei einem Wasserdruck im Felsen von ca 0.8 MPa abgedichtet. Die Bohrlöcher wurden mit einer elektrischen Handbohrmaschine gebohrt.

In einer anderen Zone wünschte der Auftraggeber anstel-

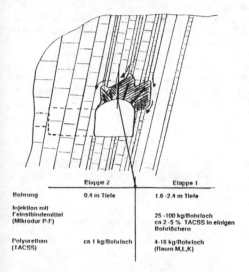

	Etappe 2	Etappe 1
Bohrung	0.4 m Tiefe	1.6 -2.4 m Tiefe
Injektion mit Feinstbindemittel (Mikrodur P-F)		25 -100 kg/Bohrloch ca 2 -5 % TACSS in einigen Bohrlöchern
Polyurethan (TACSS)	ca 1 kg/Bohrloch	4-16 kg/Bohrloch (Raum M,L,K)

In einigen Bohrlöchern wurde der Verbrauch von Feinstbindemittel mit Polyurethan begrenzt, indem man Polyurethan (TACSS) direkt an der Manschette in die Suspension mit Mikrodur P-F pumpte. Diese Methode wurde auch bei dem Außfließen der Suspension zur Oberfläche angewendet.

Ein Wasserabpreßversuch wurde für alle Bohrlöcher ausgeführt. Niedrige Wasserverluste (teilweise dichte Löcher) wurden gemessen. Die Lugeonwerte lagen bei ca 0.2 - 0.8.

Figur 7 Svedjefortet, Boden, Nordschweden. Sektion der Kaverne.

Figure 7 Svedjefortet, Boden, North Sweden. Section of cavern.

le des reparationsbedürftigen Schutzdaches eine Abdichtung der Klüfte. Die Länge der Bohrlöcher wurde auf 4.0 m vergrößert, um das stark einströmmende Wasser mit einem höheren Druck von ca 1.7 MPa abzudichten (Figur 6).

Die Arbeiten wurden im Herbst 1992 mit guten Erfolg beendet. Die Kosten der Injektionsarbeiten waren niedriger als das Anbringen eines neuen Schutzdaches.

3.3 Svedjefortet, Boden Nordschweden

Das Swedjefort ist eine alte Verteidigungsanlage, die heute als Museum dient. Einsickernes tropfendes Felsenwasser in den Kavernen waren nicht nur lästig für den Besucher, sondern erhöhten auch die relative Luftfeuchtigkeit in den Ausstellungsräumen mit Mögelbildung an den Austellungsgegenständen zur Folge. Trotz der Installation von Luftbehandlungsapperaten konnten die Probleme nicht bewältigt werden. Die Abdichtungsarbeiten starteten nach der Frostperiode im Mai 1993.

Da die Felsendeckung über der Firste nicht größer als 5-6 m ist, sind die Leckagen aufgrund des Sickerwassers stark abhängig von der Intensität der Niederschläge. Die Felsenoberfläche in der Firste ist mit Spritzbeton verdeckt, so daß die Richtung und Neigung der Klüfte nur von den Wänden aus zuersehen war (Figur7).

Zum Unterschied zu den vorher beschriebenen Arbeiten wurde als Injektionsmaterial Suspensionen mit Feinstbindemittel verpreßt. Um eine Begrenzung des Injektionsbereiches zuerhalten und den Ausfluß der Suspension an der Felsenoberfläche zustopfen, wurde der Suspension das aktive Injektionsmittel TACSS direkt an der Manschette zugegeben.

Diese Methode hatte aber nur einen Erfolg bei den Arbeiten im östlichen Teil der Festung. In diesem Kluftsysten bestehend aus feinen Gesteinsrissen mit direktem Kontakt zum Bohrloch (anisotropes Gebirge) konnte die Feinstbindemittelsuspension sehr gut eindringen und der Injektionsablauf vereinfachte sich mit der Zugabe des "aktiven " Zusatzmittels TACSS. Bei Bedarf wurden die Ausflüsse an der Oberfläche gestopft, und der Injektionsbereich wurde begrenzt. Nur 3-5% TACSS im Verhältnis zur Feinstbindmittelmenge wurde der Suspension zugegeben.

In dem westlichen Teil der Festung besteht die Felsenstruktur aus großblockigen Klüften (isotropes Gebirge). Die einzelnen Gesteinsrisse haben zum Bohrloch eine flächenmäßigen Kontakt. Dies bedeutet, daß die Injektionsrate beim Pumpen der Suspension zu hoch war im Verhältnis zur Aufnahmefähigkeit des Kluftkörpers. Obwohl auch alle Eigenschaften der Suspension optimiert wurden, so floß die Suspension nur langsam in die gröbsten Klüfte. Die Öffnungen der feineren Gesteinsrisse wurden aber aufgrund der Abnahme der Bewegung der Suspension im Bohrloch blockiert Aus diesem Grunde wurden in diesem Teil der Festung die Injektionsarbeiten hauptsächlich mit TACSS ausgeführt.

137

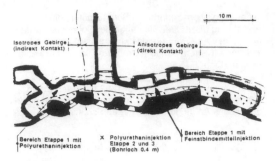

Figur 8   Svedjefortet. Übersicht der ausgeführten
           Arbeiten. Etappe 1 und 2.

*Figure 8   Svedjefortet. Survey of executed work.
            Step 1 and 2.*

Bei den Abdichtungsarbeiten dieser ersten Etappe
wurden systematisch die Bohrlöcher entlang der Kaverne
in die Firste gebohrt und verpresst. Das Ziel war außer
einer Abdichtung der Firste, das rinnende Sickerwasser
im Gebirge von der Kaverne wegzupressen.
In der zweiten Etappe wurden die kürzere Bohrlöcher
gebohrt. Die noch vorhandenen lokalen Leckagen
wurden mit TACSS abgedichtet (Figur 8).
Neue Leckagen, die nach einem regnigen Sommer
1993 auftraten, wurden mit der gleichen Methode abgedichtet. Aufgrund des einfachen Verfahrens konnten die
Ausstellungsgegenstände im Museum stehen gelassen
werden.

AUSBLICK

Seit dem Jahre 1979 ist die Methode mit der "aktiven
Eindringung mit veränderlichen Druck unter dem
Injektionsablauf" für Injektionsarbeiten im Gebirge,
Lockergestein und Beton angewendet worden. Sehr oft
kam diese Methode zum Einsatz, wenn die konventionel-
len Injektionsmethoden zum Beispiel bei stark rinnendem
Wasser versagten.
In diesem Vortrag wurde nur die Nachinjektion zur
Abdichtung des Felsens behandelt. Diese Methode
wurde auch mit Erfolg bei schwierigen
Vorinjektionsproblemen im Tunnelbau sowie auch bei
Abdichtungsarbeiten für Dämme angewendet.

*Warum war diese Methode erfolgreicher gegenüber
anderen mehr gebräuchlichen Methoden?*

Leider muß man feststellen, daß auch hervorragende
Injektionsexperten mit langjähriger praktischer Erfahrung,
das Injektionsverfahren vom dem Gebrauch von heimli-
chen Rezepten für die Suspensionen oder speziellen
chemischen Injektionsmitteln abhängig machen. Dringt
die Suspension nicht in das Gebirge ein oder wird diese
weggespült, dann versucht man mit neuen Mitteln, wobei
viel Arbeit für die Optimierung der Suspension aufgebracht wird, zum Erfolg zukommen. Zum Schluß wird
beispielsweise die Viskosität der Suspension sehr
beschleunigt oder man schäumt mit Polyurethan, so daß

die Ausbreitung des Injektionsmateriales in dem Kluftsystem blockiert wird. Neue Bohrlöcher müssen gebohrt
werden, wieder eine Injektionsrunde wird ausgeführt, und
die Kosten sind nicht mehr überschaubar.

*Der Gedanke liegt nahe, es ist unmöglich,
Feinabdichtung des Gebirges mit der Nachinjektion
auszuführen.*

Es ist doch besser, mit neuen Überlegungen, einen
anderen Weg zu beschreiten.
Im Jahre 1986 weckte der Ausdruck "aktive
Eindringung" Interesse bei den Forschern an der
Chalmers Technischen Hochschule in Göteborg,
Schweden. Das Resultat dieser Forschung wurde in dem
Rapport B 90:1 "Fintätning av berg" (Feinabdichtung von
Felsen) , Ingvar Bogdanoff veröffentlicht. In diesem
Rapport sind die Hintergründe des Begriffes "aktive
Eindringung" erläutert und an Hand von Testen im
Laboratorium nachgewiesen.
Diese Forschungergebnisse waren sehr wertvoll für die
weitere Entwicklung der Injektionsmethoden in der
Praxis. Eine besondere Methode wurde an der Chalmers
Technischen Hochshule entwickelt - FLEX-MIX-Methode -
, wo Suspensionen mit Polyurethanen mit den oben
genannten Eigenschaften als aktives Zusatzmittel in die
Injektionslöcher mit einer einfachen Spezialmanschette
gepumpt wurden.
Mit dieser Entwicklung und Forschung gibt es gute
Voraussetzungen , daß die Injektionstechnik mit der
neuen Dimension "aktive Eindringung mit veränderlichem
Druck" seinem Platz in dem " Kochbuch der Injektion" in
der Zukunft haben wird.

Litteratur:

Rapport B 90:1 "Fintätning av berg" (Feinabdichtung von
Felsen), Ingvar Bogdanoff, Chalmers Tekniska Högskola,
Göteborg. 1990.

"The individual groutability of rocks", F.K. Ewert, Water
Power & Dam Construction. January 1992.

Observations – Grouting of rock fissures

Beobachtungen bei der Injektion von Klüften im Fels

Edward D.Graf
Foster City, Calif., USA

ABSTRACT: Observations of rock grouting from over forty year years of experience are presented with the intent of improving the practice. Although not "technical" in the spirit of new ideas or mathematics, it touches upon the subjects of particle size, stable grouts, grout strengths, control of radius of permeation, drilling and stage grouting, and grouting pressures. 5 brief illustrative case histories are cited.

ZUSAMMENFASSUNG: Nach mehr als 40 Jahren Erfahrung mit Injektionen im Fels legt der Autor hier eine Reihe von Anmerkungen vor, um künftige Anwendungen zu erleichtern. Diese Anmerkungen sollen nicht "technisch" im Sinne von neuen Ideen oder Formeln sein, sondern nur die Einflußfaktoren Korngröße, stabile Suspensionen, Festigkeit des Injektionsmaterials, Injektionsdrücke und Kontrolle der Reichweite behandeln. Der Autor zieht aus seinen Erfahrungen folgende Schlußforderungen:
Die maßgebende KORNGRÖSSE (D85) in Zement-Suspensionen sollte 1/5 der Kluftweite nicht überschreiten. Die kleinste Kluftweite für Normalzement liegt daher bei 0.5 mm, für Mikrozement bei 0.05 mm. Der Einfluß der DRUCKFESTIGKEIT des Injektionsmaterials wird meist überschätzt, da unter den Bedingungen in Klüften die Festigkeit bei behinderter Querdehnung maßgeblich ist. STABILE SUSPENSIONEN sind aus heutiger Sicht vorzuziehen, da sie die bessere Verfüllung von Hohlräumen gewährleisten.
SCHLAGBOHRUNGEN mit kontinuierlicher Spülung können auch in kleinklüftigem Fels erfolgreich angewendet werden.
Die Festlegung eines INJEKTIONSDRUCKes gemäß der "europäischen Daumenregel" kann in vielen Fällen als konservativ angesehen werden. Vielmehr sollte der Injektionsdruck auf Grund von Beobachtungen vor Ort begrenzt werden.
An Hand von fünf Ausführungsbeispielen werden diese Themen untermauert. Acht Schlußfolgerungen für "gute" Felsinjektionen werden vom Autor gezogen:
(1) Geringste Kosten für den Bauherrn, um den technischen Zielvorstellungen zu genügen.
(2) Schlagbohrungen mit kontinuierlicher Spülung können zweckmäßig sein.
(3) Insbesondere bei Bohrwasserverlusten sollte gleichlaufend mit dem Abteufen der Bohrungen injiziert werden.
(4) Nicht immer führt das billigste Injektionsmaterial zu den niedrigsten Gesamtkosten.
(5) Die Korngröße des Injektionsmaterials ist von entscheidendem Einfluß.
(6) Stabile Mischungen (ohne Bluten) sind vorzuziehen, wenn möglich mit leichter Volumsvergrößerung während des Abbindens.
(7) Kontrolle der Reichweite durch optimale Kombination von Druck, Topfzeit, Fließeigenschaften und Pumpgeschwindigkeit.
(8) Keine Sandzugabe bei zu großen Aufnahmen, da damit auch andere Klüfte in der Umgebung des Bohrloches verstopft werden. Sandzugaben sind nur zur Verfüllung großer Hohlräume annehmbar.

Rock grouting involves placing a permanent material of adequate strength to fill voids in rock to a predetermined zone, usually for the purpose of shutting off water and/or strengthening a rock formation. "Good" grouting accomplishes the projekt goal for the lowest end cost.

This discussion is not meant to be an in-depth technical paper, but presents only some of the highlights of the author's experiences and observations during more than forty years of grouting, mostly as a specialty contractor.

1 GENERAL OBSERVATIONS

1.1 Particle size

Grout particle size is of major importance when using particulate grouts. Studies and tests (Ref.1) of past years have demonstrated that (1) the D_{85} (85% passing) of the grout particle sieve analysis must be one fifth of the fissure width to assure permeation and that (2) permeation is marginal with a D_{85} of one fourth of the fissure width. Although the referred study is based upon permeation of soils, the numbers are readily extrapolated to rock fissures and the extrapolations, with over 30 years of observations, seem to hold up well.

Depending upon the type and manufacturer, it is generally accepted that standard portland cement will permeate fissures down to 0.5 mm; smaller fissures may or may not be permeated by portland cement depending upon the water:cement ratio, fineness of the cement grind, roughness of the fissure surfaces, and other factors. The practice of adding sand because of high "take" will plug fissures less than 5 times the D_{85} of the sand and will plug small fissures immediately adjacent to the hole.

Unpublished analysis (Ref.2) based on standard U.S. Type I-II portland cement grinds show that, idealized, a liquid+filler:cement ratio of 6:1 by volume (4 l/kgm) yields a cement particle distribution that allows the cement particles to enter a fissure one at a time which will avoid filter plugging of the particles within marginal fissures. By using liquids plus proper filler materials finer than portland cement to achieve that ratio, such as bentonite, fly ash, etc., successful cement grouting has been accomplished under less than marginal conditions. Note that the liquid plus filler volume is calculated as absolute volume.

Ultrafine cement grouts are now commercially available that will permeate fissures down to 50 microns and finer depending upon the D_{85} of the cement.

Chemical grouts (solutions without particles in suspension) have proven themselves to be cost effective in microfractured rock.

1.2 Strength

Many civil engineers demand concrete strengths for rock grouting with portland cement. Due to their experience and expectations with concrete design, they mistakenly apply the same standards to rock grouting. They do not analyze the mechanics of what is actually necessary to accomplish a rock grouting goal.

Two items in particular are often overlooked when specifying strength for rock grouting work:

1. The mass involved because grout is rarely placed in sections less than four meters (13 feet) in diameter; a significant amount of cement grout placement results in diameters of 20 to 30 meters (66 to 100 feet) and much more.

2. Unconfined compressive strength is valuable in designing structures, but is actually meaningless in most rock grouting work. If the set cement grout is locked into voids and fissures, what force is required to extrude or compress it? Ultimate bearing strength, approximately nine times the unconfined compressive strength, is of more value in geotechnical analysis but, still, it is usually overly conservative. Successful dam grout curtains have been installed using chemical grouts with an unconfined compressive strength of 70 kPa (10 psi). There are a number of papers and articles that describe the use of acrylamide grouts, 10% or less solids in solution, for dam curtains that achieve, hopefully, that strength. Stable cement grouts with

unconfined compressive strengths of 350 kPa (50 psi), and less, have been used successfully for both water shutoff and for strengthening foundations by the author.

Only two failures of water shutoff due to low strenght are known to the author. Both failures were using acrylamide chemical grout of about 70 kPa (10 psi) unconfined compressive strength under ground water pressures of about 7,000 kPa (1,000 psi) in rock fissures. The problem was the set acrylamide chemical grout is a weak plastic material and the grout was extruded from the rock fissures in both cases *after* shooting the rock with dynamite. If a similar low strength but non-plastic grout had been used, those failures probably would not have occurred.

Of course, there are the rare occasions where high strength is imperative.

1.3 Radius of permeation

Assuming that the grout material has the capability of permeating the desired voids and fissures, the radius of permeation in any particular rock formation is controlled by the following factors:

1. Pressure. Pressure is a primary control of the radius of grout permeation. Some of the exceptions are the effect of gravity on very fluid grout materials and pumping a very fluid slow setting grout material.

2. Set time. The shorter the set time, the shorter the distance the grout will flow under a given pressure and pumping rate, especially when the set time is in seconds or a very few minutes.

3. Viscosity/cohesion. The more viscous or cohesive a grout mix, the shorter the distance it will flow under a given pressure and pumping rate.

4. Pumping rate. A slower pumping rate will usually result in a shorter radius of permeation.

1.4 Stable grouts

Much has been written about "stable" grouts and this author is a firm believer that stability is of the utmost importance for "good" grouting unless you are pumping the grout with enough pressure to force out all of the bleed water. Standard practice in deep mining of using pressures in excess of 20,600 kPa (3,000 psi) does force the bleed water from the grout but does not give radius of permeation control.

Despite the fact that many thousands of successful projects in the past used unstable grouts, today's knowledge does not give credence to their use today (historically, farm land was plowed by hand).

From observations of projects as simple as filling voids below foundations to the nearly catastrophic mine incident described in a case history below, it is obvious that, for most grouting work, settlement of the grout solids will leave (1) a weak zone subject to compression and/or erosion and (2) a void, or (3) a filter plug that is subject to failure upon formation movement caused by tectonics, explosives, or by loading of the formation. Fluid stable grouts are readily available and are usually more economic by the use of small particle size fillers such as bentonite, pozzo-, etc., together with admixtures.

The basic test for sedimentation (bleeding) that allows two hours for the test should be used only if the twenty four hour test has proven that sedimentation is complete at two hours with the materials being used.

An ideal grout material will show slight unrestrained expansion after setting into its final position. This can be accomplished easily with cement grouts by the proper use of aluminum powder if the rest of the mix is correct. *Note that proper choice of the aluminum powder for grind and type is imperative and that the dosage is critical, usually in the range of 5 to 15 grams per 50 kilograms, and must be tested for each particular cement to be used on a project.* Correct aluminum admixtures are available from admixture suppliers but these must again be tested with the particular cement to be used on each project.

Pre-drilling grout holes at intervalls that lead the grout "communication" between holes results in wasted drilling. Grout communication between bore holes results in the second hole being filled with grout under reduced pressure resulting in only plugging the smaler fissures immediately adjacent to the hole.

1.4 Drilling and stage grouting

Fortunately, in recent years percussion drilling with continuous water flushing (no air flushing allowed) has become proven and accepted as a standard for drilling grout holes even in microfractured rock. A few engineers still demand core drilling despite the substantial increase in cost in terms of both direct cost and in delays due to the slow drilling rate. With experienced percussion drillers, it has been the author's experience that the air flush handle must often be removed from their machine because of their trained reaction to blow the hole frequently.

It is the author's opinion that stage up grouting is not as effective as stage down grouting because of the potentials of (1) grout permeation upward outside of the packer and (2) plugging open fissures with the continued drilling below groutable fissures. It has been his practice to install a pipe collar at each hole with a V-notch weir or other means to measure the water return from the hole during the drilling. As soon as the water return is visibly reduced, drilling is stopped and the hole is grouted at that point. This technique can result in very short or very long stages, depending upon the formation, but has consistently produced excellent results.

1.5 Pressure

American pressure limitations of 23 kPa/m (1 psi/ft) of depth versus the much higher pressure limitations of European practice have been a source of argument in the United States for many years. It is the author's opinion that even the European practice can be considered overly conservative under certain conditions.

Given a formation that readily accepts grout (open voids/fissures) until the grout starts to fill the fissures and voids the pressure is used in the velocity of the grout travel with relatively little pressure transmitted to the formation, comparable to filling a tank. Only when a "hydraulic jack" with enough horizontal area has been developed will surface heave occur. Tightening of a formation by pressure dilation during grouting is often desirable.

Successful grouting without surface uplift has been accomplished many times with grout pressures that substantially exceed overburden pressure by each of the following methods:

1. By using a fluid slow setting grout at pressures as much as five times over the overburden pressure, pumping at a conservatively high and constant rate, and continuously observing the pressure at the collar of the hole; when the pressure rises by about 35 kPa (5 psi), the pressure is then reduced to a "safe" pressure by reducing the pumping rate. The pressure rise indicates the fissures and voids are becoming full. This technique can only be used safely when the formation will readily accept grout.

2. By using a grout with a set time of seconds or a very few minutes together with rational pumping rates, the possible development of a "hydraulic jack" effect on the formation is minimal because the set grout will not effectively transmit the pressure to the formation. In one of the following case histories, "Chemical Grout Dam Curtain", grout pressures used near the surface were as high as twenty times the overburden pressure and no surface heave was observed. In the chemical grouting of clean dune sands, it has been the author's practice to use pressures of 800 kPa (120 psi) or more at depths of 2 m (6 ft) and surface uplift has never been observed.

1.6 Cement Mixers

Althought international authorities such as Houlsby and Weaver insist that colloidal (high shear) mixers are vitally necessary to ensure a that cement particles are not agglomerated, this view is not held by the author. His experience has been that the use of surfactants such as lignin sulfonate together with adequate paddle mixer mixing, at least 5 minutes, has resulted in satisfactory grouting of marginal fissures. Using mixing tanks of ample size for the particular projekt, 5 minutes or more of mixing is easily achieved. The use of colloidal mixers does result in reducing bentonite requirements about 50% if the bentonite is not premixed.

The simplicity of paddle mixers over colloidal mixers is readily apparent. Simplicity means lower maintenance costs and much higher reliability.

2 SHORT CASE HISTORIES

Following are five case histories illustrating various experiences and observations. Only one of these cases has been published.

2.1 Water tunnel portal

The practice of pumping high water:cement ratio unstable cement grouts, even at relatively low pressures, yields an uncontrollable radius of permeation. This was demonstrated to the author about thirty five years ago when doing a small contact grout project at a water tunnel portal.
At a maximum pumping pressure of 100 kPa (15 psi), grout was observed coming to the surface in the canyon below the tunnel portal, approximately 1.2 km (0.75 mi) away. It should be mentioned that the engineer inspector, an experienced government grouting engineer inspector, was delighted and instructed the grouting crew to keep pumping with no changes. This experience triggered the author's resolve never to do uncontrolled grouting again because of the waste; there were a number of subsequent contracts lost because of this principle.

Many old-time cement grouters tell stories of cement slurry grouts plugging wells as far as 6.5 km (4 mi) away from the point of injection. When a properly designed grout program can effectively hold the grout to a radius of 4.5 m (15 ft), and even much less, grouting beyond the necessary zone is a waste of the client's (or the taxpayers') money.

2.2 Penstock tunnel high pressure grouting

A new penstock tunnel, with its portal at the top of a ridge above a new powerhouse, required high pressure grouting to strengthen the surrounding rock for the penstock pressures. A 2.7:1 water:cement ratio (4:1 by volume) had been used and approximately 250 metric tons (6,000 sacks) of cement had been pumped into one section without building any back pressure. The grout was surfacing at the bottom of the canyon about 600 m (2,000 ft) below. Using standard paddle-mixer tanks and one doubleacting pump, a system was devised to use a 0.67:1 (1:1 by volume) water:cement ratio grout with calcium chloride to bring the set time down to three minutes. 6.3 metric tons (150 sacks) of cement were used to "pressure off" the section followed by low "take" contact grouting. Now, 25 years later, it is still performing with no problems.

2.3 Mine ore bin

A mine was being developed at a depth of about 825 m (2,700 ft). During the development of the mine, the section for the underground ore bin had been grouted before mining with over 170 metric tons (4,000 sacks) of cement using a water:cement ratio of 2.7:1 (4:1 by volume) by a specialty contractor experienced in using conventional cement slurry. Ground water pressure was about 3,240 kPa (470 psi). Following the excavation of the ore bin, the water infiltration was about 2,000 l/m (500 gpm) which was not acceptable.

Using the following technique, the inflow was reduced to 20 l/m (5 gpm) using only 11.3 m³ (400 cf) of grout:
1. A two pump proportioning grout rig pumping a stable cement:bentonite and silicate grout with a set time of 45 to 60 seconds.
2. Grout holes were 3 m (10 ft) deep, wet-head percussion drilled, angled to intercept the fissures at 2.3 m (7.5 ft) with mechanical packers set at 1.5 m (5 ft).
3. Grouting pressures used were 4,000 to 4,300 kPa (570 to 620 psi) which were 700 to 1,000 kPa (100 to

150 psi) above the ground water pressure.

4. Pumping rates were up to 45 l/m (12 gpm), intermittent when necessary to stop grout leakage from the face with the fast setting grout.

The typical radius of grout travel observed on the faces of the ore bin was 9 m (30 ft) and, on occasion, up to 14 m (45 ft).

The first remedial grouting attempts used a liquid+filler:cement ratio of 4.5:1 by volume (3 l/kgm) and caused almost immediate plugging of the holes. This problem was eliminated by increasing the liquid+filler:cement ratio to 6:1 by volume (4 l/kgm).

It must be noted that the characteristics of the cement and silicate grout have not been published. It is subject to severe syneresis if not properly admixed and there are other serious problems that are only learned with experience. On this project, a five year performance check showed no increase in the water inflow into the ore bin above the 20 l/m (5 gpm) at the completion of the work.

2.4 Mine flooding

In the same mine development as the previous case, the mining and haulage drifts were being advanced under a grout cover using the same water:cement ratio of 2.7:1 (4:1 by volume) by the same experienced specialty contractor. As the mining drift was being advanced, a sudden inflow of 30,000 l/m (8,000 gpm) was encountered. The reserve pumping capacity at that time was 3,000 l/m (800 gpm). The major catastrophe of a new mine development full of water was imminent. Luckily, the inflow was at the upper mining level which allowed the lower haulage level to act as a reservoir. It was quickly calculated that there was little more than 24 hours before the haulage level would fill causing the electrical dewatering pumps to be shut off by the water inflow.

By installing additional pumps and implementing unusual grouting techniques (including the use of ladies' sanitary napkins for the grout mix), the inflow was overcome in 5 days when the water level in the haulage drift was within only 4 cm (1- 1/2 inches) of the back (ceiling). Conservatively, a full concrete plug was placed before mining was resumed.

The source of the inflow was a nearly vertical fissure about 20 cm (8 inches) wide almost normal to and on the back (ceiling) of the drift. When the round was pulled and the area opened, the fissure was exposed solidly filled with cement grout. The failure of the cement grout and the sudden inflow of water occurred during the mucking of the shot rock shortly after the shot.

2.7:1 (4:1 by volume) cement grout, by tests, yields about 25% set solids and 75% bleed water. It is the author's firm opinion that the set grout only extended a few centimeters up into the rock above the back of the drift and was overlain by bleed water. Some distance above, the fissure had narrowed enough to make a small cement plug that allowed the grout to "pressure off"; then the settled cement left bleed water filling a tall vertical fissure. Due to the dynamite shot, that small plug had been damaged allowing the full ground water pressure of about 3,240 kPa (470 psi) to develop on the inadequate grout plug above the back of the drift. A stable grout would not have allowed this to happen. Minor tectonic movement or the filling of a dam can cause the same type of failure when an unstable grout is used.

It was also observed during mining that many nearly horizontal grout holes were less than half filled with grout, further demonstrating the inadequacy of unstable grout mixes.

2.5 Chemical grout dam curtain

Two small diversion dams above a sanitary landfill required grout curtains in microfractured granodiorite to a depth of 12 m (40 ft) (Ref.3). Tests had indicated a permeability coefficient of 10^{-4} cm/sec. The grout curtain permeability was specified to be 10^{-7} cm/sec or less. Drilling was stage down, wet-head percussion, with grouting a stage whenever drill water loss was observed. A dilute urea-formaldehyde grout, 2-3 cps, was used with set times of about 2 minutes and with pumping rates up to 23 l/m (6 gpm) at pressures up to 700 kPa (100 psi). This procedure was used at all grouting depths in the effort to limit permeation to about a 1.5 m (5 ft) radius. A three row splitspaced curtain was constructed. The required reduction in permeability, to less than 10^{-7} cm/sec, was achieved.

This was one of the very few dam curtain projects where the author was able to negotiate a contract and accomplish the work in what he felt was the best and most economic method wit a cooperative client's engineer. The goals were accomplished at a total cost to the client of US$130/vm^2 (US$12.15/vsf) of curtain. A very similar project in size, scope, and geology was accomplished a few months later but the different design engineer required other procedures; the same goals were realized at a unit cost of two and one-half times the cost of this project and required almost three times the time per unit to accomplish the work.

3 SUMMARY

To summarize, the author firmly believes that "good" rock grouting involves:

1.   Least cost to the client to achieve the completed project requirements.
2.   Wet-head percussion drilling where feasible, no air flushing allowed.
3.   Stage down grouting; grout whenever drill water loss is observed.
4.   Grout material choice that will yield the desired results at the lowest *end cost* which is frequently not the cheapest grout material.
5.   Grout particle size is of the utmost importance.
6.   Stable (no bleed) grouts with, preferably, slight expansion when setting.
7.   Control of the radius of grout permeation using various combinations of pressure, set time, viscosity/cohesion, and pumping rate.
8.   No addition of sand to a mix if a hole is taking "too much" because large sand particles will plug fissures immediately adjacent

to the drill hole. Sand is acceptable when confident that there is a large void to fill.

The author is very much indebted to the science he has learned from the grouting professionals and to the common sense of the field grouting laborers he has worked with during his forty two years of being involved in grouting.

REFERENCES

(1) King, J.C. and Bush, G.W., "Grouting of Granular Materials", Proceedings, ASCE Symposium on Grouting, Vol. 87, No. SM2, Paper No. 2761, April, 1961 and extrapolations by Polivka, M. and Graf, E., 1962, not published.

(2) Polivka M., Graf, E., and Kuhn, J., 1963, not published; based on portland cement data to achieve a particle separation equal to or greater than the median diameter of cement particles using (1) the number of particles per unit weight and (2) the median diameter of the those particles.

(3) Graf, E., Rhoades, D., and Faught, K., "Chemical Grout Curtains at Ox Mountain Dams", Proceedings, ASCE "Issues in Dam Grouting", 1985.

Ultra deep grout barriers

Extrem tiefe Injektionsschürzen

W.F. Heinz
RODIO (Pty) Ltd, Midrand, South Africa

ABSTRACT

Certain grouting techniques currently used in the South African mining industry lend themselves to the execution of deep and ultra deep barriers. In particular these techniques are –
1.	Precementation of deep shafts up to 2400m.
2.	Long distance conveyance of cement–sand slurries (5km).
3.	The successful impermeabilisation of rock masses with "thin, unstable" cement grouts.

We refer, rather arbitrarily, to deep and ultra deep grout barriers as barriers beyond 500 and 1000m respectively.

The successful design and execution of deep grout barriers requires comprehensive geotechnical investigations. The determination of fracture and fissure patterns and the corresponding geohydrological parameters, the overall stressfield, its directions and other stress parameters are very important characteristics of the formation which must be determined with reliability..

ZUSAMMENFASSUNG

Verschiedene Injektionstechniken, welche zur Zeit erfolgreich in dem sudafrikanischen Bergbau eingesetzt werden, eignen sich besonders fur die Erstellung von tiefen und ultra–tiefen Injektionsschleiern. Ins besondere:
1.	Vorinjektion tiefer Schachte (2400m Teufe).
2.	Transport von Zement–Sand–Wasser Mischungen uber lange Abstande (5km).
3.	Die erfolgreiche Abdichtung des Gebirges mit "dunnen, unstabilen" Zementinjektionen.

Der Entwurf und die erfolgreiche Ausfuhrung dieser tiefen Injektionschleier benotigen umfassende, geotechnische Untersuchungen. Die Bestimmung des Risse–Systems und die entsprechenden geohydrologischen Parameter, das ubergeordnete Spannungsfeld, dessen Richtungen und andere Spannungsparameter mussen mit hinreichender Genauigkeit bestimmt werden.

## 1.	INTRODUCTION

Cement grout curtains are typically associated with the impermeabilisation of dam foundations to a depth which varies between 50 – 150m; in rare cases to 250m. The design and execution of grout curtains are reasonably well understood. In most cases curtains are success–fully executed and fulfil their required function. Deep grout barriers of 500m and deeper and ultra deep barriers beyond 1000m present special problems and have not been executed except on a small scale in the Republic of South Africa i.e. if one can regard the pre-cementation of deep shafts as deep grout barriers. The author is not aware of any barriers of significant depth and extent that have been executed elsewhere to depths well beyond 500m. Within the context of the South African mining industry three techniques relating to grouting are utilized, the knowledge and experience of which can form the basis of the design and execution of deep grout barriers.

2. PRECEMENTATION OF DEEP SHAFTS

2.1 Introduction

Pregrouting or precementation of deep shafts prior to sinking has been applied in South Africa since the fifties, with consider-able success. Nevertheless, the financial and technical benefits are still very much under discussion primarily because of the difficulty of determining conclusively once a precementation has been executed whether the sinking time was in fact shortened by the pre-grouting operation. Geological and geohydrological considerations are decisive parameters deter-mining success or failure of a precementation, indeed its desirability.

A comprehensive site investigation is vital. Prior to any precementation and shaft sinking project an exhaustive collection of information and data relevant to the project should be under-taken. Specifically the following data are of importance:

a) Logs of all boreholes in the region.
b) Water table levels and changes of the levels in the region.
c) Quality of the water in the area.
d) An analysis of aerial photographs and magnetic surveys to detect possible faults and dykes. Infrared photog-raphy in dolomitic terrain.
e) Stressfield, direction and strength parameters, in situ modulus of elasticity.
f) The presence of prominent joint and fissure patterns.
g) Natural drift of boreholes in the area.

2.2 Precementation of deep shafts

The brevity of this paper does not allow any detailed descriptions; these can be found in several publications given in the references. (Heinz, 1988). A short summary of the decided benefits of deep shaft precementation (Fig. 1) is given here:
1) It increases the safety of the sinking operation;
2) It minimizes the inflow of water and gas;

Fig. 1   Precementation of a Shaft
Bild 1   Vorinjektion eines Schachtes

3) It minimizes the time lost due to additional grouting operations during sinking and hence minimizes standing time of shaft sinking crews and equipment;
4) It provides improved rock strength for excavations in the immediate vicinity of the shaft area (grouted fissures have been found up to 60m from the pregrouted shaft);
5) It provides detailed information of the geology of the proposed shaft site and possibly information on ore grades in the shaft vicinity.
6) It reduces the number of intermediate underground grout stations during shaft sinking operations.

2.2.1 Drilling Techniques

Generally precementation of deep shafts is done prior to sinking but can also be done during sinking if the cementation operation leads

148

Fig. 2   Precementation of Dual Shaft System (Six Rigs, 1700m Deep)
Bild 2   Vorinjektion eines Doppelschachtes (Sechs Bohrmaschinen, 1700m Teufe)

the sinking by several hundred metres. A certain number of cover holes will always be drilled whether or not a precementation has been executed from surface. Cover drilling and grouting may be executed from the bottom of the shaft and / or from bays excavated for this purpose. In both cases the drilling and grouting operation interferes with the shaft sinking operation.

Drilling is usually done by slim hole wireline drilling, generally in B or N sizes. In some cases H size was added for improved flexibility. Typically boreholes are engineered as follows (fig. 2)

a)   Drill and case overburden in a large size e.g. 4–9/16" to approx. 30m.

b)   Drill HQ or similar from 30 – 300m.

c)   Case hole to 300m.

d)   Drill CHD–76 or similar from 300–600m.

e)   Case hole to 600m.

f)   Complete hole with BQ to 1070m.

Drilling Equipment
Conventional drill rigs are utilised equipped with hydraulic chuck systems, separate wireline winches and normal triplex pumps commonly used in drilling practice (Heinz, 1985).

For deeper boreholes up to 2400m the Joy Sullivan 45 has proved to be adequate. Prime movers are turbo charged diesel engines as the

South African gold fields are situated at high altitudes, in some cases up to 1700m.

Wedging and Deflections
Directional deflections are an essential part of surface precementation projects.
Conventional wedging techniques are time consuming and cumbersome, nevertheless have proved very successful in hard rock. Specifications usually require the boreholes to spiral around the shaft in an annulus between 8 and 25m. Shaft diameters are between 6 and 10m.
Between 1 to 3 wedges per 100m may be required.
Good grouting practice requires that groutable fissures are intersected normal to the plane of the fissure or at least as close to normal as possible. Therefore, wedges are introduced to improve this angle of intersection and hence increase the chances of intersecting more fissure per metre drilled. Conventional wedges facilitate a deflection of less than 2 degrees.

2.2.2   Grouting Technique

Cement is still the primary grouting material due to its easy handling, low price and naturally its pozzolanic and environmentally friendly characteristics. The grouting technique in general is described elsewhere (Heinz, 1988); however,

149

some aspects of particular pertinence in precementation are mentioned below:

a) The high pressure used in cement grouting, results in the early development of strength due to:

    (i) the accelerated hydration of the cement because of high temperatures at depth in the borehole and because of high shear mixing techniques generally used.

    (ii) the packing of the cement grains under pressure as a result of differential settling of the cement grains and the elastic response (back pressure) of the surrounding rock mass.

b) Due to the cement hydrating under near autoclaved conditions and high pressures; (water is squeezed out) (fig. 3), the hydration is incomplete in the short time available before redrilling cement. Hence the redrilling of this cement is a hazardous exercise as the cement grains produced by the diamond drilling process are very fine, well wetted and because of the higher surrounding temperatures and friction heat from the high rotation speed of the wireline rods, are highly active. Therefore, intense flushing and slow advance are a prerequisite for redrilling cement successfully and in order to avoid cementing the rods in case of water loss.

Grouting Equipment

A standard grouting unit utilised for a typical deep shaft precementation comprises the following:
Large 50 to 100 ton silos for cement and fuel ash (PFA) and a number of back–up silos per unit up to a total storage capacity of 300 tons may be required.
An electric mixing plant consisting of one primary high shear mixing vessel and two secondary storage mixing vessels, transfer pumps and control panels.

The mixing plant has a capacity of 15 tons of solids/hour which is equivalent to 20,000 litres of 1:1 grout mix per hour.
Two air–driven grout pumps with capacities to deliver 5000 litres of grout mix/hour at pressures in excess of 20MPa; compressors.

Grout Mix

The constituent materials of the grout mix that have been used are cement,
PFA (pulverised fuel ash) and bentonite, typically mixed in the following proportions:

    PFA   75%
    Cement 20%
    Bentonite 5%

The addition of PFA to the grout mix is done for economic considerations as it is less expensive than cement with an acceptable loss in final strength of the grout and a more economical redrilling operation.
Coloured dyes have been added to the solid portion in small quantities to trace cement paths. The ratio solid to liquid of a grout mix is determined by the results of the water pressure test before commencement of the grouting. Starting mixes can be as thin as 6:1, W:C by mass and may be thickened to 1:1 W:C by mass or even denser.

Grouting Procedure

Grouting is carried out when the following conditions occur:

– excessive water acceptance during water pressure tests. Water pressure tests are conducted at regular intervals and when a noticeable loss of drilling fluids occurs.

– total water loss condition.

– after an ungrouted interval has been drilled, typically not more than 100m.

The borehole is flushed with water prior to grouting to remove remnants of the drilling fluids, then a water pressure test is carried out. The solid to liquid ratio of the starting grout mix is determined.

The injection starts by pumping a certain number of mixes of water and bentonite followed by injection of the selected starting mix.
The consistency of the mix is gradually changed – according to the pressures obtained – in order to pump a grout mix as thick as the fissure can absorb without choking.
The unset grout mix is flushed out of the hole until hardness increases and redrilling of the solid grout starts.
On completion of the redrilling a water pressure test is carried out to assess the effectiveness

CEMENT GRAIN

HYDRATION (incomplete)

WATER CEMENT RATIO W:C = 0,20

CEMENT GRAIN

Fig. 3   *Cement : Incomplete Hydration*
Bild 3   *Zement : Unvollständige Hydratation*

Fig. 4   Surface Batching Plant
Bild 4   Zentrale Hauptmischanlage

of the grouting. This test will be undertaken up to a pressure of not more than 2,5 times the hydraulic head of the area. A more logical sealing criterium would be to grout to approximately 30% below sustaining pressure in hydrofracturing terminology.

Generalisations of grout absorption are difficult to make and often unreliable. Nevertheless, it is interesting to note that the overall absorption for the Karoo sediments of the O.F.S. gold fields was approximately 50kg cement per metre whereas the equivalent figure for the Central Group rocks (quartzite) in the same area was almost 2000kg cement per metre.

2.2.3 Recent Projects

	Depth m	Time M	Wedge No	Cement Pumped Tons
EDPC 1**	2210	–	27	278
EDPC 4**	2200	10	47	680
EDPC 3**	2000	13	35	840
EDPC 2**	2100	19	68	1785
NSR	2348	18	27	
Joel 3&4*	7231	10	39	3431
Avg/hole	1206		7	572
Joel 1&2*	8542	12	57	1298
Avg/hole	1424		10	216

** (Smit et al, 1984)

* Sum of six boreholes, one rig per hole. (Heinz, 1988).

Although drilling rates may reach up to 35m (N–size) in a 24 hour shift, the overall average drilling rate (including grouting, water testing possibly some fising operations, etc.) is closer to 5m per 24 hour shift.

3. CEMENT SLURRY CONVEYANCE

3.1 Introduction

Grouting in South African deep mines serves several important functions : (Dietz, 1982).

1. To cover–grout any excavation made into rock for development, stopes, etc., and to combat water in–rushes during these routine excavations.

1. For packgrouting for supporting systems. This method requires the conveyance of large volumes of cement–sand mortars (Relative Density = 1,9).

3. To seal fine fissures by cement and chemical grouting.

4. For precementation of shafts as described in this paper.

Actual operational procedures are as follows:

151

Fig. 5   Underground Relay Station
Bild 5   Verteilungs – und Mischanlage vor Ort.

1.   Determine conveyance parameters such as: distance and time required to transport slurry, calculate friction losses in pipe ranges, usually done empirically for the grout required; determine operation time available.

2.   After drilling has been completed, install high pressure packers, connect grouting pipe and pump until expected pressure is reached.
      The grout is prepared, mixed, etc., at a central surface batching plant (fig. 4) usually, semi–automatic and transferred to the underground relay station (Fig. 5). At the relay station mix can be further reticulated to the various operational areas. Dense mortars are conveyed in the same manner to the actual supporting grout packs in the stopes.

Materials

The primary component for high pressure fissure grouting is cement and for pack grouting (fig. 6) cement and filler, usually sand or tailings (slimes). Other pozzolanic fillers such as fly ash are also used. The large quantities required necessitate a compromise between a low cost material component and the required strength at the correct

time. Continual control of the quality of the mix at the surface plant is imperative.

Plant and Equipment

Plant and equipment used underground must be simple, robust and reliable. Plunger pumps capable of reaching pressures up to 40 MPa and higher are typically used; mixing devices at the underground relay station are slow rotating double or single drum mixers.
The surface batching system uses a better quality grout with minimum bleeding and hence better fluidity and flow characteristics.

Drilling Techniques

Similar to the grouting equipment the drilling equipment is simple and robust. In most cases air driven, screw feed machines are used capable of drilling to 200m (AX) or more.
Rigging of these machines is simple but special attention is given routinely to safety features such as anchoring of the machine, etc.

4.   HIGH PRESSURE "THIN UNSTABLE" CEMENT GROUTING

4.1   Introduction

In a somewhat simplified manner, cement

grouting philosophies can be described as high pressure grouting of thin, unstable grouts and low pressure grouting of thick, stable grouts. The former technique allows deformation of the lattice, indeed requires these, whereas in the latter technique, deformation of the formation is strictly taboo.

In the creation of deep grout barriers—the precementation of shafts—almost every
principle of low pressure grouting, accepted in the civil engineering industry in some countries, is violated (Houlsby, 1982). Unstable suspensions are used, the rock is deformed at high pressures, long stages of boreholes (e.g. 90m) are grouted with the down-stage method (packer at the top), yet these techniques have been successfully appplied, therefore, explanations of the related phenomena are required.

This section of the paper endeavours to explain some of these differences and phenomena and describes these techniques based on the high pressure grouting techniques utilised in the South African mining industry.

4.2   Basic Grouting Philosophy

Possibly the most intriguing, convincing and basic principle supporting the "thin" cement grouting technique is that if it is water that requires control (and it is mostly water) then the grouting, cementing liquid or slurry should in the dynamic phase have the properties of water as closely as can be achieved. That was the attraction of AM-9. The result is that the grouting liquid will follow the water paths and create similar pressure distributions and deformation patterns as water; any other slurry with different properties will create different behaviour patterns in rock masses.

The difference between the two grouting philosophies is best illustrated by describing two extreme cases of grouting applications.

The best and clearest example of low pressure "permeation" grouting is the rocrete or colcrete process. In this process a concrete-like material is created by grouting (filling) under low pressure large voids in a lattice or structure made from coarse rock or gravel. What are the main characteristics of this method:

1.   Only low pressures are required, indeed allowed.

Fig. 6   Pack Grouting Support System (Cement filled Packs)
Bild 6   Zementmörtelstützsäulen

2.   The lattice of the permeated structure is not deformed or rather, the deformations are negligible or hardly measureable.

3.   Although water is used as the medium of transport the water reacts at the point where the cement is "deposited". Therefore, no bleeding can be allowed.

4.   The grout is pumped into the formation for only a short distance. Penetration is limited due to the low pressure.

5.   As no bleeding can be allowed grout mixes have to be relatively thick and stable under gravity (W:C 2:1, 1:1, etc) (assuming that thin grouts usually bleed and are unstable under gravity).

6.   Pulsating flow and high velocities are not required (short distance, slow settling velocities) indeed, would deform the structure and are, therefore, not allowed.

High pressure fissure grouting in its extreme case is significantly and fundamentally different. The technique is applied successfully in Europe but in its extreme form possibly only in the mines at large depths in South Africa.

1.   High pressure is required to overcome the high water pressure underground in order to replace the water against the gravitational forces.

2.   Thick grouts choke boreholes, therefore when a next round is blasted

a few metres into rock, water can rush in again. Therefore, the water has to be "flushed back" considerable distances. Only thin grout can do that effectively. (W:C 8:1, 6:1, 4:1). Sometimes these grouting techniques are even referred to as "Water grouting".

3. Deformation of the rock, i.e. opening of the fissures is desirable, indeed necessary, in order to obtain a larger distance and a more comprehensive grouting of the fissure network. This requires high pressures.

4. The sound but fissured rock is not normally damaged by high velocities. If some clay is washed out it can only be beneficial.

5. High pressures and velocities are required to obtain long distances.

6. High velocities will prevent the settling of relatively "thin, unstable under gravity" grouts as the applied forces are higher than the gravitational forces.

7. It has been established in slurry transportation (coal slurries, etc.) that pulsating flow is more efficient in transporting particulate slurries than any other way i.e. the percentage of material that can be transported in suspension is highest for pulsating flow.

   Hence the human heart uses pulsating flow for very effecient flow into "fine fissures" with a frequency of between 60 to 80 pulses per minute, similar to the normal plunger pump frequencies as used in the mining industry.

8. With pulsating flow high shear mixing is desirable but not as critical as for low pressure grouting, as the water is used primarily for transporting the cement to where it is deposited; the finer particles of cement move further, the finest cement particles are pushed even further until finally only water and bentonite remains.

   Therefore, most of the water does not remain where the cement is deposited. The author has termed this segregation, SELECTIVE SEDIMENTATION or FORCED SEDIMENTATION. Forced sedimentation occurs at very low pressures already, as is shown by the filter cake development in piling and diaphragm wall construction and during drilling with bentonite muds.

9. The back pressure of the formation once the grouting has been completed, presses (squeezes) the water out and compresses the cement grain structure so that the cement in the fissure not only gains strength from hydration but also from compaction of the grout.

4.3   Thick or Thin, "Stable or Unstable" Grouts

It is important to realise that "stable" grouts are really grouts stable under gravitational forces. Therefore, stable means either sedimentation is so slow that it is almost negligible or thixotropic action, hydration or other reactions and possible forces prevent sedimentation.

It is helpful, indeed necessary, to distinguish between

   STATIC PHASE grouting and
   DYNAMIC PHASE grouting of particulate suspensions.

The ideal static phase of cement grouting is the measuring cylinder where sedimentation is predominantly influenced by:

   particle interference, gravity, very low particle velocity, stationary continuous phase.

Hence low pressure, low velocity grouting as described earlier is so similar to the "ideal" static phase that it can be categorised as static phase grouting; in fact, plug low or low velocity laminar flow would fall into this category.

In contrast, in the dynamic phase of cement grouting the sedimentation process is predominantly regulated by:

   High velocity resulting from high pressures, forces which change the resultant force on the particles in contrast to gravity only, different velocities between the suspended particles and suspending phase, selective sedimentation.

Both phases require control and manipulation. It is incorrect to assume as is often done that if the static phase is "stable" the dynamic phase is also "stable". Stable in the dynamic phase requires the properties of the grout to remain similar before and after moving through the rock mass.

A typical example of static/dynamic phase manipulation or control is the drilling fluid (thin, unstable) circulation system. The dynamic phase must carry the particles to surface, some losses will occur due to losses to the formation and selective sedimentation (filter cake, permeation) and the static phase must enable the drilling cuttings (particles) to settle rapidly in the settling tanks on surface before the drill fluid is pumped back into the borehole again. Further proof that "stable" grouts such as bentonite grouts are in fact unstable under pressure is clearly demonstrated by the filter cake formation in boreholes and bentonite piles and diaphragm walls. Particularly in diaphragm walls where the pressure on the slurry is very small, 200 – 400 KPa, a filter cake is formed on the sides of the wall, in fact, resulting in selective sedimentation.

The degree of sedimentation and the sedimentation velocity of particles in a fluid is dependent on

1. gravitational forces.
2. the size and shape of the particles, hydration changes the size of the particles (Fig. 7), floculation.
3. the density of the suspended particles relative to the density of the suspending liquid (continuous phase)
4. the motion of the liquid
5. the properties of the liquid such as viscosity.
6. Surface interaction, adhesion electrostatic forces, etc. between the suspended particles and between these particles and the suspending phase.

The corollory of the above statements, therefore, is that by manipulating the above-mentioned properties e.g. increasing the pressure and velocity or size of particles of a particulate grout one can achieve selective sedimentation; in essence this is the "philosophy" of high pressure grouting.

The addition of bentonite is a prime example of the abovementioned manipulation of particle characteristics. The known characteristics are as follows:

1. Bentonite particles are smaller than 5 microns, hence the particles static forces influence their behavior.
2. Bentonite particles have an extremely large surface; they have very aptly been described as cornflake shaped.

(a) CEMENT GRAIN   (b) CEMENT GRAIN AFTER ADDING WATER   (c) HYDRATION COMPLETE

Fig. 7   Cement : Hydration
Bild 7   Zement : Hydratation

3. Bentonite particles have large electrostatic imbalances.

Smaller particles, including water dipoles, will adhere electrostatically to larger bentonite particles thereby creating flocs. These low density, large sized flocs prevent cement particles from settling, hence sedimentation (bleeding) is delayed, which in turn prevents choking of boreholes and fissures. The artisan "sees" this delayed sedimentation as improved lubrication as lower pressures are required at the pumps; sedimentation will constrict the fissure and boreholes hence require more pressure for a similar ground and equipment configuration. This is experienced and can be measured as a higher cohesion of cement–bentonite grouts.

High pressures, high velocities and pulsating flow will partially destroy these flocs again, as the electrostatic forces are relatively weak. An interesting example of static and dynamic phase manipulation. The above description highlights the potential of possible conflicts of interpretation of these phenomena. In dense cement–bentonite grouts, the effect of bentonite is less pronounced as physical interference of particles already prevents sedimentation.

5. DEEP GROUT BARRIERS

Based on the present knowledge and experience in deep grout barriers, particularly deep precementation of shafts and other mining related grouting techniques more extensive deep grout barriers below 500m and even below 1000m seem technically possible under the following conditions:

1. Comprehensive site investigation including stress field determination, hydrofracturing tests to determine upper grout pressures.
2. Comprehensive fissure pattern determination including frequency, width and direction.
3. Comprehensive geohydrological investigation.
4. Determination of an approximate constitutive equation for the rock mass to be grouted based on in situ parameters and not laboratory measurements. An UCS of 300MPa of a sound rock sample measured in the laboratory is rather meaningless for grouting purposes in situ. Similarly the

limitation of the pressure often used in civil engineering based on the weight of the overburden is equally meaningless in deep barrier grouting, hence the requirement of determining the stress field and other parameters such as the sus–taining pressure to function as operational guide–lines during grouting.

5.	An acceptance of the basic principle that to impermeablise rock masses against infiltration of water, liquids similar to water in their dynamic phase should be used, "Thin, unstable" grouts seem to be a possibility. In view of the high cost of chemical grouts which exhibit these required characteristics (AM9 non toxic equivalents) micro fine cements and silica fume slurries seem to offer exciting alternatives.

6.	The economic feasibility of such projects would dictate large borehole spacings, possibly not closer than 30 to 50m. Hence penetrability is vital, therefore, again only water–like "thin" grouts can succeed, see point 5.

7.	It is important not to hydrofracture the formation as it is difficult to close by grout all the open fissures again, hence the requirement to determine the stress characteristics of the rock mass.

6.	CONCLUSION

The description of the above techniques as practiced in the mining environment in South Africa and their judicial application should enable engineers to execute deep and ultra deep grout barriers. The methods described have been applied with success in the mining industry. The most important conclusions can be summarised as follows:

1.	The acceptance of the principle that the grouting material should have the characteristics of water in the dynamic phase.

2.	Thin, unstable grouts" utilising micro–fine cement and silica fumes offer exciting possibilities to obtain water–like grout slurries.

3.	The high pressure thin grouting technique is significantly different to the low pressure thick grouting methods. The two extreme methods described in the paper highlight the differences, nevertheless, all these methods are tools to an end and wil find their application where required especially in the hands of the specialist who understands grouting in all its facets.

4.	The execution of deep grout barriers necessitates comprehensive in situ investigations of the stress and deformation characteristics of the rock masses as well as their fissure patterns.
Parameters determined in a laboratory can be rather meaningless in this context.

5.	Possibly the most significant disadvantage of high pressure, dynamic phase grouting is the difficulty to control the process.

6.	A careful study of the behaviour of particles at microscopic level can assist significantly in the understanding of grouting processes.

7.	The requirements of the various philosophies determine the complete grouting process from mixing to placing and controlling as well as equipment and plant specification. (mixers, pumps, pipe ranges, packers).

Finally, the author hopes that highlighting these different grouting techniques will improve the general understanding of cement grouting and its various aspects.

7.	REFERENCES

Baker, W.F. (ed) (1982) Grouting in Geotechnical Engineering, Proc. Conf. of ASCE, New Orleans.

Baker, W.F. (ed) (1985) Issues in Dam Grouting, Proc. of Sem. of ASCE, Denver.

Cope, R (1981) American Rock Foundation Grouting, Practice Procedures and Practices, Short Course on Fundamentals of Grouting, Univ. of Missouri, Rolla.

Dietz, .H.K.O (1982) Grouting Techniques used in deep South African Mines, Conference on Grouting in Geotechnical Engineering, ASCE, New Orleans.

Du Bois, H.L.E. (1978) High Pressure Grouting in Deep Gold Mines, SIAMOS–78, Granada.

Heinz, W.F. (1985) Diamond Drilling Handbook, South African Drilling Association, Johannesburg

Heinz, W.F. (1988) Precementation of Deep Shafts. The Third International Mine Water Congress, Melbourne, Australia.

Heinz, W.F. (1987) The Art of Grouting in Tunneling. Proceedings, South African National Committee on Tunnelling (SANCOT). Seminar on the Management of Underground Construction.

Heinz, W.F. (1983) "Tube-a-Manchette: Description and Applications". Proceedings, Grouting Symposium, JHB.

Houlsby, A.C. (1982) Cement Grouting for Dams Proceedings, Conference on Grouting in Geotechnical Engineering, ASCE, New Orleans.

Smit, N.J. and Lain M.J. (1984) Pre-cementation boreholes, Drilling News.

Weyermann, W.T. (1977) Rock Conditions improved through Pressure Grouting, published by RODIO.

– (1963) Grouts and Drilling Muds in Engineering Practice. Conf. of ISSMFE, London, Butterworths.

– (1982) Zement Taschenbuch, Bauverlag GmbH, Wiesbaden–Berlin, 48th edition.

(1984) Grouting Technology, Engineering and Design, Corps. of Engineers, Dept. of the U.S.A. Army.

157

Influence of grout pressure on capacity of boreinjected piles and anchors

Der Einfluss des Druckes auf die Tragfähigkeit von Bohrlochinjizierten Pfählen und Ankern

Igor M. Kleyner, Raymond J. Krizek & Stanley F. Pepper
Department of Civil Engineering, Northwestern University, Evanston, Ill., USA

ABSTRACT: Boreinjected piles and anchors are widely used in construction for foundation reinforcement and tieback walls. The installation of such devices includes boring a hole, filling it with cement grout, inserting reinforcing bars, and pressurizing (compressing) the grout. Experience has shown that the friction forces along the soil-grout interface and the resulting load capacity of the pile or anchor increase as the grout is compressed, but hydrofractures are undesirable. Using the similitude in the behavior of the soil surrounding a borehole when subjected to pressure from either a pressuremeter or cement grout, a mathematical model based on the diffusion equation and the Mohr-Coulomb failure criterion is developed to relate the external pressure applied to the grout, the dissipation of the resulting excess pore water pressure in the grout, the strength and permeability of the surrounding soil, and the increase in load capacity of the pile or anchor. Relationships are suggested for determining from pressuremeter data the appropriate incremental parameters for grouting the pile or anchor. Test results from different types of installations indicate load capacity increases of at least 150% and deformation reductions up to 40% for piles grouted under incremental pressure steps relative to those grouted by other techniques.

ZUSAMMENFASSUNG: Bohrlochinjektierte Pfähle und Anker sind in die Bauwerken fur die Stärkerung von Unterbauen und Ankerwänden weit benutzt. Die Installierung solcher Geräte besteht aus dem Bohren eines Loches, dem Füllen des Bohrloches mit Zementmörtel, dem Einsetzen der Stahlarmierung und dem Einpressen des Zementmörtels unter Druck. Aus Erfahrung weiss man dass die Reibungskräfte entlang der Boden-Zementmörtel Schnittstelle und die resultierende Lastungsfähigkeit des Pfähles oder Ankers grösser werden während der Zementmörtel verdruckt wird, aber "Hydrofraktur" sollte vermieden werden. Mit Hilfe der Ähnlichkeit zwischen dem Verhalten des Bodens um das Bohrloch herum unter dem Druck von Druckmessern oder von dem Druck des Einpressmörtels, wurde ein mathematisches Modell entwickelt. Dieses Modell benutzt die Diffusionsgleichung und das Mohr-Coulomb Fehlerkriterium und die Beziehungen zwischen den folgenden Variablen zu bestimmen: der äussere Druck am Einpressmörtel; die Verteilung des Porenwasserdruckes im Vergussmörtel; die Stärke und Durchlässigkeit des Bodens in der Umgebung des Bohrloches; die Zunehme der Belastungsfähigkeit der Pfähle oder Anker. Beziehungen werden vorgeschlagen sein um die verwendbahren zunehmenden Parameter fur eingepresste Pfähle oder Anker auf Grund von Druckmessungen zu bestimmen. Ergebnisse mit verschiedenen Typen von Einrichtungen weisen auf eine Zunahme der Belastungsfähigkeit von mindestens 150% und eine Reduktion der Deformationen von ungefähr 40% für mit Mörtel verpresste Pfähle unter zunehmendem Druck im Vergleich zu anderen Methoden mit eingepresste Pfählen hin.

1 INTRODUCTION

Boreinjected piles and anchors are widely used in construction projects to reinforce foundations and support tieback walls. Such installations are characterized by large length-to-diameter values (often 100 or more) and small diameters (on the order of 100 to 200 mm). They are constructed by drilling a hole, filling it with cement grout (usually Portland cement and water at a water-to-cement ratio of about 0.5 to 0.7), inserting reinforcing bars, and pressurizing (compressing) the grout. Data by Hobst and Zajic (1977), Lapshin (1979), Egorov et al (1982), Mishakov (1984), and others have shown that unit friction forces along the soil-grout interface increase from about 100 to 150 kPa when the grout is simply placed in the borehole and not compressed to about 300 to 400 kPa when the grout is compressed. Of course, compressing the grout increases the volume of grout needed for a given installation and extends the time required to complete the job (Smorodinov, 1983).

Various schemes are used to construct boreinjected piles or anchors. In one common type of installation, termed the "fixed collar" method, a borehole about 150 mm in diameter is drilled about 2 m deep and filled with cement grout; then, a metal casing is inserted in the grout-filled hole and left for a few days to harden, after which the grout is drilled from inside the casing (thus leaving a "fixed collar" near the surface) and the hole is advanced with the help of drilling mud to its desired depth. Next, reinforcing bars are inserted in the borehole (if desired) and the cement grout is placed, starting at the bottom of the borehole and forcing out the drilling mud. Finally, a cover is attached to the top of the "fixed collar" and pressure is applied to the grout. In another type of installation, termed the "removable casing" method, a borehole is drilled, cased, and fitted with reinforcing bars (if desired); then, the grout is pumped into the hole under pressure as the casing is withdrawn.

2 ANALYSIS

The methodology used to analyze the installation of boreinjected piles and anchors is adapted from the physical process of consolidation in saturated soils, where the grout is the consolidating medium from which water is squeezed and the surrounding soil serves as an impeded drainage boundary. In their study of the behavior of cement grout subjected to pressure, Albakidthe (1971), Ahverdov (1981), and others have determined that, for grouts with a water-to-cement ratio in excess of 0.45 or 0.50, the applied pressure is, for all practical purposes, transmitted entirely through the pore water. As the pore water is squeezed from the grout, the pressure is transferred to the cement particles and the grout becomes a viscoplastic sol at a water-to-cement ratio ranging from 0.2 to 0.4. The degree of grout consolidation (densification) depends on the magnitude and duration of the applied pressure.

From the solution to the Lamé problem, the effective radial stress, $\overline{\sigma}_r$, and effective tangential stress, $\overline{\sigma}_t$, in the soil surrounding a borehole of radius r_0 filled with a cement grout subjected to a pressure, p, can be written as (Figure 1):

$$\overline{\sigma}_r = p\left[\frac{r_0^2}{r^2}\right] - u \quad \text{and} \quad \overline{\sigma}_t = -\left[p\frac{r_0^2}{r^2}\right] - u \quad (1)$$

where u is the excess pore water pressure. Immediately after applying the pressure, p, to the grout in the borehole, the excess pore water pressure, u, in the soil near the borehole wall ($r \approx r_0$) may be approximated by p and Equations (1) may be expressed as:

$$\overline{\sigma}_r = p - p = 0 \quad \text{and} \quad \overline{\sigma}_t = -p - p = -2p \quad (2)$$

After the excess pore water pressure in the grout has dissipated, these stresses become:

$$\overline{\sigma}_r = p \quad \text{and} \quad \overline{\sigma}_t = -p \quad (3)$$

The stresses given by Equations (3) are analogous to those which exist during a pressuremeter test. From the foregoing equations it can be seen that

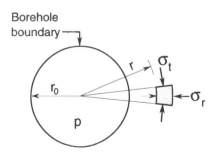

Figure 1. Notation for element in surrounding soil

the maximum tensile stresses occur in the soil immediately after the grout has been pressurized, and, if the pressure in the grout is sufficiently high, cracks will form in the surrounding soil and the "liquid" grout will penetrate into the resulting hydrofractures. However, the probability of developing hydrofractures decreases as consolidation of the grout takes place due to (a) reduced tensile stresses in the soil and (b) increased viscosity of the grout due to hydration.

2.1 Soil Strains

To compare the strains in the soil surrounding a borehole filled with pressurized grout with those caused by a pressuremeter, a series of laboratory experiments were conducted in a cylindrical tank (550 mm in diameter) filled with fine sand by using models (35 mm in diameter) of the pressuremeter probe and a borehole filled with cement grout. After the test in each case, the sand was removed layer by layer in four layers and the diameter of the deformed hole was measured in four directions on each of the four levels. A statistical evaluation of the sixteen measurements for each test showed no meaningful difference between the two types of test. To investigate volume changes, the tank was filled with concentric circular regions of colored sand and similar experiments with models showed that (a) the strain was less than 0.005 at a distance of more than five borehole radii from the borehole wall and (b) the results were similar for both types of test.

2.2 Hydrofractures

Construction experience has shown that hydrofractures cause an additional expenditure of grout without a corresponding increase in the bearing capacity of the piles or anchors. Moreover, hydrofractures (especially in the horizontal plane) may be very dangerous. To establish a safe limit for the pressure applied to the grout, the Mohr-Coulomb failure criterion was used to establish a relationship between the critical pressure, p_{cr}, to cause a hydrofracture and the proportional limit, p_{pl}, measured in a pressuremeter test (see Figure 2). For axisymmetric conditions, the Mohr-Coulomb failure criterion can be written as:

$$\sigma_t + c \cdot \cot \phi = K_a(\sigma_r + c \cdot \cot \phi) \qquad (4)$$

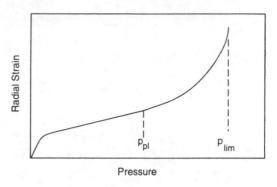

Figure 2. Schematic response from pressuremeter test

where c and ϕ are the cohesion and angle of internal friction of the soil, respectively, and $K_a = \tan^2 [(\pi/4) - (\phi/2)]$ is the coefficient of active earth pressure. If the at-rest horizontal earth pressure in the soil mass is designated as σ_h, Equations (2) and (3), respectively, become:

$$\bar{\sigma}_r = \sigma_h \quad \text{and} \quad \bar{\sigma}_t = \sigma_h - 2p \qquad (5)$$

and

$$\bar{\sigma}_r = \sigma_h + p \quad \text{and} \quad \bar{\sigma}_t = \sigma_h - p. \qquad (6)$$

Substituting Equations (5) into Equation (4) gives:

$$p_{cr} = \frac{1 - K_a}{2} [\sigma_h + c \cdot \cot \phi] \qquad (7)$$

Similarly, the substitution of Equations (6) into Equation (4) will give the proportional limit pressure in the pressuremeter test as:

$$p_{pl} = \frac{1 - K_a}{1 + K_a} [\sigma_h + c \cdot \cot \phi] \qquad (8)$$

Finally, combining Equations (7) and (8) and the indicated expression for $K_a = f(\phi)$ gives:

$$p_{cr} = \frac{1}{2} \left[1 + \tan^2 \left(\frac{\pi}{2} - \frac{\phi}{2} \right) \right] p_{pl} \qquad (9)$$

where the coefficient of proportionality between p_{cr} and p_{pl} is 1.00, 0.85, 0.75, 0.67, and 0.61 for ϕ values of 0, 10°, 20°, 30°, and 40°, respectively. The applicability of Equation (9) has been

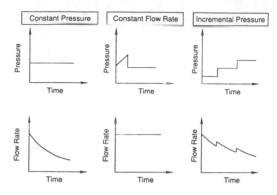

Figure 3. Typical time-dependent relationships for different grouting procedures

Table 1. Limit (maximum) pressure for pressuremeter test

Soil Type	Liquidity Index	p_{lim} (MPa) for a void ratio of		
		0.45	0.70	0.90
Clayey Soils	Negative	1.2 - 1.6	1.0 - 1.2	0.6 - 0.9
	0 to 0.25	0.9 - 1.2	0.6 - 1.2	0.4 - 0.7
	0.25 to 0.75	0.5 - 0.8	0.3 - 0.6	0.2 - 0.4
Sands		1.3 - 1.7	0.7 - 1.0	-

$$\text{Liquidity Index} = \frac{\text{Water Content} - \text{Plastic Limit}}{\text{Liquid Limit} - \text{Plastic Limit}}$$

confirmed by field investigations. Hence, there is a similitude between the soil response caused by a pressuremeter and pressurized cement grout provided (a) water can be squeezed from the grout into the surrounding soil and (b) the pressure does not cause fractures in the soil.

For boreinjected piles and anchors, the bearing capacity is determined primarily by the stresses in the soil near the borehole and the concrete strength. Accordingly, it is desirable to pressurize the grout as much as possible to compress the adjacent soil and densify the grout. In practice the grout is usually placed under constant pressure (Figure 3a) or at a constant rate of flow (Figure 3b). Under these conditions the final water-to-cement ratio of the grout is normally between 0.3 and 0.4. Lower values cannot be obtained because the duration of grouting is not coordinated with the duration of the grout consolidation and the infiltration of its pore water into the soil. In a very fluid grout the pressure is essentially transmitted completely to the pore water. Hence, the maximum tensile stress acts on the soil and hydrofractures form when $p > p_{cr}$ (about 200 to 300 kPa in most practical cases). Grouting with a constant rate of grout flow (which depends on the pump capacity) causes rapid increases in pressure and induces hydrofractures.

Experimental data indicate that, in the majority of soils, the value of p_{cr} is 30% to 50% of the limit (maximum) pressure, p_{lim}, in the pressuremeter test. Therefore, to achieve p_{lim} and avert hydrofractures, the grout pressure may be increased only as fast as the tensile stresses are reduced. Because the latter must be accompanied by the dissipation of pore water pressures, the increase in pressure is governed by the consolidation rate of the grout. Therefore, it follows that an incremental gradual (step-by-step) increase in pressure (Figure 3c) is most appropriate, where the pressure increment, Δp, is taken less than the critical value (i.e., $\Delta p < p_{cr}$). Recommended pressure increments are 0.10 to 0.15 MPa in loose sands and soft clayey soils and 0.3 to 0.7 MPa in dense sands and firm to hard clayey soils. Following this procedure the total pressure, $p = \Sigma \Delta p$, can equal the limit pressure in a pressuremeter test (see Figure 2). Approximate values for p_{lim}, as obtained from the literature and more than 200 tests, are given in Table 1.

2.3 Consolidation

The duration of each pressure increment is dictated by the consolidation rate of the grout in the borehole, which is, in turn, controlled largely by either the permeability of the grout or the permeability of the adjoining soil. The equation governing the consolidation of the grout can be written as:

$$\frac{\partial u}{\partial t} = c_v \left[\frac{\partial^2 u}{\partial r^2} + \frac{1}{r} \frac{\partial u}{\partial r} \right] \quad (10)$$

where r ($0 > r > r_0$) is the radius and c_v is the coefficient of consolidation for the cement grout (which, for portland cement, is about $0.5 \times 10^{-4} m^2/s$). The initial and boundary conditions around the periphery of the borehole are $u(r,0) = p_0$ and $u(r_0,t) = 0$, which, when incorporated into the general solution of Equation (10), yields:

$$u(r,t) = 2p_0 \sum_{m=1}^{\infty} \frac{e^{-c(\mu_m/r_o)^2 t} J_0(\mu_m r/r_0)}{J_1(\mu_m)\mu_m} \quad (11)$$

where J_0 and J_1 are Bessel functions and μ_m are roots of $J_0(\mu_m) = 0$. Considering only the first term of the series and taking $u(0,t) = 0.01\, p_0$, the time for 99% consolidation of the grout, t_g, can be written as:

$$t_g = 0.89\, r_0^2/c_v \quad (12)$$

For boreholes with diameters of 150 mm, 200 mm, and 250 mm, the grout consolidation times are 1.7 min, 3.0 min, and 4.3 min, respectively.

The time, t_s, for the excess water in the consolidating grout to infiltrate into the soil can be determined from the following modified Dupuit equation:

$$t_s = \frac{V_w \gamma_w \ln(R/r)}{2\pi u k_s L} \quad (13)$$

where V_w and γ_w are the volume and weight density of the infiltrating water, L is the grouted length of the borehole, k_s is the permeability of the soil, and R is the radius of influence of the borehole (typical values range from about 1 to 3 meters as the soil permeability increases from 10^{-7} to 10^{-4} cm/sec). The volume of water, V_w, squeezed from the grout into the surrounding soil is given by:

$$V_w = \frac{W_c}{\gamma_w}\left[(w{:}c)_0 - (w{:}c)_f\right] \quad (14)$$

where W_c is the weight of dry cement in the grout and $(w{:}c)_0$ and $(w{:}c)_f$ are the average initial and final water-to-cement ratios of the grout in the borehole. Typical calculations using Equations (12) and (13) show that t_g is many times larger than t_s for sands, while the opposite is true for clays; for intermediate soils the values are of the same order of magnitude. Hence, in sands the duration of the pressure increment is determined by the consolidation of the grout, whereas the soil permeability controls this time in the case of clays. When the permeability of the soil is about 10^{-6} cm/sec, the time for complete consolidation is approximately equal to the set time of the cement

grout and additives must therefore be used to extend the set time. If the incremental pressure procedure described above is used, the total time to complete the installation will be the sum of the times required to achieve consolidation for each pressure increment.

3 EXPERIMENTAL PROGRAM

Two series of tests were performed to document the effect of the grouting regime on the deformation and bearing capacity of a pile or anchor. In the first series twenty-five model piles were installed by one of four different techniques in either a clayey soil or a sandy soil and then loaded in tension. In the second series twelve field-size piles were installed (eight by the constant pressure method and four by the incremental pressure method) and loaded in either compression or tension.

3.1 Model Tests

Model tests were conducted on boreinjected piles in a cohesive moraine soil (void ratio = 0.53; plasticity index = 7; water content ≈ plastic limit) and in a sand (void ratio = 0.80; degree of saturation = 13%). The test piles were 35 mm in diameter and either 1.2 or 2.0 m long. Four different grouting techniques were employed; these were (a) placing the grout with no applied pressure, (b) grouting under a constant pressure of 0.2 to 0.3 MPa for 2 to 3 minutes, (c) grouting at a constant flow rate of 0.7 liters/min, and (d) grouting with incremental pressure according to the following system:

Clayey soil:-
$\Delta p = 0.25$ Mpa with $\Delta t = 10$ min to $p = 0.75$ MPa

Table 2. Average grout volume (liters) in borehole for each grouting procedure

Soil Type	Grouting Procedure			
	No Applied Pressure	Constant Pressure	Constant Flow Rate	Incremental Pressure
Clayey	2.6 (5)	3.1 (3)	5.0 (2)	4.7 (6)
Sandy	1.6 (4)	2.5 (2)	-	2.8 (3)

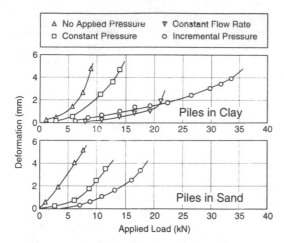

Figure 4. Load-deformation response for model piles in tension

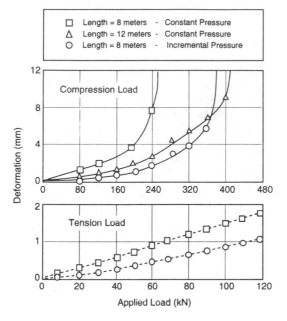

Figure 5. Load-deformation response for full-scale piles in tension and compression

Sandy soil:-
Δp = 0.20 MPa with Δt = 2 min to p = 0.60 MPa

For each grouting procedure, Table 2 gives the average volume of grout placed in the borehole (the number of tests conducted is shown in parentheses) and Figure 4 shows the average load-deformation response in tension for the piles in either the clayey soil or the sandy soil. These data indicate that, for a given deformation (say 4mm), the average load taken by the piles installed by the incremental pressure method in the clayey soil is 1.5 to 4 times that on any of the other piles; for the piles in sand this factor ranges from 1.5 to 3. Unconfined compression tests on the hardened cement grout placed by the incremental pressure technique gave values 1.2 to 1.4 times higher than those of hardened grout placed under no applied pressure.

3.2 Field Tests

The influence of the grouting regime was investigated by testing in either compression or tension a series of piles 132 mm in diameter and either 8 or 12 meters long installed in a sandy loam. Steel bars with a diameter of 36 mm were placed in each of the boreholes, and the "fixed collar" method was used to place the grout. Eight of the piles were grouted in about 2 to 3 minutes under a constant pressure of 0.2 to 0.3 MPa, and four were grouted by the incremental pressure technique in increments of 0.20 to 0.25 MPa to a final pressure of 0.7 to 0.8 MPa, with each pressure increment being maintained for 7 minutes. The typical load-deformation curves given in Figure 5 show that, under a compressive load, the 8-meter pile installed by the incremental pressure technique has a bearing capacity about 50% higher than the one installed at constant pressure and only slightly less than the 12-meter pile installed at constant pressure. Under tension, a pile installed by the incremental pressure technique manifests about 60% of the deformation experienced under comparable load by a pile installed at constant pressure.

4 CONCLUSIONS

Theoretical and experimental investigations have established a similitude in the mechanical behavior of the soil surrounding a borehole when acted upon by either a pressuremeter probe or the pressure exerted by a cement grout. The required conditions for this similitude are that (a) the soil provides for drainage of the water squeezed from the grout during the process of consolidation and (b) the grout pressure does not cause fractures in the soil. Based on this similitude, the incremental pressure technique for grouting boreinjected piles and anchors was developed by taking into account the strength and permeability of the soil. The pressure increment in each case must be less than the critical pressure to cause hydrofractures in the soil; the maximum total pressure is equal to the limit pressure in the pressuremeter test; and the time duration for each pressure increment is determined by either the consolidation properties of the grout or the permeability of the soil surrounding the borehole. Relationships have been established for determining an appropriate pressure increment from pressuremeter data. The results of static load tests on different types of boreinjected piles indicate that the load capacity of piles grouted by the incremental pressure technique is at least 1.5 times greater than for those grouted by other techniques and the deformation under a given load can be reduced by up to 40%.

REFERENCES

Ahverdov, I.N. 1981. *Foundations of concrete physics*, Stoiizdat, Moscow.

Albakidthe, M.G. 1971. *Compressing and vibrocompressing of cement grout and concrete*, Book entitled Izvestia NISIEI, Moscow, p. 21.

Egorov, A.I., Lvovich, L.B., and Mirochnik, N.S. 1982. *Practice of design and construction of foundations of cast-in-place injected piles*, Journal of Bases, Foundations and Soil Mechanics, Moscow, Vol. 6, pp. 18-21.

Hobst, L., and Zajic, S. 1977. *Anchoring in rock*, Elsevier Scientific Publishing Company, Amsterdam.

Lapshin, F.K. 1979. *Piles bearing capacity calculations*, Izdatelstvo Saratovskogo Universiteta, Saratov.

Mishakov, V.A. 1984. *Determination of boreinjected pile bearing capacity*, Book entitled Rational Use of Manpower, Material, and Energy Resources in Construction of Transportation Facilities, Moscow, pp. 23-28.

Smorodinov, M.I. 1983. *Anchor arrangement in construction*, Izdatelstvo Stroiizdat, Moscow.

Aufsprenginjektionen im Schluff – Neue Anwendung im Wiener U-Bahn-Bau

Soil-fracturing in silt and clay – New applications in the Vienna Underground

Lothar Martak
Magistratur der Stadt Wien MA 29 – Fachbereich Grundbau, Österreich

Helmut Liebsch
Wiener Stadtwerke - Verkehrsbetriebe, Gruppe MD BD-U-Bahn-Bau, Österreich

ZUSAMMENFASSUNG: Aufsprenginjektionen in veränderlich festen Gesteinen, zu denen auch die stark überkonsolidierten Ton-Schluffe des Wr.Tertiärs zählen, sind so alt wie die Injektionstechnik selbst. Waren sie in der Vergangenheit durch die daraus entstehenden plötzlichen und weitgehend unkontrollierten Hebungen in der Regel störend, so werden sie nun gezielt im städtischen Tunnelbau eingesetzt um Setzungen zu reduzieren. Am Bauabschnitt U3/14, Schweglerstraße, der Wr. U-Bahn wurden sie als Bauhilfsmaßnahme zum Tunnelvortrieb nach der Neuen Österr. Tunnelbaumethode angewandt und halfen mit, die Setzungen an der darüber befindlichen Bebauung im Bereich weniger Millimeter zu halten. Es wurden stabile Zementmischungen verwendet, die z.T. auch aus den Tunnelvortrieben heraus mit geringen Drücken verpreßt wurden. Die Problematik der Aufsprenginjektionen, ihre Zielsicherheit, ihre zeitlich begrenzte Wirkung im tonigen Schluff, sowie ihr Einfluß auf die Steifigkeitsverhältnisse des veränderlich festen Gesteins werden im Rahmen des Tragsicherheitskonzeptes der NÖT dargestellt.

ABSTRACT: Soil-fracturing in stiff rock and hard soil is as old as grouting itself. Especially in overconsolidated tertiary clay the suddenly and widely uncontrolled occuring uplift created by the fractures is a troublesome problem of the grouting technique. This principle handicap is now made use as a tool against settlements coming from tunnelling particularly in urban area. At the working lot U3/14 "Schweglerstraße" of the Vienna Underground soil-fracturing was employed as soil improvement for tunnelling by the New Austrian Tunnelling Method (NATM). It helped to minimize the settlements of the lodging area above the station tunnels and the running tunnels to an account of a few millimetres. Stable cement-water compounds were grouted under low pressure conditions from the partially excavated tunnel surface. The grouted fractures located outside of the tunnel benches prestress the overconsolidated clay. They anticipate the stress-strain distribution of the later on tunnelling excavation and are a part of the stability concept of the NATM. The previous heaves by soil-fracturing, their accuracy in magnitude and location and the relaxation of the saturated clay are discussed. Soil-fracturing in the pressure zones of a tunnel appears as a very economic mean of regulating the deformations according to the principles of NATM.

1. EINLEITUNG

Im innerstädtischen Bereich ist die Begrenzung der Setzung an der Oberfläche bei der Herstellung von Tunnel für die Schadensvermeidung an der Bebauung von größter Bedeutung. In einem konkreten Fall bei der U-Bahnstation Schweglerstraße der Linie U3 war die Aufgabenstellung bei der Herstellung der Stationsröhren, die mit einem Ausbruchsquerschnitt von 65 m2 und gleich großer Querschläge dergestalt, daß sie Anlaß zu einer Weiterentwicklung der

Aufsprenginjektionen (soil-fracturing) unter Ausnützung der Gesichtspunkte der Neuen Österr. Tunnelbaumethode (NATM) gab.

Die grundlegende Definition der NATM lautet: "Der Tunnelbau nach der Neuen Österr. Tunnelbaumethode ist dadurch charakterisiert, daß der bei der Herstellung des Hohlraumes sich einstellende Kraftumlagerungsprozeß über die gesamte Funktionsdauer durch dissipative Regelkreise so gesteuert wird, daß er in allen Phasen bis zum ruhenden Gleichgewicht

BEEINFLUSSUNG DER	SETZUNGSURSACHEN	
ÄNDERUNG DER	BODENEIGENSCHAFTEN	BODENVERFORMUNGEN
BEEINFLUSSUNG DURCH	o VERFESTIGUNGSINJEKTION o HDBV o ENTWÄSSERUNG o USW	o AUFSPRENGINJEKTION o (DRUCKLUFT)
ERGEBNIS	BESONDERE BODENEIGENSCHAFTEN	VORVERFORMUNG DES BODENS

das umgebende Gebirge selbsttragend
gestaltet, oder im Verein mit dem Tunnel-
ausbau zum Mittragen heranzieht und dadurch
die Standsicherheit des entstehenden Hohlraumes gewährleistet."

Die gezielte Verbesserung des Kraftumla-
gerungsprozesses durch Aufsprenginjektionen
brachte eine interessante Neuentwicklung
über die im folgenden berichtet werden
soll. Diese wurde als Alternative zur
Setzungsminderung mittels Druckluft oder
anderer Bodenverbesserungsmaßnahmen eingesetzt.

2. GEOLOGISCHE UND HYDROTECHNISCHE
VERHÄLTNISSE

Die Stationsröhren liegen in Wiener
tertiären Sedimenten, die vorwiegend aus
tonigen, z.T. sandigen Schluffen bestehen,
in die untergeordnet Sandschichten,
abschnittsweise auch Kiesschichten eingelagert sind. Die Quartärschichten darüber
sind hauptsächlich als schluffige sandige
Sedimentablagerung ausgebildet und reichen
in den oft eng begrenzten, tief einge-
schnittenen Rinnen der Tertiäroberfläche
bis ins Tunnelprofil hinein. Die Lockergesteine sind zufolge Scher- und Kluftflächen intensiv zerlegt und tiefgründig
verwittert. Die weitverzweigten
Grundwasserkommunikationen im Quartär,
aber auch in den jungtertiären Fein-
sandschichten bedingen sehr lange Vorlauf-
zeiten bei Grundwasserabsenkung und
Entspannung. Die Tunnelsohle liegt 10-21 m
unter Grundwasser.

3. MINIMIERUNG DER SETZUNGEN ALS
BODENMECHANISCHE AUFGABE

Bei der Herstellung von Hohlräumen im Boden
treten Beanspruchungsumlagerungen auf,

welche an der Oberfläche Setzungen verursachen. Die Größe dieser Setzungen ist von
der gewählten Baumethode, der Arbeitsweise
und den Festigkeitseigenschaften des Baugrundes abhängig. Bei der NATM wird der
Boden dergestalt zum Mittragen herangezogen, daß während der freien Standzeit
der Tunnelortsbrust, die eintretenden
Umlagerungskräfte vollständig vom Boden
und danach ein Teil derselben von der
Tunnelauskleidung aufgenommen werden.
Der Umlagerungsprozeß und seine Beeinflussung sind für das Ausmaß und die
Verteilung der Setzungen an der Oberfläche die maßgeblichen Faktoren.

Um die Setzungen und ihre Ausbreitung z.B.
im dicht verbauten Stadtgebiet klein zu
halten wurden bisher die Festigkeitseigenschaften des Bodens im tunnelnahen
Bereich durch Bodeninjektionen, Pfahlwände,
HDBV-Schirme, oder auch durch Entwässerun-
gen global beeinflußt. Eine gezielte
Vorwegnahme der späteren Beanspruchungs-
verhältnisse im Boden vor der Herstellung
des Hohlraumes erfolgte jedoch nicht.
(Tabelle 1)

Auch mit dem bislang üblichen Einsatz des
Soilfracturingverfahrens im Tunnelbau bei
dem eine Aufsprenginjektion zwischen dem zu
schützenden Objekt und dem Hohlraum angeordnet wurde, erfolgte die Setzungs-
reduktion durch die zugehörige Anhebung
nach den eingetretenen Setzungen. Der
nunmehr vorauseilenden Beanspruchumlage-
rung liegt der Gedanke zugrunde, bereits
jene Spannungsumlagerung zu erzwingen,
welche sonst erst nach Herstellung des
Hohlraumes entstehen. Durch diese vorweg
bewirkten Umlagerungen treten beim Aushub
des Hohlraumes nur noch geringe Verformungen auf. Die Beeinflussung der Umlage-
rung erfolgt an den Orten der zukünftigen
Druckspannungskonzentration, nämlich im
Regelfall in den Ulmenbereichen und zwar

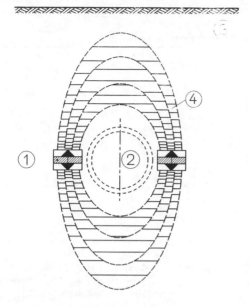

Abbildung 1   Wirkungsweise der Aufspreng-
               injektion
               1. Injektionszone
               2. Künftiger Tunnel
               3. Oberfläche
               4. Druckgewölbe

Figure 1   Efficiency of soil-fractu-
               ring
               1. Soil-fracturing zone
               2. coming tunnel
               3. street level
               4. arch of soil pressure

dadurch, daß Volumensvergrößerungen durch
Feststoffeinbringungen vorgenommen werden
(Abbildung 1)

Zweierlei wird hiedurch bewirkt:

* Entkoppelung der Untergrundbeanspru-
  chungen im Sinne der NATM dadurch, daß
  ein Druckgewölbe dem Vortrieb voreilend
  hergestellt wird.

* Druckspannungsentlastung des künftigen
  Hohlraumes, sodaß eine raschere Vor-
  triebsgeschwindigkeit für den Tunnel
  gefahren werden kann, da die Beanspruchung der frisch hergestellten
  Spritzbetonschale geringer wird.
  Auch bei den Stützmitteln vor Ort,
  z.B. bei den Dielen, sinken aus dem-
  selben Grund die Aufwendungen.

Betrachtet man den Energieaufwand für die
Gesamtherstellung so ist zu bemerken, daß
das erforderliche Injektionsvolumen für die

Herstellung der Beanspruchungumlagerung
geringer ist, als eine Abdeckplatte oder
eine großflächige Injektion oberhalb des
Tunnels.

4. SOIL-FRACTURING IM LOCKERGESTEIN

Aufsprengungen bei Injektionen sind so alt
wie die Injektionstechnik selbst, sind aber
bei Injektionenarbeiten, die die Penetration des Bodens zum Ziel haben unerwünscht.
Aufsprenginjektionen ermöglichen die Einbringung großer Feststoffmengen in vergleichbar kurzer Zeit und bewirken spontane
Hebungen. Diesen Effekt kann man für die
planmäßige Verteilung und Größe von Boden-
hebungen im Tunnelbau unter sensibler
Bebauung nutzen. Im Sinne der Abb. 2 lassen
sich Überlegungen anstellen, wo in
Abhängigkeit vom Aufbau des Untergrundes
soil-fracturing wirksam eingesetzt werden
soll und von wo die erforderlichen
Injektionsbohrungen hergestellt werden
können oder müssen.
Besonders Großprofile, die die vortriebs-
technische Unterteilungen erfordern,
bieten die Möglichkeit Überlegungen
anzustellen, nur einen Teil der
Injektionsbohrungen von der Geländeoberfläche aus herzustellen um Bohrmeter zu
sparen. Auch ist es nicht erforderlich,
einen Hebungsberg sofort für das Gesamt-
profil vorweg zu schaffen, was die darüber
befindliche Bebauung unnötig stark beansprucht. Es genügt in der Regel Hebungen
auf Teilflächen zu realisieren und aufgrund
der Steifemoduli der Bodenschichten die
notwendigen Injektionsgutmengen für die
Vorhebung abzuschätzen. Die Reaktionen des
Untergrundes auf die ersten Aufsprenginjektionen lassen eine Eichung der
Injektionstechnik und der Injektionsmischung zu. Im Zuge der Profilvergröße-
rungen kann dann die Auswahl für die
weiteren soil-fracturing Maßnahmen
erfolgen. Die in der Abbildung strichliert
begrenzt eingetragenen Hebungssektoren,
die etwa nach der Ausbreitung des passiven
Bodenwiderstandes geformt sein können,
lassen die für die erwartete Breite der
Setzungsmulde notwendigen soil-fracturing
Körper planen.

Über soil-fracturing Maßnahmen gibt es in
unterschiedlichen Lockergesteinen einge-
hende Erfahrungen. Grundsätzlich muß der
Hebungserfolg nach bindigen und nichtbindigen Böden unterschieden werden.
Wie die Abbildung 3 schematisch zeigt
reagieren kohäsive, wassergesättigte
Böden auf Aufsprenginjektionen schon
bei kleinsten Injektionsgutmengen sehr
spontan, was bei Setzungskorrekturen

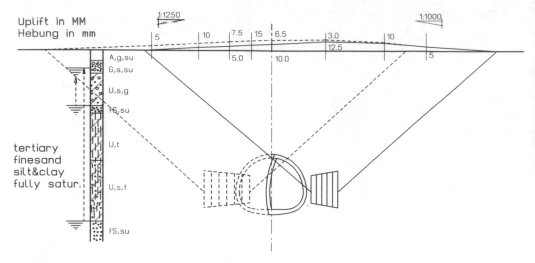

Uplift In MM
Hebung in mm

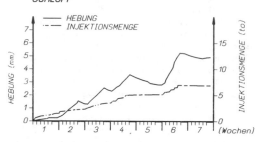

tertiary
finesand
silt&clay
fully satur.

**Abbildung 2   Geplante Hebungen aus Auf-
sprenginjektionen seitlich
der Tunnelröhre**

Figure 2   Designed uplift by soil-
fracturing acting sidewise
the tunnel bench

HEBUNGSVERLAUF IN WASSERGESÄTTIGTEM SCHLUFF

HEBUNG
INJEKTIONSMENGE

**Abbildung 3   Charakteristik der Auf-
injektionen in unter-
schiedlichen Böden**

HEBUNGSVERLAUF IN ENTWÄSSERTEM ODER NICHTBINDIGEM BODEN

HEBUNG
INJEKTIONSMENGE

Figure 3   Nature of soil-fracturing
in different soils

sehr erwünscht ist. Bedingt durch den
Abbau des Porenwasserdruckes, der vom
Sandgehalt des Schluffes oder Tones sehr
abhängig ist, gehen die anfänglichen
Hebungen wieder etwas zurück. In den
nichtbindigen Böden ist für eine
merkliche Hebung des Untergrundes bei
gleicher Feststoffeinbringungs-
geschwindigkeit wesentlich mehr Zeit
erforderlich. Allerdings bleibt die
einmal eingetretene Hebung, abgesehen
von dem Volumensverlust der nicht gänzlich stabilen Injektionsmischung vollständig erhalten. Diese charakteristischen Bilder, entnommen aus einer fach-
lichen Veröffentlichung von Raabe und
Esters (1986) zeigen aber auch die
Anwendungsgrenzen des soil-fracturing
für eine gezielte Vorwegnahme von

Druckbeanspruchungen rund um einen
geplanten Tunnelquerschnitt.

Bevor mit einem soil-fracturing Projekt
begonnen wird, sollte man sich die Frage
nach möglichen unerwünschten Begleiterscheinungen stellen. Die dabei denkbaren
Scenarien sind sowohl für die Aufspreng-
technik des soil-fracturing als auch für
den späteren Tunnelbau konzeptionell zu
untersuchen.

a) Die Hebungen treten nicht dort ein,
so sie gewünscht werden.

Die Abbildung 4 zeigt die Möglichkeit
auf, daß die Hebungen durch im Untergrund vorgegebene Zonen großer Durchlässigkeit abseits des geplanten

170

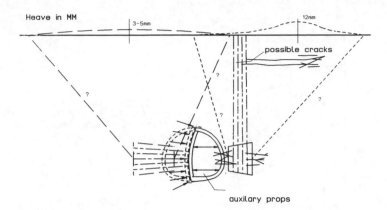

Heave in MM   3-5mm   12mm

possible cracks

?   ?   ?   ?

auxilary props

Abbildung 4   Unerwünschter Hebungsver-
              lauf durch mögliche Un-
              gleichmäßigkeiten im
              Untergrund und durch
              Nachgiebigkeit des bereits
              bestehenden Tunnelteilaus-
              bruches

Figure 4   Unsatisfying uplift caused
              by heterogenities in the
              soil and by yielding of
              the partially excavated
              tunnel

Hebungszentrums entstehen und bei
gleichbleibender Verpressgeschwindig-
keit wesentlich größer ausfallen als in
anderen, scheinbar gleichartigen Unter-
grundbereichen. Der stets mit Heteroge-
nitäten behaftete Untergrundaufbau und
die in Wien besonders häufigen tektonisch bedingten Störungszonen können
trotz gewissenhafter, kleinräumiger
Untergrundaufschließung für solche
Überraschungen sorgen.

Soil-fracturing Arbeiten aus flach
geneigten, relativ kurzen Bohrungen können
auch bei gut ausgebauten Manschettenrohren
mit Zementummantelung zu bevorzugten,
bohrungsparallelen Auspaltungsflächen
führen, die statt zu Hebungen des
gesamten überlagernden Bodens zu druckbeaufschlagten Flüssigkeitsfilmen auf der
zur Bohrung genutzten Tunnelfläche führen.
Verformungen der Tunnelschale mit Injek-
tionsmischungsaustritten können die Folge
sein. Auch Tunnelaussteifungen unter
beengten Platzverhältnissen müssen in
Erwägung gezogen werden.

b) Die durch soil-fracturing bewirkten
   Hebungen verringern sich mit der Zeit.

Überkonsolidierte sandige, tonige Schluffe,
die nahezu vollständig wassergesättigt
sind, reagieren auf gesteigerte Druckbean-
spruchung mit Porenwasserüberdrücken.
Die dabei eingetretenen Volumsvergröße-
rungen und Oberflächenhebungen reduzieren
sich im Ausmaß eines nachfolgenden Poren-
wasserdruckabbaues. Läuft der Tunnelbau mit

Setzungen aus dem Vortrieb parallel zum
soil-fracturing, sind weitere Aufsprenginjektionen nötig, um eine bestimmte
Oberflächen- oder Fundamenthöhe einer
darüber befindlichen Bebauung einzuhalten.
Eine solche Situation ist in Abbildung 5
skizziert.

c) Mögliche Festigkeitsverminderung des
   aufgesprengten Bodens.

Zu den Volumenverlusten durch Austreibung
des Porenwassers, die verglichen mit der
Tunnelvortriebsgeschwindigkeit eher
langsam vor sich gehen, kommt noch ein
weiterer setzungswirksamer Effekt.
Die für solche Arbeiten eingesetzten
stabilen Injektionsmischungen auf Zementbasis geben dennoch immer eine bestimmte
Menge an Filtratwasser ab, das hohe
Alkalität besitzt (pH-Werte zwischen 8.5
und 12). Dieses Filtratwasser dringt an
den Aufsprengungsflächen und vorgegebenen
Kluftflächen anscheinend einige Millimeter
in den Schluff ein und dürfte zu, wenn auch
geringen Änderungen des natürlichen Wassergehaltes, der Fließgrenze und der Ausroll-
grenze führen. Wie einige durchgeführte
Klassifikationsversuche an den Schluffen
entlang der Aufsprengungen zeigten, sind
deutlich verringerte Konsistenzen gefunden
worden (z.B. $I_c = 0.78$ also steif plastisch
zu $I_c = 0.63$ was einer weichen Konsistenz
entspricht), die auch Veränderungen der
Gesamtsteifigkeit des aufgesprengten
Schluffpaketes erwarten lassen
(Abbildung 5).

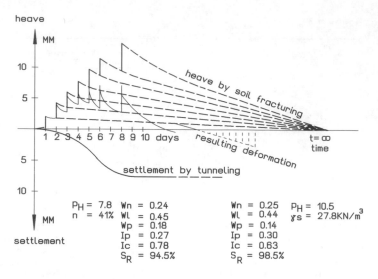

heave

MM

heave by soil fracturing

resulting deformation

t = ∞
time

settlement by tunneling

settlement

MM

	before	soil-fracturing	after
$P_H = 7.8$	$W_n = 0.24$	$W_n = 0.25$	$P_H = 10.5$
$n = 41\%$	$W_l = 0.45$	$W_l = 0.44$	$\gamma_s = 27.8 KN/m^3$
	$W_p = 0.18$	$W_p = 0.14$	
	$I_p = 0.27$	$I_p = 0.30$	
	$I_c = 0.78$	$I_c = 0.63$	
	$S_R = 94.5\%$	$S_R = 98.5\%$	

Abbildung 5   Zeitliche Verringerung der
Hebungen im tonigen Schluff
durch Zunahme des Wasser-
gehaltes und des pH Wertes.
Der Sättigungsgrad des
Schluffes nimmt zu, der
Konsistenzindex wird kleiner.

Figure 5   Relaxation of the grouted
heave in clay created by
an increasing watercontent
and pH-valve of groundwater
changing the index of
consistence and the degree of
saturation.

Besondern bei mehrfachen, zeitlich kurz
hintereinander erfolgenden Aufsprengphasen
können parallel zur Bodenverbesserung durch
die Feststoffeinbringung auch zumindest
zeitlich begrenzte Festigkeitsverschlechte-
rungen mit erhöhten Hebungsverlusten nicht
ausgeschlossen werden. Ein zum soil-
fracturing gleichzeitig verlaufender
Tunnelvortrieb sollte daher vermieden
werden um den zu verdichtenden Untergrund-
bereich neben dem Porenwasserdruckabbau
auch die Zeit zur chemischen Bindung der
zementhältigen Filtratwässer zu geben.

Wenn ein solcher Bauzeitplan eingehalten
werden kann, ist für die durch die Aufsprenginjektionen betroffenen Untergrundbereiche langfristig mit globalen Festig-
keitsanstiegen zu rechnen. Für die nachfolgenden Tunnelvortriebe bedeutet dies
eine zusätzliche Verringerung der
Setzungen darüber.

5. SOIL-FRACTURING AUF DER BAUSTELLE U3/14

EIGNUNGSVERSUCH: Entsprechend der
genannten Überlegungen wurde wie folgt
bei einem Großversuch auf der Baustelle vorgegangen:

Von einer bestehenden Streckenröhre
(Ausbruchsquerschnitt 36 m2), wurde
seitlich und von oben injiziert und
die folgenden Annahmen geprüft:
* Bleiben die Spannungsumlagerungen im
  Boden bestehend bzw. bauen sich ab;
  ab wann wird der gewünschte Vorspanneffekt wirksam?
* In welchem Abstand zum Tunnel sollen
  vorauseilende Hebungen gemacht werden,
  um unzulässige Beeinflussungen der
  Tunnelaußenschale zu vermeiden?
* Ist die angewandte Injektionstechnik
  (durch zielgenaue Bohrungen und Ausbau
  mit Manschettenrohren) für diese Art der
  Feststoffeinbringung geeignet?

Aus umfangreichen Messungen in Form von
Nivellements (Oberfläche und Tunnelfirste)
Gleitmikrometer, Konvergenzen der Tunnelschale sowie Druckmeßdosen konnte ein
einheitliches Bild der Bodenreaktionen
durch das soil-fracturing gewonnen werden,
sodaß die aufgeworfenen Fragen eindeutig
positiv beantwortet werden konnten:

* Die erzwungenen Spannungsumlagerungen
  bleiben ausreichend lange bestehen.
  Der gewünschte Vorspanneffekt entsteht
  bereits, sobald eine kleine Hebung

Abbildung 6   Herstellung der Injek-
               tionskörper in der Ulme
               durch die Ortsbrust des
               Tunnels.

Figure 6   Soil-fracturing from the
               tunnel surface

der Oberfläche erfolgt und hält über
die Zeit der Tunnelherstellung an.
* Bei Injektionen vom Tunnel aus ist mit
  einer starken zusätzlichen Beanspruchung
  der Tunnelschale zu rechnen.
* Als Injektionsgut ist eine hochviskose
  möglichst stabile Zement-Bentonit-
  mischung vorzusehen.

AUSFÜHRUNG: Nach diesem Versuch wurde
an 3 Stellen der Station Schweglerstraße,
die Aufsprenginjektionen zur vorauseilenden
Spannungsumlagerung angewandt. Die Station besteht aus 2 Tunnelröhren mit
je 65 m2 Ausbruchsfläche, sowie 2 gleich
großen Querschlägen und einem Schrägschacht für Rolltrepppen.
An insgesamt 3 Stellen wurden die Ulmen
mit soil-fracturing behandelt. Das Auffahren der Tunnel und Querschläge erfolgte mit Ulmen- und Restausbruch.
Die Injektionskörper wurden so nahe als
möglich an die Tunnelaußenschale gelegt.
Sämtliche Injektionen konnten von unter-
tage ausgeführt werden (Abbildung 6);
dies war besonders deshalb wichtig, da

für eine Ausführung von obertage die
entsprechenden Servitute in den Häusern
nicht erwirkt werden konnten. Die Abfolge
der Injektionen wurde nach einem Raster-
verdichtungsverfahren ausgeführt, wobei
in mehreren Phasen ein Injektionsvolumen
von etwa 10 bis 15 % des Bodenvolumens des
anstehenden Schluffes bzw. Feinsandes
verpreßt wurden. Als Mischung wurde verwendet: Wasser 1000 l, Bentonit 45 kg,
Zement HOZ 600 kg, Wasserglas 10 l.
Es wurde mit geringen Injektionsgeschwindigkeiten (unter 400 l/pro Stunde) und
Injektionsdrücken (unter 15 bar) gearbei-
Die angestrebte Vorhebungen betrugen im
Maximum 1 cm bis 1,5 cm.
An einer Stelle wurde von einem Ulmen-
stollen aus die Injektion durchgeführt.
Hiebei zeigte sich, daß im Ulmenstollen
selbst eine Absteifung zur Aufnahme der
Drücke, die aus den Injektionsarbeiten
entstanden, erforderlich war.
Der Abstand des Injektionskörpers von der
zugewandten Ulme betrug 2 m. Die durch die
Nähe des Tunnel bedingten vertikalen
Aufsprengungen zeigten eine geringere
Hebungsreaktion in Relation zur verpreßten

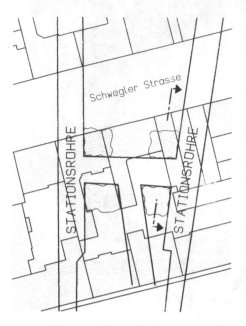

Menge des Injektionsgutes, als die an den
anderen Stellen erzielten horizontalen
Aufsprengungen. Bei den anderen beiden
Stellen konnte von einem Querschlag bzw.
von einem Schacht aus der erforderliche
Bereich vor Beginn der Ausbruchsarbeiten
behandelt werden, sodaß deutlich
günstigere Randbedingungen zur Verfügung
standen. (Abbildung 7 und Abbildung 8).
Wesentlich war, daß, wie die Setzungslinien
der Abbildung 8 zeigen, eine Setzungsminderung in der Größe von etwa 50 % der
prognostizierten Werte festzustellen ist.
Als positiver Effekt ist zu vermerken,
daß in den Stellen der größten
Beanspruchung durch das soil-fracturing
die schichtenweise Vorverdichtung des
Bodens zu höherer Steifigkeit erreicht
wurde.

6. SCHLUSSFOLGERUNGEN:

* Wird vom Tunnel aus in den Ulmenbereich
  injiziert, so muß ein "Respektbestand"
  zur Tunnelschale eingehalten werden. Dies
  kann zu unliebsamen Beanspruchungen der
  Tunnelschale führen.
  Sinnvoller ist es daher, von der Oberfläche zu injizieren, da dadurch auch
  näher an den künftigen Ausbruch heran-
  gegangen werden kann, und somit der
  Bereich der Aufsprenginjektionen
  kleiner gehalten werden kann. Bereits

Abbildung 7   Station Schweglerstraße,
                Lage der Injektionskörper

Figure 7   Undergroundstation "Schweg-
                lerstraße"
                Positions of soil-fracturing

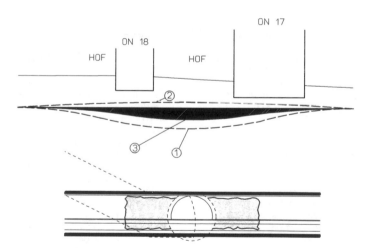

Abbildung 8   Station Schweglerstraße,
                Injektionserfolg.
                1. Setzungen ohne Auf-
                   sprenginjektion
                2. Vorhebung
                3. Setzungen mit Auf-
                   sprenginjektionen

Figure 8   Undergroundstation "Schweg-
                lerstraße", effect of soil-
                fracturing
                1. presumtive settlements
                   without grouting
                2. vertical predeformation
                3. final settlements with
                   soil-fracturing

die Planung hat darauf Rücksicht zu nehmen.

Die Wirkung der aufgebauten Druck-spannungsumlagerungen ist dann gegeben, wenn auf der Oberfläche die ersten Hebungsreaktionen eintreten. Es ist jedoch zu beachten, daß über Porenwasser-drücke Hebungen erfolgen, die die Spannungsumlagerung nur kurzzeitig beeinflussen können und die mit der Zeit verschwinden, sodaß ein geringeres Maß an Vorhebung im Boden zurückbleibt. Weiters ist sicherzustellen, daß Reaktionsnebenprodukte aus den Aufspreng-injektionen, wie Filtratwasser, durch seinen Chemismus bei den vorhandenen Drücken keine bleibenden Konsistenzver-schlechterungen der bindigen Unter-grundverhältnisse, insbesondere bei überkonsolidierten Schluffen und Tonen bewirken.

* Zweckmäßig ist es vor den Arbeiten, Prognosen über die Verteilung und die Größe der Setzungen aus dem Tunnelbau zu erstellen und diese laufend mit Bereichen wo keine Aufsprengarbeiten erforderlich sind, zu eichen. Nur durch konsequente Verwendung von methodischer und verfahrenstechnischer Rückkoppelungen zwischen Tunnelvortrieb, Setzungen, Bodenverbesserungen und möglichen uner-erwünschten Begleiterscheinungen können gezielte setzungsmildernde Maßnahmen wirtschaftlich optimal eingesetzt werden.

* Ein umfangreiches Meßprogramm zur Über-prüfung der Wirksamkeit ist unumgäng-lich erforderlich.

Die vorgeschlagene Methode des gezielten Einsatzes von Aufsprenginjektionen ist sicher kostengünstiger als großflächige Hebungsinjektionen. Dies gilt besonders dann, wenn wie in den vorgestellten Beispielen komplizierte Tunnelabzweigungen hergestellt werden sollen und vor allem dann, wenn die Setzungsprognosen ein Überschreiten der für die darüber befind-liche Bebauung verträglichen Grenzwerte erwarten lassen. Auch für den Regelvortrieb eines Tunnels werden die Aufsprenginjek-tionen wirtschaftlich sinnvoll, wenn eine mäßige Überschreitung der Grenzwerte prognostiziert wird.

LITERATUR

Ettel,E. 1991. Vorauseilende Spannungs-umlagerung - eine Weiterentwicklung der Spritzbetonbauweise im Lockergestein. Vortrag der STUVA - Tagung 1991 in Düsseldorf.

Flunck, G., Wilms, J.1987. Kontrollierte Steuerung von Senkungen durch Feststoff-Einpreßtechnik am Beispiel U-Bahn-Essen, Baulos 31 Vorträge der STUVA - Tagung 1987 in Essen

Liebsch, H. 1988. Injektionen beim Wiener U-Bahn-Bau - Erfahrungen und Zukunfts-aussichten Zement und Beton 4/88

Liebsch, H. 1992. Weiterentwicklung der Neuen Österreichischen Tunnelbaumethode im Wiener U-Bahnbau Österreichische Ingenieur- und Architekten-Zeitschrift 5/92

Raabe, E.W., Ersters, K. 1986. Injektions-techniken zur Stillsetzung und zum Rück-stellen von Bauwerkssetzungen. Tagungsband der Deutschen Baugrundtagung 1986 in Nürnberg, Seiten 337 - 366.

Ruppel, G. 1970. Ausführung von Injektionen in Lockergestein. Bergb.Wiss. 17 (1970), Heft 8, 285 - 290.

Samol, H., Priebe, H. 1985. Soilfrac - ein Injektionsverfahren zur Bodenverbesserung. Proceedings of the 11th Conference on Foundation Engineering. San Francisco 8/1985

A non-pressurized grouting method using clay for controlling groundwater around crude oil storage caverns

Druckfreies Abdichtungsverfahren unter Verwendung von Ton zur Kontrolle des Grundwassers in direkter Nachbarschaft von Untergrund-Speicherkavernen für Rohöl

Yoshiharu Miyanaga & Masahiko Ebara
Electric Power Development Co., Ltd, Japan

Abstract: Rock caverns for crude oil storage were constructed at three sites in Japan. A water barrier system was employed to prevent leakage of oil and vapor in these caverns. As one of these caverns was located in fractured and highly permeable granite, excessive groundwater discharge to the cavern was anticipated. Therefore non-pressurized clay grouting method was employed in order to reduce an amount of seepage flow. Before applying this method to the construction, indoor and in-situ tests were performed to confirm clay clogging mechanism and effects of clay grouting. Based on these tests, it was successfully applied to the cavern mentioned above. Outline of clay grouting and results is described in this paper.

Zusammenfassung: In Japan wurde der Bau von Untergrund-Speicheranlagen für eine nationale Ölreserve im Jahr 1987 in Angriff genommen. Bei der Speicheranlage am Standort Kuji, einer von drei solchen Einrichtungen, besteht das Festgestein aus Granit, der jedoch aufgrund einer erheblichen Zahl von Spalten auch einen sehr hohen durchschnittlichen Durchlässigkeitskoeffizienten aufweist. Die Kavernen wurden unterhalb der aktiven wasserundurchlässigen Zone ausgehoben, um ein zu starkes Absinken des Grundwasserspiegels auszuschliessen. Darüber hinaus war eine Reduzierung der Sickerwassermenge erforderlich, um nach der Fertigstellung der Anlage die Betriebskosten für das Abwasserbeseitigungssystem zu senken.
Üblicherweise versteht man unter "Abdichten" das Einpressen von aushärtendem Material (z.B. Zement) in Spalten beispielsweise an Bohrlöchern. Das Abdichten mit Ton im Sinne der vorliegenden Abhandlung dagegen ist ein druckfreies Verfahren, das sich den Sickerwasserfluss in Richtung von Kavernen zunutze macht. Tonpartikel, die durch das Sickerwasser in kleine oder grosse Spalten getragen werden, verschliessen diese Sickerwege allmählich und in weitem Umfang. Auf diese Weise lässt sich eine gleichmässige Undurchlässigkeit von extensiven Gesteinsmassen (einschliesslich Kavernen) erreichen. Zunächst wurde durch Laborversuche der Mechanismus des Abdichtens mit Ton geklärt. Die anschliessend durchgeführte Erprobung vor Ort bestätigte die Praktikabilität des Ton-Abdichtverfahrens. Schliesslich wurde dieses Abdichtverfahren erfolgreich bei dem Projekt zum Bau einer Untergrund-Ölspeicheranlage am Standort Kuji angewendet, wo auch die wirkungen des Ton-Abdichtverfahrens bestätigt wurden.

INTRODUCTION

In Japan, the construction of underground crude oil storage plants (3 sites-5 million kiloliters) for national petroleum stockpiling was started in 1987. At present, civil works have been completed and oil installation started in January 1993. The water barrier system is employed to prevent leakage of oil and petroleum gas at these three storage plants. The bedrock of the Kuji site, one of the three sites, consists of granite, but this is considerably fractured, and the average permea-bility coefficient is very large, $(7.6 \times 10^{-8} \text{m/sec})$. Accordingly, difficulty in the construction to excessive water leakage had been anticipated. However, in case extensive

drawdown of groundwater level around caverns occurred during construction, it might be difficult to raise the groundwater table again after completion.

Furthermore, remained unsaturated zone around caverns has possibility to invalidate functions of the artificial water barrier. Therefore, the caverns were excavated under the active water barrier system in order to prevent the excessive drawdown of groundwater level. In addition to this, it was necessary to reduce the amount of seepage in order to decrease the operation cost for wastewater disposal system after completion.

Considering the above-mentioned requirements, the non-pressurized clay grouting method was employed, using the seepage flow from water barrier system to

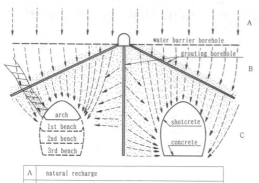

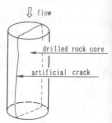

A	natural recharge
B	natural recharge and clear barrier water
C	natural recharge, clear barrier water and clayey water

Fig. 1   Clay grouting system

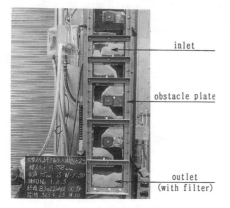

Photo 1.   Slit model

the caverns. Since the effect of the clay grout-ing has been confirmed, an outline of the process and its results are reported in this paper.

1. CLAY GROUTING

Generally, grouting means pressurized injection of hardening materials such as cement into cracks around boreholes. On the other hand, clay grout-ing stated hereafter is a non-pressurized grouting utilizing seepage flow directed toward caverns. Clay particles carried into fissures or fractures by seepage flow plug those flow paths slowly and widely. As a result impermeability of extensive rock mass including caverns can be uniformly improved.

Concretely speaking, clay grouting is accom-plished in the following way. First, water barrier tunnels and water barrier boreholes are excavated prior to excavating the storage caverns.

Photo 2.   Clay clogging in the artificial crack by splitting test

Then, while the caverns are being excavated, water is poured into the water barrier facilities in order to prevent the groundwater level from falling excessively. In the clay grouting method, clay is mixed with this water and applied to the rock through the water barrier facilities.

The sequence of cavern excavation is as follows: (1) the top heading, (2) arch enlarging, (3) first stage bench cut, (4) second stage bench cut, (5) third stage bench cut. The flow of groundwater varies in its path according to each of these stages. In this way, clay is provided to the fissures and fractures of the rock area in large, and the impermeability is improved (Fig. 1).

Since clay grouting targets a wide area and requires a period of several months to several years, clay grouting must be performed by a low pressure and a low concentration so that the clay does not clog up the grouting boreholes or their surroundings without reaching more inner areas of the rock.

For a large amount of local discharging water during cavern excavation, ordinary cement injection from inside of the cavern is also necessary to reduce the velocity of water. Thus, at the Kuji site, cement injection from the caverns was employed simultaneously besides clay grouting.

2. INDOOR EXPERIMENTAL TESTS

Success of this grouting method depends on whether or not the impermeability of a whole rock can be improved without clay clogging around boreholes.

Thus, before performing in-situ tests, various indoor tests were executed to confirm the possi-bility described above.

Two different indoor tests were performed.

One is a test that employs a large slit model, which consists of transparent acrylic plates and obstacle plates that look like a stairway, between a slit. This facility was intended to model relatively wide cracks. The width of the slit is 0.2 - 0.8 mm. The other test is a permeability test, which employs a test piece that is a drilled rock core split into two. In this case, the width of the crack, which is artificially made by the

splitting test, is about 0.03 - 0.07 mm. Based on
these two tests, the mechanism of clogging of
clay in cracks was understood, and the possibility
of clay grouting was confirmed.

1) In relatively large cracks of which width
is over 0.2 mm, that can be reproduced by the
slit model, clogging started from the filter part
at the end of cracks (corresponding to shotcrete
surface of real caverns) without any clogging
in the middle of the paths, and then clogging
expands to all cracks. In this case, most parts
of the opening of the cracks were filled with
clay particles (Photo - 1).

2) In the case of minute cracks, of which width
is smaller than 0.1 mm, that is reproduced by the
split core model, by decreasing the clay concentration, a relatively distributed piling-up pattern
can be obtained. Therefore, it is possible to
improve the impermeability of the wide area if the
clogging phenomenon proceeds by a long-term clay
pouring. On the other hand, if the clay concentration is large, the clay may be piled upnear
boreholes (Photo - 2).

Suitable clay for this grouting method has the
following three conditions. First, it consists
of particles with a small diameter. Second, its
solution has low viscosity. Third, it has low
sedimentation rate. For the grouting work, Kuji
clay that is obtained near the construction site
was used. This clay is kaolinite and does not
swell. The ratio of the 50-percent-finer-than size
(D_{50}) of the clay is 3 micron meter, which is very
minute (Fig. 2). It can keep muddy condition for
a long time, and has a low sedimentation rate.
Thus, Kuji clay satisfies the three conditions
described above (Fig. 2).

3. IN-SITU TEST

An in-situ test was performed at a part of a tunnel
of the Kuji site. In the test, clayey water
(clay / water = s/w = 1/200) was recharged from
water barrier boreholes of the upper tunnel for
four months. Changes in the supply rate of clayey
water, amount and quality of discharging water
at the lower tunnel were observed during this
period. In addition, permeability tests of water
barrier boreholes before and after clay grouting,
and resistivity tomography tests were performed.

The possibility of applying clay grouting
method to the caverns was confirmed as described
below through those tests (Fig. 3).

1) The reduction of the amount of recharging
and discharging water was confirmed by evaluating
changes of those rates.

2) It was confirmed that clay did not clog at
around the boreholes and the water barrier effect
was not hindered by clogged clay through the
observation of wet condition at the drain holes
drilled at the surface of the lower tunnel and
the results of the resistivity tomography.

3) It was confirmed that the clogged clay was

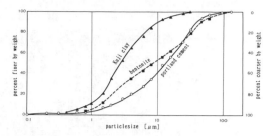

Fig. 2 Particlesize accumulation curve of Kuji clay

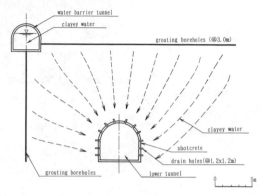

Fig. 3   General section of in-situ test

stable and did not flow out under changes of
hydraulic gradient through the injection test
using fifty times larger hydraulic gradient.

4) It would be better to dilute the clayey
water more.

4. CLAY GROUTING WORK

4-1. Extents of Grouting and its Preparation

Based on the result of the indoor and in-situ
tests, clay grouting was started in August 1990
for the real caverns. The effectiveness of clay
grouting and harmlessness to the water barrier
were confirmed, as described in the preceding
section. However, considering that clay grouting
had never been performed before, a section where
groundwater level was the lowest was chosen as
the first target for clay grouting and we decided
that the clay grouting area would be extended by
reconfirming the effectiveness and harmlessness
of the clay grouting method. The first targeted
caverns were southern half of the four rows (101
B, 102A, 102B, and 102C), and the water barrier
tunnels were 2S and 3S (Fig. 4).

A pipe method was adopted for a grouting work.
While providing water to the horizontal water
barrier boreholes with pipes, inclined grouting
boreholes (76 mm diameter, 50 m length) were

179

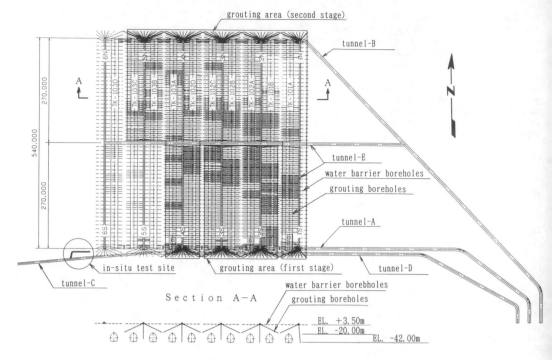

Fig. 4　General plan and grouting area

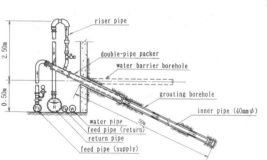

Fig. 5　Pipe system

Photo 3.　Pipe system for grouting

drilled at intervals of 10 meters (Fig. 1).

An internal pipe with a diameter of 40 mm was inserted inside boreholes, and a double-pipe packer was set at the inlet. The internal pipe of the packer was connected to a supply pipe, and the external pipe was set to a riser pipe (H = 2.5 m) for controlling pressure (Fig. 5). Clayey water was supplied to the boreholes via a supply pipe that had been settled inside the tunnel.

Then, the clayey water flew from the tip of the pipe toward the mouth, and returned to the return pipe, stabilizing the pressure by the riser pipe.

If there existed continuos cracks from boreholes to caverns and if the water flew along the cracks, the clay was supplied to cracks with water. If ther was no water flow, the clay was returned to the fee tank with water and then, was recycled and reused.

The layout of the clay grouting system of the water barrier tunnel is shown in Fig. 6 and Photo -3. At the enlarged section, three pairs of a feed tank (4 m^3 each) for supply and a return tank (1 m^3 each) for recycling from the return pipe are installed for controlling amount of pouring water for left, right, and central systems (Photo-4).

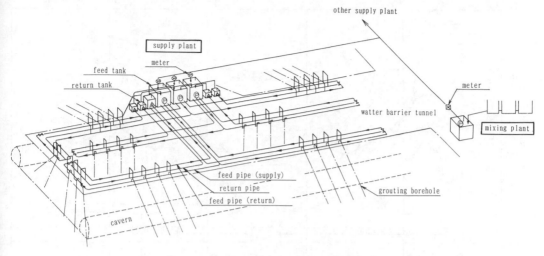

Fig. 6   Clay grouting system

4-2 Grouting

The clay obtained near the site was transferred to the production plant and then, concentrated solution (s/w = 1/2.5) was produced by removing coarse clay, using a moisture crush and classification facility. The concentrated solution was transferred to the mixing plant in the tunnel by a tank truck, and then, modified to s/w = 1/600, and transferred to the feed tank of each system.

As the clayey water penetrated from grouting boreholes into the rock, the water level of the feed tank lowered. When the water level reached a certain level, the clay water of which concentration is 1/600, was automatically supplied.

Clay grouting for the 2S group was started on August 21, 1990, and for the 3S group it was started on September 16. Although for the 3S group grouting was finished in September 1991, for the 2S group it was continued till August 1992 for 24 months.

Change of the amount of recharging water for the 2S and 3S groups during this period is shown in Fig. 7. At the starting point, the amount of pouring water is large, and is stabilized after 2 to 3 months, and then decreased slowly.

The initial concentration of clayey water was 1/600, considering the result of the in-situ test.

In the final stage, the concentration was increased gradually as 1/400, 1/300, 1/200.

By reconfirming the effects and harmlessness of clay grouting at the 2S and 3S groups (the No. 1 work), the No. 2 work for the other area except three sections of 6N, 6S , and 5S, where the drawdown of water level is small, was started and extended in August 1991. Eighty percent of the total area was targeted for clay grouting.

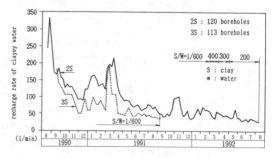

Photo 4.   Feed tanks and clayey water

Fig. 7   Recharging clayey water in 2S and 3S

181

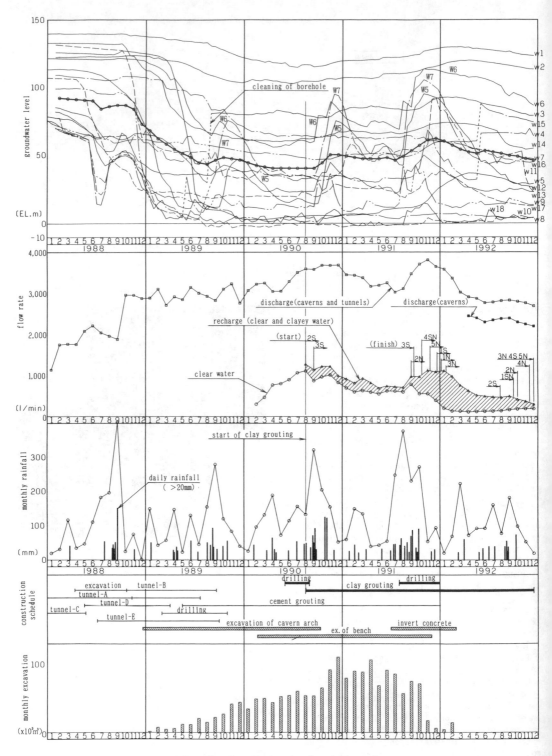

Fig. 8 Groundwater level and flow rate

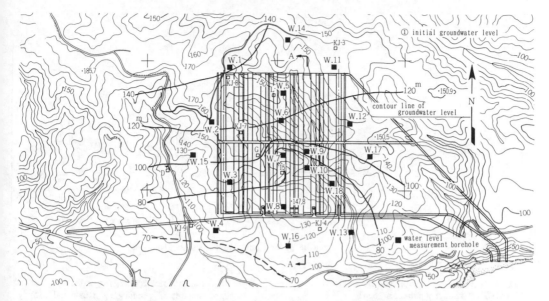

Fig. 9 Arrangement of water level measurement boreholes and initial groundwater level

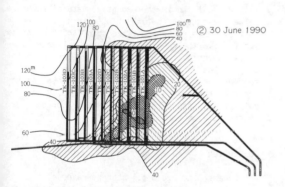

Fig. 10 Groundwater level just before commencement of grouting

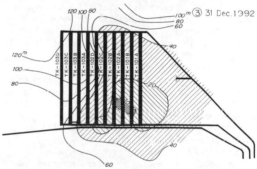

Fig. 11 Groundwater level after completion of grouting

The excavation of the caverns was completed in November 1991 and placement of invert concrete was finished in February 1992. Clay grouting was continued until December 1992 and was finished with the period of 28 months.

4-3 Confirmation of Grouting Effects

Since clay grouting is a method which demonstrates its effect by recharging low concentrated clayey water into an extensive rock mass slowly, continuous measurements of groundwater level, amount of discharging and recharging water were performed.

The level of groundwater was measured by using the 18 water level measurement boreholes arranged around the caverns (Fig. 9). The change of water

level during construction was recorded by an automatic supervision system by a telemeter.

Discharging water of the whole site was gathered at one place, and measured in early morning of every Monday, when discharge was not hindered by the construction. Since April 1992, when the excavation of the caverns and placement of invert concrete (t = 25 cm, in contact with rock) were completed, amount of discharging water of 10 caverns were measured separately with the tunnels.

Amount of recharging water was recorded for every water barrier tunnel divided into 12, such as 1~6N, S.

Figure 8 summarizes the change of the groundwater level, amount of discharging and recharging water described above, monthly and daily rainfall, and construction schedule. Effects of clay grouting,

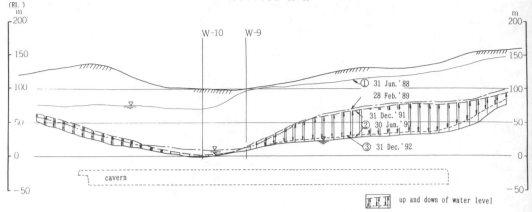

Fig. 12 The change of groundwater level in section A-A

based on Fig. 8, are enumerated as follows.

1) Raise of the groundwater level

Groundwater level raised from July 1990, when grouting was started, to December 1992, when grouting was finished, even if the recharge due to rainfall is ignored. Especially, the ground water level around the observation boreholes W-10 and W-18, where the groundwater level had been low initially, raised and stabilized significantly. (Fig. 10 and Fig. 11).

2) Decrease of discharging water

The amount of discharging water decreased from 3,600 l/min in July 1990 to 2,800 l/min in October 1992. In addition to the effect of clay grouting, the waterstop effect of shotcrete and concrete in contact with the excavated rock surface seems to be included.

3) Decrease of recharging barrier water

The amount of recharging water decreased from 1,200 l/min in July 1990 to 400 l/min in October 1992. It is assumed that clay particles in barrier water narrow the width of seepage paths of groundwater and reduces the amount of recharging and discharging water, and thus resulted in the grouting effect of raising the groundwater level.

In addition to the above-mentioned effects of the clay grouting, the following was observed.

1) The amount of clay used during the total grouting period is 790 tons. The volume of one ton of dry clay becomes 1.1 m³ in water. It is assumed that these clay particles permeated into cracks of the extensive rock mass around the storage caverns and improved the impermeability.

2) Reduction and muddiness of discharging water from the caverns were measured, and at about 30 locations, clayey discharge was confirmed.

3) In Fig. 8, changes in the amount of discharging and recharging water are almost parallel and their relationship is

$$Qg - Qp \doteqdot 2,400 \text{ l/min (constant)}$$

where

Qg : amount of discharging water,
Qp : amount of recharging clear and clayey water.

This 2,400 l/min is the amount of natural recharge.

The reason that the amount of natural recharge is very large is that surface runoff of the stream right above the caverns infiltrates entirely.

The amount of the natural recharge is constant, despite the fact that there are rainy and dry seasons, because the fractured zone located north-east of the caverns (W5, W6, and W7) plays a role of underground reservoir, which stores water in the rainy season (Fig. 12).

CONCLUSION

For underground crude oil storage caverns, the amount of discharging water must be decreased without lowering the level of groundwater. Thus, clay grouting, which improves the impermeability of the whole rock mass slowly, was needed.

First, indoor experimental tests were performed and the mechanism of the clay grouting was understood. Then, the in-situ test was performed and possibility of the clay grouting method was confirmed. Finally, this grouting method was successfully applied to the construction project of underground crude oil storage plants in Japan, and the effects of the clay grouting were confirmed.

The conceptual origin of this grouting method lies in the natural clogging phenomena, such as (1) the amount of discharging water of a tunnel decreases slowly in years, and (2) a large amount of water leakage of a dam after impoundment decreases rapidly by muddy water in the reservoir, caused by melted snow or flood. This grouting method can be used as a countermeasure to dam leakage with the combination of curtain grouting as well as to underground caverns.

Geräte für die moderne Injektionstechnik

Equipment for today's grouting technology

R. Müller
Häny & Cie, AG, Meilen, Schweiz

ZUSAMMENFASSUNG: Die moderne Injektionstechnik stellt nicht nur erhöhte Anforderungen an Ingenieure und ausführende Unternehmen sondern auch an die Geräte. Die Wahl der geeigneten Misch-, Verpress- und Registriergeräte spielt eine wichtige Rolle zur erfolgreichen Durchführung eines Injektionsprojektes. In den letzten Jahren wurden erhebliche Fortschritte in der Entwicklung besserer und umweltfreundlicher Injektionsmaterialien erzielt. Diese bedingen aber auch Geräte, welche an diese moderne Technik angepasst sind.

Von entscheidender Bedeutung sind hier sicher die Mischer, welche die Aufgabe haben, die Injektionsmaterialien so aufzubereiten, dass diese ihre optimalen Eigenschaften entfalten können. So ist es z.B. absolut sinnlos, qualitativ hochstehende Feinstbindemittel in Paddel- oder Durchlaufmischern aufzubereiten, in denen die Aufschliessung der einzelnen Partikel völlig unzureichend ist. Zur Aufbereitung dieser Materialien wie auch für Suspensionen auf der Basis von gewöhnlichem Portlandzement, sollten hochtourige Turbomischer (Kolloidalmischer) verwendet werden. Damit wird eine vollständige Hydratation aller Partikel gewährleistet und Klumpenbildungen wirkungsvoll verhindert. In diesen Mischern können homogene und stabile Suspensionen mit ausgezeichneten Verarbeitungseigenschaften hergestellt werden, welche nicht zum Absetzen neigen und sich im Grundwasser kaum entmischen.

Gleichermassen wichtig ist auch die Injektionspumpe, welche in der Lage sein muss, die unterschiedlichen Injektionsmaterialien von der dünnen Suspension bis hin zum Injektionsmörtel, bei genau vorgegebenen Werten bezüglich Maximaldruck und Menge, zu verpressen. Es gibt eine Vielzahl von Pumpsystemen, wobei jedes System seine Vor- und Nachteile hat. Wichtig jedoch ist die Wahl eines Systems, welches den Anforderungen bezüglich der Druck- und Mengenregulierung gerecht wird.

Ein weiterer wichtiger Punkt ist die Erfassung von Daten über ein Injektionsprojekt, welche zur späteren Analyse über Erfolg oder Misserfolg einer Methode erforderlich sind. Die Möglichkeit, auf Erfahrungswerte eines anderen, gleichartigen Projektes zurückgreifen zu können, sind unschätzbar. Die Datenerfassung kann sowohl manuell oder über entsprechende Geräte erfolgen, welche vom einfachen Druckschreiber bis hin zur vollcomputerisierten Erfassung aller wichtigen Parameter reicht.

ABSTRACT: The state of the art in grouting technology not only demands more expertise of consulting engineers and contractors but also equipment that can meet these requirements. The choice of the suitable mixing, pumping and recording systems plays an important part in the successful execution of a grouting project. In the past few years major progress has been achieved in the development of new and better grouting materials that are more compatible with the environment. These materials however, require equipment which is adapted to the new grouting technologies.

Of significant importance are the mixers which have to prepare the grouting materials in such a way to reach their optimum characteristics. E.g. it does not make sense to mix Microfines in a paddle or continuous mixer as the resulting hydration of the cement particles in such mixers is insufficient. For these materials as well as for ordinary Portland Cement based grouts, high-shear mixers (colloidal mixers) should be used to assure full hydration of all particles and to prevent lumps of dry cement. In these mixers, homogenous and stable suspensions can be produced which do not tend to settle out of suspension and which displace free water in ground rather than getting diluted by it.

Equally important are the grout pumps which must be capable of pumping different grouts, from thin suspensions to fairly thick sanded grouts, at exactly predefined maximum pressure and flow rate. There are a variety of pump systems available, each having its advantages and disadvantages. Important however is the choice of a system which complies with the demand for precise pressure and flow control.

A further important factor is the recording of data about a grouting project which are essential in the analysis of success or failure of a specific method. The ability to look into records of comparable projects are invaluable. The data collection can be done manually or with equipment which ranges from simple pressure recorders to fully computerized data logging of all important parameters.

1 MISCHER

Die Zusammensetzung des Injektionsgutes muss sehr sorgfältig den gegebenen Bodenverhältnissen angepasst werden. Dazu sind seriöse geologische Untersuchungen des zu injizierenden Bodens sowie Wasserdurchlässigkeitstests durchzuführen. Basierend auf diesen Untersuchungen kann nun das Anforderungsprofil an das geeignete Injektionsgut erstellt werden. So ist es z.B. wenig sinnvoll, grössere Hohlräume mit einer dünnen Zementsuspension zu verfüllen und doch wird dies immer wieder praktiziert. Die Gründe hierfür sind vielfältig, sei es dass die Injektionsmannschaft allfälligen Problemen bei der Aufbereitung oder Förderung dickflüssiger Injektionsmedien von vornherein aus dem Weg gehen will oder, dass die verwendeten Geräte tatsächlich für eine solche Anwendung ungeeignet sind. Eine Zementmischung hoher Viskosität oder Mörtelmischungen können in einzelnen Fällen bessere Resultate erzielen und sich obendrein als kostengünstig erweisen.

Auf der anderen Seite muss bei feinen Rissen der Wasser/Zement Wert dem erwarteten Fliessverhalten angepasst sein. Die groben Richtlinien bezüglich der anzuwendenden Mischungen sind normalerweise in den Spezifikationen festgehalten und werden vom Ingenieur überwacht und in Anpassung an die momentanen Verhältnisse geändert. Der ausführende Injektionsspezialist kann mit seiner Erfahrung und Beurteilung jedoch auch einen grossen Beitrag zur erfolgreichen Injektion leisten.

1.1 Mischverfahren

Die Qualität der Mischung ist ebenso wichtig wie ihre Zusammensetzung. Wir unterscheiden grundsätzlich zwischen 2 Mischverfahren:

Mischen durch Umrühren
Mischen durch Erzeugung hoher Scherkräfte

1.2 Mischen durch Umrühren

In diese Kategorie fallen die herkömmlichen Paddel- (Fig. 1) sowie die Durchlaufmischer (Fig. 2). In beiden Mischertypen werden Wasser, Zement und allfällige Zusätze solange bewegt bis diese sich langsam vermischen.

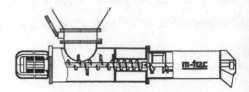

Fig. 2: Durchlaufmischer.
         Continuous Mixer.

Die in diesen Mischern erzeugten Scherkräfte sind jedoch sehr gering und kleinere Zementklumpen werden kaum aufgeschlossen. Das Resultat ist eine instabile Mischung mit einzelnen schwereren Partikeln und Klumpen mit unbenetztem Zement. Diese Zementklumpen werden durch die Oberflächenspannung des Wassers zusammengehalten und sind demzufolge grösser und schwerer als einzelne Zementpartikel. Deshalb neigt eine instabile Mischung sehr stark zum Absetzen, was zu frühzeitiger Verstopfung feiner Risse durch den Filtrationseffekt sowie zu Problemen in Pumpen und Rohrleitungen führen kann. Bei Injektionen im Grundwasser wird eine instabile Mischung sehr schnell durch das reichlich vorhandene Wasser entmischt resp. verdünnt und ändert somit ihre Eigenschaften.

Paddelmischer sind sogenannte Chargenmischer, bei welchen die Wasser/Zement Faktoren genau kontrolliert werden können. Bei den Durchlaufmischern ist die Kontrolle des W/Z Wertes wesentlich schwieriger, da der Wasserdruck auf der Baustelle erfahrungsgemäss stark schwankt und zugleich die Zufuhr des Zementes in den Mischer nicht immer gleichmässig ist.

Von beiden Mischertypen erfolgt die Entleerung in einen Stapelbehälter (Rührwerk) über Schwerkraft. Dies bedeutet, dass entweder der Mischer über dem Stapelbehälter angeordnet werden muss, was eine bestimmte Bauhöhe bedingt (Fig. 3A) oder dass bei Paddelmischern zwei im Parallelbetrieb eingesetzt werden (Fig. 3B).

Diese Mischer eignen sich für gewisse Arbeiten, bei welchen die Qualität der Mischung eine untergeordnete Rolle spielt.

Fig. 1: Paddelmischer.
         Paddel Mixer.

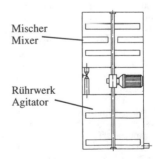

Mischer
Mixer

Rührwerk
Agitator

Fig. 3A: Paddelmischer für kontinuierliche Injektion
         Paddel Mixer for continuous Grouting.

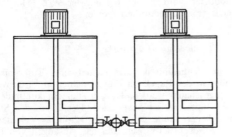

Fig. 3B: Paddelmischer im Parallelbetrieb.
Parallel operation of Paddel Mixers.

1.3 Mischen durch Erzeugung hoher Scherkräfte

Für dieses Mischverfahren eignen sich die hochtourigen Mischer, bekannt unter den Bezeichnungen Kolloidal- oder Turbomischer. Nachstehend wird die Bezeichnung Turbomischer verwendet. Diese bestehen normalerweise aus einem Mischbehälter und einer Misch- oder Zirkulationspumpe (Fig. 4). Bei diesen Mischern wird zuerst die genaue Menge Wasser in den Mischbehälter eingefüllt. Durch die Mischerpumpe wird die Flüssigkeit vom Boden des Behälters angesaugt und in dessen oberen Bereich mit hoher Geschwindigkeit wieder eingespiesen. Die Trockenkomponenten werden von oben zugegeben.

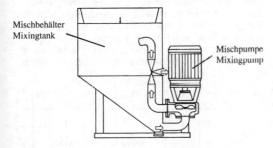

Mischbehälter
Mixingtank

Mischpumpe
Mixingpump

Fig. 4: Turbomischer (Kolloidalmischer).
High-Shear Mixer (Colloidal Mixer).

Im Behälter selbst findet nur eine Vorbenetzung der Trockenkomponenten statt. Der eigentliche Mischeffekt wird jedoch durch die hohen, in der Mischerpumpe erzeugten, Scherkräfte erzielt. Diese hohen Scherkräfte führen zu einer Trennung der einzelnen Partikel, was bedeutet, dass jedes Teilchen individuell benetzt wird und dass sich keine Klumpen bilden können. Das Resultat ist eine voll hydratisierte stabile Suspension mit ausgezeichneten Verarbeitungseigenschaften. Eine solche Mischung neigt sehr wenig zum Absetzen und dringt dank der guten Dispersion der Partikel in feinste Risse und Klüfte ein. Gleichzeitig handelt es sich um eine homogene Suspension, welche sich bei Injektionen im Grundwasser kaum entmischt. Eine stabile Mischung

verdrängt das Wasser ohne sich zu entmischen und ohne ihre Eigenschaften zu verändern.

Ein einfacher Test zur Veranschaulichung des Unterschiedes zwischen einer instabilen und einer stabilen Mischung kann wie folgt durchgeführt werden:

Es werden dieselben Komponenten (z.B. W/Z = 0,5) einmal in einem Paddelmischer und einmal in einem hochtourigen Turbomischer gemischt. Anschliessend wird eine Gegenstand (z.B. Maurerkelle) zuerst in die entsprechende Mischung und dann in einen Behälter mit klarem Wasser eingetaucht. Bei der instabilen Mischung wird das Wasser sofort stark getrübt und die Suspension vom betreffenden Gegenstand abgewaschen. Bei der stabilen Suspension findet nur eine leichte Trübung statt und die Mischung bleibt zum grössten Teil haften.

Ein guter Turbomischer sollte folgende Eigenschaften aufweisen:

1. Die Drehzahl der Mischpumpe sollte zwischen 1500 und 2000 min^{-1} liegen.

2. Die Kapazität der Mischpumpe sollte im Minimum eine dreimalige Umwälzung des gesamten Behälterinhalts pro Minute erlauben. Der Druck der Mischpumpe spielt eine untergeordnete Rolle und ist höchstens für die anschliessende Förderung des Mischgutes zum Stapelbehälter von Bedeutung.

3. Die Mischpumpe sollte hohe Scherkräfte erzeugen, entweder durch enge Toleranzen zwischen Laufrad und Pumpengehäuse (Fig. 5) oder durch Erzeugung sehr hoher Turbulenzen im Pumpengehäuse (Fig. 6). Konventionelle Zentrifugalpumpen sind

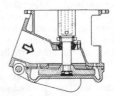

Fig 5: Mischerpumpe «System Craelius»; Erzeugung der Scherkräfte durch enge Toleranzen.
Mixing Pump «Craelius System»; The shear forces are created by close tolerances between impeller and casing.

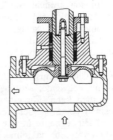

Fig 6: Mischerpumpe «System Häny»; Erzeugung der Scherkräfte durch hohe Turbulenzen im Gehäuse.
Mixing Pump «Häny System»; The shear forces are created by high turbulence in the casing.

für diesen Zweck ungeeignet, da diese dafür ausgelegt sind, ein Medium ohne oder nur mit wenig Feststoffen - bei einem hohen Wirkungsgrad - zu fördern. Dieses Ziel kann nur durch eine möglichst turbulenzarme Strömung durch die Pumpe erreicht werden. Dadurch ist der Mischeffekt einer solchen Pumpe minimal. Am besten eignen sich Pumpen mit Laufrädern, welche eine hohe Turbulenz erzeugen. Es sind dies sogenannte Wirbelräder, welche zurückgesetzt im Pumpengehäuse angeordnet sind. Diese Pumpentypen haben deutlich weniger Verschleiss als solche mit engen Toleranzen und lassen grössere Partikel passieren ohne zu verstopfen oder Schaden zu nehmen. Gleichzeitig können diese Wirbelradpumpen grössere Mengen von Sand im selben Arbeitsgang verarbeiten was den Einsatz eines zweiten, speziellen Sandmischers erübrigt.

4. Der Mischbehälter sollte so geformt sein, dass der Wirbel eines tangentialen Eintritts gegen den Behälterboden hin gebrochen wird, um das Ansaugen von Luft zu verhindern. Zum Teil werden Prallbleche oder sonstige Schikanen eingebaut, um diesen Wirbel zu brechen. Solche Einrichtungen sind wenig geeignet, da diese auf der Strömungsrückseite Rückstände aufbauen, welche mit dem normalen Reinigungsvorgang schwer zu entfernen sind.

Turbomischer eignen sich zur schnellen Aufbereitung homogener Suspensionen bis zu W/Z Faktoren von 0,35. Nur 30 bis 60 Sekunden Nachmischzeit ist erforderlich, nachdem die Zementzugabe beendet ist. Es ist jedoch zu beachten, dass diese Mischer mit einem hohen Energieeintrag sehr wirkungsvoll arbeiten. Deshalb sollte die Mischzeit kurz gehalten werden. Bei einer zu langen Mischdauer wird zuviel Energie eingetragen, das Mischgut erwärmt sich zu stark und führt zum "todmischen" einer Suspension. Das Mischgut soll also unmittelbar nach Ablauf der optimalen Mischzeit einem Stapelbehälter mit langsamdrehendem Rührwerk zugeführt werden. Da diese Turbomischer bereits mit einer Pumpe ausgerüstet sind, lässt sich das Mischgut problemlos in einen gleichhoch oder höher gestellten Stapelbehälter fördern.

Keine anderen Mischertypen eignen sich in gleicher Weise zur Aufbereitung von Bentonitsuspensionen. Mit einer Mischzeit von 4 bis 6 Minuten lassen sich Resultate erzielen, welche mit herkömmlichen Mischern nur durch eine zusätzliche Quellzeit von 12 bis 18 Stunden erreicht werden. Für Zusätze von Bentonit in der Injektion bedeutet das, dass dieses direkt in einem Arbeitsgang in die Suspension eingemischt werden können und sich deshalb die voluminösen Stapelbehälter erübrigen.

Beide Typen von Mischern, die Paddel- sowie die Turbomischer sind Chargenmischer, d.h. ein kompletter Mischzyklus muss beendet sein bevor das Mischgut verwendet werden kann. Für eine kontinuierliche Injektion ist deshalb ein Stapelbehälter oder Rührwerk erforderlich. Hingegen kann der W/Z Faktor präzise eingehalten werden, da immer eine gegebene Menge Zement einer darauf abgestimmten Menge Wasser gegenübersteht.

Die Durchlaufmischer benötigen nur einen sehr kleinen Stapelbehälter, da laufend neues Mischgut

produziert wird, hingegen kann der W/Z Faktor nicht präzise kontrolliert werden.

1.4 Stapelbehälter/Rührwerke

Dem Chargenmischer wie Paddel- oder Turbomischer sollte in der Regel ein Stapelbehälter mit Rührwerk nachgeschaltet werden. Dies erlaubt ein kontinuierliches Injizieren der Suspension. Der Inhalt des Stapelbehälters sollte das 1,2- bis 1,5-fache Volumen des Mischers aufweisen, um das Ansaugen von Luft durch die Injektionspumpe zu verhindern. Ein langsamdrehendes Rührwerk verhindert das Absetzen der Mischung während deren Verweilzeit und befreit die Suspension gleichzeitig von allfälligen Luftblasen, welche durch den intensiven Mischprozess im Turbomischer eingetragen werden.

2 INJEKTIONSPUMPEN

Um beste Injektionsresultate zu erzielen, ist die Wahl des optimalen Druckes sowie die der optimalen Fördermenge von zentraler Bedeutung.

2.1 Fördermenge

Hohe Fliessgeschwindigkeiten des Injektionsgutes in weichen Böden können zu Ausspülungen führen. Dies bedeutet, dass sich eine grössere Menge des Injektionsgutes einen Weg in Zonen bahnt, wo dieses nicht erwünscht ist und gleichzeitig nicht in die gewünschten Bereiche eindringt. Es gilt zu beachten, dass je höher die Fliessgeschwindigkeit ist, desto höher auch die Reibungsverluste sein werden (Fig. 7). Höhere Reibungsverluste wiederum verlan-

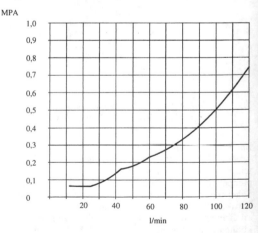

Fig. 7: Druckverluste einer Injektionssuspension; W/Z Faktor = 1, +3% Bentonit in 100 m Stahlrohr 1".
Fricton loss curve of a grout suspension; w/c ratio = 1, +3% bentonite in 100 m of 1" steel pipe.

gen höhere Drücke, welche nicht in jedem Fall wünschenswert sind. Eine Injektion mit geringeren Fliessgeschwindigkeiten hat somit die besseren Chancen gute Resultate zu erzielen. Es liegt in der Natur der Sache, dass die ausführenden Firmen nicht unbedingt eine langsame Injektion vorziehen, besonders wenn die Abrechnung rein über die Quantität des injizierten Zementes erfolgt.

2.2 Druck

Ebenso wichtig oder gar wichtiger als die korrekte Strömungsgeschwindigkeit ist der Druck. Zu hoher Druck kann zu Hebungen des Bodens oder zur Zerstörung von Bodenstrukturen führen. Zu niedriger Druck resultiert in ungenügender Füllung der Hohlräume. Für gewisse Anwendungen (z.B. Kontaktinjektionen oder Hinterfüllung von Tübbings) kann der Maximaldruck relativ leicht errechnet werden. Für Bodenkonsolidierungen und Schirminjektionen bei Dammbauten ist es aufgrund der unterschiedlichen Bodenstrukturen wesentlich schwieriger, eine genauere Druckvorgabe zu geben. Zusätzlich gibt es verschiedene Ansichten bezüglich des optimalen Injektionsdruckes. Die nordamerikanische Philosophie geht dahin, die Bodenstrukturen so intakt wie möglich zu belassen, wogegen die europäische Philosophie versucht, die ohnehin schwachen Strukturen zu durchbrechen und zusammen mit dem umgebenden Gebirge zu konsolidieren. Als Faustregel gilt für Nordamerika ein Druck von ca. 0,22 bar je Meter Injektionstiefe, wogegen der europäische Wert bei 1 bar je Meter liegt.

Gegensätzliche Ansichten bestehen besonders auch darin, ob bei möglichst gleichmässigem Druck und Fliessgeschwindigkeit injiziert werden soll oder ob Schwankungen toleriert werden können. Es ist absolut klar, dass Druckspitzen, welche über dem maximal zulässigen Höchstdruck liegen, nicht toleriert werden können (Fig. 8). Anderseits haben Erfahrungen gezeigt, dass kurze Druckabfälle innerhalb des vorgegebenen Druckbereiches dem Injektionsvorgang

durchaus förderlich sind, d.h. dass bei gleichem Druck und gleicher Durchflussmenge, mehr Material injiziert werden kann. Der Grund liegt wahrscheinlich darin, dass die Partikel sich in den Hohlräumen verkeilen und zu Brückenbildungen führen. Diese Brückenbildung beschleunigt den Filtrationseffekt und stoppt den Fluss. Bei gleichbleibendem Druck wird dieser Effekt beschleunigt, da die Partikel keine Möglichkeit haben, sich in ihrer Position zu verändern. Eine Erhöhung des Druckes bringt keine Verbesserung, da lediglich der Filtrationseffekt verstärkt wird (Fig. 9). Eine kurze Druckentlastung hingegen ermöglicht den Partikeln, sich neu zu

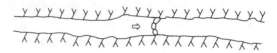

Fig 9: Brückenbildung bei konstantem Druck.
Bridging effect at constant pressure.

formieren, d.h. Brücken werden zerstört und somit der Filtrationseffekt verlangsamt (Fig. 10). Aus dieser Sicht ist die Verwendung von Kolben- oder Plungerpumpen vorteilhaft, bedingt aber, dass diese mit einer präzisen Druckregulierung ausgerüstet sind,

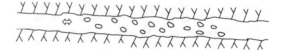

Fig 10: Neuformierung der Partikel durch kurze Druckentlastungen.
Reformation of particles through short pressure relaxation.

welche auch keine Druckspitzen zulässt. Mit solchen Pumpen kann auf die, ausserhalb Europa's noch weit verbreitete, Zirkulationsleitung verzichtet werden. Diese Art Pumpen erlauben auch eine Druckhaltung über beliebige Zeit am Ende eines Injektionszykluses.

2.3 Auswahl der Pumpe

Bei der Auswahl der geeigneten Injektionspumpe sind folgende Punkte zu beachten:
• Der maximale Injektionsdruck sollte an der Pumpe stufenlos einstellbar sein. Alle Regelorgane sollten sich antriebsseitig (pneumatische und hydraulische Antriebe) befinden, da Regelventile mediumseitig schlecht - und nur für kurze Zeit funktionieren. Das Druckregelventil sollte so konstruiert sein, dass der eingestellte Maximaldruck über beliebige Zeit gehalten werden kann. Druckschalter mit EIN-AUS Funktion sind ungeeignet, da diese die Pumpe in kurzen Abständen ein- und ausschalten und dadurch kein eigentlicher Druckhalteeffekt erzielt werden kann.

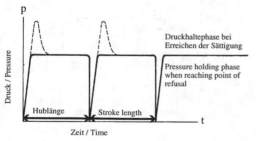

Druckspitzen von Pumpen mit ungenügender Regelung.
Pressure peaks from pumps with inadequate pressure control.

Druckverlauf von Pumpen mit präziser Regelung.
Characteristics for pumps with precise pressure control.

Druckhaltephase bei Erreichen der Sättigung
Pressure holding phase when reaching point of refusal

Hublänge   Stroke length

Druck / Pressure

Zeit / Time

Fig. 8: Druckverlauf von Kolben- und Plungerpumpen.
Characteristics of Piston and Plunger Pumps.

- Die Pumpe sollte bezüglich Fördermenge von 0 - 100% voll regelbar sein.
- Pumpen mit Direktantrieben haben in der Regel keine Druckeinstellmöglichkeiten und weisen nur selten variable Fördermengen auf. Solche Pumpen sind für Injektionen ungeeignet.
- Das abrasive Injektionsgut verlangt nach einem verschleissarmen Pumpsystem (z.B. Plungerpumpen).
- Einen hohen Stellenwert haben einfache Reinigungs- und Wartungsmöglichkeiten.
- Von wesentlicher Bedeutung sind auch die Ventile. Diese sollten einen grossen Durchgang haben damit auch hochviskose- und eventuell sandhaltige Injektionsmedien störungsfrei gepumpt werden können.

Die heute am häufigsten angewandten Pumpsysteme sind:

2.4 Exzenterschneckenpumpen

Die Exzenterschneckenpumpen (auch bekannt unter den Namen Mono-, Moyno- oder Moineau Pumpen) haben den Vorteil geringer Anschaffungskosten (Fig. 11). Bedingt durch den hohen Verschleiss an Stator und Rotor sind jedoch die Unterhaltskosten relativ hoch. Der maximale Injektionsdruck ist geringer als bei andern Pumpsystemen, was den Einsatz dieser Pumpen stark einschränkt. Die meisten Modelle haben weder eine Druck- noch eine Mengenregulierung. Druck und Menge werden über eine Zirkulationsleitung zum Stapelbehälter durch Absperrhahnen manuell reguliert. Dies ist kostenintensiv da diese Funktion durch eine zusätzliche Person ausgeführt wird. Die Sicherheit bezüglich Druck hängt völlig von der Aufmerksamkeit dieser Person ab. Dieses Pumpsystem ist vor allem in Nordamerika noch weit verbreitet.

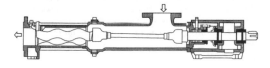

Fig 11: Exzenterschneckenpumpe.
Progressive cavity pump.

Zusammenfassung:
- Niedrige Investitionskosten
- Hohe Unterhalts- und Personalkosten
- Limitierte Anwendung (kleiner Druckbereich)

2.5 Kolbenpumpen

Kolbenpumpen (Fig. 12), welche in der Injektion Anwendung finden, sind meistens doppeltwirkende, oszillierende Pumpen. Die Anschaffungskosten liegen in der Regel höher als bei Exzenterschneckenpumpen. Es gibt Kolbenpumpen für praktisch jeden Druck- und Mengenbereich. Druck- und Mengenregulierung sind erhältlich bei einigen pneumatisch- oder hydraulisch angetriebenen Modellen. Sandhaltiger Injektionsmörtel mit kleineren Korngrössen kann gefördert werden. Durch die präzise Kolbenführung im Zylinder wird der Verschleiss bei der Förderung abrasiver Medien stark erhöht. Ein weiterer Nachteil liegt darin, dass bei der Förderung geringer Mengen von relativ dickflüssiger Suspension oder Injektionsmörtel - bedingt durch die kleine Fliessgeschwindigkeit - die Schwerkraftventile stark zu Funktionsstörungen oder Verstopfungen neigen.

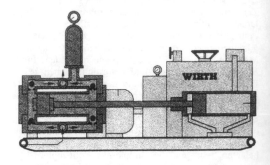

Fig 12: Kolbenpumpe.
Piston Pump.

Zusammenfassung:
- Höhere Anschaffungskosten
- Niedrigere Unterhalts- und Personalkosten
- Breite Anwendung (grosser Druckbereich)
- Neigung zu Verstopfungen

2.6 Plunger Pumpen

Plungerpumpen sind die am universellsten einsetzbaren Injektionspumpen (Fig. 13). Sie sind als einfach-

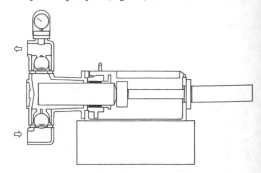

Fig 13: Plungerpumpe.
Plunger Pump.

wirkende , doppeltwirkend oszillierende (Fig. 14) und doppeltwirkend parallel arbeitende Pumpen erhältlich (Fig. 15). Die Anschaffungskosten liegen etwa bei denjenigen der Kolbenpumpen. Plungerpumpen decken praktisch alle Druck- und Mengenbereiche ab. Einige Fabrikate weisen sehr präzise Druck- und Mengenreguliersysteme bei pneumatisch- oder hydraulisch angetriebenen Pumpen auf.
Die grossen Vorteile der Plungerpumpen sind:
- Extrem niedriger Verschleiss
- Einige Fabrikate bieten die Möglichkeit, leicht auswechselbare Plungereinheiten verschiedener Durchmesser in dieselbe Pumpe einzubauen. Dies verleiht diesen Injektoren ein grösstmögliches Anwendungsspektrum bei einer geringen Antriebsleistung.

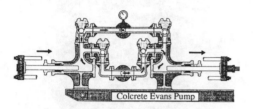

Fig 14: Doppeltwirkende, oszillierende Plungerpumpe. Double acting, reciprocating Plunger Pump.

Fig 15: Plungerpumpe, doppeltwirkend (Parallelmodule). Plunger Pump, double acting (side by side).

- Es gibt Plungerpumpen, einfach- und doppeltwirkende (Parallelmodule), welche den Rück- oder Saughub, unabhängig von Druck oder Fördermenge, immer mit derselben hohen Geschwindigkeit ausführen. Dies führt zu einer relativ hohen Fliessgeschwindigkeit des Injektionsmediums im Bereich der Saugventile, was bewirkt, dass diese bei jedem Saughub intensiv gespült werden.

Verstopfungen oder das Aufschwimmen von Ventilkugeln werden dadurch selbst bei Injektionen mit kleinsten Mengen oder hohen Drücken wirkungsvoll verhindert. Fliessmörtel mit Körnung bis zu 8 mm Durchmesser kann mit diesen Pumpen problemlos gefördert werden.

Zusammenfassung:
- Höhere Anschaffungskosten
- Niedrigste Unterhaltskosten
- Universellstes Pumpsystem
- Verstopfungsfreie Modelle mit schnellem Saughub erhältlich.

Weitere, aber nicht sehr verbreitete Pumpsysteme sind ferner: Membran- und Schlauchquetschpumpen sowie Druckbehälter.

3 QUALITÄTSKONTROLLE

Um den Erfolg eines Injektionsprojektes sicherzustellen, sollte ein Qualitätssicherungsprogramm vorhanden sein. Ein solches Programm umfasst die Injektionsversuche, die Kontrolle während der Injektion und die Überprüfung des Injektionserfolges.
Die auszuführenden Tests sowie die zu sammelnden Daten sollten aus den Spezifikationen klar hervorgehen und umfassen u.a. Wasserabpressversuche (Lugeon-Test) vor und nach der Injektion. Wichtig ist, dass diese Tests immer unter denselben Bedingungen durchgeführt werden, damit Vergleiche und Rückschlüsse möglich sind.
An dieser Stelle konzentrieren wir uns auf die Erfassung der wichtigsten Daten wie Zusammensetzung der Injektionsmischung, Injektionsdruck, Injektionsmenge je Zeiteinheit und total injizierte Menge. Diese Parameter sollten je Bohrloch und Stufe erfasst werden.

3.1 Zusammensetzung der Injektionsmischung

Einer der schwierig zu erfassenden Parameter ist die genaue Zusammensetzung der Injektionsmischung sofern diese nicht über eine automatische Dosieranlage aufbereitet wird. Moderne Dosieranlagen (Fig. 16) drucken automatisch Chargenprotokolle aus, welche alle notwendigen Informationen über die Zusammensetzung des Mischgutes enthalten. Manuell hergestellte Mischungen verlangen ein hohes Mass an Disziplin vom Bedienungspersonal. Oft ist es einfacher, die Suspension etwas dünner anzumischen, um möglichen Verstopfungen aus dem Weg zu gehen.
Ein, auf die spezifischen Bedürfnisse ausgearbeiteter Wartungs- und Unterhaltsplan, ist Bestandteil der Qualitätskontrolle und lässt der Bedienungsmannschaft wenig Freiraum für nicht fachgerechte Eigenkreationen. Der Einsatz von Densometern oder regelmässige Mischproben können einen wesentlichen Beitrag zur Einhaltung der vorgegebenen Zusammensetzung leisten.

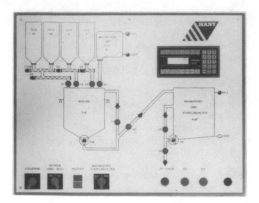

Fig 16: Steuerung einer vollautomatischen Mischanlage mit Gewichtsdosierung.
Control of a fully automated mixing plant with weigh-batching

3.2 Datenerfassung von Druck und Menge

Die heutige Messtechnik erlaubt das genaue Erfassen der Durchflussmenge sowie des Druckes. Für die Messung der Durchflussmenge werden vorzugsweise induktive Durchflussmesser eingesetzt, welche bei vollem Querschnitt keinerlei bewegliche Teile aufweisen. Die Auswahl des geeigneten Durchflussmessers sollte jedoch einem Spezialisten überlassen werden. So ist es z.B. wichtig, dass das Gerät nicht nur im Bereich des Messbereichendwertes genau misst, sondern ebenfalls bei kleinen Durchflussmengen von 1 bis 5 l/min. Ein zweites Kriterium ist die Fähigkeit des Durchflussmessers, den stark schwankenden Geschwindigkeiten des Mediums zu folgen. Dies ist besonders wichtig, wenn Kolben-, Plunger- oder Membranpumpen eingesetzt werden. Bei geeigneten Durchflussmessern liegt deshalb die Genauigkeit bei ca. ± 1% bei einem Durchflusswert von über 5% vom Messbereichsendwert. Bei niedrigen Durchflusswerten zwischen 1% und 5% des Messbereichsendwertes beträgt die Genauigkeit immer noch ca. ± 5%. Die Zusammensetzung des Mischgutes hat keinen Einfluss auf die Messgenauigkeit solange dieses eine minimale Leitfähigkeit (ca. 10 µS) aufweist.

Der Druck sollte so nahe an der Injektionsstelle wie möglich gemessen werden. Dadurch bleiben die Messresultate durch die Reibungsverluste in längeren Injektionsleitungen unbeeinflusst. Die Druckmessinstrumente sollten durch eine Membrane und Trennflüssigkeit vom Injektionsmedium getrennt sein. Es gibt auch Druckaufnehmer, welche den effektiven Druck am Packer im Bohrloch messen. Die Handhabung solcher relativ heikler Instrumente bedarf jedoch besonderer Sorgfalt, welche bei durchschnittlichen Injektionsmannschaften erfahrungsgemäss nicht immer gegeben ist. Für Wasserabpressversuche, welche normalerweise durch einen erfahrenen Ingenieur überwacht werden, ist die Anwendung solcher Imloch-Druckaufnehmer wesentlich weiter verbreitet.

3.3 Registriereinrichtung /Datenverarbeitung

Durch die weit verbreitete Anwendung von induktiven Durchflussmessern und elektronischen Druckaufnehmern können Menge und Druckverlauf kontinuierlich auf einen Linienschreiber aufgezeichnet werden. Die Totalmenge wird auf einem rückstellbaren Zähler aufaddiert (Fig.17).

Fig 17: Einfaches Druck- und Mengenregistriergerät.
Simple pressure and flow recorder.

Die heutige Datenverarbeitungstechnik erlaubt zudem unzählige Möglichkeiten der Verarbeitung und Ausgabe der Daten. All diese Möglichkeiten können bei der Analysierung eines Injektionsprozesses sehr hilfreich sein. Es darf jedoch nicht ausser Acht gelassen werden, dass diese elektronischen Hilfsmittel wertvolle Werkzeuge für den Ingenieur sein sollten und dass kein Computer das Einschätzungs- und Beurteilungsvermögen eines erfahrenen Ingenieurs ersetzen kann. Manchmal wird man jedoch den Eindruck nicht ganz los, dass Computer und Monitoren in klimatisierten Räumen nicht vorhandene oder nicht genutzte Erfahrungen kompensieren sollten.

Es ist wichtig, dass die Registriereinrichtungen den rauhen Baustellenbedingungen angepasst konstruiert sind. Der Betrieb sowie der normale Unterhalt sollten durch geschultes Fachpersonal und nicht durch Computerspezialisten durchgeführt werden können.

4 SCHLUSSBEMERKUNG

Der heutigen Injektionstechnologie angepasst ist die Verwendung von entsprechenden Geräten unerlässlich. Zu oft werden aber noch selbstgebastelte, nicht dem Stand der Technik angepasste Geräte eingesetzt und man wundert sich, weshalb die erzielten Resultate nicht den Erwartungen entsprechen. Es sollte das Ziel der Injektionsindustrie sein, ihren Ruf weiter zu verbessern und die Methoden von "Kunst" in Richtung "Wissenschaft" weiterzuentwickeln. Dazu gehören auch alte Ausschreibungsunterlagen, welche seit Jahren oder gar Jahrzehnten immer wieder verwendet werden, ohne Rücksicht auf die technischen Fortschritte der Industrie. Die Schwerpunkte in den Ausschreibungen sollten sich in Bezug auf die Geräte vermehrt mit dem zu erzielenden Resultat als mit gerätespezifischen Eigenschaften wie minimale Mischdauer, Druck der Mischpumpe oder Typ der Injektionspumpe befassen.

REFERENCES

Houlsby, A.C. 1990. Construction and Design of Cement Grouting. New York, John Wiley & Sons.
ISRM. Commission of Rock Grouting 1992. Interim Report. ISRM.
Kutzner, C., 1991. Injektionen im Baugrund. Stuttgart Enke.
Müller, R.E. 1984. Monitoring of Pumping Tests and Grouting Operations. Springer Verlag

Grouting in Rock and Concrete, Widmann (ed.) © 1993 Balkema, Rotterdam, ISBN 90 5410 350 7

Gezielte Hebung durch SOILFRAC-Injektion

Targeted raising by SOILFRAC-injection

P. Stockhammer & E. Falk
Keller Grundbau GesmbH, Wien Österreich

ZUSAMMENFASSUNG:

Anhand der grundlegenden Problemstellungen im Rahmen von Hohlraumbauten und Bauwerksgründungen werden die hauptsächlichen Anwendungsfälle für das Verfahren in Lockergestein und Fels gezeigt.

Die Durchführung von Bodenverbesserungen mittels SOILFRAC wird als Zusammenwirken von installiertem System, stufenweiser Injektion und begleitender Beobachtung beschrieben.

Für die effiziente Steuerung des Verfahrens sind genaue Kenntnisse über den Einfluß einzelner Ausführungsparameter erforderlich. Aufgrund vorliegender Erfahrungen wird über Ansätze zur systematischen Verfahrensabstimmung auf bestimmte Bodenverhältnisse berichtet. Erfahrenes Personal, das sein Wissen über die Reaktion des Bodens ständig erweitert, kann auf das Injektionskonzept anhand kontinuierlich erfaßter Daten Einfluß nehmen.

Die schematische Darstellung erfolgreicher Anwendungen zeigt das weite Einsatzgebiet des Verfahrens in der Verbesserung von Problemzonen des Untergrundes.

ABSTRACT:

Considering the fundamental problems in the area of underground excavation and foundation construction it will be shown the principal applications for the treatment of alluvium and rock.

The successful ground improvement to such zones by means of SOILFRAC will be described as a combination of the installation system, stage injection and accompanying observation.

For the efficient control of the technique exact knowledge is necessary about the influence of each individual performance parameter. From current experiences it will be reported about the emergence of a systematic method for making adjustments to the technique to suit particular ground conditions.

Experienced personnel with their constantly widening knowledge about the reaction of the ground, can influence the injection process with the help of continually recorded data.

The schematic presentation of successful applications confirms the wide uses of the technique in the improvement of problem zones in the underground.

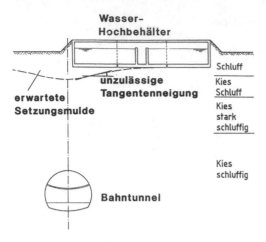

Abb. 1: Unterfahrung eines Trink-
wasser-Hochbehälters durch
einen Eisenbahntunnel

Fig. 1: Tunneling for a railway under
a drinking water holding
tank

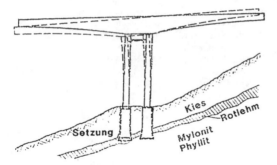

Abb. 2: Schiefstellung eines Brückenpfeilers während der
Errichtung des Tragwerks

Fig. 2: Differential settlement of a
bridge pier during the
errechen of the
ouperstructure

1. GRUNDLEGENDE PROBLEMSTELLUNGEN

1.1 Der Bedarf an Unter-
grundverbesserungen besteht vor allem
im wenig tragfähigen Gebirge, das
aufgrund von Schwächezonen durch
Verwitterung, Lösungsvorgänge oder
starke Klüftung Probleme für die
geplante Errichtung von Bauwerken
aufwirft.

1.1.1 Wenn durch geplante Hohlraumbauten unzulässige Setzungs-
bewegungen für bestehende Bauwerke
erwartet werden, besteht im
allgemeinen der Wunsch nach einer
setzungsabhängigen Nachstellmöglich-
keit. Ausgeprägte Setzungsmulden,
wie sie z. B. durch die Errichtung
oberflächennaher Tunnel entstehen,
stellen aufgrund ihrer ungünstigen
Tangentenneigungen die Nutzbarkeit
höherwertiger Gebäude in Frage. Oft
sind hohlraumnahe setzungsmindernde
Maßnahmen während des Vortriebs zur
Verminderung von Bewegungen im
Tunnelprofil selbst geeignet, ihr
Einfluß auf die Bewegungen an der
Oberfläche bleibt jedoch begrenzt.

1.1.2 Die Gründung von Bauwerken kann
nicht immer in optimalem Untergrund
erfolgen. Wenn die Felsoberkante mit
wirtschaftlichen Maßnahmen nicht
erreicht werden kann, stellt eine
Verbesserung der darüberliegenden
Schichten eine überlegenswerte
Alternative dar. Eine wichtige
Erweiterung bestehender Baumethoden
ist die Möglichkeit zur Rückstellung
von bereits eingetretenen Setzungen.

1.1.3 Hangbewegungen an Übergangszonen
mit verschiedenen Materialeigenschaften werden in der Regel
steife Verbaumaßnahmen entgegen-
gesetzt. Mit der Verbesserung der
bewegten Hangbereiche selbst steht ein
alternativer Ansatz zur Problemlösung
zur Verfügung.

1.1.4 In klüftigem Gebirge sind
Abdichtungsmaßnahmen dann proble-
matisch, wenn eingeschaltete,
undurchlässige Zonen die Reichweite
herkömmlicher Injektionen ein-
schränken. Wenn die natürliche
Erreichbarkeit von offenen Hohlräumen
nicht gegeben ist, müssen künstliche
Fließwege geschaffen werden.

1.2 Randbedingungen

1.2.1 Im Fall von Hohlraumbauten sind
die erwarteten Bewegungen bekannt und
ihr absolutes Maß ist rechnerisch
abschätzbar. Wenn unzulässige
Oberflächensetzungen oder Setzungs-
differenzen erwartet werden, ist ein

Ausgleich der auftretenden Bewegungen wünschenswert. Die aktive Einstellung auf auftretende Setzungen kann dann erfolgreich sein, wenn die selektive Rückstellung einzelner Gebäudebereiche gelingt (Kompensationsinjektion).

1.2.2 Wenn Setzungen bereits eingetreten sind oder sich noch in Gang befinden. sind die Sicherheitszustände in komplexen Systemen schwer abschätzbar. Eine dosierte Vorgangsweise mit kontinuierlicher Verbesserung der Bodenkennwerte und damit der Gesamtsicherheit ist oft die einzige Möglichkeit, die Sicherung des Gesamtsystems schrittweise zu erhöhen.

1.2.3 Der zur Verfügung stehende Arbeitsraum ist besonders bei Sanierungen nicht immer für Großgeräte ausreichend. Teilbereiche von Produktionsprozessen oder Verkehrswegen dürfen nicht unterbrochen werden.

Das Bohren eines Manschettenrohrsystems erfolgt mit Geräten verschiedener Größe. Somit kann die Flexibilität des Verfahrens den Ausschlag für die Anwendung geben.

2. BODENVERBESSERUNG

Allgemeines Ziel von Bodenverbesserungen ist die Erhöhung der Tragfähigkeit von Problemzonen des Untergrunds. Im günstigsten Fall kann damit auf die Weiterführung der Belastungen in andere, besser tragfähige Bereiche verzichtet werden. Durch die Einpressung von Feststoffmengen unter hydraulischem Druck können die Richtung der Hauptspannungen und das Seitendruckverhältnis in gewünschter Weise beeinflußt werden.

Soilfracturing ist aus dem Erdölbergbau zur Erhöhung der Lagerstättenkapazität schon seit Jahrzehnten bekannt. Die Anwendung im Erd- und Grundbau erfolgt ebenfalls seit vielen Jahren mit allerdings begrenzten Einsatzmöglichkeiten. Mit der Weiterentwicklung der Injektionstechnik und Meßtechnik haben wesentliche Verfahrensbestandteile des

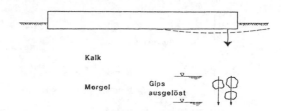

Abb.3:   Setzungsschäden an einem Industriegebäude hervorgerufen durch Lösungsvorgänge im Untergrund

Fig.3:   Settlement demage to a industrial building caused by erosion in the underground

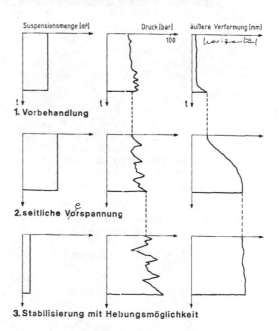

Abb. 4:   Zunehmende Einschränkung der Beweglichkeit im Untergrund durch vielfache Injektion bei steigendem Druck. Die seitliche Verspannung des Bodens ist Bedingung für eine angestrebte Bauwerkshebung.

Fig. 4:   Increase in the rigiclaty of the subsoil through multiple phase injections and higher pressures. The lateral stressing of the sort is a condition for the raising of the building

197

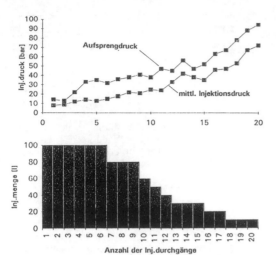

Abb. 5: Mit zunehmender Anzahl der Injektionsdurchgänge steigt mit den Injektionsdrücken die Effektivität der eingepreßten Mengen.

Fig. 5: The increase in the number of grouting phases leading to higher grouting pressures greatly increases the effectiveness of the injected material

Soilfrac-Verfahrens zu komplizierten Anwendungsmöglichkeiten geführt.

Die Weiterentwicklung betrifft insbesondere:

1) Hohe Meßgenauigkeit aufgrund neuer Meßverfahren
2) Unmittelbare Verfügbarkeit komplexer Meßdaten mittels EDV für den leitenden Ingenieur.
3) Hohe Bohrgenauigkeit und Wirtschaftlichkeit der Bohrverfahren
4) Vollautomatische Injektionsanlagen und rasche Reaktionsmöglichkeit durch neue Kommunikationsmittel
5) Verfügbarkeit von Injektionsmischungen mit
   - hohem Feststoffanteil
   - geringer Viskosität
   - Steuerbarkeit der Abbindegeschwindigkeit
6) Setzungsminimierung durch verbesserte Mantelmischungen
7) Neue Injektionstechnik in Form von speziellen Manschettenrohren, Packeranordnungen und Dosiersystemen für die injizierten Medien

2.1 Verfahrensbeschreibung SOILFRAC

2.1.1 Bohrung:

Nach einer Festlegung jener Bodenbereiche, auf die sich die Verbesserungsarbeit konzentrieren soll, erfolgt die Installation von Manschettenrohren. Mittels verrohrter Bohrung werden Manschettenrohre mit einem Durchmesser von 1 – 2 Zoll eingebaut, die für eine vielfache Injektion unter wachsendem Druck geeignet sind. Die Bohrarbeiten können sowohl von der Oberfläche als auch von Schächten oder Gräben aus durchgeführt werden. Injektionsmatten können im Boden anteilhaft die Funktion einer Bewehrung übernehmen, sodaß nach Möglichkeit eine Ausrichtung der Rohre entsprechend den auftretenden Belastungsrichtungen angestrebt wird.

Der Hohlraum rund um das Manschettenrohr wird während des Ziehens der Bohrverrohrung kontinuierlich mit einer rasch ansteifenden Mantelmischung verpreßt, um Auflockerungen des Untergrunds zu minimieren.

2.1.2 SOILFRAC-Injektion:

Die Injektionsarbeiten selbst werden von einer stationären Anlage aus bedient. Über Druckleitungen gelangt die Suspension von der automatischen Misch- und Pumpanlage zu den im Injektionsprogramm vorbestimmten Injektionsstellen. Jede einzelne Ventilöffnung des gesamten Injektionssystems kann mittels Doppelpacker gezielt beaufschlagt werden. Mit der Fortdauer der Arbeiten und in Abstimmung auf den anstehenden Boden erfolgt die Anpassung der verwendeten Mischungen auf der Basis von Wasser, Zement, Kalksteinmehl und Bentonit an den jeweiligen Verwendungszweck. Die Variation des eingepreßten Feststoffanteils folgt den vorgegebenen Zielen:

- Verfüllung
- Verfestigung
- Stabilisierung
- Herstellung des Kraftschlusses als Vorbereitung zur Hebung
- Bauwerkshebung

Durch eine Steuerung der Abbindegeschwindigkeiten und der in einem Zug eingepreßten Mengen wird eine Reichweitenbegrenzung für das Eindringen der Suspension in den Baugrund sichergestellt.

Anfangs erfolgt die Ausbreitung der Mörtellamellen ("Fracs") relativ unbehindert. Mit der Zunahme der Verpreßdurchgänge und einem Anstieg des allgemeinen Druckniveaus verstärkt sich die Bildung immer feinerer Verästelungen. Die vielfältige Geometrie dieser Mörtelflächen und deren sich ändernde Fließrichtung erzeugen im Boden ein dichtes Mörtelgeflecht, das einzelne Bodenbereiche zu Einheiten zusammenschließt. Insgesamt wird die Verschieblichkeit von Böden und Kluftkörpern durch den Feststoffeinbau unter hohem Druck behindert bzw. unterbunden.

2.1.3 Steuerung und Dokumentation

SOILFRAC als Verbesserungsmethode für definierte Bodenbereiche bedarf der Unterstützung durch eine kontinuierlich durchgeführte messende Beobachtung. Ziel ist die Erstellung eines räumlichen Modells des bearbeiteten Bodens, in dem sämtliche Daten aus vorangegangenen Erkundungen, den Injektionsbohrungen selbst, Injektionsparametern und äußeren Bewegungen laufend berücksichtigt werden. Aus der Entwicklung der Parameter wie Druck, Druckverlauf und Verformung in Abhängigkeit des Arbeitsverlaufs können Rückschlüsse auf den Festigkeitszuwachs im Boden gezogen werden. Das System aus Baugrund, Bauwerk und aktiv wirkender Injektion wird auf jene Bereiche hin untersucht, wo gegenseitige Beeinflussungen effizient möglich sind.

Je nach geforderter Genauigkeit und möglicher geometrischer Anordnung kommen folgende Systeme zum Einsatz:

Präzisionsnivellement, Schlauchwaagensysteme, elektronische Neigungsmesser, vertikale und horizontale Inklinometer, Setzungspegel, Extensometer, Inkrementalextensometer, Fissurometer, optische Distanzmessung...

Zumindest die Schlauchwaagenmeßwerte werden über Datenwandler einem PC-Programm zugeführt, um in zwei- oder dreidimensionaler Darstellung jederzeit als Grundlage für ingenieurmäßige Entscheidungen über den Injektionsverlauf zur Verfügung zu stehen.

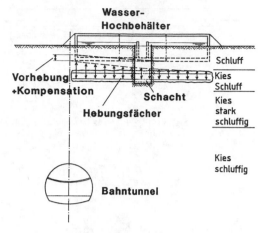

Abb.6: Kontinuierliche Kompensation von Setzungen während des Tunnelvortriebs durch einen Hebungsfächer.

Fig. 6: Continuous compensation of the settlement during the driving of the tunnel by fan of injection peruts

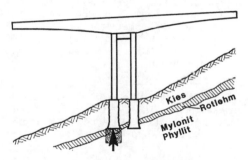

Abb.7: Gezielte Rückstellung der Differenzsetzungen von hoch belasteten Fundamenten (ca. 1200 kN/m2)

Fig.7: Targeted recovery of the differential settlement for highly loaded fondations (ca. 1200 kN/m2)

2.2 Hinweise zu Erfahrungen mit SOILFRAC

Entscheidend für den Erfolg von Injektionsmaßnahmen ist die Genauigkeit, mit der ein theoretischer Verbesserungsbereich im Boden tatsächlich hergestellt werden kann. Grundlagen für jede Sanierung sind die örtlichen Bodenaufschlüsse und Erkun-

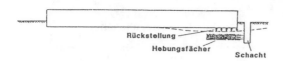

Rückstellung
Hebungsfächer
Schacht

Abb.8:   Örtliches   Ausgleichen   von
Setzungen mit späterer Nach-
stellmöglichkeit

Fig.8:   Localized   balanc   of   the
settlements and possibility
for later adjustment

dungsergebnisse.   Darüberhinaus   sind
vor   und   im   zuge   der   Ausführung
Versuche zweckmäßig und nötig, um die
wichtigsten Anmerkungen zu überprüfen.

2.2.1 Lamellenausbildung:

Stark   heterogene   Fels-   und   Bodenbereiche   können   bei   geeigneter   Abstimmung von Druck, Menge und Pumpgeschwindigkeit   verbessert   werden.
Trotz   der   Füllung   von   Hohlräumen   im
gebrochenen Fels mit tonigem Schluff
erfolgt   die   Ausbildung   der   Mörtellamellen   größteils   unabhängig   von
Materialgrenzen.

2.2.2   Oberflächennahe   SOILFRAC-
Injektionen   sind   bei   entsprechend
dosierter Verfahrensanwendung möglich
und   führen   nicht   zwangsläufig   zu
Austritten.

2.2.3 Die Ausbildung von Mörtellamellen im Torf nach dem SOILFRAC-
Verfahren ist möglich. Der Verbes-
serungseffekt   im   Boden   wird   nur
teilweise durch eine Zusammendrückung
bewirkt.   Die   dauerhafte   Verbes-
serungswirkung wird wesentlich durch
die   Ausbildung   eines   Verbundkörpers
aus Mörtel und eingeschlossenem Boden
erzielt.

2.2.4 Klüftiger   Mergel   kann   zur
Abdichtung   mit   SOILFRAC   behandelt
werden. In Bereichen, wo weder eine
Hochdruckbodenvermörtelung   noch   eine
herkömmliche Injektion einen Erfolg
versprechen, können die Wasser-
wegigkeiten mittels Aufsprenginjektion
unterbunden werden. Freilegungen erga-

ben, daß die nur wenige Millimeter
starken   Fracs   aufgrund   ihrer
intensiven   Verästelung   den   Wasser-
durchfluß wirksam behindern.

2.2.5 SOILFRAC-Injektionen können auch
in Schluff und Ton erfolgreich zur
gezielten   Hebung   von   Gebäudeteilen
eingesetzt werden.

2.2.6 Sowohl die Anhebung von Platten
als   auch   die   Einrichtung   von
Einzelfundamenten kann gezielt durchgeführt werden.

2.2.7   Durch   die   Einstellung   der
Suspensionsmischungen   auf   den
angetroffenen   Boden   wird   auch   in
rolligen   Schichten   eine   Penetration
der Porenräume weitgehend verhindert.
Die eingepreßte Feststoffmenge kann
auch in ausgeprägten Wechsellagerungen
ihre   Wirkung   durch   Aufsprengung
entfalten.

3. PRINZIP DER SOILFRAC-HEBUNGSWIRKUNG

Bei unsachgemäßer Feststoffinjektion
unter belasteten und nicht ausreichend
gegründeten Fundamenten können Setzungen und im Extremfall Grundbruchs-
bewegungen stattfinden.

Die   Rückstellung   von   bereits
eingetretenen   Bauwerkssetzungen   ist
nur unter Berücksichtigung von drei
grundsätzlich   verschiedenen   Phasen
möglich:

a) Umfassende Behandlung des gesamten
betroffenen Bodenbereichs

b)   Selektive   Verminderung   der
seitlichen   Verformbarkeit   von
Schwächezonen

c) Hebung im gegenläufigen Sinn zur
eingetretenen Setzung

Die sehr zeitraubenden Tätigkeiten der
Phasen a) und b) sind Grundlage für
den Erfolg von Phase c), welche in der
Regel die kürzeste ist.

4. ZUSAMMENFASSUNG

Die aktuelle Entwicklung der SOILFRAC-
Methode   stellt   eine   flexibel
einsetzbare Erweiterung der techni-

schen Möglichkeiten des Spezial-
tiefbaus dar. Der umfassende Einsatz
von Meßmethoden bildet die Grundlage
für die ingenieurmäßigen Entscheidungen, die laufend für den Erfolg des
Verfahrens zu treffen sind.

Das Verfahren bietet sich im Fels vor
allem für die Behandlung von Zonen
reduzierter Festigkeit an, wo die
konventionellen Verfahren der Injektionstechnik zur Verfestigung nur
begrenzt erfolgreich sein können.
SOILFRAC ist insbesondere auch im
Zusammenhang mit dem Hohlraumbau für
das gezielte Anheben von Boden- und
Gebäudebereichen vor oder nach dem
Auffahren von Hohlräumen geeignet.

SOILFRAC ist mit dem heutigen Stand
das einzige Bauverfahren der Injektionstechnik, welches einen gezielten,
rechtzeitigen und kontinuierlichen
Ausgleich von Setzungen während des
Eintretens der Bewegungen ermöglicht.

Literaturverzeichnis:

1) Brandl, H., (1991).
Stabilization of excessively settling
bridge piers.
  Proceedings of the tenth European
conference on soil mechanics and
foundation engineering. Florenz 1991

2) Brandl, H., (1992).
Fundamentsicherung einsturzgefährdeter
Brücken.
  Vortrag der Deutschen Baugrundtagung
1992 in Dresden.

3) Falk, E., Kühner, W., Bell, A.L.,
(1992).
Soilfrac und Soilcrete zur
Baugrundsanierung bei Brückenpfeilern.
  Vortrag der Deutschen Baugrundtagung
1992 in Dresden.

4) Gabener, H.-G., Raabe, E.-W.,
Wilms, J., (1989).
  Einsatz von Soilfracturing zur
Setzungsminderung beim Tunnelvortrieb.
Tunnelbautaschenbuch 1989

5) Meißner, H., Petersen, H., (1990).
Einpreßtechniken zur Erddruckerhöhung
und zum Anheben von Bauwerken.
  Bauingenieur 65, 1990

6) Moseley, M. P., (1993).
Ground Improvement.
  Blackie Academic & Professional

7) Priebe, H., Samol, H., (1985).
Soilfrac - Ein Injektionsverfahren zur
Bodenverbesserung.
  Proceedings 11th ICSMFE, San
Francisco 1985

8) Raabe, E.-W., Wehmeier, H.-J.,
Sondermann, W., (1990).
  Moderne Injektionstechniken für
Vortriebssicherung, Bebauungs- und
Grundwasserschutz.
  Vortrag der Deutschen Baugrundtagung
1990 in Karlsruhe

9) Ruppel, G., (1070).
Ausführung von Injektionen in
Lockergestein.
  Bergbauwissenschaften 17, Heft 8,
1970

Reinforcement of a stratified fracture zone in rock abutment using high pressure water jet technology

Verfestigung einer geschichteten Bruchzone im Felswiderlager mit Hochdruck-injektionen

Wang Zhiren
Foundation Treatment Company of Water Conservancy Ministry of China, People's Republic of China

Zhou Weiyuan
Tsinghua University, Beijing, People's Republic of China

ABSTRACT: In this paper a test to check the technology concerning the strengthening of a stratified fracture zone in a rock abutment of a dam is decribed. High water pressure jet grouting tests, geophysical explorations and large drillings were used to quantify the strengthening.

Following results were acchieved after reinforcement by high pressure water jet grouting. The dynamic elasticity modulus of the fractured rock zones was increased by 30 - 40 %; the elastic wave speed was increased from 1542 m/s to 2941 m/s. The later is almost two times larger than originally.

ZUSAMMENFASSUNG: In diesem Beitrag werden die Ergebnisse eines Versuches zur Prüfung eines Verfahrens zur Verfestigung einer schichtförmigen Bruchzone im Felswiderlager einer Talsperre dargestellt. Die Untersuchungen im Testfeld erfolgten vor und nach der Injektion mit dem Hochdruck-Wasserstrahlverfahren, mit geophysikalischen Verfahren und Großloch-Bohrungen, um die Verfestigung quantitativ zu erfassen.

Im einzelnen wurden folgende Ergebnisse festgestellt. Der dynamische Elastizitätsmodul konnte durch die Hochdruck-Wasserstrahl-Injektion um 30 - 40 % erhöht werden. Die Geschwindigkeit der elastischen Wellen wurde von 1 542 m/s auf 2 941 m/s vergrößert, also fast auf das Doppelte des ursprünglichen Wertes.

1 GEOLOGICAL CONDITIONS

In the dam site, the rockmasses are mainly of Dyas basalt, which were erupted five times in their formation process. Owing to different time interval after each eruption, it is obvious that five large layers of rock are presented. ($P_2\beta_1$, $P_2\beta_2$, $P_2\beta_3$, $P_2\beta_4$, $P_2\beta_5$) Especially, between the forth and fifth layers, the heavily fractured rocks C_5 could be found owing to their formation movement. The C_5 layer is formed as a fractured zone which dips about 6–8 °. In the C_5 layer, there are many faulted joints among which the main joints were heavily fractured and melonitized. Their structures were thoroughly destroyed, so that they have the least strength in the zone. As shown in picture, the white zone is C_5 layer, which is of 10–50cm in width, and has different degrees of hardness.

2. ARRANGEMENT OF GROUTING HOLES AND WASHING HOLES

As shown in Fig.1, the washing and grouting holes were arranged

Totally, there are ten holes, which are numbered from G_1 to G_{10}. with 38m in depth, 2.6m in row spacing and 3.0m in hole spacing.

There are about 5 geophysical exploration holes denoted by WT_{-1}–WT_{-5}, respectively, which are arranged in two orthogonal intersecting lines.

They are used for seismic wave penetration tests. The measuring points were arranged with spacing of 1m. The measuring instruments with the mark E_s 1210 were used for tests and amplification seismic wave producers

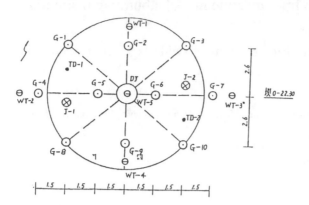

1. • Upward mobile monitor
2. ⊖ Geophysical hole
3. ⊙ Grouting hole
4. ⊗ Investigation hole
5. ○ Big hole
6. ----Acoustic wave

1. • anfwartses beweqliches Instrument
2. ⊖ Geophysikaks lock
3. ⊙ Einprepöffnung
4. ⊗ Untersuchungsloch
5. ○ gropes lock
6. ----klustische welle

Fig.1   The Plane Arrangement of Test Holes

Bild 1   Die Ebenenverteihung von Testlucher

were made in U.S.A. The seismic waves pre and post washing grouting were measured and recorded.

Two holes TD_1-TD_2 with 40m in depth, were arranged in order to moniter earth crust upward movement. Two other holes were drilled to investigate the rock seepage capacity.

After all these tests, a big hole with one meter diameter was drilled at the center of the test zone. It is 31.8m in depth. Some samples were taken and photographs, and video records were used to investigate the effects of high water pressure jet grouting.

Radial sound wave investigation holes were arranged in two sides along C_5. The thickness between them is 1 meter and 16m in depth for each hole. Another two rows of radial holes in C_5 are set up with row spacing of 20cm and depth of 80cm.

In these four rows of holes, the SYC-2 accoustic wave producer was used to measure the dynamic elasticity. By using the above described method, the effect from high water pressure grouting for basalt foundation and reinformcement of C_5 have been made clear.

3. USING HIGH PRESSURE WATER JET IN WASHING ROCK FRACTURE ZONE C_5

The diameter of groute holes is 56mm. When the grouting holes reached the depth of 10-15m, the fracture zone C_5 may be found, and then the high pressure jets were inserted into the holes. The hydraulic pressure was used about 30-40MPa, the air pressure 0.6MPa, the capacity of water 100 l/mis. Generally, the duration for washing holes was about ten hours, and sometimes 30 hours.

Soil clayey blocks and fragmencts were wept out by air blows and water jets. For each segment, the largest volume of materials washed out is $1.0m^3$. However, for squeezed hard rockmasses, it was difficult to wash them out even by 50MPa hydraulic pressure jets.

The diameter of grouting hole is Φ56mm, and the diameter for water jet pipe is Φ42mm, so that those blocks with diameter larger than 10mm, could not be washed out, and be detained in the hole. They consititute the aggregates of concrete when grouting afterwards. Generally, the groute pressure is 3.0MPa.

4. THE GROUTING EFFECT INVESTIGATION OF WEAK ZONE C_5

Three devices have been used to investigate the high pressure grouting effect of weak zone C_5.

The first is the seismic test technology, the second is the supersonic sounding and the third is to directly investigate the walls of large diameter drill holes.

In order to study the effect of high pressure grouting, the elasticities of rock C_5, corresponding to two state: pre and post–groulting were observed and compared.

As in Fig.1, WT_{-1}–WT_{-5} are shown five geophysical holes. In order to study the variation of rock character in the interfacial rock fracture zone, in the large diameter hole, four rows of horizontal holes were drilled for accoustic surveying. Now the detail are given below.

4.1 Seismic measurement

Seismic wave penetrating study was carried out from hole to hole. There were four penetrating surfaces namely: WT_{-1} to WT_{-2}; WT_{-2} to WT_{-4}; WT_{-4} to WT_{-3} and WT_{-3} to WT_{-1}

Measuring steps, for seismic operation are given below. In each hole, in turn the detonation operations were conducted, and the wave meters were set up to measure the wave velocities in other holes. The detonation tests were conducted from low to high spots with each segment of one meter long. Total measuring points were 496 and 520 respectively referred to pre and post grouting.

The wave velocities V_p were measured from the tests, then dynamic Elasticity Ed could be calculated.

$$Ed = \rho - V_p^2 \frac{(1 + \mu)(1 - 2\mu)}{(1 - \mu)}$$

where ρ is the rock density.
μ is the poisson ratio.

As shown in Tab.1, the V_p were varied largely after grouting. They have been increased clearly, in the range 3000–4000m / s. In the four orthogonal intersecting lines, there were two points with $V_p < 2000$m / s before grouting, whereas for all points $V_p > 2000$m / s after grouting. Before goruting, there are 20 points with 2000m / $s < V_p < 2500$m / s, whereas after grouting, only one point remained. For v_p 2500m / s, 3000m / s, there were 49 points, whereas after grouting, only 26 points remained in such level.

For $V_p > 3000–3500$m / s, 20 points and 45 points are referred to pre and post states respectively. For $v_p > 3500$m / s, 18 points and 37 points are referred to pre and post states respectively.

Elastic Wave Isoline Analysis

In order to make the results clear, and figurative, figures were made to show the isoline of elastic wave along the point in the periphery of test zone. Fig.2 and Fig.3 were referred to pre and post states respectively. From them, the area of the test rock zone with $V_p < 3000$m / s reaches 95cm^2, and it is equal to about 42.1% of total area. After grouting the area decreases to 14cm^2, it is equal to 6.2% of the total area. The area where $V_p > 4000$m / s is 30cm^2, is 13.1% of the total test zone, before grouting. After grouting, it reaches 54cm^2, about 23.5% of the total area.

The Variation of Dynamic Elasticity Ed

As in Tab.2, the variation of dynamic elasticity along the orthogonal intersecting lines and periphery line, are shown.

Table.1 Elastic wave velocity V_p pre and post grouting

Measure Line	Measure distance m	Elastic wave velocity V_p (K_m / s)													Σ point
		<2		2–2.5		2.5–3		3–3.5		3.5–4		4–4.5		>4.5		
		pre	post	pre	post	pre	post	pre	post	pre	post	pre	post	pre	post	
WT–1, WT–5	4.25	0	0	5	0	18	10	0	13	2	0	2	4	0	0	54
WT–2, WT–5	6.74	0	0	5	0	8	3	9	12	2	7	3	0	0	5	54
WT–3, WT–5	5.815	0	0	4	0	9	4	1	7	2	12	0	2	2	2	54
WT–4, WT–5	3.905	2	0	6	1	14	9	1	13	2	2	3	3	0	0	56
WT–1, WT–2	8.1	0	0	0	0	13	4	9	15	0	3	4	3	0	1	52
WT–2, WT–4	8.1	0	0	0	0	8	1	13	3	1	17	6	0	0	7	56
WT–4, WT–3	6.69	0	0	0	0	6	0	12	5	8	15	2	6	0	2	56
WT–3, WT–1	7.1	0	0	0	0	18	2	3	11	2	6	2	2	0	4	50
Σ	Σ	2	0	20	1	94	33	57	79	19	62	22	20	2	21	432

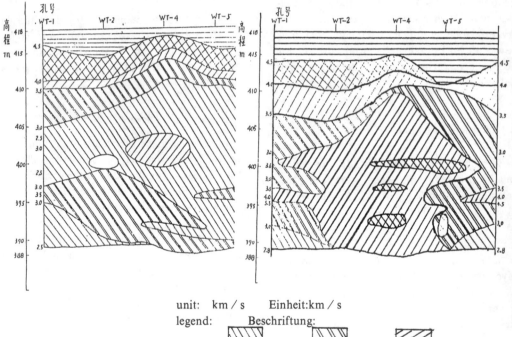

unit:　km / s　　　Einheit:km / s
legend:　　　Beschriftung:

2.5～3.0　　　3.0～3.5　　　3.5～4.0

4.0～4.5　　　>4.5

Fig.2　Isoline of V_p　before grouting　　Fig.3　Isoline of V_p　after grouting
Bild.2　Isotache　von V_p　vor der　　　Bild.3　Isotache von V_p　nach der
　　　　Eimpressung　　　　　　　　　　　　　　Einpressung

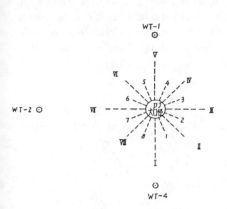

Fig.4 Arrangement of holes for acoustic wave

Bild.4 Die Anordnung der Löchern

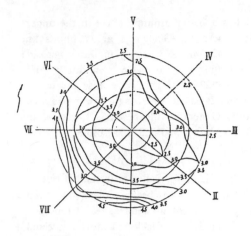

Fig.5 V_p potential Isoline in C_5 by Percolation tests

Bild.5 V_p potentiallinie im C_5 nach Test

Tab.2 Dynamic Elasticity pre and post grouting

Measure line		Dynamic elasticity Ed(10^4kg / cm^2)		Variation (%) Ed' / Ed
		pre Ed	post Ed'	
+ Line	WT−1, WT−5	13.65	19.21	40.73
	WT−2, WT−5	19.56	24.30	24.23
	WT−3, WT−5	19.07	25.16	91.93
	WT−4, WT−5	15.52	19.13	23.26
periphery	WT−1, WT−2	18.42	22.07	19.82
	WT−2, WT−4	22.10	27.60	24.89
	WT−4, WT−3	21.21	27.66	30.41
	WT−3, WT−1	18.79	25.19	34.06
	Mean	18.54	23.79	28.66

Table 3. V_p in the upper side of C_5

Point	Elastic wave velocity V_p (Km / s)					Σ
	< 2.5	2.5− 3.0	3.0− 3.5	3.5− 4.0	4.0− 4.5	
I	2	1	2	1	1	7
II	0	3	4	0	0	7
III	0	2	1	4	0	7
IV	7	0	0	0	0	7
V	0	0	3	2	2	7
VI	0	0	0	1	2	3
VII	0	0	2	1	4	7
VIII	0	0	0	3	4	7
Σ	9	6	12	12	13	52

Superacoustic Measuring Results

In order to investigate the grouting effects in C_5, in the large investigation hole, four rows of horizontal radial test holes were set up. They are 32 holes totally. Holes I−VIII were set in the upper and lower sides of C_5. They are spacing 1.0cm with 1.6m in depth. Where-

as 1−8 holes were set in the upper and lower side of C_5. They are spacing 20cm with depth 0.8m, as shown in Fig.4.

The investigation results of elastic velocity V_p were shown in Tab.3 and Tab.4.

From these Tables, the weak rock interlayer

207

C_5 has a better grouting effect in the upper side, where $V_p > 3000m / s$ at 37 points and $V_{Pmax} = 4545m / s$, whereas, in the lower side $V_p > 3000m / s$ at 26 points and $V_{Pmax} = 3922m / s$. This is because the upper side rock is basalt, whereas the lower side rock is tuff. Generally, the besalt is harder than tuff. As shown in Fig.5, the isoline of V_p is given.

In the Fig.5, V_p is higher in the left side of II-VI holes. This is because there have been wahsed out a lot of weakened and loosened rock fragments, and a lot of mortar injected afterward. This can be seen in Picture 1, 2, 3. For the fragemented rocks, they have been replaced by concrete.

Table 4. V_p in the lower side of C_5

Point	Elastic wave velocity V_p (Km / s)					Σ
	< 2.5	2.5– 3.0	3.0– 3.5	3.5– 4.0	4.0– 4.5	
I	0	1	2	4	0	7
II	3	3	1	0	0	7
III	2	5	0	0	0	7
IV	1	2	3	0	0	6
V	0	2	3	1	0	6
VI	0	3	4	0	0	7
VII	2	1	4	0	0	7
VIII	1	0	1	3	0	5
Σ	9	17	18	8	0	52

In 1-8 holes, the super acoustic tests result are given. From thc table 5, it can be drawn that the C_5 interlayer has been reinforced by high presure grouting. Their maximum value is 3571m / s, minimum value is 1667m / s, and average 2443m / s.

In Tab.6, there are shown V_p from other kinds of drilling cores from large diameter hole.

5. CONCLUSION

The effect is good for the treatment of weak interlayer using high pressure washing technology, the softer the interlayer, the easier and better results of the process. The washing pressure will be raised in case of encountering rock mass with higher hardness.

The influence diameter, and washing pressure, will be varied proportionally with the time used in grouting. Generally, it lies be-

Table 6.   Super sonic tests

Lithologic category	Elastic wave velocity V_p m / s
Mortar concrete	4166
Grouted basalt	4545
Grouted Tuff basalt	3286
Grouted faults fillings	2306
Grouted C_5	2941
C_5 in original state	1542

Table 5.   Super sonic test in C_5

Holes		1	2	3	4	5	6	7	8	Σ
Elastic wave velocity V_p m / s	upper	2703 2547	2050 2173	2564 1667	2197 2030	2740 2532	3030 1786	2272 2197	2858 3571	2552 2313	2433
Dynamic Elasticity Ed 10^4 kg / cm^2	lower	14.6 12.6	5.9 9.6	11.8 4.3	8.7 7.4	13.1 12.3	16.5 6.3	9.0 8.7	14.7 19.8	11.7 10.1	10.9

tween 1–1.5m. It will be invarsely, propotionally with rock hardness.

From the big diameter investigation hole, it counld be drawn that the dynamic elasticity V_p have a maximum 4343m / s, and a minimum 1667m / s, they are all satisfied with engineeing usage.

ACKNOWLEDGEMENT

This paper is supported by The State Foundtion Committee, which is acknowledged by the authors.

REFERENCES:

Witherspoon, P.A. et al: Hydraulic and Mechanical properties of natural fractures in low permeability rock ISRM (1989) Montreal.

ZHOU W.Y. Yang Ruozieng: Finite element analysis of water seepage in fractured rock masses of dam abutments. Intern. Symposium on Rock seepage. Xian, May 1993

Stephansson, O.: Fundamental s of Rock Joints. Proc. of the Inter. Symposium of Rock joints. Bjorkliden 1985

Oliveara, R Engineering geological Investigation of Rock masses for civil engineering project and mining operations. Proc. of 5th Intern. Congress on Engineering geology, Buenos Aires (Argentina) 1986.

Grouting in Rock and Concrete, Widmann (ed.) © 1993 Balkema, Rotterdam, ISBN 90 5410 350 7

Selection of grout mixes – Some examples from US practice

Auswahl der Injektionsmischungen – Einige Beispiele aus der Praxis in den USA

K. D. Weaver
Woodward-Clyde Consultants, Oakland, Calif., USA

ABSTRACT: U.S. grouting practice has been dominated by the practice of the Bureau of Reclamation, which entails use of thin, unstable grouts and an intuitive process of gradually thickening the grout during injection. However, there has been a movement toward use of grouts with lower water:cement ratios, incorporating stabilizing or fluidif ying additives. Use of lugeon water tests for selection of grout mixes as well as for evaluation of grouting effectiveness is gaining acceptance. Examples of past and recent U.S. practice for selection of grout mixes are presented. These examples include government agency practice and consultant practice. They also include displacement grouting and controlled hydrofracture grouting in unconsolidated materials as well as permeation grouting of bedrock.

ZUSAMMENFASSUNG: Verfahren mit Einpressarbeiten in den USA sind von dem Bureau of Reclamation angewandte Methoden dominiert. Diese Methoden schliessen die Benutzung von dünnen, instabilen Vergussmassen und ein intuitives Verfahren mit allmählicher Verdichtung der Vergussmasse während der Injection ein. Eine andere Methode beruht auf der Verwendung von niedrigen Wasser/ Zement Verhältnissen mittels stabilisierender oder flüssigmachender Zusätze. Die Anwendung der Lugeon Wasserprobe für die Auswahl der Vergussmischungen, sowie die Beurteilung der Wirksamkeit von Einpressarbeiten werden mehr und mehr benutzt. Diese Beispiele enthalten Verfahren von Regierung- und Ingenieurbüros. Einpressarbeiten mittels Verdrängung und kontrollierter "Hydrofraktur" in nicht-konsolidierten Böden sowie durchsickernde Einpressarbeiten im Grundstein sind auch dargestellt

1. INTRODUCTION

U.S. grouting practice has, for many years, been dominated by the practice of the Bureau of Reclamation. In general, t his practice has been characterized by the us e of "neat" portland cement grouts with high water: cement (w:c) ratios and by an intuitive approach to thickening the grout mix. W:c ratios as high as 10:1 to 20:1 by volume have been used in the past, but 7:1 grout was commonly being used as the standard starting mix by the mid 1950's. The rationale for the use of these thin mixes was that they would facilitate entry of cement particles into narrow fractures. It was assumed that the excess water would somehow go away, leaving the fractures filled with grout. The intuitive approach to thickening the grout mix

entailed attempting to maintain a constant injection rate as measured in bags of cement per hour. If a grout hole was found to accept grout freely at a w:c ratio of 7:1, the grout would be thickened to 6:1. Further thickening would be done gradually after intuitively selected volumes of grout were injected at each w:c ratio. The grout would be thinned if a point was reached at which the grout acceptance rate as measured in bags of cement per hour became abruptly slower than was experienced prior to thickening the mix. The decision to grout a hole or interval commonly has been based on the results of a "water test" (actually a simplified permeability test), with grout being applied if the rate of water "take" equaled or exceeded 0.2

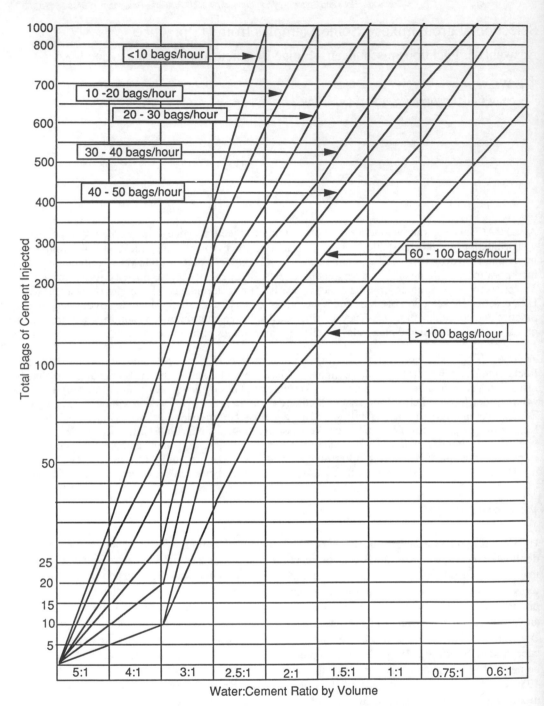

Figure 1. Grout thickening guide used at Castaic Dam, California.

212

cubic feet per minute (5.7 liters/min.). The applied pressure was not considered, and the same starting grout mix was used regardless of the rate of water take.

There has been a gradual movement away from extremely thin grout mixes, toward use of guidelines for the rate of thickening the mix, and toward the use of additives. This movement was accelerated as a result of papers presented at a grouting conference sponsored by the American Society of Engineers in 1982 (Baker 1982). A presentation by Dr. Don Deere (1982) on the use of bentonite to reduce bleed has led to wide use of that material. A.C. Houlsby's paper on optimum water:cement ratios, presented at that same conference, also had an important influence on U.S. grouting practice. A subsequent paper by Deere and Lombardi (1985) and a companion paper by Houlsby (1985) further influenced a trend away from use of highly diluted grouts. U.S. grouting practice also has been influenced by Dr. Deere's advocacy of using no more than three grout mixes (i.e., grouts with no more than three water:cement ratios) rather than the large range of mixes used in Bureau of Reclamation practice. There also has been a gradual movement toward use of a lugeon criterion for selection of starting mixes, and toward use of fluidifiers and superplasticizers.

Most grouting of fractured rock is done with cement-based grouts, formulated with type I or type II portland cement. Type III portland cement is more commonly used for its relatively rapid set time than for its finer grain size. Ultrafine cement grouts are being used on an increasing number of projects. Usage of these grouts for treatment of fractured rock is expected to increase significantly with the entry of more manufacturers and suppliers into the U.S. market. On-site grinding of conventional portland cement to produce ultrafine cement grout is now being done on a limited basis in the U.S., and may help to expand the use of ultrafine cement grouts on large projects. Chemical grouts, although acknowledged to offer the advantage of a full-volume (zero bleed) set, are rarely used in dam foundation grout curtains in rock except as a supplement to cement grouting.

2. GROUTING PRACTICE ON CALIFORNIA WATER PROJECT DAMS

Bureau of Reclamation grouting practice was adopted, essentially without change, for numerous dams designed by the California Department of Water Resources (DWR) during the 1950s and 1960s. For example, the grouting program at Antelope Valley Dam in northern California entailed starting all holes accepting water at a rate of 0.2 cfm or more with a neat cement grout having a w:c ratio of 7:1 by volume. Progressively thicker grout mixes were injected on the basis of the inspector's judgment if the thinner mixes were accepted freely by the grout holes. Thick, sanded mixes were injected to seal surface leakage through sheet joints in the granitic bedrock.

Grout with a 5:1 w:c ratio by volume was used for nearly all of the grouting at the DWR's Del Valle Dam, in coastal central California. However, grout as thick as 0.66:1 was used to seal surface leaks. The grouting program at this site differed from typical DWR practice in that the grout curtain consisted of two rows of holes having a uniform depth throughout. Bedrock at the site consists of sandstone and shale.

Grout with a 7:1 w:c ratio by volume was used as a starting mix for most of the foundation grouting at Castaic Dam, in southern California. However, a 5:1 mix was sometimes used. The grouting inspectors were provided with guidelines regarding the volume of grout to be injected at each w:c ratio before thickening the mix. As shown on Figure 1, the volume to be injected depended upon the rate of grout take. A totally different approach was taken for consolidation grouting of a landslide mass that was left in place beneath the dam embankment. This mass was grouted with a stable (zero bleed) grout formulated with portland cement, natural pozzolan, and bentonite. The bentonite retarded the setting time sufficiently that it was possible to resume pumping grout in a hole after having discontinued injection for periods up to 16 hours. Subsequent partial excavation of the slide mass revealed that this grout had thoroughly

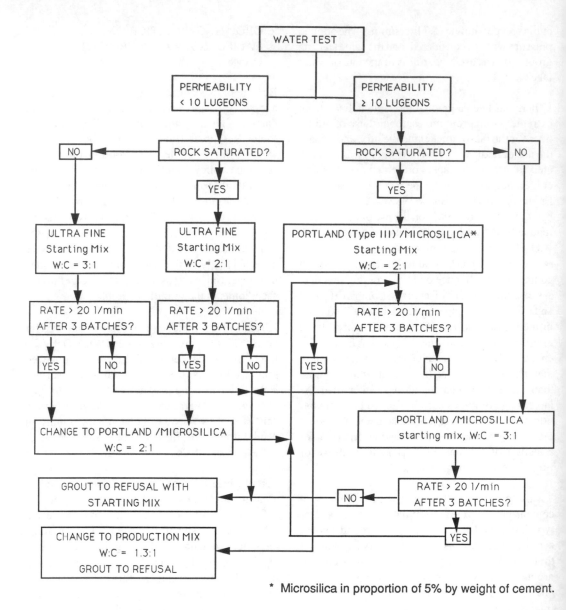

* Microsilica in proportion of 5% by weight of cement.

Figure 2. Preliminary flow chart for selection of grout mixes.

permeated the mass, completely filling fractures of all sizes and bonding equally well to sandstone and shale.

3. GROUTING PRACTICE ON OTHER U.S. DAMS

Grout formulated with a pozzolanic portland cement blend was used for constructing a grout curtain at Davis Creek Dam, in northern California. The pozzolanic blend was used due to the presence of sulfate and sulfide minerals in the foundation rock. The bedrock at the site is sandstone and shale, most of which strongly sheared as a result of its position adjoining an ancient subduction (fault) zone. The starting grout mix for all holes had a 3:1 w:c ratio by volume. Except where

214

problems requiring special treatment were experienced, only three basic grout mixes were used. A "production mix" with a 2:1 w:c ratio was used after injecting about 10 bags of the starting mix. Grouting with the "production mix" ordinarily was continued to refusal. However, a 1:1 mix was employed in a few holes that freely accepted relatively large volumes of grout. Grout was injected though drill rods extended to the bottoms of holes in an area where caving of the hole walls was so severe that circuit grouting was ineffective. A superplasticizer was added to the grout in a proportion of 1% by weight of cementitious material, improving the ability of grout to permeate the caved material in the hole annulus and enabling it to enter the bedrock formation. A further improvement was accomplished by using grout formulated with a pozzolanic microfine cement and a superplasticizer.

Type I-II portland cement was the basic ingredient used in formulating grout for a tailings retention dam in eastern Washington state. The foundation rock at this site consists principally of sandstone and shale. In general, it was intended that the production grout mix have a bleed of 5% or less. Bentonite was added to the mix to achieve this objective, but - due to lack of facilities to prehydrate the bentonite - testing of the grout yielded erratic results. In order to achieve the desired bleed in the absence of bentonite, a production mix of 1.5:1 or 1:1 by volume was employed. However, a starting mix as thin as 5:1 was used when grouting dry, porous rock. Following a finding that the "bi-modal" grout mixer apparently was not producing a satisfactorily blended and dispersed grout, a superplasticizer was added to the grout. This was found to enhance the properties of neat cement grout, but to increase the viscosity of grouts containing bentonite.

Most of the dam foundation grouting for the Merrill Creek Project, in New Jersey, was done with grout formulated with type III portland cement, with the objective of reducing the set time. Injection customarily was started with grout having a w:c ratio of 3:1 by volume. Ultrafine cement grout was used in an area where

water continued to seep from fractures in the granitic gneiss foundation following completion of grouting with the type III portland cement grout. Neither this type of grout nor subsequent injections of polyurethane chemical grout proved to be effective for stopping the seepage.

Current plans for construction of a two-row grout curtain for a new dam on a sandstone and shale foundation in California call for use of two basic types of grout, depending upon the results of permeability tests to be made immediately prior to injection of grout into any stage in a grout hole. Blends of type III portland cement and silica fume will be used for moderately permeable to highly permeable conditions, and microfine cement grouts will be used to treat intervals having a relatively low permeability. A superplasticizer will be used in the formulation of both types of grout. Selection of the grout mixes will be guided by reference to a flow chart, such as that shown on Figure 2. This flow chart will be modified on the basis of results obtained during the course of the work.

4. GROUT SELECTION FOR A HAZARDOUS WASTE SITE

Grouting operations to reduce underflow along bedding planes in dolomite underlying a hazardous waste site in New York state were preceded by a laboratory test program to verify that the grout would be compatible with the waste material (Weaver and others 1990). Various formulations containing portland cement and fly ash were tested, as were grouts formulated with two types of ultrafine cement. Some of the grouts were formulated using contaminated groundwater from the site, and some grouts were allowed to harden in a contaminated water "bath". All of the grouts tested proved to be compatible, so selection of the appropriate grout for field use was based upon viscosity and bleed characteristics.

Selection of starting grout mixes for construction of a grout curtain at the hazardous waste site was based upon the results of a simplified lugeon test of each interval that was to be grouted. In

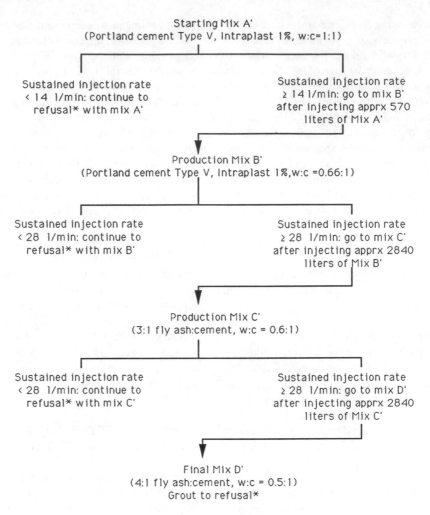

Starting Mix A'
(Portland cement Type V, Intraplast 1%, w:c=1:1)

Sustained injection rate
< 14 l/min: continue to
refusal* with mix A'

Sustained injection rate
≥ 14 l/min: go to mix B'
after injecting apprx 570
liters of Mix A'

Production Mix B'
(Portland cement Type V, Intraplast 1%, w:c =0.66:1)

Sustained injection rate
< 28 l/min: continue to
refusal* with mix B'

Sustained injection rate
≥ 28 l/min: go to mix C'
after injecting apprx 2840
liters of Mix B'

Production Mix C'
(3:1 fly ash:cement, w:c = 0.6:1)

Sustained injection rate
< 28 l/min: continue to
refusal* with mix C'

Sustained injection rate
≥ 28 l/min: go to mix D'
after injecting apprx 2840
liters of Mix C'

Final Mix D'
(4:1 fly ash:cement, w:c = 0.5:1)
Grout to refusal*

Note: *Grouting not continued beyond 4 hours or total
injected volume of approximately 11.3 cu. meters.

Modified from Weaver et al., 1992

Figure 3. Grout mix flow chart for permeabilities ≥ 50 lugeons.

general, intervals with permeabilities in excess of 50 lugeons were grouted with fly ash/cement grouts, and those with lower permeabilities were treated with ultrafine cement grouts. Flow charts were used for guidance in thickening the grout mixes. (See Figure 3.) Details regarding the project are presented in a paper by Weaver and others (1992).

5. GROUT SELECTION FOR UNUSUAL FOUNDATION CONDITIONS

Separation of a Dresser coupling in a penstock on a steep slope in the Sierra Nevada mountain range in California was inferred to be the result of settlement of talus underlying the penstock. Compaction grouting, using soil-cement mortar grout with a slump of 5 cm or less, was selected as a means of consolidating the talus within a treatment zone approximately 18 m thick. The supposed talus mass refused to accept this thick grout, so most of the consolidation work was done with a viscous grout having a slump in the range of 12 to 15 cm. Bentonite was used as a minor additive to facilitate pumping through long hoses, pipes and casings. Large grout takes, together with other data, led to the conclusion that the supposed talus mass is occupying a crevasse-like "sackung" feature (Weaver 1989). Due to the extremely open conditions within and beneath the treatment zone, sodium silicate was used to accelerate the set of cement grout placed in the annulus of slope indicator holes.

Several grout types and grouting methods were used at another site in the Sierra Nevada mountains, where a small power plant was to be constructed on landslide material underlain by boulder gravel and sandy alluvium in a stream channel. The objectives of the grouting program included prevention of loss of foundation material as a result of subsurface erosion during high-velocity stream flows (Weaver et al. 1993). High-slump mortar grout was injected in closely-spaced holes around the perimeter of the site, to create a curtain in the boulder gravel. A second row of closely-spaced perimeter holes was injected with a fast-setting blend of sodium silicate and ultrafine cement grout, with the objective of creating an impermeable curtain in the finer-grained materials. Two types of grouting were done within the perimeter curtain: compaction grouting, and controlled hydrofracture grouting. Most of the controlled hydrofracture grouting was done with a blend of portland cement and silica fume, together with sufficient water to achieve a Marsh viscosity of about 36 seconds. The silica fume was used to reduce bleed and to facilitate permeation of sand and gravel lenses intersected by grout moving along paths created by hydrofracturing the alluvial and landslide mass. Ultrafine cement grout also was used for hydrofracture grouting, in the hope that fine-grained sands might be permeated.

Pre-grouting ahead of a sewer tunnel beneath the Rocky River, in Ohio, employed portland cement grouts and ultrafine cement grouts as a means of reducing the permeability of the sedimentary bedrock. Simplified lugeon test data were used as a basis for selection of the grout formulation. Holes exhibiting permeabilities greater than 10 lugeons were injected with a formulation consisting of type III portland cement, silica fume in a proportion of 10% by weight of cement, superplasticizer, and water in a proportion of 2:1 by weight of cement plus silica fume. Ultrafine cement grouts with a w:c ratio of 3:1 by weight were used in holes with permeabilities of 10 lugeons or less. As the rock cover between the river and the tunnel was as little as 1 meter, the grouting was done through a reinforced concrete slab anchored to the bedrock of the river bed. Acrylate grout and sodium silicate-ultrafine cement grouts were specified for use in alluvial materials in the river bed as a contingency in the event that bedrock could not be reached, but a need for these grouts did not arise.

REFERENCES

Baker, W.H., ed. 1982. Proceedings of conference on grouting in geotechnical engineering: A.S.C.E., New Orleans, Louisiana, Feb. 1982.

Deere, D.U. 1982. Cement-bentonite grouting for dams. In W.H. Baker (ed.), Proceedings of conference on grouting in geotechnical engineering, A.S.C.E., New Orleans, Louisiana, Feb. 1982. 279-300.

Deere, D.U. and G. Lombardi 1985. Grout slurries - thick or thin? In W.H Baker (ed.), Issues in dam grouting: A.S.C.E., 156-164.

Houlsby, A.C. 1985. Cement grouting: water minimizing practices. In W.H. Baker

(ed.), Issues in dam grouting:
A.S.C.E., 34-75.

Houlsby, A.C. 1982. Optimum
water:cement ratios for rock
grouting. In W.H. Baker (ed.),
Proceedings of conference on
grouting in geotechnical engineering,
A.S.C.E., New Orleans, Louisiana,
Feb. 1982, 317-331.

Weaver, K.D. 1989. Consolidation
grouting operations for
Kirkwood penstock: Proceedings
of foundation engineering congress,
A.S.C.E., 342-353.

Weaver, K.D., R.M. Coad, and
K.R. McIntosh 1992. Grouting
for hazardous waste site
remediation at Necco Park,
Niagara Falls, New York. In
Grouting, soil improvement and
geosynthetics; A.S.C.E.
Geotechnical special publication no.30.
2: 1332-1343.

Weaver, K.D., J.C. Evans, and
S.E. Pancoski, 1990. Grout
testing for a hazardous waste
application. In Concrete
international, 12:7,45-47.

Weaver, K.D., T.R. Kolbe, and
S.J. Klein 1993. Foundation
grouting for the Forks of
Butte Powerhouse. In Proceedings of
third international conference on case
histories in geotechnical engineering.
June 1 - 6, 1993, St.Louis, Missouri,
U.S.A.

Grouting in Rock and Concrete, Widmann (ed.) © 1993 Balkema, Rotterdam, ISBN 90 5410 350 7

A new grouting technique using stiff cement pastes in macroporous formation

Eine neue Injektionstechnik unter Verwendung steifer Zementpasten in makroporösen Medien

Xiaodong Yang, Huaiyou Zhang & Jinjie Zhang
Institute of Water Conservancy and Hydroelectric Power Research, People's Republic of China

ABSTRACT:The reducing cost of cement grouts by adding mixtures and solving difficult problems in grouting of macroporous formation by using the properties of stiff cement pastes has been studied based on results of laboratory and in-situ experiments in this paper. The experiment results show that rheological properties of fresh stiff cement pastes can be modeled by Bigham substance and the solid properties of the grouts that will control the final infiltration distance of injecting enhance progressively with grouting accordingly. It is convenient for grouting using stiff cement pastes in macroporous formation that use vertical mixer with inject pipe in the grout compartment and the final grout ending standard with pressure or discharge of inject given primarily. Three engineering examples using stiff cement pastes show the simple in using, easy controlling, higher solid strength, full filling in macroporous formation, high efficiency and low costs.

1 INTRODUCTION

There are some needs in dam engineering such as back filling in tunnel, seepage prevention and stabilization in Karst, rockfall and macroporous soil. Grouting is often used there.

The speeded-solidify grouts or cement sand grouts were often injected for infiltration control in macroporous formation in past. The composition of the speeded-solidify grouts is complex and their costs is higher. If water glass was used in grouts, the problems of corrosion resistance would arise. The cement sand grouts, as the separation of compositions, either block the pipelines of injecting or diffuse too far in gravity to fill fully. It is needed to make the grouts to be controlled easily and lower their costs.

2 GROUTING TECHNIQUE USING STIFF CEMENT PASTES

The stiff cement pastes are generally composed of cement and some additives such as fly ash, clay and mine, and a small amount of admixtures. Additives are mainly chosen from local materials. Admixtures are chosen according to engineering demands, the power of grout equipment and grouting technique. The representative grout proportion are H-5,F-9 and L-6 in Table 1. Parts of grouts and solid natures are shown in Table 2. From Table 2, it is known that a lot of admixtures can be added to paste to reduce costs under some certain condition.

Laboratory test shows that the rheologic behavior of freshly-mixed stiff cement paste basically accords with Bigham fluid. It shows:

$$\tau = \tau_0 + \eta dv/dy \qquad (1)$$

in formula (1):

τ---shear stress;
τ_0---plastic yield strength;
η---plastic viscosity;
dv/dy---gradient of velocity.

The values of τ_0 and η of freshly-mixed slurry are shown in Table 2. Normally, when τ_0 is higher than 10-20 Pa, the colloid rate of slurry reaches to 100%. By the Bigham fluid property, when slurry flows radially in a circular pipe or a plane fissure of uniform width, the theoretical final infiltration distance under a certain pressure can be determined approximately by the formula:

$$L = P \times h/2\tau_0 \qquad (2)$$

in formula (2):

L---infiltration distance;

Table 1. Grout proportion series (by weight)

No.	Water/ solid	Cement	Fly ash	Clay	Water	Water glass	Calcium Chloride	Disperse agent
C-2	0.37-0.5	100	20	0-10	50-60	-	0-4.3	0.5-1.2
H-5	0.5-0.67	100	100	10-20	110-140	0-3	0-3.0	0-0.5
F-9	0.75-1	100	300	0-200	300-600	0-3	-	-
L-6	0.75-1	100	-	200-400	300-400	0-3	-	-

Table 2. Parts of grout and its solid property

No.	Colloid ratio (%)	Initial setting time (h)	Terminal setting time (h)	Tunnel viscosity (s)	Plastic yield strength (pa)	Plastic viscosity (pa.s)	Flexural strength (M pa)	Compressive strength (M pa)
C-2	99-100	3-5.5	3.8-9.5	85.1-drip	15.5-40.3	0.11-0.41	1.1-1.2	21.8-36.7
H-5	96-100	12.4-24	16.7-32.5	26.2-drip	5.6-49.5	0.15-0.38	0.5-0.6	6.7-9.7
F-9	71.2-100	15-33.5	28-76.5	19.2-drip	0.2-174.0	0.01-1.24	0.1-0.4	0.9-4.3
L-6	100	8.8-20.3	21-52	drip	>88	>0.34	0.03-0.13	1.1-1.6

p---infiltration pressure;

h---the radius of circular pipe or the opening of uniform width fissure.

From formula (2), when the pressure and the opening width of fissure are certain, the plastic yield strength of slurry affects the slurry flow and controls its final infiltration distance. By changing the value of plastic yield strength τ_0, grout can be controlled in a macroporous formation, and the slurry infiltration radius can be kept in a certain extent. Because stiff cement paste has a larger plastic yield strength, when shear stress of slurry is lower than its plastic yield strength, slurry does not flow and displays pastes. Since gravity has little effect on its flow, so slurry infiltration can be completed by adjusting mainly grouting pressure, and the paste filling will be well finished.

In-situ grouting engineering, to adapt some given pumping and injecting technique, the plastic yield strength of paste must be in a certain extent. So there must be some free water in fresh stiff cement paste, which is surplus than what cement hardening needs. This water makes the slurry has liquid characteristics. And it is thought that flow pressure is conveyed basically by water in slurry. But, under general cases, in the calculation of the infiltration distance of grout using formula(2) with the plastic yield of fresh stiff cement paste in Table 2, a larger result can be got, which is inconsistent with in-situ grouting engineering. It is because the slurry itself has thixotropic property which has not been taken account in formula(2), and with the injecting in process, cement hardens and moisture in slurry separates because of pressure, the property of slurry changes progressively. All the factors must lead the increase of solid property of paste gradually. If we consider additionally paste friction angle particulate loose material shear property before slurry hardened, let

$$\tau=\tau_0+\eta\times dv/dy+p\times tg\varphi \qquad (3)$$

then the final infiltration distance of paste is

$$L=h/2tg\varphi\times\ln(p\times tg\varphi/\tau_0+1) \qquad (4)$$

in the formula: φ is the friction angle of freshly-mixed paste after some time.

In the grouting process, paste infiltration can be controlled by choosing the paste with a certain plastic yield strength, adopting suitable pressure and according to special technological procedure. Considering the property of paste changes with time, and the grouted materials are non-inhomogeneous, the slurry density can be simply changed from thin to thick. Then the ending standard of grouting can be determined by jointly consider given pressure and volume.

Normally, when the plastic yield strength of slurry is bigger than 30 pa, ordinary mixers and grouting equipments will not work well. According to our initial engineering experience, slurry making can be fulfilled by forced stirring with large power, for examples, using old drilling or simply made vertical mixer with large power.

Grouting pumps are screw type, such as LB or LG mono screw pump. The grouting intake is under grouting outlet of mixers. Grouting pressure is 1.2-2.4 Mpa, flow rate is 100-400 l/min, the yield strength of paste can be controlled not bigger than 150 pa which will meet the need of pumping transport by adjusting the rotation speed. For large discharge pumping transport, other types of screw pumps or concrete pumps can be used. Besides, boring diameters are Φ76 or Φ91, grouting pipe-lines are soft or hard pipe with 38mm or 50mm in diameter and valves are sphere valves. All the transporting pipe-lines should be unblocked. Because of the good stability of stiff cement pastes, pure pressed grouting can be used. To avoid slurry free and paster uncompacted in the formation, the inject pipe should be set at the bottom section of grouting hole.

3 ENGINEERING UTILIZATION

3.1 Curtain grouting test in the rock-fill dam of Hongfeng hydroelectric power plant

Hongfeng hydroelectric power plant is the first cascade of hydroelectric station at the Maotiaohe River, stands upstream the Jichang Bridge of Qingzhen county 35 km southwest of Guiyang City. Its reservoir has a basin area of 1551 km^2 and a gross capacity of 6.01×10^8 m^3. In its construction, sloping wood anti-seepage wall was used in the dam. Since its beginning of storage in 1960, the reservoir has overrun for many years. It is needed to build a grouting curtain as a permanent impervious core to replace the wood sloping wall. The designed curtain is in loose-stone dam body having a 21.5-25.6% air voids. The rock-fill dam body at its down stream has a 37.0-38.6% air voids.

Curtain grouting in the macroporous rock-fill dam is rather difficult and few experience could be refereed. Since 1985, we have begun laboratory experiment. Based on experiment results, we made an in-situ grouting test together with the designer and constructor with setting accelerated slurry and stiff cement paste in 1986. The proportion of paste and property of grout are C-2 and H-5 shown in Table 2.

Because of the complexity of setting accelerated slurry's proportion and the high precision need of its batch amount, its field stirring operation is difficult. If the slurry is thinner than needed, the infiltration controlling becomes difficult, and if it is too thick the ordinary mixer and transport device will not be suitable, so thinning work is needed. Besides, the quality of in-situ setting accelerated

slurry is not good enough to end grouting though measures such as limiting grouting discharge and intermittent grouting are adopted. In many cases the consumption of dry materials is as high as 15t/m in macroporous formation and there is still no pressure. So if the slurry quality is not controlled very well, effective grouting controlling in macroporous formation will be very difficult.

To solve the controllability of grouting, some constructors once adopted cement-water glass grout. In this way though grouting infiltration can be controlled well and the pressure could rise rapidly, its cost was high and its grouting compaction and long time stability against permeability is of problem. In some section though crushed stones were filled, the consumption of dry materials was still as high as 21.4t/m. Apart from this, it usually took a long working time and led to pipes blocking easily which affected grouting effects.

By reforming slurry-making equipment and using LG screw pumps, constructors chose the H-5 series proportion of stiff cement grouts adding local fly as and clay referring laboratory tests. It could effectively decrease slurry waste caused by gravity in macroporous formation, then the grouting rate could be controlled by adopting different grouting pressures and the grouting infiltration could be controlled further. In tests, they succeeded in controlling the average dry materials consumption at 2.5t/m by regulating pressure. If the air porosity of loose stone is 23.6%, the grouting infiltration radius will be about 1.4m.

Though formula (2) represents the influence of plastic yield strength of slurry to the final infiltration distance, let the plastic yield strength of fresh slurry be 25.8 pa, grouting pressure be 0.2 Mpa, opening radius in formation be 20mm, the calculated final infiltration distance will be 77.5m. Such a result is obviously not in accord with the actual situation. If it is supposed that the plastic yield strength of slurry increase to 50 pa with grouting process accordingly, and the slurry has a 2° friction angle only, the calculated final infiltration radius L will be 1.42m by formula (4). It is known that this value is more reasonable. This result shows the influence of solid property of slurry injected into formation to grout infiltration with grouting time continuing. So, it is very important to choose suitable grouting pressure, slurry and change slurry technology based on the state of grouted formation for controlling grouting infiltration distance.

In the grouting test of stiff cement

pastes, we used the method of hole sealing, descending stage, pure pressed grouting. Its efficiency was more than 2 times than that when cement-water glass is used after grouting a certain quantity of thick slurry. Core samplers show that macroporus are filling fully by calculus of pastes which bond hard with loose-stones and the compressive strength reaches 7.0 Mpa asked by designers. Now the construction of dam curtain with stiff cement paste grouting is undertaking (1990).

3.2 Reinforcing grout treatment of the dam base in Longtang turbine-pump station

Longtang turbine-pump station situates in lower reaches of the Nandu River of Hainan Province. The masonry rind diversion dam is 142m long and 7.5m high. The station mainly serves lifting for irrigation. Because the dam base was not cleaned completely in construction, most part of the dam body stands on the first layer limestone about 3.16m thick and a layer of medium-fine sand about 2.17m thick. The hard rind of dam was full of block stones and river sand. During its initial operating period in 1971, leakage was found at the dam foundation. Afterwards the leakage amount increased gradually. Each year, filling with crushed stones in the dam body and leakage sealing was necessary. In 1980, not a few leakages were found caused by solution caves in the limestone layer. Until 1985, more than 30 leakage holes had been found with considerably large hole openirg and leakage amount. Though measures of filling with crushed stones in the holes and the dam body were adopted, the dam body stability was badly lowered since the caved foundation, damaged dam bottom plate and appearance of several crevices. The big one of these crevices was 40m long and 2cm wide. Serious settlement of the dam body occurred. And the tendency was obvious that all solution holes, caves were being enlarged continuously by erosion and corrosion of ground water. It was urgent to reinforce the dam foundation. Ordinary cement grout once was made on trial, but it was suspended because of shortages at controllability and cost.

Based on cases mentioned above and special property of local materials, L-6 stiff cement-clay paste series in Table 1 was choose as grouting slurry. Three sets of mixers were made by reforming old drilling machines. Grouting pumps were LB mono-screw pumps. A pre-stir slurry station was set to supply with mud. Most pores were Φ91 or Φ76 in diameter. The grouting method was hole sealing, no stage, pure pressed grouting.

The injection pipe was set at the hole bottom. Grouting pressure was 0.2-0.3 Mpa. Average dry materials consumption was 1.6 T/m.

Drilling and grouting at dam foundation was begun in February 1989. The period of accomplishing construction was less than two months. The price of the paste was low and the technique was simple easy to be controlled. Holes and caves were fully filled. Even in the condition of flow pressure, grouting can still be fulfilled.

3.3 Anti-seepage and reinforcing grout treatment of the earth dam of Yulong reservoir

Yulong reservoir stands in upper reaches of the Cantang River in Yichun City, Jiangxi Province. The dam is a homogenous earth dam of 166.5m long and 32m high. It situates on intensely weathered sandstone rock. There is an overburden of sand and gravel 4-6m thick at the river bed. Because of low construction quality, the compatibility of dam body was low, the sand and gravel at dam base were not excavated off and the leakage in dam body and dam base was serious. In the water pressure test, when discharge was 80 l/min, about 80% of drilling holes did not return water.

To increase the compatibility of the dam body and capability of anti-seepage for the safe operation of dam, grouting curtain construction was begun in 1989. A clay core wall was built in the dam body above water surface, an anti-seepage and reinforcing grout was constructed in the dam body under water surface and base.

Cement-clay slurry was used for grouting, the proportion of mixture (water vs. solid materials) was 0.33-0.45 in macroporous formation. Because the specific gravity of mud was low and there was no other additives in the locality, cement dose had to be increased to raise the plastic yield strength of pastes for meeting the demands of grout control.

Large power mixers made by ourself were used in slurry making. Because grout seal was difficult in earth dam, for preventing grout flowing backward, the grouting method was descending stage, hole sealing, pure pressed with injecting pipe set at the bottom of grout compartment. Grout pump was LB mono-screw pump. The ending standard of grouting considered both given pressure and injecting amount. Ending grout pressure was 0.1-0.2 Mpa. Average dry materials consumption was 0.8-1.0t/m. If grouting amount was large, it was needed to adopt the high limit of ending pressure. The impervious curtain was made up of single row

222

drill holes with 3 stage. The depth of hole was 37.1m and the distance between holes was 2.5m. The quantity of seepage at the dam back decreased to about 10% of what was before.

If we adopted general grouting method, it was difficult to control grout lost and reach at normal ending grout standard. Surplus water in the grout could affect the safe of the dam. The period and cost of grout were also not be able to accept. The utilization of stiff cement pastes solved these problems economically and efficiently.

4 CONCLUSION

Utilization of stiff cement pastes and relevant technique grouting leads to a good effect in engineering of anti-leakage, reinforcing and back filling in macroporous formation. The filling is full in macroporous cave and has a high solid strength. This technique has the advantages of low cost, simple technology, high controllability and work efficiency.

Fresh stiff cement paste can usually be simulated with Bigham fluid model. Its solid property of paste become obvious gradually with grouting process which controls the final infiltration distance of grout at certain grouting pressure.

It is convenient and feasible for grouting using stiff cement pastes in macroporous formation that to adopt vertical mixer with large power, screw pump to transport, pure pressed grouting with injecting pipe into the grout compartment, and ending grout standard considering given pressure or discharge.

Discussion of some problems about curtain grouting in karstic ground
Diskussion einiger Probleme bei Dichtungsschirmen in karstigem Untergrund

Xushujun
Yangtze River Scientific Research Institute, Wuhan, People's Republic of China

Synopsis

On the basis of data acquired in monitoring and quality control of curtain grouting for the Geheyan project, method including "Small diametrical diamond drilling, non — washing, non — waiting for concretion, closureof the borehole mouth, drilling and grouting by turns from top to bottom" adopted for curtain grouting in karstic formation grbund was successful. Effect of grouting was very well. Test sections of inspection borehole for seepage prevention were 100% satisfactory. According to the groutingpractice of Gcheyan Project and other engineerings, some problems for discussion were presented. Such as: washing is unnecessary for cement grouting under high pressure, waiting for concretion is unnecessary or the waiting time should be shortened; Though simple single grade water pressure test is useful there is not fixed relationship between water absorptivity and volume of cement injection; So water pressure tests of multiplegrades of pressure and multiple stages of rising and reduction of waterpressure are suitable. Nevertheless, grouting test can not be substituted by them; Each has its own strong point for closure of borehole mouth or blocking by section using high pressure grouting plug; Initial water cement ratio for grout curtain 1 : 1 is generally adoptable; Cyclic injection time of borehole closure by mechanical pressure grouting method is no good being too long.

1. Introduction

The Geheyan project, located on the mainriver of Qingjiang, the tributary of the Yangtze River, is another large — scale hydro — electrical station built in the karstic formation in China after the Wujiangdu project. It is a concrete gravity arch dam, 151m in height. Its total reservoir volume is $3.4 \times 10^3 m^3$, maximum water head 121.5m and installed capacity 1200MW.

The foundation rock in the dam site areais mainly the Cambrian Shilongdong limestone, 145k 195 m in thickness. There are harst andalong — river fractural structure developed init. There are more than 20 faults which have strong effect on grout curtain, such as, F10, F25, and shear zones such as 201#, 301#. Grout curtain is totally 1490m long and total design drilling footage is 220 km. Accordingto the construction program, the yearly highest intensity amounts to 100km, and the monthly highest intensity is 9.5km. There has been no precedent in China yet.

Principles of design for seepage prevention grout curtain are: "methods including small diametrical diamond borehole, borehole mouth closure, grouting by section from top to bottom should be adopted priori, Maximum grouting pressure 5 MPa is adopted. After grouting, unit absorption $w \leqslant 0.01$ L/min. m. m in the curtain bodybelow elevation 160m, while in the curtain body above elevation 160m, $w \leqslant 0.03$ L/min. m. mare required.

Construction of the curtain grout began in June 1990 and 150km of curtain were finished from then to the end of 1992. According to water pressure test data of inspection boreholes, in the total 2641 test sections of 218 water pressure inspection boreholes, the total footage of which is 12660m, w value of 2625 test sections is small than the designstandard of seepage prevention and is 99. 4%of the total test sections; For the 16 unsuited test sections, complement grouting and inspection were conducted, and all sections reached the design requirement. Core recoveryof the inspection borehole is apparently higher than that in the grouting borehole. As viewed from drainage holes and uplift observation holes constructed, under the conditon of upstream reservoir elevaiton 120m, there were no seepage in most drainage hole mouthes and a little seepage in a few holes; There were basically no seepage pressure in the most uplift observation holes and there were uplift only a few hundredths of MPa in a few holes. Seepage flow was very smal. The observation is being continued now. Practically, the grout technology adopted is correct andits effect is apparent.

Based on the author's experience of monitor work of curtain grout for Geheyan Projectand knowledge of other engineerings, the following problems are presented for discussion.

2. Discussion

2. 1 Borehole washing problem

In the traditional construction of cement grouting, it is required to wash the borehole until the return water being clear. In curtain grout for karstic ground or rock foundation with fault broken zones and weak intercalations, views on whether washing is needed is different. In practice, washing out karstic mud and fractural filler thoroughly istime—consuming and strenuous. To clear a borehole section in several days and nights by washing is difficult, even if adding additivein the washing water is not often effective. In the view of the author, adopting high pressure cement grouting technique, it is not necessary to wash the borehole. Requirement ofcurtain grout is reduction of seepage and increasing of stability of seepage prevention, but is not strengthening. Therefore, many countries adopt cement — clay grout to do curtain grouting. Weight of additive clay is more than 50% of the weight of cement. Since daycan be mixed in cement grout for curtain grout, why karstic mud and filler in cracks must be washed out? As viewed from the practice of the curtain grout engineering for the Geheyan hydro—electric station in constructionand the Wujiangdu project, nonwashing for borehores but increasing grouting pressure to 5—6 MPa to conduct cement grouting under high pressure, good effect has been obtained. According to water pressure test in inspection horehole, 100% of test sections reach the standard of seepage prevention in the Geheyan water power station and 98% in the Wujiangdu water power station; Failure by pipinghas not been occurred for karstic mud underhydraulic grade $J = 400 \sim 700$ in the Wujiangduwater power station; The clay injected has not been disintegrated in the water; Clay block after being injected by cement grout has an elastic modulus 2500 MPa and compressive strength 2. 4MPa; Seepage water quantity for the whole curtain (area of seepage proof is 185000 m²) is only 68 m³/24h.

2. 2 Water pressure test

2. 2. 1 Limitations of water pressure test

Owing to that the seepage path is not straight and its length is unknown, absorptivity obtained by water pressure test is not the absorptivity in the meaning of Darcy's Law. Nevertheless, unit absorption w is still considered as a numeral valve expressing the permeability of rock. So the permeability of rock is generally detected by water pressure tests so as to detect the possibility of grouting and possible grout consumption for rock. But data of crack number and width can not be directly obtained by water pressure tests and the main direction of seepage flow can not be given from pressure and absorptivity readings too. If absorptivity of a large number of fine cracks was equal to that of afew wide cracks, the seepage quantity and grouting possiblity might be greatly different. Up to date, the relationship between absorptivity and cement volume injected is not clear. So, in some engineering constructed abroad recently, water pressure tests for prediction of grout consumption are given up. But water pressure tests are useful yet. During inspection or evaluation of grouting effect, results of water pressure tests before and after grouting are often used to compare and evaluate.

2. 2. 2 Necessity of water pressure tests of multiple grades and multiple stages

Nowadays, the general program of water pressure tests is: At the geological investigation stage, simple water pressure tests using single grade of pressure 0. 3 MPa are used. According to results of water pressure tests, depthes of relative impermeable strutum and curtain bottom line are determined; During construction stage of grouting, water pressure tests are executed according to «Technological standard of cement grouting construction for hydraulic structure» (SDJ210—83). Simple water pressure tests of single gradepressure 1. 0 MPa are adopted for most borehole sections. But water heads before the damare different during operation of the engineering. For example, the maximum water head before the dam of the Geheyan water power station is 121. 5m, i. e. about 1. 2 MPa. The problem of pressure disagreement like this exists in other engineering with different degree.

Practically, the w values are different, when water pressure tests are conducted in the same borehole section with various pressure. Generally, $\omega_{0.3} > \omega_{1.0}$. When Huanghe WaterConservancy Committee Investigate and DesignInstitute conducted grouting tests under thedam of Guxian Project, contrast tests underdifferent pressures 0. 3MPa and 1. 0MPa (the equivalent water head before dam is about 100m) were done in the same borehole section and $\omega_{0.3}/\omega_{1.0} = 2 \sim 2.5$ was obtained. It was mainly dependent on flow form of ground water, fracture development condition, different charge of filler under different pressure and so on. Simple water pressure test of single grade is without response of such variation. In order to grasp the effect of curtain grout exactly and the varying regularity of curtain permeability under different pressure. it is necessary to do water pressure tests with rising and lowering pressure in multiplestages under multiple pressures.

2. 2. 3 Water pressure tests in multiple stages under multi ple grades of pressure

At Guxian Dam, water pressure tests in five stages under three grades of pressure were adopted, namely, pressures of 0. 3, 0. 6, 1. 0, 0. 6, 0. 3 MPa were adopted. After water pressure tests were conducted at each sectionin inspection boreholes, seven types of P k Qcurve were obtained. Among them, the fifth type of curves constitute the majority. Of this kind seerage flow is relatively small during pressure rising stage. By comparison with the pressure rising stage, seepage flow in the pressure lowering stage under the samegrade of pressure is reduced more. Appearingthis type of curves, the causes may be two: The first may be in such cases that the cracks are short and small, their communicating property is bad and part of cracks is a state of closure; The second may be that fillers in cracks are loose and they may move under water pressure, plug cracks and fill themdensely so as to reduce the seepage flow atthe pressure lowering stage. The second cause is true, the curtain body is corroded under the action of high water head in a longperiod after operation of the dam. Whether fillers in cracks can be stable in a long

time should pay attention to.

It is easy to see that stable standard must be reached at each stage, so the time of water pressure test is longer. Therefore, this kind of water pressure test method is suitable for the following cases: Some engineerings where the seepage regularity in the strutum is wanted to know under the action ofdifferent pressure or where requisition forthe accuracy and correctness of results of water pressure tests is high such as pilot hole and inspection hole. In order to shorten time of water pressure and to quicken construction progress, time interval of taking flow reading is changed from 10 min to 5 min. After the reading period is shortened, corresponding measures must be adopted so as to guarentes the accuracy of water pressure test. For example: water supply under automatic flow and stable pressure is adopted for water pressure test, the upper and lower limits of the measuring range of the flowmeter are deducedby w valus and so forth.

2. 2. 4 Necessity of grout test

There is not a fixed relationship between absorptivity and grout consumption of water pressure test, therefore grout tests can not be substituted by water pressure test. Whether grout is needed and successful is notonly a problem of rock permeability (w value), but also involving grouting possibility problem. Owing to the different natural conditions for various engineering, especially a great disparity in geological condition, parameters assumed according to precedent engineering analogy are not accurat in numerical quantities. So it is suggested that in the meantime of doing water pressure tests, grout tests are conducted too in the investigate stage. Results of tests can be used as data basis for determination of borehole distance, allowable pressure and estimation of cement injection.

2. 3 Grouting technology

As mentioned above, in the curtain grout of constructed Wujiangdu water power station and Geheyan water power station in construction in the karstic ground, technology of adopting small diametric diamond horehole, nonwashing, not waiting for concretion, closure of borehole mouth and cyclic cement grouting by section from top to bottom under high pressure is successful. Good effect of grouting is abtained. Owing to limited space, only the following several links are discussed.

227

2.3.1 Water cement ratio of grout

In the construction of curtain grout in China, 9 grades of ratio with initial ratio 8 : 1 (by weight) or 6 grades of ratio with initial ratio 5 : 1 are adopted for water cement ratio of grout. That is, grouting is started from thin grout and become thicker by grade. Thus part of water will separate out from the thin grout so as to make the continuity of cement filler discontinuous and form cavities, bubbles and seepage passages. With the lapse of time, seepage water may make cement grout calculus produce chemical corrosion, because calcium in calculus is solved out by seepage water. It is common occurrence that grout curtains are eroded and corroded so as to fail in China and abroad. In order to make cement grout curtain have good durability, firm calculus of cement grout and grout fully filling cracks and empties should be guaranteed. But the above requisity can not be obtained by using thin grout. Even if grout with water cement ratio 1 : 1 is used, separated water may be 30% of grout produced. Only 70% of space are filled by cement unless the water separated is drained out under grouting pressure. So, it is suggested that for common cracks, initial water cement ration 1 : 1 (by weight) is used and for smaller and wider cracks initial water cement ratio is risen to 2 : 1 or decreased to 0.8 : 1 or thicker. In the dam grouting engineering of the Yitaipu project built recently in Brazil, grout of water cement ratio 1 : 1 is adopted wholly. For United States Bureau of Reclamation, grout of water cement ratio 12 : 1 or thinner was always used formerly, but using thicker grout has been practical trend in grouting practice of the recent years.

2.3.2 Closure of borehole mouth

In view of the common rubber plug being difficult to bear grout pressure more than 4MPa, installing hole collar pipe and closing hole collar adopted in curtain grout of the Geheyan water power station are suitable methods. Grout quality can be guaranteed by closure of hole collar, not waiting for concretion and cyclic boring and grouting from top to bottom. If a grouting hole is divided into n sections, the first section will be bored and grouted repeatedly n times, the second, n—1 times, the third, n—2 times ..... Thus, though grouting quality may be guaranteed, time consuming, strenuous and grout waste are caused.

Along with development of grauting technique, Chinese Water Conservancy and Hydro—electric foundation Engineering Department developed a grouting plug under high pressuer, which can safely construct under grouting

pressure 8 MPa. So, method of plug grouting by section from top to hattom or from bottom to top can be adopted, and shortcomings of repeated boring and grouting and grout waste for hole collar closure method will be avoided. Especially, under the condition of good rock property of strutum, boring a hole to its bottom in one time and plug grouting by section from bottom to top so as to avoide to move the boring machine and contradiction between boring and grouting. Thus effect of grouting will greatly increased.

2.3.3 High pressure grouting and time of waiting for concretion

During the application of cement grouting under high pressure and closure of hole collar, using methods of not waiting for concretion and cyclic grouting and boring, good grouting effect has been obtained and phenomenon of influencing grouting quality owing to not waiting for concretion has not been discovered. When plug grouting by section from top to bottom is adopted, concretion may not waite for also, because calculus is mostly formed under the action of pressure seepage water, and the nearer to grouting point is, the greater the pressure may be sujected and the higher the strength of calculus will be. According to results of laboratory, 28d compressive strength of calculus, formed under pressure 0.1 kg/cm^2 and separated water, is as high as about 1 MPa. Therefore, high pressure grouting should be adopted as far as possible, where conditions are good. And measures of not waiting for concretion or shortening the time of waiting for concretion are suggested so as to raise construction effect of grouting.

2.3.4 Hole closure

For curtain grout engineering, Geheyan water power station is as same as other waterpower station. After finishing of grouting, the gouting hole was closed by mechanic pressing grout method. The end section of grouting hole is generally deepened below the relative impermeable strutum to certain depth, so the permeability and the grout consumption are very small. Therefore, temperature of grout rises sharply in cyclic injection even as high as 43 ℃ and grout behavior may be effected: How long the grout is laid in the transfer station where the grout is stirred continuously, is not grasped by the crew, so they still abandon the grout according to the principle of abandoning grout after or over 4 hours, and the practical time has usually been over 4 hours. Besides, too long stirring time in the process of mixing and injectionhas influence on grout behavior and hole closure quality also. So cyclic injection timeduring hole closure using mechanic pressing grout method is suggested within an hour.

3. Conclusion

Curtain grout is a covert engineering and conditions of nature and giology are different in various engineering. So the groutingmethod of curtain may not and should not be thoroughly the same. The above mentioned method including small diametric diamond boringhole, nonwashing, not waiting for concretion, hole collar closure and cyclic boring and grouting by section from top to bottom* adopted in karstic ground is effective. And it is successful in curtain grout engineering of Wujiangdu water power station and Geheyan water power station. But grouting technology will be further perfect if thicker grout, plug grouting by section using high pressure grouting plug and shortening hole closure injection time are adopted.

For water pressure test, test with multiple grade of pressure and multiple stage is suitable, even so, it is not suitable to substitute grouting test by it. For engineerings which have good conditions, in the meantime of doing water pressure test, grouting test should be conducted too.

Grouting is a science but mainly a experimental technique. In arder to guarantee thatgrouting is applied, in the treatment process of an engineering, successfully, economically and reasonably, we should depend on the emergency guiding of in-situ engineers who not only are familiar with and understand the central point of general specification and principle but also have rich practical experience, but not depend on bookishness.

References

[1] 高钟璞. 国内外水利水电基础处理施工技术现状. 《1993 年基础处理技术交流会》论文集（上册）P. 1~P. 9. 水利部建设开发公司，中国水利水电基础工程局合编.

[2] 魏龙斌. 三级压力五个阶段压水试验方法应用研究. （文献[1]P. 232~P. 237）.

Grouting in Rock and Concrete, Widmann (ed.) © 1993 Balkema, Rotterdam, ISBN 90 5410 350 7

Evaluation of foundation improvement after grouting

Beurteilung der Verbesserung des Gründungsbereiches nach dem Injizieren

P.T.Yen
Bechtel Corporation, USA

B.J.Gutierrez
Department of Energy, USA

ZUSAMMENFASSUNG: An einem bestehenden Atomkraftwerk in den USA wurde ein Programm für die Verdichtung und Einspritzvergiessung des Bodens durchgeführt. Der Grund für das Programm war den Boden unter Teilen der Anlage zu verbessern, um gegenwärtigen seismischen Anforderungen zu genügen. Frühere Einspritzungen während der ursprünglichen Bauarbeiten resultierten in das Einspritzen von grossen Mengen von Sandvergiessungs Material in den Boden. Die Vergiessung war von der Oberfläche durch Löcher durchgeführt, die in einem Gittersystem gebohrt waren. Aufzeichnungen zeigen, dass Bohrlöcher die grosse Mengen von Vergussmaterial anfnahmen in Gruppen auftraten und dass das Vergussmaterial meistens unter Schwerkraftdruck eingebracht wurde. Basierend zum Teil auf das Ergebnis von vollständigem Wasserverlust und Freifallen des Bohrgestänges, war ursprünglich angezeigt dass Hohlräume, weiches Material und kanalähnliche Zonen varhanden waren, die die Bohrlöcher untereinander und über grosse Strecken verbinden. Die gegenwärtige Erfahrung bestätigt, dass dies der Fall ist. Daher wurde eine Ueberprüfung der Qualitätskontrolle der Vergiessung der Fundamente Anfang 1992 durchgeführt. Das Projekt schloss die Ueberprüfung von Informationen von über 800 Penetrationsprüfungen der Fundamente einschliesslich Vergussbohrungen, Kerntestbohrungen, Standard Penetrations Tests, Marchetti Dehnungsmeter Tests, Piezokegelmeter Tests, Bodenproben, geophysikalische Vermessung und Tomographie ein. Eine Reihe von Hebungs– und Neigungsmessern wurden benutzt um Oberflächenverschiebungen zu messen. Messergebnisse schliessen die Bestimmung der Lage, Höhe, Dicke und seitlichen Ausdehnung der Vergussmasse ein. Drei–dimensionale Computer Modelle wurden hergestellt und Modellversuche (Simulationen) im Laboratorium durchgeführt. Dieses war das weitestgehende Anlagenmessprogramm für die Bestimmung der Lage, Dicke under Flächenausdehnung von in–situ Vergusskörpern.

ABSTRACT: A compaction grouting program was conducted at an existing nuclear plant in the United States. The objective was to improve the soils beneath portions of the site in order to meet current seismic standards. Earlier grouting conducted during original plant construction resulted in the injection of a very large quantity of sanded grout into the subsurface. Grouting was conducted from the surface through holes drilled in a grid pattern. Records showed that holes having large grout takes occurred in clusters, and that most of the grout was introduced at gravity pressure. Partial to complete water losses and drill rod drops that were previously believed to have indicated voids, soft zones and channel like areas, resulted in grout hole interconnection over great distances. The recent experience confirmed that this was occurring. Therefore, a quality control assessment of the foundation grouting was conducted beginning in 1992. The project included assessment of data from over 800 penetrations into the foundation comprised of grout holes, core holes, standard penetration tests, Marchetti dilatometer tests, piezocone penetrometer tests, soil sampling, geophysical surveying and tomography. An array of heave gauges and tiltmeters was utilized to monitor surface displacement. Results included the determination of the location, elevation, thickness, and lateral extent of grout plumes. Three dimensional computer models were prepared and laboratory simulations were conducted. This was the most comprehensive field program undertaken in North America to determine the location, thickness and the areal extent of grout bodies in situ.

1 INTRODUCTION, BACKGROUND AND PURPOSE

The principal element of this grouting project was to improve the soils beneath the foundation of an existing nuclear facility. This paper summarizes significant technical findings and puts forward conclusions that were reached about the grouting and potential soil settlement.

Initial geologic work at the site was performed in the early 1950's during construction. Surface depressions were observed at the site and soft zones (low resistance to penetration) were encountered at depth during exploratory drilling. The surface

ABB.1: ZUSAMMENFASSUNG DER ERGEBNISSE   FIGURE 1: SUMMARY OF RESULTS

PROGRAM	Total Holes Drilled	Total Holes Grouted	Total Grout Take (cubic meter)	Average Take per Hole (cubic meter)	Take per Area (cubic meter / square meter)	Maximum Individual Hole Take (cubic meter)
1991 - 1992	276	264	707	2.7	0.06	54
1951 - 1952	398	91	4,672	11.8	0.09	483

ABB. 2: BOHRUNGEN   FIGURE 2: DRILLING

PROGRAM	Method	Fluid Type	Pump Pressure (kPa)	Hole Depth (m)	Primary Hole Pattern	Primary Indicators of Soft Zones
1991 - 1992	• Conventional open–hole wash boring technique using drag bit. (Vertical) • Casing–advancer system. (Angle holes.)	• Bentonite (initial holes) • Revert (majority of holes)	172 - 2068	49	Parallel lines of holes spaced 7.6 m on center. Alternating staggered 7.6 m.	• Fluid losses during drilling
1951 - 1952	• Conventional open–hole wash–boring technique using fishtail bit. (Vertical holes only.)	• Bentonite	• Not known. Expect similar to 1991-1992 program.	64	Parallel lines spaced 7.6 m apart with holes on 15.2 m spacing. Alternating lines staggered 7.6 m.	• Fluid losses during drilling • Rod drops or unusually fast penetration rate during drilling

ABB. 3: EINSPRITZUNGEN

FIGURE 3: GROUTING

PROGRAM	Mix	Batching and Pumping Method	Procedure	Maximum Applied Pressure (kPa)	Acceptance Criteria	Hole Communication
1991 - 1992	• Sanded Mix cement-sand-bentonite-water • Neat Mix cement-bentonite-water	• Primarily on-site batching using paddle mixers and mobile batch plant • Moyno 3L-8 pumps used	• Grout injected through NQ size grout pipe (tremie) & lowered to within 1 meter of hole bottom. Vertical holes partially cased; angle holes entirely cased. Grouted also through 10.2 cm I.D. casing as casing was withdrawn in increments. • Pressure was applied at selected intervals.	138	Grouting complete at any interval if one of following occurs: • No significant take (less than 0.01 cubic meters per minute) recorded over 5–10 minutes at maximum allowable pressure • 142 cubic meters of grout is injected • Grout returns to the surface	No grout communication noted. Temporary venting of water from one hole was observed. Observed fluid communication during drilling in several cases.
1951 - 1952	• Sanded Mix cement-sand-bentonite-water	• On-site batch plant with drum-type mixers • 15 x 15 cm Duplex pumps used	• Grout continuously injected through 5 cm tremie pipe lowered into open hole to just above zone of significant fluid loss. All holes were uncased. • Following initial grouting, some holes were flushed with water and the grouting process repeated.	• Initially 172 kPa. Later increased to 276 kPa. Most holes grouted at 276 kPa. • For several tight holes, gauge pressure was recorded at 827 kPa.	• Initially a maximum of 142 cubic meters per hole. • Later changed to little to no grout take at maximum allowable pressure.	Fluid or grout communication observed in 34 holes. Some travel noted in excess of 30 m distance.

233

depressions, then believed to have been caused by the dissolution and collapse of carbonate–rich horizons at depth, and the presence of the soft zones led to the decision to inject grout into the subsoil strata to ensure the stability of the facility. Subsequent studies in 1991-1992 indicated that no detectable voids exist in the subsurface and that the original surface depressions are not associated with soft zones.

This study is based on over 850 soundings performed at the reactor site, including the data from the original site investigation and foundation grouting. In particular, the following data was evaluated.

- 75 exploration holes
- 662 grout injection holes
- 106 cone penetration test holes
- 9 Marchetti dilatometer test holes

Drilling and grouting data are summarized in Fig. 1, 2 and 3.

2 TECHNICAL INFORMATION

2.1 Geology

The marine sediments underlying the site are sands, silts, and clays. The sediments were deposited along with small carbonate banks of oyster shells. These are thought to play a role in the development of soft zones because much, or most, of the calcareous material has been dissolved. The soft zones are anomalous zones in the subsurface that are underconsolidated; they currently are not subject to the full pressure, or weight, of the overburden. Soft zones were identified by the sinking of drill rods under their own weight, by penetration devices having little resistance, and by drilling fluid losses. Soft zones do not occur everywhere beneath the site. Their extent, as determined from grout takes in boreholes, is less than 10 percent of the total area of the site.

2.2 Grouting

A total of 264 holes were grouted during this program. The total amount of fluid grout placed was 7062 m³, or an average of about 27 m³ per hole. This average includes about 7 m³ of grout needed to fill the drilled volume of each hole. The set, in–situ grout is approximately 50% of the injected fluid volume.

The largest amount of grout placed in one hole was 537 m³. More than 80 percent of the holes took less than 30 m³ of grout and the great majority of holes accepted little more than the amount needed to refill the hole. Only 17 holes had takes exceeding 85 m³, and these accounted for 45 percent of the total grout take. These results indicate that the soft zones deep in the subsurface are of small size and discontinuous distribution.

Most of the grout selectively entered locations a short distance above the soft zones at depths from 37 to 44 m. In a few instances, a small amount of grout was injected at about 27 m in a thin zone. Using a minimum dry density of soft zone material of 721 kg/m³, the grouting caused an increase in dry unit weight of about 48 kg/m³ for an assumed 2.4 m soft soil column. This also represents a corresponding decrease in void ratio from about 2.7 to about 2.5. This program indicated that the character of the subsurface materials and hydrofracturing, or hydraulic jacking, were the primary influence on grout take. In general, the amount of grout that a hole accepted was considered to be influenced by 1) the physical characteristics of the subsurface materials, 2) the fluid pressures to which these materials were subjected during drilling and grouting, 3) the properties of the grout, and 4) the procedures used to inject the grout. However, the relatively minor changes in grout properties and grouting procedures during the program had little effect on the amount of grout injected.

Two types of grout mixtures were used in the program, a sanded mix and a neat cement mix (without sand). Field observations indicated that the rate of grout take did not change significantly when the grout was changed from neat to sanded, nor was it obvious that neat grout, which is more flowable, penetrated soft zones more readily than sanded grout. The viscosity, weight, age, and temperature of the grout were varied within limited ranges, but these changes had no observable effect on grout takes.

Similarly, changes in the methods of introducing grout in the hole and interruptions of 20 minutes or less during the grouting procedure had little effect on the volume of grout take. Grout was injected at different gauge pressures of 0 to 276 kPa with no notable differences in the grout volume injected. Most holes that took large amounts of grout took it under gravity with no pressure added at the ground surface.

Grout holes were drilled using either a bentonite clay– or an organic polymer–based (Revert) drilling fluid. The type of fluid that was used did not influence the amount of grout placed. Some holes were advanced using conventional open–hole drilling methods while others were drilled using the casing advancer drilling system. The casing advancer system can be a source of drilling fluid pressures greater than 1379 kPa because of the small annular space between the casing and borehole wall.

Drilling fluid losses, as distinct from grout takes, were due to hydrofracturing and, to a lesser degree, by permeation. Losses generally were larger in holes drilled with the organic–polymer fluid because this fluid has a lower viscosity. Fluid losses were also larger in holes drilled with the casing advancer due to the higher fluid pressures associated with this method.

Experience gained through drilling and grouting indicated that grout take would be low for a hole that had no significant fluid loss and that the grout takes would be higher in holes which completely lost fluid return during drilling. High drilling fluid losses did not always correspond to high grout takes, however, and the total quantity of fluid lost during drilling could not be used to estimate potential grout take. These characteristics were demonstrated by conducting field tests at selected holes.

2.3 Heave Monitoring

To verify that no potentially damaging strains occurred at the locations of buried and above–ground facilities, instruments monitored ground heave during the grouting of each hole. A maximum allowable vertical surface movement was set conservatively at 2.5 mm because that was a measurable quantity that could be detected by monitoring instruments but that would not harm existing facilities.

Three types of monitoring systems were used. These were precise surface surveys, Borros–type stations which required deep, drilled holes, and tiltmeters, which measure the tilt parallel to the base of the instrument. Tiltmeters, which could be installed quickly, were used in conjunction with the Borros type heave–monitoring station in several areas. After field results indicated tiltmeters provided data comparable to the results of the Borros stations and surface surveying options, they were used as the primary monitoring system in the remainder of the program.

Ground heave as measured by the tiltmeters was always less than the allowable limit of 2.5 mm and trended to zero after the completion of grouting. Subsurface pressures imposed during both the drilling and grouting operations exceeded existing in situ stresses, but no permanent deformation of the ground surface was recorded.

2.4 Exploration and Testing

1. Introduction. The exploratory drilling and testing phase of the grouting program was designed to investigate the in–place grout thickness, the travel distance, and the effect on soft zones. Cone penetrometer (CPT) soundings, standard penetration test (SPT) borings, core holes, dilatometer test (DMT) soundings, seismic tomography, borehole geophysical logging, and laboratory tests were the techniques used.

Three areas selected for testing included one that had the largest grout take of any one area and also the largest single hole take of the program, a second that had the largest single hole take of the construction grouting program and a third that had the largest number of holes that took more than 85 m^3 of grout in the recent program.

2. Cone Penetration Test Soundings. The Cone Penetration Test uses a device that is pushed into the ground and records cone tip pressure, frictional resistance, and induced pore pressure. More than 100 CPTs at the site were used to identify soft zones and grouted layers.

CPT tests generally were effective for identifying soft zones. However, there was little difference in tip pressures between soft zones that were adjacent to grout holes with high takes and those that were not. Correlations between pre– and post–grout CPTs were not easily made due to variations in stratigraphy and thicknesses of soft zones over short distances. The CPT tests are consistent with the conclusion that the grouting, as conducted, caused no observable improvement to the soft zones, and demonstrates that soft zones remain after grouting.

3. Standard Penetration Test Borings. A split spoon sampler was used to secure both subsurface samples and at the same time, provide an index of the in–place strength and density of the materials. In SPT testing, the resistance to penetration is measured as the number of blows (N–value) of a weight required to produce a standard depth of penetration.

During the post–grout exploration phase, SPT data were collected in all or part of five sample holes. The grout had no apparent effect on penetration, because N–values obtained from boreholes where grout was found were no different from those in which it was not found. The tests in all three areas indicated that most of the grout entered soil above the softest zones at different levels between 37 and 43 m. The SPT tests also indicated that soft zones remained both above and below levels where grout entered and grouting made no significant improvement to the soft zones.

4. Core Holes. Core samples were obtained from 22 holes. Continuous sampling generally began somewhere between 18 and 27 m and continued to the bottom of the hole. Core recovery averaged more than 70 percent in the sampled interval.

Grout was found in thin layers interfingered with formation material. The layers ranged from less than 2.5 mm to 152 mm in thickness. Maximum cumulative thickness of all grout layers in any one hole was 310 mm and this occurred 4.6 m from a grout hole that had the largest grout take of this project. Most of the grout layers were found from depths of 39 to 40 m, but thin grout layers also were found at about 27 m in five holes. Average cumulative grout thickness per sample hole in a test area was about 76 mm.

Sanded grout injected during construction was recovered in all the holes in the test areas. Multiple layers of grout from 13 to 559 mm thick occurred from 37 to 40 m and the maximum cumulative thickness of the layers in any one hole was 1.0 m. The grout was found to be interbedded with sand and clayey sand. Permeation into the soil matrix was not observed.

5. Dilatometer Test Soundings. A dilatometer test is performed by pushing a thin, wedge–shaped blade with a circular membrane attached to one

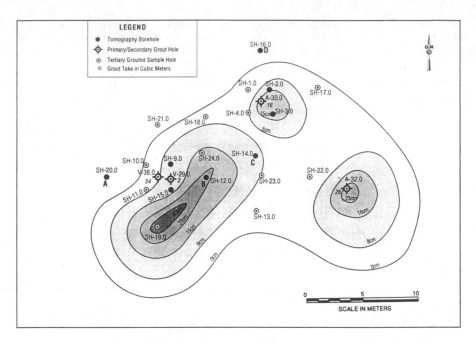

*Abb. 4 - LAGEPLAN MIT INTERPRETIERTER AUSDEHNUNG
DER EINSPRITZUNGEN*

Fig. 4 - PLAN VIEW WITH INTERPRETED PLUME EXTENTS

side into the soil. The blade is stopped at predetermined intervals and the membrane expanded. The resistance to expansion and the thrust required to advance the blade are used to develop indices that correlate to soil types, strengths, and pressures.

Nine DMT soundings were made for the primary purpose of evaluating the in situ stress and compressibility of the soft zones. The DMT results confirmed that the matrix material within the critical layer are acting as an arch which transfers overburden loads around soft zones. The values of tangent modulus obtained were used to calculate a best estimate of surface settlement.

6. Seismic Tomography. Seismic tomography creates images by measuring the speed and amplitude of sound waves traveling along paths within a slice of earth between two boreholes. It is a state–of–the–art geophysical tool that provides information on the properties of the materials.

Crosshole seismic tomographic surveys were conducted in the test areas to image the structure of grout layers between boreholes. Grout layers more than 75 mm thick were detected using the survey.

Tomography effectively imaged grout plumes and detected plume changes in before–and–after surveys. The images indicate that grout traveled up to 12 m along selective pathways from the boreholes, an observation that corroborates other evidence about grout dissemination. The results of tomography testing are shown in Fig. 4 and 5.

7. Borehole Geophysical Logging. In borehole geophysical surveys, a probe or special tool is lowered in a hole and various physical parameters of the borehole and surrounding formations are recorded. These logs provided information about the physical characteristics of the subsurface materials and formation water. They proved useful for identifying soil types and properties and for correlating stratigraphic horizons. The logs, used in conjunction with core descriptions, SPT soundings, and CPT results, provided detailed information about the occurrence and thickness of soft zones.

8. Laboratory Testing. Only a limited amount of laboratory testing was performed to provide information on moisture content, grain size, Atterberg limits (plasticity), and unit weight (bulk density). Holes were sampled with a split spoon, conventional rotary core barrel, or thin–walled sampler. Visual descriptions were made of the materials using the Unified Soil Classification System and laboratory tests were performed on selected samples.

Eighteen samples obtained from continuous core borings were analyzed using X–ray diffraction, the scanning electron microscope, and other laboratory instruments and techniques. The objectives of these analyses were to evaluate the differences in the prior and recently injected grout, the presence of multiple grout injections, and the contact between the grout and natural materials. Selected

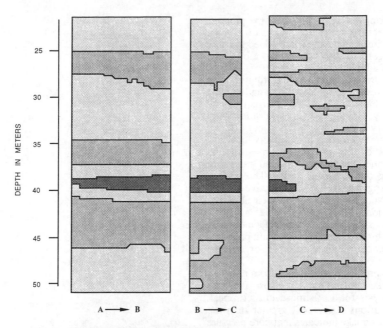

DEPTH IN METERS

A ⟶ B   B ⟶ C   C ⟶ D

Abb. 5 - SEISMICHE TOMOGRAPHY ABBILDUNG A-B-C -D
GESCHWINDIGKEITS PROFILE

Fig. 5 - SEISMIC TOMOGRAPHY PANEL A-B-C-D VELOCITY PROFILE

physical properties also were measured on a few samples. Fourteen of the samples were grout and the remaining four samples were formation materials.

A series of laboratory tests were performed to investigate grout penetration under pressure both in uniform cohesionless sand and in sand with one or two clay inclusions simulating site materials. Both soft and stiff clays were examined.

Tests showed that in a uniform deposit of wet sand, the grout simply compacted the sand around the grout hole. When soft clay layers were added, grout in most cases preferentially penetrated the materials at the boundaries of the layers. In some tests the grout did displace the soft clay. Grout also preferentially entered zones of weakness that had been introduced into soft clays prior to the tests.

3 DEFORMATION ANALYSES

The investigations demonstrated that soil arches exist above the soft zones and that the soft zones are underconsolidated. These soft zones could compress should the soil arches break down, resulting in surface settlement. A state–of–the–art finite difference method incorporating elasto–plastic soil behavior was employed to obtain an indication of strains that would be induced by such compression. The analytical method that was employed permitted the inclusion of such

important factors as stress dependent elasto–plastic soil behavior, soil dilatancy, and both plane strain and axisymmetric geometries. The effect of earthquake–induced pore water pressures was also studied.

4 CONCLUSIONS

The original construction grouting program and the recent program revealed that soft zones, the principal targets for grouting and stabilization, are not interconnected, average about 1.8 m in thickness, and have an average lateral extent of less than 15 m in any one direction. Their areal extent, based on grout takes, is less than 10 percent of the area grouted during the previous and present programs.

The post–grout investigation indicated that about 97 percent of the grout in this program was placed between the depths of 37 to 43 m. Less than one percent was detected at a depth of about 27 m. Eighty percent of the injected grout occurs in thin seams less than 50 mm thick, although seams ranged less than 13 mm to 150 mm in thickness.

Most of the grout entered thin intervals between strata from the top of the soft zones to about a meter above the top. Grout penetration by intergranular void permeation was not observed. In most cases, high drilling fluid pressures and fluid losses probably exploited preferential paths of weakness in the natural deposits by hydrofracturing (hydraulic jacking). During grouting, the pressure of grout injection, rate of injection, grout flowability, and casing depth all combined to cause grout to travel by hydrofracture along the drilling–enhanced pathways.

Exploratory drilling and in situ testing showed that soft zones remained after grouting and that the injected grout had only a minimal effect on the compressibility of the zones and additional compressibility remains. The benefit of densification due to grouting is extremely localized.

The results of the post-grout investigation program confirmed the presence of load transfer or arching around the soft–zone soils. The drilling and grouting program also was an effective method for testing the supporting strata and their ability to resist deformation. Both the 1951-1952 and 1991-1992 programs subjected the subsurface soils to fluid pressures higher than existing in situ pressures with no evidence of arch collapse or surface settlement.

Induced pore pressures anticipated from a design basis earthquake thus will not collapse the arch material. Analysis for a seismic event with peak ground accelerations 50 percent greater than the design basis earthquake indicates that pore pressure induced from such an event also will not collapse the soil arch above the soft zone.

Information gained from this program showed that soft zones represent only a small (less than 1%) percentage of the total volume of material to a depth of 49 m. These zones were not compressed by grouting to any measurable degree. The grout injected probably densified the foundation soils by a small amount. Test data indicate the foundation is stable without grouting.

New technology for grouting behind shaft walls in aqueous sand strata

Neues Injektionsverfahren für wasserführende Sandschichten hinter Schacht-Wänden

Zeng Rongxiu
China National Coal Corporation, People's Republic of China

ABSTRACT: The report deals with a recently developed grouting procedure which was applied at the first time in 1989 for aqueous sand strata behind the lining of a ventilation-shaft for a coal mine. By means of grouting cement suspensions a consolidation of the sand strata and a strengthening of the shaft lining was achieved. With this method essential improvements of the impermeability as well as the stability of the shaft were attained. This method was applied several times since then and got the chinese state patent in 1991 and was awarded China science and technology invention price.

ZUSAMMENFASSUNG: Im Bericht wird ein für die Injektion wasserführender Sandschichten entwickeltes Verfahren behandelt, das erstmals 1989 bei einem Ventilationsschacht im Kohlebergbau angewendet wurde. Durch die Injektion von Zement-Suspensionen werden die Sandschichten verfestigt und die Schachtauskleidung verstärkt. Damit konnten entscheidende Verbesserungen sowohl der Stabilität als auch der Wasserdichtigkeit erreicht werden. Dieses inzwischen mehrfach ausgeführte Verfahren erhielt 1991 das staatliche Patent in China und wurde 1992 mit dem chinesischen Preis für Erfindungen auf dem Gebiet von Wissenschaft und Technologic ausgezeichnet.

INTRODUCTION

Over a long period of time, it was regarded as a regulation in China that prohibits grouting through shaft wall in aqueous sand strata. In Coal Mine Safety Regulations of PRC, it was prescribed clearly that if the shaft wall pass through aqueous strata, the depth of grouting hole at least less 200 mm than the thickness of the shaft wall. If the shaft has double side shaft wall, the grouting hole can cross inside wall and reach up to 100 mm into the outside wall. The shaft wall must have enough strength to bear the maximum grouting pressure. If not, grouting work can't be applied.

In February 1989, in dealing with the accidents of the shaft wall failure of west ventilation shaft of Panji No3 Coal Mine, the New Technology For Grouting Behind Shaft Wall in

Aqueous Sand Strata (be called breaking shaft wall for grouting for short) developed by the author was adopted which met with success on the first field trial, and in the same year it was immediately introduced for widespread trail successfully in the west ventilation shaft of Haizi Coal Mine and the auxiliary shaft of Zhangshuanglu Coal Mine.

In 1990, it was used respectively in the main shaft, auxiliary shaft of Haizi Coal Mine, and was also used in the main shaft Zhangshuanglu Coal Mine, auxiliary shaft of Lin huan Coal Mine for widespread application. This technology has been adopted in China Coal Mine underground construction constitution and acceptance check specification in 1990, and won state patent in 1991. In 1992, it was awarded China science and technology invention prize.

HISTORICAL BACKGROUND OF BREAKING SHAFT WALL FOR GROUTING

Since 1987, 17 vertical shafts have been constructed in China using the freezing method. Subsidence of the tower, the shaft wall as well as the surface ground the shaft has been caused due to large volumes of mine water discharged resulting in drastic lowering of water head in the aquifer at the bottom of top soil and unconsolidated rock waste. This has led to serve local damage or deformation of the shaft wall and facilities such as the shaft guides, drainage piping, etc., with the concrete wall falling off up to a depth of 250 mm, and the clearance between the shaft guide and wall being reduced to a minimum of 10 mm as a result of deformed reinforcing bars inside the wall, over exceeding the stipulation requires limits of 150 mm. For instance, a 12-in dia, drainage pipe was bent over a length of 763 mm, and the consequential water inrush through the wall reached up to 66 cu.m/hr containing 5.3 % sand with the biggest particle size of 20 mm. Mine safety was seriously endangered, mining operation and construction had to be repeatedly stopped for an emergency remedy, resulting in considerable financial losses. It was unprecedented in the history of mine construction and operation in terms of the damage it had created.

To prevent the damage from deteriorating the shaft and for effective remedy, the leaders of China National Coal Corporation had convened experts conference many times, worked out comprehensive handling methods to reinforced the shaft wall with U-steel rings inside and with shotcreting, while grouting was applied to the leaking sections of the wall. However, deformations and damage were deteriorating in the 8 previously remedied shafts had suffered more severe damage, mining operations had to be stopped again, if it not be effectively remedied at this time, may be lead to be abandoned. The practice had proved that reinforcing the shaft wall with U-steel rings and concrete totalling thickness of 200 - 300 mm can't halt the occurrence of the shaft wall damage, can't assure sustained stability of the shaft wall (see Fig. 1).

Under these circumstances, the New Technology For Grouting Behind Shaft Wall In

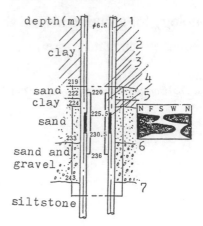

1_ 1.2 m double-side concrete shaft wall;
2_ 27.5 m distance of grouting hole between top line and bottom line;
3_ 16 m reinforced height with U-steel rings 200 m in height;
4_ 5 m vertical failure height of the shaft wall damaged in the first time;
5_ the plan view of the shaft wall damaged in horizontal direction;
6_ 250 mm maximum damaged thickness;
7_ 763 mm maximum bent length of the drainage pipe.

Fig. 1   The cross-section of the failure and reinforcement of the shaft wall of Zhangshuanglu auxiliary shaft.

Aqueous Sand Strata developed by the author was adopted, which met with success on the site trial.

THE PRINCIPLE AND TECHNOLOGICAL PROCESS OF BREAKING SHAFT WALL FOR GROUTING

PRINCIPLE

The author has studied the key problems for a long time which was concerned with by experts to break shaft wall for grouting, and reached following conclusions:

(1) Breaking shaft wall for grouting can bring the shaft wall under permanent control.

The substantial failure reason of the shaft wall was that the join external force was bigger than

Table 1.

Shaft dia. of hole (m)	0.005	0.01	0.015	0.02	0.025	0.03	0.035	0.04	0.045	0.05
water inrushing rate (cu.m/hr)	1.32	2.64	3.96	5.38	6.59	7.91	9.23	10.55	11.87	13.19

itself resistance. But the main reason of disequilibriums' force acting on the shaft wall was that the sand strata structure behind the wall was damaged after the shaft sinking by freezing, and the conditions of hydrogeology and engineering geology around the shaft wall had been caused tremendous change due to mine water discharged. Therefore, the main purpose of breaking the shaft wall for grouting was to grout to sand strata around the shaft wall, make it consolidation from porous state, change its physical mechanics properties, improve its strength and stability, reduce its permeability, compressibility and swelling property, reduce vertical compressive stress and lateral compressive stress, and make itself resistance bigger than join external force, keep the stability of the shaft wall, and thus the shaft wall can be brought under permanent control.

(2) Breaking shaft wall with a small-diameter drilling machine can assure a small volume of water inrush through wall, and can control water inrush through wall using its inrushing intermittent regularity.

Base on the underground water dynamics principle, it is known that the volume of inrushing water depends on the following three factors: the permeability of sand; the level of water head; and the water inrushing section area.

In the light of water inrushing conditions in hole, water inrushing rate can be calculated according to the following formula:

Q = 2prsk

in the formula,

Q_ water inrushing rate, cu.m/hr
r_ half diameter of hole, m;
k_ permeability coefficient, m/hr
s_ the level of water head, m.

When k and s are fixed, 2prsk is a constant, Q is in direct proportion to r.

According to the information about the west ventilation shaft of Panji No3 Coal Mine,

k =1 m/hr; s = 42 m.

The calculated value is close to the really value of the west ventilation shaft of Panji No3 Coal Mine, but 1 - 2 times less than empirical value, and therefore the water inrushing rate can be controlled by controlling the diameter of hole.

Because Q is in direct proportion to r, the more the sand inrush through the wall, the smaller the effective half diameter, and therefore the water inrushing rate could be reduced. Due to the veries of diameters of sand particle, the drilling hole can be blocked easily resulting in water and sand inrush through wall intermittently, and these will be advantageous to the control of water and sand inrush.

(3) The shaft wall commonly can't be damaged if the shaft wall was broken to grout to sand strata.

According to sand strata structure distribution features, as well as the water permeability regularity and the hydraulic transmission principles, sand strata is a no-lateral limited loose materials, the percentage of porosity is 20 - 40 % with bigger pore in it which uniformal distributed, and it have a good permeability and compressibility. When grou- ting pressure was transmitted to loose material, it was transmitted to all direction again through loose material and liquid medium. If the technological process was reasonable selected, and the safety measures were taken at the same time, safety construction can be assured, even if there was a higher was a higher grouting pressure.

(4) Concrete curtain could be formed to consolidate aqueous sand strata use of single-fluid cement grout.

Sand strata, particular sand and gravel strata, have a good and uniformal permeability, grouting work can be easy applied. A proper thickness of concrete curtain can be formed if

Table 2.

shaft	grouting time (y/m/d)	grouting length (m)	time limit (d)	concrete curtain thickness (m)	remarks
PJ NO 3 w ven.sh.	89.2.17~89.2.27	12	11	0.6	3 shifts per day
HZ w ven.sh.	89.6.15~89.7.22	23	37	0.6~1.2	3 shifts per day
ZSL aux.sh.	89.8.18~89.10.21	27.5	42	0.8~3.3	3 Shifts per day
HZ aux.sh.	89.9.10~89.11.11	34	42	1.4~2.0	2 shifts per day
HZ cen.ven.sh.	90.2.14~90.3.23	36	27	1.5~2.0	3 shifts per day
ZSL mai.sh.	90.4.17~90.6.8	40	44	1.1~1.8	3 shifts per day
HZ mai.sh.	90.5.12~90.7.9	37.5	39	1.5~2.0	2 shifts per day
LH aux.sh.	90.8.19~90.10.31	41	25	1.5~2.0	1 shift per day

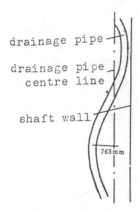

Fig. 2 The drainage pipe deformation in diameter of Zhangshuanglu auxiliary shaft

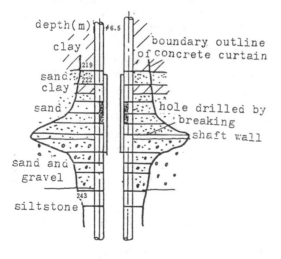

Fig. 3 The concrete curtain section of Zhangshuanglu auxiliary shaft formed by breaking shaft wall for grouting.

effective measures in technology processing were taken.

TECHNOLOGICAL PROCESS

The main features of the technology was: when single-fluid cement grout was grouted in aqueous sand strata, the sand strata around shaft wall was consolidated, and it connected with shaft wall which was remedied use of grouting, thus the shaft wall could be brought under permanent control and maintains stability.

The main technology process was: drilling collar, fixing collar pipe and mounting high-pressure valve--drilling pressure leakage hole, fixing collar pipe and mounting high-pressure valve--breaking shaft wall--grouting behind shaft wall--sweeping hole--grouting.

Primary materials: cement
Primary equipments:

ordinary pneumatic drill
2TG2-60/210 double-fluid speed control grout jump.

BREAKING SHAFT WALL FOR GROUTING

1. The engineering projects which had been complicated use of breaking shaft wall for grouting.

2. The construction example of Zhangshuanglu auxiliary shaft.

In remedied 8 vertical shafts, Zhangshuanglu auxiliary shaft, Panji No3 west ventilation shaft, Haizi west ventilation shaft were three serious damaged shafts, and were first three shafts put to trial. Here only to Present the failure and remedy cases of Zhangshuanglu auxiliary shaft.

(1) Brief introduction

The production of Zhangshuanglu auxiliary shaft was designed 1 2 Mt with thickness of 243 m of quaternary period alluvial deposit, and thickness of 24 m of sand strata in bottom, containing silt, fine sand medium coarse sand, coarse sand and gravel, with 15 kgf/sq. cm water pressure head. With 6.5 m clean diameter and 565.3 m in depth, the shaft was constructed using the freezing method with 345 m freezing depth. The shaft wall structure was double-side 300 - 500 reinforced concrete with a total thickness of 1200 mm (each side 600 mm).

(2) The damage circumstances of the shaft wall.

In July 29, 1987, 229.3 - 230.6 m length of the shaft wall was damaged with the damaged are length of 1. 5 m and the concrete wall falling off up to a depth of 250 mm. The maximum water inrush through the wall reached up to 66 cu.m/hr containing 5.3 % sand.

In May 20, 1988, the drainage pipe seat cushion was fractured in four corners, the South and North side of the shaft wall was bent over a length of 10 mm and 15 mm each other.

In July 13, 1989, the shaft guide in depth of 216 - 224 m was severe twisted, the drainage pipe had been breaking away and the clearance between the shaft guide and wall which was previously reinforced being reduced to a minimum of 17 mm and the water inrush through the wall reached up to 4.1 cu.m/hr.

In September 30, 1989, four drainage pipes were all bent in the depth of 236 - 266 m and a severely one was bent over a length of 763 mm in diameter (see fig. 2).

(3) The main reason of the shaft wall damage and resolve methods.

The Main Reason of The Shaft Wall Damage

In 243 m overburden strata which the shaft crossed, the top clay strata account for 69.5 % (169 m), the bottom was 24 m sand strata in which containing rich water. Subsidence of the surface around the shaft has been caused due to large volumes of mine water discharged resulting in drastic lowering of water head in the aquifer at the bottom of top soil, the great supplemental stress was applied on the shaft wall at the top thick clay strata, and thus the abutment area of the shaft wall certainly would engender a great deformation and damage. In July to September 1987, the subsidence of the tower reached up to 116 mm.

Resolve Methods

There were two distinct methods to be used successively to remedy the damaged shaft. In July to September 1987, the shaft wall was reinforced with U-steel rings 200 mm in height in close spacing and shotcreting in the depth of 220 - 236 m, but the stability of the shaft wall was not resolved causing further stoppage of mine operation once again (see fig. 1). In September to October 1989, the New Technology For Grouting Behind Shaft Wall In Aqueous Sand Strata was adopted, as a result the sand mortar consolidated concrete curtain was formed in a thickness of 0.8 - 3.3 m behind the shaft wall, water leakage through the shaft wall was sealed immediately, and through 16 months' observation it was found the water leakage had been eliminated and the shaft stabilized (see fig. 3).

TECHNICAL AND FINANCIAL EFFECTIVENESS OF BREAKING SHAFT WALL FOR GROUTING

1. Direct effectiveness

(1) Fast, effective and inexpensive, when sand strata behind the shaft wall was consolidated, the water leakage through the shaft wall was eliminated immediately. For example, in the west ventilation shaft of Panji No3 Coal Mine, when grouting work was being applied only two days (more than 10 tons of cement was grouted), both water leakage places were stocked. In the west ventilation shaft of Haizi Coal Mine, 48 cu.m/hr water leakage had been eliminated at 10 days when grouting work being applied. Up to now, through four years' observation it was found the water leakage was eliminated permanent in 8 vertical shafts which were treated with this technology.

(2) When cement grout being grouted to sand

strata behind the shaft wall, the clearance between the inside and outside shaft wall was also stowed at the same time, resulting in the resulting in the remedied inside and outside shaft wall were connected in one body which improved the strength and stability of the shaft wall itself.

(3) After grouting work was applied, some thickness of consolidation concrete curtain was formed behind the shaft wall, strength and stability of the shaft wall was improved, permeability and compressibility was reduced, and vertical compressive stress working on the shaft wall was reduced due to the connection of the shaft wall with consolidation sand strata.

Through long times' observation it was found that the shaft wall not only had no water leakage but also had a quit little deformation keep the stability of the shaft wall, and it was brought under permanent control. The measured deformation curve contrast of the shaft wall of Haizi Coal Mine centre ventilation shaft which was initially reinforced with U-steel rings and later was reinforced by breaking shaft wall for grouting as follow.

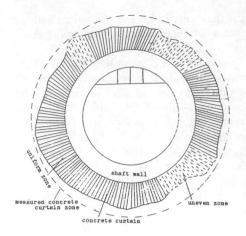

Fig. 5(a) Plan view of concrete curtain.

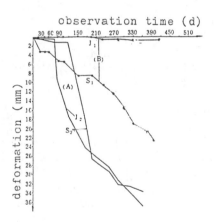

J1, J2_ deformation in dia.;
S1, S2_ vertical deformation.

Fig. 4 The measured deformation curve of the shaft wall of Haizi Coal Mine centre ventilation shaft which was initially reinforced with U-steel rings (A) and later was reinforced by breaking shaft wall for grouting (B).

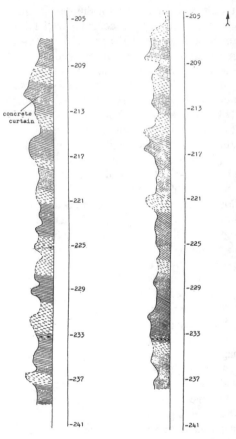

Fig. 5(b) Cross-section view of concrete curtain.

Fig. 5 was the plan view and cross- section view of concrete curtain zone measured use of high stress pulse of the auxiliary shaft of Lin huan Coal Mine formed by grouting behind the shaft wall.

Fig. 6 was the specimen photograph of unconsolidated and consolidated sand fetched from a drilling hole in different depth behind concrete curtain of the auxiliary shaft of Zhangshuanglu Coal Mine.

2. Technical, financial and social effectiveness.

In the 8 severe damaged shafts which were treated with this technology, the total grouting length only 251 m, and only took 9 months for grouting work. This indicated that the technology has fast effective, good quality, security and inexpensive features.

The use of the technology, bringing the 4 mines with a total annual capacity of 7.5 million tons of run-of-mine coal back to operation, and save more than 10 millions for country.

3 Grouting procedures – Case histories
Ausführung von Injektionen – Beispiele

Final grouting Warm Springs Dam

Abschließende Injektion bei der Talsperre Warm Springs

William J. Clarke
Geochemical Corporation, Ridgewood, N.J., USA

Millard D. Boyd
US Army Corps of Engineers, Sacramento, Calif., USA

ABSTRACT: In 1990, a test grouting program was performed at the control shaft for the Warm Springs Dam, Dry Creek, California. Final grouting the Warm Springs Dam by Jensen Drilling for the Corps of Engineers was accomplished from May through December, 1992. Portland type II cement was used for primary and secondary grouting and ultrafine slag/portland was use for permeation grouting. Because of early problems in drilling a very straight hole, 4.5 inch diameter drill steel was used for most of the holes. Straight hole percussion drilling was accomplished by installing a long surface nipple which came up through the drill table, plumbing the nipple and the drill steel, grouting the nipple into the hole and then cutting the nipple off to the desired length.

The last seven holes were grouted with a combination of M1 ultrafine slag and M3 ultrafine portland because of the unavailability of M5 slag/portland. Viscosities were satisfactory and no changes in formulation or mixing procedure were required for these holes. Properties of the ultrafine cements are outlined in the ASCE references by the author.

Placement of over 14300 cubic feet of grout indicates that many voids near the control shaft accepted large amounts of grout. The ultrafine cement filled many very fine openings and achieved the required seal around the shaft. During the last few weeks of the contract, there was a rainfall accumulation of several inches which raised the level of the reservoir into the flood pool zone and the control shaft still remains dry.

AUSZUG: Im Jahre 1990 wurde ein Probe-Zementierungsprogramm am Steuerschaft für den Warm Springs Damm in Dry Creek, Kalifornien ausgeführt. Die Schlußzementierung des Warm Springs Dammes bei Jensen Drilling für das Ingenieur Korps wurde von Mai bis Ende Dezember 1992 durchgeführt. Zement der Portland Sorte II wurde für das primäre und das sekundäre Zementieren gebraucht und ultrafeine Schlacke/Portland wurde für das ausfüllende Zementieren gebraucht. Weil es anfangs Probleme gab ein sehr gerades Bohrloch zu bohren, wurde ein Bohrmeißelstahl, der 4,5 Zoll im Diameter war, für die meisten Bohrlöcher gebraucht. Die Durchführung des Schlagbohrens gerader Bohrlöcher bestand indem man einen Langoberflächennippel installierte, der durch den Bohrtisch heraufkam, den Nippel und den Bohrmeißelstahl lotete, den Nippel in das Bohrloch zementierte und dann den Nippel nach gewünschter Länge abschnitt.

Die letzten sieben Bohrlöcher wurden mit einer Kombination von M1 ultrafeiner Schlacke und M3 ultrafeinem Portland zementiert, weil die M5 Schlacke/Portland nicht vorhanden war. Die Viskositäten waren zufriedenstellend und Änderungen in der Formulierung oder in dem Mischverfahren waren für diese Bohrlöcher nicht nötig. Die Eigenschaften der ultrafeinen Zemente sind in den ASCE Referenzen beim Autor dargestellt.

Das Legen von mehr als 14300 Kubikfuß des Vergußmörtels zeigt, daß viele Hohlräume neben dem Steuerschaft große Mengen von Vergußmörtel hinnahmen. Der ultrafeine Zement füllte viele sehr kleine Öffnungen und vollbrachte die erforderte Abdichtung um den Schaft. Während der paar letzten Wochen des Kontrakts, gab es eine Regenansammlung von mehreren Zoll, die das Niveau des Reservoirs in die Zone des Überschwemmungsbassins erhob, weil der Steuerschaft immer noch trocken ist.

1.0 BACKGROUND FOR FINAL GROUTING WARM SPRINGS DAM

In 1990, a test grouting program was performed at the control shaft for Dry Creek (Warm Spings) Dam in California. The purpose of the test grouting program was to determine whether leakage into the shaft could be effectively controlled by grouting around the shaft. The design concept was to provide an outer ring grout curtain constructed with Portland cement grout, and then grout the inner area with ultrafine cement grout. The purpose of the outer ring was to provide containment of the more expensive ultrafine cement grout, keeping it within the rock mass between the control shaft and the outer ring of grout holes. The test grouting program permitted construction and evaluation of a portion of the final area to be grouted, and provided information on grout takes, time required for grouting, grout mix designs and materials, effect of grout injection on the fish hatchery water supply and stream, and effectiveness of grouting. The test grouting program utilized subsurface surveying technology to monitor and control the location of the drill holes, and used very fine grained cement grout (ultrafine cement) to provide a relatively tight and impervious barrier around the control shaft. The aspect of the work involving monitoring of the hydrogen-ion concentation (pH) of the water downstream in the in the outlet works was to be sure that the grouting work did not raise the pH enough to kill fish or interfere with the normal operations of the hatchery.

To accurately determine the position of the drill holes at any depth so that the holes did not penetrate the control shaft, and so that they remained in nearly the same plane, subsurface surveying was necessary. A tolerance of 2.5 feet from the horizontal coordinates of the drill hole at

the surface was established as the allowable deviation from a vertical line. The surveying instrument operated on a non-magnetic principal and the surveying was accomplished within a cased hole to protect the surveying tool from loss or damage. In addition to the casing, the surveying equipment needed to operate on a non-magnetic principle due to the presence of steel surface pipe (nipples) and of rockbolts and anchor bars throughout the depths of the drill holes, making magnetic measurement devices unacceptable.

Although the specifications permitted some latitude in drilling equipment and technique, the equipment had to be capable of drilling through steel anchor bars. With the tolerance specified for drilling the grout holes, it was unreasonable to assume that anchor bars would not be encountered.

Evaluation of the results of the test grouting indicated that completion of the grout curtain around the control shaft would provide significant reduction of water inflow into the shaft. A design for the final grout curtain configuration was prepared and submitted for review and comments and the final package went to bid.

1.1 Drilling

Before the drilling and grouting program commenced, the system to monitor the pH of the water coming through the tunnel was reactivated. This system, which was installed for the test grouting program, provided both a record of any change in the pH of the water as the result of the drilling and grouting operations, and an alarm to alert the personnel at the adjacent fish hatchery so that steps could be taken to protect the fish hatchery from any change in pH. The recording meter was located at the downstream outlet works and the probes were located

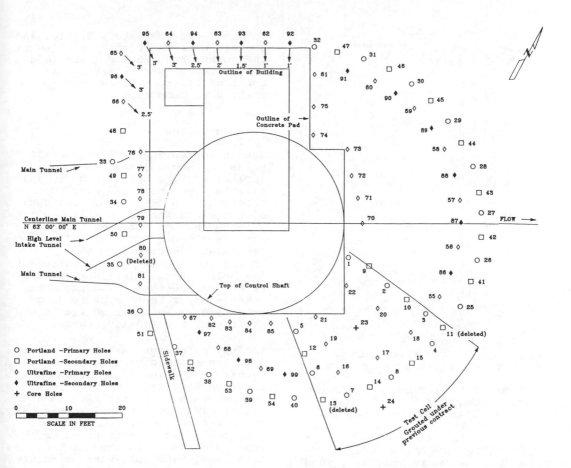

Figure 1.0   Grout Hole Layout

Abbildung 1.0   Grundriß eines Bohrloches mit Vergußmörtel

on the wall of the outlet works just upstream from the water intake for the fish hatchery. During the course of the drilling and grouting no changes in the pH were recorded.

On 13 May 1992, the first load of equipment was mobilized onto the site and mobilization continued sporadically over the next two weeks. On 28 May, the first primary hole # 36 was started (see Figure 1.0 for hole layout). As the holes were drilled, downhole or subsurface surveying of the hole was performed to determine the location of the drill bit to prevent drilling into the control shaft . A tolerance of 2.5 feet at the bottom of each hole was specified in the contract documents. Since recovery of core was not required, the Contractor

chose to use a down-the-hole hammer system which uses a fairly large diameter, thick walled rod providing rigidity and resulting in less deflection of the hole from vertical. Initially, the Contractor chose to use a 6 inch diameter drill bit. Then he proposed using a drill string con-figuration which included an in-line packer to eliminate having to pull the drill string in order to perform water pressure tests.

This configuration lasted through only part of one hole before the in-line packer was completely destroyed. He than went to a 4.5 inch diameter drill bit with which the remaining drilling was completed. In several of the holes drilled early in the program, obtaining correct alignment was

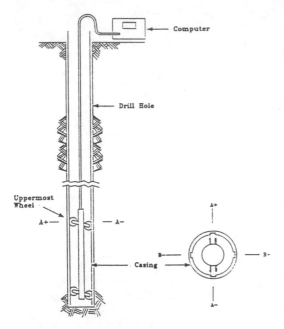

Figure 2.0   Subsurface Survey
Configuration and Axes Orientation

Abbildung 2.0   Überblick einer Schicht
unter der Oberfläche
Struktur und Orientierung der Achsen

an area were completely drilled and grouted were the ultrafine holes in the interior of the cell and long the northwest side started. A total of 74 grout holes were drilled to completion.

1.2 Surveying

Subsurface surveying of the drill holes was accomplished using an inclinometer, a spiral sensor, and plastic pipe which had four trackslots in the walls. One set of the slots was physically oriented by the surveyor to the north and south and the inclinometer was inserted into these tracks and lowered to the bottom of the hole. As the inclinometer was raised in the hole, readings were taken every 2 feet. The readings were taken a second time with the instrument rotated 180 degrees.

The spiral sensor was then inserted into the plastic pipe in the same manner as the inclinometer and the readings were taken every 5 feet. The recording computer then used the readings from both instruments, corrected for spiral changes and gave corrected offsets which were true readings correlated to the north-south and east-west principle axes. The survey of each hole was taken at every 40 foot increase in depth as the hole was drilled. See Figure 2.0 for configuration and axes orientation for subsurface surveying.

It is noted that in some of the holes the deviation from vertical exceeded the 2.5 foot tolerance. Early in the program it was determined that unless the deflections were going to be large, it would be neither cost nor time effective to hold strictly to this requirement providing the deflection was not toward the shaft. These deviations from the tolerance do not appear to have affected the results of the grouting program. During the final phase of the grouting program, tolerances were adhered to as closely as possible and did not for the most part, exceed those allowed during the test grouting program.

difficult. The method used to remedy this was to install a long surface nipple which came up through the drill table, plumb the nipple and the drill steel, grout the nipple into the hole, and then cut the nipple off to the desires height. Because of the previously grouted test cell and the location of ultrafine holes around the northwest side of the grout curtain, the grouting could logically be divided into three areas each independent from the others. These three areas are as follows. The first extends from the test cell counterclockwise to the northeast corner of the building. The second extends from the test cell clockwise to the west side of the building. The third area contains the ultrafine holes located between the other two areas. The primary holes around the perimeter of each area were drilled and grouted before the secondary holes were drilled and grouted. Only when all of the Portland cement perimeter holes in

252

1.3 Washing and Pressure Testing

During and after completion of the drilling in each hole and prior to starting of the grouting, washing and pressure testing were performed (see Table 1.0 for pressure used).

The first pressure test was to be performed using no pressure but conducted as a static head test. In subsequent tests a packer was attached to a pipe, inserted into the hole to a predetermined depth, and inflated with compressed gas. Water was then pumped into the hole through the pipe. The water was metered as it was pumped into the hole and the test was performed for a timed interval. The result of pressure testing gives an indication of what the grout take will be for the interval tested. Where bypass was known to occur, the packer was moved to a different depth and the test run again. Originally it was intended that pressure tests be taken at intervals of no more than 40 feet as the hole was drilled. The Contractor submitted a VE proposal, which was accepted, to modify this to the same intervals at which the holes were to be grouted. The first test was the interval from grout surface to a depth of 100 feet and the second test was the interval from 100 feet to the bottom of the hole.

1.4 Grouting

The test grouting involved the use of two types of cement. The first was Portland cement Type II and the second was an ultrafine cement. The Portland cement grout was used in the primary and secondary holes which made up most of the outer ring of the test cell (Holes 25 - 54 on Figure 1). The ultrafine cement was used in all the interior holes of the test cell and the exterior holes along the northwest portion of the grout ring (holes 55 - 99). In grouting the primary and secondary Portland cement holes, most were started with either a 3:1 water to cement grout mix consisting of 3 cubic feet of water, one bag of cement (94 pounds), 3 percent (by weight)

Table 1.0   Water Pressure and Grout Pressures Related to Depth

Tabelle 1.0   Wasserdruck und Verguβmörtel-drücke in Beziehung zur Tiefe

MAXIMUM ALLOWABLE PRESSURES AT THE HEADER
(GAGE PRESSURES IN PSI)

DEPTH OF GROUT OR PRESSURE TEST (FT)	WT OF WATER 62.4 (pcf)	WT. OF GROUT MIXES IN POUNDS/CUBIC FOOT (pcf)						
		0.6:1 120	1:1 105	1.5:1 94	2:1 88	3:1 80	4:1 76	5:1 74
0 - 20	0	0	0	0	0	0	0	0
20 - 40	15	0	0	10	11	12	13	13
40 - 60	29	0	18	21	22	24	26	26
60 - 80	44	0	21	31	33	37	38	39
80 - 100	59	0	21	35	43	49	51	52
100 - 120	73	0	21	35	43	54	59	62
120 - 140	78	0	21	35	43	54	59	62
140 - 160	78	0	21	35	43	54	59	62
160 - 180	78	0	21	35	43	54	59	62
180 - 200	78	0	21	35	43	54	59	62
200 - 220	78	0	21	35	43	54	59	62
220 - 240	78	0	21	35	43	54	59	62
240 - 260	78	0	21	35	43	54	59	62
260 - 280	78	0	21	35	43	54	59	62
280 - 300	78	0	21	35	43	54	59	62

Table 2.0   Recommended Mix to Begin Grouting

Tabelle 2.0   Emphohlene Mischung um das Zementieren anzufangen

WATER LOSS (GPM) (1)	GROUT MIX	
	WC RATIO	FLUIDIFIER (2)
0-2	3:1	3
2-4	3:1	2
4-6	3:1	1
6-8	2:1	2
8-10	2:1	1
10-15	1.3:1	2
15-20	1.5:1	1
20-30	1:1	2
30-40	1:1	1
40-50	0.6:1	1.5
Over 50	0.6:1	1

(1) At maximum allowable pressure from pressure test

(2) Percent of cement by weight

Bentonite, and 1 to 3 percent (by weight) fluidifier or a 2:1 water to cement grout mix consisting of 2 cubic feet of water, one bag of cement, 2 percent bentonite, and 1 or 2 perent fluidifier (see Table 3.0 for Grout Mix Quantities). As the grouting of each hole progressed, the grout mixture was modified to suit the conditions indicated in the hole (see Table 1.0 for grouting pressures used). During grouting, many of the holes initially took large amounts of grout. In order for the grout mixing to keep up with grout placement, it was found that a

Table 3.0   Grout Mix Quantities

Tabelle 3.0   Mengen der Vergußmörtel Mischung

QUANTITY FOR EACH BAG (94 LBS) PORTLAND CEMENT

WATER LOSS (GPM) (1)	WC RATIO	WATER (LBS)	WATER (GAL)	BENTONITE (LBS)	FLUIDIFIER (LBS)	YIELD (CF)
0-2	3:1	187	22.4	3	3	3.54
2-4	3:1	187	22.4	3	2	3.53
4-6	3:1	187	22.4	3	1	3.525
6-8	2:1	125	15	2	2	2.525
8-10	2:1	125	15	2	1	2.52
10-15	1.5:1	94	11.3	1.5	2	2.02
15-20	1.5:1	94	11.3	1.5	1	2.02
20-30	1:1	62	7.5	1	2	1.52
30-40	1:1	62	7.5	1	1	1.51
40-50	0.6:1	37	4.5	0.6	1.5	1.23
Over 50	0.6:1	37	4.5	0.6	1	1.2

QUANTITY FOR EACH BAG (44 LBS) ULTRAFINE CEMENT

WATER LOSS (GPM) (1)	WC RATIO	WATER (LBS)	WATER (GAL)	BENTONITE (LBS)	FLUIDIFIER (LBS)	YIELD (CF)
0-2	3:1	87.5	10.5	1.4	1.4	1.65
2-4	3:1	87.5	10.5	1.4	0.9	1.65
4-6	3:1	87.5	10.5	1.4	0.5	1.65
6-8	2:1	58.5	7	0.9	0.9	1.18
8-10	2:1	58.5	7	0.9	0.5	1.18
10-15	1.5:1	44	5.3	0.7	0.9	0.95
15-20	1.5:1	44	5.3	0.7	0.5	0.95
20-30	1:1	29	3.5	0.5	0.9	0.7
30-40	1:1	29	3.5	0.5	0.5	0.7
40-50	0.6:1	17.3	2.1	0.3	0.7	0.6
Over 50	0.6:1	17.3	2.1	0.3	0.5	0.6

Table 4.0   Thinning Table

Tabelle 4.0   Verdünnungstisch

REQUIRED MIX

INITIAL MIX	0.60	0.70	0.80	0.90	1.00	1.25	1.50	1.75	2.00	2.50	3.00	3.50	4.00	5.00
0.60	0.00	0.09	0.19	0.28	0.37	0.60	0.83	1.07	1.30	1.76	2.23	2.69	3.15	4.00
0.70	----	0.00	0.08	0.17	0.25	0.47	0.68	0.89	1.10	1.53	1.95	2.38	2.80	3.58
0.80	----	----	0.00	0.08	0.16	0.35	0.55	0.74	0.94	1.33	1.72	2.11	2.50	3.23
0.90	----	----	----	0.00	0.07	0.25	0.44	0.62	0.80	1.16	1.52	1.89	2.25	2.93
1.00	----	----	----	----	0.00	0.17	0.34	0.51	0.67	1.01	1.35	1.69	2.03	2.67
1.25	----	----	----	----	----	0.00	0.14	0.29	0.43	0.78	1.01	1.30	1.59	2.14
1.50	----	----	----	----	----	----	0.00	0.13	0.25	0.51	0.76	1.01	1.26	1.75
1.75	----	----	----	----	----	----	----	0.00	0.11	0.34	0.56	0.78	1.01	1.44
2.00	----	----	----	----	----	----	----	----	0.00	0.20	0.40	0.60	0.80	1.20
2.50	----	----	----	----	----	----	----	----	----	0.00	0.17	0.33	0.50	0.83
3.00	----	----	----	----	----	----	----	----	----	----	0.00	0.14	0.29	0.57
3.50	----	----	----	----	----	----	----	----	----	----	----	0.00	0.13	0.38
4.00	----	----	----	----	----	----	----	----	----	----	----	----	0.00	0.22
5.00	----	----	----	----	----	----	----	----	----	----	----	----	----	0.00

NOTE: Values are multiplied by the number of cubic feet of initial mix to be thinned.
EXAMPLE: You have 10 cubic feet of 1:1 - how much water do you add to thin to 2:1 ?
Read 0.67 from the chart - water needed is 0.67 x 10 = 6.7 cubic feet water.

pumping rate of approximately 2 CFM was needed. To facilitate mixing, a separate unit was used to prehydrate and store the bentonite.

This was done by mixing a 100-pound bag of bentonite with 9.5 CF of water (resulting in 10 CF of mix) and allowing the mixture to hydrate with constant agitation. A minimum

time of 1-hour was usually allowed for this process. Varying amounts of this mix could then be used depending upon the percentage needed for the grout mix. Thinning of the grout mixes was accomplished by using the procedures outlined in Table 4.0.

Ideally, all of the primary holes would be drilled to full depth before any grouting started. After grouting the primary holes, the secondary holes would be drilled and grouted, and finally, the ultrafine holes would be drilled and grouted. However, because of experience gained during the test grouting, each hole was grouted upon completion of drilling and pressure testing (see Table 5.0, column 2 for grouting sequence). With the layout of the holes and the test cell previously completed, the grouting could logically be divided into three areas which could be treated as separate portions of the grout curtain. These three areas are from the test cell to the northwest corner of the building, from the northwest corner of the building to the west side of the building, and from the west side of the building to the test cell. Two of these three areas could be completed with the Portland cement primary and secondary holes being grouted first followed by the ultrafine primary and secondary holes. The third area which consisted only of ultrafine holes could be grouted in the primary and secondary hole sequence. The sequence of completing primary holes before starting secondary holes and then ultrafine holes was, except for two holes, maintained. Grouting of the deep holes was accomplished by setting the packer at the 100 foot depth and grouting the bottom of the hole to refusal. The packer was then raised, set in the nipple and the grouting resumed until the upper zone also went to refusal and the hole was complete. Whenever grout bypassed the packer at the 100 foot depth, it was usually not known until grout began to return through the nipple. When this occured, the packer was raised and set in the nipple. The reason for this was to prevent grouting of the

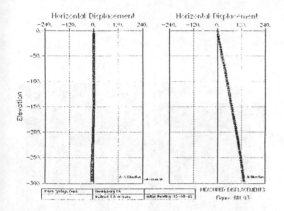

Figure 3.0   Horizontal Displacement of Hole # 93

Abbildung 3.0   Horizontale Verstellung des Loches # 93

packer into the hole resulting in the loss of the packer and the grout pipe.

Several factors combined to affect the takes in grout holes including weathering and the proximity of the hole to the control shaft excavation. Highly to moderately weathered and naturally fractured material of the near surface (upper 60 to 80 feet) resulted in some fairly large but slow grout takes. At depths below the weathering and where only natural fracturing occurred, grout takes were usually small. Near the shaft, where the rock surrounding the shaft has been fractured and existing fractures opened by the blasting and excavation, takes were larger and in some cases very large. In addition, many voids were encountered adjacent to the shaft and these also accounted for large amounts of grout take.

In the following paragraphs, a short description of grouting the last seven holes is presented. These holes were grouted with a combination of 75 percent M1 ultrafine slag and 25 percent M3 ultrafine portland in place of M5 slag/portland.

Hole 93 (UF-Sec)   Total depth 300 feet. Anchor bars were drilled at 230 feet, 261 feet and 277.2 feet. The packer was set at 103 feet and

Table 5.0   Grout Quantitities Placed

Tabelle 5.0   Gelegte Mengen des Vergußmörtels

HOLE #	ORDER GROUTED	GROUT PLACED IN CUBIC FEET	GROUT TO FILL HOLE	GROUT OVER AMT. TO FILL HOLE
25-Primary	(1)	85	53	32
26-Primary	(5)	182	30	152
27-Primary	(2)	90.5	30	60.5
28-Primary	(7)	58	30	28
29-Primary	(4)	191.5	33	158.5
30-Primary	(8)	92.4	33	59.4
31-Primary	(6)	107.5	33	74.5
32-Primary	(3)	186.8	33	153.8
33-Primary #1	(16)	368	33	135
34-Primary	(33)	81.5	30	51.5
35-Primary #2	(34)	117.5	13	104.5
36-Primary	(18)	259	29.5	229.5
37-Primary	(25)	203	33	170
38-Primary	(15)	141	33	108
39-Primary	(11)	216.5	33	183.5
40-Primary	(22)	134	33	101
41-Secondary	(9)	108.5	28.5	80
42-Secondary	(13)	116	30	86
43-Secondary	(17)	81.5	30	51.5
44-Secondary	(12)	298	33	265
45-Secondary	(14)	51.5	33	18.5
46-Secondary	(23)	94	33	61
47-Secondary	(19)	470	33	437
48-Secondary	(21)	225.5	33	192.5
49-Secondary	(38)	438	30	408
50-Secondary	(20)	361	9	352
51-Secondary	(10)	390	30	360
52-Secondary	(30)	195	33	162
53-Secondary	(35)	115.5	33	82.5
54-Secondary	(28)	73.5	33	40.5
55-UF-Prim.	(24)	309	29.5	279.5
56-UF-Prim.	(29)	163.5	29.5	134
57-UF-Prim.	(36)	90	29	61
58-UF-Prim.	(42)	62	29.5	32.5
59-UF-Prim.	(47)	54	33	21
60-UF-Prim.	(53)	70	33	37
61-UF-Prim.	(45)	103.5	33	70.5
62-UF-Prim.	(39)	526	33	493
63-UF-Prim.	(50)	189	33	156
64-UF-Prim.	(54)	165.5	33	132.5
65-UF-Prim.	(44)	520.5	33	497.5
66-UF-Prim.	(51)	271	33	238
67-UF-Prim.	(40)	101.5	33	68.5
68-UF-Prim.	(58)	62	33	29
69-UF-Prim.	(61)	67.5	33	34.5
70-UF-Prim.	(27)	41.5	22	19.5
71-UF-Prim.	(31)	35	22	13
72-UF-Prim.	(37)	104.5	22	82.5
73-UF-Prim.	(43)	206	23	183
74-UF-Prim.	(26)	65	11	54
75-UF-Prim.	(32)	109	25	84
76-UF-Prim.	(60)	40.5	29.5	11
77-UF-Prim.	(66)	148	29.5	118.5
78-UF-Prim.	(72)	26	9	17
79-UF-Prim.	(55)	21	9	12
80-UF-Prim.	Deleted	--	--	--
81-UF-Prim.	(48)	161.5	29.5	132
82-UF-Prim.	(46)	277.5	25	252.5
83-UF-Prim.	(49)	293	23	270
84-UF-Prim.	(52)	67	22.5	44.5
85-UF-Prim.	(41)	72	22	50
86-UF-Sec.	(56)	32	29.5	2.5
87-UF-Sec. #3	(73)	24.5	29.5	--
88-UF-Sec.	(57)	34	29.5	4.5
89-UF-Sec.	(69)	43	32	11
90-UF-Sec.	(63)	107	33	74
91-UF-Sec. #4	(74)	21	33	--
92-UF-Sec.	(59)	63	33	30
93-UF-Sec.	(64)	117.5	33	84.5
94-UF-Sec.	(68)	278	33	245
95-UF-Sec.	(62)	60	33	27
96-UF-Sec.	(70)	51	33	18
97-UF-Sec.	(65)	60	33	27
98-UF-Sec.	(71)	43	33	10
99-UF-Sec.	(67)	39	33	6
				8555.7

the lower zone grouted with 82.4 CF of 2:1 mix. The upper zone was then grouted using 2.1 CF. Total placement was 84.5 CF. Horizontal displacement of hole 93 is shown in Figure 3.0.

Hole 94 (UF-Sec)   Total depth 300 feet. The packer was set at 100 feet and the lower zone grouted using 242 CF of 2:1 mix. When moving the packer up to the nipple, it was found that some of the grout had bypassed the packer. The upper

255

Photograph 1.0 Drilling Rig at Angle for Drilling Angle Hole

Photographie 1.0 Bohrgestell unter einem Winkel zum Bohren eines Winkel-Loches

Photograph 3.0 Readout of Survey Coordinates on Computer

Photographie 3.0 Ablesung der Begutachtung koordiniert mit einem Computer

Photograph 2.0 Lowering Plastic Tube into Drill Hole for Surveying

Photographie 2.0 Hinablassung eines plastischen Rohres in das Bohrloch zur Begutachtung

Photograph 4.0 Addition of M5 Ultrafine slag/portland to mix tank

Photographie 4.0 Zugabe von M5 ultrafeiner Schlacke/Portland in den Vermischungstank

zone was then grouted using only an additional 3 CF of 2:1 mix. Total placement for this hole was 245 CF.

Hole 95 (UF-Sec) Total depth 300 feet. Anchor bars were drilled at 234 and 280.6 feet. The packer was set at 60 feet because of cave in the hole and the lower zone grouted using 3.5 CF of 2:1 mix. The upper zone was then grouting using 23.5 CF of grout. Grout had communi-

256

cated to this hole during the grouting of Hole 92. Total placement 27 CF.

Hole 96 (UF-Sec)   Total depth 300 feet. At 55 feet a 3-foot void was encountered. Because of cave in the hole, the packer was set at 60 feet and the lower zone grouted using 1.5 CF of 2:1 mix. The upper zone was grouted using 16.5 CF of 2:1 mix. Total grout placement was 18 CF.

Hole 97 (UF-Sec)   Total depth 300 feet. An anchor bar was drill at 205 feet. The packer was set at 100 feet and the lower zone was grouted using 2:1 mix. The upper zone was then grouted accepting no additional grout over that required to backfill the open hole. Total grout place 27 CF.

Hole 98 (UF-Sec)   Total depth 300 feet. A packer was set at 100 feet and the lower zone grouted using 6 CF of 2:1 mix. The upper zone was grouted with 4 CF of 2:1 mix. Total placement was 10 CF.

Hole 99 (UF-Sec)   Total depth 300 feet. The lower zone of this hole was grouted to refusal using 2:1 mix. The upper zone was then grouted but accepted no grout over that needed to backfill the hole. Total grout placement was 6 CF.

Grout takes in some of the primary and secondary holes were large. It is believed that most of the large grout placements went into areas of overbreak which were left unfilled behind the concrete liner of the control shaft and tunnels. It is also felt that the takes of portland cement and particularly ultrafine were governed more by where the hole was located in relation to the shaft excavation than by the filling of the numerous very small openings in the adjacent bedrock. This is not to say that the ultrafine was not needed. The ultrafine cement certainly filled many of these very small openings and achieved the required seal around the shaft.

The last seven holes (hole # 93 to 99) where grouted with a combination of M1 ultrafine slag and M3 ultrafine portland because

Photograph 5.0   Salt Deposits on Shaft Wall from Previous Seepage

Photographie 5.0   Saltzablagerungen auf der Schaftwand von vorhergehender Versickerung

of the unavailability of M5 slag/ portland. Viscosities were satisfactory and no changes in formulation or mixing procedure were required for these holes.

1.5   Conclusions for Final Grouting Warm Springs Dam

Personnel on the grouting contract, workers assigned to the dam staff, and county personnel working with the power plant in the control shaft have all made observations and comments that the shaft appears to be significantly drier now than before the grouting started. Whether this is entirely due to grouting or partially related to the current drought conditions, the placement of over 14,300 feet of grout (combined total of test and final grouting) indicates that water filled voids and abundant fractures existed in the area immediately adjacent to the control shaft and that these voids accepted the large amounts of grout injected during both the test grouting and the final grouting programs. It can be noted that during the last few weeks of the contract and the month following completion of the contract, there was a rainfall accumulation of several inches which raised the level of the reservoir into the flood pool zone and the control shaft still remains dry.

REFERENCES

American Concrete Institute Committee Report #226, (1987), "Ground Granulated Blast-Furnance Slag as a Cementitious Constituent in Concrete", American Concrete Institute Materials Journal, July-August, pp 1-15.

Clarke, W.J., (1884) "Performance Characteristics of Microfine Cement", Preprint 84-023, ASCE Geotechnical Conference, Atlanta, GA, May 14-18, pp 1-14.

Clarke, W.J. (1989) "Alkali Activated Slag and Portland/Slag Ultrafine Cements", Material Research Society Symposium on Specialty Cements with Advanced Properties, Boston, MA, November 27-29, pp 219-232

Clarke, W.J., Boyd, D.B. and Helal, M. (1992) "Ultrafine Cement Tests and Dam Test Grouting", ASCE Geotechnical Grouting Conference, New Orleans, LA, Feb. 25-28, pp 626-638

Clarke, W.J., Boyd, D.B. and Helal, M. (1993) "Ultrafine Cement Tests and Drilling Warm Springs Dam", ASCE Geotechnical Practice in Dam Rehabilitation Conference, Raleigh, NC, April 25-28, pp. 718-732.

Helal, M. (1991) "Microstructure of Microfine Cement Grouted Sand and its Anisotropic Behavior", PhD Dissertation, Northwestern University, December, pp. 1-129.

Houlsby, A.C., (1990) "Construction and Design of Cement Grouting", John Wiley & Sons, Inc., New York, pp. 61-62.

Karol, R.H., (1990) "Chemical Grouting", Marcel Dekker, Inc., New York, pp 429-432.

Krizek, R.J. and Helal, M., (1992) "Anisotropic Behavior of Cement-Grouted Sand", ASCE Geotechnical Conference, New Orleans, LA, Feb. 25-28, pp. 541-550.

Schwarz, L.G. and Krizek, R.J., (1992) "effects of Mixing on Rheological Properties of Microfine Cement Grout", ASCE Geotechnical Conference, New Orleans, LA, Feb. 25-28, pp 512-525.

Abdichtung des Stausees von Salanfe (Schweiz) – Injektionsschleier

Sealing of the Salanfe Reservoir (Switzerland) – Grout curtain

Philippe Dawans
EOS, Lausanne, Schweiz

Michel Gandais
SIF-BACHY, Paris, Frankreich

Toni R. Schneider
Beratender Geologe, Uerikon, Schweiz

Jean-Paul Waldmeyer
ETHZ, SIF-GROUTBOR, Renens, Schweiz

ZUSAMMENFASSUNG : Das Stauwerk von SALANFE befindet sich im Kanton Wallis in der Schweiz. Der Bau des hydroelektrischen Kraftwerks wurde 1952 abgeschlossen. Seither wurde die Füllung des Sees, bis zur ursprünglich geplanten maximalen Höhe, wegen einer im Bereich des Rückhaltebeckens vorhandenen Sickerzone mit grossen Verlusten, verunmöglicht. So wurde ein Programm von Abdichtungsarbeiten, verteilt zwischen 1992 und 1994 beschlossen, um die volle Kapazität der Stauanlage ausnützen zu können. Die Arbeiten beinhalten zuerst das Erstellen eines Inejektionsstollens von 600 m Nutzlänge in der vorher genannten Verlustzone. Die eigentlichen Abdichtungsarbeiten werden von diesem Stollen aus durchgeführt. Sie umfassen die Ausführung einer einreihigen Injektionsschürze, welche oberhalb und unterhalb des Stollens liegt. Die untere Schürze senkt sich bis 150 m unter den Stollen ab. Der behandelte Fels hat sehr unterschiedliche Strukturen und enthält unter anderem Dolomiten und Rauhwacken, deren Behandlung spezielle, stark eindringungsfähige Injektionsgüter erforderlich macht. Der Artikel beschreibt das Problem der Abdichtung des Stausees von Salanfe und die Erarbeitung der angewandten Methode. Sämtliche Injektionsarbeiten wurden mittels Computer und EPICEA gesteuert, überwacht, ausgewertet und aufgezeichnet. Eine solche Entscheidungshilfe ist heute unumgänglich.

ABSTRACT : The hydroelectric scheme of SALANFE is located in the Valais, in Switzerland, on the left bank of the Rhône. The gravity dam creates a lake at an altitude of 1925 meters above sea level with a nominal capacity of 40 million cu.m. The construction was completed in 1952. Due to very high leakage located in an area on the left bank of the reservoir, the lake was never filled up to its capacity. Since then, the reservoir has been used to only 50 % of full capacity. After locating the leakage area, it was decided to undertake work to make it watertight in order to use the full capacity of the lake. The work, started in 1992, and will extend into 1994. It consists of excavating a grouting gallery over a length of 600 m into the leakage area. From this gallery, a grout curtain is constructed with one part located above the gallery and one part below. The grout curtain is monolinear. The area includes many different types of rock from extremely hard quartzite to triasic weathered and eroded dolomies and cornieules. Grouting in these last types of rock cails for special grouts with very high penetrability characteristics. The paper describes the different aspects of the leakage problem of the Salanfe lake and gives a description of the beginning of the work. Focus is made on grouting methods and on their evolution due to the discovery of geological features more complex than expected. All the grouting process is steared, controlled, evaluated and recorded with a computer and EPICEA. This type of process control ist indispensable in current conditions.

1. EINLEITUNG

Die Gewichtsmauer Salanfe (Länge : 616 m, Höhe : 52 m, Betonvolumen : 230'000 m2) wurde in den Jahren 1948 bis 1952 gebaut.

Das Stauziel auf Kote 1925 m ü.M., dem ein Nutzvolumen von 40'000'000 m3 entsprechen würde, wurde bis anhin nie erreicht. Das Becken verliert ab der Staukote 1890 m ü.M., in zunehmendem Masse Wasser. Die

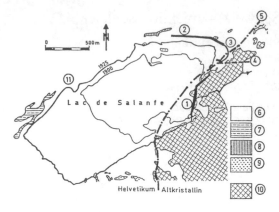

Fig. 1 : Situation, geologische Skizze

1 : Gewichtsmauer, 2 : Injektionsstollen
3 : Zugangsstollen 4 : Unterirdische
Standseilbahn mit Baufenster
5 : Stratigraphischer Kontakt :
Altkristallin des Aiguilles Rouges-
Massivs - Nordhelvetische Sedimente
6 : Lockergesteine, 7 : Alttertiär,
8 : Jura, 9 : Trias, 10 : Altkristalline
Gneise, 11 : Stauziel

Verlustmengen erreichen bei der Kote von
1915 m ü.M. ca. 1 m3/s, was in Anbetracht
des Nutzgefälles von ca. 1400 m ein
beträchtliches Energiepotential darstellt.

Einer ersten Expertengruppe, die unmittelbar nach dem Erkennen der Verluste
beauftragt wurde, gelang es in den Jahren
1954 bis 1956 nicht, ein klares Bild über
die Verluststellen zu gewinnen. Die einzige sichere Erkenntnis war, dass die be-
trächtlichen Verlustmengen oberflächlich
nirgends zu Tage treten. Mit Beginn der
80-er Jahre wurde die Suche nach den Verluststellen aufgrund der sich abzeichnen-
den kritischen Situation auf dem Energie-
sektor erneut aufgenommen.

2. GEOLOGISCHE SITUATION

Die Sperrstelle Salanfe ist auf Gneisen
des Altkristallins des Aiguilles Rouges -
Massivs fundiert. Ihre Fundation und die
nähere Umgebung, d.h. das Altkristallin,
können als weitgehend dicht bewertet werden. Für die Stauhaltung ist jedoch charakteristisch, dass sie in die autochthone
und parautochthone Sedimentbedeckung des
Kristallins übergreift (Fig. 1). Die Oberfläche des Kristallins fällt relativ flach
mit 10 - 20 Grad NNW-wärts ein (Fig. 2).

Die autochthone Sedimentbedeckung besteht
im wesentlichen aus Quarziten und Tonschiefern der unteren Trias, sowie Dolomiten der mittleren Trias. Beide Serien sind
zusammen rund 25 m mächtig und bedecken
relativ ungestört den Kristallinkörper. Im
Hangenden folgen die intensiv verschupp-
ten z.T. auch verfalteten Serien des
Parautochthons. Sie bestehen aus Dolomiten
und Rauhwacken der Trias, wobei der
oberste Bereich der Dolomite z.T. auch
quarzitische Einschaltungen enthalten
kann. Darüber folgen Quarzite des Lias
sowie dunkle, z.T. oolithische, eisen-
haltige Kalke des Doggers und dunkle
Kalkschiefer des Malm. Die tektonische
Beanspruchung dieser Serien ist ausge-
sprochen intensiv, wobei insbesondere die
Dolomite und Rauhwacken der Trias z.T.
tektonisch angehauft auftreten.

Typisch für die geologischen Verhältnisse ist, dass einzig der Kristallin
oberflächlich ansteht. Sämtliche helve-
tischen Serien sind im Beckenbereich von
mächtigen Lockergesteinsserien bedeckt,
wobei der Glazialkolk der Ebene von
Salanfe über 80 m Tiefe erreicht.

3. HYDROLOGISCHE SITUATION

Aus Wassermessungen vor dem Bau der Staumauer Salanfe geht hervor, dass die Fels-
schwelle aus Kristallin, die den Ueberlauf
des Beckens enthält, auch in wasserarmen
Perioden immer Wasser führte. Es heisst
dies, dass mit grosser Wahrscheinlichkeit,
das Felsbett unterhalb der Ueberlauf-
schwelle (Kote ca. 1880 m ü.M.) als dicht
zu bewerten ist. Die Wasserverluste müssen
deshalb durch denjenigen Felsbereich erfolgen, der durch den Einstau ab. ca. Kote
1890 m ü.M. neu benetzt wird. Die zu
lösende hydrogeologische Aufgabe bestand
somit darin, unterhalb der vorhandenen
Lockergesteinsdecke die Verlustzonen geologisch stärker einzugrenzen und mittels
entsprechender Untersuchungen die Verluststellen nachzuweisen. Aus grundsätzlicher
geologischer Sicht kamen für die gross-
räumigen Verluste einzig Karsterscheinun-
gen im Bereich der triadischen Serien in
der Grenzzone Autochthon - Parautoschthon,
d.h. im speziellen in den als karstanfäl-
lig bekannten Dolomiten und Rauhwacken in
Frage. Diese Serien begrenzen den Felsuntergrund des Beckens hauptsächlich in den
Sektoren NNW - N - NNE der Uferzone. Es
musste somit der Nachweis erbracht werden,
dass die Verluste effektiv in dieser
Region auftreten.

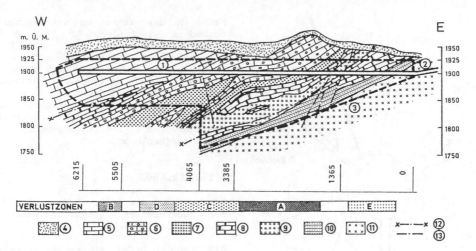

Fig. 2 : Geologischer Längsschnitt

1 : Injektionsstollen, 2 : Zugangsstollen, 3 : Ausdehnung des Dichtungsschirmes,
4 : Quartär, 5 : Malm, 6 : Dogger, 7 : Lias, 8 : Trias ; Dolomite, 9 : Trias ;
Rauhwacken und Gips, 10 : Trias ; Tonschiefer und Quarzite, 11 : Gneise des
Aiguilles Rouges-Massivs, 12 : Ueberschiebungen, 13 ; Verwerfungen

4. ERFASSUNG DER VERLUSTZONE

Das Ueberwachungssystem zur Erfassung der
Verlustzonen basiert im wesentlichen auf
36 Bohrungen, die entlang des "verlustver-
dächtigen" Uferbereiches durch die Lockergesteinsserien hindurch bis in den Fels
vergetrieben wurden. Die Bohrungen wurden
durchwegs mit geschlitzten 4"-Rohren bestückt welche eine genauere Beobachtung
der Grundwasserverhältnisse erlaubten. In
Abhängigkeit vom Seespiegel wurden folgende Parameter erfasst :

- Depression der Grundwasseroberfläche im
Bereich der Verlustzonen
- Temperatur des Grundwassers
- Leitfähigkeit des Grundwassers
- δ 0 18
- Radiohydrometrische Messungen in den
Bohrlöchern zur Erfassung der Sickerungs-
richtung und -Geschwindigkeit

Dieses redundante Informationssystem
erlaubte eine recht eindeutige Gliederung
des untersuchten Uferbereiches in einzelne
Verlustzonen. Insgesamt wurden fünf verschiedene Bereiche ausgeschieden. Wie sich
später zeigte, liegt das Zentrum der Verluste im Bereich einer stärkeren Anhäufung
von Rauhwacken und Dolomiten an der Basis
des Parautochthons. Die Verlustzonen sind
von A bis F mit abnehmender Bedeutung in
Fig. 2 aufgezeichnet.

5. KONZEPT DER DICHTUNGSMASSNAHMEN

Die Lockegesteinsbedeckung im Bereich der
zu dichtenden Felsoberfläche war durchwegs
so gross (über 20 - 30 m), dass eine Abdichtung von der Oberfläche aus nicht in
Betracht gezogen werden konnte. Dies insbesondere auch aufgrund der Höhenlage und
der damit verbundenen nur kurzen Saison
für Bauarbeiten, wie auch der Lawinenge-
fährdung des ganzen Uferbereiches. Aufgrund dieser Tatsache, musste die Abdichtung aus einem Stollen heraus, der in
einem vernünftigen Abstand von der Felsoberfläche verläuft, erstellt werden. Als
Kote der Stollensohle wurde 1900 m ü.M.
gewählt, womit der Stollen selbst aufgrund
seiner Auskleidung einen Teil der Dich-
tungsmassnahme darstellt. Zudem erlaubte
er eine Zweiteilung des Schirmes in einen
untern und einen obern Teil. Zur Ueberprüfung der Lage des Stollens wurde zusätzlich eine Detailgeophysik zur Erfassung der genauen Situation der Felsoberfläche, die ihrerseits durch einige Kurz-
bohrungen verifiziert wurde, durchgeführt.
Detaillierte strukturelle Untersuchungen
des zu dichtenden Felskörpers liessen den
Schluss zu, dass der Tiefenschirm vertikal
erstellt werden kann. Dies insbesondere
auch unter Berücksichtigung der relativ
engständigen Bohrungen mit in der vierten
Phase nur 1,5 m Abstand.

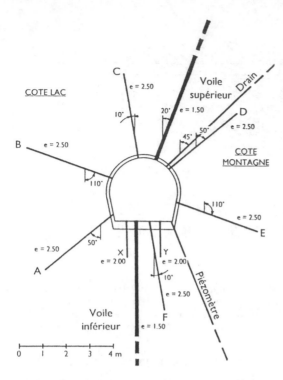

COTE LAC

C e = 2.50

Voile supérieur

Drain

10° 20° e = 1.50 D

50° e = 2.50

45°

B e = 2.50

COTE MONTAGNE

110°

110°

e = 2.50

E

50°

e = 2.50 X e = 2.00 Y e = 2.00

A

Piézomètre

10°

e = 2.50

Voile inférieur F e = 1.50

0 1 2 3 4 m

Fig. 3 : Injektionsstollen-Querschnitt

Bereits in der Zeit der Voruntersuchungen wurde erkannt, dass dem Becken während der Schneeschmelze und bei Starkregen viel Hangwasser zufliesst. Aus diesem Grunde musste der bergseitigen Drainage des Schirmes grosse Beachtung geschenkt werden. Vorgesehen ist die Ueberleitung dieses Hangwassers mittels Pumpen in das Seebecken.

6. GEOMETRIE DER SCHUERZE

Die Lage der Dichtungsschürze ist auf der Fig. 1 schematisch dargestellt. Die Injektionsarbeiten sind zur Zeit im Gang.

Die ursprünglich vorgesehene Abdichtungsbehandlung musste im Verlauf der Arbeiten aufgrund von neu auftauchenden Erkenntnissen angepasst werden, sowohl in der Methodologie wie in der zu erreichenden Tiefe.

Neue Tatsachen waren unter anderem :

- Vorhandensein einer Störzone um den Injektionsstollen, welche insbesondere unter der Sohle aufgelockert war.

- Vorhandensein von stark tektonisiertem

Fels in der Tiefe im unteren Schleier besonders zwischen den KP 1365 und 4065. Die Schichten fallen viel schneller als vorgesehen ab, den Kontakt mit dem Gneis fanden wir zum Teil bis über 150 m unter der Sohle des Stollens. Die Rauhwacken- und Dolomite- Schichten sind viel mächtiger und die sehr schlechte Qualität diesr Böden erzwingt wesentliche Veränderungen und Anpassungen der Behandlungsmethode.

6.1 Dichtungsschürze

Sie ist auf dem Längsschnitt (Fig. 2) und dem Querschnitt (Fig. 3) schematisch dargestellt.

Die Schürze wird ab einem Injektionsstollen aus erstellt, welcher auf der Kote 1900 vorgetrieben wurde. Die Länge des Stollens beträgt 630 m und hat ein Hufeisenquerschnitt von 3 m x 3 m.

Sie besteht hauptsächlich aus :

- einer einreihigen oberen Schürze von ca. 16'000 m2 welche mit einer Neigung von 20 Grad talwärts ausgeführt wird. Diese Bohrungen steigen bis zur Kote 1925 an.

- einer einreihigen unteren Schürze mit einer Fläche von 41'000 m2, welche ab senkrechten Bohrungen injiziert wird, zwischen dem Stollenende (PK 6215) und den PK 4065 werden diese Bohrungen bis zur Kote 1840 abgeteuft. Diese Tiefe von 60 m übertrifft damit den im ursprünglichen Plan vorgesehenen mittleren Wert um 20 m. Zwischen dem PK 4065 und dem Stollenbeginn erfuhr das ursprüngliche Schleierprojekt wesentliche Veränderungen.Es musste beim PK 4065 die Kote 1750 erreichen was eine Tiefe von 150 m unter dem Stollen entspricht um allmählich gegen den Stollenbeginn abzunehmen.

- eine fächerförmige "Verbindungsschürze" am Stollenende. Ihre Fläche beträgt ungefähr 3000 m2.

- eine Verbindungsinjektion der oberen und unteren Schürze um den Stollen herum

Die Verbindungsinjektion wird mit Hilfe von senkrechten Fächern zur Stollenachse ausgeführt. Jeder Fächer besteht aus 6 radialen Bohrungen von 4 m Tiefe mit einem Fächer-Abstand von 2,50 m.

- zusätzliche Behandlung unter der Sohle

Da der Fels unter der Stollensohle stark

zerklüftet ist und da diese Zone durch die aufeinanderfolgenden Injektionsphasen sehr beansprucht sein wird, ist eine zusätzliche Dichtungs- und Konsolidierungsbehandlung notwendig. Dies wird durch 2 m tiefe Bohrungen, die alle 2 m beiderseits der Stollenachse liegen, erreicht.

6.2 Zusätzliche Massnahmen

Die Massnahmen umfassen :

- ein Entwässerungsnetz, bestehend aus 26 m tiefen steigenden Entwässerungsbohrungen im Abstand von 6 m, welche 45 Grad zur Senkrechten geneigt sind und talwärts von der oberen Schürze liegen.

- ein Netz von Piezometern, im Abstand von 12 m, 50 m tief ab Stollensohle, mit einer Neigung von 20 Grad zur Senkrechten und talwärts von der unteren Schürze gelegen.

7. INJEKTIONSGUETER

Die Bodenbehandlung erfolgt mit 2 klassischen Bentonit-Zement-Suspensionen und 2 speziellen Injektionsmischungen

7.1 Klassische Bentonit-Zement-Suspension

a) Grundmischung

Es handelt sich um eine stabile Mischung bestehend aus :

Für 1 m3 :   CLK Zement :   350 kg
             Bentonit :   38 kg
             Wasser   885 Liter

Wichtigste Eigenschaften :

Viskosität : 33 bis 36 " Marsch

Druckfestigkeit (ohne Seitenbehinderung) nach 28 Tagen : 15 bar.

Diese Mischung war für die Anfangsphase der Behandlung der gesamten Schürze vorgesehen.

b) Injektionsgut für Verfüllungen

Es handelt sich um eine feststoffreiche Mischung. Sie ist für eventuelle Verfüllungen von sehr durchlässigen Schichten gedacht.

Dosierung für 1 m3 :

CLK Zement :   860 kg

Bentonit :   30 kg
Wasser :   715 Liter

7.2 Spezielle Injektionsmischungen

a) C3S

Es handelt sich um eine Bentonit-Zement-Suspension welche mit speziellen Zuschlagsstoffen behandelt ist um eine gute Wassserzurückhaltung (Wasserretention) unter Injektionsdruck zu erreichen. Diese Eigenschaft verbessert das Eindringungsvermögen der Mischung in den zermalmten oder fein zerklüfteten Zonen. Sie wird oft für die letzten Behandlungsphasen vorgesehen.

Dosierung für 1 m3 :

CLK Zement :   350 kg
Bentonit :   38 kg
Dispersionszusatz :   4 kg
Wasser :   885 Liter

b) Rheosil S

Es handelt sich um ein Gut, das zum Erreichen einer grossen mechanischen Festigkeit stark zementhaltig ist. Es wird spezielle mit Zusatzmitteln behandelt um eine einwandfreie Stabilität zu erreichen (kein freies Wasser bei einer Stauzeit von 2 Std.) obwohl es kein Bentonit enthält. Es hat eine extrem niedrige Fliessgrenze und eine sehr gute Wasserretention.

Dieses Injektionsgut dient der Zusatzbehandlung der Sohle und ist für die Behandlung von zermalmten und zerklüfteten Zonen geeignet, welche nur unter leichtem Druck injiziert, werden können.

Dosierung für 1 m3 :

CLK Zement :   675 kg
Zuschlag für Stabilisierung : 150 kg
Dispersionsmittel :   3 kg

8. INJEKTIONS- UND KONTROLLMETHODE

8.1 Bohrlochabstand

Die obere und untere Schürze haben den gleichen Bohrlochabstand. In der primären Phase liegen die Bohrungen 18 m auseinander. Diese Bohrungen begrenzen Zonen welche ab dem Stolleneingang von 1 bis 34 nummeriert wurden und so als Bezugspunkte während den Arbeiten dienen.

In dieser Phase wird jede zweite Bohrung

als Kernbohrung abgeteuft, je nach Komplexität der Geologie wurden zusätzlich Kernbohrungen angesetzt.

In der sekundären Phase liegen die Bohrungen je 6 m auseinander, also je 2 solcher Bohrungen zwischen 2 primären Bohrungen.

In der tertiären Phase vermindert sich der Abstand auf 3 m.

In der quarternären Phase, falls durchgeführt, beträgt der Abstand der Bohrungen 1,50 m. Die sekundären, tertiären und quarternären Bohrungen werden im Drehschlagverfahren ausgeführt.

8.2 Bestimmung der Anfangsdurchlässigkeit

Sie wird in den primären Kernbohrungen durch einen Abpressversuch (Lugeonversuch) 10 bar während 10 min. bestimmt. Eine Messung erfolgt alle 5 m während dem Fortschreiten der Bohrung.

8.3 Injizierte Mengen und Druckbereiche

8.3.1 Behandlung um den Stollen

Diese Bohrungen sind mit C3S injiziert, mit einem maximalen Volumen von 2000 l und einem Höchstdruck von 5 bar. Ausser in Zonen mit schlechter Standfestigkeit wo Anpassungen nötig sind, insbesondere dort wo die Rauhwacken bis zum Stollen aufsteigen, erfolgt die Injektion dieser 4 m Bohrungen mit Kopfanschluss.

8.3.2 Zusätzliche Behandlung der Sohle

Die 2 m tiefen Bohrungen X und Y werden mit Rheosil per Kopfanschluss injiziert. Die Menge des Injektionsgutes ist pro Bohrung auf 500 l begrenzt und der Druck auf 8 bar beschränkt.

8.3.3 Obere Schürze

Die Bohrungen dieser Schürze werden in 5 m Abschnitten steigend injiziert. Die Injektion wird bis zur Sättigung durchgeführt, wobei die Sättigungsdrücke wie folgt beschränkt sind :

15 bar von 0 - 5 m
25 bar von 5 - 10 m
40 bar für mehr als 10 m

Die Injektion wird mit der Bentonit-Zement

Grundmischung ausgeführt. In Fällen von zu hoher Aufnahme ohne Druckanstieg wird die Verwendung des Verfüllgutes in Betracht gezogen.

8.3.4 Untere Schürze

Die Injektionsmethode ist unterschiedlich je nach Schürzenzonen.

Vom PK 4065 zum PK 6215

In diesr Zone des Stollenendes wird die gleiche Methode wie für die obere Schürze verwendet.

Einige Bohrungen, welche wegen der schlechten Gebirgsstandfestigkeit nicht steigend injiziert werden konnten, wurden in 5 m langen Abschnitten von oben nach unten ausgeführt.

Die primären, sekundären und tertiären Bohrungen werden mit Bentonit-Zement Grundmischung injiziert, die quarternären Bohrungen hingegen mit C3S.

Von PK 0 zum PK 4065

In diesem von schlechter Standfestigkeit gekennzeichneten Bereich erfolgen die Injektionen in Abschnitten von 10 m von oben nach unten.

Der erste Abschnitt von 10 m wird nach der Injektion auf der ganzen Länge verrohrt.

Die primäre Phase wird mit Bentonit-Zement Grundmischung ausgeführt, in den folgenden Phasen wird C3S injiziert.

Die Höchstdrücke sind die gleichen wie die der oberen Schürze.

8.3.5 Fächerförmige Verbindungsschürze

Diese am Stollenende gelegene Schürze wird gleich behandelt wie die obere Schürze.

8.4 Kontrollbohrungen-Enddurchlässigkeit

Die Kontrolle wird für jede Zone von 18 m Länge mittels einer diagonal geneigten Bohrung ausgeführt. Diese in der Schürzenebene gelegene Bohrung schneidet die anderen Bohrungen des Abschnittes. Die Enddurchlässigkeit wird alle 5 m durch Lugeon-Tests überprüft.

Wenn eine Kontrollbohrung beendet ist, wird sie gleich wie die quarternären

Bohrungen der oberen Schürze injiziert.

9. ABWICKLUNG DER ARBEITEN

Die wichtigste Phase der Arbeiten findet während der Abfassung dieses Artikels statt. Die Behandlung desjenigen Teils der Schürze der Dolomiten und Rauhwacken bis 150 m Tiefe wurde im Frühling 1993 begonnen und wird erst in den ersten Monaten von 1994 beendet sein.

Einige Injektionsphasen sind beendet und können zusammenfassend dargestellt werden.

9.1 Behandlung um den Stollen, Anschluss obere und untere Schürze

Die Injektion der Bohrungen B, C, D, E wurde wie vorgesehen durchgeführt.

Die Bohrungen A und F erforderten wegen der schlechten Standfestigkeit des Felsens unter der Sohle besondere Massnahmen. Sie wurden in 2 Phasen von 2 m ausgeführt.

Die Aufnahmen für diese Behandlung betrugen durchschnittlich 46 Liter C3S pro Fächer, d.h. 23 Liter pro Laufmeter Bohrung. Das Aufnahmehistogramm zeigt eine unregelmässige Verteilung der injizierten Mengen längs dem Stollen.

9.2 Zusätzliche Behandlung der Sohle

Die Behandlung der Linien X und Y wurde wie vorgesehen durchgeführt.

Für den bereits ausgeführten Teil war die durchschnittliche Aufnahme von Rheosil S 470 Liter pro KP, d.h. 117 l pro Laufmeter Bohrung. Diese Aufnahme ist fünf mal höher als bei der Behandlung des Stollengewölbes. Wenn man berücksichtigt, dass Rheosil S zementreicher ist, ergibt sich eine 10 x grössere Aufnahme von Zement pro Meter Bohrung als bei der Gewölbebehandlung.

Die Aufnahmehistogramme zeigen eine gute Proportionalität zwischen Aufnahmezonen von Rheosil S und von C3S entlang dem Stollen. Damit wird klar ersichtlich, dass die besonderen rheologischen Eigenschaften von Rheosil S, ihm ausgezeichnetes Eindringungsvermögen verleihen.

9.3 Behandlung der oberen Schürze

Die Arbeiten sind gemäss dem ursprüng-lichen Verfahren im Gang, d.h. steigende Injektionen in 5 m Abschnitten mit Bentonit-Zement. Für den Teil zwischen KP 3885 bis zum Stollenende (KP 6215) wurde die Durchführung der quarternären Phase als überflüssig erachtet. Der Durchschnitt der Aufnahmen bis zur tertiären Phase betrug 40 l Suspension per Laufmeter Bohrung, und die durchgeführten Kontrollen haben eine gute Abdichtung des Felsens nachgewiesen. Die individuellen Aufnahme-werte sind in der Tabelle 1 dargestellt.

9.4 Fächerförmige Verbindungsschürze am Stollenende

Die Behandlung wurde gemäss dem ursprünglichen Injektionsverfahren in 5 m Abschnitten steigend mit Bentonit-Zement durchgeführt.

Für den oberen Teil des Fächers betrug die Aufnahme im Durchschnitt 38 l per Laufmeter Bohrung, d.h. also praktisch gleich dem Durchschnittswert der oberen Schürze. Für den unteren Teil betrug die Aufnahme 63 l per Laufmeter Bohrung.

9.5 Untere Schürze

Sie besteht aus 2 grossen Zonen die im Verlauf der Arbeiten entsprechend den ständig genaueren Kenntnissen der geologischen Eigenschaften, definiert wurde. Diesen beiden Zonen entsprechen in der Praxis auch 2 verschiedene Behandlungs-methoden.

a) Zone PK 4065 bis 6215

Die Arbeiten begannen am Stollenende (KP 6215). Bis KP 4065 konnte das Injektios-verfahren mit steigenden Injektionen in 5 m Abschnitten verwendet werden. Die Sondierbohrungen sollten ab den vorge-sehenen Tiefen von 40 m erst in einem Boden mit weniger als 2 Lugeon-Einheiten Durchlässigkeit eingestellt werden. Mit dieser Bedingung wurde in dieser Zone die Schürze auf 60 m Tiefe verlängert. Die in den Sondierbohrungen ausgeführten Abpressversuche ergaben Durchlässigkeits-werte zwischen 0 und 14 Lugeon-Einheiten.

Bis zur tertiären Phase wurden die Injektionen mit Bentonit-Zement vorge-nommen. Die quarternäre Injektionsphase mit C3S wurde nur für die Zonen 31 bis 34 (KP 5505 bis 6215) ausgeführt.

Die in den Kontrollbohrungen durchge-führten Abpressversuche ergaben in der

TABELLE 1 - Obere Schürze - Detail der Aufnahmen in den Zonen 22 bis 34

ZONE Nr.	von KP	zu KP	INJEKTIONSGUT-VOLUMEN PRO LAUFMETER
22	3'885	4'065	63
23	4'065	4'245	87
24	4'245	4'425	58
25	4'425	4'605	12
26	4'605	4'785	39
27	4'785	4'965	27
28	4'965	5'145	49
29	5'145	5'325	23
30	5'325	5'505	28
31	5'505	5'685	42
32	5'685	5'865	10
33	5'865	6'045	32
34	6'045	6'215	44

TABELLE 2 - Untere Schürze - Detail der Aufnahmen in den Zonen 23 bis 34

ZONE Nr.	von KP	zu KP	INJEKTIONSGUT-VOLUMEN PRO LAUFMETER
23	4'065	4'245	97
24	4'245	4'425	27
25	4'425	4'605	19
26	4'605	4'785	25
27	4'785	4'965	80
28	4'965	5'145	42
29	5'145	5'325	23
30	5'325	5'505	35
31	5'505	5'685	110
32	5'685	5'865	103
33	5'865	6'045	86
34	6'045	6'215	135

Regel Werte zwischen 0 und 2 Lugeon-Einheiten, mit einigen Werten bis zu 4 L.E. Infolge der Abweichungen aus der zentralen injizierten Zone wurden diese Resultate als zufriedenstellend betrachtet.

Für diese gesamte Zone betrug der Durchschnitt der Aufnahme 78 l Mischung per Laufmeter Bohrung. Die einzelnen Absorptionswerte sind in der Tabelle 2 aufgeführt.

Das Aufnahmehistogramm in 5 m Abschnitten ist auf Fig. 4 dargestellt. Die meisten Aufnahmen sind kleiner als 1000 Liter pro Abschnitt. Die Grunddurchläs-

sigkeit ist in der Schürzenfläche also relativ gleich- mässig verteilt. Allerdings lässt sich das Vorhandensein von 2 Familien von Volumenaufnahmen feststellen, eine zwischen 2500 l und 3000 l, die andere über 4500 l je Abschnitt. Dies deutet auf das Vorhandensein von bevorzugten Durchgängen, welche unterschiedlicher Herkunft sein können, die aber bedeutende Wasserverluste verursachen können.

b) Zone PK 4065 bis PK 0

Die Arbeiten setzen sich hier in Richtung des Stolleneinganges fort. Aufgrund der Sondierbohrungen wurde die Schürze bis

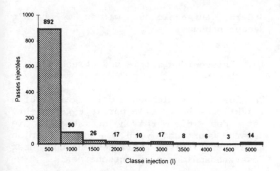

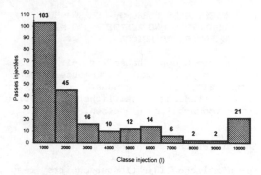

Fig. 4 : Unterer Schirm von KP 4095 bis KP 6215

Injektionsgutaufnahmehistogramm in Litern pro 5-Laufmeter-Abschnitt.
Primäre bis quarternäre Phasen zusammen

Fig. 5 : Untere Injektionsschürze von KP 0 bis KP 4095

Injektionsgutaufnahmehistogramm in Litern pro 10-Laufmeter-Abschnitt.

150 m unter den Stollen abgesenkt. Die Abpressversuche ergaben Werte zwischen 0 und 24 L.E.

Genauer ausgedrückt dringt die Schürze von KP 0 zu KP 2625, 2 m in den Gneissockel, damit die Undurchlässigkeit der Kontaktfläche Sockel / Tonschiefer gesichert ist. Zwischen KP 2625 und KP 4065 dringt die Schürze nur bis 10 m in den Tonschiefer ein. Es wurde nicht als notwendig erachtet, die Kontaktfläche Sockel / Tonschiefer in diesem Bereich zu behandeln da sie keine direkte Verbindung zum See hat.

Die sehr schlechte Gebirgsfestigkeit verunmöglicht den Einsatz von Druckluftpackern für die Injektion. Teilweise lassen sie sich nicht in die Bohrungen absenken ; sie verklemmen sich oder platzen beim Kontakt von zu weichem Boden. Man muss also von oben nach unten injizieren, die Injektionen werden mit Kopfanschluss ausgeführt und nach Abbinden des Injektionsgutes, muss die Bohrung wieder aufgebohrt werden, vor Aufbohren des nächsten Abschnittes. Als Abschnittslänge wurde 10 m gewählt.

Um das Aufstossen von Injektionsgut unter die Stollensohle zu verhindern, wurden auf den 10 oberen Metern in jeder Bohrung nach deren Injektion ein Rohr einzementiert.

Die Injektionen begannen mit dem klassischen Bentonit-Zement. Wenn man diese Suspension unter Druck setzt, lässt sich ein Teil des Wassers auspressen, das Injektionsgut wird zu hart und das Wie-

deraufbohren kann nicht mehr unter guten Bedingungen ausgeführt werden.

In weichen Böden führt das Vorhandensein von stark erhärtetem Injektionsgut zu grösseren Abweichungen der Bohrkronen und die Bohrung weicht schnell von der ursprünglichen Bohrachse ab.

Dieses Problem konnte durch Verwendung von C3S gelöst werden. Dieses Injektionsgut hält den Wasseranteil viel besser zurück, härtet dementsprechend weniger und führt nicht zu Abweichungen beim Wiederaufbohren in weichen Böden.

Die hauptsächlich mit Bentonit-Zement durchgeführten primären Injektionen dieser Zone ergaben einen Durchschnitt der Aufnahme von 291 l per Laufmeter Bohrung.

Das Aufnahmehistogramm für die Abschnitte unter den ersten 10 m ist in 10-m Schritten auf Fig. 5 dargestellt. Man findet wiederum eine Familie mit hoher Injektionsgutaufnahme welche auf das Vorhandensein von einzelnen sehr durchlässigen Bereichen hindeutet, die grössere Wasserverluste durchlassen können.

Die Bohrungen der sekundären Phase wurden zu 60 % mit Bentonit-Zement und zu 40 % mit C3S injiziert. Es ergab sich eine durchschnittliche Aufnahme von 262 l per Laufmeter.

Die Injektionen der tertiären Bohrungen waren im Zeitpunkt des Verfassens dieses Artikels noch nicht im Gang.

10. INJEKTIONSSTEUERUNG, UEBERWACHUNG AUSWERTUNG UND AUFZEICHNUNG MITTELS COMPUTER UND EPICEA

In unserer Gruppe haben wir seit mehr als 6 Jahren ein komplexes Computerprogramm entwickelt, EPICEA genannt, welches wir zur Ausführung sämtlicher grösseren Injektionsarbeiten einsetzen.

10.1 Injektionssteuerung

Für sämtliche Injektionsabschnitte (resp. Manschetten bei Lockergesteinsinjektionen) sind die zulässigen Injektionskriterien definiert und in einer zentralen Einheit gespeichert.

Der Injektionsspezialist muss nur bei Beginn jeder Abschnitts- (oder Manschetten-) Injektion diese Abrufen und der Computer steuert selbsständig die automatischen Injektionspumpen. Die maximale Injektionsgeschwindigkeit, der Maximaldruck, das maximale Injektionsgut-Volumen je Abschnitt sowie für Felsinjektionen die Sättigungs-Einpressgeschwindikeit welche den Abbruch der Injektion des Abschnittes bestimmen kann, wird vom Computer überwacht und die Injektion fehlerfrei gesteuert.

10.2 Injektionsüberwachung

Alle Injektionspumpen sind mit einem elektronischen Durchflusszähler und einem elektronischen Druckaufnahmegerät ausgerüstet. Alle Injektions-Daten werden kontinuierlich in der zentralen Einheit während der ganzen Pumpentätigkeit eingegeben und gespeichert. Gleichzeitig beobachtet der Injektionsspezialist auf einem Schirmbild die Aufzeichnungen sämtlicher Pumpen und kann die momentanen Leistungen für jede einzelne Pumpe ablesen.

10.3 Auswertung und Aufzeichnung

Die ganze Injektionstätigkeit kann auf Papier ausgedruckt werden und erlaubt eine gründliche Untersuchung aller Sonderheiten.

Das EPICEA-Programm ermöglicht aber auch die Bearbeitung dieser riesigen Datenbank, indem übersichtliche Aufzeichnungen der Injektionen in Längs- und Querschnitten, sowie Grundrisse über Injektionsmengen, mittlere Injektionsdrücke, Sättigungs-

drücke, aufgezeichnet werden, welche wir Scaner nennen.

10.4 Entscheidungshilfe und Abrechnungsbasis

Da das Aufzeichnen der geleisteten Injektionsarbeit alle par Tage geschieht, ist das die Entscheidungshilfe, die dauernd gebraucht wird, um Anpassungen an neue Verhältnisse vorzunehmen und über Zusatzbehandlungen zu entscheiden.

Heute kann man nicht mehr pflichtbewusst grössere Injektionen anders führen und beurteilen.

Zur Abrechnung können natürlich sämtliche Daten mit EPICEA ausgedruckt werden.

11. SCHLUSSFOLGERUNG

Ein wichtiger Teil der Arbeiten harrt noch der Ausführung. Es handelt sich um die Behandlung der unteren Schürze welche bis 150 m unter den Stollen abzuteufen ist.

Die untere Schürze wurde vom Stollenende her erstellt, wo die Behandlungsverhältnisse weniger kompliziert waren. Damit konnte die Injektionsmethode beim Vorrücken in schwierigeren Fels sukzessive angepasst werden. Sie ist heute gut geignet und dieser entscheidende Teil der Arbeiten kann unter guten Voraussetzungen in Angriff genommen werden.

Permeability and groutability of the Valparaiso Dam foundation, NW of Spain

Durchlässigkeit und Injizierbarkeit des Untergrundes der Talsperre Valparaiso, NW Spanien

A. Foyo & C. Tomillo
University of Cantabria, Santander, Spain

ABSTRACT: In this paper the results of the Permeability Tests, the criterious of Grouting Programme and the relations between the permeability and groutability of rock mass foundation have been analyzed, when the cleavage and closed or filled joints are the main structural discontinuities.

As the consequence, the results obtained in the analysis of the permeability and groutability characteristics of the Valparaiso Dam foundation must be analysed by means of the Hydraulic Monitoring criteria, which they are in concurrence with the results obtained in some Spanish Dams, i.e. Las Portas Dam and San Cosmade Dam, when is the cleavage the main structural discontinuity of rock mas foundation.

ZUSAMMENFASSUNG: In diesem Artikel werden Durchlässigkeitsversuche, Kriterien für Injektionsprogramme und die Beziehungen zwischen Injizierbarkeit und Durchlässigkeit von massigen Felsfundamenten untersucht, wenn die wichtigsten strukturellen Diskontinuitäten schieferhaltige Gesteine und geschlossene oder gefüllte Klüfte sind.
Zusammenfassend muß man drei voneinander abhängige Faktoren für die Notwendigkeit einer Abdichtung beachten:
1. Die Sickerwassermenge darf nicht die Nutzung des Stausees beeinträchtigen.
2. Die Unterströmung darf nicht eine progressive Erosion fördern.
3. Der Auftrieb darf nicht die Standsicherheit gefährden.
Geht man von der Hypothese aus, daß eine generelle Behandlung durch Injektionen nicht als erforderlich erachtet wird, ist eine Kontrolle des Auftriebs notwendig. Deshalb ist es unbedingt nötig, die Worte des russischen Experten E.G. Gaziev zu beachten, die er anläßlich des 2. Internationalen Ingenieurgeologen-Kongresses in Sao Paulo (Brasilien) vortrug:
"Das beste Instrument für die Kontrolle des Felsmassivs sind Dränagen und Piezometer."
Die Analyse der Beobachtungen in diesen Meßeinrichtungen, als hydraulische Untersuchung bezeichnet, führt zum Erkennen von Veränderungen, die das durchsickernde Wasser im felsigen Untergrund von Stauseen verursachen kann.
Die Ergebnisse der Durchlässigkeitsanalyse und die Injektionsdaten der Talsperre von Valparaiso, nach den Kriterien der hydraulischen Untersuchung analysiert, stimmen mit denen anderer Talsperren, wie Las Portas und San Cosmade überein, in denen durch die Schieferung die wesentliche strukturelle Diskontinuität des Felsuntergrundes der Talsperren verursacht wird.

1 INTRODUCTION

The Valparaiso Dam, NW of Spain, is a gravity dam of 67 m. hight with a reservoir capacity of 168.5 Hm^3.

The dam foundation is made of a gneisic rock called "Ollo de Sapo Formation" where the cleavage is the

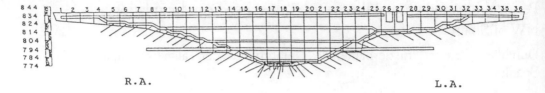

Fig. 1
Valparaiso Dam
Valparaiso Talsperre
L.A. Left abutment. R.A. Right abutment
Link rand.   Recht rand.
1 - 36 : concrete block number.
nummer von beton block.

main structural discontinuity with a disposition parallel to the river course and a subvertical dip.

For the grouting treatment, the rock foundation was divided in 36 zones or blocks, in correspondance with the block division of the wall. Fig. 1.

The sequence of grouting operations have been as follow:

a. Complete boreholes with 5 metres between every one of them and a minimun of 15 metres depth in the rock foundation.

b. Lugeon Permeability Test or Water Pressure Test, WPT, in the last 5 metres in the borehole. The water pressure levels they were 5, 10 and 5 bar.

c. If the water absorption at the bottom of the borehole it was less than 1 Lugeon, the grouting operation it has been carried out in ascending stages of 5 metres length.

d. When the water absorption was more than 1 Lugeon, the borehole was continued for another 5 metres length.

Furthermore, the grouting pressures were as follows:

0	to	5 m.	15 bar
5	to	10 m.	25 bar
10	to	15 m.	30 bar
under	15 m.		35 bar

2 ANALYSIS OF THE RESULTS

To analyse the relations between the permeability and groutability of dam foundation, the results obtained in the Water Pressure Test, WPT, carried out in the first investigation boreholes, the results of the WPT carried out at the bottom of the grouting boreholes, the cement absorption at different depth and the corresponding grouting pressures have been taking into account.

To have a representative point of view of the results along the foundation, have been selected four zones in concurrence with two blocks on the right abutment, Fig.1, and other two blocks on the left abutment as follows:

Right abutment.
Block nº 14.
Block nº 8.
Left abutment.
Block nº 23.
Block nº 30.

2.1. Block nº 14. Analysis of the results. Fig. 2

Table A show the results under the Block nº 14 foundation zone.

Fig. 2 show that a relation between the increase of grouting pressure and cement absorption could be established. However the absolute value of the cement absorption is very little, maximun absorption about 20.40 kg/ml.

Table A.

| B.14 | S-1 | | | |
D	Pi	C	Qi	Q
2.50	15	0.00	>1	>4<6
7.50	25	10.20	>1	>4<6
12.50	30	10.20	>1	>4<6
17.50	35	10.20	<1	4.94
22.50	35	20.40	<1	<4
27.50	35	20.40	<1	<4

270

VALPARAISO DAM

B.14. S-1

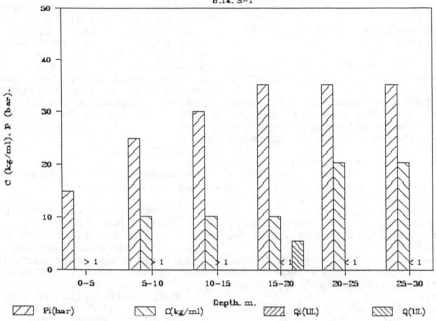

Fig. 2

Pi. Grouting pressure   Qi. Flow in grouting borehole
    Injektionen Druck   Flusswasser in Bohrung

C. Cement absorption   Q. Flow in WPT previous test
    Zement Absorption   Flusswasser in WPT test

Furthermore, Table A, the progressive increase of the cement absorption is absolutely independent of the permeability results obtained from grouting boreholes, Qi, which they are less than 1 Lugeon under the 15 to 20 metres zone and they are less than 4.94 Lugeon if the results of permeability test in the investigation boreholes, Q, are taking into account.

Consequently, whereas the increase of cement absorption could be considered as a consequence of the progressive increase of grouting pressures, the possible relations between the water absorption capacity, traditionally called permeability, and the cement absorption capacity, called groutability, they must be analysed carefully.

Table B.

```
----------------------------------
B.6   S-3
   D      Pi      C      Qi      Q
  2.50    15    10.00    >1      -
  7.50    25    11.20    >1      -
 12.50    30   190.00    >1     6.36
 17.50    35    51.00    >1     4.96
 22.50    35    81.60    >1     5.70
 27.50    35    40.80    >1     5.00
 32.50    35    10.20    <1     <5
----------------------------------
```

2.2 Block nº 6. Analysis of the results. Fig. 3

The Block nº 6 is located in the high zone of the right abutment.
Table B show the permeability test and grouting results.
Fig. 3 show that the possible relations between the grouting pressures and cement absorptions showed in Fig. 2, they have disappear. The cement absorption has been controled by the geology

271

through the structural characteristics of the rock mass.

2.3 Left Abutment. Analysis of the results.

Fig. 4A and 4B show the results under the Block nº 23 and Block nº 30, Fig.1.

Both pictures show the low cement absorption of the rock mass and the difficulties to establish a relation of proportionality with grouting pressures.

Furthermore, the water flow in the WPT previous test it was more than 1 Lugeon with a maximun about of 5 Lugeon, alike that in the grouting borehole WPT test.

As consequence of this simply analysis, two conclusion could be achieved:

1. The geological characteristics has had a important influence on the permeability and groutability of the rock mass.

2. The cement absorption has been carried out under hydraulic fracturing conditions.

Table C and Table D show the values of the different parametres used to make the pictures.

Like this, the little water absorption in the WPT previous test must be interpreted as a very low permeability of the rock mass, the little cement absorption should be translated as low groutability or cement absorption capacity of the rock mass foundation.

3. CONCLUSIONS

The conditional factors of the depth treatment are three:

1. Compatible seeepage flow with the normal operation.
2. Compatible filtration rates with regressive erosion.
3. Compatible water pressures, uplift, with the stability.

It must begin with the hypotesis that a generalized grouting treatment of the foundation is not

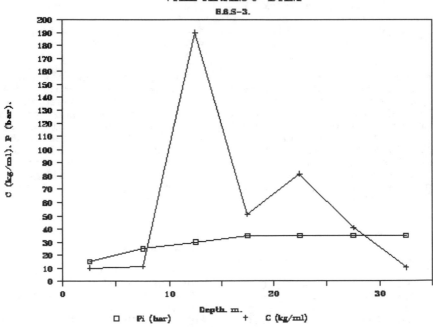

Fig. 3

Cement absorption C, and grouting pressures Pi
Zement absorption C, und injektion Drucken Pi

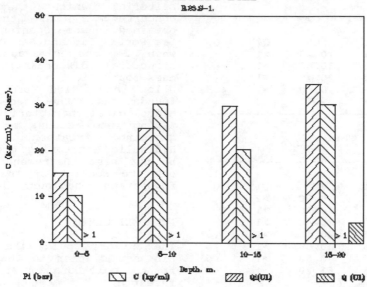

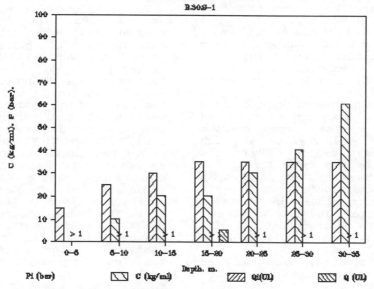

Fig. 4A. B.23 S-1
Fig. 4B. B.30 S-1

Pi. Grouting pressure   Qi. Flow in grouting borehole
    Injektionen Druck   Flusswasser in Bohrung

C. Cement absorption   Q. Flow in WPT previous test
    Zement Absorption   Flusswasser in WPT test

273

Table C.

```
----------------------------------------
B.23   S-1
      D      Pi       C        Qi      Q
    2.50     15     10.20      >1      -
    7.50     25     30.60      >1      -
   12.50     30     20.40      >1      -
   17.50     35     30.60      >1     4.58
   22.50     --      ---       <1      -
----------------------------------------
```

Table D.

```
----------------------------------------
B.30   S-1
      D      Pi       C        Qi      Q
    2.50     15      0.00      >1      -
    7.50     25     10.20      >1      -
   12.50     30     20.40      >1      -
   17.50     35     20.40      >1     5.38
   22.50     35     30.60      >1     <5
   27.50     35     40.80      >1     <5
   32.50     35     61.20      >1     <5
   37.50     --      ---       <1      -
----------------------------------------
```

neccesary, and the control of uplift pressures is necessary.

Like this, the words that the russian expert E.G. Gaziev said in the Second International Congress on Engineering Geology, Sao Paulo (Brasil, 1974), should be remembered today:

"The best instrument for rock mass appraisal are drains and piezometres. The comprehensive analysis of the results as such observations called Hydraulic Monitoring is a powerfull means of detecting the slightest changes in rock dam foundation".

As a working model, an extensive drainage pattern in depth will be arrange. This drainage pattern will be complete with piezometres at differents depths.

When in a local zone of the foundation the seepages exceed the admissible amount, calculated taking into account the conditional factors, grouting work will be done in the affected zone through the drainage that will remain clean afterwards, and with the help of others drilling holes.

As the consecuence, the results obtained in the analysis of the permeability and groutability characteristics of the Valparaiso Dam foundation must be analysed by means of the Hydraulic Monitoring criteria, which they are in concurrence with the results obtained in some Spanish Dams, i.e. Las Portas Dam and San Cosmade Dam, when is the cleavage the main structural discontinuity of rock mass foundation.

In this kind of large dam foundations, the geological and hydrological characteristics must be taking into account to desing the grouting treatment, and the Hydraulic Monitoring criteria will be the best instrument to carried out the control of the rock mass foundation behaviour.

ACKNOWLEDGEMENT

The authors would like to express his appreciation to IBERDROLA S.A. for the collaboration at the present research.

REFERENCES

Foyo,A. & Cerda,J. 1990.The critical permeability. News criterion to determination of permeability on Large Dam foundations.6th Congress of the IAEG. T-3. 1177-1185. Amsterdam. The Netherlands.

Foyo,A. & Tomillo,C & Cerda,J. 1991. The Low Pressure Test. Analysis of permeability and groutability of Large Dam foundations. XVII ICOLD. Q.66, R.5. 61-77. Vienne. Austria.

Foyo,A & Tomillo,C. 1991.Tratamiento de las cimentaciones. Ensayo de Inyecciones.Presa de Las Portas. Geogaceta. Vol. 10. 85-87. España.

Foyo,A & Tomillo,C. & Ewert,F.K. 1991. The permeability of slate rocks. Analysis of the relations between the permeability and groutability of Large Dam foundations. 7th. Int. Congress of the IRSM. Aachen. Germany.

Foyo,A. 1993. Permeability, Groutability and Hydraulic Monitoring of Large Dam foundations. The IRSM Int. Symposium EUROCK'93. Lisboa.

Untergrundsanierung beim Ausbau der Werratal-Brücke im Zuge der Bundesautobahn A7 Hannover-Kassel

Ground stabilization for the expansion of the A7 autobahn bridge across the River Werra

H.Geißler & H.Möker

Niedersächsisches Landesamt für Bodenforschung, Hannover, Deutschland

ZUSAMMENFASSUNG: Die Bundesautobahn A 7 überquert die Werra bei Hedemünden mit einer großen Talbrücke. In unmittelbarer Nachbarschaft einer in den Jahren 1986 bis 1989 gebauten Eisenbahnbrücke für die Neubaustrecke Hannover-Würzburg, begann im Frühjahr 1987 der Ausbau der Autobahnbrücke auf 6 Fahrspuren. Das Niedersächsische Landesamt für Bodenforschung war für beide Brücken mit der Gründungsberatung beauftragt. Probleme ergaben sich im Zusammenhang mit "Hangzerreißungserscheinungen" in den anstehenden Gesteinen des Mittleren Buntsandstein. Davon betroffen waren die nördlichen Widerlager. Es wird über die ingenieurgeologischen Befunde und die daraus resultierenden baulichen Konsequenzen berichtet.

ABSTRACT: Expansion of the A7 autobahn bridge crossing the River Werra near Hedemünden from 4 to 6 lanes was begun in early 1987 during the construction of a new railroad bridge next to it for a new railroad route from Hannover to Würzburg by the German Federal Railway (1986 - 1989). Work was necessary, however, on the pillar foundations and abutments of the bridge so that the expanded bridge could carry the increased load. Fissures were present in the Middle Bunter rocks at the base of the northern abutment. The load on the abutment of the old bridge had led to widening of these fissures. To avoid unacceptable movement of the rock below the abutment, the fissures were injected with concrete to obtain a sufficiently large, homogeneous rock mass. More than 12 000 tonnes of concrete were injected.

1 DAS BAUVORHABEN

Im Zuge der neuen Schnellbahnstrecke Hannover-Würzburg der Deutschen Bundesbahn und des erforderlichen Ausbaus der Autobahn A 7 Hannover-Kassel werden im Bereich des Werratales bei Hedemünden umfangreiche Brückenbaumaßnahmen durchgeführt.

Nach langwierigen Diskussionen über mögliche Varianten haben sich die Verkehrsträger unter Berücksichtigung konstruktiver, landschaftspflegerischer, gründungstechnischer und anderer Gesichtspunkte entschieden, eine neue Eisenbahnbrücke zu erstellen und die in den Jahren 1935 bis 1937 errichtete, 1945 gesprengte und zwischen 1950 und 1952 neu gebaute bestehende Autobahnbrücke unter Verwendung der vorhandenen Pfeiler sechsspurig auszubauen. Aus gestalterischen Gründen wurde der Pfeilerabstand bei der Eisenbahnbrücke auf den der Autobahnbrücke abgestimmt. Die Eisenbahnbrücke ist inzwischen fertigge-

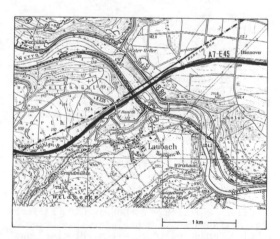

Bild 1 Trassenführungen im Bereich der Werra-Überquerung
Fig.1 Routes of the autobahn and railway where they cross the River Werra

WERRATAL - BRÜCKE

Bild 2 Autobahnbrücke, Hilfsbrücke und Eisenbahnbrücke über die Werra bei Hedemünden
Fig. 2 Autobahn bridge, temporary bridge, and railroad bridge across the River Werra near Hedemünden

stellt. Sie verläuft in einem Abstand von ca. 58 Metern westlich der Autobahn.

Mit Rücksicht auf die Verkehrsführung während der Bauzeit wurde unmittelbar westlich der bestehenden Autobahnbrücke eine Hilfsbrücke mit der bereits endgültigen Richtungsfahrbahn Kassel/Frankfurt als Überbau errichtet. Der bauzeitliche Verkehr wurde während des Rückbaus der vorhandenen Überbauten, der Sanierung der alten Pfeiler und Widerlager sowie der Herstellung eines Teilüberbaus an der Ostseite zunächst vierspurig über die Hilfsbrücke geführt. Später wird deren Überbau – jeweils aufeinmal über 416 Meter Länge – von den Hilfspfeilern auf die sanierten Pfeiler seitlich verschoben. Die Pfeiler der Hilfsbrücke werden wieder abgetragen.

Die Sanierung der alten Pfeiler und Widerlager war erforderlich, um diese für die Aufnahme der vermehrten Belastungen durch die neue Konstruktion zu ertüchtigen.

Das NLFB hatte sowohl von der Deutschen Bundesbahn als auch von der Straßenbauverwaltung den Auftrag für die ingenieurgeologische und gründungstechnische Beratung. Neben der Erkundung der allgemeinen geologischen Situation war der unmittelbare Baugrund im Bereich der Pfeiler und Widerlager zu untersuchen. Hierzu gehörte die Festlegung, die Betreuung und die Auswertung von entsprechenden Aufschlußbohrungen. Es waren Angaben über die Gründungsarten der verschiedenen Pfeiler und Widerlager zu machen und die zulässigen Belastungen zu ermitteln. Probleme ergaben sich aufgrund der ungünstigen Beschaffenheit des Untergrundes am nordöstlichen Widerlager (im folgenden auch als Göttinger Widerlager bezeichnet). Aus statischen Gründen mußte dieses als festes Widerlager mit den daraus resultierenden hohen Belastungen des Untergrundes ausgebildet werden.

2 GEOLOGISCHE SITUATION

Das Werratal ist tief in den Buntsandstein der Nordostabdachung des Kaufunger Wald-Sattels eingeschnitten. Nördlich der Werra beginnt die Dransfelder Hochfläche, in deren Bereich der Buntsandstein nach Nordosten abtaucht und vom Muschelkalk überlagert wird. Die werraabwärts unter dem Solling- Bausandstein (smSS) zutage tretenden obersten Teile der aus Sandsteinen und Schluffsteinen aufgebauten Hardegsener Wechselfolge (smH) wurden auch im Bereich der Brückenbauwerke in einigen Erkundungsbohrungen im Taltiefsten angetroffen. Die Hardegsener Wechselfolge weist im Raume Hannoversch Münden eine Mächtigkeit um 150 Meter auf.

Die im Widerlagerbereich oberflächennah anstehenden Bausandsteine der Solling-Folge sind violette bis violettbraune, mittelbankige bis dickbankige, glimmerschichtige Fein- bis Mittelsandsteine mit z.T. mäßiger bis schlechter Kornbindung. Zwischengeschaltet sind gelegentlich geringmächtige Schluffstein-Lagen.

Bei regional überwiegend flacher Schichtlagerung finden sich mehr oder weniger deutliche Abweichungen in der Nachbarschaft tektonischer Störungen. So sind beispielsweise im Bereich des Brückenpfeilers 2 die Schichten im Talboden an einer Schar von Störungsbahnen bis zu 60^0 steil aufgerichtet. Deutliche Schichtverbiegungen zeigten sich auch in der Baugrube des nordöstlichen Hilfswiderlagers.

Eine bestimmende Rolle hinsichtlich der Standsicherheit der Hänge in den Widerlagerbereichen spielen Großklüfte, die bei

276

NE   SW

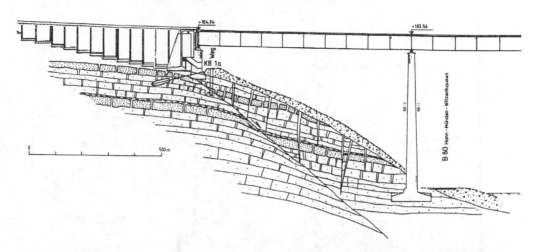

Bild 3 Werratal-Brücke, Nordosthang und "Göttinger Widerlager" (Entwurf J.HANISCH)
Fig. 3 Cross section of the northern approach of the autobahn bridge across the River
Werra (by J. Hanisch)

Tab 1 Ausbildung der Haupttrennflächen am Göttinger Widerlager
Table 1 Characteristics of the bedding planes and main fractures in the rock at the
northern abutment

	SCHICHTUNG	HAUPTKLUFTRICHTUNGEN		
	ss	K 1	K 2	K 3
Streichen u.Fallen (Mittelwerte)	168/13–85 W	23/70 SE	121/69 NE	97/78 S
Erstreckung Abstand	>10 m dm bis m	m bis>10m dm bis 3m	dm bis ca.3m dm	dm bis 3m m (dm)
Öffnungsweite	--	--	bis >5cm	bis >8cm

hangparallelem Streichen zur Spaltenbildung neigen. Als Hinweis auf einen solchen
"Hangzerreißungseffekt" ist die Beobachtung zu werten, daß das alte Widerlager
Göttingen um einige Zentimeter nach Süden
abgekippt ist. Vermutlich haben die Wider-
lagerlasten und die sehr großen Belastungen aus der langen Brückenrampe zu einem
Talzuschub geführt.
    Das Trennflächengefüge wurde auf der
Gründungssohle für das nordöstliche Hilfs-
brückenwiderlager eingemessen (Bild 4).
Der Bausandstein war hier stark geklüftet
und bereichsweise sogar kleinstückig zerbrochen. Neben der Schichtung ließen sich
drei Hauptkluftrichtungen unterscheiden
(Tab 1).

Die annähernd hangparallel streichenden
Hauptkluftrichtungen K2 und K3 sind als
"Hangzerreißungsklüfte" wirksam. Diese
Klüfte öffnen sich in bergauswärtiger
Richtung, wo dem Gebirge durch den
Taleinschnitt das Widerlager entzogen
wurde. Die Richtung K1 ist eher recht-
winkelig zum Hangstreichen orientiert.
Infolge der seitlichen Einspannung der
Kluftkörper blieben die Fugen der K1-
Klüfte geschlossen.
Der Effekt der "Hangzerreißung" ist durch-
aus nicht auf den unmittelbaren Hangbe-
reich beschränkt. Er ließ sich nördlich
und südlich der Werra in den Vorein-
schnitten des Rauhebergtunnels bzw. des
Mündener Tunnels (DB-Neubaustrecke

277

Werratal – Brücke   BAB A7
Widerlager Göttingen

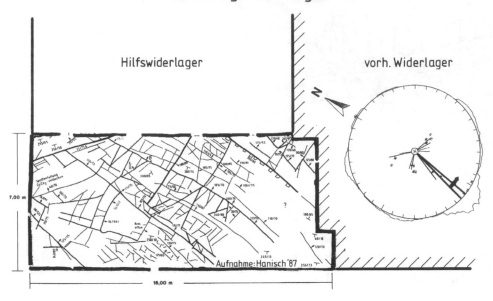

Bild 4 Trennflächen in der Gründungssohle des nordöstlichen Widerlagers (Sohlkartierung J.HANISCH)
Fig. 4 Fractures and bedding planes at the base of the foundation of the northern abutment; rose diagram of the fractures is shown at the right (mapping by J.Hanisch)

Hannover-Würzburg) sowie auch im Mündener Tunnel selbst beobachten.

3 UNTERGRUNDSANIERUNG

Zur Vermeidung unzulässiger Bewegungen an den nordöstlichen Widerlagern der Hilfsbrücke und der endgültigen Autobahnbrücke war eine Sanierung des Untergrundes erforderlich. Es wurde angestrebt, die vorhandenen Spalten mittels Zementinjektionen zu verfüllen und so einen ausreichend großen, möglichst homogenen Gesteinsblock unterhalb der Widerlager zu schaffen. Da von vornherein nicht davon auszugehen war, daß durch derartige Verbesserungsmaßnahmen der gesamte Bereich der "Hangzerreißung", der nachweislich einige hundert Meter bergeinwärts reicht, erfaßt werden kann, wurde der Gebirgskörper direkt unter dem Fundament zusätzlich mit Felsankern fixiert.

Für die Abschätzung der erforderlichen Verpreßmengen wurde aufgrund der Befunde im Hangbereich davon ausgegangen, daß ein Kluftvolumen von ca.10% zu verfüllen wäre. Diese Größenordnung hatte sich beim Bau der Eisenbahnbrücke als realistisch erwiesen. Dort wurden ca. 1000 t Feststoffe

in den Untergrund eingebracht. Aufgrund der größeren Fundamentabmessungen und der offenkundig stärkeren Zerrüttung des Gebirges bis zum Talboden war am Widerlager der Autobahnbrücke mit entsprechend höherem Zementverbrauch zu rechnen.

Die Injektionsbohrungen wurden in mehreren Reihen vor und neben dem vorhandenen Widerlager angeordnet (Bild 5). Da ein Durchbohren der Fundamentvorsprünge vermieden werden sollte, wurden die Injektionsbohrlöcher teilweise gegen das Bauwerk geneigt.

Die Verpreßarbeiten wurden anfänglich im Manschettenrohrverfahren durchgeführt. Bereits in den ersten Bohrlöchern waren sehr große Aufnahmen an Verpreßgut zu beobachten. Teilweise waren weit mehr als 100 t Zement injiziert, ohne daß sich ein Abschluß, d.h. eine Sättigung, andeutete. Da über die Manschettenrohre nur eine begrenzte Menge pro Zeiteinheit eingebracht werden kann, war abzusehen, daß eine vollständige Vergütung des Gebirges einen Zeitraum von mehr als einem Jahr in Anspruch nehmen würde. Das Verfahren wurde umgestellt: Der Bausandstein vor dem Widerlager wurde in einem Streifen von mehreren Metern freigelegt. Die Klüfte und

Spalten wurden ausgeblasen, z.T. aufgebohrt und dann drucklos, wie im Leistungs-
verzeichnis vorgesehen, mit Zementsuspension verfüllt.

Mit dem Ziel, das Abwandern der Suspension nach Möglichkeit zu verhindern bzw.
zu begrenzen, wurde die Zusammensetzung
des Verpreßgutes mehrfach geändert. Die
optimale Mischung hinsichtlich der Verarbeitbarkeit und des Verpreßerfolges bestand aus einer reinen Zementmischung mit
einem Wasser-Feststoff-Quotienten von ca.
0,6, d.h. 6 Gewichtsteilen Wasser auf 10
Gewichtsteilen Zement. Auf die Zugabe
grober Zuschlagstoffe wurde verzichtet, um
Entmischungen der Suspension zu vermeiden
und eine lückenlose Verfüllung zur Tiefe
hin sicherzustellen.

Die Aufnahme pro Verfüllbohrung bzw. pro
Verfüllstelle betrug bei dieser drucklosen
Einbringung zwischen ca. 1 t und mehr als
500 t.

Nach Abschluß der drucklosen Verfüllung
wurde noch eine Manschettenrohrverpres-
sung durchgeführt, bei der die Aufnahmemengen bereits deutlich geringer waren.

Das Verpreßgut bestand zu je 50 % aus
Zement (HOZ 35 L) und EFA - Füller. Für
die Herstellung der Mantelmischungen wurden Wasser-Feststoff-Quotienten zwischen
W/F= 0,9 bis W/F= 1,2 gewählt. Das Verpreßgut bei der eigentlichen Gebirgsvergütung (Sekundärmischungen) hatte Wasser-
Feststoff-Quotienten zwischen W/F= 0,5 bis
W/F= 0,6.

Um Ankerkräfte in der gewünschten
Größenordnung von 0,64 MN bis 1 MN in den
hinteren Bereich einleiten zu können, war
auch hier eine vollständige Verfüllung
aller Spalten und Klüfte erforderlich. Die
Ankerbohrungen wurden mit einer Neigung
von ca. 20° gegen die Horizontale abgeteuft. Die Bohrungen ließen sich infolge
Nachfalls oft nicht in einem Zuge herstellen. Sie mußten dann mehrfach zementiert
und wieder aufgebohrt werden. Die injizierten Feststoffmengen bei den Ankerbohrungen lagen zwischen ca. 20 t und 100 t.
Im Normalfall reichen wenige hundert Liter
Suspension aus, um einen Anker zu zementieren.

4 SCHLUßFOLGERUNGEN

Der Verbrauch an Verpreßgut bei der
Sanierung des Untergrundes am Göttinger
Widerlager der Werrabrücke zeigt, daß
Material auch in Bereiche abgewandert sein
muß, die außerhalb des Spannungsbereiches
des Widerlagerfundaments liegen. Aus der
Literatur und Erfahrungen bei anderen
Baumaßnahmen ist bekannt, daß die "Hang-
zerreißung" bis mindestens zum Niveau des

Werratal – Brücke   BAB A7
Widerlager Göttingen, Fundamentdraufsicht
Untergrundvergütung (Kluftverfüllung)

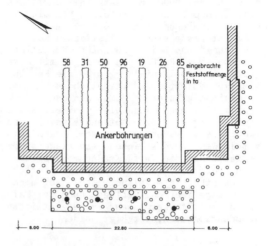

o Verpreß- bzw. Verfüllbohrungen
O Verpreßbohrungen, Aufnahme >300 to
● geneigte Kontrollbohrungen

Bild 5   Anordnung der Verpreßbohrlöcher
bzw. Einpreßstellen am nordöstlichen
Widerlager der Werratal-Brücke
Fig. 5 Location of the injection holes and
anchors around the northern abutment of
the autobahn bridge across the River Werra

Talbodens hinabreicht. Aus diesem Grunde
sind für die Ausbreitung des Injektionsgutes zunächst vertikale Richtungen anzunehmen. Damit ergibt sich zwangsläufig
eine Verfüllung auch der unteren Hangbereiche. Die horizontale Ausbreitungs-
komponente dürfte aufgrund der quasi
hangparallelen Orientierung offener Spalten (Hauptkluftrichtungen K2 und K3)
bevorzugt zu den Seiten und weniger in den
Berg gerichtet sein (K1- Klüfte überwiegend geschlossen).

Bei der Sanierung des Untergrundes am
Göttinger Widerlager der Werrabrücke
wurden im Zuge der Verfüll- und Verpreßarbeiten insgesamt über 12 000 Tonnen
Feststoffe verbraucht. Dies ist auch im
Nachhinein ein Beleg für das Ausmaß der
Gebirgszerrüttung und damit der Nachweis
der Notwendigkeit derartiger Maßnahmen,
auch wenn sich diese nicht auf den unmittelbaren Gründungsbereich beschränken
lassen: In dem unter dem Widerlagerfunda-
ment zu vergütenden Gebirgskörper mit

$$V_{Gebirge} = 32\ 500\ m^3$$

(Auflagerfläche F = ca. 650 m^2, Höhe über dem Talboden h = 50 m)

wurden anstelle der einem maximal vorstellbaren Kluftvolumen von 15% entsprechenden ca. 5000 m^3 tatsächlich mit den o.g. 12 000 Tonnen Feststoffen Volumina in der Größenordnung von 10 000 m^3 verfüllt.

Im Interesse einer realistischen Kostenschätzung und Zeitplanung für Verfüllarbeiten im großklüftigen Gebirge mit ausgeprägter Spaltenbildung sind die Massenansätze entsprechend reichlich zu wählen. Injektionstechnische Maßnahmen oder Mengenlimitierungen sind hier in der Regel kaum geeignet, den Aufwand sinnvoll zu reduzieren. Wären beim beschriebenen Beispiel des nordöstlichen Widerlagers der Werrabrücke die Verfüllmaßnahmen abgebrochen worden, bevor ein Sättigungseffekt (Druckaufbau,Materialaustritte) erreicht war, wäre zwar das Gebirge nahe dem Talboden, nicht aber oben, im eigentlichen Gründungsbereich, vergütet worden.

Literatur

HOLZ,H.-W.(1966): Talzuschub an flachen Hängen – Geol.Mitt.,6 (Breddin-Festschrift), S.87-114, 16 Abb., Aachen

KRAUSE,H.: Oberflächennahe Auflockerungserscheinungen in Sedimentgesteinen Baden-Württembergs –Jb.Geol.L.A.Bad.Württ.,8, S.269-323,Abb.56-69,Taf.15-23,Tab.51, Freiburg 1966

GEISSLER,H. & KUNERT,N.(1976): Injektionen im Druckstollen des Pumpspeicherwerkes Langenprozelten/Main – Technische Akademie Wuppertal,Berichte 12, S.80-87, 3 Bilder, 2 Taf.,Vulkan-Verlag, Essen

GEOLOGISCHE ÜBERSICHTSKARTE 1:200 000, CC 4718 Kassel, Hannover 1979

Technique and technology to grout silt stratum at Xidan Subway Station, Beijing

Technik und Technologie der Injektion zur Verfestigung der feinkörnigen Sandschicht bei Xidan Untergrundbahnstation, Beijing

Hu Fafu, Fu Qiang & Xu Wenxue
The Sixteenth Construction Bureau, Railway Ministry, People's Republic of China

ABSTRACT: The Xidan Subway Station in Beijing, designed by the Third Railway Surveying and Designing Institute, Railway Ministry of PRC and constructed by the Sixteenth Construction Bureau, Railway Ministry of PRC, is a 2-storeyed structure with 3 arches and 2 columns. It is the first large subway station in busy down-town area of China, located in silt stratum, with only 6 m covering depth, and constructed by subsurface excavation method. During the whole construction period, the ground settlement should not exceed 30 mm, no interruption of city traffic is allowed, and all the underground pipelines should work normally. To meet the above requirement a construction method of 'double spectacles' is adopted to combine big pipe-shed and small mortar spouting ducts in an advanced shedframe system. In order to consolidate surrounding soil stratum, the composition of cement mortar is determined by physical property tests of soil specimens in advance. The construction technology and operation specifications are followed exactly, with proper cement mortar compostion and arranging the grouting radius to 30 ~ 45 cm, a consolidation strength of 0.5 ~ 0.7 MPa is obtained successfully.

ZUSAMMENFASSUNG: Die Xidan Untergrundbahnstation in Beijing, vom 3. Planungsinstitut des Ministeriums für Eisenbahn der V.R.China geplant und vom 16. Konstruktionsbüro des Ministeriums für Eisenbahn der V.R.China gebaut, ist ein zweistöckiger Bau mit drei Gewölben und zwei Säulen. Sie ist die erste große Untergrund-Bahnstation, die in einem lebhaften Straßenviertel der chinesischen Innenstadt in einer Sandschicht mit einer Überdeckung von nur 6 m gebaut wurde. Während der gesamten Bauzeit sollte die Bodensenkung 30 mm nicht überschreiten. Man mußte gewährleisten, daß der Stadtverkehr nicht unterbrochen und unterirdische Rohrleitungen normal benutzt werden dürfen. Um diese Forderung zu erfüllen, wurde ein Bauverfahren mit "doppelten Brillen" angewendet. Eine große Schale wurde über die Injektion durch kleine Mörtelrohre verfestigt und so ein voreilendes Rahmensystem geschaffen. Damit konnte das Konstruktionsziel eingehalten werden. Um die Konsolidierung der Bodenschichte zu erreichen, wurden zuerst Bodenproben entnommen und die Zusammensetzung des Injektionsmaterials durch physikalische Versuche optimiert. Auf der Baustelle wurden die Ausführungsrichtlinien genau beachtet, sodaß die vorgesehene Wirkung der Verfestigungs-Injektionen erreicht wurde. Der Radius der Verfestigung lag bei 30 ~ 45 cm, die Festigkeit von 0.5 ~ 0.7 Mpa wurde erreicht.

1. SURVEY

The Xidan Subway Station in Beijing, located to the east of Xidan crossroads, is under the Changan Street, where traffic is most heavy. It is the first large subway station in busy down-town area of China, located in the place with heavy traffic on the ground surface and dense pipe network under the ground, and constructed by subsurface excavation. The main body of the station is a 2-storeyed island-like platform, with 3 arches and 2 columns.

The station is 26.04 m wide and 13.5 m high, with only 6 m covering depth. The area of excavation surface is 340 m^2. Engineering geological features are as follow: It is alluvium of Quaternary System located in the crestal pyrt of alluvial fan. Its main components are fine silt, medium and coarse sand, and unsaturated or saturated clay. The arches of the station are located in fine silt, and its surrounding rock is friable and selfstability is poor. See figure 1.

During the construction period, the ground set-

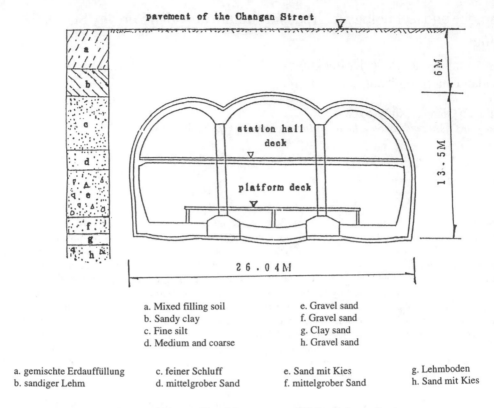

pavement of the Changan Street

station hall deck

platform deck

6 M

13.5 M

26.04M

a. Mixed filling soil
b. Sandy clay
c. Fine silt
d. Medium and coarse

e. Gravel sand
f. Gravel sand
g. Clay sand
h. Gravel sand

a. gemischte Erdauffüllung
b. sandiger Lehm

c. feiner Schluff
d. mittelgrober Sand

e. Sand mit Kies
f. mittelgrober Sand

g. Lehmboden
h. Sand mit Kies

Figure 1. The profile of the stratum of Xidan Subway Station
Bild 1.   Profil der Schichten bei der Xidan Untergrundbahnstation

tlement should not exceed 30 mm, no interruption of ground traffic is allowed, and the underground pipelines should not seep or be broken. So 'double spectacles type excavation method', by using big pipe-shed and small duct as the advanced shed frame system, was developed successfully to achieve the desired engineering goal.

In order to meet the above needs, chemical grouting is used to consolidate the silt stratum of the arches and strengthen selfstability.

2.   CHEMICAL GROUTING TEST

2.1 Determining the stratum parameter
   (see table 1)

From table 1, it can be known that the stratum the station passing through is silt, medium sand, and gravel sand, etc. The station arches are mainly in the silt, and its permeability coefficient is 3×10^2

cm/s. Normal cement grouting can not consolidate the silt, so chemical grouting must be used.

2.2 Selecting grout materials

The selected grout materials must have the following properties:

1. Good groutability, easy to grout into the stratum

2. Good stability, that is, the materials can be stored for a long time and its properties do not change at normal temperature, nor would there be chemical reaction taking place.

3. No poison, no odour, no decay, no pollution to environment, and no harm to human body.

4. Easy to control the time of consolidation. The consolidated body must have the proper compression strength, especially during the early stage.

5. Wide range of material sources and low lost.

Table 1.   The characters of different statum of Xidan Subway Station
Die Eigenschaften der verschiedenen Bodenschichten

	A	B	C	D	E	F	G	H
Natural moisture content (%)	3.17	2.9	4.7	4.2	20.5	10.5	26.0	4.5
Stone content (%)		4.8	13.0	19.0				20.5
Silt content (%)		4.6	2.7	2.3		7.8		1.7
Natural unit weight	1.97	1.59	1.76	1.56	1.87	1.53	1.83	1.75
Specific density	2.73	2.66	2.68	2.69	2.33	2.59	2.75	2.09
Dry unit weight		1.55	1.68	1.50	1.50	1.32	1.44	1.61
Fineness modulus		1.53	2.48	3.51		1.67		2.36
Porosity ratio (%)		50.0	39.2	42.0		49.0		40.0
Natural slope		54.9	48.3	47.7		55.5		46.5
Relative density		0.685	0.563	0.538		0.704		0.580
Liquid limit	36.6				36.2		33.2	
Plastic limit	21.9				25.0		20.3	
Index number	14.7				11.2		13.2	
Cohesive force	0.18				0.31		0.38	
Slope angle (*)	26.0				33.2		33.8	
pH vlaue		8.3	8.2	8.3		8.3		8.3
Soil name	A	B	C	D	E	F	G	H

A. Silty clay
B. Fine silt
C. Medium sand (sampling of No. 2 Shaft)
D. Gravel (sampling of No. 1 Shaft)
E. Sandy clay (or loam)
F. Finest silt
G. Sandy clay (or loam)
H. Medium sand

A. Schluffiger Lehm
B. Feinkörniger Schluff
C. Mittlerer Sand
D. Kies
E. Sandiger Lehm
F. Feinster Schluff
G. Sandiger lehm
H. Mittlerer Sand

6. Easy to make up grout, the equipments are simple and easy to operate.

In the light of the geological condition of the Xidan Subway Station and the results of innumerable tests, the mixed sodium silicate grout is made up to consolidate the silt stratum, with the sodium silicate as the main content, and sulphuric acid and other materials on the subsidiary.

2.3 The character of the grout

According to figure 2, when adding acid (acid firming agent) to the sodium silicate liquid (the SiO_2 content of it is 4 ~ 10 %) with proper density, the pH value of the mixed liquid will be decreased and the liquid will get the ability of jellification. With the increasing of the acid, the jellification time will be shortened. When pH value is in the range of 8 ~ 9, the liquid will be jellified quickly. At this time, adding more acid, the jellification time will be

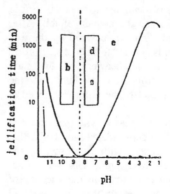

a.   Alkali
b.   alkalischer
c.   saurer Bereich
d.   neutral
e. Nichtalkali

a.   alkali
b.   alkaline zone
c.   acid zone
d.   neutral
e.   non-alkali

Figure 2.   The regulation of acid-based reaction
Bild 2.   Regulierung der sauren und basischen Reaktion

Table 2

No.	Sand size	pH value	Consolidation time (min)	The strength after 1 hour (Mpa)	The strength after 8 hours (Mpa)
1	medium	3.2	7	4.10	5.90
2	fine	3.2	8	3.09	5.10
3	medium	3.5	5	4.88	6.60
4	fine	3.5	7	3.60	6.10

prolonged. When pH value is in the range of 1 ~ 2, the time will reach the peak value.

According to the above regulation, the sodium silicate with different Baume (Be), the sulphuric acid with different density, and the different subsidiary materials are selected to make up the grout liquid with different jellification time to adapt the grouting of different stratum.

2.4 Consolidation test

The compression strength of the consolidated body is the main index of grouting quality. The index is decided by many factors, including the soil granularity, permeability coefficient, the ingredient of jell liquid, the characters of the grout (i. e. density, viscosity, and modulo), and the jellification time, etc.

The strength of the consolidated sand body is related to the size of sand, also to the total surface area of sand grain and the amount of jell. After grouting, the sand surface is covered by the jell, and the sand grain is jellified to each other to form the 'framwork' for bearing heavier load. Two sand samplings brought from field, medium and fine sand, are tested in the lab, the results are listed in table 2.

Table 2 indicates the reason why the sand size affect the strength of consolidated sand body, that is, the sand grain contact with each other by the points, the large sized grain has the larger vacancy, the grout can flow into the vacancy of large sized grain more easily than that of small sized ones, so the amount of grout flowing into the medium sand vacancy is more than that of fine sand, the strength of consolidated medium sand is stronger than that of the fine sand.

This fact indicates that the sizes (coarse, medium, fine) of sand grain must be determined first when determining the strength of

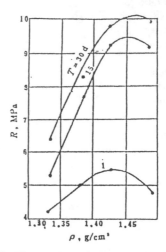

Figure 3. The relation of the strength of consolidated sand body and the density of the sodium silicate

Bild 3. Das Verhältnis zwischen der Dichte des Wasserglases und der Festigkeit des konsolidierten Sandes R

consolidated sand body. Another factor that affects the strength of consolidated sand body is the density of sodium silicate, figure 3 shows the relation of the strength of consolidated sand body (R) and the density of sodium silicate.

From figure 3, it can be noted that when the strength (R) reaches the maximum value, the density (P) is 1.4 ~ 1.45 g/cm*. If the density increases further, the viscosity will also increase. If the grout permeate or diffuse into the sand unevenly, the strength of the consolidated sand body will decrease. When the density is constant, the modulo of sodium silicate will also affect the strength, see figure 4.

The strength of the consolidated sand body increase with the increasing of the sodium silicate

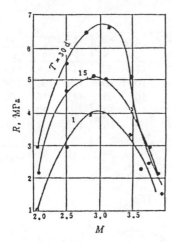

Figure 4. The effect of sodium silicate modulo to the strength of consolidated sand body

Bild 4. Der Einfluß des Wasserglasmodul M auf die Festigkeit des konsolidierten Sandes

modulo before the modulo reaching 2.75 ~ 3.10. On the contrary, the strength decreases when the value increase further, this is due to the reason that the density is constant, the viscosity increase with the increasing of the modulo, the permeability will decrease and the grout diffusive range will become small.

3. DETERMINATION OF GROUTING PARA-METER

3.1 Determining the depth of horizontal grout hole and the angle of elevation

To prevent the flowing collapse of fine silt stratum with poor selfstability when excavating the surrounding rock, chemical grout is grouted through the pipes into the periphery of cave room to increase the selfstability, see figure 5. According to field test, the optimum length of inserted duct is 3 ~ 4.5 m, the angle of eleration is 15 ~ 20°, in normal condition, the effective duct length is 3 ~ 3.5 m, the lap joining length can not be less than 1.0 m, the concrete sizes are shown in table 3.

3.2 Grouting pressure

The stratum that the Xidan Subway Station passes through contains less water, and almost no static water pressure exists. The only resistance to the grouting comes from the grout flowing resistance. Considering the thin covering stratum, the grouting pressure can not be too high. The value of 0.3 ~ 0.5 Mpa is determined as the grouting pressure, and the maximum value can not exceed 0.6 Mpa.

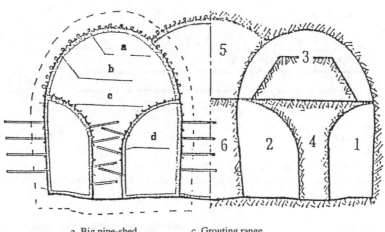

a. Big pipe-shed   c. Grouting range
b. Small duct   d. Anchor bolt

a. große Rohrschale   c. Injektionsbereich
b. kleines Führungsrohr   d. Anker

Figure 5. Double spectacles type excavation method

Bild 5. Baumethode mit 'doppelten Brillen'

Table 3

a	L=320	D=200			L=370	D=250			L=420	D=300			L=450	D=300		
	l	h	c	b	l	h	c	b	l	h	c	b	l	h	c	b
10					364	64	114	43	414	73	114	52	443	78	113	57
13	312	72	112	45	360	83	110	56	409	94	109	67	438	101	108	74
15	309	83	109	52	357	95	107	65	406	109	106	78	435	116	105	85
17	306	94	106	58	354	108	104	73	402	123	102	88	430	132	100	96
20	301	109	101	68	348	127	98	86	395	144	95	103	423	154	93	113
25	290	135	90	85												

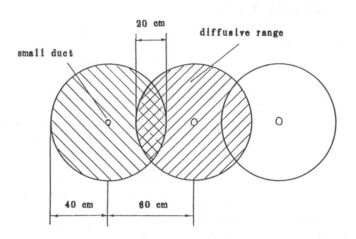

Figure 6. The diffusive range of small duct
Bild 6. Ausbreitungsbereich der kleinen Führungsrohre

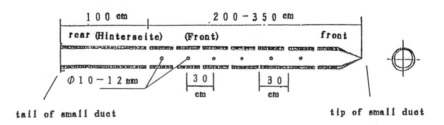

Figure 7. The sketch of processing small duct
Bild 7. Kartenskizze der Verarbeitung für kleines Führungsrohre

3.3 Grout diffusive radius

One of the preconditions for chemical grouting is that the stratum to be grouted must have permeability. The permeability coefficient derived from the Darcy Law can show the groutability of the grout exactly. The permeability coefficient of the fine silt stratum that the Xidan Subway Station passes through is 3×10^2 cm/s. The field tests show that the diffusive radius of small duct grouting is about 40 cm, so the distance between the ducts can be decided from the diffusive radius, see figure 6.

286

3.4 The amount of grout for one grouting hole

To ensure all the vacancy of sand in the range of diffusive radius, each grouting hole should be grouted until it can not take any more grout. The following formula give the amount of grout for one grouting hole in the fine silt stratum.

$$Q = r^2 \times \pi \times K \times e \times L$$

r: Diffusive radius
e: Porosity ratio
K: Filling coefficient
L: The depth of borehole

4. GROUTING MACHINES AND TOOLS AND GROUT TECHNOLOGY

4.1 Grouting machines and tools

1. Grouting pumps: Pneumatic grouting pumps are used, the type is QZH-50/60.
  2. Boring hole machines and tools: The borehole can be made by the steel pipes with high-pressure blast and coal-electric drills.
  3. Grout retaining cap: Two kinds of grout retaining cap were developed to fit for the field condition, one is a retaining cap with simple structure and outer cover, the other is a rubber grout retaining cap whose inner part can expand.
  4, Grout mixer: This mixer can uniformly mix a great amount of grout at one time, and pH value can be controlled easily.

4.2 Grouting technology

4.2.1 Making grouting dues

Figure 8 shows the structure of small grouting duct, the front of the duct should be cone shaped and sealed up. The holes for grout overflowing are set on the duct body, and arranged in the shape of plum blossom, the distance between the holes is 30 cm, no holes should be set on the rear segment of 1 m, the duct end should be reinforced with a ring, and kept straight.
  When boring the grouting hole, the cross piece should be set first to determine the direction, location and the eleration angle of the hole, then the holes are made by the coal-electric drill. The direction of the grouting hole should be kept straight, no crooked or blocked holes are allowed, in order that the grouting duct can be inserted smoothly. The duct should be aimed at the direction and the angle of the hole, and be pushed inside by hydraulic thruster. The ends of the grouting ducts are best in the same section, with the optimum length of 30 cm on the outside. The heading face must be sealed with shotcrete with the thickness of 5 ~ 8 cm, especially the face around the ducts must be sealed tightly, so that the grout can not overflow.

4.2.2 Making up chemical grout

Chemical grout is sensitive to chemical reaction, apart from being sensitive to the amount of grout, the order of feeding materials, and the speed of mixing, it is also sensitive to the grout temperature, so that the grout of different materials must be made up to the proper density, and stored in the container with distinctive marks for later use. The diluted sulphuric acid liquid can not be used until cooled down in the normal temperature. When making up grout, the amount of feeding materials must be controled strictly, the liquid must be mixed in time and the pH value must be tested constantly.

4.2.3 Grouting

Before grouting, the number and the length of the grouting ducts and geological parameter of the grouted segment must be known clearly and recorded carefully. Then the grout proportion shoud be determined according to different stratum parameter, and the amount of raw materials should be calculated. When grouting, it is necessary for a special person to control the pressure meter of the grout retaining cap and adjust the grouting pressure constantly. After grouting, the grouting machines and tools should be checked up and the ducts should be kept clean and unblocked.

5. INSPECTION OF GROUTING EFFECT

The borehole should be bored directly to the designed depth, the prepared grout should be grouted by pneumatic grouting pump. When grouting the segments of the stratum, each segment should be divided into three groups, 9 blocks (3 blocks from each group) should be tested for natural permeating and consolidating sand effect by using modified sodium silicate. Inspection should be carried at the time of age, 1 hour, 3 hours and 24 hours during the process of grouting. Excavation should begin after grouting the whole ducts. Actual radius of grouted sand should be checked in the period of excavation, and the strength of the samplings on the point should be tested. According to these actual feedback information, the grout proportion and grout pressure could be adjusted further.

6. CONCLUSION

Through a long period of field test, a consolidation strength of 0.5 ~ 0.7 Mpa and a consolidation radius of 30 ~ 45 cm are obtained, the requirement of design and construction security are guaranteed successfully. From October, 1989 to February, 1992 the total sodium silicate of 6656 T was used to grout, and the total sand stratum of 86000 m^2 was consolidated, the successful rate is 100 % and construction security was guaranteed perfectly.

REFERENCES

Du Jiahong   Foreign chemical grouting
  Liaoning   Publishing House

He Xiuren   Technique of chemical grouting
  Liaoning   Publishing House

Pont de Normandie – Stabilisierungsinjektionen für unverrohrte Pfahlbohrungen mit ø 2,10 m

Pont de Normandie – Stabilization grouting for uncased piles with ø 2,10 m

Frank Huppert & Jörn M. Seitz
Bilfinger + Berger Bauaktiengesellschaft, Tiefbauabteilung, Mannheim, Deutschland

ZUSAMMENFASSUNG: Bei den Pfahlarbeiten zur Gründung der weltgrößten Schrägseilbrücke über die Seine bei Le Havre mußten zur Sicherung der unverrohrt mit Suspensionstützung hergestellten Pfahlbohrungen besondere Maßnahmen zur Stabilisierung ergriffen werden. Durch Bodeneinbrüche in einer ca. 30-35 m tief liegenden Kiesschicht war die Standsicherheit der unverrohrt abgeteuften Suspensionsbohrung nicht mehr gewährleistet. Die Stabilisierung der Bohrlöcher erfolgte für die tiefliegenden Kiesschichten durch gezielte Hohlraumverfüllungen und Injektionen um den späteren Pfahlschaft. Die Injektion zur Stabilisierung der Kiesschicht hatte folgende Kriterien zu erfüllen: Sie sollte einmal die am Bohrlochrand liegenden einzelnen Steine beim Aufbohren halten und zum anderen aus Gründen der Bohrleistung (Rollenmeißel - Werkzeugabnutzung) keine zu großen Festigkeiten aufweisen. Desweiteren sollte eine möglichst wirtschaftliche Verfüllung erreicht werden. Erschwernisse ergaben sich hier durch die hohen Durchlässigkeiten der Kiesschichten. Vor Beginn der Maßnahmen wurden umfangreiche Baustellenversuche zur Optimierung der Geometrie der Injektionsbohrungen, der Verpreßdrücke und Verpreßmengen und der Mischungen durchgeführt. Begleitet wurden die Baustelleninjektionen von zahlreichen Qualitätssicherungsmaßnahmen, wie z.B. Dokumentation, Laborversuche, Pressiometer-versuche, Versickerungsversuche.

ABSTRACT: The foundation for the world's largest cable stayed bridge to date, the "Pont de Normandie" near Le Havre in France, consists of 2 x 28 piles ($\emptyset$ 2,10 m, length up to 60 m) for the two main pylons and 124 piles ($\emptyset$ 1,50 m length up to 50 m) for the approach viaducts. Piles are installed by a combination of the patented Hochstrasser-Weise method and air-lift drilling under bentonite slurry. The ground consists of approx. 25 m sand/silt overlying a 15 m granular quaternary deposit. The gravels are underlain by heavily consolidated hard clay. Limestone sequences in which the piles are seated are present at depths of approx. 50 m and 60 m. During initial stages of pile construction it became clear that stability in the gravel horizon was not sufficient for the large diameter boreholes without danger of collapse. It was decided to pre-grout the gravel horizon from the surface for the piles of the two main pylons. Two limiting conditions had to be met by the grouting operation: The gravel needed to be stabilized to an extent that large blocks ($\emptyset$ 0.7 - 1.0 m) were held in place or could be cut during drilling, but strength and hardness of the layer should be held at the necessary minimum to enable reasonable drilling progress and minimum wear of tools. Extensive investigations were carried out in the laboratory to arrive at suitable mix-designs for the grout. In various trials on the building site the layout and exact method of the grouting operation were tested to arrive at a scheme wich produced the desired results at minimum cost. It could be shown that aimed void-filling and stabilization is possible by grouting even in a material wich is not normally the object of classical grouting. The grouting operation was accopanied by numerous measures of quality control and documentation.

1 EINLEITUNG

Bei der Stadt Le Havre an der Atlantikküste Frankreichs entsteht zur Zeit im Mündungsgebiet der Seine die größte Schrägseilbrücke der Welt, die "Pont de Normandie". Bauherr des Jahrhundertbauwerks ist die Industrie- und Handelskammer von Le Havre, die das Bauvorhaben über ein Bankenkonsortium frei finanziert und auch Eigenmittel aus Mauteinnahmen ein-

PONT DE NORMANDIE

4 spurige Autobahnbrücke mit einem Mittel - und zwei Rand-
streifen

Brückenlänge	: 2141 m
Brückenbreite	: rd. 24 m
mittlere Spannweite	: 856 m davon 624 m aus Stahl, bestehend aus 33 Einzelstücken à 180 to und 2 x 116 m Betonüberbau
Freie Durchfahrtshöhe	: 52 m, bei Jahrhunderthochwasser
Pylone	: umgekehrtes Y, rd. 215 m hoch 40.000 Tonnen ständige Last
Schrägseile	: 8 x 23 = 184 sechseckige Bündel aus Parallel-litzenseilen ø 160 - 173 mm bestehend aus 31 - 53 Litzen ø 15 mm, insgesamt 2.300 Tonnen
Vorlandbrücke Süd	: Widerlager und 11 Pfeiler Spannweite 43 m Kragarm 116 m
Vorlandbrücke Nord	: Widerlager und 15 Pfeiler Spannweite 43 m Kragarm 116 m

Gründung

Pylone	: 56 Pfähle ø 210 cm mit Bohrtiefen von 45 - 60 m. Insgesamt rd. 3.000 lfdm Traglast des Einzelpfahles 3.300 to
Vorlandbrücke Süd	: 52 Pfähle ø 150 cm mit Bohrtiefen von 38 - 55 m. Insgesamt rd. 2.200 lfdm Traglast des Einzelpfahles 1.700 to
Vorlandbrücke Nord	: 72 Pfähle ø 150 cm mit Bohrtiefen von 40 - 55 m. Insgesamt rd. 2.700 lfdm mit rd. 1.000 to verlorene Stahlrohre ø 150 cm Traglast des Einzelpfahles 1.700 to

Bild 1   Hauptdaten der Brücke;
Fig. 1   Dimensions of the brigde.

setzt, die von der in den 50er Jahren erbauten Hängebrücke "Pont de Tancarville" stammen. Fachliche Beratung für Ausschreibung, Vergabe und Bauleitung obliegt der DDE 76, einer ministeriell gesteuerten Administration. Das Brückenbauwerk wurde vom Bauherrn in zwei Lose aufgeteilt: Los 1 - Beton; Los 2 - Metall. Auf Wunsch und in Abstimmung mit dem Bauherrn wurde BILFINGER + BERGER, Spezialtiefbau, mit der Ausführung der Gründungsarbeiten von einer französischen Arbeitsgemeinschaft - GIE Pont de Normandie - beauftragt. BILFINGER + BERGER ist als sogenannter "Named Subcontractor" eingesetzt und wird vom Bauherrn direkt bezahlt (Arz, Seitz;

1992). Der Gesamtpreis der Brücke einschließlich künstlicher Insel für den Pylon Nord, Hilfsbrücke und Anschlußdämme betrug zum Zeitpunkt der Vergabe im Mai 1990 ca. 430 Mio DM. Davon machte seinerzeit die Gründung ca. 18 Mio DM aus. Die Eröffnung der Brücke ist für 1994 vorgesehen.

Neben der Weltrekordspannweite von 856 m zwischen den 215 m hohen Pylonen sind in Bild 1 weitere Hauptdaten der Brücke dargestellt. Die bis zu 40 MN reichenden Lasten pro Pylon werden über je 28 Pfähle (Ø 2,10 m, Länge bis 60 m) abgetragen. Die Pfeiler und Widerlager der Vorlandbrücken stehen auf insgesamt 124 Pfählen (Ø 1,50 m, Länge bis 50 m).

2 BAUGRUND

Für die Gründungsmaßnahme lagen umfangreiche Schichtenprofile von Probebohrungen aus den schon 1973 begonnenen Untersuchungen vor. Die Erkundung des Baugrundes erfolgte mit Kern- und Pressiometerbohrungen sowie Pegeluntersuchungen zur Beobachtung des Grundwassers. Ergänzt wurden die Voruntersuchungen durch Laborversuche, in denen neben der Bodenklassifizierung, Scherfestigkeit, ein- und dreiaxiale Druckfestigkeit und Zusammendrückbarkeit dokumentiert wurden. Kurz vor Beginn der Gründungsarbeiten beauftragte der Bauherr für die Vorlandbrücke Nord zusätzliche Bohrungen in den nun endgültig festliegenden Standorten der Pfeiler (je Pfeiler eine Bohrung), um die genaue Lage der Gründungsschichten einzugrenzen.

Die für die Gründung maßgebenden Schichten lassen sich in 2 Epochen einteilen: Quartär und oberes Jura. Als Gründungshorizonte wurden vom Bauherrn für die Vorlandbrücken und Widerlager ein Kalksteinhorizont in ca. 45 m Tiefe (Calcaire à Harpagodes) und für die Pylone eine Kalksteinbank in ca. 55 m Tiefe (Banc de Plomb) vorgegeben (Bild 2).

Der erste ca. 3 m mächtige Gründungshorizont (Calcaire à Harpagodes) besteht aus einer Wechselfolge von dünnen Kalksteinbänken und zwischengelagertem mergeligem Ton. Die Stärke der individuellen Lagen reicht von 5 cm bis einige dm. Die Pfähle der Vorlandbrücken binden ca. 20 cm in diesen Horizont ein.

Der zweite Gründungshorizont (Banc de Plomb) besteht aus zwei harten Kalksteinbänken von 60 bis 80 cm Stärke, die durch eine 40 bis 60 cm mächtige Tonlage getrennt sind. Der gesamte Horizont hat nur eine Mächtigkeit von ca. 1,80 m. Die Pfähle der Pylone sollten auf Wunsch des Bauherrn nicht mehr als 20 cm in die obere der beiden Kalkbänke einbinden.

Der über dem oberen Gründungshorizont (Calcaire à Harpagodes) anstehende jurassische Ton (Argile Kimmeridgienne) ist durch eine hohe Vorbelastung von ca. 300 m überkonsolidiert. Als Kern gewonnen ist er hart bis sehr hart, sobald er aber entlastet wird und mit Wasser in Berührung kommt, verwandelt er sich in eine plastische und schwer zu bohrende "Kittmasse".

Über dem Ton liegt eine 12 bis 15 m mächtige quartäre Kiesschicht. Hauptbestandteil des Kieses ist der sehr harte Feuerstein (Silex) aus der Quarzart Chalcedon. Den Abschluß bilden ca. 25 m mächtige quartäre bis holozäne Schluffe und Feinsande.

Das Grundwasser wird im Baubereich vom Wasserstand der Seine beeinflußt. Der Tidehub der Seine liegt im Mündungsbereich bei 7 bis 9 m. Der höchste Grundwasserstand wurde mit 3 m unter Geländeoberkante gemessen.

3 PFAHLGRÜNDUNG

Die Pfahlbohrungen für die Vorlandbrücke Süd ($\varnothing$ 1,50 m) und die für die beiden Pylone ($\varnothing$ 2,10 m) wurden in ihrem oberen Drittel mit einer temporären und wiedergewinnbaren Verrohrung (18 bis 25 m) hergestellt. Darunter wurde unverrohrt mit einer stützenden Bentonit-Suspension weitergebohrt. Während der Pfahlbohrarbeiten traten unvorhergesehene Schwierigkeiten auf, die im Bereich der Pylone zu den in diesem Beitrag beschriebenen Injektionen zur Stabilisierung des Untergrunds führten. Die im Wasser stehenden Pfähle für die Vorlandbrücke Nord wurden aus Stabilitäts- und Sicherheitsgründen für die unmittelbar daneben stehende Hilfsbrücke bis zur Oberkante des Tons (30 bis 35 m unter Geländeoberkante) mit verlorener Verrohrung hergestellt. Das Konzept von BILFINGER + BERGER sah für die Ausführung der Bohrungen eine Kombination des firmeneigenen Hochstrasser-Weise-Verfahrens (HW: Pylon Nord, Vorlandbrücken Süd und Nord) bzw. des hydraulischen Verrohrungsverfahrens (Pylon Süd) mit dem Lufthebebohrverfahren mit Rollenmeißel vor (Arz, Seitz; 1992).

4 PROBLEME

Da die provisorische Verrohrung für die Pfähle der Pylone oberhalb der Kiesschicht enden sollte, mußte erschütterungsarm gebohrt werden, um Nachfall des Bodens ins offene Bohrloch zu verhindern bzw. gering zu halten. Das war ein technischer Entscheidungsgrund für die Auswahl des von BILFINGER + BERGER vorgeschlagenen Bohrverfahrens. In Bohrprofilen und

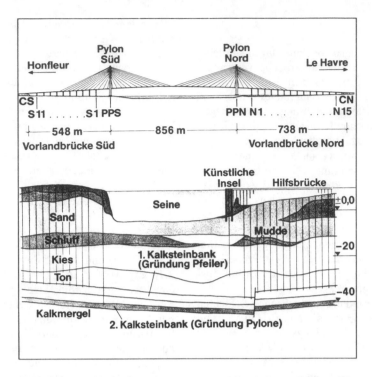

Bild 2   Geologie;   Fig. 2   Soil profile.

Baugrundbeschreibung war für die 12 bis 15 m mächtige Kiesschicht eine maximale Korngröße von 150 mm angegeben. Schon zu Beginn der Arbeiten am Pylon Nord stellten sich Schwierigkeiten beim Bohrvorgang ein. Diese konnten auf das Vorkommen von Gesteinsblöcken zurückgeführt werden, deren Dimensionen die im Baugrundgutachten angegebene maximale Korngröße weit übertrafen. Diese unregelmäßig verteilten und mit Kantenlängen von 700 bis 1000 mm geborgenen Blöcke ließen sich mit dem Rollenmeißel nur schwer zerkleinern, da sie in der Kiesschicht lose eingelagert waren. Mittlere und große Blöcke wurden beim Bohrvorgang aus der Bohrlochwandung gerissen und vor dem Rollenmeißel nach unten in den Ton mitgenommen, wo sie wie ein Rührwerk wirkten. Das hatte erhebliche "Ausräumungen" im Ton zur Folge, sodaß dort ein für diese Bodenart unüblicher hoher Betonmehrverbrauch registriert wurde. Darüber hinaus wurde der Ton während des Bohrens durch die vor dem Meißel mitgeführten Blöcke durchgeknetet. Es entstand ein Konglomerat aus Blöcken, Steinen und weich gekneteten Ton, das nicht mehr die guten

bohrtechnischen Eigenschaften des ursprünglich harten Tons besaß und deshalb vom Rollenmeißel nicht mehr sauber geschnitten und gefördert werden konnte. Durch den hohen Zeitaufwand für das Bohren im Ton wurde weiterhin durch die Einwirkung der Bohrsuspension das Quellen und Aufweichen des Tons noch begünstigt.

Diese Vorgänge und Probleme waren zu Beginn keinem der am Projekt Beteiligten bekannt und auf Grund des Bodengutachtens auch nicht zu erwarten gewesen. Sie konnten erst später nach der Bergung größerer Blöcke und weiterer Untersuchungen des Tones durch externe internationale Fachleute nachvollzogen werden. Ausgelöst wurden diese weitergehenden Untersuchungen durch ein zunächst unglückliches Ereignis: ein partieller Einbruch einer Bohrung mit $\varnothing$ 2,10 m am Pylon Nord im Bereich der Kiesschicht begrub das noch im Bohrloch befindliche Bohrwerkzeug in einer Tiefe von ca. 55 m. Obwohl das Werkzeug geborgen und das Bohrloch für die Herstellung des Pfahls gerettet werden konnte, wurde klar, daß die Instabilität der Kiesschicht ein zu hohes Risiko nicht nur

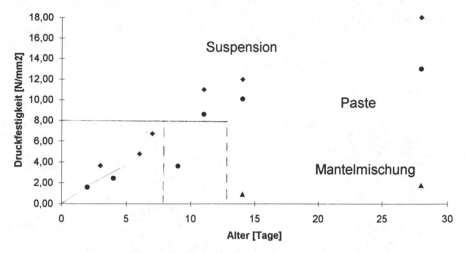

Bild 3   Einaxiale Druckfestigkeit der Injektionsmischungen;
Fig. 3   Uniaxial compressional strength of the grouting mix.

für die Bohrarbeiten (Tagbrüche, Verlust des Bohrwerkzeugs), sondern auch für das gesamte Gründungskonzept darstellte. Durch weitere Einbrüche und der damit verbundenen Boden-auflockerung würde die Gesamtstabilität der Pylongründung in Frage gestellt. Aus Platz-gründen, besonders auf der künstlichen Insel des Pylon Nord, wäre die Anordnung von Zusatz-pfählen beim Verlust einer Bohrung nicht ohne weiteres möglich gewesen.

5 KONZEPT DER STABILISIERUNG

Die Arbeiten wurden zunächst eingestellt, um über ein neues Bohrkonzept und eine Stabili-sierung des Baugrunds nachdenken zu können. Schließlich entschloß sich der Bauherr nach langer Überlegung zu einer der Pfahlherstellung vorauseilenden Injektion der bis zu 15 m mäch-tigen Kiesschicht, jeweils im Bereich der vorge-sehenen Pfahlbohrungen an beiden Pylonen. Die Schicht sollte insbesondere gegen das Aus-brechen von größeren Blöcken aus der Bohr-lochwandung stabilisiert werden. Andere Lösungen - wie z.B. Verrohrung bis in den Ton - wurden zum einen aus Zeitgründen (Beschaffung entsprechend großer Rohre, Werkzeuge und Maschinen) verworfen. Zum anderen waren bereits 12 der 28 Pfähle am Py-lon Nord fertiggestellt und es wäre bei einer Permanentverrohrung der noch zu bohrenden

Pfähle das unterschiedliche Last-Setzungs-Ver-halten in der gesamten Pfahlgruppe zu über-denken gewesen.

Nach der Entscheidung des Bauherrn die in-stabile Kiesschicht zu verfüllen und zu injizie-ren wurden Vorversuche begonnen. An die In-jektion des Kieshorizontes wurden folgende An-forderungen gestellt:

• Die am Bohrlochrand und im Bereich der Bohrung liegenden einzelnen Steine und Blöcke sollten in eine Matrix eingebunden werden, um beim Aufbohren zerkleinert werden zu können bzw. gegen Einfall in die Bohrung gesichert zu sein.

• Die injizierte Schicht sollte aus Gründen der Bohrleistung und der zu erwartenden Werk-zeugabnutzung keine unnötig große Festig-keit aufweisen.

5.1 Laborversuche

Aus den vor Ort verfügbaren Ausgangsstoffen (Zement CPJ 55 PM, Kalksteinmehl "Calcaire brayé D", Bentonit Clarsol, Verflüssiger Addiment IH1) wurden mehrere Rezepturen für verschiedene Einsatzvarianten zusammenge-stellt:

• eine Mantelmischung aus Wasser, Zement und Bentonit zur Ausfüllung des Hohlraums zwischen Injektionsrohr und Bohrlochwan-dung;

293

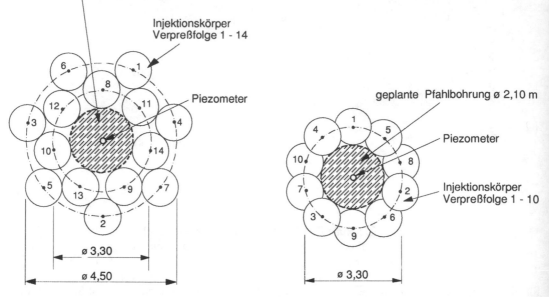

Bild 4   Anordnung der Injektionsbohrungen.
                 a: Ausgangsvariante Versuchsinjektion (14 Bohrungen in 2 Ringen);
                 b: Endgültige Anordnung (10 Bohrungen in einem Ring).

Fig. 4   Position of the grouting boreholes.
                 a: Initial position of the trial injection (14 bores in 2 circles);
                 b: Final position (10 bores in one circle).

- eine dickflüssige Paste aus Wasser, Zement, Kalksteinmehl, Bentonit und etwas Verflüssiger zur primären Verfüllung großer Hohlräume;
- eine dünnflüssige Suspension aus Wasser, Zement, Bentonit und Verflüssiger zur Nachinjektion von kleinen Hohlräumen.

Mit der dickflüssigen Paste sollte erreicht werden, daß sich das Injektionsgut auch bei dem zu erwartenden großen Porenvolumen nur im engeren Bereich der Bohrung ausbreitet, mit dem Ziel, die Injektionsbohrungen gezielt plazieren und die Injektionsmenge auf das notwendige Maß beschränken zu können. Mit der dünnflüssigeren Injektionssuspension sollten die Poren verfüllt werden, die von der Paste nicht erreicht wurden.

Für jede dieser Anwendungen wurden mehrere Mischungen angesetzt und im Labor von BILFINGER + BERGER auf Absetzmaß, Zähigkeit, Schwindmaß und Druckfestigkeit untersucht. Nach Auswertung der Ergebnisse

wurden folgende Mischungen für die Injektionsarbeiten empfohlen:

- **Mantelmischung**

      100,00 kg Wasser
       50,00 kg Zement
        3,00 kg Bentonit

- **Suspension**

      100,00 kg Wasser
      104,00 kg Zement
        4,25 kg Bentonit
        1,04 kg Verflüssiger

- **Paste**

      100,00 kg Wasser
       93,30 kg Zement
     109,40 kg Kalksteinmehl
        9,40 kg Bentonit
        0,97 kg Verflüssiger

Die Entwicklung der einaxialen Druckfestig-

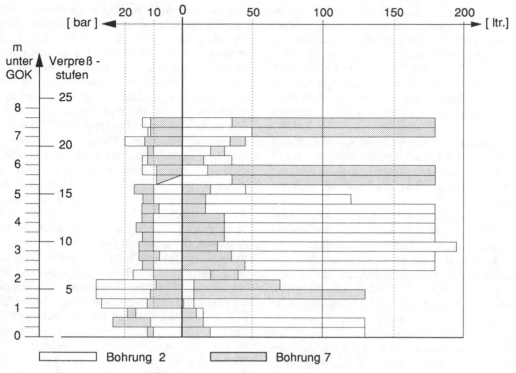

Bohrung 2   Bohrung 7

Bild 5   Verpreßdruck und Menge in zwei benachbarten Injektionsbohrungen;
Fig. 5   Grouting pressure and volume in two neighbouring boreholes.

keit der Injektionsmischungen ist in Bild 3 aufgezeichnet. In-situ war eine vergleichbare Festigkeitsentwicklung zu erwarten, da das betonähnliche Gemisch Kies-Injektionsgut zwar eine höhere Festigkeit als der reine Zementstein entwickeln würde, diese aber aufgrund der niedrigeren Temperatur zu einem späteren Zeitpunkt erreichen würde. Bei einer im Zementstein erreichten Druckfestigkeit von 8 - 10 N/mm^2 wurde davon ausgegangen, daß im Kieshorizont ein ausreichender Zusammenhalt zwischen Steinen und Injektionsgut existieren würde, um größeren Nachfall in die Bohrung zu vermeiden. Es durfte aber auch nicht zu lange mit dem Beginn der nachfolgenden Pfahlbohrarbeiten gewartet werden, damit Bohrleistung und Werkzeugabnutzung durch zu hohe Festigkeiten nicht nachteilig beeinflußt würden.

5.2 Feldversuche

Zur Überprüfung des Konzeptes für die Stabili-

sierung des Kicshorizontes durch eine Injektion und zur Dimensionierung der Injektionsbohrungen wurden am Pylon Nord und Pylon Süd bei je 4 vorgesehenen Pfahlbohrungen Versuchsinjektionen durchgeführt.
Ausgangspunkt war das in Bild 4.a dargestellte Konzept:
  In 2 Ringen mit je 7 Bohrungen in einem Abstand von 0,6 m bzw. 1,2 m von der Pfahlperipherie entfernt wurden insgesamt 14 Bohrungen bis 1,0 m in den Ton niedergebracht. Zum Bohren wurden das Raupenbohrgerät Klemm KR 806 und eine Verrohrung mit $\varnothing_a$ 108 mm verwendet. Nach Erreichen der Endtiefe wurden in die mit Mantelmischung gefüllten Bohrungen im Schutz der Verrohrung Manschettenrohre ($\varnothing$ 42,5 mm, Manschettenabstand 33 cm) eingebaut. Beim Ziehen der Verrohrung wurde die Kiesschicht über die Bohrrohre in 2 m Stufen Paste mit einem Druck von max. 16 bar verpreßt. Dazu war das Bohrgestänge an einen mit 3 Spezialpumpen ausgestatteten Injektions-Container angeschlos-

295

sen, in dem Verpreßdruck und -menge automatisch aufgezeichnet wurden. Nach Ansteifen der Paste (ca. 24 Std.) wurde über die Manschettenrohre mit Hilfe von Doppelpackern in 33 cm-Stufen mit Suspension nachinjiziert. Kriterium für das Beenden der Injektion bei einer Manschettenstufe war das Erreichen eines Verpreßdruckes von 20 bar oder eines injizierten Volumens von 180 l pro Stufe.

Der äußere Ring wurde immer vor dem inneren Ring abgebohrt und injiziert und die Bohrreihenfolge war so ausgelegt, daß benachbarte Bohrungen nie direkt nacheinander injiziert wurden.

Bei den restlichen Versuchsinjektionen wurden Anzahl und Position der Bohrungen, Einsatz der 3 vorgeschlagenen Mischungen, Ausführungsart und Injektionsparameter variiert, um den notwendigen Aufwand einzugrenzen und gleichzeitig zu optimieren. Eine minimierte Versuchsinjektion sah einen Ring mit 10 Bohrungen im Abstand von 0,6 m von der Pfahlperipherie vor (Bild 4.b). Die Bohrungen wurden beim Ziehen der Rohre mit Injektionsgut verpreßt. Auf den Einbau von Manschettenrohren und eine Nachinjektion wurde ganz verzichtet.

Der Verlauf der Injektion und die aufgezeichneten Verpreßdrücke und -volumen wurden analysiert, um die Wirksamkeit der Maßnahme abzuschätzen. Auf Bild 5 ist die

Entwicklung des injizierten Volumens und Druckes für jede Manschettenstufe einer Bohrung dargestellt. Bild 6 zeigt die daraus errechneten Durchmesser mit den mittleren Abmessungen der Injektionssäulen um eine Pfahlbohrung. Obwohl die gemittelten Werte nur idealisierte Säulen darstellen und unregelmäßige Ausläufer unberücksichtigt lassen, eignen sie sich zum Vergleich der injizierten Volumina. Zur endgültigen Überprüfung der Wirksamkeit wurden Versickerungsversuche durchgeführt (Bild 7). Dazu wurden in den mit einem Injektionskörper umgebenen Pfahlbohrungszentren Pegel gebohrt und bis zur Oberkante mit Wasser aufgefüllt. Über eine festgelegte Zeitdauer (> 10 min.) wurde die Wassermenge gemessen, die notwendig war, um den Wasserstand im Pegel zu halten. Derselbe Versuch wurde zur gleichen Zeit an einem außerhalb des injizierten Bereichs liegenden Nullpegel durchgeführt, um Verfälschungen durch den Tideeinfluß auszuschließen. Mit den gemessenen Werten wurde ein fiktiver Durchlässigkeitsbeiwert errechnet. Die so ermittelten Werte entsprechen zwar nicht den in der Wasserhaltung verwendeten Durchlässigkeitsbeiwerten, dienen aber dennoch als Grundlage eines Vergleiches zwischen injiziertem und jungfräulichem Kies.

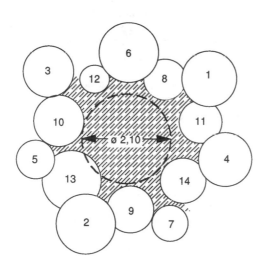

Injektionen um Pfahl II SE

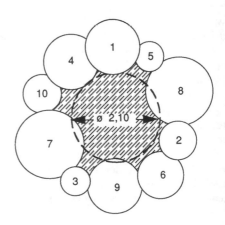

Injektionen um Pfahl 10 SE

Bild 6   Gemittelte Dimensionen der Injektionskörper;
Fig. 6   Averaged dimensions of the grouted soil body.

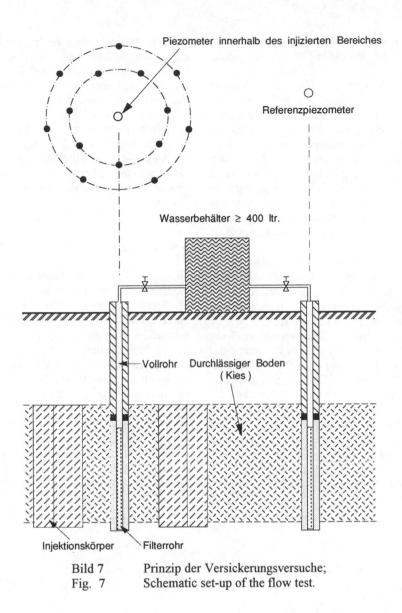

Piezometer innerhalb des injizierten Bereiches

Referenzpiezometer

Wasserbehälter ≥ 400 ltr.

Vollrohr   Durchlässiger Boden
(Kies)

Injektionskörper   Filterrohr

Bild 7   Prinzip der Versickerungsversuche;
Fig. 7   Schematic set-up of the flow test.

5.3 *Ergebnisse der Versuche*

Obwohl der in seiner Kornzusammensetzung sehr inhomogene Kieshorizont für eine Zementinjektion im klassischen Sinn nicht sehr gut geeignet ist, ergab sich aus der Auswertung der Versuchsinjektionen, daß eine ausreichende Verfestigung des Kieses in der Peripherie der Pfahlbohrungen möglich ist. Die in den Versickerungsversuchen ermittelten k_f-Werte im injizierten Bereich waren durchgehend um ca. 2 - 3 Zehnerpotenzen geringer als im nicht injizierten Bereich. Daraus konnte auf eine ausreichende Wirksamkeit der Injektion geschlossen werden. Die Versuche zeigten außerdem, daß der Boden direkt über die Bohrrohre ohne Verwendung von gerichteten Manschetten injizierbar ist und mit dieser Verpreßmethode der gewünschte Effekt zu erreichen war. Die Abdichtung der Bohrrohre zur Atmosphäre war durch die relativ lange Strecke zum Kies gewährleistet. Aus diesem Verfahren ergaben sich zeitliche und wirtschaftliche Vorteile gegenüber einer Injektion über Manschettenrohre. Die Gesamtaufnahme an Injektionsgut variierte beträchtlich (40 m³ bis 105 m³/Pfahl), war jedoch nicht abhängig von der Anzahl der Injektionsbohrungen (10 oder 14). Auf der Nordseite wurden in den einzelnen Verpreßstufen sehr große Unterschiede in der Mengenaufnahme festgestellt. Auf der Südseite war die Aufnahme im allgemeinen gleichmäßiger. Die in einzelnen Verpreßstufen aufgenommenen, sehr großen Mengen an Injektionsgut führten wahrscheinlich zu großen, in benachbarte Injektionsfelder hineinreichende Ausbreitradien. Es konnte davon ausgegangen werden, daß sich das injizierte Volumen bei nachfolgenden Injektionen verringern würde.

Aufgrund der in den Versuchsinjektionen gewonnenen Erkenntnisse wurde für die Injektion der Kiesschicht im Bereich der restlichen Pfähle folgendes Vorgehen als technisch und wirtschaftlich optimierte Lösung festgelegt:

Um jeden Pfahlansatzpunkt werden in einem Ring mit Abstand 0,60 m von der Peripherie des Pfahles 10 Bohrungen bis ca. 1,0 m in die Tonschicht niedergebracht. Beim Ziehen der Verrohrung wird das Injektionsgut direkt über das Bohrgestänge in 2 m Stufen verpreßt. Als Anhaltspunkt für das Beenden der Injektion auf einer Stufe gilt das Erreichen eines Druckes von 16 bar oder eines injizierten Volumens von 1800 l. Die Entscheidung, ob eine ausreichende Injektion erreicht ist, wird jedoch vor Ort und

unter Berücksichtigung der im direkten Umfeld erreichten Drücke und Volumina gefällt. Ab Oberkante Kiesschicht wird drucklos verfüllt. Aufgrund des festgestellten Aufnahmeverhaltens des Bodens wird am Pylon Nord mit Paste verpreßt. Bei zu schnellem Erreichen der Begrenzungskriterien wird auf die dünnflüssigere Suspension umgestellt. Auf der Südseite wird zunächst mit Suspension injiziert, bei fehlendem Druckaufbau wird auf Paste umgestellt. Die Anordnung der Injektionsbohrungen ist auf Bild 4.b dargestellt.

6 AUSFÜHRUNG DER INJEKTION

Nach den Versuchsinjektionen wurden die verbleibenden Pfahlzonen (9 Stück auf Pylon Nord, 24 Stück auf Pylon Süd) mit zwei Geräteeinheiten im 24-Stunden Schichtbetrieb injiziert. Bei jedem 4. Pfahl wurden aus dem Injektionsgut 12 Probezylinder hergestellt, die nach 3, 7, 14 und 28 Tagen zur Bestimmung der Druckfestigkeit abgedrückt wurden. Ein Pfahlansatzpunkt wurde für die Pfahlbohrarbeiten freigegeben, wenn bei den Probezylindern eine Druckfestigkeit von 8 - 10 MPa festgestellt wurde. Dies war in der Regel nach ca. 2 Wochen der Fall. Zusätzlich war zu berücksichtigen, daß die Herstellung eines Pfahles weniger Zeit in Anspruch nahm als die vorherlaufende

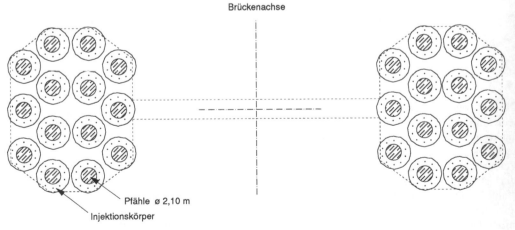

PONT DE NORMANDIE, PYLON SÜD

Brückenachse

Pfähle ø 2,10 m

Injektionskörper

Bild 8   Aufsicht des Fundaments am Pylon Süd;
Fig. 8   Plan view of the foundation, Pylon South.

Bild 9   Injektionsarbeiten am Pylon Süd;
Fig. 9   Grouting in progress, Pylon South.

Injektion inklusive Wartezeit für das Erreichen der notwendigen Druckfestigkeit. Aus diesen Randbedingungen konnte für die Injektions- und Pfahlbohrarbeiten ein Zeitplan erstellt werden, der eine Abwicklung ohne Unterbrechung und Auflaufen ermöglichte.

Der Verlauf der Injektionsarbeiten bestätigte die in den Versuchsinjektionen gewonnenen Erkenntnisse. Auf Pylon Nord wurde vergleichsweise mehr Injektionsgut verpreßt als auf Pylon Süd, was auf eine inhomogenere Lagerung des Kieshorizontes mit größeren Wegigkeiten zurückgeführt wurde. Im Verlauf der Injektionen ging sowohl innerhalb einer Pfahlzone bei aufeinanderfolgenden Injektionsbohrungen als auch bei nacheinanderfolgenden Pfahlzonen der Verbrauch an Injektionsgut zurück.

Die nachfolgenden Pfahlbohrarbeiten bestätigten die Wirksamkeit der Injektionsmaßnahme. Die Bohreigenschaften des Tons waren erheblich verbessert. Nachfall beim Durchteufen der Kiesschicht wurde nicht mehr festgestellt. Der Mehrverbrauch an Beton hielt sich im üblichen Rahmen.

Insgesamt wurden zur Stabilisierung der Kiesschicht unter beiden Pylonen 1 760 m³ Injektionsgut über 15 932 lfm Injektionsbohrung verpreßt. Davon entfielen auf Pylon Nord 813 m³ Injektionsgut und 6 015 lfm Bohrung,

auf Pylon Süd 947 m³ Injektionsgut und 9 917 lfm Bohrung. Für die Injektionsarbeiten wurden auf Pylon Nord 13 Wochen, auf Pylon Süd 22 Wochen benötigt, wobei sich die Bauzeit nicht nur nach der möglichen Injektionsleistung, sondern aus Gründen der Koordination auch nach den nachfolgenden Pfahlbohrarbeiten richtet. Der injizierte Bereich des Pylon Süd ist zur Veranschaulichung auf Bild 8 dargestellt.

7 QUALITÄTSSICHERUNG

Das gesamte Bauvorhaben unterlag während aller Bauphasen einer streng gehandhabten Qualitätssicherung, vergleichbar den jetzt auch für die deutsche Bauindustrie relevant werdenden Grundsätzen der Normen DIN ISO 9000 bis 9004. Um sicherzustellen, daß die Qualität der Ausführung den hohen an sie gestellten Anforderungen entsprach, wurde vor Beginn der Arbeiten ein Qualitätssicherungsplan (Plan d'Assurance Qualité, PAQ) erarbeitet und mit dem Bauherrn abgestimmt. Darin waren für jeden Arbeitsgang Verfahrensweise, Verantwortlichkeiten, Art der protokollarischen Abnahme und Haltepunkte, bei denen vor Weiterführung der Arbeiten bestimmte Kriterien erfüllt sein mußten, dargelegt. Der Qualitätssicherungsplan diente als verbindliche Richtlinie für die Fremdüberwachung und für die Eigenüberwachung, die von einem eigens auf der Baustelle stationierten Ingenieur für Qualitätssicherung wahrgenommen wurde.

In den Qualitätssicherungsplan wurden auch die Injektionsarbeiten in Form einer Verfahrensbeschreibung mit detaillierten Arbeitsanweisungen eingearbeitet.

Neben der Dimensionierung der Injektionsmaßnahme (Tiefe und Anordnung der Injektionsbohrungen, verwendetes Gerät) wurden darin die maßgeblichen Grenzwerte für die Injektion (Druck, Volumen) beschrieben. Es war festgelegt, auf welche Weise und wie oft das angemischte Injektionsgut zu überprüfen war (Zähigkeit, Dichte) und wie oft Proben entnommen wurden, um über die Druckfestigkeit den Beginn der Pfahlbohrarbeiten bestimmen zu können.

Für jede Bohrung wurden Injektionsdruck und -volumen automatisch aufgezeichnet und zusammen mit Diagrammen, die Druck- und Volumenverteilung über die Tiefe zeigten (Bild 5), und Tagesberichten, in denen angelieferte Mischungskomponenten, angemischtes und verbrauchtes Injektionsgut, Bohrmeter etc. vermerkt waren, zur Dokumentation der Maßnahme abgelegt.

LITERATUR

Arz, P. & Seitz, J.M. 1992. Besonderheiten der Pfahlgründung für die Schrägkabelbrücke Pont de Normandie bei Le Havre in Frankreich. Vorträge der Baugrundtagung 1992 in Dresden, Deutsche Gesellschaft für Erd- und Grundbau Essen.

Grouting in Rock and Concrete, Widmann (ed.) © 1993 Balkema, Rotterdam, ISBN 90 5410 350 7

Repair grouting of a sudden increase of leakage during initial impounding

Sanierung einer plötzlichen Steigerung der Wasserdurchtritte während des Aufstaues durch Injektionen

R. Iida
Japan Dam Engineering Center, Japan

I. Shibata
IDOWR Engineering Co., Ltd, Japan (Formerly: Japan Dam Engineering Center)

ABSTRACT: During initial impounding of the reservoir, if a sudden increase of leakage is happened, it is required to confirm the function of water barrier and to repair it, if necessary. This paper paper reports the design of repair grouting on such case.

1. INDISPENSABILITY TO OBSERVE LEAKAGE

To monitor the dam safety, it is efficient to observe the deflection of dam and/or the quantity of leakage.

The deflection of dam is a integral of the strain at each part of dam, and is expressed as following,

$$y = \iint \frac{M}{E\,I}\ dx\ dx \quad\cdots\cdots\cdots\cdots\cdots\cdots (1)$$

where, x, y; coordinate (displacement), M; bending moment, EI; rididity.

As long as the structure behaves elastically, we can confirm the dam safety if the observed deflection is in proportion to loads.

Similarly, seepage through the structure is governed by Darcy's low, an individual velocity of seepage at each part of it is expressed as following,

$$v_n = k_n \cdot i_n \quad\cdots\cdots\cdots\cdots\cdots\cdots\cdots (2)$$

where, v_n; velocity of seepage at a part n, k_n; permeability coefficient at a part n, i_n; pressure gradient at a part n. Then,

$$q_n = a_n \cdot v_n = a_n \cdot k_n \cdot i_n \quad\cdots\cdots (3)$$

where, q_n; a quantity of leakage through a part n, a_n; an area where the seepage pass through a part n.

A quantity of leakage through the structure is derived by a integral of eq.(3) as following,

$$Q = \sum_n a_n \cdot k_n \cdot i_n \quad\cdots\cdots\cdots\cdots\cdots\cdots (4)$$

In eq.(4), a chracteristic of the structure is k_n only. Therefore, as long as Q is proportional to a change of reservoir water level, we are able to confirm the dam safety because k_n is kept to be a constant. But if the passage of seepage isenlarged by wash out of fine particles, a sudden increase of Q is recognized caused by the increases of k_n. Teton dam failure during initial impounding proves the phenomenon mentioned above and suggests the especial indispensability to obseve a quantity of leakage during intial impounding.

However, it is not clear that how to treat the problem if a sudden increase of the leakage is observed while inintial impounding.

This paper gives a good example at Simoyu dam.

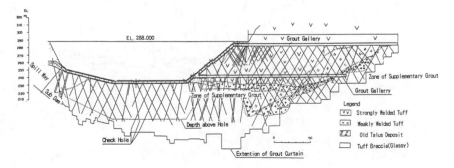

Fig. 1   Geology at right bank of addle dam

2. PROFILE OF DAM AND GEOLOGY

The Shimoyu Dam was completed on the Tsutsumi River by Aomori Prefecture Government in 1988. A 70 meter high, 366 meter long rockfill main dam with central claycore was built on the main stream, and a 52 meter high, 255 meter long saddle dam was built as an inclined core type rockfill dam, and the spillway was provided between the maindam andthe saddle dam.

The foundation of the dam was mainly composed of pyroclastic rock and shale in the Miocene formation. As shown Fig. 1, the right bank contained buried river deposit, welded tuff, buried river deposit andwelded tuff from the bottom up respectively. The lower welded tuff layer which was sandwiched by two gravel layers was weakly welded and was therefore not highly permeable. But the upper welded tuff layer was extremely permeable because it was strongly welded and almost vertical openings and fissures were developed. In a densely jointed portion, openings and fissures were observed at 0.5 to 1 meter intervals. The maximum opening width was 30 centimeters, but the width of most of the openings was 5 or so centimeters.

3. CURTAIN GROUTING

In spite of highly permeable zones above 25 Lugeon were densely developed, apreferable grout curtain was accomplished with grout holes at intervals of 0.75mas shown Fig.2. Grouting was carried out under the maximum injection pressure of 15kgf/cm. Bentnite, 3 percent of the cement, was added to grout.

Prior to grout into strongly welded tuff cement grout with a mix of C : W = 1 : 1 and cement mortar with a mix of C : S : W = 1 : 1 : 0.8 were poured into the openings under no pressure.

At first, litile improvement was expected to seal the buried old river depositlayers by grouting. However, the results of the grout works in three rows were considered almost satisfactory chiefly because the permeabilities of 5 Lugeon orless were obtained to over 90% of the treated area and it corresponded to Darcy's coefficient of 5×10^{-5} cm/s, and the grouting operations were terminated.

4. MONITORING SYSTEM FOR LEAKAGE

Eight partition walls with a pipe line which let water flow into observation spot were provided at the base of downstream filter and rock fill in order to check where a large leakage happened. Furthermore, wells were drived into the right bank foundation rock in order to gather the leakage through it.

5. LEAKAGE FROM RIGHT BANK

When the impounding of the constructed dam began and the water level of the reservoir reached at elevation 261 meter, a sudden increase of leakage from the foundation on the right bank was observed as shown Fig. 3. The

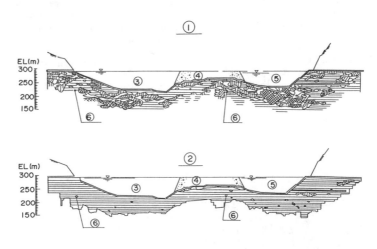

① Before grouting, ② After grouting,
③ Main dam, ④ Spillway, ⑤ Saddle dam, ⑥ Diversion tunnel,
⑦ Below 5 Lugeon, ⑧ 5-10 Lugeon, ⑨ 10-25 Lugeon, ⑩ Above 25 Lugeon

Fig. 2 Lugeon map before and after grouting

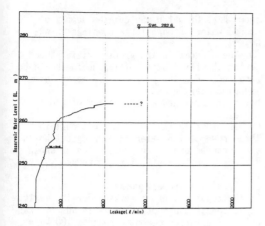

Fig. 3 Relations between reservoir water level and leakage at right bank during first stage of impounding

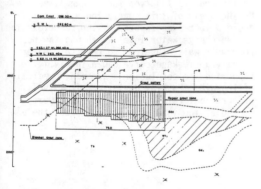

Fig. 4 Hydraulic geology

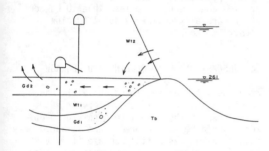

Fig. 5 Extent of repair works in grout curtain

permeability of the grout curtains was checked immediately, and it was found that the grout curtain in the welded tuff showed the same permeability as that obtained at the time of the completion of the construction. Most of the buried old river deposit layer showed 5 or so Lugeon, but some portions showed as high as

20 Lugeon. It suggested that the grout curtain at an extent of containing high potential due to reservoir water was broken partially and windows were shaped.

By the checked results of the grout curtain in the buried old river deposit, Gd_2 layer shown in Fig.4 was a subject of our investigation.

6. HYDRAULIC GEOLOGY

According to the relations between reservoir water level and quantity of leakage, the quantity of leakage increased suddenly after when the water level raised up to EL. 261 m which agreed with the elevation where the highly permeable Quaternary welded tuff overlay impervious bed rock.

As shown Fig. 4, it is able to introduce a hydraulic model as followings ;

① When the reservoir was filled up to the level of EL. 261 m , stored water begin to spill over the impermeable Neogene pyroclastic bed rock into the buried old river deposit layer (Gd_2) through large openings developed in the overlay strongly welded tuff (Wt_2).

② The supplied water into Gd_2 flows through the windows in grout curtain and springs up into the downstream-side large openings in the Wt_2.

③ Then the leakage flows out ground surface and is gathered up from downstream filter of dam.

Further to add, it was full of axiety about to wash out fine particles of old river deposits through piping holes into large openings developed in the strongly welded tuff .

7. REPAIR WORKS

(1) Zone and method

According to the results of permeability tests, it was decided to repair the grout curtain in Gd_2 for the extent of 100 m as shown Fig. 5, where was deduced the existence of windows in the grout curtain.

Under the circumstances, it was obliged to choose a grouting for repair works.

(2) Design

As mentioned before, quantity of seepage through the grout curtain in Gd_2 is shown as following,

$$Q_2 = A_2 \cdot k_2 \cdot i_2 \qquad (3')$$

It is useful to reduce each facter in this equation as a means to design. So, they should be examined.

A_2 is unchangeable facter.
k_2 may be reduced by clogging the windows in grout curtain.

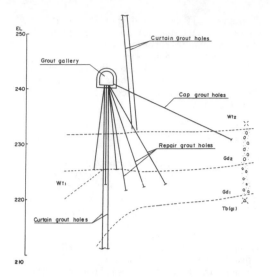

Fig. 6 Arrangement of cap grout holes and repair grout holes

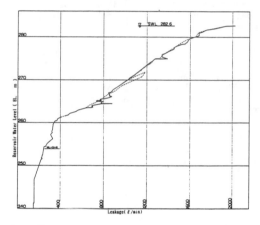

Fig. 7 Relations between reservoir water level and leakage at right bank during resumed impounding

shown Fig. 6, which may cut the supply of stored water through large openings in Wt_2 to Gd_2.

② The windows in the grout curtain shaped during the first stage of impoundingshould be repaired. For the purpose, 1.5 meter interval supplementary grout holes are arranged at the center of grout curtain.

③ In order to thicken the grout curtain, three rows of 1.5 meter interval grout holes are arranged under cap grout holes as shown Fig. 6. They are efficient to reduce i_2 and/ or k_2.

2) Results of repair works

On the repairing grout works under the maximum injection pressure of $30kgf/cm^2$ in Gd_2, it was only 5 % absorbed cement more than 100 kg/m, 11 % absorbed cement50 through 100 kg/m, and remaining 84 % absorbed cement less than 50 kg/m.

But, permeability coefficient of repair grouted zone reduced to less than 10 Lugeon after careful grouting by split spacing closure method.

9. RESUMPTION OF IMPOUNDING AND LEAKAGE

From the works as mentioned above, it was concluded that even if the reservoir water level rose further, no more highly-permeable zones would be encountered, and it would only result in increasing the seepage flow in the buried old river deposit layer according to steepening in the hydraulic gradient. And it was expected that the grouted zone in Gd_2 would keep its function for a long time. On the basis of this conclusion, it was decided to resume impounding the reservoir to the surcharge water level which was the peak reservoir water level. The reservoir water level v.s. leakage curve shown in Fig.7 supported the conclusion concerning prediction of leakage. Partial deviations from the proportional relation shown in Fig.7 were caused by the thawing of snow around the dam.

i_2 may also be reduced by thickening the grout curtain.

If k_2 and i_2 both can be reduced to 90 % respectively, quantity of seepage through the grout curtain in Gd_2 may decrease to $0.9 \times 0.9 \fallingdotseq 0.8$.

(3) Repair grouting

1) Arrangement of grout holes

① It was made clear by preceded curtain grouting that large openings developedin Wt_2 can be filled up with cement grout to be water -tight.

So, in order to reduce i_2 by increasing pass length, 1.5 meter interval cap grout holes in Wt_2, just above Gd_2, are arranged as

Sanierung der Kölnbreinsperre – Injektionskonzept und Ausführung

Kölnbrein Dam repair works – Grouting concept and report on execution

Kurt Kogler
Insond Gesellschaft mbH, Neumarkt, Österreich

ZUSAMMENFASSUNG: Die Kölnbreinsperre ist das zentrale Bauwerk der Kraftwerksgruppe Malta der ÖSTERR. DRAUKRAFTWERKE und liegt in Kärnten im hinteren Talende des Maltatales.
Bei der Kölnbreinsperre handelt es sich um eine bogengewölbte Staumauer mit einer Höhe von 200 m, einer Kronenbreite von 626 m und einem Betonvolumen von 1,6 Mio. m3.
Nach dem Überschreiten der Staukote von 1860,00 m traten die ersten Schäden am Bauwerk auf, sodaß in den Jahren 1979 bis 1985 der Kraftwerksbetrieb nur eingeschränkt aufrecht erhalten werden konnte.
Neben mehreren Sanierungsvorschlägen entschied man sich zu der von Prof. Lombardi vorgeschlagenen Lösung mit der Errichtung eines luftseitigen Stützgewölbes und umfangreichen Injektions-maßnahmen, die einen wesentlichen Bestandteil dieses Sanie-rungskonzeptes ausmachten.
Die Injektionsarbeiten sind Gegenstand dieser Darstellungen und beschreiben die erfolgreiche Anwendung neuer Injektions-kriterien für Suspensionen und Harze.

SUMMARY: Kölnbrein dam is the main structural component of the Malta Hydroelectric group of ÖSTERR. DRAUKRAFTWERKE in Austria. The 200 m high double curvature arch dam had some leakage problems since its first impoudings, and a major repair concept was developed by Prof. Lombardi in the 80ies.
For the 1,6 Mio m3 concrete dam, a 500.000 m3 supporting structure using more than 600 Nos Neoprene pads as load transferring elements was conceived. Together with a major cement and epoxy resin grouting program. This program is the subject of this paper and is described in detail. New grouting criteria where introduced and a close monitoring program established to control the grouting operation.

Leakage was reduced by two orders of magnitude and the reservoir is about to be filled to maximum level, after last years sucessful performance upto 7 m below final target.

1. EINLEITUNG

Die Kölnbreinsperre ist das zentrale Bauwerk der Kraftwerksgruppe Malta der Österr. Draukraftwerke und liegt in Kärnten im hinteren Talende des Maltatales.

Bei der Kölnbreinsperre handelt es sich um eine bogengewölbte Staumauer mit einer Höhe von 200 m, einer Kronenbreite von 626 m und einem Betonvolumen von 1,6 Mio. m3 und er-zeugt somit einen Speicherinhalt von 206 Mio. m3 Wasser.

Mit diesen Abmessungen gehört die Kölnbreinsperre zu einer der höchsten Sperren Mitteleuropas.

2. GEOLOGISCHE LAGE

Geologisch liegt das Kraftwerksgebiet in 1700 - 1900 m Seehöhe im massiven Zentralgneis, dem sogenannten Tauernfenster.

Die Sperrenstelle liegt in einem u-förmigen Talquerschnitt, an der rechten Flanke dominieren kluftarme Granodioritgneise, die linke Flanke besteht aus wenig geklüfteten Blattengneisen, im Bereich der Aufstandsfläche befinden sich zwei starke Schieferzonen mit hohem Glimmeranteil, sowie dünne schieferungsparallele Mylonitlagen.

3. BAUPHASEN UND SCHADENSBILD

Die Anlage wurde in den Jahren 1971 bis 1978 errichtet, wobei die Betonierung der Kölnbreinsperre von 1974 bis 1978 erfolgte.

1979 konnte das erstemal der Vollstau mit einer Staukote von 1902,00 m erreicht werden.

Nach dem Überschreiten der Staukote von 1860,00 m traten die ersten Schäden am Bauwerk auf und boten folgendes Bild:

- Wasserverluste erreichten Werte über 200 l/sec.
- Anstieg der Sohlwasserdrücke bis zu 100 % im Bereich der Aufstandsfläche der Mittelblöcke.
- Ausbildung von zwei schalenförmigen Rißgruppen von der Wasserseite der Sperre gegen die Aufstandsfläche bis in den Felsuntergrund, mit einer max. Öffnungsweite von bis zu 30 mm bei Vollstau.
- An der Wasserseite bildete sich ein vertikaler Riß zwischen Beton und Felsvorland, die sogenannte Bewegungsfuge,
- ein luftseitiger Rißbereich bis zu 10 m in den Sperrenbeton mit einer max. Öffnungsweite von 3,5 mm bei leerem Becken.

In den Jahren 1979 bis 1985 konnte der Kraftwerksbetrieb nur eingeschränkt aufrecht erhalten werden, dazu wurde eine Reihe von ergänzenden Baumaßnahmen ausgeführt, wie:

- Ergänzende Zementinjektionen im Dichtschirmbereich im Felsuntergrund, sowie Kunstharzinjektionen im Sperrenbeton.
- Errichtung eines Gefrierschirmes.
- Errichtung eines wasserseitigen Vorbodens mit einer dauerelastischen Dichtungsfolie.

4. SANIERUNGSKONZEPT UND AUSFÜHRUNG

Neben mehreren von verschiedenen Experten ausgearbeiteten Sanierungsvorschlägen entschied man sich zu der von Hr. Prof. Lombardi vorgeschlagenen Lösung mit der Errichtung eines luftseitigen Stützgewölbes.

Der Stützkörper mit einer Basislänge von ca. 60 m und einer Höhe von 70 m ergibt ein Betonvolumen von ca. 450.000 m2 und somit ein starres Widerlager zur Aufnahme der Sperrenkräfte.

Als Kraftübertragung zwischen Stützkörper und Sperre wurde ein System von insgesamt ca. 600 Neoprenlagern in 9 Lagerhorizonten angeordnet. Dieses System ermöglicht die Verformung der Sperre während des Aufstaues, wobei die einzelnen Lagerelemente von unten nach oben nach einem vorberechneten Schema angelegt werden, wobei die Verformung der Sperre auf 1/3 reduziert werden sollte. Zusätzlich zu dieser Stützkonstruktion sind umfangreiche Injektionsmaßnahmen vorgesehen, und bilden somit auch einen wesentlichen Bestandteil des Sanierungskonzeptes.

5. INJEKTIONSMASSNAHMEN

Injektionskonzept:
Das Injektionskonzept kann in zwei wesentliche Bereiche unterteilt werden.
- Im Felsuntergrund werden Injektionen mit Zementsuspensionen ausgeführt zur Wiederherstellung des ursprünglichen Dichtungschirmes sowie zur Konsolidierung der Sperrenaufstandsfläche.
- Im Beton sind Kunstharzinjektionen vorgesehen, die nach Möglichkeit eine kraftschlüssige Injektion der aufgetretenen Rißstrukturen bzw. Abdichtung dieser Zonen ermöglichen sollen.
- Weiters soll durch die Errichtung eines neuen Drainageschirmes das Auftreten von Sohlwasserdrücken verhindert werden,
- und schließlich wurden die bestehenden Meßeinrichtungen um weitere Extensometer, Gleitmikrometer und Piezometer ergänzt.

Beschreibung der Injektionsetappen:
Die einzelnen Injektionsmaßnahmen wurden im wesentlichen in folgende Bereiche eingeteilt, wobei die einzelnen Abschnitte jeweils stauabhängig zu den entsprechenden Zeiten behandelt wurden.

In Kurzform zusammengefaßt waren dies:

1990 Bereich "S" - Wiederherstellung des Dichtschirmes bis ca. 50 m mit Hammerbohrungen und Zementinjektionen.
Bereich "N1" - Konsolidierungsinjektionen mit Hammerbohrung und Zementinjektionen bis ca. 25 m.
Bereich "M" - Konsolidierungsinjektion mit Zement im Bereich der Aufstandsfläche des Stützkörpers. Die Injektion erfolgte bei Staulagen von 60 - 100 m unter Vollstau

1991 Bereich "N2" - Zementinjektion im
Fels unter dem Vorboden zur Konsolidierung des wasserseitigen
Widerlagers für den Lastfall
"leeres Becken".
Bereich "BF" - Injektion der Bewegungsfuge zwischen Sperre und
Felsvorland mit Feinzement-
mischungen.
Bereich "P u. Q" - Primärinjektion der Rißgruppe 1 und 2 mit
Kunstharzen verschiedener Viskositäten in jeweils getrennten Ar-
beitsgängen mit Hilfe von ver-
lorenen Packern.

Die Ausführung der Bohrungen für diese
Bereiche erfolgte als Rotationskernbohrung mit 46 mm Bohrdurchmesser, wobei in einer Bohrung bis zu 4 verschiedene Arbeitsgänge ausgeführt wurden, die Injektion erfolgte bei einer
Staulage von 40 - 60 m unter Vollstau.

1992 Bereich "P u. Q" - Sekundärinjektion der beiden Rißgruppen im
Beton mit Kunstharz mit niederer
Viskosität und höheren Injektionsdrücken.
Bereich "S2" - Sekundärinjektion
des Anschlusses Dichtschirm an
den Sperrenbeton mit Kunstharz
in Form eines zweireihigen
Schirmes, beidseitig der Dichtungsebene bis ca. 10 m in den
Felsuntergrund. Auch diese Bohrungen wurden als Rotationskern-
bohrungen Durchmesser 46 mm ausgeführt. Diese Injektionen wurden bei relativ hoher Staulage
von 20 - 40 m unter Vollstau
ausgeführt.

1993 - Injektion der Sperrenblockfugen in den Blöcken 10 - 20
mit Kunstharz.
- Nachinjektion der Rißgruppe "2"
in den Blöcken 19 und 20 über
bis zu 40 m lange Bohrungen vom
Kontrollgang IV aus.
- Verschwenken des Dichtschirmes
in den Blöcken 15 - 17 zur Luftseite und Nachinjektion mit
Kunstharz.
Die Austeilung der Bohrgeometrie erfolgte bei den Zementinjektionen mit

6 m bei einem seitlichen Abstand von
3 m, die Packerpositionen wurden in 5 m
Passen geteilt.
Für die Kunstharzinjektionen wurde
ein Bohrlochraster von 3 x 3 m bezogen
auf die Rißfläche gewählt.
Insgesamt wurden in den 4 Bausaisonen
von 1990 bis 1993
- 18.000 m Hammerbohrung,
- 29.000 m Rotationskernbohrung,
- 1.300 m Bohrungen für die Meßtechnik,
- 300 to Zementinjektionen und
- 200.000 kg Kunstharzinjektionen ausgeführt.

6. BAUSTELLENORGANISATION UND GERÄTE-
EINSATZ

Die Ausführung der Arbeiten erfolgte
bedingt durch die oft extreme Witte-
rungssituation und dem vorgegebenen
Staubetrieb pro Bausaison jeweils ca.
6 Monate von April bis September im
Tag- und Nachtschichtbetrieb.

Die Bauleitung war neben dem Bauleiter
mit
1 Außenbauleiter,
2 Technikern für die Arbeitsvorberei-
tung und Abrechnung,
2 Technikern für Dokumentation und Auswertung der Injektionsergebnisse,
1 Baukaufmann und Sekretariat besetzt.

Auf der Baustelle vor Ort waren
1 Polier, je 2 Schichtführer für Bohr-
und Injektionsarbeiten, 6 - 8 Mann für
Infrastruktur, wie Werkstatt, Magazin
und Versorgung, sowie bis zu 60 Mann an
Bohr- und Injektionsspezialisten im
Einsatz.
Zur Ausführung der Hammerbohrungen
wurden 4 Stück SIG-Hammerbohrgeräte mit
hydraulischen Bohrhämmern eingesetzt.
Der Bohrdurchmesser betrug 51 mm bei
Bohrtiefen bis zu 60 m.
Die Rotationsbohrungen wurden mit
insgesamt 6 Stück hydraulischen Bohrge-
räten DIAMEC ausgeführt.
Die Bohrdurchmesser betrugen hauptsächlich 46 und 56 mm. Bei Bohrtiefen
zwischen 15 und 40 m konnten mit Hilfe
der hohen Drehzahlen von ca. 2000
U/min. und entsprechenden Diamantkronen
Bohrleistungen von ca. 25 lfm. pro
Schicht im Schnitt inkl. Umstellungen
erreicht werden.
Zur Ausführung der Meßbohrungen und
diverser Bohrungen standen noch
2 Stück Rotationsbohrgeräte für Bohrungen mit Bohrdurchmesser bis zu
200 mm zur Verfügung.
Eine Anzahl von über 5000 Stück Bohrungen erforderte eine effiziente
Transportmöglichkeit der Geräte unter
den beengten Platzverhältnissen. Dazu
wurde im gesamten Arbeitsbereich eine
Kranschiene an die Kontrollgangfirste
montiert.
Für die Bohrgeräte wurden eigene Ar-
beitsplateaus angefertigt, welche mit
der Kranschiene transportiert werden
und an jeder Stelle des Kontrollganges
rasch bohrbereit fixiert werden konnten.

Die Zementinjektionen wurden von
einem zentralen Injektionscontainer mit
4 Stück Obermann-Pumpen vom Sperrenfuß
aus über Injektionsleitungen bis zu
250 m Entfernung ausgeführt. Die Versorgung mit Zement erfolgte über eine
zentrale Mischanlage an der Sperren-
krone mit 2 Stück Vorratssilos und
einem Injektionscontainer mit 2 Stück
Förderpumpen über eine Versorgungslei-
tung durch den Sohlgang der linken
Flanke. Die Kunstharzinjektionen wurden
mit insgesamt 8 Stück Kunstharzinjek-
tionseinheiten, bestehend aus hydr. ge-
steuerter Kunstharzpumpe, Wiegeeinrich-

tung und Druck-/Mengenschreiber, ausgeführt.

7. INJEKTIONSMATERIALIEN

Für die Sanierung der Kölnbreinsperre wurden prinzipiell zwei Injektionsmaterialien verwendet und zwar Zementsuspensionen im Felsuntergrund und Kunstharz für die Rißinjektionen im Sperrenbeton.

Bezüglich der Wasserzementwerte setzte sich auch bei den Injektionsmischungen die Erkenntnis der Betontechnologie durch, daß die Festigkeit umgekehrt proportional zum Wasserüberschuß ist. Es wurden daher nur stabile Zementsuspensionen mit einem WZ-Wert von 0,7 verarbeitet, wobei erst durch Beigabe von Additiven die Fließeigenschaften der Mischungen wesentlich verbessert wurden und dadurch die geforderten rheologischen Eigenschaften einer stabilen Zementmischung erreicht werden konnten. Die Feinzementmischung wurde nur in dem vorhin erwähnten Bereich der Bewegungsfuge angewendet.

Sämtliche Injektionen innerhalb des Sperrenkörpers im Beton sowie die Kontaktinjektion Sperre / Fels wurde ausschließlich mit Kunstharz der Marke "RODUR" ausgeführt. Dabei kamen hauptsächlich drei verschiedene Produkte mit abgestuften Viskositätswerten zwischen 4.000 und 48.000 m PAs. und Gammawerten zwischen 1,45 und 2,20 zur Anwendung. Die drei verschiedenen Injektionsharze wurden jeweils in einer anderen Farbe eingefärbt, sodaß bei der Auswertung der Kernaufnahmen eventuelle Rückschlüsse auf den Verlauf der Injektionsarbeiten getroffen werden konnten.

8. INJEKTIONSKRITERIEN

Gegenüber den herkömmlichen Methoden solcher Injektionsmaßnahmen, wie Wasserabpreßversuche, LUGEON-Wertbestimmung, Injektion mit wässrigen Zementsuspensionen und Kontrolle der Injektion über Kontrollbohrungen, wurden in Kölnbrein neue Wege beschritten.

Maßgeblich für die Beurteilung des Injektionsablaufes ist hier keinesfalls der an der Pumpe oder am Packer gemessene Injektionsdruck, sondern die gesamte Kraft, die sich aus der Integration des Druckverlaufes über die injizierte Fläche ergibt.

Aus dieser Betrachtung wurde von Hr. Prof. Lombardi ein Grenzwertkriterium abgeleitet bei dem das Produkt aus Injektionsdruck und eingepreßten Injektionsvolumen eine Größe ergibt, die als "Injektionsintensität" bezeichnet wird. In der graphischen Darstellung ergibt dies eine Hyperbel mit der Beziehung P x V ist konstant. Für die Bestimmung des Druckes für das Einhalten dieser Injektionsregel wird der Injektionsdruck bei Rate O

gemessen. Für die Praxis bedeutet dies, daß die Injektionspumpe nach einem konstanten Injektionsvolumen abgestellt wird und der sich einstellende Ruhedruck auf dem Schreibstreifen des Injektionsdiagrammes beobachtet wird.

Dieser Ruhedruck dokumentiert somit den effektiv herrschenden Injektionsdruck im Riß bzw. im Kluftsystem und läßt Rückschlüsse auf die Rißgeometrie sowie auf den Sättigungsgrad der Injektionsbereiche zu.

Im vorliegenden Fall wurde für die Behandlung der Felszonen für die Injektion mit Zementmischungen ein Energiekriterium von P x V = 7.500 für 15 bar max. Injektionsdruck bzw. P x V = 12.500 für max. 30 bar Injektionsdruck ausgewählt

Für die Injektion der Rißflächen im Sperrenbeton wurde das Injektionskriterium für die jeweilige Rißgruppe im einzelnen festgelegt. Durch die hohe Viskosität des Injektionsmateriales wurden für die Kunstharzinjektionen wesentlich höhere Injektionsdrücke als bei der Zementinjektion angewendet. So erfolgte die Primärinjektion mit einem Ruhedruck von 30 bar, die Sekundärinjektionen wurden mit einem Ruhedruck von 60 bar ausgeführt. Als Arbeitsdrücke wurde der zwei- bis dreifache Wert des jeweils festgelegten Ruhedruckes zugelassen, d.h. Arbeitsdrücke von 120 bis zu 150 bar.

9. ERFASSUNG UND AUSWERTUNG DER INJEKTIONSDATEN

Die Erfassung der Injektionsdaten erfolgte über firmenintern entwickelte Druckmengenschreiber, welche die Zementinjektion im Injektionscontainer am Sperrenfuß zentral angeordnet waren. Für die Kunstharzinjektionen wurde jede Injektionseinheit vor Ort mit einem Injektionsdatenschreiber versehen.

Die Datenerfassungseinheit besteht aus einem Fünf-Bandschreiber, auf dem die Werte für Injektionsdruck, Injektionsvolumen, Rate sowie zwei Verformungswerte parallel gegen die Zeit aufgezeichnet werden können. Diese Daten werden vor Ort auf den Schreiberstreifen aufgezeichnet und können über Datenübertragung auf dem Bildschirm des Zentralrechners in der Bauleitung "One Line" mitverfolgt werden. Zusätzlich besteht die Möglichkeit, diese Daten über eine Modemleitung bis nach Klagenfurt in die Hauptverwaltung der Österr. Draukraftwerke zu übertragen, wo diese im Zentralrechner abgespeichert und jederzeit wieder abgerufen werden können.

Aus den gesamten Injektionsdaten konnten somit relativ einfach Mengenbilanzen über die verschiedenen Injektionsbereiche erstellt werden und die Unterschiede in den Aufnahmen für Primär- und Sekundärinjektion dokumentiert werden.

Zur Kontrolle der Injektionsergebnisse wurden noch eine Reihe von

Kontrollbohrungen mit Kerngewinn und
Fernsehsondierung vorgenommen.
  Kernbeispiele.

Resümee:

Im Herbst 1992 konnte ein Stauziel von
1895 m, also 7 m unter Vollstau, erreicht werden, und die Sickerwasserverluste betrugen zu diesem Zeitpunkt
10 l/sec.

Für Herbst 1993 ist der Vollstau mit
einer Staukote von 1902 m vorgesehen.
Nach den bisher vorliegenden Auswer-
tungen der Meßergebnisse verläuft das
Verhalten der Sperrre ziemlich genau
den errechneten Werten.

Das Beispiel der Kölnbreinsperre zeigt,
daß eine erfolgreiche Sanierung von so
komplexen Bauwerken nur durch eine kooperative Zusammenarbeit von Bauherrn,
Statikern, Spezialisten und Produkther-
stellern mit dem Einsatz ihres gesamten
"KNOW-HOW" zum Erfolg führen kann.

Grouting in Rock and Concrete, Widmann (ed.) © 1993 Balkema, Rotterdam, ISBN 90 5410 350 7

Grouting in underground excavations in Carpathian Flysch

Injektionsverfahren in Hohlräumen im Karpaten Flysch

R.Łukaszek
Chemkop-Geowiert GmbH, Cracow, Poland

K.Thiel
Institute of Hydroengineering, Polish Academy of Sciences, Gdansk, Poland

ABSTRACT: The paper deals with description of grouting performed in the site of pumped-storage power plant. The object was constructed in polish Carpathian Flysch in Porabka-Zar. It is composed of many underground excavations such as pressure and non-pressure tunnels, caverns, shafts, etc. All the excavations were lined with reinforced concrete or with steel lining. As the geologic conditions of the rock mass built of sandstone and weak and slaking clay shale were complex and relatively difficult, and there were many overbreaks and voids at the contact between the rock mass and lining, it was necessary to improve generally the quality behind the lining. Grouting in two stages was therefore performed, i.e.contact grouting (at the contact between rock and lining) and consolidation grouting (in the zone of the rock mass up to about 3.0m from external surface of tunnel support). The paper presents two systems of grouting (hole and without-hole methods) as well as their results.

ZUSAMMENFASSUNG: Das Referat stellt die Beschreibung eines Injektionsverfahrens vor, das in den Hohlraumbauten des Pumpenspeicherkraftwerkes durchgeführt wurde. Dieses Obiekt war in polnischem Karpatenflysch in Porabka-Zar gebaut. Es gibt viele Hohlräume, die in dem ausgeführt wurden, wie z.B. Tunnels, Druckstollens, Kavernen, Wasserschloss, usw. Die waren mit der Stahlpanzerung oder mit dem Stahlbeton ausgekleidet. Die geologische Verhältnisse waren sehr komplizierte als auch verhältnismäßig schwere, weil das Felsmassiv ist mit dem Sandstein und Tonschiefer gebaut, der kann leicht löschen. Viele Ausbrüche und grössere Löcher entwickelte während des Vortriebes, und der Kontakt zwischen Ausbau und Felsmassiv nicht dicht war. Es war erforderlich diesen Kontakt als auch Felsmassivseigenschaften zu verbessern. Injektion in zwei Stuffen wurde zu diesem Zweck gemacht. Im ersten Stuff s.g. Kontaktinjektion (auf dem Kontakt zwischen Ausbau und Fels) und im zweiten Stuff s.g. Konsolidationsinjektion durchgeführt wurde (in der 3 m tiefe Zone des Felsmassivs). Das Referat beschreibt zwei Methoden der Injektion (die Injektion mit und ohne Löcher) als auch die Ergebnisse und seine Vergleichung. Es wurde gezeigt, daß die zweite Methode, in der die Injektionsrohren hinter dem Ausbau, parallel zur Längsachse der Tunnels installiert waren und der Zement durch diese Rohren gepumpt war, ist mehr effektiv und billig. Das Pumpenkraftwerk in Porabka-Zar ist schon seit ungefähr 20 Jahre im Betrieb und keine grössere Unfälle stattgefunden haben. Man kann also sagen, daß die Injektion ausreichend und fehlerfrei war.

1 INTRODUCTION

In time of construction of pumped-storage power plant in Porabka-Zar in Beskid Slaski Mountains grouting works were carried out in great extent. The scope was to fill caverns, overbreaks, voids and fissures in the rock mass behind the lining of the pressure tunnels and other underground openings.

Underground excavations were performed by blasting in sandstone--shale Carpathian Flysch (Thiel 1989). Rock mass was strongly jointed and fissured and shale was very sensitive to water and its slaking was often observed. Two basic flysch complexes are distinguished in the region (fig.1), i.e. Kg3 and Kg4 series. Their lithologic and petrographic properties differ significantly. The Kg3 series, in which almost all openings are located, consists of approximately the same volumes of sandstones and shales. The Kg4 series is built of thick-bedded sandstones (practically more than 90%) with thin shale interbeddings. Two zones of weaker rock, few meters wide, are located along the longitudinal side walls of the electro-station cavern as well as in tunnels in its vicinity. Second zone, about 100m wide is situated in the region of lower reservoir (Łanocha 1977).

In these conditions numerous overbreaks as well as irregularities developed in tunnel. Although blasting was carefully performed, these took place due to layered structure of the medium. It was very difficult to fill these spaces during installation of preliminary support and it was thus decided to liquidate them by grouting in later stage of the construction, i.e. after installation of final lining.

Grouting was performed using two techniques:

a) "hole method", i.e. by introducing grout mixture to the rock mass by the hole bored in the lining; this method was used first of all in slightly inclined tunnels,
b) "without-hole method", i.e.without making the holes in the lining; this method was used first of all in deeply inclined tunnels.

In both of cases the purpose of grouting was to liquidate water leakages to the openings, to stiffen the rock mass, to improve uniformity of loadings acting on the lining and to provide appopriate joint-work of the system, composed of lining and rock mass (by filling the gap at the rock - concrete interface).

System of underground openings in constructed object is shown in the fig 1. Grouting was performed in investigation drift 5, in concrete--lined section of tunnel 3 as well as in excavations 2, 4 and 6-8. Careful grouting was also carried out in foundation of turbo-generators and suction pipes.

Great variability of dimensions, objectives and constructional details of excavations, caused differences in grouting technique. Hole method was employed in most of cases (slightly inclined tunnels, shafts), and without-hole method in upper pressure tunnels as well as in foundation of turbo- generators.

2 HOLE METHOD

Example is described here, showing the grouting procedure in concrete--lined section of outlet tunnel and technical effects of grouting measures..

Tunnel cross-section and scheme of grouting is presented in fig 2. It must be mentioned that thicknes of reinforced concrete lining was determined with taking into account

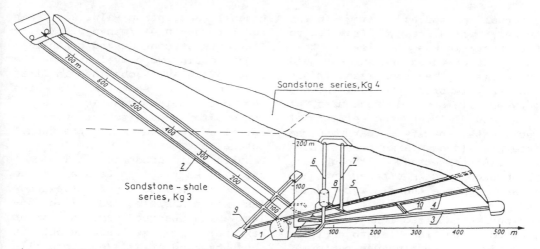

Fig. 1. Scheme of the pumped-storage power plant in Porabka-Zar
1 - power station cavern; 2 - upper pressure tunnels; 3 - outlet tunnel;
4 - service tunnel; 5 - investigation drift; 6 - surge tank; 7 - ventilation shaft; 8 - ventilation tunnel; 9 - cross-cut to upper pressure tunnels; 10 - cross-cut to the outlet tunnel

Abb. 1. Plan des Pumpenspeicherkraftwerkes in Porabka-Zar
1 - Kraftkaverne; 2 - Oberwasser Druckstollen; 3 - Unterwasser Druckstollen; 4 - Zugangstollen; 5 - Versuchstollen; 6 - Wasserschloß; 7 - Bewet-
terungschaft; 8 - Bewetterungtunnel; 9, 10 - Verbindungstunnel

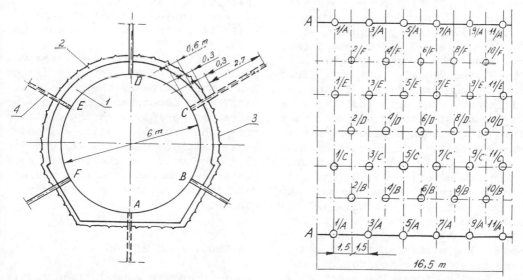

Fig. 2. Distribution of grouting holes in concrete-lined outlet tunnel
1 - permanent lining (reinforced concrete); 2 - preliminary support; 3 -
rock mass; 4 - boreholes

Abb. 2. Anordnung der Injektionsbohrlöcher im mit Stahlbeton ausgekleidet
unterwasser Stollen
1- Stahlbetonauskleidung; 2 - Verbau; 3 - Felsmassiv; 4 - Bohrlöcher

313

internal pressures acting in the tunnel, which change from 0.1 to 0.6 MPa. It was also assumed that the part of these pressures should be carried by the rock mass. It was thus very important to seal the gap between lining and rock mass, as this could lower the effects of rock mass capacity in carrying the internal loads. Some prestressing was also expected, although this phenomenon was not taken into account when the tunnels were designed.

Experimental section was first grouted with the length of 16.5m (see fig 2). Grouting was performed in this section in following layout:
- odd holes 1/A, 9/A, 5/A, 3/A, 7/A and analogically in rows C and E,
- even holes 2/B, 10/B, 6/B, 4/B, 8/B and analogically in row F,
- holes in row D (in roof), similarly as in row B.

This layout was changed in next grouting sections, i.e. holes in one ring (A-F) were bored first (in one stage) in order A, E, C or B, F, D and immediately grouted. Rings of holes were divided into first, second and third order rings.

Each individual hole was first bored, crossing final lining and preliminary support and reaching the depth of 0.3m in the rock mass (1-st zone). Water absorption test (Lugeon test) was carried out and depending on its results an extent of contact grouting was determined. Grout mixture was then pressed into the contact zone, until all the voids and fissures were filled. Then the hole was bored to the depth of 3 meters in the rock mass (2-nd zone) and consolidation grouting was performed in similar way to above described.

Water absorption tests and grouting works were carried out with maximum pressure equal to 0.4 MPa in 1-st and to 0.6 MPa in the second zone. As a criterion of tightness, absorption value was taken equal to $q <= 0.01 l/min*m*0.01MPa$.

Effects of contact and consolidation grouting were evaluated on the base of results obtained in first four grouting sections. In these sections seismoacoustic measurements before and after grouting were also carried out in the tunnel bottom zone.

Results are presented in Tables 1 and 2. As it can be seen in the Table 1, only some part of all holes was capable to grout (i.e. holes which absorbed water and grout mixture). In few cases grout was not absorbed at all (e.g. consolidation grouting in IV-th section) or absorption was relatively low (e.g. I-st section). It proves, that the initial contact between the lining and rock as well as surrounding rock mass zone were relatively tight and compact. However, few zones were met strongly decompressed and loosened.

Consumption of grout mixture was relatively high, i.e.:
1.4 t/m, in 1-st (contact) zone,
0.48 t/m, in 2-nd (consolidation) zone.

Consumption in the 1-st zone was about 3 times greater than in the 2-nd one. It means that the loosening and decompression phenomena decrease rapidly with the distance from the tunnel walls, although some part of mixture was used to fill the voids on the support/rock contact which resulted from not accurate performance of preliminary support.

3 WITHOUT-HOLE METHOD

Upper pressure tunnels (see fig 1) were supported with rockbolts and shotcrete during excavation stage. In weaker rock mass zones, steel arches were additionally installed, and concrete beams were placed as blocking between rock walls and ar-

Table 1. Consumption of grout mixture in concrete-lined section of outlet tunnel - hole method

Grouted section	Ring	Grouting order	Contact grouting % of effective holes	Grout consumpt.,kg min.	max.	total	Consolidation grouting % of effective holes	Grout consumpt.,kg min.	max.	total
I	1-11	---	30	17.5	8100	24113	30	1	5341	8909
II	12-23	---	69	10	7280	28421	48.5	16	1101	2948
III	24-33	I-st	100	12	5950	10167	100	30	1110	1990
		II-nd	100	26	1400	2497	50	200	6450	6950
		III-rd	60	3	3050	9753	53	150	750	2800
IV	34-44	I-st	77	6	3000	5800	22	1800	2100	3900
		II-nd	44	40	6900	7040	0	0	0	0
		III-rd	66	6	1325	4300	60	50	2400	4680
SUM						92091				32170

Table 2. Velocities of longitudinal seismic waves before and after grouting

Medium	Depth [m] from wall of lined tunnel (from-to)/aver.	Velocity of longitudinal seismic wave [m/s] (from-to)/average before grouting	after grouting
Concrete in permanent lining	(0.0-1.2)/1.0	(3400-3900)/3600	(3500-4000)/3700
Concrete in preliminary support + + voids	(0.8-2.5) / (1.0-1.8)	(2000-4200)/3700	(2000-4300)/3800
Voids in rock mass, 0.0-0.3m	(1.5-2.2) / (1.7-2.1)	(2000-3600)/2800	(3500-4200)/3900
Fissures and and secondary cracks in rock mass, 0.3-4/5m	(1.5-4.0) / (1.9-3.1)	(2000-4400)/3600	(3100-4700)/4000
	(2.3-5.0) / (2.7-4.4)	(3700-4500)/4000	(3800-5000)/4200

ches. Such kind of support did not provide close contact between rock and support and many empty spaces were observed. One supposed that it would be impossible to fill all these voids by hole grouting. It was thus initially decided to remove preliminary support before the installation of permanent lining

However, this process was practically very difficult and collapses appeared in few cases. The idea was thus revised. Instead of removing the support, modified without--hole grouting technique was proposed and employed (Łukaszek 1980).

Voids and fissures in the contact zone were filled with grout mixture

315

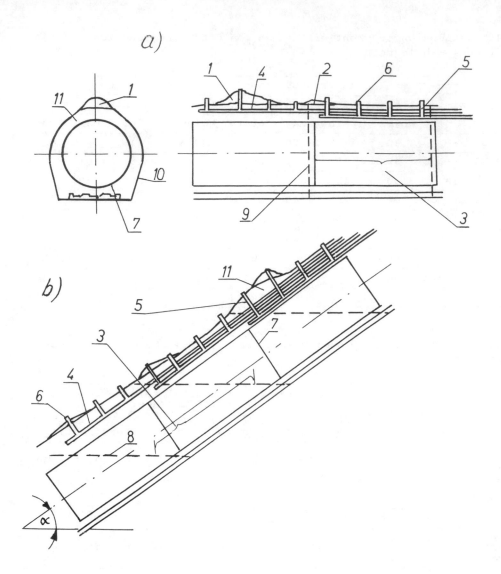

Fig. 3. Distribution of grouting conduits in pressure tunnels
a) horizontal section; b) inclined section; 1, 2 - voids, overbreaks; 3 -
grouting section; 4 - lateral pipes; 5 - injection radial pipes; 6 - end
of injection pipe wrapped with rubber hose; 7 - steel lining; 8 - level
of filling concrete; 9 - shuttering; 10 - rock mass wall; 11 - permanent
lining

Abb. 3. Anordnung der Injektionsrohren in Druckstollen
a) horizontal Segment; b) schräg Segment; 1, 2 - Löcher und Kavernen
zwischen Felsmassiv und Auskleidung; 3 - injeziert Sektion des Stollens;
4 - seitliche Rohren; 5 - radiale Injektionsrohren; 6 - Anschluß des Injektionsrohres mit Gummimuffe; 7 - Stahlpanzerung; 8 - Niveau des Betons
hinter der Panzerung; 9 - Schalung; 10 - Wand des Felsmassivs; 11 - Auskleidung

having C/W ratio equal to 1:1 - 1:0.75. Grouting was performed in sections (fig 3) and the mixture was pressed to the contact zone by injection pipes (5), placed in tunnel roof, side walls and bottom ,in places where voids and caverns were observed, and connected with pipes (4), set parallel to the longitudinal axis of the tunnel. Each end of injection pipe (5) was perforated and wrapped with rubber hose (6), operating as non-return valve. Each grouting section had separate installation. All the pipes were set between steel lining and preliminary support, before this space was filled with concrete.

After installation was arranged in the section, the space between steel shell and preliminary support was filled with concrete. In horizontal sections shuttering was made, preventing concrete outflows and concrete was then pumped to the space, whereas inclined sections were filled gravitationally. Grouting pipes and conduits were elongated upwards, successively with advance of the concrete works. Grouting in any section was carried out after few next sections were completed. Thanks to this procedure, concrete between steel shell and support composed reaction plug and it was thus possible to realize grouting with relatively high pressure.

The method was also successfully used to eliminate the gap between steel lining and concrete.

Filling degree in grouted zones was inspected by water absorption tests and the same tightness criterion was assumed as in case of hole method, i.e. $q \leq 0.01 l/min*m*0.01MPa$.

Consumption of grout mixture during the filling of caverns, overbreaks and voids was many times higher in comparison to the volume of the mixture used for filling the gap between shell and concrete.

4 FINAL REMARKS

The object described was successfully completed about 20 years ago. The tunnels operate until now without any greater defects. It proves appropriateness of grouting in every cases.

The choice of grouting technique (i.e. hole or without-hole method) should be done on the base of economic effects: the second method was about 6 times less expensive than the first and thus more effective.

There are many more or less sophisticated grouting systems (e.g. Seeber 1975). It seems however, that these described above are well designed for Carpathian Flysch rock mass.

REFERENCES

Łanocha, R. 1977. Underground hydroelectric constructions in Porabka-Zar pumped-storage power plant. Wiadomosci Gornicze 1:1-7 (in polish).
Łukaszek, R. 1980. Tightening and strengthening of soils and rocks by grouting. Proc.of Conference on Found. Eng.:263-281. Warsaw: Inst.of Meteorology and Water Management (in polish).
Łukaszek, R., Nosal M. Koehsling J. & Gutwiński L. The method of grouting of caverns and voids behind the lining of mining excavations and grouting apparatus. Patent No. 105659 (in polish).
Seeber, G. 1975. Neue Entwicklungen fuer Druckstollen und Druckschaechte. Oesterreichische Ingenieur- -Zeitschrift 5:140-149.
Thiel, K. 1989. Rock mechanics in hydroengineering. Warsaw-Amsterdam: PWN-Elsevier.

Injektionsmassnahmen an der Staumauer Vermunt
Grouting of the Vermunt Gravity Dam

E. Pürer
Vorarlberger Illwerke Aktiengesellschaft, Schruns, Österreich

ZUSAMMENFASSUNG: Die in den Jahren 1928 bis 1930 errichtete, 53 m hohe und etwa 400 m lange Gewichtsmauer Vermunt wurde im Zuge von Erhaltungsmaßnahmen in den Jahren 1987 bis 1989 am wasserseitigen Mauerfuß umfangreichen Dichtinjektionen unterzogen. Als Injektionsgut wurden Suspensionen aus Portlandzement PZ 475 bzw. hochfeinem Zement, unter Zusatz von Betonit und Verflüssigern, angewandt. Aufgrund der hohen Durchlässigkeit und Injektionsgutaufnahmen von flächenhaft ausgedehnten, nestrigen und porösen Zonen waren mehrere Injektionsdurchgänge mit PZ 475 und einem W/Z-Wert von 0,8 bis 2,0 erforderlich. Der endgültige Abdichtungserfolg konnte jedoch erst durch Injektionen mit hochfeinem Zement erreicht werden. Um die Standsicherheit der während der Bauarbeiten im Staubetrieb stehenden Sperre nicht zu gefährden, wurden Mengen- und Druckbegrenzungen vorgenommen und kontinuierliche Meßeinrichtungen installiert. Mit den durchgeführten Injektionsmaßnahmen und der vorgesehenen Überwachung konnte das Ziel einer sicheren Abdichtung des wasserseitigen Mauerfußes erreicht werden. Die Sickerwässer des gesamten Sperrenkörpers betragen derzeit 0,2 l/s und sind damit auf ein Dreißigstel des ursprünglichen Wertes reduziert worden.

SUMMARY: The Vermunt dam is a 53 m high and 400 m long gravity dam and has been constructed during the years 1928/30. The increase of seepage water made comprehensive remedial measures necessary. 60 cm of facing concrete were applicated on the upstream side and the upstream heel of the dam was sealed by grouting measures. Six chambers had to be excavated in the highest blocks of the dam in order to provide space for the drilling machine and the grouting equipment. Cement, ultrafine cement, bentonite and liquifiers from different suppliers were put to several qualifying tests. Due to the high permeability of the flatspread character of the zones of discontinuity, respectively the concrete nests, several grouting sequences with Portland cement PZ 475 and with a water cement ratio between 0,8 and 2,0 were made. A final and sufficient sealing effect was obtained by grouting with ultrafine cement. In order to avoid a reduction of stability of the dam during the grouting process special attention was paid to the observation of a limitation of pressure and the take of the grout. Therefore a special monitoring system – tiltmeters and jointmeters – was installed. Due to the grouting measures and the supervision system a safe sealing of the upstream heel of the dam has been obtained. The seepage water has been reduced from former 6 l/sec to 0,2 l/sec.

1 STAUMAUER VERMUNT

Die Staumauer Vermunt ist eine 53 m hohe und rund 400 m lange Gewichtsmauer, welche in den Jahren 1928 bis 1930 errichtet wurde. Die Art und Weise der seinerzeitigen Einbringung, Verteilung und Verdichtung des Betons hat, wie umfangreiche und wiederholt durchgeführte Untersuchungen zeigten, zu einem quasi schichtförmigen Aufbau des Sperrenkörpers geführt. Er ist durch einen Wechsel von Beton höchster Qualität und dazwischen-liegenden, flächenhaft ausgedehnten Nestern geringerer Festigkeit gekenn-zeichnet. Wiewohl dem Projektanten seinerzeit diese begrenzten Möglichkeiten der Betontechnologie bekannt waren und daher die Sperre auch mit einem Dichtputz versehen wurde, haben im Laufe der Jahrzehnte die Durchsickerungen des Sperrenkörpers selbst - vorzugsweise in den poröseren Zonen geringerer Festigkeit - zugenommen.

Aus diesem Grund wurde in den Jahren 1987 bis 1989 ein umfangreiches Erhal-tungs- und Erneuerungsprogramm durchge-führt. Dieses bestand im wesentlichen aus einer zuverlässigen Abdichtung der Wasserseite durch Errichtung eines Vorsatzbetons und durch Dichtinjektionen am wasserseitigen Mauerfuß, bei gleich-zeitiger Erhaltung der Durchlässigkeit des Betonkerns der Sperre. Die Erhal-tungsmaßnahmen wurden von G. Innerhofer und E. Pürer (1988) beschrieben. Im nachfolgenden wird auf das Projekt, die Injektionen und die Sperrenüberwachung während der Durchführung der Injektions-maßnahmen näher eingegangen.

2 PROJEKT DER INJEKTIONSMASSNAHMEN

Das Projekt sah vor, den wasserseitigen Mauerfuß, welcher durch den Vorsatzbeton allein nicht zuverlässig abgedichtet werden konnte, mittels Injektionen abzudichten, ohne den Speicher zu lange entleeren zu müssen.

Die Injektionen sollten daher größten-teils von der Sperre und zum geringeren Teil vom wasserseitigen Mauerfuß aus durchgeführt werden. Um die Injektionen von der Sperre aus durchführen zu können, wurden in den Mauerblöcken großer und mittlerer Höhe insgesamt 6 Bohr- und Injektionskammern mit einer Grundriß-fläche von 3,5 x 3,5 m und einer Höhe von 3,5 m mit kuppelförmiger Kalotte vorge-sehen.

Von diesen Injektionskammern und in geringem Ausmaß auch vom vorhandenen Kontrollgang sollten für mehrere Injek-tionsdurchgänge die Injektionsfächer gebohrt und die Injektionen (bei um 10 m abgesenktem Stauspiegel) durchgeführt werden.

Als Injektionsgut war normaler Port-landzement, hochfeiner Zement und Kunst-harz vorgesehen.

Die Eignungsprüfung für das Injektions-gut wurde in 3 Versuchsabschnitten durchgeführt:

Der erste Versuchsabschnitt diente zum Vergleich der Zemente verschiedener Lieferanten sowie der generellen Ermitt-lung des Einflusses der Bentonitzugabe auf das Injektionsgut.

Im zweiten Versuchsabschnitt wurden die Auswirkungen verschiedener Wasser-/Zement-werte, verschiedener Bentonitbeigaben und die Wirksamkeit von Zusatzmitteln (Ver-flüssiger) verschiedener Lieferanten miteinander verglichen.

Im dritten Versuchsabschnitt wurden die optimalen Beigaben von Bentonit und Verflüssiger für den hochfeinen Zement bestimmt.

Aufgrund dieser Versuche wurde folgende Zusammensetzung des Injektionsgutes festgelegt:

- Portlandzement PZ 475, Bentonitbeigabe 1,5 %, bezogen auf die Masse des Wassers, Verflüssiger 1 %, bezogen auf die Masse des Zements, W/Z = 1,5.
- Hochfeiner Zement mit einem Blainwert von 8800, Bentonitbeigabe 1 %, bezogen auf die Masse des Wassers, Verflüssiger 2 %, bezogen auf die Masse des Zements, W/Z = 2,0.

Für die Injektionen wurden Mengen- und Druckbegrenzungen festgelegt:

Aufgrund der Wasserabpreßversuche, bei welchen trotz hoher Wasseraufnahmen (Wasserverluste) kein nennenswerter Druck erreicht werden konnte, und aufgrund des Erscheinungsbildes des Betons beim Ausbruch der Injektionskammern wurde die Ausdehnung der flächenhaften, porösen Zonen geringerer Festigkeit und die damit zusammenhängende, mögliche Injektionsgut-aufnahme vorweg abgeschätzt. Über stati-sche Sicherheitsbetrachtungen wurden dann die zulässigen Drücke in Abhängigkeit von der Injektionsgutaufnahme festgelegt. Bei druckloser Aufnahme wurde die Menge mit 500 l pro Passe festgelegt. Bei einem Injektionsdruck von 5 bar durften bis zu maximal 500 l pro Passe injiziert werden und bei Injektionsaufnahmen bis zu 200 l pro Passe wurde der Injektionsdruck mit 7 bar begrenzt. Dazu wurde bei der Injek-tion in den hohen Mauerblöcken die Absenkung des Wasserspiegels um 10 m, bei der Injektion der Mauerblöcke mittlerer

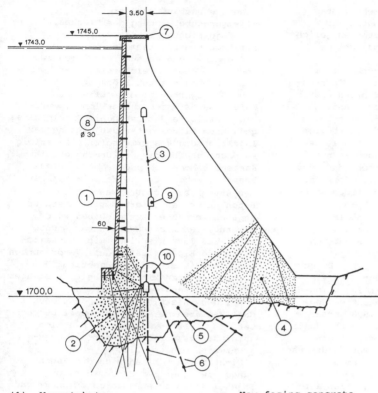

(1)	Vorsatzbeton	– New facing concrete
(2)	Dichtinjektion	– Sealing grouting
(3)	Drainagebohrung	– Relief boring
(4)	Injektion ohne Dichtanspruch	– Grouting without sealing
(5)	Nicht injizierter Bereich	– Area not grouted
(6)	Entlastungsbohrung	– Relief boring
(7)	Kronenplatten neu	– New crest slab
(8)	Dübel Ø 30 mm	– Anchor bar Ø 30 mm
(9)	Kontrollgang	– Gallery
(10)	Bohr- und Injektionsraum	– Grouting chamber

Höhe um 7 m unter Stauziel festgesetzt. Neben der Überwachung der Injektionsdrücke und der Injektionsgutaufnahme wurde die Sperre einer laufenden Überwachung unterzogen, welche in Punkt 4 beschrieben wird.

3 INJEKTIONEN DES WASSERSEITIGEN MAUERFUSSES

Der Großteil der Dichtinjektionen am wasserseitigen Mauerfuß wurde 1987 durchgeführt.

Wie aufgrund der Wasserabpreßversuche zu vermuten war, gestalteten sich die ersten Injektionen im Beton des wasserseitigen Mauerfußes schwierig. Häufig wurden, ohne daß nennenswerte Injektionsdrücke erreicht worden wären, schon vor der Mengenbegrenzung Übertritte in benachbarte Bohrlöcher, aber auch der Austritt von Injektionsgut in die Bohr- und Injektionskammern längs der porösen Schichten festgestellt. Fallweise wurde auch festgestellt, daß bei vorgängig ausinjizierten Bohrlöchern Injektionsgutaustritte stattfanden. Um überhaupt einen sinnvollen Fortschritt zu erzielen, mußten zwei Bohr- und Injektionskammern mit Spritzbeton wasserseitig abgedichtet werden. Weiters mußte der W/Z-Wert der Injektionen bis auf 0,8 herabgesetzt werden, um überhaupt einen ersten Erfolg zu erzielen.

Aufgrund dieser Schwierigkeiten und der Notwendigkeit, Injektionsgut mit niedrigeren W/Z-Werten zu verwenden, war es nicht möglich, mit dem ursprünglich projektierten Injektionsschema das

Auslangen zu finden. Vielmehr mußten Zwischenschirme und horizontal zur Wasserseite bzw. sogar nach oben gerichtete, zusätzliche Injektionsschirme angeordnet werden.

Die Injektionen wurden zunächst mit Injektionsgut auf Basis PZ 475 durchgeführt. Die danach durchgeführten ersten Wasserabpreßversuche zeigten dann, daß mit der normalen Zementinjektion bei Einhaltung der Mengen- und Druckbegrenzung keine befriedigende Dichtheit des Sperrenkörpers erzielt werden konnte. Die entnommenen Bohrkerne zeigten, daß zwar ein Großteil, aber nicht alle Poren mit Injektionsgut verschlossen werden konnten.

Aus diesem Grunde wurden weitere Injektionsdurchgänge mit hochfeinem Zement nachgeschaltet. Zur leichteren Überprüfung der Injektionserfolge wurde der hochfeine Zement eingefärbt. Durch dieses Einfärben konnte dann auch vielfach die Wirksamkeit dieser Injektionen an Bohrkernen sichtbar gemacht werden.

Im darauffolgenden Jahr 1988 wurden dann bei entleertem Speicher noch zusätzliche Injektionen vom wasserseitigen Mauerfuß zuerst wieder mit Portlandzement 475 und dann mit hochfeinem Zement durchgeführt. Im allgemeinen konnte mit vergleichsweise geringem Injektionsaufwand die gewünschte Dichtheit des Betons am wasserseitigen Mauerfuß erreicht werden. Vereinzelt mußten jedoch Nachinjektionen mit hochfeinem Zement durchgeführt werden. Die Injektion wurde als erfolgreich abgeschlossen betrachtet, wenn pro 5-m-Passe bei einem Abpreßdruck von 3 bar und einer Haltezeit von 5 min der Wasserverlust nicht mehr als 1 l/min betrug (entspricht o,67 Lugeon). Vielfach wurden überhaupt keine Wasserverluste gemessen.

Insgesamt wurden am wasserseitigen Mauerfuß 110 t PZ 475 und 30 t hochfeiner Zement injiziert.

Am luftseitigen Mauerfuß wurden zur Verbesserung der Festigkeit des Betons schematisch Injektionen mit PZ 475 durchgeführt. Die Erzielung einer Dichtheit war hier nicht erwünscht.

4 ÜBERWACHUNG DER SPERRE WÄHREND DER INJEKTIONEN

Da der Großteil der Injektionen im Sperrenkörper bei eingeschränktem Staubetrieb erfolgen mußte, sind neben der Absenkung des Speicherspiegels um 10 m bzw. 7 m unter Stauziel folgende, kontinuierlich arbeitende Überwachungssysteme eigens eingebaut worden:
- Neigungsüberwachung des Blockes, in dem die Injektionen stattfanden, mittels elektronischer Klinometer
- Fugenbewegungsmessung der Fugen zwischen dem Block, welcher injiziert wurde und den benachbarten Blöcken.

Um Grenzwerte für die zulässigen Neigungsänderungen zu bestimmen, wurde die elastische Verdrehung im Umfeld eines gedachten Risses unter der Wirkung des Injektionsdruckes ermittelt. Dies erfolgte nach der Theorie der Bruchmechanik – Kantenriß Mode 1 – und nach der Biegetheorie, wobei für die Materialkennwerte des Betons eine größte Bandbreite von Betonkennwerten aufgrund von Literaturangaben verwendet wurde. Dabei wurde erkannt, daß sich sowohl unter Ansatz hoher wie auch niedriger Rißzähigkeitswerte und Elastizitätsmoduli Verdrehungen ergaben, die mit den eingesetzten Neigungsmeßgeräten mit Sicherheit beobachtet werden konnten.

Bei den Injektionen wurde dann deutlich das Reagieren der Neigungsmesser beobachtet. In 3 Fällen wurde die Injektion aufgrund des Erreichens der Neigungsgrenzwerte eingestellt.

Nicht bewährt haben sich die Fugenmessungen zwischen dem Block, der den Injektionsmaßnahmen unterzogen wurde und den benachbarten Blöcken. Wiewohl im injizierten Block deutlich Neigungsänderungen festgestellt wurden, konnte kein Fugenspiel zwischen den "unabhängig gedachten Blöcken" beobachtet werden. Dies wird darauf zurückgeführt, daß die Gewichtsmauer, solange sie im geschlossenen Verband steht, praktisch wie ein Monolith wirkt.

Nach Abschluß der Injektionsarbeiten wurden neue Entlastungsbohrungen und Sohlwasserdruckmeßstellen eingerichtet. Die Summe der gesamten Sickerwässer beträgt derzeit 0,2 l/s und ist damit auf ein Dreißigstel des Wertes vor den Sanierungsmaßnahmen zurückgegangen.

REFERENCES

Innerhofer, G. & E. Pürer 1988. Ageing and remidial measures for the sixty years old Vermunt dam. Q. 65, 17th ICOLD-Congress, Vienna.
Stefko, E. & G. Innerhofer 1967. Condition of Vermunt dam after more than 30 years of operation. Q. 34, 9th ICOLD-Congress, Istanbul.

Kraftschlüssige Rißinjektion an der Gewölbemauer Zillergründl
Force-locking crack injection at Zillergründl Arch Dam

P. Schöberl
Tauernkraftwerke Aktiengesellschaft, Salzburg, Österreich

H. Huber
Tauernplan Prüf- und Meßtechnik GmbH, Strass, Österreich

H. Döpper
Tauernplan Consulting GmbH, Salzburg, Österreich

ZUSAMMENFASSUNG: Ein lokal begrenzter Riß im wasserseitigen Aufstandsbereich der Sperre wurde mittels einer kraftschlüssigen Injektion mit Kunstharz bei mittlerer Stauhöhe und anschließender Behandlung bei leerem Becken erfolgreich abgedichtet. Es wird über die Planung und Ausführung der Arbeiten und die Bewältigung von dabei zutagegetretenen, für die Beteiligten in diesem Ausmaß neuen Problemen berichtet.

ABSTRACT: In the 186 m high Zillergründl dam, which was completed in 1985, a crack in one of the dam blocks occurred at a storage level of approx. 175 m in autumn 1987. The crack running more or less horizontally extended from the upstream side of the dam as far as the inspection gallery at base level, it covered an area of approx. 350 m² and had a maximum opening width of 1 mm. Via a measuring chamber situated upstream of the inspection gallery at base level and via the elevator shaft built into this block water entered at a rate of 160 l/s.

The planned crack-sanitation works essentially consisted of force-locking injection of the crack with sythetic resin carried out in two stages according to the sanitation concept developed by the company Insond.

After the injection method had been tested in two test fields, the crack area was opened up with grout holes in a bore screen of 2 m x 2 m from the galleries in the dam. The primary injection of the crack was carried out at a storage level of approx. 120 m. The two component epoxy resin Rodur was used as grout. In the process high grouting pressures of up to approx. 200 bar were applied with crack widenings of 0,4 mm being measured.

After the primary injection, a secondary injection of the crack was carried out through additional grout holes with the reservoir being empty. In this process 4 additional vertical cracks occurred due to the high grouting pressure applied (up to approx. 170 bar). These cracks were also caused by simultaneous horizontal tensile stresses on the upstream side of the dam blocks in dead load. These tensile stresses were increased even further due to the low temperatures prevailing at the time.

Further injections were continued at a pressure between 60 to 80 bar. A total of approx. 20 t of Rodur was grouted. Inspection borings showed that force-locking grouting covering the entire area of the main crack had been achieved. The block joints adjacent to the crack were also partially grouted. The crack traces on the upstream side of the dam were sealed off with plastic foils. In addition to that a series of measures inside the dam were taken so as to improve the stress conditions at the upstream dam toe.

After providing a number of drainage holes in the crack and after carrying out a subsequent grouting in the year 1989, maximum water level in the Zillergründl reservoir could be reached without further problems in the years 1990 and 1992.

1 BESCHREIBUNG DER ANLAGE

Die 186 m hohe Bogenmauer Zillergründl wurde in den Jahren 1979 bis 1985 durch die Tauernkraftwerke Aktiengesellschaft errichtet. Die Betonsperre besteht aus 26 Blöcken zu je ca. 20 m Breite. Die größte Dicke der Mauer beträgt rd. 42 m an der Felsaufstandsfläche. Unmittelbar auf der Felsoberfläche wurde im wasserseitigen Drittel ein Sohlgang ausgebildet. Die Sperre weist darüber hinaus ein neuartiges Abdichtungskonzept sowie einige konstruktive Besonderheiten auf, die im folgenden kurz beschrieben werden (Abb. 1).

Am wasserseitigen Sperrenfuß ist der soge-

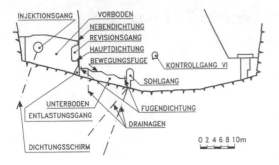

INJEKTIONSGANG
VORBODEN
NEBENDICHTUNG
REVISIONSGANG
HAUPTDICHTUNG
BEWEGUNGSFUGE
KONTROLLGANG VI
SOHLGANG
UNTERBODEN
ENTLASTUNGSGANG
FUGENDICHTUNG
DRAINAGEN
DICHTUNGSSCHIRM
0 2 4 6 8 10m

Abb. 1: Widerlager - Regelquerschnitt
Fig. 1: Abutment - typical cross section

nannte Vorboden angeordnet, das ist eine
massive Betonplatte, die 3 Gänge aufweist.
Oberwasserseitig liegt der Injektionsgang,
von dem aus der Injektionsschirm zur Untergrunddichtung hergestellt wurde. An der
Bewegungsfuge zur Sperrenwasserseite hin
befindet sich der Revisionsgang, der die
Hauptdichtung enthält und im normalen
Staubetrieb geflutet ist. Die Hauptdichtung dichtet die Bewegungsfuge zwischen
dem an sich in Ruhe befindlichen Vorboden
und der unter Wasserlast bis zu 10 mm talwärts verschobenen Sperre gegen einen Wasserdruck von rd. 185 m ab. Unterhalb davon
befindet sich der sogenannte Entlastungsgang, der im Bereich der Talmitte aus sta-
tisch-konstruktiven Gründen als nach Fertigstellung der Sperre nicht mehr zugäng-
licher Hohlraum mit Trapezquerschnitt ausgebildet wurde. Zweck dieser Vorbodenkon-
struktion war es, den Dichtungsschirm so
weit oberwasserseitig der Sperre zu situ-
ieren, daß dieser durch vom Sperrenkörper
verursachte Untergrundverformungen nicht
abreißen würde. Die wasserdichte Anbindung
an die Sperre machte die Vorbodenplatte
und die Hauptdichtung notwendig.
Die zweite wichtige Besonderheit der
Staumauer ist die Fuge Unterboden-Sperre.

Wie dem Regelschnitt zu entnehmen ist,
wurde im Bereich der Talmitte zwischen der
Wasserseite der Sperre und dem Sohlgang
diese weitere Bewegungsfuge angeordnet.
Sie verläuft ca. 2 m oberhalb der Felsaufstandsfläche parallel zu dieser in Trep-
penform im Beton und trennt somit den sogenannten Unterboden vom übrigen Sperrenbeton. Diese Fuge, die unterhalb der
Hauptdichtung verläuft, wurde ausgebildet,
um allfällige Zugspannungen und das Auftreten von Rissen am wasserseitigen Sperrenfuß zu verhindern.
Die erste Teilfüllung des Speichers erfolgte erst nach vollständiger Betonierung
des Sperrenkörpers im Jahr 1986 bis Kote
1820 m ü.A. Für 1987 war der Aufstau bis
Kote 1840 m ü.A. geplant, das ist 10 m unter dem Stauziel von Kote 1850 m ü.A.

2 RISSEREIGNIS

Am Abend des 28.9.1987 wurde von der automatischen Bauwerksüberwachungsanlage zunächst eine Überflutung des Liftschachtes
im Sperrenblock 10 und wenige Minuten später eine Überschreitung des Grenzwertes
für die Sickerwassermenge aus dem rechten
Teil des Sohlganges gemeldet. Zu diesem
Zeitpunkt, um ca. 20.34 Uhr lag der Stauspiegel auf Kote 1838,8 m ü.A., die Stauhöhe betrug somit rd. 175 m. Die übrigen
fernübertragenen Meßwerte zeigten keine
abnormalen Veränderungen.
Neben dem Liftschacht, der vom untersten
bis zum obersten horizontalen Kontrollgang
reicht, beinhaltet der Block 10 als einer
der wesentlichen Meß- und höchsten Sper-
renblöcke eine sowohl nahe der Felsaufstandsfläche als auch nahe der Wasserseite
liegende Meßnische (Abb. 2, 3).
Bei der sofort veranlaßten Begehung durch
das Fachpersonal der TKW wurde ein auf den
Block 10 beschränkter Wassereintritt in
einer Größenordnung von rd. 160 l/s fest-

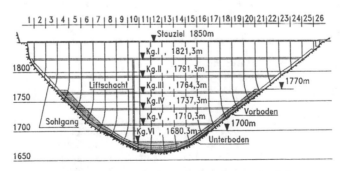

Abb. 2: Längenschnitt, luftseitige Ansicht
Fig. 2: Longitudinal section, downstream
       view

324

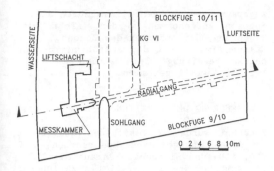

Abb. 3: Bl. 10, Horizontalschnitt 1680 m
Fig. 3: Bl. 10, horizontal section 1680 m

gestellt. Vor Eintritt des Schadens hatte die Gesamtsickerwassermenge rd. 17 l/s betragen. Der Wassereintritt erfolgte über einen zur Luftseite geneigten Riß, der in die Meßkammer, den Zugang zum Lift sowie den Liftschacht mündete (Abb. 4). Im Sohlgang waren lediglich nässende Haarrisse an der wasserseitigen Wand gegen den Block 9 hin zu erkennen.

Abb. 4: Wassereintritt im Zugang zur Meß-
        kammer
Fig. 4: Seepage in the access to the
        measuring chamber

3 SOFORTMASSNAHMEN

Als erstes wurde eine Gesamtmessung aller vorhandenen Meßeinrichtungen und gleichzeitig eine optische Rißkontrolle in sämtlichen Gängen der Sperre durchgeführt. Die Auswertung der Meßergebnisse zeigte lediglich in der unmittelbaren Umgebung des gerissenen Bereiches Veränderungen. Alle übrigen Meßergebnisse sowie die Rißinspektion zeigten ein normales Verhalten der Sperre an. Trotzdem wurde vorsorglich eine Absenkung des Stauspiegels im Speicher um zunächst rd. 14 m durch einen verstärkten Einsatz der Maschinen des Kraftwerkes Häusling eingeleitet. Diese nach 4 Tagen erreichte Spiegelabsenkung bewirkte eine Verringerung der Wasserdurchtritte um mehr als 50 %.

Als nächstes wurde eine Reihe von Sondierbohrungen festgelegt, um Ausmaß, Lage und Verlauf des Risses zu bestimmen. Mit der Ausführung dieser Bohrungen wurde bereits am 30.9. begonnen. Am 1.10. fand eine Begehung und Besprechung mit den Sachverständigen der Wasserrechtsbehörde statt, als deren wesentliches Ergebnis festzuhalten ist, daß die Gesamtstandsicherheit der Sperre zweifelsfrei gegeben ist. Neben der Klärung der Frage nach der Rißursache war umgehend ein Sanierungskonzept für den gerissenen Bereich zu erstellen, auf das im folgenden näher eingegangen werden soll.

4 RISS-SANIERUNGSKONZEPT

Allen Beteiligten war klar, daß der Riß möglichst kraftschlüssig zu verkleben, also zu injizieren ist, ein Vorhaben also, das bei Betonsperren ähnlicher Dimension bereits erfolgreich durchgeführt wurde (Berchten 1985, Wuhrmann 1986). Viele Fragen, etwa nach dem Injektionsgut, dem Verfahren und insbesonders den Randbedingungen, wie etwa Höhe des Speicherstauspiegels etc. waren zu lösen. Darüber hinaus ergaben die Sondierbohrungen sowie die anläßlich des Rißereignisses gemachten Beobachtungen ein insoferne kompliziertes Bild, daß der Riß in tangentialer Richtung gesehen an der Wasserseite des Blockes relativ horizontal verlaufen dürfte, somit an der Blockfuge 10/11 im Stauraum beginnend, in den Revisionsgang verlaufend, die Hauptdichtung querend bis zur Fuge Unterboden-Sperre im Entlastungsgang bzw. bis zur Blockfuge 9/10 reicht. In radialer Richtung ergaben die Sondierbohrungen eine Rißerstreckung ungefähr bis zur Ebene des Sohlganges. Die Rißfläche betrug somit rd. 350 m², die Öffnungsweite an der Wasserseite rd. 1 mm (Abb. 5).

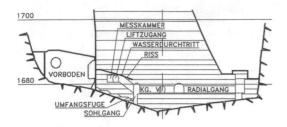

Abb. 5: Block 10, Radialschnitt
Fig. 5: Block 10, radial section

Damit wurde klar, daß für die Ausführung einer soliden Sanierung der Speicher entleert und der Riß auch unmittelbar von der Wasserseite her behandelt werden mußte. Damit war der Terminplan vorgegeben: die Sanierungsarbeiten, zumindest jene an der Wasserseite, mußten im Frühjahr 1988 abgeschlossen sein. Darüber hinaus mußte selbstverständlich für den Aufstau im Jahr 1988 der Speicher - und somit natürlich auch die Sperre - zumindest teilweise wieder zur Verfügung stehen.

Zur Ausarbeitung eines Injektionsprojektes wurden Konstruktionsaufträge an 3 namhafte Bohr- bzw. Injektionsfirmen (Sonderbau, Consonda und Insond) erteilt. Der von Insond ausgearbeitete Sanierungsvorschlag war mit Abstand am ausgereiftesten. Darüber hinaus konnte Insond über die Rodio-Gruppe Referenzen über Rißinjektionen in Betonsperren vorweisen und bot eine Gewährleistung für die kraftschlüssige Verbindung des Rißbereiches.

Das technische Konzept von Insond sah im wesentlichen folgende Ausführungsschritte vor:
- Ausführung von 2 Testinjektionsfeldern;
- Installation von 5 Extensometern;
- primäre Rißinjektion mit "Rodur", einem 2-Komponenten-Epoxydharz von Rodio bei einer Stauhöhe von 90 bis 120 m, die bis Ende 1987 abgeschlossen sein sollte;
- Sekundärinjektion während der Entleerungsphase bzw. in der Aufstauperiode 1988;
- Verklebung der oberwasserseitigen Rißspur, Blockfugeninjektion sowie Verfüllung der im untersten Teil des Blockes 10 vorhandenen Hohlräume mit Beton während der Entleerungsphase im Winter;
- Drainage- bzw. Kontrollbohrungen.

Das Ziel aller Maßnahmen war die Wiederherstellung des Sperrenblockes durch die Injektion von Rodur entsprechender Viskosität mit einer Haftfestigkeit auf nassem Beton, die größer ist als die Zugfestigkeit des Betons bzw. 2 N/mm². Allfällige Zugspannungen sollten allein durch die Injektion aufgenommen werden, nicht durch Bewehrung oder Ankerung des Betons.

Insond wurde mit der Ausführung der Arbeiten beauftragt, sie wurden am 17.11. 1987 in Angriff genommen.

5 AUSFÜHRUNG DER ARBEITEN

Zunächst wurden 2 Testfelder ausgeführt, in denen verschiedene Rodursorten (510, 520, 600) und Mischungen daraus mit Injektionsdrücken im Extremfall bis zu 180 bar, im Regelfall zwischen 50 bis 100 bar bei unterschiedlichen Bohrlochabständen verpreßt wurden (Abb. 6). Um einen besseren Kontrast des an sich betongrauen Rodur im Hinblick auf die Bohrkernentnahme zu erreichen, wurde das verwendete Rodur rot gefärbt. Die wichtigsten Materialeigenschaften von Rodur sind:

	Rodur 510	Rodur 520
Dichte/20°C (g/cm³):		
Komponente A	≈ 2,25	≈ 2,17
Komponente B	≈ 1,02	≈ 1,35
Komponenten A+B	≈ 2,00	≈ 1,82
Viskosität/20°C (mPa.s):		
Komponente A	22 000	62 000
Komponente B	60	pastös
Komponenten A+B	2 400	hoch thixotrop
Potlife/20°C/1 kg:	≈ 50 Min.	≈ 50 Min.
Haftzugfestigkeit/20°C (N/mm²):		
auf nassem Beton	2,5 - 3,5	2,5 - 3,5
auf trockenem Beton	> 3	> 3

Die Stauhöhe im Speicher lag dabei auf Kote 1802 m ü.A., d.h., der Wasserdruck im Riß betrug rd. 120 m. Erste Bohrproben wurden bereits 3 Tage nach Beendigung der Injektion entnommen, sie zeigten eine vollflächige Rißverfüllung im Bereich beider Testfelder. Die tatsächliche Rißweite lag zwischen 0,5 und 1,9 mm, auch ein feiner Nebenriß mit einer Weite von 0,2 mm war verfüllt. 2 Wochen nach der Injektion wurden je Testfeld 2 Bohrkerne ø 83 mm entnommen. Die in der Materialversuchsanstalt Strass durchgeführten Zugversuche ergaben Festigkeiten von 1,1 bis 1,6 N/mm². Bei allen 4 Proben trat der Bruch im Beton auf. Somit war die Kraftschlüssigkeit der Rodurinjektion und die Ausführbarkeit des Injektionsprojektes nachgewiesen. Nun konnten die eigentlichen Sanierungsarbeiten beginnen.

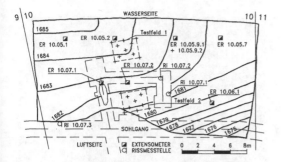

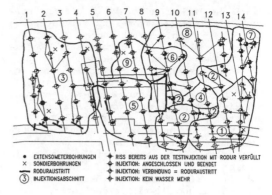

Abb. 6: Draufsicht Rißfläche mit Test-
 feldern und Extensometern
Fig. 6: Section through crack with test
 fields and extensometers

5.1 Primärbohrungen

Über die Rißfläche wurde zunächst ein
Bohrlochraster mit einem Achsmaß von 2 x
2 m gelegt. Die Rasterachsen verliefen einerseits parallel zum radialen Kontrollgang (Primärschirmachsen 0 bis 15), andererseits parallel zur Wasserseite der
Sperre (Abb. 7). Die Schirme 1 bis 4 wurden vom Sohlgang bzw. der Meßkammer aus,
die Schirme 5 bis 8 vom Zugangsstollen zum
Lift aus, die Schirme 9 bis 14 von den
Kontrollgängen VI bzw V aus gebohrt. Insgesamt wurden 106 Injektionsbohrungen (⌀
46 mm, insges. 1228 lfm, größte Einzel-
länge rd. 35 m) und 5 Extensometerbohrun-
gen (⌀ 56 mm, insges. 99 lfm) hergestellt.
Sämtliche Bohrungen wurden als Rotations-
kernbohrungen mit durchgehendem Kerngewinn
ausgeführt. Dazu standen 2 hydraulische
Bohrgeräte vom Typ "Diamec 250" zur Verfügung.
 Die wasserseitigste Bohrlochreihe wurde
mit einem Mindestabstand von 2 m zur Sperrenwasserseite konstruiert. Das Durchörtern des Risses, dessen Lage nur aus einigen Erkundungsbohrungen ungefähr bekannt
war, konnte durch den verstärkten Wasserandrang relativ genau festgestellt werden. Sicherheitshalber wurde der Riß im
Normalfall um rd. 1 m überbohrt. Weiters
wurde getrachtet, die Rißfläche mit den
Bohrungen möglichst normal zu durchörtern.
 Die gewonnenen Bohrkerne wurden einer
eingehenden Kernbeschau unterzogen. Sämtliche, aus den Primärbohrungen gewonnenen
Informationen hinsichtlich der Beschreibung des Risses lassen sich wie folgt zu-
sammenfassen. Der Riß ist von seiner Ausdehnung her auf den Bereich zwischen Wasserseite der Sperre und Sohlgang beschränkt. Es handelt sich im wesentlichen
um einen Einzelriß, der, im Radialschnitt
gesehen, an der Wasserseite relativ flach,

Abb. 7: Primärbohrraster und Ergebnis der
 Primärinjektion
Fig. 7: Primary bore screen and result of
 primary injection

also horizontal beginnt, gegen die Ebene
des Sohlganges hin fällt und im luftsei-
tigsten Abschnitt etwas versteilt.

5.2 Primärinjektion

Sämtliche Bohrungen wurden mit Injektions-
packern ausgerüstet. In Bohrungen bis 12 m
Länge wurde der Setzkolben in einer Tiefe
von rd. 0,5 m, also sehr nahe dem Bohrlochmund plaziert. In längeren Bohrungen
erfolgte der Setzkolbeneinbau auf einer
Tiefe von rd. 5 m oberhalb des Risses. Die
wasserseitigste und luftseitigste Bohrlochreihe sollte nur zur Beobachtung der
Injektionsgutausbreitung dienen und nicht
verpreßt werden. An der Wasserseite bestand einerseits die Gefahr der Beschädigung der Hauptdichtung, andererseits sollten unkontrollierbare Injektionsgutaustritte in den Speicher vermieden werden.
An der Luftseite bestand die Gefahr der
Rißausbreitung infolge der Injektion. Vom
generellen Injektionsvorgang her war geplant, in radialer Richtung die Injektion
von der Rißwurzel in der Ebene des Sohlganges gegen die Wasserseite der Sperre
hin, in tangentialer Richtung von der
Blockfuge 10/11 gegen die Fuge 9/10 hin
voranzutreiben. Im Detail richtete sich
der Injektionsverlauf nach den sich einstellenden Verbindungen und Übertritten.
Die im unmittelbaren Nahbereich der beiden
Blockfugen befindlichen Schirmachsen 0 und
15 waren zur Abdichtung des Risses gegen
die Blockfugen hin gedacht und sollten
vorrangig behandelt werden. Als Injektionsgut war eine Mischung aus Rodur 510/520
vorgesehen. Zu Beginn der Injektion waren

sämtliche Packer geöffnet, um die Ausbreitung des Injektionsgutes an Veränderungen der Schüttungen der Bohrlöcher sofort erkennen zu können. Die anzuwendenden Injektionsdrücke wurden mit Werten von 50 - 100 bar angeschätzt und waren nach oben eigentlich nur durch die Leistungsfähigkeit der Pumpe, einer Graco-Kolbenpumpe, die einen Maximaldruck von rd. 220 bar erbrachte, beschränkt. Bei der Primärinjektion wurde wegen der zu erwartenden großen Aufnahmen auf eine Einfärbung des Rodur verzichtet. Die sichtbaren Rißränder im Bereich der Meßkammer, des Zugangsstollens und des Liftschachtes wurden vor Beginn der Injektion teilweise abgedichtet.

Die Rißaufweitungen sollten über bis zu 30 m lange Einfachstangenextensometer, die mit Meßuhren bestückt wurden, gemessen werden. Als Grenze für die Ein- bzw. Umstellung der Injektion wurde eine Rißaufweitung von 0,2 mm festgelegt. Der Injektionsdruck war an einem Manometer an der Injektionsstelle abzulesen. Die Protokollierung des Injektionsdruckes sowie des Injektionsgutverbrauches erfolgte manuell. Diese Parameter, sowie die Zeit, besondere Beobachtungen und die Extensometermeßergebnisse wurden in handgeschriebenen Injektionsprotokollen festgehalten.

Die Primärinjektion wurde im Zeitraum vom 14. bis 18.12.1987 durchgeführt. Der Stauspiegel im Speicher wurde ziemlich konstant auf Kote ca. 1800 m ü.A. gehalten. Bezogen auf die Rißfläche betrug der mittlere Staudruck somit rd. 120 m Wassersäule. Wie geplant wurde als Injektionsgut hauptsächlich eine Mischung aus Rodur 510 und 520 im Verhältnis 1:1 verwendet. Die maximalen Injektionsdrücke lagen zwischen 30 bar im Minimum und 220 bar im Maximum. Bei rd. einem Drittel der angeschlossenen Bohrlöcher (25 von 72) lag der angewendete Maximaldruck über 100 bar. An dieser Stelle ist festzuhalten, daß der Sperrenbeton von der Endfestigkeit her etwa einem B 300 entspricht. Die erst nach Beendigung der Injektionsarbeiten publizierte Regel, wonach der maximale Injektionsdruck (in bar) kleiner als ein Drittel der Betongüte (in kp/cm²) sein sollte, wurde somit in etlichen Fällen mißachtet (Asendorf, 1988).

Von den 91 zu injizierenden Bohrlöchern wurden 70 aktiv verpreßt, in 21 Bohrlöcher konnten Injektionsgutübertritte erreicht werden. Die festgestellten Übertritte und Austritte sowie der ungefähre zeitliche Ablauf der Injektion (Bereiche 1 bis 9) sind den Abb. 7,8 zu entnehmen. Der gesamte Injektionsgutverbrauch betrug 5600 kg, das ist ein Volumen von rd. 2950 l. Das Bohrlochvolumen betrug rd. 2000 l, sodaß abzüglich diverser Austritte und Mani-

Abb. 8: Injektionsgutaustritt im Liftschacht

Fig. 8: Resin seepage from elevator shaft

pulationsverluste rd. 900 l für das Auffüllen des Rißvolumens verblieben. Bei einer Rißfläche von rd. 350 m² errechnet sich somit die mittlere Injektionsgutstärke im Riß zu rd. 2,5 mm. Die mittels der Extensometer gemessene Rißaufweitung betrug jedoch nur maximal 0,4 mm, sodaß man auf Bohrkerne mit der tatsächlichen Rißstärke gespannt sein durfte. Weiters zeigte sich, daß knapp talseitig des Liftschachtes ein feiner Vertikalriß in radialer Richtung vorhanden sein dürfte. Dieser Riß erstreckte sich vom Sohlgang, wo er aufgrund eines Injektionsgutaustrittes sichtbar wurde, wasserseitig in zunächst unbekanntem Ausmaß. Abschließend sei zur Primärinjektion erwähnt, daß sie eine Verringerung der Sickerwassermenge von zuvor 20 l/s auf 9 l/s bewirkte.

5.3 Sekundärbohrungen

Unmittelbar nach Beendigung der Primärinjektion wurde die Absenkung des Speichers fortgesetzt und die weitere Vorgangsweise bestimmt. Es wurden 13 Sekundärschirme parallel zwischen den Primärschirmen festgelegt. Parallel zur Wasserseite der Sperre, in einem Bereich also, wo eine noch unvollständige Rißverfüllung angenommen werden mußte, wurden 3 Bohrlochreihen (0, 1, 2) ausgeführt, deren Durchstoßpunkte mit der Rißebene in Abständen von 1, 3 und 5 m von der Wasserseite der Sperre zu liegen kamen (Sekundärraster somit ebenfalls 2 x 2 m). Luftseitig davon wurde die Anzahl der Bohrungen verringert und nur mehr jede zweite Bohrlochreihe bzw. Schirm ausgeführt (Bohrlochraster 4 x 4 m). Diese Arbeiten sowie das Erbohren eines Zustieges zum Entlastungsgang wurden sofort nach der Weihnachtspause in Angriff genommen.

Mitte Jänner war mit der Speicherabsenkung die Kote 1709 m ü.A. erreicht, ab welcher der unterste Teil des Revisionsganges entleert und begangen werden kann. Kurze Zeit später war auch der Entlastungsgang aufgebohrt und zugänglich. Anfang Februar war dann auch der Speicher soweit entleert, daß auch der im Stauraum oberhalb des Vorboden befindliche Teil des Risses zugänglich war. Dieser präsentierte sich als Einzelriß mit einer Weite von etwa 0,5 mm in der aus den Bohrungen abgeleiteten Höhenlage. Im Revisionsgang waren deutliche Injektionsgutaustritte von der Primärinjektion vorhanden. Weiters ist erwähnenswert, daß in der talseitigen Hälfte des Blockes oberhalb des Hauptrisses ein zunächst ca. 5 m langer, vertikaler Riß (V 1) von geringer Weite festgestellt wurde.

Insgesamt wurden 62 Sekundärbohrungen mit 1100 lfm als Rotationsbohrungen mit Kerngewinn (∅ 46 mm) ausgeführt. Die Bohrkerne zeigten, daß der Riß durchwegs satt mit Rodur in einer Stärke von 1 bis 2 mm verfüllt war. An einigen Kernen wurden in der Umgebung des Risses auch rodurverfüllte Haarrisse angetroffen. In sämtlichen Bohrungen wurden Wasserabpreßversuche durchgeführt, um vor der Injektion ein Bild über die Gängigkeit der Bohrlöcher zu erhalten.

5.4 Sekundärinjektion

Geplant war, diese beginnend am Schirm S 1 gegen die Talseite bzw. Luftseite hin voranzutreiben. Gegen die Wasserseite hin war die Verwendung von Rodur 510, ansonsten von eher dünnflüssigem Rodur geplant. Der Riß wurde an der Wasserseite nicht verschlossen, Injektionsgutaustritte sollten verstemmt bzw. durch Flämmen unterbunden werden. Ansonsten galten für die Sekundärinjektion die gleichen technischen Vorschriften wie für die Primärinjektion.

Die Injektion wurde am 11.2.1988 begonnen. Die Aufnahmen (Rodur 515) gestalteten sich unterschiedlich, es wurden Injektionsdrücke bis zu 170 bar angewendet. Bereits am Nachmittag des 11.2. wurden im bergseitigen Bereich des Blockes an der Wasserseite 2 weitere Vertikalrisse (V 2, V 3) festgestellt. Bei der Fortsetzung der Arbeiten am 12.2. trat ein weiterer Vertikalriß auf (V 4), dessen offensichtliches Längenwachstum am 13.2. zur Einstellung

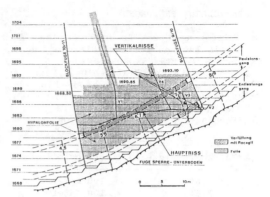

Abb. 9: Ansicht von der Wasserseite mit Rissen und Foliendichtung

Fig. 9: Upstream view with cracks and foil sealing

der Injektionsarbeiten führte (Abb. 9). Die Nebenrisse zeigten nur teilweise Injektionsgutaustritte, an den Rißmeßstellen und Extensometern waren keine nennenswerten Bewegungen zu beobachten. Während beim Nebenriß V 1 die Entstehung nicht zwingend durch die Primärinjektion verursacht worden sein muß, kann aufgrund der Beobachtungen geschlossen werden, daß die Nebenrisse V 2, 3 und 4 durch die Injektionsarbeiten verursacht wurden (Stäuble, Wagner 1991). Dazu muß jedoch bemerkt werden, daß die vertikalen bzw. die vertikal verlaufenden Risse, die vertikal bzw. parallel zu den Blockfugen der Sperre verlaufen, keinerlei Auswirkung auf die Standsicherheit der Sperre hatten. Als Mitursache für das Auftreten der Nebenrisse waren äußerst tiefe Temperaturen und dadurch hervorgerufene allseitige Zugspannungen sowie auch horizontale Querzugspannungen aufgrund der großen Vertikalspannungen im Lastfall leeres Becken in Betracht zu ziehen, denn im besagten Zeitraum war auch der außerhalb des unmittelbaren Injektionsbereiches liegende Riß V 1 um mehr als 10 m nach oben gewachsen.

Für die Fortsetzung der Sekundärinjektion wurde der maximal zulässige Injektionsdruck auf 80 bar reduziert und es wurden nur mehr jene Bohrungen verpreßt, die eine Wasseraufnahme zeigten. Jene Bereiche, in denen kein Injektionsgutaustritt aus dem Hauptriß an der Wasserseite erzielt werden konnte, wurden mit kurzen Bohrungen von der Wasserseite aus angebohrt und mit einem maximalen Druck von 60 bar injiziert ("Vernähungsinjektion").

Im Zuge dessen wurde auch ein Nebenriß im Block 11, der gleichzeitig mit dem Hauptriß entstanden sein dürfte, mitinjiziert. Bei der Sekundärinjektion wurden weitere 3100 kg Rodur verpreßt.

6 KONTROLLEN, PRÜFUNGEN

Die Lieferungen von Rodur wurden auf der Baustelle laufend auf ihre Gleichmäßigkeit überprüft (Dichte, Viskosität).

Zum Nachweis der Kraftschlüssigkeit der Rißverfüllungen wurden nach Abschluß der Injektionsarbeiten 8 Bohrkerne entnommen, bei denen bei Zugversuchen in 7 Fällen die Kraftschlüssigkeit nachgewiesen wurde:

	Haftzugfestigkeit	Bruchstelle
1	1,04 N/mm²	100 % Beton
2	1,84 N/mm²	100 % Beton
3	0,74 N/mm²	im Injektionsgut
4	1,64 N/mm²	85 % Beton
5	1,69 N/mm²	90 % Beton
6	1,12 N/mm²	100 % Beton
7	1,30 N/mm²	100 % Beton
8	1,58 N/mm²	100 % Beton

7 ZUSATZMASSNAHMEN

Anschließend wurden die den Hauptriß seitlich begrenzenden Sperrenblockfugen in den untersten Abschnitten über das Rohrsystem sowie über Bohrungen mit Rodur 510 bzw. 600 unter einem maximalen Druck von 40 bar injiziert. Die Nebenrisse wurden ebenfalls über kurze Bohrungen mit Rodur 600 unter einem maximalen Druck von 40 bar injiziert. Eine vollständige Injektion dieser statisch unbedeutenden Nebenrisse wurde nicht angestrebt.

Sämtliche Rißspuren an der Wasserseite der Sperre wurden, soweit zugänglich, mit einer hochwertigen Kunststoffdichtungsfolie (Hypalon) verklebt. Die nicht zugänglichen Rißbereiche zwischen der Sperre und dem Vorboden wurden mit einem quellfähigen Acrylharz (Rocagil) aufgefüllt (Abb. 9).

Trotz härtester äußerer Bedingungen konnten die Arbeiten an der Wasserseite der Sperre Mitte April abgeschlossen und somit der Speicher zum Wiederaufstau freigegeben werden. Im Zuge der Speicherfüllung wurden insgesamt 56 Entlastungsbohrungen in die Rißfläche hergestellt.

Neben den Injektionsarbeiten im Block 10 wurden eine Reihe von Maßnahmen durchgeführt, die das Auftreten ähnlich gelagerter Risse verhindern sollten, wie etwa das Ausbetonieren sämtlicher nahe der Wasserseite gelegenen Meßnischen, das Verkleinern des Sohlganges in gewissen Abschnitten, das Abteufen von Drainagebohrungen im Beton in allen hohen Sperrenblöcken und insbesondere das Aufbringen eines begrenzten Wasserdruckes in der Bewegungsfuge Unterboden-Sperre, der dafür sorgt, daß bei geöffneter Fuge vertikale Druckspannungen im Sperrenbeton vorhanden sind.

Im Herbst 1988 wurde die Staukote 1840 m ü.A. erreicht, die Entlastungsbohrungen aus dem Rißbereich brachten insgesamt eine Menge von rd. 1,3 l/s, wobei eine Konzentration in der Gegend des Vertikalrisses V 1 festzustellen war. Im Winter/Frühling 1989 wurde dieser Bereich nachinjiziert. Trotz äußerst geringer Injektionsgutaufnahmen konnte eine deutliche Verringerung der Wasserzutritte (20 l/min bei Staukote 1845 m ü.A.) erreicht werden. Eine weitere Nachinjektion von 4 rinnenden Entlastungsbohrungen im April 1990 mit einem maximalen Druck von 40 bar erbrachte keine Aufnahme und auch keine Verringerung der Schüttungen.

8 SCHLUSSBEMERKUNG

Der in einem Block der Sperre Zillergründl aufgetretene Riß wurde in mehreren Etappen mit insgesamt rd. 20 t Kunstharz kraftschlüssig injiziert. Die durch anfängliche Anwendung hoher Injektionsdrücke entstandenen sekundären Risse wurden ebenfalls saniert.

Die bei der Ausführung der Injektionsarbeiten aufgetretenen offenen Fragen bewirkten unter anderem die Gründung eines Arbeitskreises, der sich mit Injektionen im allgemeinen und Kunstharzinjektionen im Beton im besonderen befaßt.

Abschließend sei erwähnt, daß in den Jahren 1990 und 1992 ohne Probleme der Vollstau des Speichers Zillergründl erreicht werden konnte. Eine im April 1993 entnommene Kontrollbohrung aus dem Riß im Block 10 lieferte einen einwandfrei verfüllten, kraftschlüssig verbundenen Kern aus dem Rißbereich. Die Sanierung des Risses im Block 10 ist somit erreicht.

LITERATURHINWEISE

Wuhrmann, E. 1986. Regenerierung von schadhaftem Beton hinsichtlich Festigkeit und Dichtigkeit. Schriftenreihe des Deutschen Verbandes für Wasserwirtschaft und Kulturbau e.V. Heft 77.

Berchten, R.A. 1985. Repair of the Zeuzier Arch Dam in Switzerland, 15th Congress on Large Dams, Q 57, R 40, Lausanne.

Asendorf, K. 1988. Das Strömungs- und Ausbreitverhalten von Injektionsharzen im Riß. Bautenschutz und Bausanierung 2

Stäuble, H., Wagner, E. et al 1991. Cracking of Boreholes in Arch Dams Due to High Pressure Grouting. Proc. Int. Conference on Dam Fracture, Boulder.

Quality control of a rockfill dam foundation treatment

Qualitätskontrolle der Behandlung des Untergrundes eines Steinschüttdammes

A. Silva Gomes

Foundations Division, Geotechnical Department, Laboratório Nacional de Engenharia Civil (LNEC), Lisboa, Portugal

ABSTRACT: The main aspects of the quality control of the foundation treatment of a 48 m high rockfill dam with a bituminous concrete upstream facing, are reported, following the presentation of the main results of the geotechnical investigation performed in the foundation rock mass (essentially formed by greywackes and shales), and of the guidelines governing the water testing and grouting.

Mention is also made to the analysis of the results obtained during the foundation treatment and immediatly after its ending, which made it possible to set up a series of grouting curtain verification holes, which were afterwards used to install pressure take chambers, located along each hole, according to criteria which are also reported. The corresponding piezometers, together with two other piezometers, one located upstream of the grouting curtain, and the other downstream, will make it possible to evaluate the grouting curtain effectiveness during the first filling of the reservoir and during the following dam operation phase.

Some final remarks complete the report, stressing the importance, as regards the efficiency of the works, its economy, and the dam safety, of performing an adequate quality control of the dam foundation treatment.

ZUSAMMENFASSUNG: Im Anschluß an die Vorstellung der hauptsächlichen Ergebnisse der geotechnischen Untersuchung des, im wesentlichen, aus Grauwacken und Tonschiefern bestehenden Gründungsgebirges eines 48 m hohen Steinschüttdamms mit oberwasserseitiger Asphaltbetonverkleidung, und der die dort ausgeführten WD-Versuche und Injektionen bestimmenden Richtlinien, wird über die Hauptaspekte der Qualitätskontrolle der Fundamentbehandlung berichtet.

Die Fundamentbehandlung bestand aus der Injektion von 3 Bohrlochreihen, die von dem, im oberwasserseitigen Dammfuß angelegten Injektionsstollen aus aufgebohrt wurden, wobei die 2 äußeren Reihen als Begrenzungsreihen dienten, mit der Absicht einen übermäßigen Verbrauch an Injektionsgut zu vermeiden, und die mittlere, und tiefste, Reihe die nötige Wasserabdichtung gewährleistete.

Die Analyse der während der Fundamentbehandlung und sofort nach ihrer Vollendung erhaltenen Ergebnisse erlaubte es, eine Serie von Injektionsschirmüberprüfungsbohrlöchern festzusetzen, die später dazu benutzt wurden, Druckaufnahmekammern zu installieren, die, in Übereinstimmung mit Kriterien, über die auch berichtet wird, entlang jedes Bohrlochs angelegt wurden. Die entsprechenden Piezometer, zusammen mit 2 anderen Piezometern, das eine flußaufwärts vom Injektionsschirm und das andere flußabwärts gelegen, werden es erlauben, die Injektionsschirmwirksamkeit während der ersten Füllung des Speichers und während der anschließenden Dammbetriebsphase abzuschätzen.

Einige Schlußbemerkungen schließen den Bericht ab und betonen die Wichtigkeit, für den Erfolg der Arbeiten, ihre Wirtschaftlichkeit und die Dammsicherheit, eine angemessene Qualitätskontrolle der Dammfundamentbehandlung durchzuführen.

1. INTRODUCTION

Apartadura dam is a 48 m high rockfill dam with a bituminous concrete upstream facing and a grouting, drainage, and inspection gallery in the upstream toe (fig. 1).

The dam foundation is essentially formed by greywakes and shales. The geotechnical investigation showed that the rock mass is, in general, highly to completely weathered (W_{4-5}) down to a depth of 5-6 m in the left bank, 14 m in the medium part of the right bank, and 7 m in the higher zone of this bank. In the lower part of the valley, the very weathered zones reach about 3 m in depth. With increasing depth, the rock mass is less weathered, generally fresh to slightly weathered (W_{1-2}). Regarding the fracture spacing, the rock mass presents close to very close (F_{4-5}) fractures up to a depth of 15 m in both banks, and 10-15 m in the lower part of the valley; further down, the rock mass maintains the type of fracturation, with wide (F_2) to moderately distanced (F_3) fractures (Vazquez Gonzalez et al., 1988).

Lugeon tests, performed during the investigation studies, indicated absorptions generally ranging from 20 to 60 Lugeon (Lu), down to depths of about 15 m. In general, the Lugeon values decreased significantly with the increase of the depth.

2. FOUNDATION TREATMENT

According to the dam project, COBA (1983), the foundation treatment consisted of injections with cement based grouts in three rows of holes, drilled from the grouting gallery (fig. 2). The two external rows were to act as contention rows, aiming at avoiding an excessive grout consumption, and the central and deepest row, was to guarantee the required watertightness.

The first row to be executed was the downstream contention row (F3), followed by the upstream contention row (F1) and, finally, the central row (F2). In each row, the grouting procedure followed the split-spacing method (in general, only up to the secondary holes). The primary holes were 9 m apart, and the secondary holes halfway between the two adjacent primary holes.

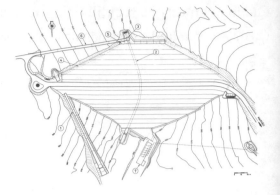

(1) Spillway · (1) Überlauf
(2) Diversion gallery · (2) Umlaufstollen
(3) Intake tower · (3) Entnahmeturm
(4) Drainage gallery · (4) Entwässerungsstollen
(5) Upstream toe · (5) Oberwasserseitiger Fuß
(6) Access bridge · (6) Zugangsbrücke
(7) Pumping station · (7) Pumpanlage

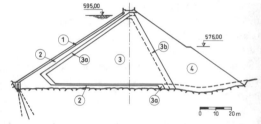

(1) Watertight facing
    (1) Wasserdichte Verkleidung
(2) Upstream rockfill
    (2) Oberwasserseitiges Steinschüttmaterial
(3a)+(3b) Transition rockfill
    (3a)+(3b) Übergangssteinschüttmaterial
(3) Central rockfill
    (3) Mittleres Steinschüttmaterial
(4) Downstream rockfill
    (4) Unterwasserseitiges Steinschüttmaterial

Fig.1 — Apartadura dam. Plan and cross-section (Staudamm Apartadura. Grundriß und Querschnitt)

Both the dam project and the specifications included guidelines to be followed during the treatment. However, and as usual in this type of work, those guidelines were somehow modified, in order to take into due account the experience

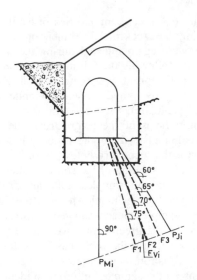

Fig.2 — Apartadura dam. Contention and watertight curtains, and piezometric hole-fans (Staudamm Apartadura. Begrenzungs- und Wasserabdichtungsschirme, und Piezometerbohrlochfächer)

obtained during the water tests and grouting of two test blocks (one in the upper zone of each bank), and during the injection of the first holes of the grouting curtain.

Some of the most relevant aspects regarding the above mentioned modifications, will be mentioned below.

In accordance with the modified guidelines, the following methods were used, generally, for the drilling and performing of water absorption tests in each hole:

- down to 8 m depth (from the concrete-rock mass interface), the descending stage method was used; the first stage was 3 m long, and the second one 5 m;

- following the water absorption tests of these 2 stages, the hole was drilled to the expected depth (28 m for the contention rows and 33 m for the central row);

- in the last 5 m of the hole, a Lugeon type test was then performed, aiming at establishing the hole's final depth; when the test result was lower than 2 Lu (for the central row) or 4 Lu (for the contention rows), the hole was considered com-

pleted, else 5 m more were drilled;

- finally, Lugeon type tests were performed in the remaining 5 m stages of the hole, following the up-stage method;

- in all Lugeon type tests, the pressures (MPa) held during 10 min each, were:

- 0.10, 0.20, and 0.10, for the first stage (0--3 m);
- 0.20, 0.40, and 0.20, for the second stage (3-8 m);
- 0.25, 0.50, and 0.25, for the next 2 stages (8-18 m);
- 0.35, 0.70, and 0.35, for the next 2 stages (18-28 m);
- 0.50, 1.00, and 0.50, for the deeper stages (more than 28 m).

Also according to the same guidelines, the methods generally used for the hole grouting were:

- grouting by the descending stage method down to 8 m;

- grouting by the up-stage method, from the bottom of the hole up to 8 m.

The applied grouting pressures (MPa) were:

- 0.20, for the first stage (0-3 m);
- 0.40, for the second stage (3-8 m);
- 0.60, for the third stage (8-13 m);
- 1.00, for the deeper stages (more than 13 m).

This last pressure (1.00 MPa) (instead of the 1.50 MPa preconized in the dam project and the original specifications) was used because the analysis of the results previously obtained showed that hydraulic fracturing frequently occurred when the pressure exceeded about 1.00 MPa.

The new guidelines also refer to the type of grouts to be successively used in any stage of each hole, and to the grout volumes not to be exceeded. According to the guidelines, the injection of any stage must begin with the thinnest, less viscous, usable grout (water/cement ratio - - 3:1).

Schematically, 3 different situations may occur when injecting the 3:1 grout mix:

i) the grouting pressure tends to rise too rapidly;

ii) the grouting pressure tends to rise slowly, when keeping, as much as possible, a constant grout flow rate;

iii) the grouting pressure tends to stabilize or drop.

In case (i), as soon as the specified maximum pressure has been reached, the pressure has to be kept constant during the next 10 min, independently of the flow rate.

In case (ii), there are 2 possible procedures: if the pressure limit is attained before injecting 1 m^3 grout, this pressure has to be kept constant during the next 10 min; if 1 m^3 grout is injected before attaining the pressure limit, the grout has to be thickened (2:1 grout mix).

In case (iii), again the grout has to be thickened (2:1 grout mix).

During the injection of the 2:1 grout mix, the procedures to be adopted are essentially similar to those previously reported, except as regards case (i), the volume limit, and the thicker grout.

In fact, if the grouting pressure rises too rapidly, it is necessary to return to the 3:1 grout mix. Then, if the specified maximum pressure is attained before injecting 0.5 m^3 of the 3:1 grout mix, this pressure has to be kept constant during the next 10 min; else, it is again necessary to pass to the 2:1 grout mix.

The volume limit for the 2:1 grout mix is 2 m^3, instead of the 1 m^3 limit, established for the 3:1 grout mix.

If a thicker grout is needed, a 1:1 water/cement ratio is used afterwards.

The injection of the 1:1 grout mix also develops essentially in a similar way (passing to a thinner grout, if the pressure rises too rapidly, or continuing to inject this type of grout), the maximum allowed volume of grout being now 3 m^3. If 3 m^3 grout are injected without reaching the specified pressure limit, the injection shall be suspended until it is sure that the grout has settled. Then, it shall be reinitiated, again with a 1:1 grout mix, after a washing of the hole. If the pressure still keeps constant or rises slowly, a thicker grout mix (1:2) shall be injected during a maximum of 20 minutes, but not exceeding 3 m^3.

The injection of the downstream contention rows followed a slightly different method. In fact, a pre-grouting of the first 3 stages (0-13 m) was performed, using a thicker grout mix (1:1) since the beginning, and lower pressures and volumes in each injection, in order to create a "ceiling", and thus avoiding a longer penetration of the later grouts, especially towards downstream, and minimizing the risk of having them reaching the rockfill. The mentioned first 3 stages were again injected during the up-stage grouting, following the full drilling of the hole.

The applied grouts were cement grouts, for the contention rows, and cement-bentonite grouts (2% of bentonite), for the central row.

In most holes of the central row, the grouts were coloured (red for the primary holes, and green for the secondary holes), in order to enable the identification of the source of the grouts appearing in the samples collected from the grouting curtain verification holes. However, only a few samples clearly showed the presence of red grouts; the green grouts were not apparent, probably because the rock mass is of the same tonality. It is foreseen to perform some chemical analysis on the samples, in order to identify the eventual presence of coloured grouts, especially the green ones.

3. ORGANIZATION OF THE QUALITY CONTROL

In all dams owned by the Directorate General of Natural Resources (DGRN), a body of the Portuguese Central Public Administration, the supervision of all works concerning the dam construction, including the foundation treatment, has been entrusted to teams of surveyors from the DGRN.

In the case of fill dams, these surveyors were usually assisted by a technical team of the National Laboratory of Civil Engineering (LNEC), which, in close collaboration with the DGRN surveyors, performed the earth- or rockfill placement tests for the fill quality control. The surveyors decided on the acceptance or rejection of each placed layer, according to the results obtained by the LNEC technicians. It is recognized that this kind of collaboration has given good results concerning the serviceability and safety of the large DGRN fill dams. Prior to the construction of the Apartadura dam, this collaboration, however, had never included, in a systematic way, the works concerning the foundation treatment.

As is well known, the treatment of a dam foundation always involves very specific aspects, requiring not only a specialized knowledge but also a thorough continuous technical assistance. In fact, contrarily to what happens with other dam construction works, is it very difficult to assess the quality of the foundation treatment after its execution, because the results, usually, can not be visually assessed, and the verification tests are, generally, not very representative.

For the Apartadura dam, an agreement was reached between the DGRN and LNEC, establishing the conditions for a technical collaboration, with similar terms to those above mentioned for the quality control of the fill placement. As a consequence, the LNEC technicians (usually 2) assisted the local surveyors, essentially by performing the following activities:

- assessment of the conformity of the works with the above mentioned guidelines, regarding both the water tests (pressures and flows) and the grouting procedures (pressures, flows, and grout types and volumes);

- testing of the used grouts, in order to assess their main characteristics, namely, their Marsh viscosity, sedimentation, density, and strength;

- recording, in appropriate forms (fig. 3), data regarding the water absortion tests and the grouting operations;

- processing the data at the site on a microcomputer, in order to get diagrams (fig. 4).

Despite the good overall results obtained with the performed quality control, namely, regarding the compliance of the works with the specified guidelines, and the reliability of the collected data, some difficulties arised, which must be overcome in future similar works. In fact, while it was relatively easy to follow the execution of the water absorption tests, with the recording of the corresponding pressure values from the manometer, and the water flow from the flowmeter, both located next to the hole top, even when only 1 LNEC technician was present at the site, there were some difficulties regarding the holegrouting. When the 2 LNEC technicians were present at the site, it was possible to control and record the grouting pressures, measured with the manometer located near the hole, as well as the grout mixes used during the hole injection (at the grouting plant). But, when only 1 technician was present, obviously following the operations de-veloping at the hole,

WATER ABSORPTION TEST LOG
APARTADURA DAM
Hole no: *E3-55*   Work area: *GGD*
Stage from: *13* to *18*
WATER TEST
No of pressure stairs: *3*
Started testing: *15:15* (HH:MM)   Date: *92-02-04* (YYMMDD)
Ended testing: *15:45* (HH:MM)   Dist. Man.-Water: — (m)
                                     Dist. Man.-Water: — (m)

MIN	1st Stair		2nd Stair		3rd Stair	
	PRESS	VOL	PRESS	VOL	PRESS	VOL
1	2,5	34	5,0	60	2,5	33
2	"	35	"	59	"	33
3	"	34	"	60	"	33
4	"	35	"	59	"	33
5	"	35	"	59	"	33
6	"	35	"	60	"	34
7	"	35	"	60	"	33
8	"	35	"	60	"	33
9	"	35	"	60	"	32
10	"	35	"	60	"	33

Units: Press (kgf/cm²); Vol (l)
COMPUTER DATA
Volume : *348*   Volume : *591*   Volume : *330*
Time   : *10*   Time   : *10*   Time   : *10*
Pressure : *2,5* Pressure : *5,0* Pressure : *2,5*

GROUTING LOG
APARTADURA DAM
Hole no: *E3-55*   Work area: *GGD*
Stage from: *13* to *18*
GROUTING
Date: *92-02-04* (YYMMDD)
Started grouting: *16:00* (HH:MM)   Dist. Man.-Water: — (m)
Ended grouting : *19:00* (HH:MM)   Dist. Man.-Water: — (m)

W/C - 3/1			W/C - 2/1			W/C - 1/1			W/C - 1/2		
C	P	T	C	P	T	C	P	T	C	P	T
300	1,0	10									
			850	1,5	70						
						650	1,5	20			
									2250	10,0	80

Units: C - Cement (Kg); P - Pressure (Kgf/cm²); T - Time (min)
COMPUTER DATA
Mix (W/C) :   *3/1*   *2/1*   *1/1*   *1/2*
Cement   :   *300*   *850*   *650*   *2250*
Pressure :   *1,0*   *1,5*   *1,5*   *10*
Time   :   *10*   *70*   *20*   *80*

Fig. 3 — Water test and grouting logs (WD-Versuchs- und Injektionsprotokolle)

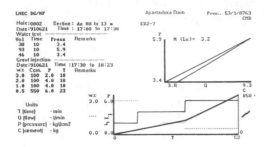

LNEC DG/NF   Apartadura Dam   Proc. 53/1/8763
                                                                CMD
Hole:0002   Section : de 88 to 13 m   ED2-7
Date:910621   Time : 17:00 to 17:30
Water test
Vol   Time   Press   Remarks
 30   10   3.4
 93   10   5.9
 46   10   3.4
Grout injection
Date:910621   Time :17:30 to 18:23
wcc   Ccm.   P   T   Remarks
3.0   100   2.0   10
2.0   100   4.0   10
1.0   100   4.0   10
0.5   550   6.0   23

   Units
T [time]   - min
Q [flow]   - l/min
P [pressure] - kg/cm2
C [cement]   - kg

Note: Snapshot of computer display (slightly retouched)

N.b.: Schnappschuß eines Komputerbilds (leicht retuschiert)

Fig.4 — Typical water test and grouting diagrams (Typische WD-Versuchs- und Injektionsdiagramme).

he was only able to control and record the grouting pressures, having to rely upon the information given by the contractor, re-garding the grout mixes used.

The quality control would have been easier to perform, and have given more detailed information with an automatic scheme (which was not available at the Apartadura dam, where all data were manually recorded). However, the existence of such a scheme does not avoid the presence at the site of a team (whose constitution must be adequated to the importance of the works), in order to periodically check and analyse the automatically recorded information, and to help the surveyors in taking the necessary decisions, when facing situations that have not been anticipated.

The experience gained by the LNEC during the foundation treatment of the Apartadura dam, will make it possible to better organize the quality control activities for similar works in the near future. In addition, this experience may also be used to modify some aspects regarding the project of foundation treatments, and the corresponding specifications, aiming at improving the quality of these important works.

4. ANALYSIS OF THE RESULTS

The results obtained during the foundation treatment, were immediately analysed, though not very thoroughly, in order to detect any situation calling for the use of a different procedure, in which case the responsibles for the work management were alerted, in order to adopt the adequate modifications to the established guidelines.

A more detailed analysis of the results made it possible to understand the most important characteristics of the overall response of the dam foundation to the treatment, and to establish, in accordance with the methods presented in item 5, from the identification of the zones where doubts about the grouting efficiency may subsist, the location, orientation, and depth of the grouting curtain verification holes (Silva Gomes, 1993).

The general characteristics concerning the response of the foundation to the treatment, are the following:

- the permeability decreases with the depth;
- the permeability decreases from the primary to the secondary holes (exceptions to this trend are some cases where the absorption in a secondary hole was higher than the absorption in the adjacent primary hole(s); these cases are relatively frequent for the upper injection stages, mainly for the 0-3 m stage);
- the permeability decreases from the F3 to the F1 contention rows, and from these to the central (F2) row;
- the cement grout takes do not exhibit a clear correlation with the absorptions obtained from the corresponding water tests (this fact is essentially due to the different rheologic behaviour of the water and the grout, to the characteristics of the rock mass jointing, mainly the joint spacing and conductivity, and to the different pressures used in the water tests and in the grout injections); however, as happened with the water tests, in general, the grout takes decrease with the depth and from the primary holes to the secondary holes, as well as from the F3 to the F1 row, and from these to the F2 row.

This analysis shows that the treatment was, in general, adequate; a special remark must be made about the excelent results obtained, for depths of more than 18 m, with the grouting pressure limit of 1.0 MPa, instead of the initially preconized limit of 1.5 MPa, thus avoiding the hydraulic fracturing that would have been responsible for higher grout consumptions without any profit.

Despite the already mentioned permanence of relatively high absorptions in the upper zones of the holes (which is probably due to the fact that the excavation of the trench for the drainage gallery took place long before the beginning of the gallery construction, thus causing superficial stress release), it was decided not to apply higher grouting pressures (which would guarantee a better grout penetration) than the specified ones, due to lack of sufficient overload (as said before, the gallery is perimetral).

The thorough analysis of the results stored in the microcomputer files, in conjunction with the analysis of the results obtained from the hydraulic piezometers, will allow the establishment of a conceptual model, regarding the hydraulic behaviour of the dam foundation. This model will be especially useful during the first filling of the reservoir. The thorough analysis will be made soon, because the treatment has just been finished, and the first filling will possibly begin next winter.

5. EVALUATION OF THE GROUTING CURTAIN EFFICIENCY DURING THE FIRST FILLING AND OPERATION STAGES

In the final phase of the foundation treatment, a series of grouting curtain verification holes have been set up, aiming, as said, at the assessment of the treatment effectiveness in the zones where doubts still subsisted. In addition, and depending on the results obtained along these holes, the possibility of installing piezometers, with pressure take chambers located in appropriate stages, was envisaged.

The methods used in order to define a verification hole, and to establish the subsequent procedures, are summarized below:
- the location, orientation, and depth of a verification hole were set up taking into account only the results from the central row, as this row was the last to be executed; the hole was located

in the neighbourhood of central row holes in which high water and grout absorption values had still been obtained;
- almost all verification holes were executed along the intersection of the plane containing approximately the nearest central row holes, with a plane normal to the gallery; the holes were drilled by descending stages (the first one, down to 3 m from the concrete-rock interface, and the others, 5 m long);
- in each stage, a Lugeon type water test was performed, using 3 pressure stairs (the same ones as those used in the central row hole tests, and again maintained during 10 min each);
- in the stages where the absorptions exceeded previously established limits (5 Lu, for the stages between 0-8 m; 6 Lu, for the stages between 8-18 m; 7 Lu, for the stages between 18--28 m; and 8 Lu, for the deeper stages), injections were made, the grouting pressure limits exceeding by 0.1 MPa the equivalent limits of the foundation treatment; the injections were carried out in accordance to procedures similar to those already reported, starting with a 3:1 grout mix, which was successively thickened up to a 1:1 grout mix, if the specified pressures could not be reached before a consumption of 500 kg of cement, for each used grout mix (3:1 and 2:1); for the 1:1 grout mix, if the refusal pressure was still not attained until the injection of 1000 kg of cement, the hole was redrilled and reinjected, according to the same methods;
- on the contrary, if the absorption was lower than the mentioned limit values, the next stage was drilled and tested.

The above mentioned procedure was based on the following reasons:
- if, after the conclusion of a treatment, involving the execution of 2 contention rows and 1 watertight row, the water absorption value obtained in a given stage, is still higher than the established limit, this means that the injection of the adjacent holes did not reach the neighbourhood of the verification hole (for instance, because the joints to be injected were very thin, or because the grout percolation was essentially orientated in other directions); in any case, it will be necessary to inject thinner grouts, with slightly higher pressures than those previously used, which can now be applied, because the

foundation treatment has already been made;

- although the project established lower absorption limits (as said, 4 Lu, for the contention rows, and 2 Lu, for the central row) than the new proposed limits, for the verification holes, it was considered advisable to adopt these higher values, as they approach, still by defect, the values which have more recently been suggested by well known experts (Houlsby, 1985, 1990; Kutzner, 1991), for dams similar to the Apartadura dam.

Pressure take chambers for hydraulic piezometers were set up in all grouting curtain verification holes. The hole diameter (0.086 m) allowed the installation of up to 2 pressure chambers in each hole, their location depending on the Lugeon values obtained along the hole, and being established, in principle, in accordance with the following rules:

- from the Lugeon value N_i, obtained for the stage i (i being the number of the stage going down the hole, i.e, 1 for the 0-3 m stage, 2 for the 3-8 m stage, etc.), the theoretical value N_{ia} was computed as $N_{ia} = N_i - i$;

- the first chamber was located in the stage with the highest N_{ia} value; if several stages presented that value, the stage chosen was the one situated further up in the hole;

- the second chamber was located using the same criterion, but disregarding the stage with the first pressure chamber and the adjacent one(s).

These rules consider that the water will pass through the zones of higher permeability, and, when choosing between 2 different stages with the same permeability, more through the upper stage than the lower one, as the way through the deeper stage is longer. The rules were followed in almost all holes, but, when special circumstances occurred, some adaptations were made.

Together with the majority of the verification holes, piezometric hole-fans were set up (fig. 2). For that, in the plane normal to the gallery including a given verification hole, 2 other, 0.066 m diameter holes were drilled (with a roto--percussion hammer drill), one normal to the gallery floor, and the other one forming an angle of 5°, towards downstream, with the F3 contention row plane; their depths were established in accordance with the results obtained in the verifica-

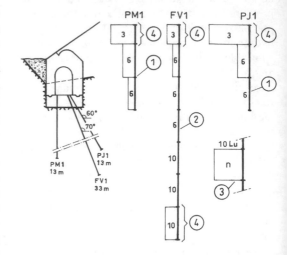

(1) Hammer-drilled hole
                (1) Schlagbohrer-Bohrloch
(2) Rotary-drilled hole
                (2)Diamantbohrer-Bohrloch
(3) Pressure of the water test (n)
                (3) Druck des WD-Tests (n)
(4) Pressure take chamber
                (4) Druckaufnahmekammer

Fig.5 — Hole-fan with the FV1 verification hole (Bohrlochfächer mit dem FV1 Überprüfungsbohrloch)

tion hole tests.

In order to set up the respective pressure chambers, following rules similar to the ones already indicated, but having also in mind the pressure chambers of the corresponding verification hole, water tests were made along these 2 holes. An example of a hole-fan, which includes the FV1 verification hole, is presented in fig. 5.

The type of hydraulic piezometers installed in the Apartadura dam foundation, is schematically represented in fig. 6.

6. FINAL REMARKS

As is well known, the component "foundation" of the structural whole "dam-foundation" plays a very important role, regarding the safety of

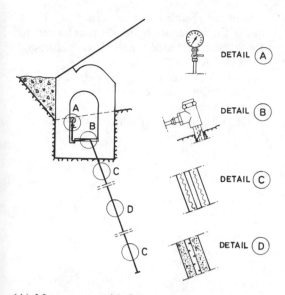

DETAIL (A)

DETAIL (B)

DETAIL (C)

DETAIL (D)

(A) Manometer with 3-way tap
     (A) Manometer mit 3-Wegverschluß
(B) T-junction with plug
     (B) T-Verbindung mit Spund
(C) PVC tube in grout
     (C) PVC-Rohr in Injektionsgut
(D) PVC tube with small clefts in sand
     (D) PVC-Rohr mit kleinen Schlitzen in Sand

Fig. 6 — Piezometer installed in the Apartadura dam foundation. Scheme (In der Gründung des Staudamms Apartadura installierter Piezometer. Schema).

that whole. On the other hand, the improvement of the rock mass foundation characteristics usually causes significant costs in relation to the overall expenses of the development, those costs generally exceeding (in some cases even to a great degree) the initial budget.

The difficulties related with the inspection of the treatment works (lack of surveyors with a due technical training in this field; frequent involvement of the surveyors in other work areas, jeopardizing the necessary assistance to the works; constitution of the surveyor team made with economic criteria prevailing over the technical criteria; etc.) originate cases in which the costs increase in an unexpected and apparently unjustified way, or in which, immediately after an extensive treatment, the hydraulic behaviour of the foundation puts severe problems of serviceability or even of safety.

In our opinion, both the owners and the public Authorities concerned with the safety of dams, must have in mind the 2 following considerations, when deciding about the procedures to adopt, and the human means to affect to the dam construction inspection:

- it is not admissible that activities of the construction process, which have a great influence on the safety of developments with a high or significant potential risk, such as the activities regarding the inspection of the works, depend exclusively upon economic factors; in fact, the possible lack of serviceability conditions of a dam or, even more, its possible failure, which could have been avoided by an efficient inspection during the construction stage, will correspond to social and economic consequences which no argument will be able to justify;

- the presence of an efficient team of surveyors will invariably carry out a global economy of the works, because it favours an acceleration of the decision procedures, and guarantees the necessary conditions for the adoption of safe and economic procedures, especially when unexpected situations occur (which is very frequent, as said, in the case of foundation treatment works); on the contrary, and, as happens in the majority of cases, a saving in the inspection activities, in addition to the above mentioned consequences, is followed by direct and indirect further costs (due to increasing delays, additional works, etc.) which are much higher than the costs of an efficient inspection.

In addition to a convenient organization of the inspection activities, it is obviously necessary to have permanently the designer's assistance, in order to decide about the modifications or adaptations that the treatment project usually calls for.

The adoption of the necessary control procedures will give the following benefits to the concerned entities:

- the owner is able to justify the costs of the foundation treatment, which may be higher than those anticipated in the initial budget, based on the duly documented imponderables, occurred during the treatment;

- the designer, by closely accompanying the works, will be able to point out the necessary adaptations or alterations to the project, thus ensuring the aims of the treatment;

- the team of surveyors, using their own capabilities (with the designer's assistance), and having other technical consultants (namely, in Portugal, those of the LNEC), may control the quality of the works, and intervene in the construction process, protecting the interests of the owner and, due to the importance of the developments, also the national interest;

- finally, the contractor, by satisfying, under the permanent supervision of the surveyor team, the project and the specifications (or the provisions resulting from approved alterations or adaptations), does not incur in the risk of provoking suspicions as regards his work, even when the costs significantly exceed the initial budget, or when unexpected behaviour anomalies occur.

AKNOWLEDGEMENTS

The permission of the Directorate General of Natural Resources (DGRN) to publish this paper, is gratefully acknowledged.

Thanks are also due to Mr N.F. Grossmann, principal research officer of the LNEC, for his revision of this text and for the translations into German.

REFERENCES

COBA, Consultores para Obras, Barragens e Planeamento, SA (1983) - Hydroagricultural Development of Marvão. Project (in Portuguese). Lisbon PORTUGAL

Houlsby, A.C. (1985) - Design and construction of cement grouted curtains. R60, Q58. 15th ICOLD Congress. Lausanne SWITZERLAND

Houlsby, A.C. (1990) - Construction and design of cement grouting. A guide to rock foundations. John Wiley & Sons. New York NY USA

Kutzner, C. (1991) - New criteria for rock grouting in dam engineering. R18, Q66. 17th ICOLD Congress. Vienna AUSTRIA

Silva Gomes, A. (1993) - Complement of the observation plan of the Apartadura dam. Location of the grouting curtain verification holes and of the foundation hydraulic piezometers (in Portuguese). Internal report. LNEC. Lisbon PORTUGAL

Vazquez Gonzalez, J.; S. Fernandes Rodrigues (1988) - Bituminous concrete faced rockfill dam. A Portuguese case. R41, Q61. 16th ICOLD Congress. S. Francisco CA USA

Grouting in Rock and Concrete, Widmann (ed.) © 1993 Balkema, Rotterdam, ISBN 90 5410 350 7

Die Anwendung der SOILFRAC-Technik zur Gebäudehebung beim Wiener U-Bahnbau, Baulos U 3/13

The use of the SOILFRAC-technique to raise a building during the construction of the new Vienna U-Bahn Section U 3/13

G. Sochatzy
Fachbereich Grundbau, Wien, Österreich

P. Stockhammer
Keller Grundbau GmbH, Wien, Österreich

ZUSAMMENFASSUNG: Anläßlich des Wiener U-Bahnbaues war im Zug der Herstellung des Bauloses U 3/13 in der Nähe des Wiener Westbahnhofes ein Verwaltungsgebäude der Österr. Bundesbahnen zu unterqueren.
Das auf Streifenfundamenten gegründete 5-geschoßige Bauwerk wurde auf einem Drittel seines Grundrisses durch das Stationsbauwerk der U-Bahn unterquert. Der gesamte Ausbruchquerschnitt von insgesamt 224 m2 wurde nach der "Neuen Österreichischen Tunnelbauweise" aufgefahren. Die rechnerischen Gesamtsetzungen von 8 cm und die rechnerische Tangentenneigung von 1:350 machten eine Zusatzmaßnahme vor und während der Tunnelherstellung erforderlich, da diese Tangentenneigungen nicht ohne wesentliche Beschädigung des Bauwerkes aufgenommen hätten werden können.
Die Auswahl fiel auf ein modifiziertes SOILFRACTURING-Verfahren, womit bei fachgerechter Anwendung eine gezielte Hebung des Gebäudes und ein laufender Setzungsausgleich während der Vortriebsarbeiten möglich war.
Über 3 Injektionsschächte wurde eine nahezu horizontale und 2 m starke Injektionsmatte zwischen Tunnelfirste und Fundamentsohle hergestellt.
Für einen Zeitraum von mehr als 1 1/2 Jahre wurde eine gezielte SOILFRAC-Injektion über die Manschettenrohre durchgeführt, um die durch den Vortrieb eintretenden Setzungen schadensfrei zu kompensieren.
Die Steuerung der Injektionsmaßnahmen erfolgt über kontinuierliche Setzungsbeobachtung an 36 Schlauchwaagenmeßstellen, welche über PC kurzfristig auswertbar waren und die Entscheidungsgrundlage für das jeweilige Injektionsprogramm bildeten.

ABSTRACT: In the course of the construction of section U3/13 of the Vienna underground, in the vicinity of the West Vienna railway station, a tunnel was driven under an existing Austrian Railway administration building.
The 5 Storey building was founded on strip foundations and the underground tunnel works would traverse under 1/3 of its floor area. The complete excavation cross-section of 244 m² was to be driven using the "New Austrian Tunnel" construction method. With the calculated total settlement of 8 cm and tangent slope of 1 : 350 the building could have experienced significant damage, therefore ist was nesessary to introduce additional measures to reduce the settlments.
The choice fell to a modified SOILFRAC technique, whereby the aim was, by application of precise calculationl, to raise the building prior to tunnelling and equalize the settlements during the drive of the tunnel.
From 3 grouting shafts an almost horizontal injection matt, 2 m thick, was constructed between the roof of the tunnel and underside of the building foundations.
The Soilfrac injections were performed throughout the construction period of the tunnel of 1 1/2 years in order to compensate for settlements ans thereby maintain the building damage free.
Control of the injection steps was by means of continuously monitoring 36 N° electronic Settlement measurement points which were connected to a computer, allowing short term plotting. It was on the basis of these results that decisions were taken for the respective injection programmes.

1.) Allgemeines

Der Bauabschnitt U 3/13 beinhaltet ein klassisches Kreuzungsbauwerk zwischen den Linien U3 und U6 samt den dazugehörenden Stationen. Hergestellt wurden die Baulichkeiten im wesentlichen in offener Bauweise und in Deckelbauweise. Das zuvor erwähnte Bürogebäude der ÖBB und eine unter Naturschutz stehende Grünfläche wurden in geschlossener Bauweise nach der "Neuen Österreichischen Tunnelbaumethode" aufgefahren.
Mit insgesamt 224 m² Ausbruchsfläche war der zu bauende Stationsquerschnitt der größte Tunnelquerschnitt, der jemals in Wien aufgefahren wurde. Aufgrund rechnerischer Abschätzungen waren für das Gebäude zufolge des Tunnelvortriebes ohne Berücksichtigung der Wasserhaltung maximale Setzungen und Tangentenneigungen (wie zuvor erwähnt) zu erwarten gewesen, die im Grenzbereich der Gebäudeverträglichkeit lagen und somit die Gebrauchssicherheit eines Bürogebäudes mit hunderten Beschäftigten gefährdeten. Dies führte bereits im Zuge der Ausschreibung zur Suche nach wirkungsvollen setzungsmindernden Zusatzmaßnahmen.

2.) Untergrundverhältnisse

Der Untergrund im Bereich des zu unterfahrenden ÖBB-Gebäudes besteht, abgesehen von einer kulturellen Deckschicht und Resten des Quartärs, im wesentlichen aus feinkörnigen Sedimenten des unterpannonen Tertiärs.
Die jungtertiären Schichten, die gegenständlich etwa ab 6 m unter Gelände anstanden, bestanden dabei größtenteils aus Tonen und tonigen Schluffen mit vereinzelt zwischengelagerten sandigen Schluffen bis hin zu schluffigen Feinsanden. Im Nahbereich des Anfahrschachtes erreichten diese Feinsande im Ausbruchsquerschnitt allerdings Schichtmächtigkeiten bis zu 1,5 m.
Das erste und freie Grundwasser lag im Objektsbereich in etwa am Tertiärrelief auf. Die Grundwässer im Tertiär waren dagegen mit zumeist stark gespannten Druckverhältnissen als Kluft und Schichtwasser anzutreffen.

3.) Tunnel - Querschnitt/Vortrieb

Die U3 Station ist als mehrröhriger Tunnel konzipiert und beinhaltet zwei Stationsröhren und einen mittig gelegenen Passagentunnel.
Die Schienenoberkante war mit etwa 23 m

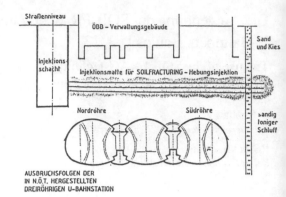

AUSBRUCHSFOLGEN DER
IN N.Ö.T. HERGESTELLTEN
DREIRÖHRIGEN U-BAHNSTATION

Abb.1
Ausbruchsfolgen der in N.Ö.T. hergestellten dreiröhrigen U-Bahnstation.

Breakout cross-section by the N.Ö.T. technique to form triple tube underground station.

unter Gelände vorgesehen, so daß der Tunnelquerschnitt zur Gänze im Tertiär liegt, und zum ÖBB-Verwaltungsgebäude hin eine 10 m mächtige Tertiärüberdeckung aufweist.

Aufgrund des großen Querschnittes wurde der Tunnel in sieben Einzelvortrieben aufgefahren. Begonnen wurde der Tunnelausbruch mit den Vortrieben der zwei Pfeilerstollen (je 37 m² Querschnittsfläche). Anschließend erfolgte der Einbau der Betonpfeiler und der Vortrieb der beiden Seitenstollen (je 26 m² Querschnittsfläche) der außenliegenden Nord- und Südröhre. Danach erfolgte der Mittelausbruch des Passagentunnels (ca. 40 m² Querschnittsfläche) und zuletzt die beiden Restausbrüche der Nord- und Südröhre (je 29 m² Querschnittsfläche).

4.) Setzungsproblematik
Planungsüberlegungen Problemlösung

Ausgehend von der bereits eingangs erwähnten Problematik, daß die im mindesten prognostizierten Verformungen im Grenzbereich der Verträglichkeit des ÖBB-Gebäudes lagen, wurden in der Planungsphase verschiedenste Möglichkeiten und Maßnahmen zur Problemlösung diskutiert. Im wesentlichen umfaßten diese Überlegungen:
a.) Maßnahmen von den Vortrieben aus
b.) Maßnahmen von fertigen Einzelvortrieben oder Vortriebsabschnitten aus
c.) Maßnahmen vom Gelände, vom Keller oder

von Schächten aus
d.) Maßnahmen am Gebäude selbst
e.) Temporäre Büroabsiedelungen mit
abschließender Sanierung des Hauses.
Entsprechend dem Ernst des Problems sollte
die bereits in der Planungsphase
angesiedelte intensive Diskussion der
Problemlösung jedenfalls für die
Ausführungsphase auch späteres
Wunschdenken, Zufälligkeiten, oder Umstände
der Not und dem Zeitzwang gehorchend,
gerade noch praktikable Lösungen erfinden
zu müssen, weitgehend ausschalten.

Baugrund, Technologie, Sicherheit und
Wirtschaftlichkeit waren dabei die
maßgebenden Beurteilungsprameter. So
schieden z.B. jegliche Injektionen von der
Ortsbrust mit dem Ziele der Bodenverfestigung aus, da sie, nicht nur den im
feinkörnigen Lockergestein so notwendigen
raschen Ringschluß verzögert hätten,
sondern bei den gegebenen Tonen und tonigen
Schluffen auch nicht durchführbar gewesen
wären.
Hochdruckbodenvermörtelungen, Pfähle,
Vereisung, Aufsprengungsinjektionen etc.,
durchgeführt von fertigen Einzelvortrieben
oder Vortriebsabschnitten aus, zeigten
wieder im wesentlichen neben den hohen
Kosten der Baumaßnahme selbst, meist nur
eine begrenzte Wirkung und zum Teil die
Gefahr unkontrollierter Bodenbewegungen
und/oder die Gefährdung der fertigen
Außenschalen.
Zudem bedeuteten diese Maßnahmen
grundsätzlich enorme Zeitverzögerungen bei
den Tunnelvortrieben und beim Einbau der
Innenschale, was einerseits hinsichtlich
des generellen Setzungsverhaltens negativ
zu sehen, und andererseits aber vor allem
auch mit hohen zusätzlichen zeitgebundenen
Kosten verbunden gewesen wäre.
Das Gebäude zur Gänze auf Pressen zu
stellen schied infolge hoher Kosten und
wegen hohen Risikos aus, von einer
temporären Absiedelung der Büros, verbunden
mit einer Teilsanierung des Gebäudes nach
der Tunnelherstellung, mußte aus
Kostengründen Abstand genommen werden.
Letztendlich als wirtschaftlichste und für
das ÖBB-Gebäude risikoärmste Baumaßnahme
gewählt, der Ausschreibung zugrundegelegt
und ohne wesentliche Variante angeboten,
wurde die Anwendung von gezielten
Aufsprengungsinjektionen, durchgeführt von
tiefen Schächten aus. Diese
Aufsprengungsinjektionen sollten dabei in
der tertiären Tunnelüberdeckung und dem
ÖBB-Haus auf die Setzungserscheinungen
reagieren und, ohne die Vortriebe
wesentlich zu beeinflussen, ausgleichende
Hebungen bzw. Setzungsrückstellungen
bewirken.

5.) SOILFRACTURING - als setzungsrückstellende, den Tunnelvortrieb begleitende Maßnahme

Die in der Tunnelüberdeckung im
wesentlichen anstehenden steif bis
halbfesten zumeist ausgeprägt plastischen
und wassergesättigten Tone und tonigen
Schluffe liegen jenseits der
Anwendungsgrenze einer konventionellen
Injektion. Für eine gezielt anzuwendende
Bodenaufsprengung und Feststoffeinpressung
konnte das vorangestellte Faktum aber
geradezu als günstige Voraussetzung
angesehen werden. Mit dem modifizierten
SOLFRACTURING-Verfahren wurde jedenfalls
durch gezielte Einpressung von
sedimentationsstabilen feststofffreichen
Zementsuspensionen mit gesteuertem
Abbindeverhalten erreicht, daß die Tone
und Schluffe örtlich begrenzt aufrissen
und sich im Boden verästelte
Feststofflamellen bildeten. Die
Möglichkeit örtlich mehrfach und durch
gesteuert-beschleunigtes Abbinden auch
örtlich begrenzt zu verpressen, stellt
dabei vor allem bei den stark geklüfteten,
von Scher- und Sedimentationsflächen
durchzogenen feinkörnigen Lockergesteinen
eine wichtige Voraussetzung für eine in
jeder Phase kontrolliert durchgeführte
Hebungsinjektion dar.

Das modifizierte SOILFRACTURING-Verfahren
bot nun die Möglichkeit die aus den
Tunnelvortrieben resultierenden
Setzungsmulden, durch mehrmalige an das
Baugeschehen angepaßte
Setzungsrückstellungen, stets innerhalb
der für das Gebäude verträglichen
Grenzwerte zu halten. Als erklärtes und
später auch erreichtes Ziel wurde der
Planung ein Maximalwert von 4 cm für die
vortriebsbedingten Setzungen und eine max.
Tangentenneigung von 1:1000
zugrundegelegt.

6.) Durchführung der Hebungsarbeiten nach dem modifzierten SOILFRAC-TURING-Verfahren

Am Rande des Gebäudes wurden ingesamt 3
Injektionsschächte mit einem lichten
Innendurchmesser von 5,5 m und einer Tiefe
von ca.13 bis 15 m hergestellt.

Aus diesen Bohr- und Injektionsschächten
wurden zunächst rund 9.000 lfm nahezu
horizontale und zielgenaue
Injektionsbohrungen mit einer Länge bis zu
38 m abgeteuft. Die Anordnung erfolgte
fächerförmig in 3 Lagen.

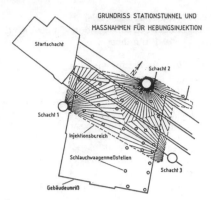

GRUNDRISS STATIONSTUNNEL UND
MASSNAHMEN FÜR HEBUNGSINJEKTION

Startschacht

Schacht 2

Schacht 1

Injektionsbereich

Schlauchwaagenmeßstellen

Schacht 3

Gebäudeumriß

Abb.2
Grundriß Stationstunnel und Maßnahmen füı
Hebungsinjektion.

Plan of station tunnels and measures to
implement raising injektion.

Zur Verwendung gelangten
Stahlmanschettenrohre und rasch abbindende
Mantelmischungen.
Die Tiefenlage der rund 2 m mächtigen
Injektionsmatte lag etwa mittig zwischen
Tunnelfirste und Kellersohle in einem
Abstand von jeweils 6 m.

Nach Herstellung der Bohrungen wurde über
die gesamte Grundrißfläche eine
Erstinjektion ausgeführt, um eine
Verspannung des Mattenbereiches zu
erreichen.
Für die Erstinjektion waren zwei versetzt
angeordnete Bohrfächer mit einem
Vertikalabstand von 1,6 m zu verwenden. Im
unteren Drittel zwischen beiden Fächern
wurde der Hebefächer angeordnet, sodaß
letztlich 3 Fächer übereinander
vorlagen. Die insgesamt für die Hebung zur
Verfügung stehende Mattenfläche betrug rund
2.700 m² und überdeckte den durch die
Tunnelherstellung beeinflußten Teil des
Gebäudes.

Insgesamt wurden für die Erstinjektion rund
ein Viertel der gesamten Mengen an
Injektionsgut verpreßt, bis die
Hebebereitschaft hergestellt war.

Für die Durchführung der Erst- und
Hebungsinjektion nach dem modifizierten
SOILFRACTURING-Verfahren kamen Suspensionen
auf Zement-Bentonit-Kalkstein-Basis zur
Verwendung. Durch die Zugabe von Wasserglas
wurde die Abbindegeschwindigkeit gesteuert.
Die Verpreßung erfolgte über automatische
Misch- und Pumpstationen, von wo aus mit
bis zu 6 Pumpen das Injektionsgut verpreßt
wurde.

Zur Verwendung gelangten folgende Mischungen:

800 - 1.000 l Wasser
 10 -   50 kg Bentonit
300 -   500 kg HOZ 275
300 - 1.200 kg Kalksteinmehl
  0 -   8 % Wasserglas

Insgesamt wurden bei den 4
Tunneldurchfahrten einschl. Erstinjektion
rund 900 m3 Suspension verbraucht, wobei
allein für den Setzungsausgleich während
der Herstellung der beiden Pfeiler
stollen ein Drittel der Injektionsgutmenge
verwendet wurde.

Die Injektionsdrücke lagen bei der
Erstinjektion noch zwischen 3 und 10 bar
und mußten mit zunehmender
Bearbeitungszeit gesteigert werden. Beim
letzten Vortrieb des Restausbruches
betrugen sie bis zu 40 bar.

7.) Eignungsversuche auf der Baustelle

Neben der Überprüfung der einzelnen
Suspensionen wurden insbesondere
Injektionsversuche zum Nachweis der
Eignung des Verfahrens unter den
Baustellenbedingungen durchgeführt.

Zunächst wurde von einem Versuchsschacht
aus ein 2 x 2 m großes Fundament
hergestellt. Darunter wurden die
Manschettenrohre für die Injektion
angeordnet und ein erster Hebeversuch zur
gezielten Bewegung des Fundamentes
ausgeführt. Nach einer Absoluthebung von
8,33 mm wurde eine bleibende Hebung von 7
mm nachgewiesen.

In einem zweiten Versuch nach 1 1/2
Monaten konnte nach einer weiteren Hebung
von 6,7 mm eine bleibende zusätzliche
Hebung von 5,7 mm erzeugt und damit die
Wiederholbarkeit der Maßnahme nachgewiesen
werden.
Nach Herstellung der Injektionsbohrungen
aus dem Schacht 1 wurde in einem 3.
Versuch eine flächenhafte Hebung unter dem
bestehenden Gebäude vorgenommen. Auch
damit konnte noch vor Beginn der
eigentlichen Hebungsarbeiten der Nachweis
erbracht werden, daß die gewählte Maßnahme
zielführend ist.

8.) Steuerung der Hebungsinjektion

Ein wesentlicher Bestandteil der
Injektionsarbeiten war nach dem
modifizierten SOILFRACTURING-Verfahren das
Meßprogramm. Zur Überwachung der
Gebäudebewegungen wurde ein

344

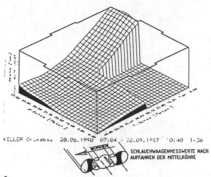

Abb. 3

Schlauchwaagenmeßwert nach Auffahren der Mittelrohre.

Values of settlement measurement points after driving of centre tube.

Schlauchwaagensystem bestehend aus 36 Meßzellen im Gebäude installiert. Mit den Schlauchwaagen waren die Relativbewegungen zu messen. Das Niveau der Meßflüssigkeit wurde in jedem Gefäß durch einen Schwimmer auf einen Weggeber übertragen. Nach Verarbeitung der Signale im PC konnten die Meßergebnisse sowohl in Meßlisten als auch in 2 oder 3-D-Bildern jederzeit dargestellt werden. Damit war es dem leitenden Ingenieur während der gesamten Dauer der SOILFRACTURING-Arbeiten an den einzelnen Manschettenrohrstufen möglich, die Gebäudereaktion zu beobachten und mit abgestimmten Maßnahmen bei der Festlegung des weiteren Injektionsprogrammes zu reagieren.

Zur Ergänzung der Schlauchwaagenmeßdaten wurden regelmäßig Oberflächennivellements an Hausbolzen im Gebäude und an den Außenwänden durchgeführt.

Das Meßprogramm wurde noch durch 11 Gleitmikrometer innerhalb und 7 Gleitmikrometer außerhalb der Injektionsmatte ergänzt. Die dargestellte Gleitmikrometermessung weist beispielhaft die Verformungen zufolge der Hebungsinjektion aus, wobei unterhalb der Matte Stauchungen und im Bereich der Injektionsmatte Hebungen nachgewiesen sind.

9.) Setzungsausgleich außerhalb der Injektionsmatte

Im südöstlichen Teil des Gebäudes traten nach dem Vortrieb des Mittelstollens Versteilungen der Tangentenneigungen außerhalb der Injektionsmatte auf. Es war daher wünschenswert, außerhalb der Matte einen sogenannten "Absenkfächer" herzustellen, um gezielte

Gebäudeabsenkungen in diesem Bereich zu ermöglichen. Zu diesem Zweck wurden Hohlraumbohrungen, Durchmesser 133 mm, ausgeführt, welche sicherheitshalber mit einem Manschettenrohr ausgerüstet aber nicht mit Mantelmischung verfüllt wurden. In regelmäßigen Abständen waren auch Hebebohrungen anzuordnen, um bei Eintritt von zu großen Setzungen mittels Hebeinjektion auszugleichen. Damit war es möglich, die Tangentenneigung der Fundamente auch außerhalb der Injektionssohle zu verbessern, ohne daß die Sicherheit des Gebäudes gefährdet war.

10.) Überprüfung des Einpreß- und Hebeerfolges

Zur Überprüfung ob gezielt - auch durch wiederholte Beaufschlagung - in räumlich begrenztem Breich Feststofflamellen aufgebaut werden können, wurden nach Auffahren der Pfeilerstollen und Durchführung der ersten begleitenden Hebeinjektionen Kernbohrungen abgeteuft. Alle diese Kernbohrungen zeigten im entsprechenden Tiefenbereich - eingebettet zwischen den unverändert gebliebenen gegenständlichen steif bis halbfesten Tonen und tonigen Schluffen - Feststofflamellen mit Stärken von wenigen Millimetern bis zu mehreren Zentimetern.

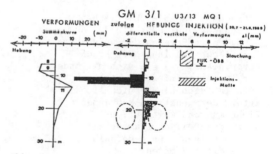

Abb. 4

Verformungen zufolge Hebungs-Injektion.

Deformation according to raising injections.

Die Auswirkungen der Feststoffeinpressung auf den Untergrund, im speziellen die differenziellen vertikalen Bodenverformungen, konnten anhand vieler Gleitmikrometerauswertungen überprüft werden. Stellt man den Verformungszuwachs, der innerhalb eines Meßintervalles einer Hebungsphase zuzuordnen ist, dar, so zeigen sich zumeist recht typische Verläufe mit großen Dehungen im Mattenbereich, nicht unwesentlichen Stauchungen unterhalb der Matte und

geringfügigen Stauchungen im Bereich
zwischen Fundament und Matte. Beim
Gleitmikrometer 3/1 bedeutet bspw. die
Auswertung eines Meßintervalles, welches
die Hebungsinjektion nach einem kleinen
Wassersandeintritt an der Ortsbrust
umfaßte, daß etwa 50 % der durch die
Feststoffeinpressung in der Matte
bewirkten Dehnungen für die Fundamente als
voll hebungsaktiv zu sehen waren.

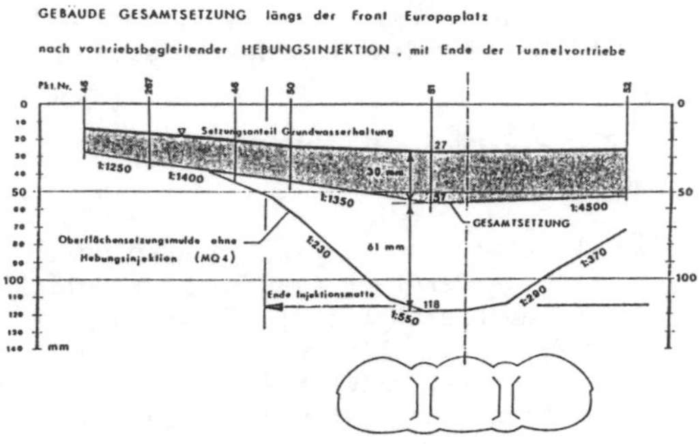

Abb. 5
Gebäudegesamtsetzung längs der Front Europaplatz.

Total settlements of building along the front
of Europaplatz.

Betrachtet man die Oberflächen-
setzungsmulden außerhalb des ÖBB-
Gebäudes im Bereich der unbebauten
Grünfläche und längs der Front
Europaplatz, wie sie sich mit Ende
der Tunnelvortriebe darstellten,
so zeigt sich klar ersichtlich der
Erfolg der gesetzten Maßnahmen.
Zieht man den weitläufigen
Setzungsanteil der Wasserhaltung
ab, so zeigen sich außerhalb der
Matte die maximalen
vortriebsbedingten Setzungen mit
90 bis 95 mm um ca. 200 % größer
als im Bereich des
Setzungsausgleiches, wo etwa 30 mm
erreicht wurden. Weiters konnten
die maximalen Tangentenneigungen,
die außerhalb des Gebäudes
letztendlich bis auf Werte von
1:230 bis 1:290 abfielen,
innerhalb des von der
Injektionsmatte beeinflußten
Bereiches, auf Werten zwischen
1:1350 bis 1:4500 gehalten werden.

Zur Problematik der Abdichtungen in der rechten Felsflanke des Dammes Feistritzbach

Difficulties of grouting works in the right rock bank of Feistritzbach dam

Ernst H. Weiss
Universität für Bodenkultur Wien, Österreich

ZUSAMMENFASSUNG: In der südlichen Koralpe wurde im steirisch-kärntnerischen Grenzgebiet zur Republik Slowenien in den Jahren 1988 bis Spätherbst 1990 der 85 m hohe Steinbrockenschüttdamm Feistritzbach von der Kärntner Elektrizitäts AG errichtet. Trotz der begleitenden und umfangreichen Arbeiten am maximal 70 m tief reichenden Injektionsschirm konnten in der rechten Gneisflanke die Sickerwasserverluste beim Erstaufstau nicht gestoppt werden. Erst durch die Ausführung eines Stollens knapp luftseitig der Schleierebene wurde eine quer stehende, 15 m breite Sekundärstörung als Verursacher der hohen Durchlässigkeiten entdeckt! Hinweisende Spuren oder Merkmale dieser Struktur an der Oberfläche waren vorher nicht zu beobachten. Die vom 108 m langen "Injektionsstollen" ausgerichteten Abdichtungsmaßnahmen führten zwar verspätet, doch zum Erfolg und am 21. Dezember 1992 wurde der Vollstau erreicht.

In der vorliegenden Arbeit werden die einzelnen Schritte der Injektionsphasen - unter Einbeziehung der geologischen Kenntniserweiterung als Folge der notwendig gewordenen Abdichtungsstrategien - und die Problematik der rechten Flanke aufgezeigt.

ABSTRACT: The 85 m-high Feistritzbach dam, which was built from 1988 to 1990, is situated in the Provinces Styria and Carinthia, near the boundary to the Slovenian Republic and is an important construction of the water power utilization "Koralpe" of KELAG, Carinthia (fig. 1). The rockfill dam with a volume of 1.700.000 m^3 is founded in old metamorphic gneisses (fig. 2, 5). The asphaltic concrete core is placed on the cutoff wall with the inspection gallery (fig. 4). From this grouting inspection gallery the grout curtain was made into a depth of max. 70 m (fig. 3, 4).

The annual storage capacity of the reservoir Soboth is 16.230.000 m^3, and its tributary is named river "Krumbach". The utilization happens through pressure tunnel and pipes to the powerhouse Lavamünd, the hydraulic gradient notes with 735,5 m.

The preliminary investigations and foundation explorations in the years 1976 - 1981 revealed that the gneiss formation was structurally very weathered, only parts were sound (fig. 5, 7) with main joints and faults. The underground rock presents no major problems to make watertight. At the damsite the river "Feistritzbach" cuts in the gneiss mass, which contains gneisses, pegmatites and marbled bindings with different thickness to max. 12 m. On the left bank there are anticline, on right syncline structures, and western of "primary fault" (fig. 10) there is the bedding unconform. The gneiss on the right bank is 20 m and more weathered, the main part of surface is overlied by mighty blocks, debris and weathered materials or loamy soils. Prospecting by borings has shown a relatively intact rock mass in the depth with specially weathered and fractured zones. During the execution of the rockfill damm the groutings began with Standard Portland cement from the inspection gallery in the dam axis. After finishing of this primary grout curtain (stage 1) a higher rock leaking in the region of valley axis and on the rigt bank was ascertained. The chemical (stage 2) in the groutings with S.P. cement and microcement types MXI and MXII followed (stages 3, 4 and 5). The sealing up started in the zone around the downstream diversion tunnel (=unwatering conduit) immediately by the concrete plug ("Stoppel") with "starlike" directions of drillings (stage 4 b). Water pressure tests, colour tests, corings and the reality of groundwater flow from reservoir to the drains (E 1 - *E 3* - E 7) in the unwatering conduit show, that the rock mass on the right bank was not watertight!

The following excavation of injection gallery on the right bank on 1013 m level with 108 m length crossed suddenly the primary fault and the big shearing zone in a 15 m extent. Mylonites, shearing zones with marbles, "red sands" (a product of an old deep weathering) and permeable spaces were observed (fig. 9). This discontinous fault was not mappable on the surface, because the shearing structure was beginning in the depth nearby the western primary major fault (fig. 10). Through the structure infiltrated reservoir water to the drains - specially to E 3 in the unwatering conduit.

In the following times groutings concentrated with cement and MX1 on the right side with the stages 6, 8, 9 and 10 in direction of the tectonical structures. During the groutings (fig. 8) corings, colour tests and continously measurements of the leakage were made. In this last injection period values of injections units to 11 tons by 33 cm pressure length were determined! After finishing the works was arrived the dammed-up water followed very slowly under controlling and the top water level in December 1992.

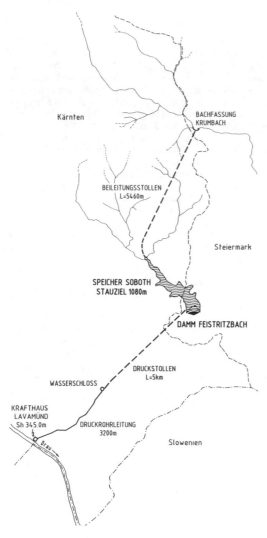

Abb. 1: Übersichtsplan
Fig. 1: General layout

1 EINFÜHRUNG

Zur energiewirtschaftlichen Nutzung der Abflüsse der südlichen Koralpe (Krum- und Feistritzbach mit E = 66,8 km^2) errichtete die KELAG den Damm Feistritzbach (I = 1,7 Mio m^3). Der Krumbach wird durch einen Stollen dem Speicher Soboth (Nutzinhalt 16,23 hm^3) zugeführt (Abb. 1). Die Abarbeitung erfolgt über den Druckstollen und die Druckrohrleitung zum Krafthaus Lavamünd in die Drau (installierte Leistung 50 MW). Die Rohfallhöhe beträgt 735,5 m.

Der Steinbrockenschüttdamm (Abb. 2) ist mit einem Asphaltbetondichtkern (D = 50 - 70 cm), auf der Herdmauer mit Kontrollgang fundiert (Abb. 3, 4), ausgestattet. Die Stützkörper setzen sich aus qualitätsunterschiedlichen Gneismaterialien zusammen, das Gestein wurde in drei Steinbrüchen des Stauraumes gewonnen, für die Filter- und Bremszonen wählte man gutes Hartgestein. Gleichlaufend mit der Dammschüttung wurden vom Kontrollgang aus in der Dichtungsebene die Injektionen mehrphasig und mehrreihig ausgeführt (Abb. 4).

2 GEOLOGISCHE GRUNDLAGEN - "ERSTER KENNTNISSTAND"

Bereits Anfang der 60iger Jahre setzten die ersten Erkundungen ein und nach Konkretisierung des Projektes "KW Koralpe" hat H. LITSCHER mit den Detailuntersuchungen 1976 begonnen und zusätzlich zur geologischen Kartierung wurden 11 Aufschlußbohrungen (Abb. 5) und seismische Prüfungen vorgenommen. Außer den Kernbohrungen erfolgten zur Prüfung der Gebirgsdurchlässigkeit und Wasserwegigkeit Abpreßversuche, die generell ein dichtes Gebirge anzeigten, in einigen wenigen Passen waren lokal Spitzenwerte von 14 Lugeon zu beobachten, in den liegenden Kalksilikat-Marmoren (Abb. 7) sprang der Wert auf 15,84 Lugeon.

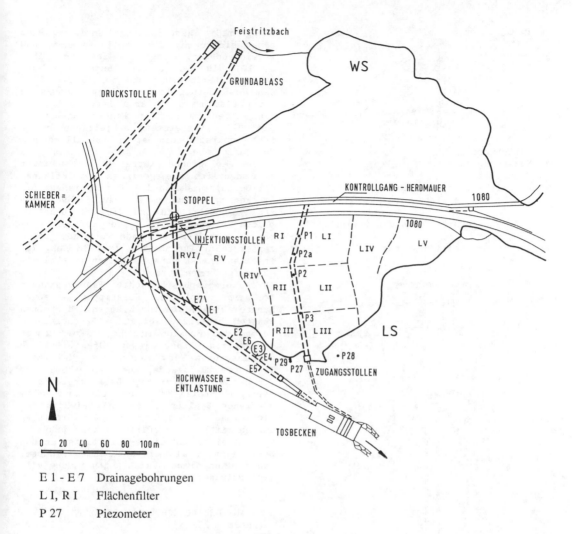

Feistritzbach

WS

DRUCKSTOLLEN

GRUNDABLASS

KONTROLLGANG – HERDMAUER 1080

SCHIEBER =
KAMMER   STOPPEL

INJEKTIONSSTOLLEN   R I P1 L I   1080

R VI   R V   P2a   L IV   L V

R IV   P2

R II   L II

E 7   P3

E 1

E 2   R III   L III   LS

E 6

E 3   E 4 P 29   • P 28

E 5   P 27   ZUGANGSSTOLLEN

HOCHWASSER =
ENTLASTUNG

N

0 20 40 60 80 100 m

TOSBECKEN

E 1 - E 7   Drainagebohrungen

L I, R I   Flächenfilter

P 27   Piezometer

Abb. 2: Lageplan des Dammes
Fig. 2: Plan of dam site

Im Vergleich mit anderen im Altkristallin und in den Zentralgneisen der Hohen Tauern liegenden Sperren wurde eine geringere Durchlässigkeit, die sich mehr auf die geländenahen Zonen beschränkte, angenommen. Obzwar keine Testfelder vor Injektionsbeginn eingerichtet wurden, erstellte man für das Gneisgebirge mit dem ausgeprägten Trennflächeninventar ein, wie wir glaubten, ausreichend abgesichertes Dichtungsprogramm.

Als dominierende Gesteinsart liegt der Disthenflasergneis vor, der mit flaserigen Ausbildungen, Glimmerschiefereinschaltungen, Pegmatiten und quarzitischen Typen den Untergrund des Sperrenbereiches auf-

baut. Unangenehm wurden von mir die Marmoreinschaltungen empfunden, da sie vorgegebene Wasserwege darstellen.

Beim Vergleich eines geologischen Projektsquerschnittes vom Mai 1977 in Dammachse mit dem Querschnitt aus dem Jahre 1990 (Abb. 7) ist festzustellen, daß im ersten Erkundungsstadium generell die Struktur des Untergrundes erkannt wurde: Im linken Einhang liegt eine Antiklinale vor, die durch die Talstörung abgeschnitten wird. Rechtsseitig herrscht ein leichter Synklinalbau, welcher sich im westlichen Schenkel in Richtung zur Hochwasserentlastung versteilt.

Diese Veränderung im Lagerungsgefüge des

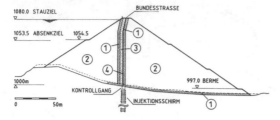

1) Filterzone
2) Steinschüttung
3) Bremszone
4) Asphaltbeton-Innendichtung
   (Asphaltic concrete core)

Abb. 3: Querschnitt durch den Steinschütt-
        damm
Fig. 3: Cross section of the rockfill dam

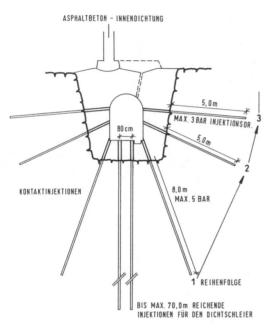

Bohrlochanordnung für die Tiefeninjektionen
(Bore holes positions of grout curtain)

Abb. 4: Schnitt durch Herdmauer und
        Kontrollgang
Fig. 4: Cross section of cutoff wall and
        inspection gallery

unregelmäßig geschieferten Gneiskomplexes hängt sicher mit der steilstehenden, NNW streichenden Störungsbahn zusammen (Abb. 5, 10) die leider bis heute wegen der starken Überlagerung und der tiefreichenden Verwitterung in ihrer Längserstreckung nicht klar zu definieren ist. Im Baustadium kamen im rechten Einbindungsbereich die völlig zersetzten, schiefrigen Gneise fast senkrecht bis steil ostfallend zum Vorschein.

Die geologische Karte (Abb. 5) zeigt sehr deutlich die geringen Ausbißflächen der Gesteine, dafür die Überlagerungen mit Blöcken bis Schuttmäntel und Bachsedimenten. Die Bohrungen und die Felsaushubarbeiten dokumentierten gerade auf der rechten Flanke tiefreichende und hochgradig zersetzte Gneiseinheiten (M-V in Abb. 7), in denen später hohe Mengen an Injektionsgut aufgenommen wurden.

In der orographisch rechten Flanke standen teilweise isolierte Felsrippen - sogenannte "Öfen", die von Klüften und breiten Spalten begrenzt, selektive Erscheinungsformen der Erosion mit hoher Kompaktheit darstellten. Einer dieser "Öfen", östlich der markanten Verwitterungszone M-V (Abb. 7), wurde im Zuge des Künettenaushubes für die Herdmauer gesprengt. Dabei traten ausgeprägte Steilklüfte und bis zu 40 cm klaffende Spalten berg- und talwärts der Blockeinheit auf. Die offenen Großfugen wurden mit Beton verfüllt, später jedoch festgestellt, daß diese weit tiefer in den Berg reichen, als man anzunehmen glaubte, daher waren diese Abdichtungsversuche völlig unwirksam!

3 ABLAUF DER ABDICHTUNGSMASSNAHMEN
  (PHASE 1 BIS 5)

Im Zuge der vom Kontrollgang ausgeführten Injektionsarbeiten, die H. LITSCHER als Geologe steuerte, waren S. JACOBS und zwei Diplomanden, R. SPREITZER und K. WALCHER auf der Baustelle im Einsatz, letztere dokumentierten vortrefflich die einzelnen Injektionsphasen, die nachfolgend kurz umrissen werden. Zur Übersicht der einzelnen Phasen dient die Abb. 6 und die folgende Tabelle.

Phase 1 - Primärschleier

Nach Ausführung der Kontaktinjektionen folgten die einzelnen Injektionsphasen in Felder aufgeteilt (F I bis F IX) vom Kontrollgang. Der Abstand der Injektionsbohrlöcher für die beiden Schleier der Tiefeninjektionen (Abb. 4) ebenso die Tiefe

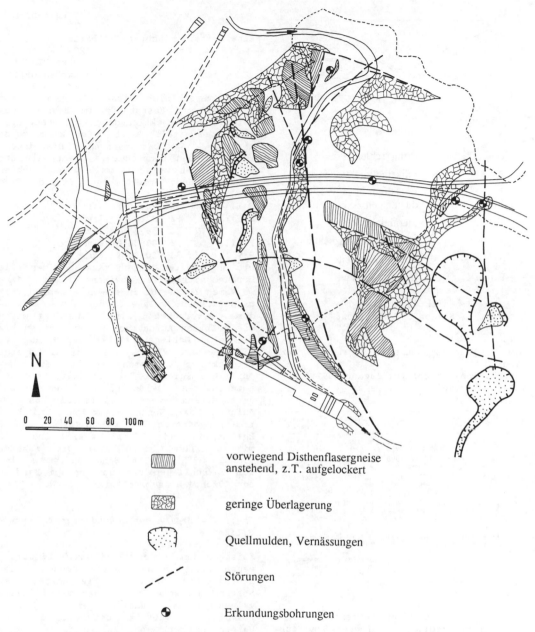

vorwiegend Disthenflasergneise
anstehend, z.T. aufgelockert

geringe Überlagerung

Quellmulden, Vernässungen

Störungen

Erkundungsbohrungen

Abb. 5: Geologische Karte der Sperrenstelle, nach H. LITSCHER, 1981
Fig. 5: Geological map of the dam site, by H. LITSCHER, 1981

hing von der geologischen Erhebung, von der maximalen Einpreßmenge und dem zulässigen Injektionsdruck ab. Die Festlegung der Abstände erfolgte nach Erfahrungswerten oder über Probeinjektionen, die vor Ort ausgeführt wurden. Über die Bestimmung der Durchlässigkeit (WPT) und der Prüfung eines Injektionsverfahrens wurde nach Abbindung des Zementgutes und der Abteufung einer Kernbohrung wiederum eine Dichtheitsprüfung vorgenommen und das Bohrloch zum zweitenmal injiziert.

Der Injektionsablauf vollzog sich nach dem "Pilgerschrittverfahren", es wurde von

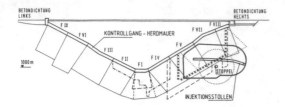

PHASEN:

1	⌞___⌟	Primärschleier
2	—·—·—	Silikatgelinjektionen
3	— — — —	Injektions-
4a	·············	dichtungen
4b	◯	Sterninjektion
5	xxxxx	Nochmalige Verdichtung
6	——	Injektionsstollen
F I - IX		Injektionsfelder

Abb. 6: Injektionsphasen 1 - 5
Fig. 6: Stages of injections 1 - 5

Tab. 1: Injektionsphasen - Damm Feistritz-
bach;
Informationsbericht vom August 1991 der
KELAG - korrigiert E.H. Weiss

Phase	Injektionsmittel	Zeitraum	Stauhöhe	Sickerwas-sermenge
1	Zement PZ 375	Juli 1988-Nov. 1990	um 1020 m	ca. 18 -20 l/sec
2	Gel-Injektion	Juni 1990-Feb. 1991	um 1020 m- 1027 m	ca. 18 -20 l/sec
3	Zement PZ 375+ Mikrozement	Feb. 1991-März 1991	um 1027 mmax. bis1030 m	14 l/sec
4	Zement PZ 375+ Mikrozement	Apr. 1991-Mai 1991	um 1030 mAnstiegauf1031,50 m	14 l/secAnstiegauf20 l/sec
5	Zement PZ 375+ Mikrozement	Mai 1991-Aug. 1991	Absenkungauf1018,00 mAnstiegauf1039,90 m	---Anstiegauf11 l/sec

unten nach oben mit Manschettenrohren in 5
m-Passen verpreßt. Am laufenden Band wurden Umläufigkeiten registriert, die auf
erhöhte Wegsamkeiten über offene Trennflächen zu interpretieren waren. Neben vertikalen Bohrungen wurden auch normal auf die
Sohlneigung des Kontrollganges ausgerich-
tete abgeteuft und in bis zu 6 Durchgängen
verpreßt. Im Feld IV und V erreichte man
Teufen bis über 70 m! Die hauptsächlich
verwendete Mischung (1) bestand aus:

220 l Wasser
230 kg PZ 375

4,6 kg Bentonit
(W/Z = 0,96)

Eine weitere Mischung (2) hatte

220 l Wasser
250 kg PZ 375
5 kg Bentonit
(W/Z = 0,80)

Im Primärschleier wurde ein Mengenlimit
von 1000 kg Feststoffe/lfm Bohrung gesetzt. Der Injektionsdruck war teufenge-
recht gestaffelt, erst unter 20 m wurde
mit 30 - 40 bar verpreßt. Die durchschnittlichen Einpreßmengen innerhalb der
Felder lagen zwischen Talung und rechter
Dammeinbindung wie folgt (Stand Ende
1990):

F I: PZ 20,00 kg/lfm
F IV: PZ 22,00 kg/lfm
F V: PZ 56,75 kg/lfm
F VII: PZ 103,68 kg/lfm
F VIII: PZ 20,08 kg/lfm

Die angeführten Mengen täuschen, denn die
Durchschnittswerte zeigen nicht jene
wegsamen Zonen auf in die hohe Mengen verpreßt wurden, welche in der Abb. 7 besonders hervorgehoben sind. Eindeutig waren
die höchsten Aufnahmen in den höheren bis
mittleren Gebirgsbereichen der Felder V,
VII und auch VIII - also auf die später
entdeckte Störungszone konzentriert. Natürlich reagierten die aufgewitterten und
aufgemürbten, geländenahen Zonen besonders. Erwähnenswert dabei ist die hohe PZ-
Aufnahme der Kontaktinjektionen im Kon-
trollgangbereich der Felder V mit 45,48
kg/lfm, VII mit 261,15 kg/lfm! und VIII
mit 76 kg/lfm. Im Feld VII wurde zusätzlich 10 m^3 Kies, 2 m^3 Sand und 4 m^3 Mörtel
zum Verschließen der Spalten im Bereich
des "Ofens" eingebracht (KELAG - Nov. 91)

Phase 2 - Beginn der Sekundärinjektionen -
Silikatgel

Zwischen den Feldern F II und F IV wurden
wegen der erhöhten Aufnahmemengen, der
Störungszone und des Tiefpunktes der
Sperre chemische Injektionen mit Silikat-
gel vorgenommen. Während der Arbeiten am
Primärschirm war festzustellen, daß der
Untergrund doch mehr zerlegt und verwit-
tert ist, sowie gewiße Bereiche auf die
Injektionen nicht angesprochen hatten.
Folglich wurde für den gesamten Bereich im
ersten Durchgang zuerst mit PZ und dann
mit Silikatgel verpreßt.

Mischung: 9 l Silikatgel
            18 l Wasser
             1 kg Kalziumchlorid (wird mit
                4 l Wasser aufgerührt)
            20 - 30 Minuten = angestrebte

Drücke:

Abbindezeit
Im Kontaktbereich
5 - 8 m Tiefe : 3 - 5 bar
Dichtschleier
je 5 m   : 5 bar
maximal   : 30 bar

Aufnahmen: PZ = 16,51 kg/lfm
                  Silikatgel = 24,84 l/lfm

Um eine höhere Dichtheit zu erhalten, erfolgte ein zweiter Durchgang mit einer
Aufnahmekapazität von Silikatgel =
32,75 l/lfm. Daraus ist zu ersehen, daß
der Fels unter der Bachwiege im 2. Durchgang noch mehr chemisches Injektionsgut
aufnahm und eine weitere Injektionsphase
erforderlich wurde.

Nach diesem Injektionsschritt erfolgte
mit Genehmigung der Wasserrechtsbehörde
(Nov./Dez. 1990) der erste Aufstau bis auf
Höhe 1020 m und es traten am luftseitigen
Dammfuß und aus dem Spritzbeton des
Grundablaßstollens, ungefähr wo die E-Mar-
kierungen in Abb. 2 eingezeichnet sind,
schlagartig Sickerwässer auf (in Summe 18
- 20 l/sec) und bei einem weiteren Stau
bis 1027 m erhöhte sich die nicht mehr exakt meßbare Gesamtmenge.

*Phase 3 - Injektionsverdichtung (Einsatz
von MX 1)*

Seit der 1. Phase war die hohe Aufnahmebereitschaft und leider auch Durchlässigkeit
des Gebirges in den oberen Feldern bis zum
Grundablaßstollen durch die Bohrungen, WPT
und Färbeversuche erkennbar geworden. Der
Stau schwankte von 1018 bis 1027 m und
wurde nur für die Injektionen im F I zu-
rückgenommen, die Sickerwassermenge pen-
delte um 14 l/sec. Die nunmehr bekannten
Bereiche erhöhter Durchlässigkeit wurden
nun mit PZ und einem Feinzement der Bezeichnung Micromix MX1 verpreßt und dabei
festgestellt, daß nach dem Einpreßen von
PZ, nachträglich noch 25 % bis fast 50 %
MX 1 vom geklüfteten und aufgemürbten Gebirge aufgenommen wurden. Die guten physikalischen Eigenschaften, in der Tab. 2 angeführt, bewirkten bei Drücken über 40 bar
ein günstigeres Einschleusen in die feinen
Fugen sowie Mikroräume und verbesserten,
durch WPT überprüft, den Abdichtungsef-
fekt.

Aufnahmemengen:
F I .... MX 1 = 11,57 kg/lfm
F IV,V,VII,VIII: PZ + MX 1 = 36,67 kg/lfm
    Als Folge dieser Maßnahme ging die
Sickerwasserschüttung um bescheidene 1,5
l/sec zurück und der Stau wurde von Mitte

Tab. 2: Auszug aus dem Prüfergebnis
Micromix MX1
(Techn. V-F-Anstalt, TU-Graz, 1993)

Art der Prüfung		Meßwert
Reindichte	(g/dm^3)	3,01
Schüttdichte	(kg/m^3)	696
Blaine-Wert	(cm^2/g)	9930
Spez.Oberfläche BET	(cm^2/g)	30200
Wasserzusatz f. Normensteife	(M%)	58
Erstarrungsbeginn	(min)	110
Erstarrungsende	(min)	160
Raumbeständigkeit		bestanden
Ausbreitmaß Mörtel	(cm)	12,4 x 12,7
Biegezugfestigkeit	(N/mm^2)	8,85
Druckfestigkeit	(N/mm^2)	61,9
mittlerer Korndurchmesser d50	(μm)	2,4
Korndurchmesser den 90% unterschreiten	(μm)	6,5
Kasumeter (lk/dk = 40; W/Z: 2,0)		
Auslaufzeit	(min)	3,5
Stagnationshöhe	(cm)	6,3

März bis Ende April 1991 auf 1030 m erhöht. Gleichzeitig wurde wegen der doch zu
hohen Sickerwassermengen, die direkt aus
dem Berg und aus dem Flächenfilter R III
stammten, im Grundablaßstollen die Drainagebohrungen E 1 bis E 7 vorgenommen (Abb.
2), um so eine Bergwasserentspannung ein-
zuleiten und die Zuflußrichtung zu erkun-
den. Als Indikator für die Mengenmessungen
und für die Ausspiegelung des aquiferähn-
lichen Bergwassers mit den Piezometern P
27, P 29 hat sich die Bohrung E 3 als er-
wichtig erwiesen, denn sie springt als erste Entlastung an. Jede Staubewegung kann
hier registriert und kontrolliert werden.
Im Juli 1992 wurden zum Zwecke der besseren Entlastung zusätzlich drei Bohrungen
abgeteuft, so daß auf 6 m Streckenlänge 4
E-Bohrungen (als Quellgruppe E 3 bezeichnet) vorliegen. Werden die Absperrventile
dieser Gruppe geschlossen, dann steigen
die Spiegel in den genannten Piezometern
und die Schüttungen in R III. Diese Entla-
stungsmaßnahme war notwendig geworden, als
man über Färbeversuche (Uranin) feststellen mußte, daß eine Verbindung zwischen
aufsteigenden Bohrungen, die aus dem Stop-
pelbereich (Grundablaßstollen) ausgeführt
wurden, eine direkte Verbindung zum Flächenfilter R III besteht! Die Färbemittel
kamen von zwei Einspeisestellen in zwei
steil zur Oberfläche weisenden Bohrlöchern
nach jeweils 2 Stunden im E 3-Bereich zum
Austritt. Zusätzlich wurden sechs Verbindungen zwischen Herdmauer-Kontrollgang zum
R III mit einer Zeitdauer von 4, 8, 22 und
30 Stunden nachgewiesen.

Damit konnten Wegsamkeiten zwischen dem
Flankenkern des Gebirges zum rechten auf
der Oberfläche aufsitzenden Flächenfilter
bzw. über Trennflächenstrukturen zum Ein-

gangsbereich des Grundablaßes aufgezeigt werden. Aus den Prüfungen, den geologischen Überlegungen und den Ergebnissen der Injektionsdurchgänge resultierte die gemeinsame Entscheidung von Bauherrn und uns Sachverständigen, daß die Durchlässigkeitsfront konzentriert eine Zone betrifft, die unbedingt durch weitere Dichtungsmaßnahmen zu bekämpfen ist. Somit war die Freigabe zu einem Weiterstauen von den Sickerwassermengen abhängig.

Phase 4 - zusätzliche Dichtungsmaßnahmen

Noch während die Arbeiten im Kontrollgang für die Phase 3 liefen, begannen die neuen Injektionen im oberen Anteil des Kontrollganges (4 a) und im Grundablaßstollen (4 b). Damit wurde der Injektionsfächer in der Dichtungsebene von Phase 3 verstärkt und "sollte" die Zone über dem Grundablaßstoppel wirksam abdichten!

Die Aufnahme von 4 a im zweiten Durchgang betrug ca. 9,55 kg/lfm. Die Bohrungen wurden vertikal und schräg gegen den Bergleib geneigt, vom Kontrollgang abgeteuft. Über den Grad der Dichtwirksamkeit konnten damals noch keine Aussagen getroffen werden und sie blieben eigentlich den Ergebnissen der Phasen bis einschließlich 10 vorbehalten.

4 b - Sterninjektionen vom Stoppel

Insgesamt 7 Injektionsschleier vom Grundablaß knapp luftseitig der primären Schleierebene wurden sternartig ausgeführt. Die zuordenbaren Bohrungen wurden 10°, 15° und 20° von der Stollenquerschnittsebene zur Wasserseite geneigt um den direkten Kontakt mit der Dichtungsfront herzustellen. Gleichzeitig konnte vom Stollen aus jene Zone mit den großen Durchlässigkeiten (siehe Abb. 7) systematisch erkundet und verpreßt werden. Als Mischungen wurden verwendet:

1) Normalzement - Stoppel, 45 l Wasser, 50 kg PZ 375, 1 kg Bentonit, trocken, W/Z = 0,9

2) Nach Aufbohren eines injizierten Bohrloches oder in der Endphase einer Injektionsreihe mit sehr hohen Aufnahmemengen wurde verwendet: MICROMIX MX 2 - Stoppel, 45 l Wasser, 25 kg MX 2, W/Z = 1,8!

Zusätzlich wurden drei Versuchsmischungen gesondert ausprobiert, erreichten nach K. WALCHER aber keinen durchschlagenden Erfolg. Gegenüber dem MX 1 hat der hier eingesetzte MX 2 eine geringere Korndichte, eine angehobene Korngröße und ist die spezifische Oberfläche geringer.

Tab. 3: Übersicht der Injektionsgutaufnahmen in den Schleierebenen der "Stoppelbohrungen"

PZ .... Portlandzement 375
MZ .... Microzement MX2
JH1 .... Injektionshilfe 1

Schleier	Injektionsgut	[t] Summe	größte Aufnahme [kg/lfm] [Bohrung, Grad, Passe]
I	PZ 38,2 t		1313,5
	MZ 11,8 t	Σ 50,0 t	[2, 145°, 5 - 7 m]
II	PZ 1,0 t		190,6
	MZ 4,5 t	Σ 5,5 t	[113A, 220°, 10 - 15 m]
III	PZ 1,0 t		147,8
	MZ 5,3 t	Σ 6,3 t	[3, 180°, 5 -9 m]
IV	MZ 0,4 t		55,0
	MZ+JH1 0,3 t	Σ 0,7 t	[13D, 200°, 0 - 2 m]
V	PZ 1,2 t		315,1
	MZ 4,1 t	Σ 10,7 t	
	MZ+JH1 5,4 t		[13B, 210°, 30 - 35 m]
VI	PZ 1,3 t		643,0
	MZ 11,2 T	Σ 15,6 t	
	MZ+JH1 3,1 t		[201A, 210°, 25 - 30 m]
VII	PZ 4,9 t		1743,0
	MZ 21,7 t	Σ 26,6 t	[200, 230°, 10 - 11 m]

In der folgenden Tab. 3 sind die Injektionsgutaufnahmen in PZ und MZ (= MX 2!) für jede Schleierebene angegeben und auf der rechten Rubrik die größten Aufnahmen der jeweiligen Bohrung (Nr 2, 145° geneigt, Aufnahmebereich 5 - 7 m als Beispiel) angeführt.

Aus der Übersicht sind die hohen Injektionsgutaufnahmen, stets in den Bohrungen talseits des Stoppels, festzustellen und erreichten Spitzenwerte von 1,3 t/2 1fm und 1,7 t/1 lfm! Bei Betrachtung des Verhältnisses PZ : MX2 wurden

im Schleier III nach der PZ-Injektion das fünffache,
im Schleier V das vierfache,
im Schleier VI das achtfache und
im Schleier VII das viereinhalbfache
an MX2 verpreßt!

Somit erwies sich das Nachinjizieren mit Feinzement, wie auch die alleinige Verwendung als sehr erfolgreich im Bereich der mürben, verwitterten und kluftdurchzogenen Gebirgsformation. Nach einer internen KELAG-Aufstellung (Stand 18.11.1991) war für den Stoppel-Grundablaß folgender Aufwand erforderlich:

Bohrungen: 5.254 lfm
PZ   : 51.766 kg
MX2   : 65.770 kg

Phase 5 - nochmalige Verdichtung

Über die gesamte Front von F IV bis zum F VIII legte man wegen der noch immer be-

achtlichen Aufnahmemengen in den vorhergehenden Phasen einen weiteren Dichtungstep-
pich an, um die erkannte durchlässige Zone
nun einzuengen. Die Kernbohrungen erschlossen eine spitz nach unten zu verlau-
fende Lamelle, die als tektonische Störung
angesprochen wurde. Über die tiefreichenden, senkrechten Bohrungen wurde damit der
Kontakt zu den übrigen Dichtschleiern hergestellt und konnte mit dem Aufstau bis
zur Kote 1039,90 m fortsetzen. Hier
schnellte die Sickerwassermenge wieder
hoch, die E3 von 7,5 l/sec auf 15 l/sec.
So kam die Erkenntnis, daß die Maßnahmen
doch nicht ausreichen und eine neue Stra-
tegie gefragt ist.

4 BILANZ DER DICHTUNGSMASSNAHMEN - "ZWEITER KENNTNISSTAND"

Die bisherigen Erfahrungen führten zu folgenden Schlüssen: In den Kernbohrungen
wurde zwar sehr geklüfteter aber kompakter
Fels angetroffen, mit PZ gefüllte Spalten
zwischen 7 und 12 cm bewiesen zwar die
Wirkung der Maßnahmen, jedoch waren lokalisierbare Bereiche mit hohen Kernverlu-
sten und sandig-tonige Mürbzonen von einigen Metern Mächtigkeit festzustellen. Das
tonreiche, sandige Material ("Roter Sand")
war nicht bis schwer injizierbar und
dürfte genetisch durch stillgelegte Ther-
mal- oder mineralisierte Wässer in einem
zerscherten Gebirgskörper entstanden sein.
Wichtig war die Beobachtung, daß die
Sickerwässer nie trüb und das eingebrachte
Injektionsgut trotz des leichten Aufstaues
abgebunden war. Ebenso fehlten Nachweise
von Sand- oder Tonausschwemmungen, die als
Folge des Injizierens gegen den Stau zu
erwarten waren. Wiederholt mußten wir bei
Injektionen unter der Nivelette von 1020 m
auf eine Stauabsenkung drängen, denn die
Kluftwässer verschleppten die Injektionsmittel sehr rasch. Vielmehr war die Meinung, die Injektionen hätten in der Dichtebene einen Höcker von unten her bis auf
Höhe 1018 m aufgebaut und übersteigt der
Stauspiegel diese Marke, dann fließt Wasser über die Kante dieses Injektionskörpers zu den Drainagen. Der Nachweis mit
dem Injektionsgut in Spalten und Klüften
vom Primärschirm und den folgenden Phasen
stammend, schloß eine solche Vorstellung
jedoch aus. Außerdem waren die Anzeichen
großer Wegsamkeiten im rechten, auch tie-
ferliegenden Flankenanteil evident. Vielmehr wurde ein Modell der rechten Flanke
hergestellt und an Hand der eingepaßten
Kernbohrungen konnte jene durchlässige und
geologisch sehr inhomogene Zone aufgezeigt
werden. Über die genaue Ausbildung, den
Entstehungsmechanismus und über die Durch-

schleusungswege der Wässer wußten wir relativ wenig, lediglich die Bohrungen und
die Aussagen durch die Injektionen ließen
den Schluß zu, einer ausgeprägten Störungszone, die im Oberflächenbereich nicht
erfaßbar war, gegenüber zu stehen.

Die Durchsickerung von Stauwässern,
ebenso reiner Bergwässer über den Untergrund der rechten Flanke zum Indikator E 3
und die Querverbindung zu P 29, P 27
führte zur Vorstellung, daß die Einschleu-
sung nicht nur über diese Zone, sondern
auch über sich kreuzende und markante
Hauptkluftscharen, die in der Herdmauerkü-
nette registriert wurden, erfolgen kann.

*Zur Frage des "Sättigungsgrades" der
Sterninjektionen (Phase 4 b)*

S. JACOBS (1992-a) versuchte das "Maß" für
den Injektionserfolg der 7 Schleier aus
dem Stoppel herauszuarbeiten und über-
prüfte vorerst die vorhergegangenen Injektionsphasen hinsichtlich ihrer Wirkung und
Überlappung. Daraus konnte er den Sättigungsgrad in "injektionsgesättigt", z.T.
injiziert, "untersättigt" (hohe Aufnahmen
noch zu erwarten) und nicht verifizierba-
rer Dichterfolg ableiten! Der Bearbeiter
kommt zum Schluß, das sehr mürbe und sehr
stark geklüftete Gebirge wurde langsam gesättigt, jedoch bestehen noch Zwickel-
räume, die als untersättigt anzusprechen
sind. Durch die EDV-gestützte Auswertung
gelang ihm dieser Nachweis über Isolinien-
karten der Injektionsgutaufnahmen, die
auch den Überlappungseffekt der Schirm-
durchgänge aufzeigten. Beispielsweise wurden nach dem 6. Schleier noch beachtliche
Mengen bei Ausführung des 7. Schleiers in
lokalisierbare Bereiche verpreßt. Wie wir
später noch sehen, haben die nachfolgenden
Phasen 6, 8 - 10 extrem hohe Einpreßquan-
ten erreicht.

5 INJEKTIONSSTOLLEN - ZUSÄTZLICHE AUS- KÜNFTE UND INJEKTIONSFRONT DER PHASE 6

Trotz des Nachweises der guten Abdich-
tungswirkung mittels MX1, MX2 und der Verringerung der Sickerwassermengen sprachen
sehr viele Faktoren für die Einrichtung
einer neuen Injektionsfront in der rechten
Flanke. Es war nämlich zu erwarten, daß
bei einer Stauanhebung die Schüttungen der
Sickerwässer progressiv zunehmen und Ero-
sionen in den Klüften eintreten würden.
Optimistische Nachweise über den vollen
Dichterfolg der Phasen 1 - 5 waren nicht
zu führen. Im Juni 1991 schlugen die Sachverständigen daher den Vortrieb eines
Stollens (Abb. 6, 8), ausgehend vom Zu-

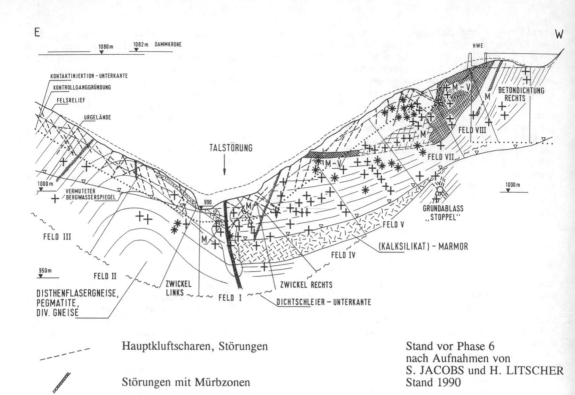

E　　　　　　　　　　　　　　　　　　　　　　　　　　　　　　　　　W

Abb. 7: Geologischer Schnitt durch die Dammachse
Fig. 7: Geological section along the dam axis

Hauptkluftscharen, Störungen

Störungen mit Mürbzonen

Stand vor Phase 6
nach Aufnahmen von
S. JACOBS und H. LITSCHER
Stand 1990

M　　Starke Verwitterung bis
　　　mürbe Zonen

M-V　Ausgeprägte Mürbzonen und völlige
　　　Verbandsentfestigungen (tiefgreifende
　　　Verwitterung)

Extreminjektionsgutaufnahmen (PZ 375 und Microzement)

+　　bis 500 kg/lfm

*　　500 - 1000 kg/lfm

gangsstollen zur Schieberkammer (Abb. 2) in Richtung Dichtebene vor, und nach Herstellung des 108,5 m langen "Injektionsstollens" (Abb. 9) erhielten wir einen aussagewichtigen Einblick in die Gefüge und Qualität der Disthengneisabfolge. Die Stollensohle steigt an und verläuft dann in einem 4 metrigen Abstand von der Achsenebene des Dammes fast waagrecht (1012,72 bis 1013,68 m Sh). Folgende Gesteinsbereiche liegen vor:

A　Zuerst steil nach West, dann sehr steil nach Ost einfallende Gneise, fast unverwittert, meist kompakt, mit scharfen Klüften, die teilweise starkes Tropfwasser führen.

B Die "sekundäre" Störungszone, die nur indirekt vermutet werden konnte, in einer Breite von 15 m (siehe das vereinfachte Detail in Abb. 9) mit dunklen Myloniten, aufgemürbten und zerscherten Gneis- und Marmorlinsen sowie die fein- bis grobsandige, schon früher erbohrte Serie der "Roten Sande".

C Hangend ist der Gneis wieder stärker verwittert, die Ost einfallende Schieferung verflacht sich zunehmend bis zu einer söhligen Lagerung. Der Gneis ist stärker von Klüften durchsetzt, die Wasser führen, tonig- lehmige Füllungen aufweisen und offene Fugen sind oftmals von PZ verfüllt. Damit konnte endlich der Gebirgszustand vor Ort gesichtet und die weiteren Maßnahmen konsequent verfolgt werden.

Phase 6 - Erkundungen - Injektionen -
Folgerungen

Der Verlauf des Bereiches B war besonders interessant und über zahlreiche Vertikal- und einigen Schrägbohrungen stellte sich die Störungszone, wie in Abb. 9 skizziert, so dar. Der Grad der Zersetzung der Gesteine (14 bis max. 25 m mächtig), die räumliche Verteilung des "Roten Sandes" und der Marmorscherlinge wurde erfaßt. Die völlige Zerlegung des unterschiedlichen Materiales wird von mir durch eine scherende Bewegung erklärt.

Mit der Steuerung der weiteren Injektionen paßte man sich an die tektonischen Strukturen und deren Materialkomponenten an, wobei die aufgelinsten Bereiche und die Säume der "Roten Sande" die höchsten Durchlässigkeiten aufwiesen.
Reihenfolge der Injektionen:

1. Vertikal von der Sohle 25 m nach unten, im talseitigen Bereich bis 35 m Aufnahmekonzentration und tiefer als Grundablaß mit Maxima zwischen 122 und 583 kg/ 5 lfm, 30-40 bar, in 200 bis 660 Minuten.
2. Fischgrätenartig vom First schräg zur Oberfläche und von der Sohle parallel zur Geländelinie, generell 40 m tief. Die Aufnahme geringer, mit vier kleineren Aufnahmestellen, ein Maximum erreichte 593,6 kg/5 lfm, 40 bar, in 1073 Minuten.
3. Schleierbohrungen von der Brust gegen die rechte Talseite und in den Gesteinsbereich A - Bergseite: unbedeutende Aufnahmemengen!
4. Vertikal von der Firste nach oben, 25 m - geringe Aufnahmen.

Aus den Ergebnissen war zu vermuten, daß nunmehr das Gebirge in der Dichtungsebene gesättigt und allmählich der Aufstau von Kote 1018 beginnen konnte. In weiser Voraussicht stimmten alle von der Gebirgsdurchlässigkeit geschockten Techniker und Experten einer Verdichtung der Injektionen zu, die vor allem jene aufnahmebereiten Bereiche der letzten Durchgänge betrafen. Zusätzlich führte der Bauherr noch gesonderte Maßnahmen aus.

6 LETZTE ABDICHTUNGSMASSNAHMEN
(PHASE 7 - 10)

Phase 7 - Linksseitige Schleierverdichtung

Diese Arbeiten erfolgten während des Injektionsstollenvortriebes und zeigten angehobene Injektionsgutaufnahmen punktuel

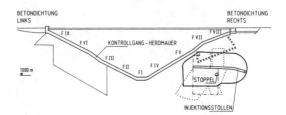

PHASEN:

6 ⌒ + —— Injektionsstollen und Mehrfachinjektionen

7 —— Linke Seite-Verdichtung

8 —·—·—·—
9 ·········· Zusätzliche Dichtungsmaßnahmen
10 xxxxxxx

Abb. 8: Injektionsphasen 6 - 10
Fig. 8: Stages of injections 6 - 10

1er Natur: Im oberflächennäheren Bereich Aufnahmen zwischen 101 und 284 kg/lfm, in einer Bohrung wurden zwischen 40 und 45 m Teufe 104 kg/lfm verpreßt, ansonsten lagen die Mengen unter 100 kg/lfm. Allgemein wird das linksseitige Gebirge als dicht eingeschätzt, obwohl über die gesamte Fläche (Abb. 8) 12,84 kg/lfm Microzement verpreßt wurden.

Phase 8 - Rechtsseitige, vertikal
ausgeführte Verdichtung

Aus dem Injektionsstollen durchgeführt, ergab sich flächenbezogen ein Wert von 13 kg/lfm MX1-Verpreßung. In der nachfolgenden Phase stiegen die Aufnahmen jedoch rapide an.

Phase 9 - Rechtsseitige Verdichtung

Hier wurden auf die beiden Flächen bezogen, PZ und MX1 im Ausmaße von 48,56 kg/lfm injiziert. Die Abdichtungen waren auf die Störungszone ausgerichtet, die Bohrlochabstände betrugen 1,0 m und es ergaben sich folgende Spitzenwerte:
1. Vertikalbohrungen vom First im direkten Ausstrich der Störungszone erbrachten Werte von 414 kg/lfm, 561 kg/lfm und an einer Stelle auf Kote 1030 m 8,5 t/1,5 lfm!
2. Schrägbohrungen von der Stollensohle:
Im talseitigen Fächer eine Stelle 339 kg PZ + 485 MX1/lfm, in einem Bereich unter Kote 980 m 2,5 t/lfm.
3. Vertikalbohrungen mit Bohrlochabständen zwischen 0,5 und 1,0 m brachten die

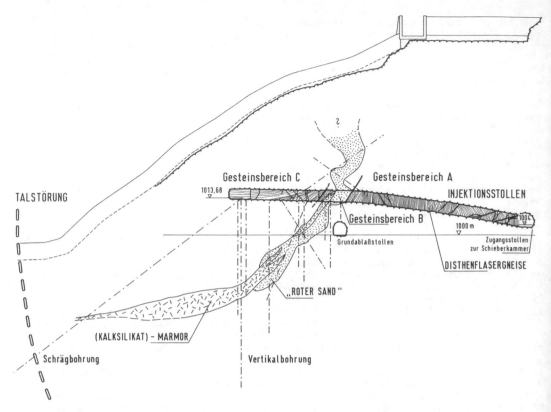

Abb. 9: Querprofil durch die rechte Flanke mit Injektionsstollen und Sekundärstörung
Fig. 9: Cross section of the right bank with injections gallery and secondary fault
        by H. LITSCHER, 1992

höchsten bisher ausgewiesenen Einpreßmengen - gerechnet zwischen den Stollenmetern 80 - 87 abwärts! Einen Wert von über 20 t/auf wenigen Laufmetern, an einer 33 cm-Passe wurden sogar über 11 t eingedrückt! Werte zwischen 1 t und 3 t/lfm im Bereich zwischen den Höhenkoten 990 und 1000 m sind von mehreren Stellen zu nennen. Alle diese extremen Eintrittsstellen betreffen die hangende Hälfte der hier minimal 6 m mächtig eingeschätzten Störungszone, die sehr steil gegen Osten abtaucht und dann sehr verflacht. Aus den Zahlenangaben ist zu ersehen, welche Aufnahmekapazität punktuell die tektonische Struktur hat und daher Mengenangaben über Injektionsflächen das Werturteil über den Sättigungsgrad sehr verfälschen. Diese Injektionsphase, in die nächste übergehend, zeigt auch auf, daß trotz 7-facher Überlappung der einzelnen Injektionsphasen noch immer hohe und extreme Mengen vom zerrütteten Streifen aufgenommen wurden.

Phase 10 - Verdichtung vom Kontrollgang

Diese ist ergänzend zur vorherigen Phase zu betrachten und wurden Schräg- und Vertikalbohrungen, gleichfalls auf die Störung und auf die Verwitterungszone (M und M-V in Abb. 7) ausgerichtet. Nach den Angaben der Bauleitung wird wieder nur ein flächenbezogener Wert von injizierten PZ und MX1 mit 32,47 kg/lfm angeführt.

Der Autor ist überzeugt, daß theoretisch nachfolgende Injektionsdurchgänge die Aufnahmebereitschaft dieses gesamten Bergabschnittes innerhalb der Dichtungsebene weiterhin unter Beweis gestellt wird.

7 DICHTUNGSERFOLG UND GEOLOGIE - "DRITTER KENNTNISSTAND"

7.1 *Weg und Ziel*

Mehrere Faktoren waren für das Erreichen des Erfolgszieles ausschlaggebend:
    Das frühzeitige Erkennen der stauabhän-

gigen Sickerwasser-Zugrichtungen, die In-
dikatordrainage E 3 und die Verbindung zu
den Piezometern 27 und 29 sowie die Konsequenz, einen Injektionsstollen zu errich-
ten, Microzement einzubringen und ziel-
strebig die über WPT, Färbeversuche und
Kernbohrungen eingeengte "Sekundärstörung"
zu suchen, zu finden und mit gerichteten,
aufwendigen Injektionen ihrer Herr zu werden. Im Laufe dieses Arbeitsweges vermehrten sich die Erfahrungen, die Kenntnisse,
auch die Zweifel jemals "dicht" zu werden.
Erst durch den massierten Einsatz in den
letzten Phasen gelang der Dichtschluß und
im Mai 1992 konnte mit dem eigentlichen
Aufstau begonnen und am 21. Dez. 1992 vor-
derhand das Ziel - der Vollstau erreicht
werden.

Im Zuge des langsamen Aufstauens mit begleitender Kontrolle wurde sofort das li-
neare, leichte Ansteigen der E 3 und der
Summenschüttung, ebenso der Flächenfilter
und fast unmerklich der Piezometer, festgestellt.

Auf Stauziel betrug dann die Summenschüttung aller Sickerwässer 28,5 l/sec,
im Injektionsstollen und im vorderen
Grundablaß traten zwar verstärkt rinnende
Wässer aus, im luftseitigen Vorland und am
Dammfuß waren keine Austritte zu beobachten. Die Schüttungen der L- und R-Flächen-
filter wurden leicht angehoben, die Piezometer P1, P2a stiegen stauabhängig an, die
übrigen verhielten sich nicht sehr aty-
pisch. In der Zwischenzeit wurde Anfang
April das Absenkziel und am 1. Juni 1993
wird wieder das Stauziel erreicht.

7.2 *Geologische Erkenntnisse*

Die während der letzten Injektionsphasen
ausgeführten Tests und Aufschlußbohrungen,
sorgfältig registriert von H. LITSCHER,
ließen erkennen, daß an den Rändern der
"sekundären" Störungszone die höchsten Injektionsgutaufnahmen in den Klüften und
Spalten vorlagen. Bis zu 20 cm lange Ze-
mentkerne wurden erbohrt! Im Nahbereich
waren viele Klüfte und kleinere Störungen
zementiert. Aus dieser parallel zu den Injektionsarbeiten laufenden Untersuchungs-
technik konnten wichtige Hinweise gegeben,
aber auch ein geologisches Bild über die
tektonische Struktureinheit vorgewiesen
werden (Abb. 9).

Die Darstellung aus den Stollen- und
Bohrkernaufnahmen zeigt ein aufgeschertes
Spektrum, das in ein tektonisches Vorstel-
lungsbild eingepaßt werden muß. Westlich
der Hochwasserentlastung stellte S. JACOBS
(1992 b) stark zersetzte Marmorausbisse
fest, im Injektionsstollen und aus den

Bohrlochaufnahmen ist der hohe Zerlegungs-
grad der förmlich in den "Roten Sanden"
schwimmenden Scherlingen bekannt. Drei
Bohrungen haben westlich der Talstörung
ein dickes (Kalksilikat)-Marmorband aufgeschlossen, welches vermutlich versetzt
ist.

Als Folge des langsamen Herantastens an
diese tektonische Struktur kam H. LITSCHER
im KELAG-Bericht 1992 zu einer Vorstellung, die geomechanisch anders zu interpretieren ist. Der sogenannte
"Schaufelbruch" als Folge einer Entspannung der rechten Felsmassen zur Talfurche
kann aus folgenden Gründen nicht möglich
sein:

1. Fallen die Zerrspalten im Bereich des
"Ofens" nicht in den Berg, sondern nach
den Aufnahmen S. JACOBS` steil und schräg
aus dem Querschnitt gegen Osten ein. Eine
Entspannung zur freien Talkerbe war wirksam und erzeugte diese Fugen und Spalten.

2. Eine Abwicklung der Bewegung nach einem Gleitkreis mit der Drehachse im Kern
der bezeichneten "Faltenmulde" - die es
nicht gibt - hätte eine andere Konstellation von Fugen und Spalten hervorgerufen,
zum anderen wären analoge Absetzungen in
der Felsböschung entstanden.

3. Das "sichelartige" Erscheinungsbild
kann nicht mit einer Rotation der Gebirgs-
masse zur Bachkerbe erklärt werden, weil
jene eingespannt zwischen Talstörung und
bergseitiger Primärstörung gar nicht die
Möglichkeit hatte mit ihrer Zehe im Tal-
grund auszuweichen.

Dem stehen nun folgende Kriterien entge-
gen, die durch geologische Beobachtungen
und Kriterien zu untermauern sind:

a) Die aufgefahrenen Marmorlinsen und
die gesamte Zone zeigt doch eine deutliche
Scherbewegung, die von Osten her entlang
des Marmorzuges gegen die steil ost-
fallende Primärstörung ansteigend, wirksam
wurde (Abb. 10).

b) Die Scherbruchstruktur ist an den
Marmorscherlingen doch gut abzulesen,
selbst das Durchreißen der Scherbahn durch
die Primärstörung ist damit zu beweisen,
daß Marmorausbisse im Liegenden dieser
Flächenschar aufscheinen.

c) Eindeutige Anzeichen für einen Fels-
grundbruch im Sinne einer Gleitkreiskon-
struktion hätte klare Spuren im Eingangs-
bereich des Grundablaßstollen hinterlassen
- sie sind bisher nicht bekannt.

d) Der ehemalige Aufstieg von Thermal-
oder mineralisierten Wässern, welche das
aufgescherte Karbonatgestein im besonderen
zersetzte, spricht auch dagegen, weil bei
einer jüngeren, entlastenden Rotationsbe-
wegung nicht vorher schon tektonisch ange-

E   W

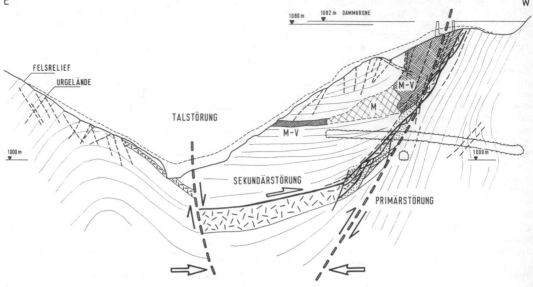

Abb. 10: Strukturgeologischer Schnitt durch die Dammachse nach
         Ausbruch des Injektionsstollens
Fig. 10: Section along the dam axis after excavation of injection
         gallery with geological structures
         E.H. WEISS, 23.5.1993

legte Aufstiegswege für das Wasser vorhanden gewesen wären.

e) Die Großlamelle zwischen den beiden
Hauptstörungen ist noch heute eingespannt
und die Bachkerbe hatte zur Endzeit der
Vereisung (Pleistozän) bestimmt 50 m und
mehr an Überlagerungshöhe!

Die von mir als "sekundäre Scherzone"
deklarierte Spur ist rein tektonomecha-
nisch angelegt und hängt mit einer horizontal wirksamen Schubspannung zusammen
(Abb. 10). Dabei glitten die hangenden
Massen aufwärts in Richtung zur gebirgs-
schwächsten Gebirgsformation - der bergseitigen Primärstörung! Als Folge dieser
Mechanik wurde teilweise der liegende,
mächtige Marmorzug zerschert und nach oben
zu aufgeschert. Hernach setzten aus der
Tiefe und von der Oberfläche einwirkend,
die physikalischen und chemischen Zerset-
zungsprozesse ein. So ist es nicht außer-
gewöhnlich, daß die Verwitterungsexzesse
gerade dort so tief reichen, wo sie sich
mit den markanten Spuren der Schertektonik
verbinden. In meinem Profil habe ich dies
anzudeuten versucht.

8 SCHLUSSBEMERKUNG

Die tektonischen Merkmale der rechten Ein-
bindungsflanke waren bis auf die Anlage

einer primären Störung unbekannt und Erkundungsbohrungen, Stollenvortriebe und
obertägige Felsaufschließungen ließen eine
zweite, sehr breite Störungsfront beweis-
kräftig nicht vermuten. Über Bohrungen,
Färbeversuche und Sickerwasserzutritte
konnte die Streichrichtung einer höchst
durchlässigen Zone fixiert aber nicht kon-
kretisiert werden. Die mehrphasigen Abdichtungsmaßnahmen bestätigten die Tatsache, waren aber im Kreuzungsbereich der
"Sekundärstörung" mit der Dichtungsebene
fast unwirksam. Erst ein Injektionsstollen
auf Höhe 1013 m, knapp luftseitig der
Schleierebene vorgetrieben, entdeckte die
15 m breite Störungszone und löste das
Rätsel der hohen Durchlässigkeit. Mit
großem Aufwand an Injektionen wurde der
Dichtschluß hergestellt und Ende 1992 der
Vollstau erreicht.

Über Ersuchen von Herrn Dipl.-Ing. Dr.
R. Widmann, Salzburg, wurde dieser zusam-
menfassende Bericht vorgelegt und als
Sachverständiger der Wasserrechtsbehörde
möchte ich mich für die kooperative Zusam-
menarbeit aller Beteiligten aufrichtig be-
danken, im besonderen den Herren Dr. H.
Wellacher, Prok. Dipl.-Ing. K. Nackler,
Dr. H. Litscher, Dr. S. Jacobs und den
Mitarbeitern der KELAG-Bauleitung, sowie
Min.-Rat. Dr. P. Hochmaier und Dipl.-Ing.
G. Innerhofer. Für die Ausführung der

Zeichnungen sei Frau H. Heller gedankt.

SCHRIFTENAUSZUG

JACOBS, S.: Injektion des Untergrundes der Sperre Feistritzbach - Diskussion der bisherigen Ergebnisse. Erstellt an KELAG und Autor, April 1992 (a)

JACOBS, S.: Geologie des Untergrundes der Sperre Feistritzbach - ein Diskussionsbeitrag. Erstellt an KELAG und Autor, Juli 1992 (b)

KELAG: Injektionsarbeiten (Untergrunddichtung des Dammes) - Informationsbericht; August 1991

KELAG: "Ein Beitrag zur Geologie der orographisch rechten Sperrenflanke". Informationsbericht; September 1992 (H.LITSCHER)

KELAG: Injektionsschleierauswertungen von HANAUSEK, 20.11.1991

LITSCHER: "KW Koralpe - Geologische Erkundung und Betreuung während der Planung und Bauausführung". Mitt. Abt. Baugeologie, Univ. für Bodenkultur Wien; 1991

RIEDMÜLLER, G.: Petrographische - mineralogische Beurteilung des vorgesehenen Schüttmaterials für den Feistritzbachdamm. Gutachten an KELAG; 18.10.1982

SPREITZER, R.: Bohr- und Injektionsarbeiten zur Untergrundabdichtung des Feistritzbachdammes. Diplomarbeit der Abt. Baugeologie, Universität für Bodenkultur Wien, 126 S., 12 Planbeilagen, Okt. 1990

Techn. Versuchs- und Forschungsanstalt, Technische Universität Graz (Leitung Prof. Dr. H. GEYMAYER): Prüfbericht über Untersuchungen an einem Micromix Feinzement, 23.4.1993

WALCHER, K.: Die Ergebnisse der Sterninjektionen vom Grundablaßstollen des Feistritzbachdammes KW Koralpe, Kärnten; Diplomarbeit der Abt. Baugeologie, Universität für Bodenkultur Wien, 101 Seiten, 10 Planbeilagen; Mai 1992

WEISS, E.H.: Kritische Anmerkungen des geologischen Sachverständigen zum Injektionsprogramm der KELAG (6 Seiten); 12.7.1989

WEISS, E.H.: Bericht über die Injektionsmaßnahmen vom Injektionsstollen der rechten Flanke - Damm Feistritzbach; Okt. 1991

WEISS, E.H.: Berichte, Gutachten, Stellungnahmen an das BMLuFW und an die KELAG; 1981 - 1993

Ferner wurden mir von der KELAG Unterlagen (Pläne bis Berichte) seit 1977 zur Verfügung gestellt.

Application of the technology of immediate blocking of water-leakage by direct grouting

Anwendungen eines Verfahrens zur unmittelbaren Abdichtung von Sickerungen durch direkte Injektionen

Zeng Rongxiu
China National Coal Corporation, People's Republic of China

ABSTRACT: The report deals with numerous successful applications of this method during the driving of tunnels, inclined and vertical shafts through aquifers in Chinese coal mines. The method was always successful very quickly, also if conventional methods have failed before. So the natural leakage up to several 1000 m^3/h was reduced to a few percent of the original value by well-aimed treatment of the aquifers. The new technique has been authorized Chinese patent by Chinese Patent Bureau.

ZUSAMMENFASSUNG: In diesem Bericht werden zahlreiche erfolgreiche Anwendungen dieses Verfahrens während des Vortriebs von Tunneln sowie geneigten und vertikalen Schächten in wasserführenden Schichten im Kohlebergbau Chinas beschrieben.. In allen Fällen führte das Verfahren zu einem raschen Erfolg nachdem konventionelle Verfahren versagt hatten. Dabei konnten die anfänglichen Wassereintritte bis zu mehreren 1000 m^3/h durch gezielte Behandlungen der wasserführenden Schichten auf wenige Prozente des ursprünglichen Wertes abgemindert werden. Dem Verfahren wurde ein chinesisches Patent verliehen.

Having worked in blocking up water grouting for many years, we've developed "immediate blocking of water-leakage by direct grouting" technology and a complete set of grouting equipment of the new technology.

The new technique has been authorized Chinese patent (CN85105852) by China Patent Bureau and obtained the forth prade of invention prize, and has been taken in China National Standards and in acceptation check specifications of underground roadway in mine.

With the characteristics of immediate blocking water-leakage from adobe rock crack by direct grouting quickly and efficiently, this new technique has been not only used in immediate blocking water emitted from aquifer for drifts, incline and vertical shafts driving, but also used as necessary construction method

in normal driving operation, resulting in that water prospecting, grouting and driving are practiced in same time. Application of the new technique can save a considerable amount of time, materials and cost in one hand and can provide an important assurance for safe construction in another hand.

Since the new technique is featured by simple process, equipment and operation; less occupied worders; high efficiency of blocking water-leakage; better quality; less material consumption and low cost, several hundreds of engineering projects have used this new technique in Chinese underground construction for blocking water in recent years. These project involved immediate blocking of water from aquifer paralleling with water prospecting and driving by direct grouting in drifts, inclines and vertical shafts in different kinds of rocks such as sandstone, gravel, limestone and fault zones.

In recent years, we have trained many special technical persons and established 9 grouting teams.

The main characteristics of finished engineering projects are: complicated in geological conditions and difficult in grouting construction.

It was failed to meet the requirements and results with conventional technique originally. Very good results have been obtained, showing very clear comparability technically and efficiently when new technique was used.

Some practical examples of grouting technique in drifts, inclines and vertical shafts are given as followings:

1. GROUTING ENGINEERING IN DRIFTS:

a) Da Yu coal mine of Ben Xu coal administration in Liao Ning province.

The roadway in -180 m level and the roadway in -220 m level are under the Taizi river (maximum discharge >10000 m^3/h). In the drift construction, it had to drive through 420m in length of Cambrian limestone with developing cracks, and aquiferous karst, the prediction of drift water discharging >1000 m^3/h, pressure head of 27 - 31 kg/cm^2.

At beginning, conventional technique was used in the mine to pregrouting the heading face, which involved in many difficulties. Then the new technique was used to grout in 420m long water-bearing strata paralleling with water prospecting and driving, resulting in a safe and quick driving in strata. The total water discharging measured 2056 m^3/h, and after grouting the water discharging remained 84 m^3/h. This technique was first successfully applicated in drift construction of Da Yu Mine.

When driving in drift with a depth of 180m, 0.3 - 0.5 fault zone with water discharging of 98 m^3/h and pressure head of 27 kg/cm^2 was run into. After grouting for 3 days with new technique the water discharging was reduced to 3.7 m^3/h, which consumed cement and other materials of 5.5 T.

Driving in the drift of -228 m level, 1.2 m fault zone with a water discharging of 135 m^3/h

had to be passed through. After grouting with new technique for 5 days, which consumed cement and other materials of 17 T, the remaining water discharging was decreased to 5 m3/h.

b) Qui Pi coal mine of Nan Piao coal administration in Liao Ning Province.

Both main haulage roadway and main return way were arranged in middle ordevician limestone with developing cracks, the widths of which range 1 - 200 mm. The Xiao Lin River on surface crosses the coal bearing formations. The river is the main source of underground water and the mine was flooded by water inrush several times. The maximum water discharging among the flood, was recorded to 120 m^3/h, pressure head reached to 28.5 kg/cm^2, which seriously harmed the life of workers and structures, delayed the mine development, which cost 500,000 Yuan of electricity annually for water drainage.

At -250 m level, the haulage roadway had to be passed through the aquifers limestone, from which water discharging was predicted more than 1000 m^3/h and pressure head was 25 kg/cm^2.

It originally planned to drill large diameter hole to reduce down water level and dewater for the roadway, then start to drive. However it was cancelled because of high cost, long periods and difficulties of construction. After then, surface and underground grouting were tried many times with conventional technique, but no good result had been obtained, so construction of the roadway had to be stopped for one and an half years.

In 1981, this project was undertaken by our corporation and the new grouting technique developed by us was used in the construction of the project. First, it took us one day to block three drain openings bared in the roadway, with the maximum water discharging of one opening being 120 m^3/h. Then in 560 m of aqueous limestone water prospecting, grouting and driving were conducted safely and smoothly. An accumulated water discharging of the zone measured in prospecting hole to be 194 m^3/h and the remaining water discharging was less than 10 m^3/h after grouting processing, with the cement consumption being 235 T.

After experiencing this project the mine bought our corporation's grouting equipment, trained their personnel in our corporation and established a "Qui Pi mine grouting team". The team has blocked 49 drain openings with a water discharging more than 20 m³/h in average and with a total blocking water discharging of 1273 m³/h, which brought a safe underground construction condition.

Besides, several other mine's grouting projects were conducted, which included shaft wall for 3 shafts, with remaining water discharging 1.5-3.0 m³/h and 5 bridges with very good results.

c) Zhang Shuanglou coal mine of Da Tun Corporation in Jiang Su Province

On the level of 538.5 m in shaft, 87 m long roadway had to drive through aquifer of quaternary sandstone fissured. When driving across this layer, the water discharging measured 95 m³/h and prediction of roadway water discharging in the aqueous zone was more than 100 m³/h.

The mine discussed and consulted our corporation with their problems, purchased equipment from our corporation and sent persons to be trained in our corporation and then set up a grouting team of seven persons. In early 1986, they firstly used the new technique in water prospecting and grouting while driving, resulting in no water appeared in 87 m of roadway, therefore the roadway driving was practiced safely and quickly with a cement consumption of 500 T, and two months of construction time saved.

d) Bai Yuan coal mine of Ping Xian administration in Jiang Xi Province

In 1985 our corporation signed the grouting contract with Bai Yuan mine for vertical shaft and main airway. Investigation and examination showed that the water discharging from one of prospecting water hole reached to 360 m³/h. We used our new grouting technique to block all water, consumed 7.5 T of cement and other materials.

The mine then bought our corporation's grouting equipment, made a technical consult with us, sent persons to be trained in our corporation, and set up a grouting team of seven people.

In August, 1986, with the new technique they spent only two days to block water discharging of 960 m³/h in main airway with 60 T of cement consumption.

e) Fang Zi coal mine in Shan Dong Province

It is a production mine where all roadways and chambers or rooms are driven in gneiss rock. The mine also bought our grouting, equipment and sent personnel to be trained in our corporation and set up a grouting team. They successfully block water discharging more than 1000 m³/h, with a pressure head of 42 kg/cm², and saved more than 5 million yuan of electrical cost for drainage.

f) Tunnel of Quing Shan to Hua guo shan in Beijin

The tunnel crossed pyrogene rock with a water discharging in working reaching to 410 m³/h. At beginning, they tried with the conventional technique to block water, but running into many difficulties. In July of 1985, they bought our grouting equipment, sent persons to be trained and set up a grouting team. It took only 16 minutes the water discharging in the tunnel to be blocked with new technique, and then they drove and prospected water at same time with grouting caused the driving speed reaching more than 100 m/month.

The examples above indicated that in drift driving there is a big difference between water discharging predicated and water discharging measured practically due to complicated geological conditions. However, with the application of new technique,, we can solve all problems of variant water discharging, resulting in that grouting and driving are progressed parallely at same time. In contradiction with the new technique, by using conventional pregrouting, the following two cases will be in fact:

(1). The real water discharging is very small, but it is treated as large one, resulting in construction delay, materials and money waste.

(2). The real water discharging was big, it is failed to be blocked, and waterflows out from some parts while driving so construction work should be stopped, and long time and a lot of materials and money should be needed.

It has approved through many construction engineering projects that blocking with our new technique would speed up drift construction and guarantee construction safety and obtain obvious economical benefits.

2. EXAMPLES OF GROUTING APPLICATION IN INCLINES

a. Do Yu coal mine of Ben Xi coal administration in Liao Ning province

The main and auxiliary inclines to be driven through andesite aquifer with developing cracks with a width of 28 mm in maximum, the pressure head of 12 - 19 kg/cm^2 and water discharging of inclines of 1000 m^3/h.

The mine adopted the conventional technique for pregrouting, but they were failed and leakage was still seen from brick walling and grout was flowed out, so the construction was delayed. Then they used our new technique to grout at same time with water prospecting and driving. They were satisfied with the result. The total water discharging measured in prospecting hole was 234 m^3/h but after grouting it was reduced down to 83 m^3/h and the construction was one year ahead.

b. Chao Hua coal mine in He Nan province

The construction of the outlet incline would be driven across quaternary loss and gravel with running sand. In 60 - 100 m long incline area there were three major problems should be considered and solved: (1). water discharging of 82 m^3/h from walling should be blocked to protect the incline from deformation as quickly as possible. (2). The rock fall above the roof should be cleared. (3). There were two famous water wells for agriculture just 40 m above the main incline, of which no water existed and needed to turn water back.

In order to solve these problems, the mine's grouting team and another research unite planned to use conventional technique to block and reinforce, for which 81 days was spent but they failed because of long period of blocking and much more difficulties.

In 1984, our corporation undertook this project, we immediately blocked water leakage by direct grouting, which took 11 days and consumed 50 T of cement and other material, water discharging downed to 6.2 m^3/h, water level of well return to what we want, the roof of incline was reinforced and 70 days was saved, compared with the original plan.

c. Xuan Dong coal mine of Hia Huayan in He Bei province

The water discharging from the main and auxiliary inclines was 66 m^3/h, after half an year's treatment by conventional methods it was unsuccessful because of complicated geological conditions.

In Nov. of 1986, our corporation signed a contract and took this project block water leakage by grouting and 12 hours water leakage from 66 m^3/h decreased to zero, consumed 42 T cement.

d. Zhang Shuanglou coal mine of Da Tun corporation in Jiang Su province

The depth of the main incline is 538.5 m, of which a part of 20.5 is sandstone and driving by the new technique. In this area rock was fractured, there were 7 small faults with 0.3 m of throw, and 0.12 of width in maximum and fracture width was 3.0 cm. The maximum water discharging of flank hole was 60 m^3/h, the maximum water discharging of full-face flank hole was 125 m^3/h and the total water discharging in the whole area was 780 m^3/h. It only took one month to drive through this section with grouting and water prospecting. The remaining water discharging was 14.4 m^3/h after it was treated by the new grouting method.

e. Yang Chun coal mine in Shi Chen county of An Hui province

The incline in this mine was 217 m long, which lined all with rock. The section of 36 m was in the quaternary gravel layer with a water discharging of 80 m^3/h.

This project was originally contracted by Wan Nan grouting company of An Hui province. But the result was not satisfied with a water discharging of 110 m^3/h because of complicated grouting conditions.

In the beginning of 1987, the project was contracted by Huai Bei grouting company of coal industry ministry. They adopted the new method which was imported from our corporation for the first time, the water discharging decreased to 6.2 m^3/h and cement of 150 T was consumed. this grouting team was trained in our corporation and their grouting equipment were purchased from our corporation.

3. EXAMPLES OF GROUTING TREATMENT IN VERTICAL SHAFT

(1) Zhang Shuanglou coal mine of Da Tun corporation in Jiang Su province

The main and auxiliary shafts would across through sandstone aquifers individually. The water discharging for the whole shaft were predicted to be 25 m^3/h, but really it was more than 300 m^3/h. It was found that water discharging still reached to 45 kg/cm^2, large water discharging, high pressure and bad quality of shaft walling construction caused many difficulties in shaft-wall grouting.

Firstly the grouting teams from mine and other companies proceed with shaft-wall grouting several times by using conventional technique, with long construction period (207 days spend for auxiliary shaft), large materials consumed (120 T for auxiliary shaft), remaining water discharging of 45 m3/h for main shaft, and no any effect could obtained and it became obstacle in the mine development and called as "Cancer".

In 1984, our corporation contracted the construction of auxiliary shaft. It only took 19 days to block water leakage by direct grouting, and consumed materials less than 40 T, and the water discharging was reduced to 6.2 m^3/h. In 1985 our corporation contracted the construction of main shaft, of which we took only 11 days to block water leakage by direct grouting and consumed materials less than 40 T, the water discharging was reduced to 4 m^3/h. It is ensure that the mine was put into production on time.

(2) Bai Yuan coal mine of Ping Xiang coal administration in Jiang Xi province

The outlet shaft had to drive through gravel layer and siliceous limestone aquifers. It is predicted that water discharge would be very small, so the conventional method was used in construction. When the shaft was sunk to 171 m depth, water discharging of 620 m^3/h came from the gravel layer and the shaft was submerged.

The treatment lasted several month with conventional technique in the mine. There was another water aquifer of 110 m in thickness within 198 - 308 m in depth. For further prediction of water discharge, the examination hole was drilled, but the speed of drilling was very slow, that is only 0.2 m per day for hard and broken rock (F= 10 - 12).

In 1984 our corporation contracted the project and adopted the new technique in the cycle of construction, making the water prospecting and grouting progressed while driving, and sinking through the difficult rock zone safely and smoothly. The real water discharging was measured 114 m^3/h and after grouting the water discharging was reduced to 5.4 m^3/h, consuming cement and materials of 50 T, saved cement of 400 T and 4 months of construction period.

(3) Chang Cun coal mine of Lu An administration in Shan Xi province

The mine was designed to produce 4 million tone/year of coal, sinking three vertical shafts of main, auxiliary and air, which was the first project using financial loan by the world bank for China Coal Industry, and was required to be finishing in 4.5 years.

The water discharging of these three shafts was big, which brought many difficult to the construction, but with the new technique, the project was fulfilled on plan and satisfied by the world bank.

a. Air vertical shaft

The depth of the shaft was designed to be 347 m with a diameter of 6.5 m, water discharging was predicted to be 127 m^3/h. Actually in depth of 0 - 4 m water discharging was zero and in depth of 44 m reached to 280 m^3/h, the shaft was submerged.

A grouting team contract the project using conventional technique to block water leakage, which consumed 315 T of cement, water discharged was reduced to 10 m^3/h, but driving work was stopped for 7 months.

With in section of 90 - 310 m in the shaft, three aquifers with the thickness of 110 m in total existed.

It was plan to grout with conventional technique and it was estimated that 4 months would be spent and 1000 T of cement would be used, so it was impossible because of long time delay of construction.

In June of 1986, our corporation contracted the mine for the project by using new techniques with water prospecting and grouting while sinking. In that case the sinking was progressed smoothly and successfully through the aquifers of 110 m. Therefore the total water discharging measured really of 480 m^3/h was reduced down to 15 m^3/h after grouting. It consumed 100 T of cement and other materials and it took less than one month.

(a) An aquifer with 52 m in thickness at 150 - 202 m in depth and with a total water discharge of 345 m^3/h. It consumed 40 T cement and other materials for grouting, caused the water discharging reduced down to 3.0 m^3/h, and the grouting period was 15 days.

In general, 52 days were spent for water prospecting grouting and shaft sinking in this section.

(b) An aquifer with 30 m in thickness was found in 240 - 270 m depth, with the water discharge measured of 50 m^3/h. It consumed cement 40 T, the remaining water after grouting was 1.9 m^3/h, and the grouting period

was 5 days. The period for water prospecting, grouting and sinking for the 30 m think aquifer was 15 days.

b. Main shaft

The depth of the shaft was 447 m and the diameter was 6.5 m by designing. In 80 - 310 m of depth, the shaft had to drive through three aquifers of 110 in thickness, the total water discharge predicted to be 203 m^3/h.

At beginning, a grouting team used conventional technique to grouting the first aquifer and predicted 19.9 m^3/h of the total water discharge. but when sinking to 113 m in depth, the water discharging of whole shaft was 134 m^3/h, it took 3 months to sink 31 m of shaft, which was too slower.

In 1980 our corporation assisted to block water for this project, it only took 12 days to pass through the remaining 11 m section of aquifer.

After that, our corporation took contract of the shaft construction for second and third aquifers, and the grouting depth was 80 m in total, prediction of shaft water discharge was 203 m^3/h, and the real total water discharge was more than 500 m3/h, for which the grouting time was 1 month, the consumption of cement and other materials was 100 T, and finally the remaining water was less than 10 m^3/h.

(a) the section of 51 m grouting between 169 - 222 m in depth, the water discharge of single prospecting hole was more than 100 m^3/h for three times, one of these was 240 m^3/h, and water ejection pillar reached to 19 m high and additionally, there were more than 100 blasting holes drilled on workings. It only took us 6 hours to block water, with a cement and other materials consumption of 3 T and then sinking without water discharge from a crack of 10 mm. We continued in Dec of 1986, under the situation of the total shaft water discharge of 50 m^3/h and we spent 30 days for water prospecting grouting and shaft sinking of 34.4 m.

In the section of 245 - 308 m in the shaft, there were three aquifers with a total thickness of 40 m, the prediction of water discharge was 60 m^3/h, by using our technique which 3.5 T of materials consumption, the remaining water was less than 10 m^3/h, and in 30 days we sunk 75 m shaft with water prospecting and grouting. But it was estimated that it would take one month and 300 T of cement by conventional technique to sink 40 meters.

With the evidence of practice, although shaft grouting for vertical shaft was more difficult than doing for incline shaft, the technical and economical efficiency of new technique has more advantages than that of conventional technique. The new technique of direct grouting and driving at same time has proved to be an effective measurement and an important guarantee for safe driving and fast construction.

According to large amount of literary material, there are big difference between prediction and real in water discharging because of variations of aqueous property, porosity, crack and fast stratum, so it is difficult to determine water discharging only by depending on a few of test holes, the existing methods of experiment and calculation.

Water source points bared can not be directly blocked by conventional technique. To determine whether to grout and how to grout for an aquifer which is plan to be driven can only depend on prediction and related parameters obtained by above methods. In that case, it could happen that big treatment just for smaller water discharging or convert, and exist much more blindness in grouting.

With application of new grouting technique, we can not only grout directly for bared water points, but also can knowing the real water discharge for each section by prospecting hole cross aquifer which need to be excavated, so we can grout just according to water discharge, that is "big treatment for large water, small treatment for small water, no treatment for no water situation". We can grout to the thickness range from a few meters to several ten meters, if it is needed. In summary we can block water leakage easily under common grouting geological conditions.

Development of rock grouting in dam construction in China

Die Entwicklung der Felsinjektionen beim Talsperrenbau in China

Zhou Weiyuan, Yang Ruoqiong & Yan Gongrui
Department of Hydraulic Engineering, Tsinghua University, Beijing, People's Republic of China

ABSTRACT : In this paper the state of the art of rock grouting in China is presented. Many large dams with complicated rock formations have been built in China.

A large amount of rock grouting was carried out: such as in carstic areas and in weak rock masses. Both cement and chemical grouting have been well developed in China. This paper deals with the experiencies at two high concrete dams.

ZUSAMMENFASSUNG: In diesem Beitrag wird der Stand der Technik der Injektionen in China dargestellt. Viele große Talsperren sind in China bei geologisch komplizierten Untergrund-Verhältnissen gebaut worden.

Viele Injektionen sind durchgeführt worden, auch in Karstgebieten und auf stark verformbarem Untergrund. Sowohl Zement-Injektionen als auch chemische Injektionen sind in China hoch entwickelt. Über diese Arbeiten bei zwei rund 170 m hohen Talsperren aus Beton wird hier berichtet.

1 INTRODUCTION

In general, foundations must have realiable permeability, efficient strength and must be adequately monolithic in case of large dams and other structures on them. In practice, a great majority of foundations do not have such characteristics. So foundation treatments are needed and rock grouling has been the main method in foundation-reinforcement and seepage control.

In China as many dams have been completed, rock grouting has been developed very fastly. Some dams or structures of which foundations are much complicated, after rock grouting and other new technical treatments, have been built up. For example, in the dam Xinanjiang and Danjiangkuo, large faults and associated fractured zones were found. In Qintongxia dam, not only faults but also karsts were widely distributed. During a score of years, these projects after grouting have been running very well.

Wujiangdu Dam and Lonyanxia Dam which well represent the development of rock grouting in China were built in the seventies and eighties respectively. Wujiang Dam is located in the South-western China where karst could be found popularly. The amount of grouting has been very large, and the performance of grouting very complicated. High-pressure grouting (3.0 MPa to 6.0MPa) and other technical measures were used. Longyangxia dam is located in the upper-reach of Yellow River and the rock masses are fractured, with low strength and high permeability which were infavourable to the Dam's stability. After careful research and grouting tests, high-pressure grouting (8.0MPa) and chemical material grouting treatment were performed.

So far they all run nomally and the grouting treatment are successfully.

This paper presents the grouting methods employed and systematized in the construction of large dams in China. forcusing on the high-pressure grouting and chemical material grouting works.

2 CEMENT GROUTING

Cement is normally used as the main material for foundations grouting in view of its cost, stability, strength and easy operation. It is widely used in consolidation grouting, contact grouting and impermeabilization curtain grouting.

During the primary design stage, grouting tests must be executed in order to determine the feasibility and economy of grouting, above all, to obtain all kinds of technical data for the design and performance of grouting treatment.

After grouting carried out, tests are needed to judge the effects of grouting.

The practice of Wujiangdu Dam grouting is exemplified to illustrated the cement grouting here.

2.1 Brief introduction of Wujiangdu Dam and its engineering-geology

Wujiangdu Dam is an arch-gravity dam with a height of 165 meters and reservior capacity of 2.41 giga-cubic meters. The amount of impermeabilization curtain grouting is 190 kilometers and consolidation grouting is 89 kilometers.

The rock in the foundation is composed of limestone. The sedimentary rock is intercrossed by many faults oriented NE and NEW, and is highly fissured and jointed. These faults and joints have been widely damaged by underground water erosion and many karst caves and cracks are formed in it. The permcability is controlled primarily by these karsts. Some karst cracks are filled with clayey-soil, karst caves are all filled with clayey-soil, fine sands and porous-breccia. The width of these infillings in joints is ranging from 1cm to 10cm. These infillings might become soft and washed away under the water pressure of 0.3MPa to 0.9MPa. It is obviously that from technical view point of dam stability most of these infillings in the karst caves and joints were unstable.

From the above facts it is clear that the seep-age control treatment of its foundation is the key work in Wujiangdu Dam.

2.2 Impermeabilization grouting curtain

Before grouting some tests had been done at three zones of Wujiangdu in the investigation adits of the right abutment in order to determine the lay-out of grouting holes, grouting materials and other technical performance of the grouting treatment.

Impermeabilization curtain is used as a main treatment to control the seepage of foundation. Cement impermeabilization wall and fissured rock excavated methods were also used on the right bank. This kind of comprehensive method was planed and used in Wujiangdu Dam. The curtain is 1, 175 meters long and with a size of 189,000 square meters.

2.2.1 Design of impermeabilization curtain

Adits on each bank and on the bottom of dam in the river bed were used to carry out the grouting. The vertical interval of the adjacent adits is 30 to 40 meters. The curtain were linked up at each adit with a intercrossed area of 5 meters high. The maximum deepth of curtain is 260 meters.

a) Layout of grouting holes

Three rows of grouting holes were planed where the water pressure head was more than 60 meters and in the abutment sections, the spacing of rows is 1.1 to 1.5 meters. Where in some sections less important, two rows of holes were planed and spacing of rows was 1.2 to 1.8 meters. In other sections there were only one row of holes. The spacing of column were mainly 2meters.

b) Depth of curtain

The depth of curtain in the bed is about 80 meters (half of the water head). It is mostly 40 to 50 meters deep in the banks of the river. If some larger karsts were found during construction, the curtain must be expanded to reach the karsts.

c) Target for improvement

In some places nearer to the dam, the specified Lugeon values: Lu<0.5; In other places: Lu< 1.0.
Target for the rtio of uplift water pressure is : ALF<0.3

2.2.2 Execution of grouting

The main characteristics of grouting perform-ance are listed as follows:

a) Grouting pressure

The grouting pressure is controlled by the depth of the section to be grouted. The data are shown in Table 2.1

When performing grouting for fissured rocks, grouting pressure is decreased rationally to prevent the rocks from too much dilation and endangering the structure.

Table 2.1   Limited grouting pressure

Depth of grouting (m)	0 to 2	2 to 3	3 to 5	5 to 10	> 10
Limite pressure (MPa)	1.0	2.5	4.0	6.0	6.0

During the grouting, the holes did not need to be washed to remove the infillings in the karste.

b) W / C value of grout

For the purpose of making the grouting more rational, the W / C value of grout is getting denser and denser in sequence of grouting steps. Grouting started with the water cement ratio of W / C = 8:1. In the final step W / C = 0.5:1. The performance was divided into 8 steps.

c) Conditions for the stop of grouting

Normally, (1) grouting is continuously per-formed for one hour after the amount of in-jecting grout is less than 0.5l / m per minute and (2) continuously performed for one and half hours under the specified pressure, till the above two conditions are all reached, the grouting could not be stopped.

2.3   Investigation of the effects of grouting

The effects of impermeabilization curtain grouting in Wujiangdu Dam can be concluded as fellows:

a) Seismic investigation and supersonic sounding tests all show that the dynamic modulus of elasticity of foundation had been increased after grouting. In places where the rock is defect,the increase is large, and occa-sionally the maximum ratio of increase amounts to 310. The rock is more monolithic after grouting. Before grouting the ratio of the maximum modulus to minimum modulus is 4.6, after grouting this ratio become 1.56.

b) The sample cores took from the large-diamter test in karst show that the infillings in karsts were injected with cement. The condensed cement masses consolidated the substances of infillings in the karst caves and joints. The core sample also show that some joints with a width over 0.1mm were in-jected very well.

c) The 113 check holes with a total depth of 6330 meters were drilled for the permeability test of the foundation. A total number of 1177 tests had been done in these holes. The results are shown in Table 2.2

Table 2.2   Lugeon value of foundation after grouting

Number of holes	Number of test	Lugeon value (Lu)			
		<0.1	<0.5	<1.0	>1.0
		Number of test and its percentage			
113	1177	1024	1137	1163	14
	100%	87%	96.6%	98.8%	1.2%

These results are lying in the extent of permis-sion and reach the target, especilly in the river bed where the maximum Lugeon value is only 0.2.

3   CHEMICAL MATERIAL GROUTING

Chemical material grouting is used when the

joints and fissures in the rock are too small to be injected by cement grout. When performing the grouting, cement grouting is carried out at first step to fill the big joints in the rock and consolidated it primarily. Then chemical material grouting is performed to inject the small fissures. In recently years it is widely used in view of its efficiency.

The main purpose of chemical material is to improve the permeability and mechnical properties of foundations. It has been used in China since the late fifties. So far there are more than 50 hydraulic engineering structures in China where this method has been used, and the results are efficiently.

We explain chemical material grouting method and its development in China with the example of Longyangxia Dam in this paper.

3.1 Brief introduciton of Longyangxia Dam

Longyangxia Dam is located on the upper reaches of the Yellow River. It is a gravity-arch dam with a height of 176 meters, the capacity of reservoir is 24.7giga-cubic meters.

The rock in the foundation is granite, the unweathered rock is hard and light. but there are a lot of large geological tectonics (such as faults F18, F120, fissured zone G4) in the foundation. The weathered rock is very thick in depth, the rock in foundation was cut into large blocks by these faults. The infillings in the faults are porous-breacia and debris.

The strength of faults and its associated fractured zones is very low, and the Lugeon values in these areas are large. For example, the modulus of deformation of fault F18 is 2.0GPa, of fault F120 is 0.86GPa, fissured zone G4 is 1.4 to 2.2GPa; the shear-strength of fault F18: $f = 0.35$, $c = 0.0$ to 0.05 MPa, of fault F120: $f = 0.30$, $c = 0$, of fissured zone G4: $f = 0.4$, $c = 0.05$MPa.

Hence the main work of the Longyangxia Dam foundation treatment is to consolidate faults and associated fissured zones, and to improve its sliding stability of abutment and permeability of foundation.

In order to reduce the excavation of weathered rock and the replacement by concrete, high-pressure grouting and chemical material grouting were used to strengthen the foundation.

The design target of improvement is laid on that the mechnical properties and permeability of the faults and fissured zones reach the following values after chemical material grouting:

Modulus of displacement: $E = 5.0$ to 7.0 MPa.

Lugeon value: $Lu = 1.0$

The chemical material grouting was achieved in 1983.

3.2 Performing of grouting

The chemical materical grouting was carried out in faults F10, F120 and fissured zone G4. At the first step cement grouting is performed, after a month as the grout condensed, chemical material grouting was executed. The main component of the chemical materialis is epoxy resin.

3.2.1 Layout of grouting holes

In all the three grouting areas, the spacing of hole rows was 2.0 to 3, 0 meters the spacing of columns is 1.0 to 1.5 meters. The holes were paralleled to the faults, the dipping angles of holes in fault F18 were 60. in F120 were 80. The diameters of holes were 75mm. 66mm and 45mm respectively. Holes were divided into three groups and they were performed one by one.

3.2.2 High pressure cement grouting

The grouting was performed in sequence from top to bottom. The sections in top 7 meters of the hole were not grouted directly to prevent the rock from cracking and blasting. The interval is 1.0 meter long at first, and as ascends it increases. At last, it reaches 5 meters. The limited grouting pressure was 8.0MPa.

Untill the two conditions--(1). Grouting is continuously performed for two hours after the amount of injecting grout is less than 1.0L / W. per Min. (2). Grouting is continuously performed for three hours under the limited pressure. --were all reached, the grouting should not be stopped.

3.2.3 Chemical material grouting

After the cement grouting had been carried out and the grout condensed, chemical material grouting would start. The length of injection pipe was 5 meters.

The performance of grouting was controlled by the amount of solution injected and its rate. The specified amount of solution injected was derived from the equation:

$$Q = 3. \ 14 * n * K * L * R * R$$

Where: Q: amount of specified solution injected,

n: ratio of joints and fissuses in the rock,

K: coefficients of evenness of rock,

L: length of injection pipe.

R: radius of solution expanding.

The specified rate of solution injected was derived from the condensing time of solution. It is 0.2 to 1.0 Litre / Minute.

During the performance, grouting pressure was 1.5 to 2.5MPa in F18 and in F120 it was 2.5 to 3.5MPa.

3.3 Results of grouting

3.3.1 Amount of grout injected

The amount of cement injected was comparativily large in these areas. In faults F18, its average was 275kg / m; in F120 790kg / m.

The amounts of chemical solution injected

were 76L / m, 140L / m and 210 to 360L / m in faults F18, F120 and fissured zone G4 respectively.

3.3.2 Permeability of rocks

The permeability of grouting foundation listed in Table 3.1

Table 3.1   Lugeon value of foundations. (Lu)

AREA	PRJOR TO GROUTING	AFTER CEMENT GROUT	AFTER CHEMICAL GROUT
F18	9.20	0.095	0.006
F120	29.19	0.09	0.03
G4	5.86	0.006	

These values show the achievement of grouting in reducing the permeability of foundation.

3.3.3   Deformative and elastic modulus of the foundation

The results of defofmative and elastic modulus of the rock are listed in Table 3.2.

The above data show that the faults and fissured zones could be consolidated and its modulus reached the design target after cement grouting. After chemical material grouting modulus increased even more and the evenness of the foundation was improved also.

3.3.4   Strength and mechanical properties of foundations

Apart from the above properties of rock, strength and other mechnical characteristics were also tested in laboratory the results are listed in Table 3.3.

Table 3.2 Results of modulus of rock in grouting areas

ITEMS	F18		F120		G4
	DEFORMATION MODULUS(GPa)		DEFORMATION MODULUS(MPa)		ELEASTIC MODULUS(MPa)
	VERTIC TO F	PARAL TO F	VERTIC TO F	PARAL TO F	
PRIOR TO G	6.7	4.2	1.9	3.1	2.2
AFTER C. G	20.5	8.3	8.0	8.5	18.5
AFTER G	31.0	24.0	9.0	14.2	25

Note   prior to g means prior to grouting.   after g means after chemical material
        after c.g. means after cement grouting.   grouting.

Table 3.3 Strength and other properties of rock after grouting

ITEMS		AREAS			NOTES
		F18	F120	G4	
COMPRESSIVE STH	MPa	73.0	40.5	42.9	
TESILE STH	MPa	3.33	2.93	3.04	
ELASTIC MODULUS	MPa		0200	40000	In laboratory
THREE–AXIAL SHEAR STH	f		0.48	0.41		Cement and chemical material all performed
	C	MPa	6.5	5.5		
MID–SIZE SHEAR STH	f		1.34	1.72		Cement and chemical material all performed
	C	MPa	0.43	0.36		
	f		2.1	1.2		Cement and chemical material all performed
	C	MPa	1.1	0.2		

3.3.5 Consolidated rock inspection

These include observations in core–sampling holes and large diameter holes (with a diameter of 1.0m) in fault F18 and F120 area and vertical well in G4 area. The results of investigation are given as fellows:

a) Condensing of grout in joints

A great majority of the joints with comparative large width were filled with the cement and chemical condensed grout. These condensed masses were linked up in meshes or lines. In these areas the rock are compact and there is no way for drainage. Sample core recovery was more then ninty precentage, in some places even to 100%.

b) Condesning of grout in small fissures

Only chemical material grout could be injected to these fissures. With the aid of microphotograph we found that the finest width of fissure that filled with condensed chemical grout was 0.006mm.

c) Condensing of grout in the clayer–soil infillings

A majority of infillings in the joints and cracks were injected by the cement and chemical grout and consolidated.

d) Durability of condened masses

Samples took from investigation holes were still compact and no cracks were found in them after half an year laying in labrotary without any protection.

4. INTERPRETATION FOR ROCK GROUTING

During the performance of grouting, grout was forced by pressure, while cracks in the rock were likely to be wideened, to be injected into the joints and fissures of rock. Then the grout was condensed and filled up the cracks. Hence the strength of the condensed mass, its cohesive strength with the surrounding rock and its erosion resistance are the main factors to be considered.

By cement grouting, the rock could be highly strengthened, however with come disadvantages in it. They are listed as below:

1) When the injected grout condensing, some water segregate from it. So fissures are formed in the condensed mass, especial along the interfaces with surrounding rock. These fissures lead to the decrease of strength of condensed mass and cohesive strength. Table 4.1 shows the ratio of segregation of cement.

Table 4.1   Relationship of cement W / C and the ratio of segregation.

W / C(water / cement)	10:1	8:1	6:1	4:1	3:1
Segregation (%)	89	87	84	76	70
W / C(water / cement)	1.5:1	1:1	0.8:1	0.6:1	0.5:1
Segregation (%)	49	35	27	20	10

Note: These data are obtained from labrotary tests.

2) Erosion resistance of cement
Sometimes, underground water was erosive to cement. This caused the condensed cement mass fractured so that the grouting become undurable.

3) The cement grout is a mixed liquid. In case the size of grains in cement grout is larger than the width of joints in rock, the grout can't be injectd into some micro−joints.

Because of these factors, the usage of cement grouting has been restricted.

However, it is well known that cement grouting is very effective in improving the deformability and permoability of soft and weathered rock. Below are given some concepts about cement grouting:

1) The elastic modulus and deformation modulus of rock could be raised after grouting. In general the ratio of increase could be 30% to 100%, it could be 150% to 200% in case of better conditions. It is well known that the worse the quality of rock, the higher the increase after grouting.

2) The permeability of foundation coul be improved after grouting, it has been found in many dams examples, such as Longyanxia Dam. Danjiangkou Dam, Liujiaxie Dam in China.

3) The grouting effects were more better in the rock with cracks and joints of no infillings than those filled with clayer−soil.

4) It is believed that cement grouting couldn't improve the shear strength of soft rock very much.

Table 4.2 shows the information of some projects in China.

5 CONCLUSIONS

So far, grouting is a kind of technolodge predominated by experiences and no foundamental theory and standard laws can be fellowed with when designing and performing the grouting. The information obtained from the previous works are needed to be analyzed in order to get systematic and experiemental concepts.

Parameters to be used to reveal and describe the physical, mechanical and permeable characteristics of rock are numerous and it is difficult to define the groutability of the rock. So it is difficult to obtain a direct relation between the geological conditions and the grouting by information summing up method.

In general, materials widely used at nowadays for grouting can be classfied as two kinds: ce-

Table 4.2 Dynamic elastic modulus of foundations before and after grouting.

NAME OF PROJECT AND TEST AREA		WAVE SPEED (M / S)		DYNAMIC ELASTIC MODULUS (GPa)		NOTES
		B.G	A.G	B.G	A.G	
WUJIANGDU DAM		4360	5160	41.6	63.7	
GUXIANG DAM		2930–4670	4100–5000	17–45.5	34–52.5	
ZHUZHUANG DAM		1920–2000	3740–4650	8.2–8.7	23.4–37.7	
GEZHOUBA DAM		5323	5647	6022	7347	
LONGYANGXIA DAM F18	PARA1 TO F	2600–3800	3900–4000 4100–5100(*)	13–29	30–37 35–55(*)	
	VERTIC TO F	3600–4300	4500–5000 4600–5700(*)	26–38	42–52 44–69(*)	(*)MEANS AFTER CHEMICAL GROUTING
	PARALL TOF	2500–3400	3600–4500 4100–4800(*)	12–23	26–42 35–43(*)	VERTIC TO F
		2400–3100	3000–4100 3900–4300(*)	7–19	29–35 30–38(*)	

Table 5.1 Properties of cement grout before and after grinding

W / C	Ratio of segregation		Compressive strength(MPa)				
	Prior to grinded (%)	After grinded (%)	Prior to grinded			Aftet grinded	
			3d	7d	28d	3d	7d
28d	2:1	60	50	0.36	1.85	4.29	1.19
4.81	9.98	1:1	29	19.5	0.68	2.14	4.48

ment and chemical material. They are all limited due to their properties.

As shown in Section 4, cement grout is a mix liquid which can not be used to improve the shear strength of rock. Chemical material grout is a solution, however it is poisonous and expensive and can not be used widely. New matericals are produced in China.

After grinding the strength of cement increases and its ratio of segregation decrease. Table 5.1 shows some results of laboratory test.

REFERENCES

[1] Shoung zhao & Li maolie: Grouting for Dam Foundation. 1986.

[2] Report of grouting treatment in Wujiangdu Dam foundation. 1984.

[3] General report of high–pressure and chemical material grouting test on the foundation of Longyanxia Dam. 1985.

4 Research works – In situ tests
Forschungsarbeiten – In Situ Versuche

Untersuchungen zur Injektion von Kunstharzen in Fugen und Rissen in Massenbeton

Investigations on grouting of resins into joints and cracks of mass concrete

R. Baban & H. Geymayer
Technische Universität Graz, Österreich

ZUSAMMENFASSUNG: Bei Instandsetzungsarbeiten an Talsperren tritt häufig die Aufgabe auf, Fugen oder Risse durch Kunstharzinjektionen möglichst vollständig und kraftschlüssig zu verfüllen, ohne durch zu hohe Injektionsdrücke neue Rißbildungen oder standsicherheitsrelevante Zusatzlastfälle zu schaffen. Die Arbeit beschreibt ein Rechenmodell zur Berechnung der Injektionsdruckverteilung im elastisch nachgiebigen Spalt sowie Versuche zur experimentellen Verifikation der Rechenwerte. Rechnung und Versuch zeigen, daß bei geringen Anfangsspaltweiten die elastische Aufweitung zufolge der Druckverteilung im Spalt in der Größenordnung der Anfangsspaltweite liegt, wodurch ein gewisser Selbstregelmechanismus entsteht, der eine Vergleichmäßigung der Druckverteilung und der gesamten Auftriebskraft auch bei stark unterschiedlichen Injektionsbedingungen zur Folge hat.

ABSTRACT: In the course of repair operations on concrete dams frequently the task arises to grout joints and cracks with resins as completely as possible, in order to restore impermeability and monolithic behaviour, but to avoid grouting pressures, that are too high and may cause additional cracks and loads, which dangerously reduce structural safety.

The paper describes an iterative calculation procedure to evaluate the hydraulic pressure distribution during the injection of a small gap in an elastically deformable halfspace by computing the pressure distribution due to radial flow from a central borehole in a gap af variable width and by subsequently calculating the elastic widening of the gap due to this pressure distribution, using Boussinesqu's relationship, and repeating the procedure untill the changes are negligible.

The paper also describes an experimental setup to verify the calculation procedure and gives examples of results that were obtained. The experimental setup consisted of 4 circular conrete slabs Ø 140 cm, which were stressed together by means of a stiff frame, 10 Diwidag prestressing rods and a flat jack Ø 120 cm, thereby providing a possibility to controll the gap width independently from the injection pressure.

Calculation and experimental results showed, that for gaps and cracks of small initial width the elastic widening caused by the hydraulic pressure during the injection is in the same order of magnitude as the initial crack width. This initiates a certain self regulating mechanism and leads to a significant reduction in the variation of pressure distributions and resultant forces due to changing injection parameters as compared to conventional calculations for hydraulic pressure distributions in gaps with rigid walls. Tests and calculations show reasonable agreement and verify, that in the early stages of a grouting procedure with constant rate a pressure maximum will develop until the injection pressure acts on an area wide enough to locally cause a significant increase of crack width i.e.of the hydraulic cross section, whereupon the grouting pressure will start to drop.

1. EINLEITUNG

Im Rahmen einer an der Technischen Versuchs- und Forschungsanstalt für Festigkeits- und Materialprüfung der TU Graz durchgeführten und von der Österreichischen Verbundgesellschaft unterstützten Forschungsarbeit wurde für die beim Injizieren von Fugen und Rissen auftretenden Strömungsvorgänge und Druckverteilungen sowie die dadurch verursachten elastischen Verformungen der Fugen- bzw. Rißwandungen ein iteratives Rechenmodell entwikkelt. Dabei wird die Berechnung der voneinander abhängigen Rißaufweitungen und Strömungsdrücke solange wiederholt, bis die Änderungen ausreichend klein sind. Parallel zu den theoretischen erfolgten auch experimentellen Untersuchungen in einem speziell dafür gebauten Versuchsstand, um die errechneten Werte mit Versuchsergebnissen vergleichen zu können. Die Arbeit sollte einen Beitrag dazu leisten, die Injektionstechnik vom bisherigen Stand einer im wesentlichen auf Erfahrung und Einfühlungsvermögen beruhenden Kunst in Richtung einer vorherberechenbaren, optimierbaren Technologie weiterzuentwickeln.

2. RECHENMODELL

2.1 Allgemeine Gleichung der nicht parallelen Strömung

Die Energieverluste (Reibungsverluste) in Leitungen beliebigen Querschnitts werden im allgemeinen durch eine auf Darcy, mitunter auch auf de Voisins und Weisbach zurückgeführte Formel beschrieben:

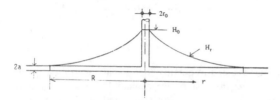

Abb. 1: Druckverteilung bei radialsymmetrischer Spaltströmung

Fig. 1: Pressure distribution for radial flow in a gap of constant width

$$J = \frac{\rho}{2} \frac{\lambda}{D_h} v_0^2 = -\frac{dp}{dr} = -\frac{dH}{dr} \quad \left[\frac{N}{cm^3}\right] \quad (1)$$

$$\text{Mit } v_0(r) = \frac{Q}{4r\pi a} \quad (2)$$

für die radialsymmetrische Strömung

$$\text{sowie } \frac{v}{g} = \frac{\eta}{\gamma} \text{ bzw. } \frac{\eta}{\rho} = vRe = \frac{D_h v_0}{v}, \ \lambda = \frac{96}{Re} \cdot c$$

$$\text{erhält man: } J = \frac{dH}{dr} = -\frac{6Qc\eta}{r\pi(2a)^3} \quad \left[\frac{N}{cm^3}\right] \quad (3)$$

bzw. nach Integration:

$$H_0 = \frac{3\eta cQ}{4\pi a^3} \ln\frac{R}{r_0} \quad \left[\frac{N}{cm^2}\right] \quad (4)$$

Der hydraulische Druck in einem Punkt zwischen r_0 und R ergibt sich mit:

$$H_r = \frac{3\eta cQ}{4\pi a^3} \ln\frac{R}{r_0} + H_a \quad \left[\frac{N}{cm^2}\right] \quad (5)$$

und die Auftriebskraft P läßt sich aus dem Volumen unterhalb der H_r- Fläche berechnen:

$$P = R^2\pi H_0 - \frac{3Qc\mu}{8a^3}\left(2R^2\ln\left(\frac{R}{r_0}\right) - R^2 + r_0^2\right)[N] \ (6)$$

Für die laminare Strömung im hydraulisch glatten Spalt ($k/D_h \leq 0,032$) gilt nach Poiseuille ein Widerstandsbeiwert $\lambda = c \ 96/Re$ mit $c = 1$.
Im Bereich $k/D_h > 0,032$ wird die allgemeine Formel

$$c = 1 + c_0\left(\frac{k}{D_h}\right)^{1,5} \quad (7)$$

verwendet, wobei zwischen $0,032 < k/D_h < 0,1$ nach Louis (1967) $c_0 = 8,8$ und darüber in Anlehnung an Lomize (1951) $c_0 = 17$ gesetzt wird (nur bei sehr kleiner Rißweite bzw. sehr großer relativer Rauhigkeit).

2.2 Radialsymmetrische Laminarströmung im elastischen Spalt

Sobald das Injektionsgut unter der Wirkung des anfänglich rasch anwachsenden Injektionsdruckes $H_{0(t)}$ eine bestimmte Reichweite überschreitet, beginnen die Rißwandungen sich infolge der Druckverteilung $H_{r(r,t)}$ stärker elastisch zu verformen, die Rißweite wird größer, wodurch der für eine konstante Injektionsrate erforderliche, zeitabhängige Injektionsdruck $H_{0(t)}$ bzw. der zeit- und ortsabhängige Spaltdruck $H_{r(r,t)}$ wieder sinkt.

2.3 Vertikale Oberflächenverformung S_r infolge einer schlaffen kreisförmigen Gleichlast (Topflast) p_a vom Radius r_a

Nach Untersuchung von Boussinesq (1885) und anderen Autoren (z.B. Ahlvin und Ulery, 1962) gelten für die Berechnung der Verformungen des elastisch isotropen Halbraumes nachstehende Beziehungen:

$$S_r = p_a r_a H \frac{1 + \mu^2}{E} \quad [cm] \quad (8)$$

H = Setzungsbeiwert

Als Setzungsbeiwert (H) für beliebige Punkte des Halbraumes gibt Boussinesq tabellarische Zahlenwerte an.

Die Setzungsbeiwerte der Oberflächenpunkte, die uns besonders interessieren, stellen eine komplizierte Kurve dar, siehe Abb. 2. Diese Kurve läßt sich mit unter einem Prozent liegenden Fehlern durch Gleichung 9 darstellen, wenn die Variablen x und y abschnittsweise gemäß Tabelle 1 verändert werden.

$$H = \frac{2}{1 + x\left(\frac{r}{r_a}\right)^y} \quad [cm] \quad (9)$$

r = Entfernung des Punktes von der Lastachse

Tabelle 1. x- und y-Werte

r / r_a	x	y
≤ 0,4	0,45	2,5
≤ 0,8	0,38	2,3
≤ 1,2	0,56	3,9
≤ 1,75	0,8	1,91
≤ 2,5	1,12	1,32
≤ 3,5	1,12	1,33
≤ 4,5	1,2	1,26
≤ 7	1,25	1,21
≤ 9	1,71	1,04
> 9	1,88	1

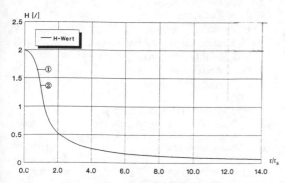

Abb. 2: 1. Kurve: Setzungsbeiwert H der Oberfläche von Boussinesq
2. Kurve: Setzungsbeiwert H der Oberfläche nach Gleichung 9
Fig. 2: Values for H: 1. according to Boussinesq
2. according to equ. 9

Durch Einsetzen der Gleichung 9 in Gleichung 8 ergibt sich die allgemeine Gleichung der vertikalen Oberflächenverformung S_r des Halbraumes infolge p_a vom Radius r_a wie folgt:

$$S_r = \frac{2 p_a r_a \left(1 - \mu^2\right)}{E\left(1 + x\left(\frac{r}{r_a}\right)^y\right)} \quad (10)$$

Damit ist eine Abschätzung der Verformung einer Fugen- oder Rißoberfläche in einem unendlich großen elastischen Halbraum möglich. Die gesamte Riß- oder Fugenaufweitung infolge des hydraulischen Druckes ist doppelt so groß, da sich an der zweiten Fugen- oder Rißwand derselbe Vorgang vollzieht.

2.4 Beliebige radialsymmetrische Schlafflast H_r vom Radius R

Eine numerische Lösung mit ausreichender Genauigkeit erhalten wir, wenn die stetige Druckverteilung H_r durch mehrere (z.B. z = 10 oder mehr) konzentrische, übereinanderliegende, kreisförmige Gleichlasten p_i, mit abgestuften Radien r_i, ersetzt wird. Die Gesamtauftriebskraft ergibt sich dann zu:

$$P = \pi \sum_{i=1}^{z} r_i^2 \cdot p_i$$

wobei die p_i-Werte direkt, dem Druckverlauf H_r entsprechend, dem Rechenprogramm entnommen werden. Demnach läßt sich die vertikale Oberflächenverformung S_r in jedem Punkt der Oberfläche des Halbraumes berechnen.

$$S_r = \frac{2\left(1 - \mu^2\right)}{E} \sum_{i=1}^{z} \frac{p_i r_i}{1 + x\left(\frac{r}{r_i}\right)^y} \quad (11)$$

2.5 Radialsymmetrische Laminarströmung in einem radialsymmetrischen Spalt mit veränderlicher Weite $2a + 2S_r$

Analog zur Gleichung (5) ergibt sich nach zweimaliger Integration die allgemeine Gleichung des Druckverlaufes im Spalt zwischen elastischen Halbräumen:

383

$$H_r = \frac{-3Q}{4\pi} \int_{r_0}^{R} \frac{\eta(1+c_0)\left(\frac{k}{4Y}\right)^{1,5}}{rY^3} \cdot dr \qquad (12a)$$

$$Y = a + \frac{2(1-\mu^2)}{E} \sum_{i=1}^{z} \frac{p_i r_i}{1+x\left(\frac{r}{\eta}\right)^y} \qquad (12b)$$

Relative Rauhigkeit

Durch die Berücksichtigung der elastischen Verformungen der Rißflächen infolge des hydraulischen Drucks ist die relative Rauhigkeit k/D_h nicht mehr konstant, sondern nimmt mit der Entfernung r vom Injektionskanal zu. Für jeden relativen Rauhigkeitsbereich bekommt c_0 einen anderen Wert.
Für sehr kleine Rißweiten $(2a+2S_r)$ kann es vorkommen, daß $k/D_h > 0,1$ ist.

Dynamische Viskosität

Da die Viskosität von Temperatur und Geschwindigkeitsgefälle abhängig ist, kann sie als Funktion von r nach folgender Gleichung

$$\eta = \eta + a \cdot r^b + c \cdot \ln\left(\frac{1}{r}\right) \qquad (13)$$

(a, b und c sind materialabhängige Zahlenwerte), ohne Berücksichtigung der Kohäsion τ_0, in das Rechenprogramm eingegeben werden.
Die Druckverteilung H_r lt. Gleichung 12a läßt sich daher nicht durch Integration, sondern nur numerisch berechnen.

Durch nachstehenden Iterationsvorgang erhält man die Druckverteilung H_r und die zugehörigen Oberflächenverformungen S_r an jeder beliebigen Stelle des Spaltes.

Rechenvorgang

1. Als Ausgangspunkt wird zuerst die Rißweite 2a (>0) angenommen.
   Nach Gleichung 5 und 6 werden H_r bzw. P berechnet. Sie werden H_{r0} bzw. P_0 genannt.
2. Die Rißaufweitung S_{r0} wird infolge H_{r0} berechnet.
   Neue Rißweite = $2a + 2S_{r0} = S_{r1}$
3. Für die Rißweite = S_{r1} werden H_{r1} und P_1 neuerlich berechnet
4. Die Rißverformung wird infolge H_{r1} berechnet.
   Aktuelle Rißweite = S_{r2}
5. Für die Rißweite = S_{r2} wird H_{r2} bzw. P_2 neuerlich berechnet.
6. Die Rißaufweitung wird infolge H_{r2} berechnet.
   Aktuelle Rißweite = S_{r3}

Bis folgende Zusammenhänge erfüllt sind:
$S_{r,i} = S_{r,i-1}$ bzw.
$H_{r,i} = H_{r,i-1}$ und $P_{ri} = P_{r,i-1}$

Mit Hilfe eines EDV-Programmes (Turbopascal) wurden Parameterstudien durchgeführt und verschiedene Injektionsversuche nachgerechnet.

Rechenergebnisse:
Im Gegensatz zur starren Spaltwand besteht bei der elastischen Spaltwand (Halbraum) keine Proportionalität zwischen einem Anstieg der Zahlenwerte für die Variablen (R, k, Q, η und 2a) und den Rechenergebnissen für die Spaltaufweitung $(2S_r)$, den Injektionsdruck (H_{0-A}, H_{0-E}), das injizierte Harzvolumen (V) etc. (Tabelle 2).

Tabelle 2. Ergebnis der Parameterstudien

Variablen	Rechenergebnisse					
	[mm] $2S_r$	[bar] H_{0-A}	[bar] H_{0-E}	[kN] P	[cm³] V	[cm/s] v°
R	++	+-	--	--	+++	---
k	++	+++	+	++	+	-
Q	++	+++	+	++	++	++
η	++	+++	+	++	+	-
2a	--	---	-	--	++	-
E	-	++	++	++	-	+

Erklärung der Zeichen
H_{0-A}: Hyd. Druck am Anfang (wenn r klein ist)
H_{0-E}: Hyd. Druck am Ende (r = R)

+: leichte Zunahme
++: Zunahme
+++: starke Zunahme
+ -: ohne Zu- oder Abnahme
-: leichte Abnahme
- -: Abnahme
- - -: starke Abnahme

Formelzeichen und Abkürzungen
2a	Weite des Risses (Anfangsrißweite)	[cm]
D_h	Hydraulischer Durchmesser = $4 \cdot Rh = 4.F/U$	[cm]
E	Elastizitätsmodul	[N/cm²]
F	Fläche	[cm²]
g	Erdbeschleunigung	[cm/s²]
H	Setzungsbeiwert des Halbraumes	[/]
$H_0=p_0$	Druckhöhe im Bohrloch (Injektionsdruck)	[bar]
$H_r=p_r$	Druckverteilung im Spalt in Abhängigkeit von r	[bar]
H_a	Hydraul. Spaltdruck vor Beginn der Injektion	[bar]
J	Druckgefälle pro Längeneinheit (Gradient)	[N/cm³]
k	äqivalente Rauhigkeit	[cm]
P	Auftriebskraft des Injektionsgutes	[N]

p_a	Topflast	[bar]
Q	Durchfluß (Fördermenge, Injektionsrate)	[cm³/s]
r	Radius	[cm]
r_a	Radius der Topflast	[cm]
r_0	Bohrlochradius (Zutrittsöffnung)	[cm]
R	Max. Reichweite des Harzes (Injektionsradius)	[cm]
Re	Reynolds-Zahl	[/]
R_h	hydraulischer Radius	[m]
S_r	elastische Verformung des Halbraums	[mm]
v_0	mittlere Geschwindigkeit	[cm/s]
V	injiziertes Harzvolumen	[cm³]
γ	Spezifisches Gewicht des Harzes = ρ g	[N/cm³]
η	Dynamische Viskosität	[N.s/m² = Pa.s]
λ	Widerstandszahl	[/]
μ	Poisson Koeffizient	[/]
ν	Kinematische Viskosität = η/ρ	[m²/s]
ρ	Rohdichte des Harzes	[g/cm³]

3. EXPERIMENTELLE UNTERSUCHUNGEN

Die in Abbildung 3 und 4 dargestellte Versuchseinrichtung bestand aus vier übereinander beto-
nierten, kreisförmigen Stahlbetonplatten mit
140 cm Durchmesser und je 30 cm Dicke, die mit
Hilfe von 10 Diwidag Spannstangen (Ø32 mm)
und einer massiven Betonbodenplatte bzw. oben
liegenden Stahlträgern vertikal zusammenge-
spannt wurden. Die oberste und die unterste Kreisplatte
dienten als Lastverteiler. Zwischen dem unteren
Lastverteiler und der Bodenplatte befand sich ein
einbetoniertes Druckkissen (Ø 120 cm, 2 mm starkes Blech, Nenndruck 200 bar, Aufweitung bis
10 mm), das zusammen mit einer Hydraulikpumpe
eine regelbare Vorspanneinrichtung bildete. Zur
Messung des Gesamtdruckes waren alle Spannstan-

gen mit DMS und das Druckkissen mit einem Manometer bestückt. Die Betonierfuge zwischen den
zwei mittleren, auswechselbaren Platten bildete die
Injektionsfläche und war mit 20 eigens dafür entwickelten einbetonierten Druckgebern (Abb. 5), 7
einbetonierten induktiven Spaltaufweitungsgebern
(Hottinger) und am Außenrand der Fuge mit drei
induktiven Weggebern instrumentiert (Abb. 6).
Alle 40 Meßstellen sowie die über die Schwim-
merabsenkung im Vorratsbehälter gemessene Fördermenge des Injektionsgutes wurden während des
Versuchs alle 4 Sekunden gruppenweise abgefragt
und mit einer Datenerfassungsanlage registriert.
Zum Injizieren fand eine Zweikolbenpumpe (Poly-
plan) Verwendung, welche mit einer normalen
Handbohrmaschine betrieben wurde. Die eingesetzte zweigängige AEG Bohrmaschine hatte zwar
eine veränderliche Drehzahl, doch blieb diese unter
Last nicht konstant, sodaß das angestrebte Ziel
einer gleichbleibenden Fördermenge nicht immer
erreicht werden konnte. Der aus einem Stahlrohr
mit Innendurchmesser 4 mm bestehende Injektionskanal endete im Mittelpunkt der Injektionsfläche.
Eine unter der Injektionsfuge angebrachte Um-
fangsrinne sammelte das aus der Fuge austretende
Injektionsgut und ermöglichte seine Wiederver-
wendung. Unmittelbar vor und nach jedem Versuch
wurde mit einem Rotationsviskosimeter (Contraves
Rheomat) die Viskosität des jeweils verwendeten
Injektionsharzes gemessen.

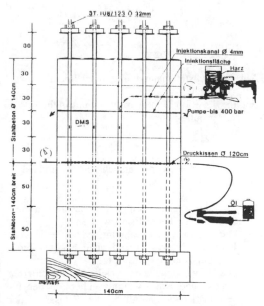

Abb. 3: Versuchsstand
Fig. 3: Experimental setup

Abb. 4: Schematische Darstellung des Versuchs-
aufbaus
Fig. 4: Schematic drawing of test rig

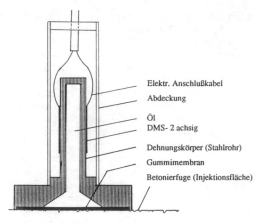

Elektr. Anschlußkabel

Abdeckung

Öl

DMS- 2 achsig

Dehnungskörper (Stahlrohr)

Gummimembran

Betonierfuge (Injektionsfläche)

Abb. 5: Schematische Darstellung des Druckgebers
Fig. 5: Schematic drawing of embedded pressure transducer

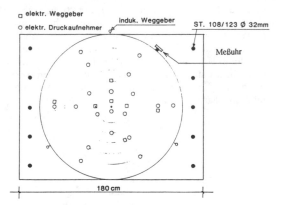

□ elektr. Weggeber

○ elektr. Druckaufnehmer

induk. Weggeber

ST. 108/123 Ø 32mm

Meßuhr

180 cm

Abb. 6: Meßstellenanordnung in der Injektions-fläche
Fig. 6: Position of measuring devices in the grouting plane

Als Besonderheiten des Versuchsstandes sind hervorzuheben:

1. Injektionsdrücke bis über 100 bar können ohne Schwierigkeiten realisiert werden.
2. Die regelbare Vorspannung erlaubt eine vom Injektionsdruck weitgehend unabhängige Steuerung der Umfangs-Rißweite.
3. Die Verwendung von speziellen, einbetonierten Druck- und Rißaufweitungsgebern in der Injektionsfläche ermöglicht die Erfassung der Druckverteilungen und der Verformungen im Injektionsspalt.
4. Für die Variation der Injektionsflächen

(Rauhigkeit der Rißflächen) sind nur die beiden mittleren Kreisplatten mit der dazwischenliegenden Injektionsfuge auszutauschen.
5. Durch Einbetonieren gespaltener Betonplatten im Mittelbereich monolithischer Versuchskörper können bruchmechanische Untersuchungen bzw. Rißfortschrittuntersuchungen vorgenommen werden.

Die bisherigen Injektionsversuche sind mit nahezu kohäsionslosen Kunstharzen (ohne Härter) mit einer dynamischen Viskosität von 350 bis ca 10000 mPas (bei 20-23°C) und Injektionsraten zwischen etwa 0,5 bis 20 cm³/sec. sowie maximalen Injektionsdrücken bis ca 120 bar (Zuleitung) durchgeführt worden. Dazu wurde nach kreuzweisem Anziehen aller Muttern der Diwidag Spannstangen zunächst das Druckkissen auf einen Druck von 10 bis 20 bar gebracht (je nach Versuchsreihe) und anschließend mit dem Injizieren bei einer vorgewählten Geschwindigkeitseinstellung der Bohrmaschine begonnen. Die Injektionsdrücke erreichten bei höheren Injektionsgeschwindigkeiten rasch anfängliche Spitzenwerte (erste Spaltaufweitung) sanken aber dann bald auf ein ziemlich konstantes, je nach Harzviskosität und Injektionsgeschwindigkeit zwischen etwa 40 und 80 bar liegendes Niveau ab, bei welchem das Injektionsgut meist ziemlich gleichmäßig aus der Umfangfuge auszutreten begann (stationärer Zustand). Am Ende jedes einzelnen Injektionsversuchs wurde zunächst der Injektionskanal abgesperrt und der zeitliche Druckabfall beobachtet, nach dessen weitgehendem Abklingen das Rückflußventil der Pumpe geöffnet und solange gewartet, bis die Fugenschließung zufolge Vorspannung abgeklungen war. Da die Rißweitenmessung keine absoluten Rißweiten, sondern nur Änderungen der Rißweite lieferte, und auch dies nur für zeitlich unmittelbar hintereinanderfolgende Messungen, konnten die absoluten Rißweiten zu Beginn jedes Injektionsversuchs, ebenso wie die Rauhigkeit der Injektionsflächen, nur geschätzt werden. Bei allen Versuchen lag die anfängliche Fugen- bzw. Rißweite im Bereich von ca. 0,05 bis 0,2 mm. Die in der Injektionsfläche gemessene Druckverteilung und die gemessenen Rißaufweitungen ließen vor allem bei kleinen Rißweiten und hohen Injektionsdrücken deutlich den Einfluß der an der Injektionsstelle beginnenden (ungleichmäßigen) Rißaufweitung erkennen und zeigten für vier durchgerechnete Versuchsserien auch eine akzeptable Übereinstimmung zwischen Messung und Rechnung.

Abb. 8 gibt ein typisches Beispiel für den zeitlichen Verlauf des Injektionsdruckes in der Zuleitung (H_0), und in der Injektionsfläche in 5 cm Entfernung von der Injektionsstelle (H_{r5}), sowie für die Riß- (Fugen-)aufweitung in 5 cm Entfernung vom Injektionskanal ($2S_{r5}$), im Zuge eines Injektionsver-

2. Versuchsreihe

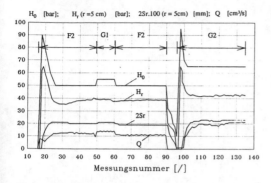

Abb. 7: Darstellung der Meßergebnisse bei einem Injektionsversuch

Fig. 7: Measuring results of a grouting test

Beispiel Nr. 2

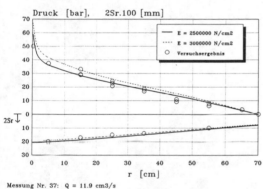

Messung Nr. 37: Q = 11.9 cm3/s

Abb. 8: Vergleich von Meß- und Rechenwerten für Messung Nr. 37 in Abb. 7

Fig. 8: Comparison of experimental and calculation results for measurement No. 37 in fig. 7

suchs mit einem Harz mit einer Viskosität von ca 830 mPa.s, wobei die Injektion nach Erreichen eines stationären Zustands kurzzeitig, bis zum weitgehenden Schließen der Injektionsfuge unterbrochen (und das Rückflußventil geöffnet) wurde, und anschließend eine Wiederaufnahme der Injektion mit erhöhter Fördermenge erfolgte. Abb. 8 zeigt eine Gegenüberstellung von Rechen- und Meßwerten für die Druckverteilung und Spaltaufweitung in der Injektionsfuge nach Erreichen eines stationären Zustandes (Meßreihennr. 37 in Abb. 7).

Zur Rechnung und ihrer hier außerordentlich guten

Übereinstimmung mit den Meßwerten ist allerdings zu sagen, daß sie eine "passende" Annahme für die meßtechnisch nicht erfaßten Einflußgrößen Ausgangsspaltweite 2a und äquivalente Rauhigkeit k voraussetzt. Natürlich war die Übereinstimmung zwischen Rechnung und Messung nicht in allen Fällen so gut, vor allem in den Randbereichen der Injektionsfläche kam es häufig zu größeren Abweichungen der gemessenen Spaltaufweitung von der gerechneten, was aber an sich nicht überrascht, da das Rechenmodell vom elastischen Halbraum ausgeht, im Versuchsstand der Halbraum aber nur unzulänglich realisiert werden kann. Im Lichte dieses Umstandes war die Übereinstimmung in fast allen Fällen zufriedenstellend.

4. SCHLUSSFOLGERUNGEN

1. Bei geringen Spaltweiten der zu injizierenden Fuge (bzw. des Risses) tritt bei einer gleichbleibenden Injektionsrate zunächst ein deutliches Maximum des Injektionsdruckes auf, welches aber nur auf eine kleine, die Injektionsstelle unmittelbar umgebende Fläche wirkt. Zufolge dieses Druckes kommt es bald zu einer progressiven, lokalen Rißaufweitung und damit zu einer wesentlichen Verringerung des hydraulischen Widerstandes in diesem "Engpaß", wodurch der Injektionsdruck (bei konstanter Fördermenge) abzusinken beginnt. Die Höhe des Maximums hängt auch von der Anfangsspaltweite, der Viskosität des Injektionsgutes etc. und natürlich von der Injektionsrate ab. Man kann die hohen Anfangsdrücke vermeiden oder wesentlich veringern, wenn die Injektion mit niedrigen Fördermengen begonnen wird.

2. Nach Überwindung dieses Anfangsmaximums, stellt sich bei gleichbleibender Injektionsrate infolge der Spaltaufweitung ein fallender Injektionsdruck ein.

3. Der Einfluß der Ausgangsspaltweite, der Rauhigkeit, der Harzviskosität, Injektionsrate und Reichweite ist infolge bzw. bei Berücksichtigung der elastischen Rißaufweitung wesentlich kleiner als es die Rechnung für eine konstante Rißweite erwarten läßt, weil bis zu einem gewissen Grad ein Selbstregelmechanismus eintritt: bei Ansteigen des Druckes kommt es zur stärkeren Spaltaufweitung mit ihrem drucksenkenden Effekt. Daher ist der Bereich in dem sich die Injektionsdrücke auch bei recht unterschiedlichen Injektionsbedingungen bewegen, verhältnismäßig klein. Dies zeigt sich sowohl im Versuch als auch im Rechenmodell, gilt aber nur für kleine Ausgangsspaltweiten im Bereich von wenigen Zehntel mm, bei denen die elastische Rißaufweitung in der gleichen Größenordnung liegt, wie die Ausgangsweite.

387

4. Die Zunahme der durch den hydraulischen Druck in der Injektionsfläche verursachten Auftriebskraft mit größer werdender Reichweite ist infolge des oben beschriebenen Regelmechanismus ebenfalls wesentlich kleiner als es die Rechnung für die Druckverteilung im Spalt mit konstanter Weite erwarten läßt.

5. Das vorgeschlagene Rechenmodell liefert für den Bereich, in dem die getroffenen Annahmen weitgehend zutreffen (Newton'sche Injektionsharze, keine turbulente Strömung, enger Einzelspalt im elastischen Halbraum) eine zufriedenstellende Übereinstimmung zwischen Rechnung und Messung, wenn für die nicht erfaßten Einflußgrößen wie Anfangsspaltweite Rauhigkeit der Injektionsfläche "passende" (plausible) Annahmen getroffen werden.

LITERATUR

Ahlvin, R.G. und Ulery, H.H. 1962. Tabulated values for determining the complete pattern of stresses, strains and deflections beneath a uniform load on a homogeneous half space. Highway Research Board. Bulletin 342, S 1-13.

Baban, R. 1992. Rißinjektion im Massenbeton mit Kunstharz. Dissertation TU Graz.

Boussinesq, M.J. 1885. Application des potentiels à l'étude de l'équilibre et du mouvement des solides élastiques. Paris. Gauthier-Villars.

Lomize, G.M. 1951. Strömung in klüftigen Gesteinen. Gosenergoizdat.

Louis, C. 1967. Strömungsvorgänge in klüftigen Medien und ihre Wirkung auf die Standsicherheit von Bauwerken und Böschungen in Fels. Universität Fridericiana Karlsruhe. Institut für Grundbau und Bodenmechanik. Heft 30.

Modeling of grout penetration at dynamic and static injection

Modellierung des Eindringvorganges bei dynamischer und statischer Injektion

Lennart Börgesson
Clay Technology AB, Lund, Sweden

ABSTRACT: The penetration of grout into fractures has been modeled. The penetration calculation is based on the derived material models for which account is taken of the influence of vibrations. The models have been checked by several tests in which bentonite-based grouts as well as cement-based grouts were injected into a 3 m long artificial fracture.

The technique has been tested in situ in granite rock in the Stripa mine at 320 m depth for the following purposes:

- Sealing of a natural fractured zone for redistribution of water flow
- Sealing of the rock around deposition holes
- Sealing of the disturbed rock around a blasted tunnel

A clear effect of the grouting was seen in the first two cases, while very little effect was achieved in the last case. The overall conclusion of the project was that the dynamic injection technique works very well in open fractures with no infillings while the sealing effect is strongly reduced by the complex nature of narrow fractures, in which infilling materials, especially chlorite, hinder grout penetration.

ZUSAMMENFASSUNG: Die Eindringung von Vergiessmörtel in Bruchflächen ist nachgebildet worden. Die Berechnung der Eindringung basiert auf den abgeleiteten Materialmodellen, bei denen der Einfluss von Schwingungen berücksichtigt wurde. Die Modelle sind bei mehreren Tests geprüft worden, bei denen Vergiessmörtel auf Bentonitbasis sowie auf Zementbasis in eine 3 m lange künstliche Bruchfläche eingespritzt wurde.

Die Technik wurde vor Ort in der Stripa-Mine an einem Granitfelsen in 320 m Tiefe getestet, und zwar mit folgendem Ziel:

- Abdichtung einer natürlichen Bruchzone zur Rückverteilung des Wasserflusses
- Abdichtung des Felsens um die Deponielöcher
- Abdichtung des Felsens um einen gesprengten Tunnel

Einer klarer Effekt des Vergiessens konnte in den ersten beiden Fällen festgestellt werden, während sich im letzten Fall wenig Wirkung zeigte. Die Schlussfolgerung des Projekts was, dass die dynamische Einspritzteknik sehr gut bei offenen Bruchflächen ohne Füllungen eingesetzt werden kann, während der Dichtungseffekt bei der komplexen Natur niedriger Bruchflächen stark vermindert ist, bei denen Füllmaterialien, besonders Chlorit, das Eindringen des Vergiessmörtels verhindern.

1 INTRODUCTION

The nearfield rock is often more fractured and permeable than the undisturbed farfield rock due to effects of blasting and stress changes. The proximity to the nuclear waste of this nearfield rock and of potential intersecting natural fracture zones, makes them very important. If these fractures can be effectively sealed, the function of the repository will be improved.

An effective sealing of these fractures puts higher demand on the technique than in normal grouting in the following respects:

- penetration into finer fractures is required ($<$100 µm)
- stiffer grouts are required

To make this possible, a new technique using oscillating grout pressure (dynamic injection) was introduced. This paper will briefly describe how the grout penetration at dynamic injection can be modeled and the results of full scale dynamic grouting. The properties of the grout materials and the effect of the vibrations are described in another paper (Börgesson, 1993).

2 GROUT MATERIALS AND GROUT TECH-NIQUE

The following two main types of grouting material and two main types of technique were tested:

Material: Cement slurry
             Bentonite slurry

Technique: Dynamic injection
             Static injection

A schematic diagram of the grouting system is shown in Fig. 1. The water and bentonite, or the water, cement and superplasticizer were mixed in the colloid mixer for about 15 minutes. Then the grout was filled in the screw pump where it, in the case of cement, was allowed to circulate for no more than 15 minutes. In the case of dynamic injection the borehole was connected and completely filled via the injection machine, while in the case of static injection the borehole could be connected and filled directly from the

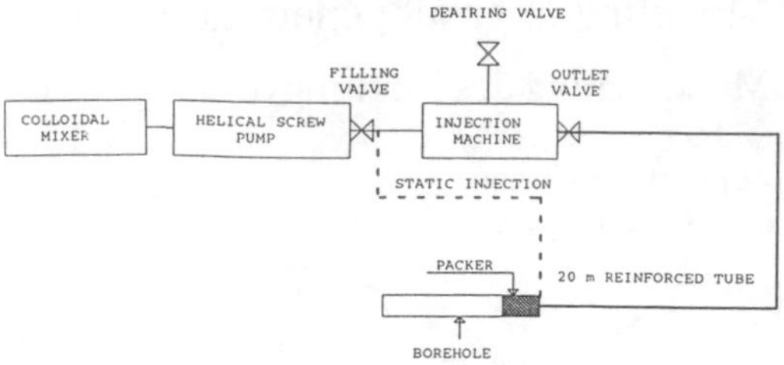

Fig. 1 Schematic drawing of the components in the grouting system
Schemazeichnung der Komponenten des Vergiess-system

Fig. 2 Picture of the dynamic injection machine taken during in situ grouting
Bild der dynamischen Einstpritzmaschine, das beim Vergiessen vor Ort aufgenommen wurde

screw pump. The average time from start of the cement mixing procedure to the grouting was 30 minutes, the upper limit being 90 minutes.

Fig. 2 shows a picture of the injection machine. It was composed of a percussion machine, which hammered at the frequency 100 Hz against a piston that moved in the grout-filled cylinder. The variable static "backpressure" was produced by a pneumatic cylinder pushing the movable percussion machine.

3 FRACTURE PENETRATION

The theories and laboratory results accounted for in the other article (Börgesson, 1993) showed that a vibrating shear strain with an amplitude $\gamma_A > 0.01$-0.1 significantly reduces the shear resistance. Such vibrations can at grouting be produced by superposing a static and a dynamic pressure. The dynamic pressure pulses will propagate from the injection machine via the grout-filled borehole into the fracture into which the grout is penetrating. In the fracture the pressure pulse will to some extent be transformed to a displacement pulse governed by the compressibility of the grout. Fig. 3 shows the amplitude of the velocity profile of the pressure pulse in a fracture. Only the oscillating part of the velocity is shown. The grout is thus vibrating backwards and forewards with no displacement close to the rock surface and maximum displacement s_{max} in the centre of the fracture. The vibrations are thus producing an oscillating shear strain amplitude

$$\gamma_A = \tan\alpha = s_{max}/(d/2) \qquad (1)$$

where d is the fracture apcrature. Since the fracture apertures are very small ($d \approx 0.1$ mm), it means that the displacement s_{max} only need to be 0.001-0.01 mm in order to produce a shear strain amplitude large enough to reduce the shear resistance of the grout.

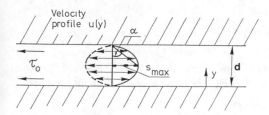

Fig. 3   Velocity profile produced by the oscillating pressure in a fracture with the aperture d

Geschwindigkeitsprofil durch den Schwingungsdruck in einer Bruchfläche mit Öffnung d

4 GROUT FLOW MODELING

In this chapter the basic equations used for calculating the flow and pressure when grout penetrates a fracture at dynamic as well as static pressure will be given.

The grout is generally assumed to be non-Newtonian with the following relation between the shear resistance τ and the rate of strain $\dot{\gamma}$ (Börgesson, 1993):

$$\tau = \left(\frac{\dot{\gamma}}{\dot{\gamma}_0}\right)^n \qquad (2)$$

or

$$\left.\begin{array}{l} \tau = m_1 \cdot (\dot{\gamma})^n \\ m_1 = m/(\dot{\gamma}_0)^n \end{array}\right\} \qquad (3)$$

which means that $m_1 = m$ at the reference rate of strain $\dot{\gamma}_0 = 1.0$ l/s.

The total pressure P and total flow Q in a thin rectangular cross-section, simulating a fracture with the aperture d, can be expressed as the sum of the static (p, q) and dynamic (p', q') components:

$$P(x,t) = p(x,t) + p'(x,t) \qquad (4)$$

$$Q(x,t) = q(x,t) + q'(x,t) \qquad (5)$$

t = time after start
x = distance from entrance (m)

The parameters in Eqns 6 to 17 are listed and explained below:

P	= total pressure (Pa)
p	= static pressure (Pa)
p'	= sinusoidal pressure pulse (Pa)
p_0'	= sinusoidal pressure pulse amplitude at the entrance (Pa)
$\|p'(x)\|$	= sinusoidal pressure pulse amplitude at shear distance x from entrance
L_1	= penetration (m)
Q	= total flow (m³/s)
q	= static flow (m³/s)
q'	= sinusoidal flow pulse (m³/s)

$|q'(x)|$ = sinusoidal flow pulse amplitude at the distance x from the entrance (m³/s)

ω = angular frequency (l/s)

ρ = grout density (kg/m³)

A = fracture area perpendicular to flow direction per meter fracture width (m)

n = rheological parameter according to Eqn. 2

d = fracture aperture (m)

m_1 = rheological parameter according to Eqn. 3

E_r = E-modulus of intact rock (Pa)

d_r = thickness of unfractured rock (m)

E = E-modulus of grout (Pa)

s_A = oscillating deformation amplitude

γ_A = oscillating shear strain amplitude

The boundary conditions according to Fig. 4 with the pressures $P_0=p_0+p_0'$ at the entrance ($x=0$), the penetration $x=L_1$ and the pressure $P=0$ at $x=L_1$, yield the oscillating pressure amplitude and oscillating flow amplitude according to Eqns. 6 and 7 if p_0' is the amplitude of a sinusoidal pressure pulse at the entrance (Pusch et at., 1988).

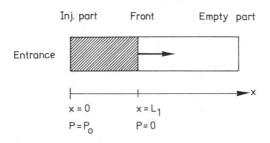

Fig. 4 Boundary conditions for the penetration modeling

Randbedingungen für die Eindringungs-nachbildung

$$|p'(x)| = p_0'\sqrt{\frac{\cosh(2f(L_1-x))-\cos(2k(L_1-x))}{\cosh(2fL_1)-\cos(2kL_1)}} \quad (6)$$

$$|q'(x)| = $$
$$= \frac{p_0'}{g\rho}\frac{C\omega}{\sqrt{f^2-k^2}}\sqrt{\frac{\cosh(2f(L_1-x))+\cos(2k(L_1-x))}{\cosh(2fL_1)-\cos(2kL_1)}} \quad (7)$$

f and k are given in Eqns 8 and 9.

$$f = \frac{CR\omega}{2k} \quad (8)$$

$$k = \sqrt{\frac{CL\omega^2}{2}+\sqrt{\frac{C^2L^2\omega^4}{4}+\frac{C^2R^2\omega^2}{4}}} \quad (9)$$

L, C and R can be solved according to Eqns 10 to 12.

$$L = \frac{4}{3gA} \quad (10)$$

$$C = \frac{gA}{v^2} \quad (11)$$

$$R = nQ^{(n-1)}\cdot\frac{1}{B^n}\cdot\frac{1}{d^{(1+2n)}}\cdot\frac{1}{g\rho} \quad (12)$$

B in Eqn 12 is

$$B = \frac{2}{m_1^{1/n}}\cdot\frac{1}{\left(\frac{1}{n}+2\right)2^{\left(\frac{1}{n}+2\right)}} \quad (13)$$

while v in Eqn 11 is the wave velocity, which can be calculated according to Eqn. 14:

$$v = \sqrt{\frac{E/\rho}{1+\frac{E}{E_R}\cdot\frac{d}{d_r}}} \quad (14)$$

The total flow through the fracture can be calculated according to the following relation

$$Q = B\left(-\frac{dp}{dx}\right)^{\frac{1}{n}}\cdot d^{\left(\frac{1}{n}+2\right)} \quad (15)$$

which also describes the non-oscillating flow q_1 if dp/dx is taken as the static pressure gradient.

The corresponding oscillatory deformation amplitude will be

$$|s_A| = \frac{|q'(x)|}{A\omega} \quad (16)$$

and the oscillating shear strain amplitude

$$\gamma_A = \frac{|q'(x)|}{A\omega \cdot d/2} \qquad (17)$$

Eqn 15 makes it thus possible to calculate the flow rate at different penetration depths and thus the total penetration during the time used for the injection. However, the flow rate is a function of the oscillating shear strain amplitude γ_A. Since γ_A varies not only with time and penetration depth L_1, but also with the distance from the entrance x, it is obvious that a theoretically "correct" penetration cannot be achieved from an analytical solution. Instead the penetration must be calculated by an iterative numerical technique. Such a technique, based on the presented theories, is described by Jönsson (1989a, 1989b) and Börgesson & Jönsson (1990).

5 SLOT INJECTION TESTS

5.1 *Artificial fracture*

In order to be able to check the theories and actual penetration of different grouts in narrow fractures, a series of laboratory tests were performed using a specially designed artificial fracture.

The device shown in Fig. 5 consisted of two, very stiff and perfectly plane steel plates. The plates were bolted together, distanced by a copper foil, to form a slot with an aperture of

Fig. 5 The artificial fracture
Die künstliche Bruchfläche

100-300 μm, 5 cm width and about 3 m length. The surfaces forming the slot were slightly roughened, and the plates were equipped with glass plugs at regular distances so that the penetration of the advancing grout could be directly viewed. Also, pressure transducers were mounted so that the pressure wave pattern could be recorded at different distances from the injection point.

5.2 *Test examples with a cement-based grout*

With the numerical iterative technique the penetration into an idealized fracture with constant aperture can be calculated as a function of time. The application will be illustrated by a calculation of the penetration into the artificial fracture with an aperture of 100 μm. The calculation corresponded to the test 900606 in which the pressure at the entrance was measured during the test as shown in Fig. 6. The average pressure was 840 kPa with a variation between about 500 and 2500 kPa yielding a pressure pulse of 1000 kPa.

The basic data were:

p_0 = 840 kPa
p_0' = 1000 kPa
ω = 251 l/s (f=40 Hz)
ρ = 1760 kg/m³
E = 2.1 ·10⁹ Pa
d = 0.0001 m

Since $E_r \gg E$ and $d_r \gg d$, the wave velocity can be written as:

$$v \approx \sqrt{E/\rho} \qquad (18)$$

The grout used in the injection was Alofix cement with SP=1.2% and w/c=0.45. The parameter m in Eqn 1 is a function of the shear strain amplitude γ_A according to Eqn. 19:

$$m = a \cdot \left(\gamma_A / \gamma_{A_0}\right)^b \qquad \text{at } \gamma_A > 0.1 \qquad (19a)$$

$$m = a \cdot \left(0.1 / \gamma_{A_0}\right)^b \qquad \text{at } \gamma_A \le 0.1 \qquad (19b)$$

393

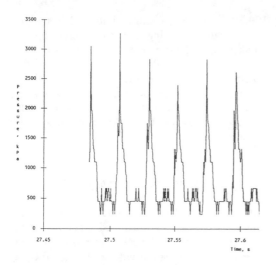

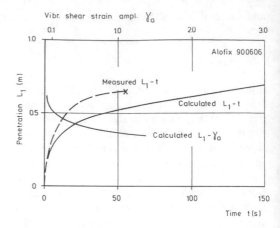

Fig. 6 Test 900606. Pressure pulses measured at the entrance of the artificial fracture. Time from the start of the test

Test 900606. Druckimpulse, gemessen am Eingang der künstlichen Bruchfläche. Zeit vom Beginn des Tests an.

with the following parameter values

$$n \quad = 1.0$$
$$a \quad = 0.2$$
$$b \quad = -1.0$$
$$\dot{\gamma}_0 \quad = 1.0$$
$$\gamma_{A_0} \quad = 1.0$$

Thus at $\gamma_A < 0.1$ the properties at $\gamma_A = 0.1$ were assumed to be valid. This assumption yielded a small underestimation of the m-value.

The calculated penetration as a function of time is shown in Fig. 7 together with the calculated shear strain amplitude at the front. In this test the measured penetration was 0.64 m after 55 s. Fig. 7 shows two main things:

1. The measured and calculated penetration agreed fairly well
2. The shear strain amplitude was very high at small penetration lengths L_1 and decreased rapidly at $L_1 > 0.35$ m.

Fig. 7 Test 900606. Calculated penetration at dynamic injection in the artificial fracture as a function of time and the vibrating shear strain amplitude at the front as a function of the penetration. The measured penetration is also shown.

Test 900606. Berechnete Eindringung bei dynamischer Einspritzung in die künstliche Bruchfläche als Funktion von Zeit und der Rüttel-Scherbeanspruchungsamplitude auf der Vorderseite in Abhängigkeit von der Eindringung. Die gemessene Eindringung wird ebenfalls aufgeführt

5.3 *Compilation of slot tests with bentonite-based grouts*

Using the developed grout flow theory and rheological models, the penetration into the artificial fracture have been calculated and compared to the measured penetration. Different bentonite-based grouts have been used and the results are shown in Fig. 8 where the grout composition is expressed as the relation between the water ratio w and the liquid limit w_L. Different types of bentonite and different salt contents in the added water were used. The figure shows acceptable agreement, considering the complex nature of the grout flow, although the measured penetration has some scatter. The liquid limit w_L is the water ratio at which transition of the material from plastic to liquid consistency takes place according to the soil mechanical definition.

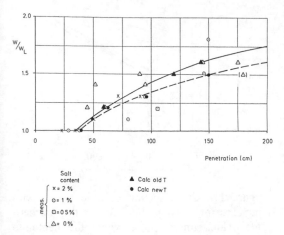

Figure 8 Comparison of measured and calculated penetration in the artificial fracture as a function of the relation between the water ratio and the liquid limit

Vergleich des gemessenen und berechneten Eindringens in die künstliche Bruchfläche als Funktion des Verhältnisses zwischen dem Wasserverhältnis und der Fliessgrenze.

6 FIELD TESTS

The technique has been tested in situ in granite rock in the Stripa mine at 320 m depth for the following purposes:

- Sealing of a natural fractured zone for redistribution of water flow
- Sealing of the disturbed rock zone around a blasted tunnel
- Sealing of the rock around deposition holes

The last test will be outlined in this article.

The test was conducted in two simulated deposition holes, their diamater and depth being 76 cm and 3-3.5 m, respectively. Careful mapping and hydraulic testing of the holes preceded the groutings. The hydraulic tests were made by pressurizing water over the entire periphery of 55 cm long sections in the holes by use of the Large borehole Injection Device (LID). It is a 55 cm high injection cylinder with a diameter that is only 4 mm smaller than that of the holes, the cylinder being surrounded by two large packers. This gave the average hydraulic tconductivity of the rock surrounding the holes at different levels, typical ranges being $3 \cdot 10^{-7}$ m/s in the upper part of hole No. 2 and $3 \cdot 10^{-10}$ m/s in the lower part of the same hole. Very close to the floor in hole No. 1 it was higher than 10^{-6} m/s.

The dynamic injections were also made by use of the LID equipment (Fig. 9). After the injections, a heater was installed in each hole and power applied to yield a rock temperature of just below 100°C. After about two months the power was turned off and the rock allowed to cool.

The hydraulic conductivity measurements were repeated after the grouting and after the heat pulse. The floor was levelled before and after the grouting as well as after the heat pulse. These measurements showed:

- that the average hydraulic conductivity had decreased to $7 \cdot 10^{-10}$ m/s in hole No. 1 and $2 \cdot 10^{-10}$ m/s in hole No. 2 after the grouting

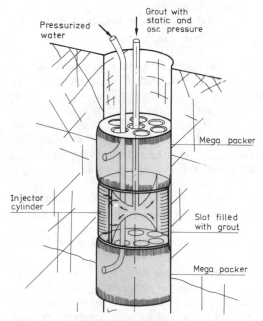

Figure 9 Large Borehole Injection Device (LID) used for hydraulic testing as well as grouting in deposition holes

Einspritzvorrichtung für grosse Bohrungen (LID) zur Verwendung bei Wasserdruckproben und zum Vergiessen von Deponielöchern.

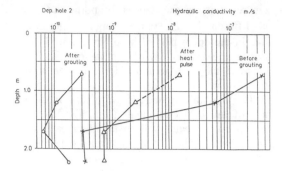

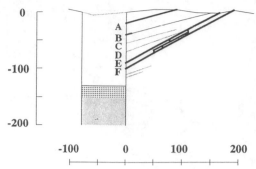

Figure 10 Measured hydraulic conductivity at four levels in hole No.2 at three different occasions

Gemessene Wasserleitfähigkeit auf vier Ebenen in Bohrung Nr. 2 bei drei verschiedenen Gelegenheiten

Figure 11 Cross section of hole No. 2 with fractures intersecting the hole on the northern side. Thick lines denote where the bentonite grout was found

Querschnitt von Bohrung Nr. 2, wobei die Bruchflächen die Bohrung auf der Nordseite schneiden. Die dicken Linien markieren, wo der Bentonit-Vergussmörtel gefunden wurde.

- that the average hydraulic conductivity had increased considerably due to the heat pulse although it was still considerably lower than before the grouting
- the residual heave of the floor around the holes was 200 μn as an average

In order to check the grout penetration into the fractures, the rock around one of the holes was excavated. The excavation showed that grout had penetrated as deep as 2 m into flat-lying fractures that were free from debris of disintegrated fracture coatings like chlorite, while it also showed that steep fractures with such debris were poorly penetrated.

The results of the hydraulic conductivity measurements in hole No. 2 is shown in Fig. 10 and the results of the excavation of the floor on the northern side of this hole is shown in Fig. 11.

The results of *the other two field tests* were that a clear effect of the grouting was seen after injection of the natural fractured zone while no reduction in total hydraulic conductivity was seen after injection of the disturbed zone around a blasted tunnel.

7 CONCLUSIONS

- The penetration of grout into fine fractures at dynamic grouting has been successfully modeled for well-defined fractures.
- Fractures intersecting large boreholes drilled through the disturbed shallow zone in blasted drifts, can be sealed with bentonite-based grout by use of "megapackers" and dynamic injection technique.
- Natural fracture zones can be sealed by using cement injected under high pressure applying dynamic injection technique. The spacing of the injection holes determines the effectiveness of the grouting.
- Shallow rock with fine fractures coated with fracture minerals like chlorite is not effectively groutable, probably because of debris produced by blasting-induced disintegration of the fracture minerals.
- The groutability of rock can be predicted by hydraulic testing and evaluation of the rock structure provided that debris is not present in the fractures.

8 REFERENCES

Börgesson, L. & Jönsson, L. 1990. Grouting of fractures using oscillating pressure. *Proc. of the Int. Conf. on Mechanics of Faulted Rock,* Vienna: Balkema

Börgesson, L., Pusch, R., Fredrikson, A., Hökmark, H., Karnland, O. & Sandén, T. 1991. Final Report of the Rock Sealing Project - Volume I - Sealing of the Near-Field Rock Around Deposition Holes by Use of Bentonite Grouts. *Stripa Project Technical Report 91-34*, SKB, Stockholm

Börgesson, L., Pusch, R., Fredrikson, A., Hökmark, H., Karnland, O. & Sandén, T. 1992. Final Report of the Rock Sealing Project - Sealing of Zones Disturbed by Blasting and Stress Release. *Stripa Project Technical Report 92-21*, SKB, Stockholm

Börgesson, L. 1993 Rheological properties of cement and bentonite grouts with special reference to the use of dynamic injection. *Proc. International Conference on Grouting in Rock and Concrete*, Salzburg: Balkema

Jönsson, L. 1989a. A simplified method for computation of very high-viscosity transient flow in a small fracture. *Internal Report. Dep. of Water Resources Engineering*, University of Lund, Sweden

Jönsson, L. 1989b. Computation of high-viscosity transient flow in small fractures. *6:th International Conference on Pressure Surges*, Cambridge, England. BHRA Fluid Engineering Centre, Cranfield, Bedford

Pusch, R., Karnland O., Hökmark, H., Sandén, T & Börgesson, L. 1991. Final Report of the Rock Sealing Project - Volume V - Sealing Properties and Longevity of Smectite Clay Grouts. *Stripa Project Technical Report,* SKB, Stockholm

Wylie & Streeter. 1978. Fluid Transients. McGraw-Hill Inc.

Der erforderliche Druck zur Einleitung von Injektionen
The pressure necessary to start the grouting procedure

G. Feder
Montanuniversität Leoben, Österreich

ZUSAMMENFASSUNG: Bei normal bis schwer injizierbarem Fels muß bekanntlich das zu injizierende Medium unter einem bestimmten Mindestdruck stehen, damit es vom Bohrloch in das
Kluft- oder Rißsystem eindringen kann. Um aber das zu injizierende Gebirge oder Bauwerk
durch den Flüssigkeitsdruck nicht in seinem Inneren aufzureißen, ist man andererseits
bemüht, insbesondere bei diesbezüglich sensiblem Gebirge, den Injektionsvorgang mit möglichst niedrigem Injektionsdruck zu konzipieren.
Bei einigen Versuchsreihen an der Montanuniversität Leoben zeigte es sich, daß der Ansprechdruck, also der für die Einleitung der Injektion erforderliche Mindestdruck, nicht
nur von der Spaltweite, sondern auch von der Art des Injiziergutes, bzw. dessen
rheologischen Eigenschaften abhängt. Generell ergab sich dabei erwartungsgemäß, daß der
erforderliche Ansprechdruck bei Suspensionen stark von der Korngröße der in der Suspension enthaltenen Partikel ebenso wie von der Ausformung des Überganges vom Bohrloch
in den Spalt abhängt und durch den Aufbau eines Filterkuchens vor dem Spalt mit der
Dauer des Injektionsvorganges weiter steigt. Bei Lösungen bzw. homogenen Medien, wie
z.B. Kunstharzen kann zunächst mit einem vergleichsweise geringerem Ansprechdruck das
Auslangen gefunden werden, der zufolge der Thixotropie nach Unterbrechungen wieder
größer werden kann.

ABSTRACT: To start the grouting procedure at normal to hard groutable rock it is
necessary that the grouting pressure exceeds a certain minimum, to enable the permeation
from the borehole into the joint. To avoid hydraulic fracturing within the rock or
structure one tries to perform the grouting procedure with a pressure as low as
possible, especially in sensitive rock or structures.
The first tests at the Montanuniversity of Leoben (Austria) were performed with an
equipment shown in Fig. 1. The tests were carried out with cement suspensions (PC 375;
w/c ratios of 0.5, 0.7, 0.9; additives at 0.5 and 1.0 %) and with epoxy resins. In spite
of the fact that the test results varied widely (Tab. 1), a tendency became very clearly
in the course of the tests. From the results it should be stressed that the so-called
"starting pressure" depends not only upon the joint width but also the grout type, the
rheological properties of the grout material (Fig. 2) and the grain size in the suspension as well as from the shape of the transition zone between the borehole and the
joint. The starting pressure increases with viscosity/cohesion and decreasing joint
width; it increases additionally with the duration of the grouting procedure due to the
development of filter cakes of suspensions (Fig. 3). This disadvantage of suspensions
may be reduced with the use of smaller grain size cement particles; on the other side
the advantage of solutions is reduced by thixotropy.
With these tests it was possible to find reference points for this starting pressure
which can be much higher without any risc than the various "allowable" grouting
pressures according to the numerous rules of thumb. The tests should be continued to
improve the knowledge and to enable a clear definition of the relationships between
starting pressure, joint width and rheological properties.

1) EINFÜHRUNG.

Der Erfolg von Injektionsarbeiten hängt unter anderem von der Wahl des optimalen Injektionsdruckes unter den gegebenen Verhältnissen ab. Aus wirtschaftlichen Gründen wird ein möglichst hoher Injektionsdruck angestrebt, der meist am Bohrlochmund gemessen wird. Ein höherer Injektionsdruck ermöglicht eine größere Reichweite des Injektionsgutes und damit einen größeren Bohrlochabstand, weniger Umstellungen des Bohrgerätes und weniger Injektionsabschnitte und damit eine wirtschaftlichere Ausführung. Um jedoch unerwünschte Auswirkungen eines zu hohen Injektionsdruckes zu vermeiden, ist eine Begrenzung des maximalen Injektionsdruckes erforderlich. Um aber auch den Mindestdruck abschätzen zu können, der im Injiziergut herrschen muß, wenn es in Spalten bestimmter Weite eindringen soll, wurden an der Montanuniversität Leoben einige Versuchsreihen gefahren. Dabei zeigte sich, daß dieser sogenannte Ansprechdruck nicht nur von der Spaltweite, sondern auch wesentlich von der Art des Injiziergutes abhängt.

Es überrascht nicht, daß reines Wasser einen vergleichsweise geringen Ansprechdruck erfordert. Suspensionen hingegen, die schwebende Partikel enthalten, die sich auch ausfiltern lassen, erfordern deutlich höhere Ansprechdrücke, die außerdem noch zeitabhängig erscheinen, da sie von der Mächtigkeit des inzwischen vor dem Spalt aufgebauten Filterkuchens abhängig sind. Lösungen, die aus einem einheitlichen Stoff bestehen, sind hier zweifellos von Vorteil, aber natürlich i.a. kostspieliger als z.B. Zementsuspensionen. Der Nachteil der Suspensionen verringert sich allerdings mit der Mahlfeinheit der festen Partikel; der Vorteil der feststofffreien Medien verringert sich, sobald strukturviskose oder thixotrope Eigenschaften zum Versteifen des Injiziergutes führen, sobald die Strömungsgeschwindigkeit absinkt oder der Injiziervorgang unterbrochen werden muß. Diese Eigenschaften des Injiziergutes werden aber noch zusätzlich von der Temperatur und der Dauer der Ruhezeit beeinflußt. Bei der Auswahl des Injiziergutes ist daher dessen vielseitige Untersuchung erforderlich.

Im folgenden sollen nun einige Einzelheiten über die Ausführung und die Ergebnisse dieser Versuche berichtet werden.

2) DER VERSUCHSAUFBAU.

Die Ausbildung des Gerätes wurde nach mehreren Vorversuchen entwickelt und ist dem Bild 1 zu entnehmen.

Das Injektionsgut wird mit einer Handpumpe unter Druck gesetzt und gelangt über eine Zuleitung zum eigentlichen Versuchsgerät und über dessen Zentralraum zum Spalt, der die einzige Austrittsmöglichkeit bietet. Der Übergang von der Bohrung in den Spalt kann durch einen auswechselbaren Ring scharfkantig oder mit abgeschrägter Kante ausgebildet werden.

Die Spaltweite kann mit Klingerit oder Cu-Blech auf die gewünschte Weite eingestellt werden. Dazu werden die beiden zunächst miteinander verschraubten Flansche voneinander gelöst und je nach der einzustellenden Spaltweite Cu-Blech (o.1 mm) oder Klingerit (0.3 mm bzw. 0.5 mm) eingelegt. Beim Anspannen der Schrauben wird die Dicke dieser Zwischenlage verringert. Die wahre Spaltweite wird dabei mit einem induktiven Wegaufnehmer auf 0.001 mm ge-

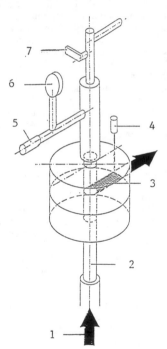

Bild 1) Versuchsgerät   Test equipment

1 .. Fließrichtung	Direction of flow
2 .. Bohrlochimitation	Boreholeimitation
3 .. Spalt	Joint
4 .. Messung der Spaltweite	Measurement of joint width
5 .. Messung des Absolut-Druckes	Measurement of absolute pressure
6 .. Kontroll-Manometer	Control-Manometer
7 .. Absperrhahn	Valve

nau gemessen, wobei die Eichung des Null-
punktes im ohne Zwischenlage fest verschraubten Zustand erfolgt. Bei den Vorversuchen hat es sich übrigens herausgestellt, daß die Spaltaufweitung während
des Injektionsvorganges bei diesem Versuch
vernachlässigbar ist.

Der Absperrhahn ermöglicht eine Druckentlastung und Zerstörung eines eventuellen Filterkuchens im Zentralraum während
des Versuches und erleichtert das Reinigen
des Gerätes durch Spülung mit Wasser.

Die Messung des Gewichtes bzw. Volumens
des aus dem Spalt austretenden Injektionsgutes erfolgt mit Hilfe einer elektronischen Waage, an deren Seiten zwei induktiv
arbeitende Wegaufnehmer angeordnet sind,
welche die durch das austretende Injiziergut verursachte Absenkung der Waagschale
messen. Zur Umrechnung dieser Absenkung
in Gewicht (jenes Injiziergutes, das
den Spalt bereits durchströmt hat) dient
eine vor Versuchsbeginn ermittelte Eich-
kurve.

Die Meßwertaufnahme erfolgte bei allen
Versuchen mit einem Vielstellen-Meßgerät
und einem PC zur Meßdatenspeicherung. Diese Rohdaten wurden mit einem Datenumfor-
mungsprogramm verarbeitet und in eine für
die Auswertung zweckmäßige Form gebracht.

Tab. 1: Versuchsergebnisse   Test results

$\frac{W}{Z}$	Addi-tiv	scharfe Kante t mm	scharfe Kante p bar	schräge Kante t mm	schräge Kante p bar
0.5	0.5	0.400	6-12	-	-
		0.395	6-12	0.381	4- 6
		0.300	15-20	-	-
		0.250	20-25	-	-
		0.220	22-30	0.224	20-30
		0.190	25-35	-	-
		-	-	0.150	45-50
	1.0	0.401	6-10	0.385	4- 8
		0.260	18-25	0.250	20-25
		0.244	15-25	0.242	25-30
		0.190	25-30		
		0.160	50-60	0.175	45-50
		-	-	0.130	55-65
0.7	0.5	0.394	4- 6		
		0.220	8-12		
		0.220	8-12		
		0.180	12-16		
	1.0	0.830	1-1.5	-	-
		0.381	6- 8	0.384	4- 6
		0.310	7-10	-	-
		-	-	0.254	6-10
		0.232	10-15	0.230	15-17

3) DURCHFÜHRUNG DER VERSUCHE:

3.1) Das Injektionsgut.

Die Versuche wurden mit dem auf einer
Großbaustelle verwendeten Injektionsgut
durchgeführt. Untersucht wurden Zementsuspensionen auf der Grundlage von PZ 375,
dessen Mahlfeinheit 4300 Blaine und dessen
maßgebender Korndurchmesser D85 0.03 mm
beträgt. Es wurden Wasser-Zementwerte von
0.5, 0.7 und 0.9 unter Zugabe von Additiven in Dosierungen von 0.5 % und 1.0 % untersucht. Zum Vergleich wurde RODUR 51o
als Beispiel für eine Lösung untersucht.

3.2) Zusammenfassung der Ergebnisse.

Wie der Tabelle 1 zu entnehmen ist, streu-
en die Versuchsergebnisse sehr stark, was
unter anderem auf die Verwendung einer
Handpumpe mit stoßweisem Druckaufbau zurückzuführen sein dürfte. Für eine erste
Darstellung der Tendenzen (Bild 2) wurden
daher allzuweit abweichende Versuchsergebnisse ausgeschieden. Aus den bisher vorliegenden Versuchsergebnissen läßt sich
ableiten, daß der Ansprechdruck

1. zunächst durch eine Filterkuchenbildung (Bild 3) bei den Suspensionen schon
für vergleichsweise großen Spaltweiten an-
zusteigen beginnt;

2. gemäß der empirischen Daumenregel
stark ansteigt, wenn die Spaltweite den
3 - 5 fachen maßgebenden Korndurchmesser
der Suspension unterschreitet.

$\frac{W}{Z}$	Addi-tiv	scharfe Kante t mm	scharfe Kante p bar	schräge Kante t mm	schräge Kante p bar
0.9	0.5	0.401	2- 5	0.381	2- 4
		0.310	10-15	0.320	4- 7
		-	-	0.25o	8-10
		0.241	6-10	0.241	5- 9
		0.220	20-25	0.220	10-13
		0.150	35-45	0.180	25-30
	1.0	0.388	2- 4	0.395	2- 3
		0.274	6-10	0.235	3- 5
		0.260	7-11	0.230	4- 7
		-	-	0.190	15-20
		0.160	15-25	0.160	25-35
RODUR 510 Kompo-nente A.		-	-	0.240	2- 3
		0.190	40-50	0.18o	20-25
		0.155	55-65	-	-
		0.140	60-75	-	-
		-	-	0.110	60-70
		-	-	0.100	80-90
		-	-	0.070	90-100

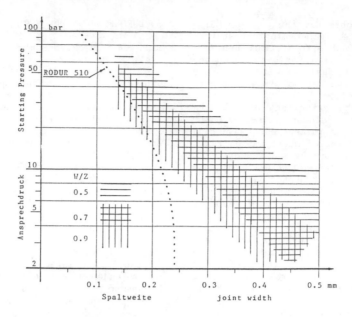

ild 2) Ergebnisse der Versuche   Test results.

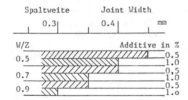

Bild 3) Beginn der Filterkuchenbildung.
       Start of Filter Cake Developing.

3. durch Verbesserung der Fließeigenschaften, sei es durch einen höheren
Wasser-Zement-Wert, sei es durch Zugabe
von Verflüssigern verringert werden kann.
So selbstverständlich diese Tendenzen
auch erscheinen, so sind doch die Versuche
deswegen von großer Bedeutung, weil damit
erstmals Anhaltspunkte für die zur Einleitung des Injektionsvorganges erforderlichen Drücke ermittelt werden. Diese
Drücke können demnach gefahrlos über den
sich aus den diversen Daumenregeln ergebenden Maximaldrücken liegen, was insbesondere für seicht liegende Injektionsabschnitte von Bedeutung ist.

4) AUSBLICK.

Eine Weiterführung der Versuche unter
Nutzung der bisher gewonnenen Erfahrungen
erscheint zweckmäßig, um mit weiterent-
wickelten Versuchsanordnungen die Zusammenhänge zwischen den rheologischen Eigenschaften der Injektionsmittel und dem
– für die Einleitung des Injektionsvorganges erforderlichen Ansprechdruck,
– Druckabfall beim Eintritt des Injektionsgutes in den Spalt für verschiedene
Injektionsraten und damit dem im Spalt
wirksamen Druck
im Vergleich zu den bei Wasserabpreßversuchen gegebenen Druckverhältnissen zu
erfassen. Damit könnten auch Vergleichs-
werte für die analytische Lösung dieser
Probleme gewonnen werden.

DANKSAGUNG.

Die vorliegende Arbeit ist ein Auszug aus
der Diplomarbeit von Herrn Dipl. Ing.
Brandl (Montanuniversität Leoben). Der
Verfasser möchte an dieser Stelle der
Österreichischen Draukraftwerke AG. dafür
danken, daß sie die Durchführung der Versuche und die Veröffentlichung der Ergebnisse ermöglicht haben, weiters den Herren
Dr. Stadler (Fa. Insond) für zahlreiche
Anregungen und Hilfestellungen sowie dem
Institutionsvorstand Herrn Prof.Dr. Golser
und seinem Team (Montanuniversität Leoben)
für die Erfassung und Auswertung der Meßergebnisse.

Fließen in Spalten – Laborversuche und Berechnungen

Flow in joints – Laboratory tests and calculations

G. Feder

Montanuniversität Leoben, Österreich

ZUSAMMENFASSUNG: Das Fließen in Spalten wurde für Wasser bereits vielfach untersucht, auch mathematische Modelle zu Beschreibung dieses Vorganges in rauhen Spalten wurden entwickelt und stimmen gut mit Beobachtungen überein. Weniger gut ist diese Übereinstimmung bei Flüssigkeiten, die nicht dem Newton'schen Ansatz entsprechen. Der vorliegende Bericht ist das Ergebnis mehrerer Versuchsreihen an der Montanuniversität Leoben, Österreich, mit dem Ziel, die Ursachen dieser mangelnden Übereinstimmung aufzuklären.

Dabei hat sich ergeben, daß diese Ursachen weniger im Aufbau der mathematischen Modelle liegen, sondern vielmehr in der Beschreibung der maßgebenden Fließeigenschaften Nicht-Newton'scher Flüssigkeiten. Zur Ermittlung dieser Eigenschaften sollte ein Kegel-Platte-Viskometer eingesetzt werden, bei dem die Schergeschwindigkeit entlang der gesamten Scherfläche konstant ist. Wesentlich ist weiters, daß am Viskometer das Drehmoment gewählt und die sich dabei einstellende Drehzahl abgelesen werden kann. Weiters wurde das mathematische Modell dahingehend abgeändert, daß die im Viskometer gemessenen Wertepaare unmittelbar in das mathematische Modell eingeführt und die sekundliche Durchflußmenge q direkt abgelesen werden können (Bild 4).

Schließlich hat sich noch herausgestellt, daß bei den verwendeten Kunstharzen die Rauhigkeitstäler und unregelmäßige Berandungen zwar aufgefüllt, aber dann nicht mehr durchflossen werden. Als Spaltweite war daher nur der Abstand der Rauhigkeitsspitzen, als Fließbreite jene der Engstellen anzusetzen. Mit diesen Parametern konnte eine gute Übereinstimmung von Rechnung und Messung erzielt werden.

ABSTRACT: Substantial research has been performed in the past investigating the flow of water in joints. Many mathematical models were developed to describe this process in rough joints which agreed very well with the test results. However, fluids which are not Newtonian fluids do not agree with those mathematical models. This report is the result of some test series performed at the Montan University at Leoben, Austria.

These investigations resulted in the fact that the reason for these differences is not found in inadequate mathematical models, but in the inadequate descriptions of the rheological properties of the fluids. To determine these properties a cone-plate-viscometer should be used, in which the shear velocity is constant upon the total shear area at a certain speed of rotation. It is essential that the moment of torque is adjustable and the rotation velocity is readable (and not reverse). Therewith it is possible, for example, to clearly determine the yield point which represents that shear stress which leads to the start of the flow. The couples of shear stress and shear velocity may depend on the structural viscosity of non-newtonian fluids and are not well described by a formula. The mathematical model was revised to enable the input of the relevant couples of shear stress and shear velocity as well as the joint width and the available gradient of pressure. From the diagram (Fig. 4), one can also see the distribution of the velocity across the joint together with the flow rate.

Finally, the investigations showed that the epoxy resin used filled the valleys of roughness as well as the bays of the borders of the flow path but there was no flow through those areas after being filled. The best conformity between analyses and test was reached with the cross-section between the peaks of the roughness and the least distance between the side borders.

1. EINFÜHRUNG

An der Montanuniversität Leoben (Österreich) wurden vergleichende mathematische
und physikalische Modelle entwickelt, um
die Ausbreitung von Injektionsgut in Spalten zu erfassen. Um dieses Ziel zu erreichen, mußten zwei grundsätzliche Fragen
behandelt werden:
- Quantitative Beschreibung der für das
Injizieren maßgebenden Eigenschaften flüssiger Medien in Form rheologischer Daten.
- Verifizierung des mathematischen Modells zur Erfassung der Ausbreitung flüssiger Medien in engen Spalten an Hand der
Ergebnisse von Laborversuchen.
Die Injektionsversuche wurden mit Wasser
und Kunstharz durchgeführt, um zwei in ihrem Fließverhalten unterschiedliche Medien
zu erfassen. Im folgenden sollen nun die
Ergebnisse dieser Arbeiten zusammengefaßt
werden.

2. DIE FLIESSEIGENSCHAFTEN

Die Fließeigenschaften von Wasser sind bekannt. Die Viskosität dieser kohäsionslosen, Newton'schen Flüssigkeit ist temperaturabhängig, was gemäß Bild 1 für den für
Injektionen infrage kommenden Bereich zwischen 5° C und 25° C berücksichtigt wurde.

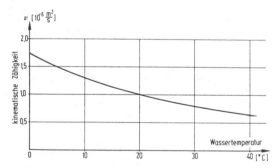

Bild 1: Temperaturabhängigkeit der
Viskosität von Wasser
Viscosity of water dependent on
temperature

Zur Bestimmung der Fließeigenschaften von
anderen Injektionsmaterialien sind zwei
Verfahren weit verbreitet:
- Der Marsh-Trichter, mit dem eine Auslaufzeit gemessen wird, die vor allem als
Funktion der beiden rheologischen Kennwerte Fließgrenze und Viskosität anzusehen
ist. Eine Trennung dieser beiden Kennwerte
ist mit diesem Gerät nicht möglich, sodaß
es vor allem für Prüfungen konstanter

Fließeigenschaften auf der Baustelle,
nicht aber zur Ermittlung der rheologischen Kenndaten im einzelnen geeignet ist.
- Die Rotationsviskometer, von denen im
wesentlichen zwei Typen zu unterscheiden
sind: jene, bei denen die Drehzahl vorgegeben wird und das zur Erreichung dieser
Drehzahl erforderliche Drehmoment gemessen
wird und jene, bei denen das Drehmoment
vorgegeben wird und die zugehörige Drehzahl gemessen wird. Letztere ermöglichen
bei langsamer Steigerung die Messung jenes Drehmomentes, bei dem die Bewegung beginnt und damit kann aus den Abmessungen
des Viskometers (Rotordurchmesser, be-
netzte Rotorfläche) die Fließgrenze, also
das Drehmoment bzw. die Schubspannung, bei
der die Bewegung anläuft, gerechnet werden.
Die Drehkörper handelsüblicher Viskometer sind entweder zylindrisch oder ko-
nisch. Die Kosten zylindrischer Viskometer
sind zwar geringer, aber die Meßergebnisse
sind ungenauer, da dort im Spalt zwischen
Rotor und Stator theoretisch bei jedem Radius eine andere Schergeschwindigkeit
herrscht. In vielen Fällen reicht aber
diese Genauigkeit aus, da das Größenver-
hältnis zwischen Spaltweite und Rotorra-
dius klein gehalten wird, sodaß die
Schergeschwindigkeit (dv/dr) als innerhalb
des Spaltes annähernd konstante Größe angenommen werden kann.
Wenn aber eine höhere Genauigkeit benötigt
wird, so verwendet man ein Kegel-Platte-
Viskometer, das allerdings kostspieliger
ist. Dort befindet sich die zu prüfende
Flüssigkeit zwischen einer horizontalen
Platte und einem kegelförmigen Rotor, dessen Spitze in der Plattenebene liegt und
dessen Achse vertikal steht. Die Kegeler-
zeugenden stehen im spitzen Winkel zur Ho-
rizontalebene. Die Schergeschwindigkeit
stellt sich dann im gesamten Flüssigkeits-
volumen zwischen Kegel und Platte, also in
allen Scherflächen, in gleicher Größe ein.
Bei Untersuchungen der Fließeigenschaften
strukturviskoser bzw. thixotroper Flüssigkeiten wurde im Rahmen dieses Projektes
ein Viskometer des letztgenannten Typs
verwendet.
In vielen Kunstharzen baut sich im Ruhezustand eine versteifende Struktur auf,
die bei späterer Durchwirbelung wieder
zerstört wird. Solche "strukturviskosen"
Medien werden durch den Injiziervorgang
allmählich dünnflüssiger. Daher ist das
Bighamsche Fließgesetz zur Charakterisierung des Fließverhaltens nicht immer ausreichend, hängen doch die Fließeigenschaften insbesodere in dem für die Reichweite
des Injektionsgutes maßgebenden Bereich
geringer Dreh- bzw. Fließgeschwindigkeiten

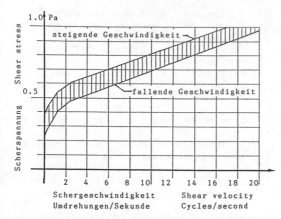

Bild 2: Fließeigenschaften von Kunstharz in Abhängigkeit von der Dreh- bzw. Fließgeschwindigkeit und der Strukturviskosität.
Flow properties of Epoxy resin dependent on the rotation resp. flow and the structural viscosity.

auch wesentlich davon ab, ob die innere Struktur bereits durch vorherige Durchmischung zerstört wurde (Bild 2).

Auch die Temperatur des Injektionsmaterials kann eine große Bedeutung für die Fließeigenschaften haben. Die Versuche mit dem verwendeten Kunstharz zeigten ein Ansteigen der Fließgrenze auf das vierfache bei einem Absinken der Temperatur von 25°C auf 5°C (Bild 3).

Bei Injiziermedien, deren Eigenschaften wesentlich von vereinfachenden Annahmen nach Newton, Bingham usw. abweichen, wird

im folgenden ein direkter Weg gezeigt, die viskometrisch ermittelten Wertepaare im mathematischen Modell zu nützen. Nochmals sei betont, daß für kohärente Medien nur solche Viskometer geeignet sind, bei denen das Drehmoment vorgegeben und die Drehzahl abgelesen wird.

Im Kegel-Platte-Viskometer wird das zwischen zwei unterschiedlich bewegten Wänden befindliche und zu prüfende Medium bei einer bestimmten Rotationsgeschwindigkeit einer konstanten Schergeschwindigkeit unterworfen. Mit Hilfe der Gerätekonstanten läßt sich dann aus dem Wertepaar Drehmoment/Rotationsgeschwindigkeit das Wertepaar Schubspannung/Schergeschwindigkeit im Medium ermitteln. Dazu wird angenommen, daß das Medium an den Wänden haftet.

Im zu injizierenden Spalt ergeben sich die Spannungsverhältnisse hingegen aus der treibenden Kraft, die vom Injizierdruck wachgerufen wird und die vom Scherwiderstand über den Spaltquerschnitt allmählich wieder aufgezehrt wird. Für die rechnerische Erfassung ist daher das Druckliniengefälle I_p des Injizierdruckes maßgebend, sodaß die Gleichgewichtsbeziehung

$$\tau = z \times I_p$$

gilt. Dabei bezeichnet z den Abstand der betrachteten Stelle des Mediums von der Mitte der Spaltweite. Die Fließgeschwindigkeit v_F erhält man durch Summation der Schergeschwindigkeit v_τ (die ja den Änderungen der Fließgeschwindigkeit entspricht) von der Spaltwand ($v_F=0$, $\tau=\tau_{max}$) aus.

$$v_F = \int_{z_W}^{0} v_{\tau,z} \cdot dz$$

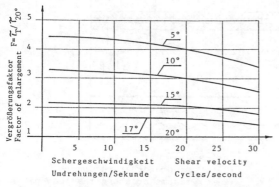

Bild 3: Fließeigenschaften eines Kunstharzes in Abhängigkeit von der Temperatur.
Flow properties of an Epoxy resin dependent on the temperature.

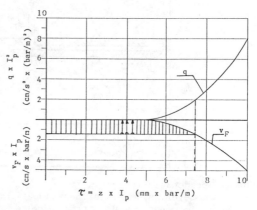

Bild 4: Darstellung der Ergebnisse eines Viskometerversuches für die direkte Berechnung des Fließens in Spalten.
Presentation of the results of viscometer tests for the direct analysis of the flow in joints.

405

Bei dieser Darstellung (Bild 4) ergibt sich auch das im Spalt zu erwartende Geschwindigkeitsprofil zwischen der Spaltmitte und der Spaltwand. Der Inhalt dieser in Bild 4 schraffierten Fläche entspricht der Durchflußmenge je cm Spaltbreite und kann mit der Linie q in $cm^3/s.cm$, also cm^2/s dargestellt werden.

Als Beispiel sei ein Punkt der Darstellung der Ergebnisse eines Injektionsversuches (Bild 7) angeführt. Bei diesem Versuch betrug die Spaltweite 0,2 mm, wovon noch die Rauhigkeit 2 x 0,06 mm abzuziehen ist, sodaß die halbe Spaltweite z_w = 0,094 mm wird. Das Druckgefälle I_p an der betrachteten Stelle sei 80 bar. Dann ergibt sich für $z_w \cdot I_p$ = 7,5,

$$\tau = 750 \text{ mPa},$$
$$v_F = 1,4 \text{ cm/min},$$
$$q = 0,09 \text{ cm}^3/\text{min (für 5 cm}$$
$$\text{Durchflußbreite)}$$

3. DAS PHYSIKALISCHE MODELL

Zum Beobachten der beim Injizieren in situ tatsächlich zu erwartenden Vorgänge wurde ein Modell mit folgenden Möglichkeiten hergestellt (Bild 5):
- einstellbare Spaltweiten in einem Größenbereich von 0,2-0,5 mm,
- Injektion von Medien in einen Wasserstrom und Messung von deren Ausbreitung,
- Imitation einer wechselnden Strömungsbreite - wie in situ - durch Kontaktinseln

der Spaltwände, sodaß das Medium wechselweise zwischen Buchten und Engstellen fließen muß.
- Anwendung von Injektionsdrücken bis zu 150 bar,
- gleichmäßige und reproduzierbare Wandrauhigkeit,
- konstante Spaltweite,
- 1,6 m langer Injizierweg,
- wechselnde Spaltbreite zwischen 50 und 100 mm.
- Zusätzlicher Anschluß im 1. Drittelpunkt des Injizierweges, um auch das Injizieren in einen wasserdurchströmten Spalt bzw. das Eindringen von Wasser in einen Injektionsstrom beobachten zu können.

Die Schwierigkeit lag im Erzielen ausreichender Präzision bei der Einhaltung einer Spaltweite, die selbst nur in der Größe weniger Zehntelmillimeter lag, bei einer Spaltbreite von 50 bis 100 mm auf einer beobachtbaren Strömungslänge von 1,6 m und einem Injizierdruck bis zu 150 bar.

Konstruktiv gelöst wurde dies durch folgende Stahlkonstruktion: Boden- und Deckplatte, je 50 mm dick, begrenzten die vom Injizierdruck bewirkte Durchbiegung auf Größen, die ohne nennenswerte Fehler durch Rechenwerte bestimmt und durch Wegmesser überwacht werden konnten. Die Abdichtung von 150 bar Innendruck wurde durch einen endlos verlöteten Quetschdraht aus Kupfer erreicht, der von den beiden 50 mm dicken Stahlplatten mit vorgespannten Schrauben

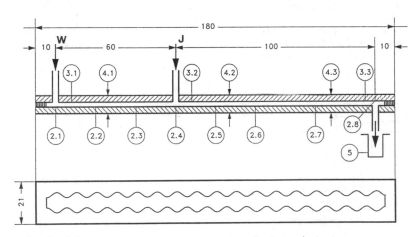

Bild 5: Versuchseinrichtung.
(1) .. Manometer an der Injektionspumpe
(2) .. " entlang des Strömungsweges
(3) .. Thermometer
(4) .. Wegmesser zur indirekten Überwachung der Spaltweite
(5) .. Zylinder zur Messung des austretenden Injektionsgutes
(6) .. Stoppuhr

Test equipment
(1) .. manometer at the pump
(2) .. manometer along the flow path
(3) .. thermometer
(4) .. indirect measurement of joint width
(5) .. glass tube for measurement of the outflow
(6) .. stop clock

so quergepreßt wurde, daß er den hohen Injizierdruck abzudichten vermochte.

Die seitliche Begrenzung des Injizier-
spaltes und zugleich den Überlastungs-
schutz für den Quetschdraht bildete jeweils eine der handelsgängigen Platten aus
Weichkupfer oder Klingerit, die in Dicken-
abstufungen von Zehntelmillimetern erhältlich sind. Aus jeder dieser Platten wurde
vor dem Einbau der für das Injiziergut
vorgesehene Weg herausgeschnitten. Sie
bildete dadurch auch die seitliche Begrenzung des Injektionsgutes. Diese Begrenzung
wurde sinusförmig ausgeschnitten, sodaß
die Spaltbreite zwischen 50 und 100 mm
wechselt. Die definierbare Wandrauhigkeit
wurde durch Sandstrahlung der Deckbleche
erreicht. Die Anordnung der Meßgeräte zur
Erfassung des Fließvorganges ist Bild 5 zu
entnehmen.

Die Ablesung aller Manometer erfolgte
stets gleichzeitig bei allen Meßstellen
(fotographisch). Nach jedem Versuch wurde
zusätzlich die eingestellte Spaltweite
durch Messen des Restdurchmessers des plastisch verformten Quetschdrahtes nachge-
prüft.

4. DIE VERSUCHE UND DEREN AUSWERTUNG

Grundlage für die rechnerischen Untersuchungen waren die bekannten Gleichungen
für die Rechteckströmung im Spalt (Louis
1970), die bei Wasser allerdings erst dann
eine befriedigende Übereinstimmung mit den
gemessenen Werten brachten, wenn
- als Strömungsbreite der Mittelwert aus
größter und kleinster Breite des Fließwe-
ges,
- die Wandrauhigkeit mit dem oberen Wert
des Streubereiches von 0,06 mm und
- die nur wenige Hundertstel Millimeter
klaffenden Zwickelbereiche über und unter
der Zwischenlage nicht berücksichtigt wurden.
Die Übereinstimmung von Rechnung und Messung ist im weiten Bereich von Druckgefällen zwischen 3,6 und 100 bar/m und
Spaltweiten von 0,2 und 0,5 mm durch eine
Varianz von 0,02 gekennzeichnet. Ein Beispiel ist in Bild 6 dargestellt.

Bei den Versuchen mit Kunstharz, das für
die Versuche ohne Härter verwendet wurde,
um die Wiederverwendbarkeit des Gerätes zu
ermöglichen, konnte eine ähnliche Übereinstimmung von Berechnungs- und Versuchsergebnissen dann erreicht werden, wenn
- die rheologischen Daten des injizierten Mediums mit einem Viskometer aufgenommen wurden, das die Ablesung der Drehzahl
bei vorgegebenem Drehmoment möglich machte
und die Berechnung der Durchflußmengen un-

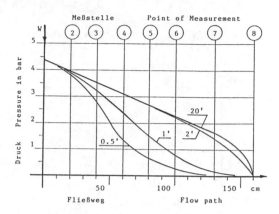

Bild 6: Injektionsversuch mit Wasser.
Grouting test with water.
Spaltweite (joint width) 0,2 mm
q=const.=0,42 cm^3/s nach 20 min

mittelbar mit den Wertepaaren Scherspan-
nung und Schergeschwindigkeit aus den Vis-
kometer-Versuchen ermittelt wurden (Bild
4),
- bei den rheologischen Daten auch die
Auswirkungen der Temperaturänderungen beachtet wurden.
Im Unterschied zu den Versuchen mit Wasser zeigte sich, daß das Kunstharz offen-
bar in den Buchten und Vertiefungen stehen
bleibt und daher
- als Strömungsbreite die Weite zwischen
den Engstellen in Rechnung gestellt und
- als Spaltweite der Abstand der Rauhigkeitsspitzen angesetzt werden muß, das dem
entspricht. Unter Berücksichtigung dieser
Gesichtspunkte ergab sich bei den Versuchen mit Kunstharz eine gute Übereinstimmung zwischen Messung und Rechnung im Bereich von Druckgefällen zwischen 22 und
125 bar/m bei Spaltweiten von 0,4 und 0,6
mm. Ein Beispiel in Bild 7 dargestellt. In
eine wassergefüllte Kluft von 0,2 mm
Spaltweite wird Kunstharz injiziert. In
den ersten 45 min wurde der Einpreßdruck
allmählich auf 100 bar gesteigert. Während
dieser Phase kam es noch zu keinem Eindringen von Injektionsgut in den vorgesehenen Sickerweg. Erst nach Überschreiten
dieses Druckes kam es zu einem langsamen
Fortschreiten des Injektionsgutes, bis
38 min später die Austrittsstelle erreicht
wurde. Eine weitere Drucksteigerung bis
auf 170 bar brachte eine gewisse Beschleu-
nigung des Durchflusses.
Als besonders interessant sei erwähnt, daß
die Rückverformung der Spaltweite bei
Absenkung des Injektionsdruckes sehr langsam vor sich ging, was mit einem langsamen
Ausquetschen des Injektionsgutes auch zu

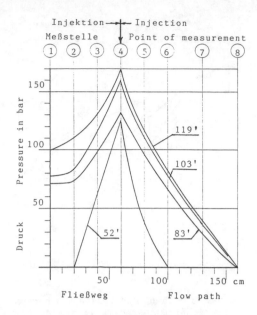

Bild 7: Injektionsversuch mit Kunstharz
       Grouting test with epoxy resin
       Spaltweite 0,2 mm joint width

einem langsameren Absinken der Strömungsgeschwindigkeit führte.

6. DANKSAGUNG

Der Verfasser möchte an dieser Stelle der
Österreichischen Draukraftwerke AG dafür
danken, daß sie die Durchführung und Veröffentlichung der Versuche ermöglicht haben, weiters Herrn Dipl. Ing. G. Lichten-
egger (ÖDK) für zahlreiche Anregungen sowie Herrn Dr. G. Stadler (Insond) für
zahlreiche praxisnahe Ratschläge, für die
Beistellung der Injizierpumpe und der verwendeten Kunststoffmaterialien.

Prognostizierung der Ausbreitung von Kunstharz bei der Injektion von Rissen in Massenbeton

Prognosis of the propagation of epoxy resin at grouting of cracks in mass concrete

Harald R.Gaisbauer
Verbundgesellschaft, Wien, Österreich

Helmut Huber
Tauernplan Prüf- und Meßtechnik GmbH, Staatlich autorisierte Materialversuchsanstalt, Straß, Zillertal, Österreich

ZUSAMMENFASSUNG: Der vorliegende Artikel stellt einen Beitrag zur Prognostizierung der Ausbreitung von Kunstharz bei der Injektion von Rissen in Massenbeton dar. Bei der Injektion von Rissen in Massenbeton stellt sich die Problematik, daß über den zu injizierenden Riß nur wenige Aussagen bekannt sind. Die Zielsetzung der gegenständlichen Arbeit ist es, durch einen vorweg durchgeführten Wasserabpreßversuch die notwendigen Rißparameter so zu bestimmen, daß in der Folge ein geeignetes Injektionsmaterial zur Rißinjektion ausgewählt werden kann.

Gestützt auf einen Wasserabpreßversuch werden wertvolle Parameter für den darauffolgenden Injektionsvorgang mit Harzen gewonnen. Der Wasserabpreßversuch wird in verschiedenen Druckstufen durchgeführt und die zugehörige Injektionsrate bestimmt. Die angewendeten Strömungsgesetze berücksichtigen sowohl die Art des Strömungsvorganges die auftritt als auch die Materialparameter des Injektionsmittels. Beim Wasserabpreßversuch ist in der Bohrlochumgebung auf grund der höheren Injektionsraten mit turbulenten Strömungsverhältnissen und Strömungsverhältnissen im Übergangsbereich zu rechnen. Bei der Harzinjektion treten vor allem laminare Strömungsverhältnisse auf. Im Unterschied zum Wasserversuch ist hier zusätzlich die Kohäsion des Injektionsmateriales zu berücksichtigen. Die gegenständliche Untersuchung beschränkt sich vorwiegend auf die Verwendung von Wasser und Injektionsharzen. Die grundsätzlichen Aussagen sind jedoch auch mit Einschränkungen für Zement/Bentonit Suspensionen gültig. Bei einer Zement/Bentonit Injektion sind andere Werte für Kohäsion und Viskosität der Injektionsflüssigkeit anzusetzen.

Die für die Strömungsberechnung maßgebenden Parameter wie äquivalente Öffnungsweite des Risses, Rißrauhigkeit, Ausbreitungsform des Strömungskörpers, Rißeinströmverlust und Nachgiebigkeit der Spaltflächen werden aus dem Wasserabpreßversuch bestimmt beziehungsweise aus Erfahrungswerten geschätzt. Mit der Kenntnis über die Rißparameter und den Fließeigenschaften des Injektionsharzes kann dann unter Anwendung der geeigneten hydraulischen Fließgesetze ein geeignetes Harz für die Rißinjektion ausgewählt werden und der Injektionsvorgang prognostiziert werden.

In einem wirklichkeitsnahen Blockversuch wurde ein Injektionsvorgang simuliert. Ziel des Versuches war es einerseits, die grundsätzlichen Probleme, die bei einer Harzinjektion auftreten zu studieren, die wesentlichen Harzeigenschaften zu bestimmen und andererseits die Zulässigkeit der angewendeten hydraulischen Strömungsgesetze zu überprüfen. Versuch und Berechnung zeigen in Anbetracht der vielen unbekannten Parameter eine zufriedenstellende Übereinstimmung.

ABSTRACT: The present article is a contribution towards the prognosis of the propagation of epoxy resins in cracks in mass concrete. When grouting cracks in mass concrete the problem arises that little is known of the properties of the crack. The scope of this investigation is to determine the necessary crack parameters with a water pressure test in such a way that one is able to choose a suitable resin for the grouting.

Based on the water pressure test valuable parameters for the following grouting are determined. The water pressure test is carried out in various stages and the corresponding injection rate is measured. The applied flow laws take into account the kind of flow that occurs as well as the material parameters of the injected fluid. At the water pressure test one has to take into account turbulent and transient flow in the vicinity of the borehole due to higher injection rates. When injecting the resin mainly laminar flow dominates. In contradiction to the water pressure test in addition cohesion has to be taken into account. The present investigation only deals with water and grouting resin. The basic findings nevertheless are also valid to cement/bentonite grouting to a certain extent. At cement/bentonite grouting different values for cohesion and viscosity of the injected material have to be used.

The important parameters for the flow calculation like equivalent crack opening, crack roughness, shape of fluid body, crack entrance headloss and stiffness of the surrounding structure are determined from the water pressure test or assessed from experience. With the knowledge of the crack parameters and the flow properties of the epoxy resin a suitable resin for the following crack injection can be chosen and the grouting procedure can be prognosed by applying the appropriate flow laws.

In a 'close to nature' block test a grouting procedure was simulated. The aim of the test was on one hand to study the basic problems that arise at resin grouting, to study the dominant resin properties and on the other hand to check the applicability of the applied hydraulic flow laws. Test and calculation show a satisfying matching when the large amount of unknown parameters is considered.

1 Problemstellung

1.1 Einleitung

Bei der Injektion von Rissen in Massenbeton stellt sich die Problematik, daß über den zu injizierenden Riß nur wenige Aussagen bekannt sind. Die ungefähre Rißlage ist aus der Vorerkundung, vor allem durch Erkundungsbohrungen und eventuell seismische Versuche bekannt. Die ungefähre Öffnungsweite des Risses kann aus Extensometer oder Gleitmikrometermessungen bestimmt werden. Nachdem diese Messungen jedoch nur Punktmessungen sind und mit einer variablen Rißöffnungsweite gerechnet werden muß, besitzt man über die tatsächliche Rißgeometrie und die Rißeigenschaften nur wenige Angaben.

Das Ausbreitverhalten eines Injektionsmaterials ist von dessen Fließeigenschaften abhängig [1]. In einer grundsätzlichen Untersuchung wurde das Fließverhalten von verschiedenen Injektionsmaterialien in Rissen untersucht. Die Zielsetzung der gegenständlichen Arbeit ist es, durch einen vorweg durchgeführten Wasserabpreßversuch die notwendigen Rißparameter so zu bestimmen, daß in der Folge ein geeignetes Injektionsmaterial zur Rißinjektion ausgewählt werden kann. Die gegenständliche Untersuchung beschränkt sich vorwiegend auf die Verwendung von Wasser und Injektionsharzen. Die grundsätzlichen Aussagen sind jedoch auch mit Einschränkungen für Zement/Bentonit Suspensionen gültig. Bei einer Zement/Bentonit-Injektion sind andere Werte für Kohäsion und Viskosität der Injektionsflüssigkeit anzusetzen.

In einem naturnahen Blockversuch wurden die Berechnungsansätze überprüft.

1.2 Injektionsmaterialeigenschaften

Die maßgebenden Parameter zur Charakterisierung der Eigenschaften eines Injektionsmateriales sind dessen Viskosität, Dichte und Kohäsion. Weitere Parameter wie z.B. Thixotropie, Adhäsion und Oberflächenspannung werden hier im weiteren nicht berücksichtigt. Die drei erstgenannten Parameter und hier im besonderen die Viskosität weisen neben einer zeitlichen auch eine starke Temperaturabhängigkeit auf. Bei der Untersuchung wurde bei verschiedenen Chargen des gleichen Injektionsmittels eine starke Streuung der Parameter festgestellt. Es dürfte sich hierbei um alterungsbedingte Einflüsse handeln.

Nachdem die Kohäsion nur bei ungefüllten Injektionsharzen nahezu Null ist, gewinnt die Bestimmung der Injektionsparameter bei gefüllten Harzen immer größere Bedeutung, da Viskosität und Kohäsion in der Berechnung gemeinsam berücksichtigt werden müssen. Dabei kann nicht mehr von dem einfachen Newton'schen Strömungsmodellgesetz ausgegangen werden, sondern es muß ein Modell höherer Ordnung (z.B. Bingham'sches Strömungsmodell) zugrundegelegt werden [2].

Die Bestimmung der Materialkennwerte einer kohäsionsbehaftenden Flüssigkeit führt auch auf große Probleme bei der Laborgeräteauswahl. In Vergleichsversuchen wurde festgestellt, daß die Bestimmung

der Viskosität für einen Newton'schen Flüssigkeitskörper ohne größere Schwierigkeiten durchzuführen ist und unabhängig von der Gerätewahl auf gleiche Ergebnisse führt. Bei der Bestimmung von Viskosität und Kohäsion für eine Bingham'sche Flüssigkeit wurden jedoch bei verschiedenen Laborgeräten verschiedene Ergebnisse unter gleichen Prüfbedingungen erhalten.

1.3 Ausbreitverhalten der Injektionsflüssigkeit unter in situ Bedingungen

In Blockversuchen wurde das Ausbreitverhalten von Wasser und Injektionsharzen in Rissen untersucht [3,4,5]. Um die natürlichen Verhältnisse möglichst genau zu simulieren, wurde ein Betonblock betoniert und der noch junge Beton zur Erzielung einer natürlichen Rißfläche gespalten. Der so entstandene Riß wurde in der Folge wie bei einem aktuellen Injektionsvorgang injiziert. Beim nachträglichen Auseinanderheben der Blockhälften konnte die Ausbreitung des Injektionsharzes ermittelt werden. Sie erfolgte bei allen Versuchen in etwa parabelförmig.

Besondere Bedeutung kommt hierbei dem Bereich am Verschnitt des Bohrloches mit dem Spalt zu. Es zeigte sich, daß hier bei höheren Strömungsgeschwindigkeiten (Wasserabpreßversuch) überraschend große Einströmverluste auftreten. Weiters ist auf die große Abhängigkeit des Druckverlaufes vom Querschnitt des Bohrloches hinzuweisen. Ein kleiner Bohrlochdurchmesser wie etwa 25 mm führt zu immensen Druckverlusten in den ersten Zentimetern des Risses in der Umgebung des Bohrloches. Dieser Druckabfall kann durch die Wahl eines größeren Durchmessers (45 mm und mehr) entscheidend verringert werden.

1.4 Rißeigenschaften

Die maßgebenden Rißeigenschaften sind Spaltweite und Spaltrauhigkeit. Beide Parameter sind nicht bekannt und müssen über die gesamte Rißgeometrie als nicht konstant angenommen werden. Für die folgende Untersuchung wurden Spaltweiten vom 1/10 mm-Bereich bis in den mm-Bereich für die Harzinjektion berücksichtigt. Aufgrund der geringen Spaltweite ist mit hohen relativen Rauhigkeiten (bis etwa 0.5) zu rechnen.

1.5 Prognostizierung des Injektionsvorganges

Unter Berücksichtigung der Anforderungen in der Praxis liegt der Prognostizierung des Injektionsvorganges folgender Grundgedanke zugrunde:

Ein vorweg durchgeführter Wasserabpreßversuch dient der Ermittlung der Spaltparameter (Spaltweite, Spaltrauhigkeit). Die Ausbreitfigur des Injektionsmateriales muß geschätzt werden und ist vom gewählten Bohrraster abhängig. Eine parabelförmige Ausbreitungsfigur hat sich für die Berechnung als zweckmäßig erwiesen. Auf dieser Grundlage wird dann mit den bekannten Harzeigenschaften der Injektionsvorgang vorberechnet, beziehungsweise aufgrund der Vorausberechnung mit unterschiedlich steifen Harzen das entsprechende Harz für einen optimalen Injektionserfolg ausgewählt.

Zunächst wird der Injektionsvorgang mit Wasser durchgeführt (Wasserabpreßversuch), wobei zu verschiedenen Druckstufen die zugehörige Injektionsrate (Q[l/min]) bestimmt wird. Aufgrund dieser so erhaltenen Punktepaare kann mit Hilfe von Bestimmungsdiagrammen eine äquivalente Spaltweite ermittelt werden, die es in der Folge erlaubt, für die nachfolgende Harzinjektion das entsprechende Injektionsharz auszuwählen. Bei den bisher durchgeführten Versuchen hat es sich gezeigt, daß die relative Rauhigkeit für Betonrisse bis 1.0mm Öffnungsweite mit Werten von 0.25 bis 0.5 hoch anzusetzen ist.

Die Auswahl des für die vorhandene Rißgeometrie günstigsten Injektionsharzes richtet sich vor allem nach der Viskosität. Die Kohäsion des Injektionsharzes ist eher als unangenehme Nebenerscheinung einzustufen, nachdem durch diese die Reichweite des Injektionsharzes und auch dessen Gängigkeit beschränkt wird. Bei etwa 1000 mPa.s ist die Kohäsion nur gering, sodaß hier näherungsweise von einem Newton'schen Fließgesetz ausgegangen werden kann. Mit zunehmender Viskosität des Harzes steigt jedoch auch die Kohäsion und es muß in der Berechnung auf ein Strömungsgesetz höherer Ordnung übergegangen werden, bei welchem die Kohäsion der Flüssigkeit Berücksichtigung findet (z.B. Bingham Körper).

2 Beschreibung der Versuchsanordnung

2.1 Versuchsaufbau

Da die Einflußgrößen auf die Strömungsverhältnisse in Klüften und Rissen theoretisch nicht exakt erfaßbar sind, wurde versucht, durch Injektion eines künstlich hergestellten Risses eine Reihe von Parametern in Abhängigkeit von Viskosität, Fließgrenze, Kohäsion, Dichte, Temperatur etc. zu ermitteln.

Herstellung der künstlichen Rißfläche:
Um eine möglichst große Rißfläche zu injizieren, wurde ein Betonblock mit den Ausmaßen L 200 x B 100 x H 100 cm in einer Betongüte B 400 hergestellt. Für den Injektionsversuch konnte eine Methode entwickelt werden, welche den Block im Alter von 48 Stunden ohne Probleme spaltbar macht.

Abb.1 - *Fig.1*
Versuchsblock - *Testblock*

Dadurch konnte eine Rißfläche von 1,00 x 2,00 m erzeugt werden.

Blockversuchsvorbereitung:
Der gespaltene Block wurde ohne Trennung der Blockhälften an 3 Außenseiten abgedichtet. Um ein starke Aufweiten des Risses während der Injektion zu verhindern, wurde der gesamte Block in einem Stahlrahmen eingespannt (siehe Abb.1). Zur Ermittlung der Drücke im Riß wurden an der Injektionsstelle im Riß und dann an 2 weiteren Stellen in der Längsrichtung des Blockes Druckmeßeinrichtungen eingebaut (siehe Abb.2).

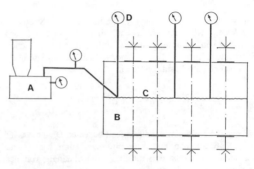

Abb.2 - *Fig.2*
Versuchsanordnung - *Test configuration*
A Injektionspumpe - *Injection pump*
B Gespaltener Betonblock - *Split concrete block*
C Riß - *Crack*
D Druckmessgerät - *Pressuremeter*

2.2 Versuchsdurchführung

Vor jedem Kunstharzinjektionsversuch wurde der Riß mit Wasser bei verschiedenen Druckstufen durchgespült und die zugehörige Injektionsrate bestimmt. In weiterer Folge wurde dann rechnerisch eine äquivalente Spaltweite und Spaltrauhigkeit ermittelt.

Von dem zu injizierenden Harz wurden vorab die Dichte, Temperatur und Viskosität (Brookfield) bestimmt. Während der Injektion wurden in Abhängigkeit der Zeit neben der mengenmäßigen Harzaufnahme auch die Drücke in den verschiedenen Rißbereichen und der Druckabfall von der Injektionspumpe bis zur Injektionsstelle im Riß bestimmt. Nach Erreichen eines stationären Fließzustandes wurden bei verschiedenen konstant gehaltenen Injektionsdrücken die Injektionsraten und die dazugehörigen Drücke im Riß ermittelt.

3 Hydraulische Nachrechnung der Versuche

3.1 Die der Berechnung zugrundegelegten Strömungsgesetze

Die in der Folge angeführten Strömungsgesetze beziehen sich auf eine Spaltströmung. Für die Wasserströmung wurde ein Newtonscher Strömungskörper zugrundegelegt. Einem laminaren Fließen liegt folgender Ansatz zugrunde:

$$\tau = \mu * \frac{dv}{dn} \quad (1)$$

In dieser Formel wird die Schubspannung zwischen zwei Flüssigkeitselementen der dynamischen Viskosität der Flüssigkeit und dem Gradienten der Geschwindigkeit normal zur Strömungsrichtung gegenübergestellt. Aus dieser Beziehung ist ersichtlich, daß die in der Flüssigkeit herrschende Schubspannung nur vom Gradienten der Geschwindigkeit abhängig ist. Die Integration dieser Gleichung mit Berücksichtigung der Randbedingungen führt zur Gleichung 3.

Die Berücksichtigung einer zusätzlich in der Strömung wirkenden Kohäsion führt zu einer Erweiterung des Newtonschen Strömungskörpers auf den Bingham'schen Strömungskörper. Der einer solchen Strömung zugrundegelegte Ansatz lautet wie folgt:

$$\tau = \mu * \frac{dv}{dn} + c \quad (2)$$

wobei c die Kohäsion der Flüssigkeit bedeutet. Dieser Ansatz zeigt, daß Strömen erst dann einsetzen kann, wenn eine bestimmte Anfangswiderstandskraft, die abhängig von der von der Flüssigkeit benetzten Oberfläche ist, überwunden wird. Das bedeutet, daß ein bestimmter Anfangsdruck vorhanden sein muß, um Fließen zu ermöglichen. Analog zum vorhergehenden Ansatz führt auch hier die Integration der Gleichung unter Beachtung der Randbedingung auf das die Strömung bestimmende Fließgesetz (siehe Gleichung 7,8).

Die absolute Rauhigkeit der Strömungsberandungsflächen, die bei großen hydraulischen Durchmessern bei laminaren Strömungsvorgängen praktisch keinen Einfluß auf den Strömungsvorgang besitzt, gewinnt mit Abnehmen des hydraulischen Durchmessers immer mehr an Bedeutung. Vor allem im Bereich geringer Rißweiten (< 0,5 mm) besitzt die absolute Rauhigkeit der Rißoberfläche einen nicht zu vernachlässigenden Einfluß auf den Strömungsvorgang. Im Grenzfall, bei dem die absolute Rauhigkeit die halbe Spaltweite erreicht, beträgt der hydraulische Gradient das 2.1-fache des hydraulischen Gradienten bei Vernachlässigung der Rauhigkeit. Anders ausgedrückt bedeutet dies, daß zur Erzielung des gleichen hydraulischen Effektes die 2.1-fache Druckhöhe notwendig wird. Dieser einfache Vergleich zeigt anschaulich den bedeutenden Einfluß der Rauhigkeit.

Im folgenden finden sich die in der Berechnung angewendeten Strömungsgesetze.

3.1.1 Laminares Strömen

Newton Fließgesetz für k/D_h < 0.032

$$Q = \frac{\gamma * I}{\mu} * \frac{2}{3} * b * a_i^3 \quad (3)$$

umgeformt:

$$I = \frac{96}{\Re} * \frac{1}{D_h} * \frac{v^2}{2g} \quad (4)$$

Newton Fließgesetz für k/D_h > 0.032

$$Q = \frac{\gamma * I}{\mu} * \frac{2}{3} * b * a_i^3 * \frac{1}{1 + 8.8 * (\frac{k}{D_h})^{1.5}} \quad (5)$$

umgeformt:

$$I = \frac{96}{\Re} * \frac{1}{D_h} * \frac{v^2}{2g} * (1 + 8.8 * (\frac{k}{D_h})^{1.5}) \quad (6)$$

Beim Bingham'schen Fließgesetz ist der Strömungsgradient nicht mehr explizit darstellbar.

Bingham Fließgesetz für k/D_h < 0.032

$$Q = \frac{\gamma * I}{\mu} * \frac{2}{3} * b * a_i^3 * (1 - \frac{3}{2} * \frac{e_c}{a_i} + \frac{1}{2} * (\frac{e_c}{a_i})^3) \quad (7)$$

mit dem Parameter

$$e_c = \frac{c}{\gamma * I} \quad (8)$$

Bingham Fließgesetz für k/D_h > 0.032

$$Q = \frac{\gamma * I}{\mu} * \frac{2}{3} * b * a_i^3 * (1 - \frac{3}{2} * \frac{e_c}{a_i} + \frac{1}{2} * (\frac{e_c}{a_i})^3) * \frac{1}{1 + 8.8 * (\frac{k}{D_h})^{1.5}} \quad (9)$$

3.1.2 Turbulentes Strömen

Vor allem beim Wasserabpreßversuch ist mit turbulenter Strömung und Strömung im Übergangsbereich zu rechnen.

Newton Fließgesetz für k/D_h < 0.032

$$Q = A * \sqrt{\frac{I}{\lambda * \frac{1}{D_h} * \frac{1}{2g}}} \quad (10)$$

umgeformt:

$$I = \lambda * \frac{1}{D_h} * \frac{v^2}{2g} \quad (11)$$

Der λ Beiwert in obriger Formel errechnet sich wie folgt.

Hydraulisch glatter Bereich k/D_h = 0 - Prandtl, Karman

$$\frac{1}{\sqrt{\lambda}} = 2 * \log \frac{\Re * \sqrt{\lambda}}{2.51} \quad (12)$$

Übergangsbereich $k/D_h < 0.032$ - Colebrook, White

$$\frac{1}{\sqrt{\lambda}} = -2*\log(\frac{\frac{k}{D_h}}{3.71} + \frac{2.51}{\Re*\sqrt{\lambda}}) \quad (13)$$

Hydraulisch rauher Bereich $k/D_h < 0.032$ - Prandtl, Karman, Nikuradse

$$\frac{1}{\sqrt{\lambda}} = 2*\log\frac{3.71}{\frac{k}{D_h}} \quad (14)$$

Hydraulisch rauher Bereich $k/D_h > 0.032$ - Louis

$$\frac{1}{\sqrt{\lambda}} = 2*\log\frac{1.90}{\frac{k}{D_h}} \quad (15)$$

3.1.3 Grenzbedingungen

Der Übergang zwischen dem laminaren und dem turbulenten Fließzustand ist sowohl von der Rauhigkeit als auch von der Reynoldszahl $\Re$ abhängig. Die Zusammenhänge ergeben sich wie folgt:

Grenzbedingung laminar - turbulent für $k/D_h < 0.0168$

$$\Re_{krit} = 2300 \quad (16)$$

Bei Nichtberücksichtigung des Übergangsbereiches zwischen laminar und turbulent nach Colebrook, White erfolgt der Übergang zwischen hydraulisch glatt turbulent (Prandtl, Karman) zu hydraulisch rauh turbulent (Nikuradse) bei:

$$\Re_{krit} = 2.552*(\log\frac{3.7}{\frac{k}{D_h}})^8 \quad (17)$$

Grenzbedingung laminar - turbulent für $0.0168 < k/D_h < 0.032$

$$\Re_{krit} = (142000*(\log\frac{3.7}{\frac{k}{D_h}})^2)^{\frac{1}{1.76}} \quad (18)$$

Grenzbedingung laminar - turbulent für $k/D_h > 0.032$

$$\Re_{krit} = (142000*(\log\frac{1.9}{\frac{k}{D_h}})^2)^{\frac{1}{1.76}} \quad (19)$$

3.2 Injektionsversuch mit Wasser (Wasserabpreßversuch)

Aufgrund der langgestreckten Rißgeometrie und aufgrund von Erfahrungen beim Auseinandernehmen der Blockteile

wird von einer parabelförmigen Ausbreitung der Injektionsflüssigkeit ausgegangen. Strömungsunter-suchungen mit zuerst radialsymmetrischer Ausbreitung und einer darauf folgenden Rechtecksströmung haben sich als nicht zielführend erwiesen und wurden nicht weiter verfolgt. Bei den verwendeten Injektionsdrücken von max. etwa 0,60 bar ergeben sich zugehörige Injektionsraten bis etwa 10 l/min. Die zugehörigen Reynoldszahlen liegen in der Größenordnung von 2000 und darüber. Aus diesem Grund ist in der unmittelbaren Bohrlochumgebung bei der Rißströmung mit turbulenten Fließzuständen zu rechnen. Mit zunehmendem Abstand vom Bohrloch geht dieser turbulente Fließzustand in den Übergangsbereich und später in den laminaren Fließbereich über. Dieser Umstand ist in der hydraulischen Berechnung berücksichtigt.

In Abb.3 ist ein Bestimmungsdiagramm für die Spaltweite auf Grundlage einer parabolischen Ausbreitungsfigur dargestellt. Die durchgezogenen Kurven entsprechen den Wertepaaren Druck im Bohrloch am Rißeintritt und zugehörige Injektionsrate. Die strichlierten Kurven entsprechen einer Druckmessung in 5 cm Abstand vom Bohrloch in der Rißebene und wiederum der zugehörigen Injektionsrate. Diese Kurven entstanden aus einer Variation der einzelnen Parameter wie Ausbreitungsform des Strömungskörpers, Ausgangsöffnungsweite des Risses, Spaltrauhigkeit, Verformungsverhalten der Spaltflächen und dem Einströmverlust. Speziell der Einströmverlust erwies sich in den Versuchen als überraschend hoch. Dies ist möglicherweise darauf zurückzuführen, daß während des Bohrvorganges Bohrklein in den Riß gedrückt wird, welches zu einer Querschnittsverminderung im Einlaufbereich des Risses am Verschnitt mit der Bohrlochwandung führt. In der Berechnung wurde diesem Umstand durch einen erhöhten Einströmverlust Rechnung getragen.

Die markierten Punkte entsprechen den Versuchspunkten. Mit der Zunahme des Druckes im Bohrloch bei höheren Injektionsraten kommt es zu einer Unterdrucksetzung der Spaltflächen, was eine zusätzliche Aufweitung des Risses bewirkt. Diese Veränderung der Spaltweite wurde in der Berechnung mit einem Federmodell simuliert und berücksichtigt. Als Ausgangsspaltweite bei Druck = 0 wurde ein Parallelspalt angenommen, welcher im Zuge der Unterdrucksetzung der Spaltflächen zu einem Spalt mit variabler Spaltweite in Abhängigkeit von der Druckaufbringung verformt wird.

Wie aus der Abbildung ersichtlich ist, ergibt sich mit den Versuchspunkten eine äquivalente Ausgangsspaltweite von etwa 1.3 mm. In dieser äquivalenten Spaltweite sind variable Rißöffnungsweiten, unterschiedliche Rauhigkeiten der Rißflächen und Abweichungen in der Strömungsausbreitungsform enthalten. Zufolge der nur geringen Druckbelastung der Spaltflächen beim Wasserabpreßversuch stellt sich praktisch keine Aufweitung des Spaltes ein. Die mit Hilfe des Diagrammes ermittelte Spaltweite dient in weiterer Folge als Grundlage zur Auswahl des Injektionsharzes für die nachfolgende Harzinjektion.

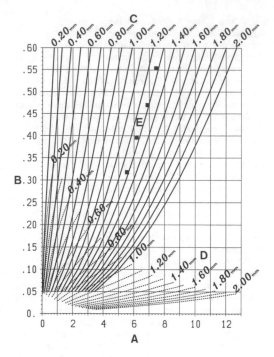

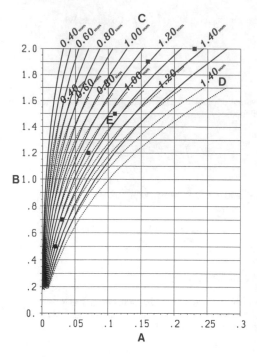

Abb.3 - *Fig.3*

Wasserabpreßversuch in Druckstufen in Funktion der äquivalenten Ausgangsspaltweite - *Water pressure test in stages as a function of the equivalent origional crack opening*

A Injektionsrate - *Injection rate* Q[l/min]
B Injektionsdruck - *Hydraulic pressure* P[bar]
C Druck am Risseinlauf - *Pressure at crack entrance*
D Druck im Riß 5cm entfernt vom Bohrloch - *Pressure in crack at 5cm distance from borehole*
E Gemessene Versuchspunkte - *measured test points*

3.3 Injektionsversuch mit Harzen

Um die Gültigkeit der angewendeten Strömungsgesetze überprüfen zu können, wurde wie folgt vorgegangen:

Für die einzelnen Versuchsharze wurde ebenfalls ein Spaltweitenbestimmungsdiagramm erstellt. Die angewendeten Injektionsdrücke bei dem hier beschriebenen Harzversuch liegen bei maximal 2 bar und die zugehörigen Injektionsraten bei etwa 0.2 l/min. Diese im Vergleich zum Wasserversuch geringen Injektionsraten ergeben Reynoldszahlen in der Größenordnung von 0.05, also eindeutig im laminaren Bereich. Bei den höheren Drücken kommt es zu einer zusätzlichen Aufweitung des Spaltes im 1/100 mm Bereich, welche sich vor allem bei den geringen Spaltweiten aufgrund des prozentuell höheren Anteiles deutlich auf den Strömungsvorgang auswirkt.

Die Spaltweitenbestimmungsdiagramme für die

Abb.4 - *Fig.4*

Harzinjektion in Druckstufen in Funktion der äquivalenten Ausgangsspaltweite - *Resin injection in stages as a function of the equivalent origional crack opening*

Beschreibung siehe Abb.3 - *Notation see Fig.3*

Harzversuche wurden mit den aus dem Wasserversuch ermittelten Parametern, wie Strömungsausbreitungsform, Spaltrauhigkeit und Einströmverlust bestimmt.

In Abb.4 ist beispielhaft das Diagramm für ein Harz mit einer Viskosität von 4800 mPa.s dargestellt. Die durchgezogenen Kurven entsprechen wiederum dem Druck im Bohrloch in der Rißebene und die strichlierten Kurven entsprechen einer Druckmessung 5 cm vom Bohrloch entfernt im Riß. Im Unterschied zu den Wasserabpreßversuchen ist hier für den Strömungsvorgang eine geringfügige Kohäsion des Injektionsmaterials (1.0 Pa) berücksichtigt (Binghamkörper Strömungsgesetz).

Die fett eingetragenen Versuchspunkte streuen um eine äquivalente Spaltweite von 1.3 mm. Die Übereinstimmung mit der Spaltweite des vorangegangen Wasserabpreßversuches ist in Anbetracht der vielen Unbekannten als durchaus zufriedenstellend zu bewerten.

3.4 Zeitliche Ausbreitung des Injektionsharzes

Unter Anwendung eines Differenzenverfahrens ist es möglich, die zeitliche Ausbreitung der Injektionsfront zu simulieren. In Abb.5 sind Injektionsrate und Injektionsweite

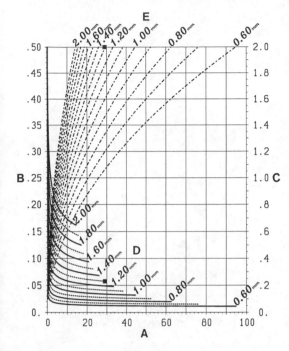

Abb.5 - *Fig.5*

Injektionsvorgang in Abhängigkeit der Zeit als Funktion der
äquivalenten Ausgangsspaltweite (2a$_i$)
für konstanten Injektionsdruck am Rißeinlauf P von 1 bar
für Harzviskosität von 4800 mPa.s
*Injectionprocedure in dependance on time as a function of
the equivalent original crack opening (2a$_i$)
for constant injection pressure at the crack entrance of 1 bar
for resin viscosity of 4800 mPa.s*

A   Zeit - *time*   T[min]
B   Injektionsrate - *Injection rate*   Q[l/min]
B   Injektionsdruck - *Hydraulic pressure*   P[bar]
C   Lage der Harzfront - *Position of resin front*   L[m]
D   Injektionsrate - *Injection rate*   Q
E   Ausbreitung der Harzfront - *Propagation of resin front*

über der Zeit dargestellt. Die hier gezeigten
Rechenergebnisse gehen davon aus, daß die Rißinjektion mit
einem konstanten Druck im Bohrloch erfolgt. Auf diese
Weise ergeben sich zu Beginn der Injektion die höchsten
Injektionsraten und die schnellste
Fortschreitungsgeschwindigkeit der Injektionsfront. Mit
Zunahme der Zeit und der Entfernung der Injektionsfront
vom Bohrloch (= fortschreitende Füllung des Risses) ergeben
sich immer kleiner werdende Injektionsraten. Der stärkste
Abfall der Injektionsrate erfolgt jedoch in der unmittelbaren
Bohrlochumgebung, was darauf hinweist, daß die
Rißinjektion vor allem von den Verhältnissen in der Nähe
des Bohrloches bestimmt wird.

Der Parameter der einzelnen Kurven ist die
Ausgangsspaltweite. Mit Fortschreiten der Injektionsfront
kommt es zufolge des ansteigenden Druckes und der
Vergrößerung der druckbeaufschlagten Fläche zu einer
stetigen Aufweitung der Spaltflächen. Auf der linken
Ordinate ist die Injektionsrate in l/min aufgetragen. Mit dem
Fortschreiten der Injektionsfront und somit Zunahme der
Zeit nimmt die Injektionsrate ab. Auf der rechten Ordinate
ist die Injektionsweite (Lage der Injektionsfront) in Metern
aufgetragen. Der Anstieg dieser Kurven bedeutet die
Fortschreitungsgeschwindigkeit der Injektionsfront. Diese ist
zum Zeitpunkt Null am größten und nimmt mit Fortschreiten
der Injektionsfront stetig ab. Zum Endzeitpunkt ist gerade
die Injektionsstrecke (Blocklänge) von 2 m zurückgelegt
worden.

Mit der Ausgangsspaltweite von 1.3 mm (abgeleitet aus dem
Wasserabpreßversuch) kann man aus Abb.5 eine Zeitdauer
von 30 min bis zum Erreichen des Blockendes ablesen. Für
die Berechnung wurde vorausgesetzt, daß der Druck im
Bohrloch mit 1.0 bar über die gesamte Injektionsdauer
konstant bleibt. Dies ist etwas im Widerspruch zum
tatsächlich durchgeführten Injektionsversuch, wo
verschiedene Druckstufen aufsteigend gefahren wurden. Der
mittlere Versuchsdruck belief sich jedoch in etwa auf 1.0 bar.
Unter Berücksichtigung der Unterschiede zwischen
Rechenannahmen und Versuch, wo etwa 35 min für die
Füllung des Spaltes benötigt wurden, aber zu Beginn
geringere Drücke im Bohrloch vorlagen, kann man
zumindest von einer Übereinstimmung der
Größenordnungen sprechen.

4   Schlußfolgerung

Der vorliegende Artikel stellt einen Beitrag zur
Prognostizierung der Ausbreitung von Kunstharz bei der
Injektion von Rissen in Massenbeton dar. Gestützt auf einen
Wasserabpreßversuch, der in verschiedenen Druckstufen
durchgeführt wird, werden wertvolle Parameter für den
darauffolgenden Injektionsvorgang mit Harzen gewonnen.
Auf diese Weise ist es möglich, aufgrund der erworbenen
Kenntnisse ein geeignetes Injektionsharz mit entsprechender
Viskosität und Kohäsion für die Rißinjektion auszuwählen.
Die angewendeten Strömungsgesetze berücksichtigen sowohl
die Art des Strömungsvorganges als auch die
Materialparameter des Injektionsmittels.

In einem wirklichkeitsnahen Blockversuch wurde ein
Injektionsvorgang simuliert. Ziel des Versuches war es
einerseits, die grundsätzlichen Probleme, die bei einer
Harzinjektion auftreten zu studieren und andererseits die
Zulässigkeit der angewendeten Strömungsgesetze zu
überprüfen. Versuch und Berechnung zeigen in Anbetracht
der vielen unbekannten Parameter eine zufriedenstellende
Übereinstimmung.

5   Danksagung

Die Versuche wurden an der Materialversuchsanstalt Straß
in Tirol durchgeführt. Die Berechnungen erfolgten großteils
an der TU Wien mit freundlicher Genehmigung des
Institutes für konstruktiven Wasserbau.

6 Literatur

1. Bérigny, C.: Mémoire sur un Procédé d'Injection propre à prévenir ou arreter les Filtrations sous les Fondations des Ouvrages Hydrauliques. Paris: 1832.

2. Lombardi, G.: Injection des massivs rocheaux. In: Mitteilungen der Schweizer Ges. für Boden- und Felsmechanik 115 (1987), S. 1 - 3.

3. Hässler, L., et al.: Simulation of grouting in jointed rock. 6. ISRM Kongreß, Montreal 1987, Vol. 2, S. 943 - 946.

4. Hässler, L., et al.: Computer simulated flow of grouts in jointed rock. Conference on Grouting, Soil Improvement and Geosynthetics, ASCE, New Orleans 1992.

5. Wittke, W.: Felsmechanik. Berlin: Springerverlag, 1984.

Notation:

$Q\left[m^3/s\right]$ ... Injektionsrate

$v\left[m/s\right]$ ... Strömungsgeschwindigkeit

$A\left[m^2\right]$ ... Strömungsquerschnitt

$\gamma\left[N/m^3\right]$ ... Spezifisches Gewicht Injektionsmittel

$g\left[m/s^2\right]$ ... Erdbeschleunigung

$I\left[...\right]$ ... Hydraulischer Gradient der Strömung

$\lambda\left[...\right]$ ... Hydraulische Widerstandszahl

$\Re\left[...\right]$ ... Reynoldszahl der Strömung

$\mu\left[Pa.s\right]$ ... Dynamische Viskosität Injektionsmittel

$\tau\left[Pa\right]$ ... Schubspannung Injektionsmittel

$c\left[Pa\right]$ ... Kohäsion Injektionsmittel

$(2a_i)\left[m\right]$ ... Anfangsöffnungsweite Spalt

$e_c\left[m\right]$ ... Parameter Spaltströmung

$k\left[m\right]$ ... Absolute Rauhigkeit

$D_h\left[m\right]$ ... Hydraulischer Durchmesser

Grouting in Rock and Concrete, Widmann (ed.) © 1993 Balkema, Rotterdam, ISBN 90 5410 350 7

The influence of flow geometry on the interpretation of Lugeon tests and the choice of grout material and grouting method

Der Einfluß der Fließgeometrie auf die Interpretation von Lugeon-Tests und die Auswahl von Injektionsmaterial und -Methode

Lars Hässler
Golder Associates AB, Uppsala, Sweden

Ulf Håkansson
ABV Rock Group KB, Jeddah, Saudi Arabia

ABSTRACT: In the article the interpretation of Lugeon tests are discussed. A simplified theory of steady state flow in 1, 2 and 3 dimensions are presented. The presentation concludes that the interpretation of conductivity from Lugeon tests are very much dependent on the actual flow dimension.

The article also discusses how to transform from continuum flow to discrete flow in joints i. e. how to transform from soil conductivity to number of joints and joint openings. With this as a base the influence of flow geometry on grout penetration is discussed.

ZUSAMMENFASSUNG: Der Bericht befaßt sich mit der Beurteilung von Lugeon-Tests. Dazu wird eine vereinfachte Theorie des stationären Fließens im ein- zwei- und dreidimensionalen Raum vorgestellt. Als Schlußfolgerung ergibt sich, daß eine Beurteilung der Leitfähigkeit auf Grundlage dieser Versuche wesentlich von den tatsächlichen Fließverhältnissen abhängt.

Weiters wird besprochen, wie aus dem Fließen im Kontinuum auf das Fließen in einzelnen Klüften geschlossen werden kann, wie also die Leitfähigkeit von Lockerböden in die Leitfähigkeit einer Anzahl von Klüften mit bestimmten Kluftweiten umgewandelt werden kann. Auf dieser Grundlage wird der Einfluß der Fließgeometrie auf das Eindringvermögen besprochen.

Hauptergebnis des Berichtes ist, daß je nach der Fließdimension bei gleichem Lugeon-Wert und gleicher Kluftdiche ein unterschiedliches Eindringvermögen des Injektionsgutes gegeben sein kann. So kann das Eindringvermögen im eindimensionalen Fall 10 mal höher als im driedimensionalen Fall sein. Das deutet darauf hin, daß die Ergebnisse von Lugeon-Tests nicht unbedingt für die Wahl des Injektionsgutes und- Verfahrens geeignet sind. Es gibt aber möglichkeiten, ähnliche, aber instationäre Versuche für die Abschätzung von Fließdimensionen und Leitfähigkeit durchzufüren. Auch gibt es Möglichkeiten, aus dem Verlauf in einem Injektionsabschnitt auf die Wirkliche Fließdimension und das Eindringvermögen zu schließen.

1 INTRODUCTION

Lugeon tests are commonly used as a base for the choice of grout and grouting method. No special evaluations are usually done and the method of choice is empirical. The choice of grout and grouting method from standard Lugeon tests does not take the flow geometry into account and can therefore be questionable.

A demonstration of the importance of the flowgeometry is easiest done in fixed geometries (note that standard Lugeon tests are evaluated as a steady state system). The discussion presented herein is based on 3D flow in a sphere, 2D radial flow in a circular-cylindrical structure and 1D flow in channel-like structures. Common for all three geometries are the inflow area, the distance from the inflow point to the outer boundary and the Lugeon value.

Grouts usually behaves as non-Newtonian fluids. This has as a result that flow calculations of grouts needs information on the actual geometries of the conduits. The comparison of grout penetration between different flow dimensions needs a transformation of conductivity to number and shape of acting conduits per unit area. This can be done in several ways. In this article rectangular conduits are used.

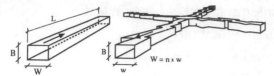

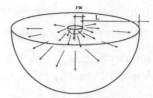

Figure 1: 1-D case with flow area independent of distance from grouthole.

Bild 1: 1-D Fall, Fließbereich unabhängig vom Abstand zum Injektionsloch.

Figure 3: 3-D case with with flow area increasing with the square of the distance from the grouthole.

Bild 3: 3-D Fall, Fließbereich vergrößert sich im Quadrat mit dem Abstand zum Injektionsloch.

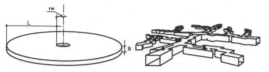

Figure 2: 2-D case with flow area increasing lineary with distance from the grouthole.

Bild 2: 2-D Fall, Fließbereich vergrößert sich linear mit dem Abstand zum Injektionsloch.

2 FORMULATION OF FLOW IN THE DIFFERENT GEOMETRIES

Below, a simple description of the different flow geometries can be found in figure 1-3.

The inflow area and the influence distance will be defined as equal in this simple demonstration. The inflow area for the three different flow geometries is defined as follows:

$A_{rw1D} = W\,B$

$A_{rw2D} = 2\,\pi\,r_w\,B$

$A_{rw3D} = 4\,\pi\,r_w^2$

where W is the width of the rectangular structure in the 1D case, B is the height of conductive area and r_w is the radius to the inflow area for the 2D and the 3D cases.

The distance of influence L, is defined as $r_L - r_w$ where r_L is the radius out to a reasonable undisturbed part of the rock mass. Note that r_w is not valid for the 1D case but that L is valid for all cases.

Common for the flow in all three cases is Darcy's law:

$U = -\dfrac{dh}{dr}\,k$

where U is the apparent velocity, dh/dr is the

gradient causing the flow and k is the conductivity [m/s] of the conductive structure.

For the 1D case the velocity and gradient is independent of the distance from the inflow area. The flow can be defined as:

$$Q_{1D} = \dfrac{h_w - h_L}{L}\,k_{1D}\,W\,B$$

or as:

$$Q_{1D} = \dfrac{h_w - h_L}{r_L - r_w}\,k_{1D}\,A_{rw\,1D}$$

where $h_w - h_L$ is the head difference acting over the distance of influence

For the 2D case the velocity and the gradient will decrease with the distance from the inflow area but the flow will remain constant at different distances. The flow can be defined according to the following:

$$Q_{2D} = -\dfrac{dh}{dr}\,k_{2D}\,2\,\pi\,r\,B$$

$$\dfrac{dr}{r} = -\dfrac{k_{2D}\,2\,\pi\,B}{Q_{2D}}\,dh$$

$$\int_{r_w}^{r_L}\dfrac{dr}{r} = -\dfrac{k_{2D}\,2\,\pi\,B}{Q_{2D}}\int_{h_w}^{h_L}dh$$

$$\ln\left(\dfrac{r_L}{r_w}\right) = -\dfrac{k_{2D}\,2\,\pi\,B}{Q_{2D}}\left(h_L - h_w\right)$$

and finally

$$Q_{2D} = \dfrac{h_w - h_L}{r_w}\,k_{2D}\,\dfrac{A_{rw\,2D}}{\ln\left(\dfrac{r_L}{r_w}\right)}$$

418

For the 3D case the velocity and the gradient will decrease with the distance from the inflow area. The decrease will be faster than for the 2D case. The flow will still remain constant at different distances. The flow can be defined according to the following:

$$Q_{3D} = -\frac{dh}{dr} k_{3D} 4 \pi r^2$$

$$\frac{dr}{r^2} = -\frac{k_{3D} 4 \pi}{Q_{3D}} dh$$

$$\int_{r_w}^{r_L} \frac{dr}{r^2} = -\frac{k_{3D} 4 \pi}{Q_{3D}} \int_{h_w}^{h_L} dh$$

$$\frac{r_L - r_w}{r_L r_w} = -\frac{k_{3D} 4 \pi}{Q_{3D}} (h_L - h_w)$$

and finally:

$$Q_{3D} = \frac{h_w - h_L}{r_L - r_w} k_{3D} A_{rw\,3D} \frac{r_L}{r_w}$$

Assume that:

$$A_{rw\,1D} = A_{rw\,2D} = A_{rw\,3D}$$

and that

$Q_{1D} = Q_{2D} = Q_{3D}$ (implies that the lugeon value is equal for all three cases)

A study of the conductivity for the three cases will give the following relations:

$$k_{1D} = \frac{r_L - r_w}{r_w \ln\left(\frac{r_L}{r_w}\right)} k_{2D} = \frac{r_L}{r_w} k_{3D}$$

From the above follows that the conductivity can not be evaluated without a known flow dimension.

3 THE TRANSFORMATION OF CONDUCTIVITY TO GEOMETRIES OF THE CONDUITS

The conductivity can be transformed to conduit geometries in several different ways. A simple way is to assume a certain number of geometri-

cally defined conduits per square meter of cross section of the permeable structures. If the conduits are assumed to be plane parallel joints the result of such a transformation can look as follows

$$k = \frac{N b^3}{12 \mu_w} \rho g$$

where N is number of joints per square meter, b is the joint opening, ρ is the density of water and g is acceleration due to gravity (see figure 4).

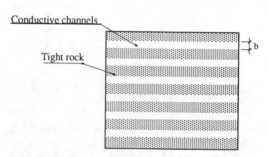

Figure 4: The conductivity describing the homogenous material is transformed to a number of plane-parallell joints.

Bild 4: Die Leitfähigkeit des homogenen Materials übertragen auf eine Anzahl planparelleler Klüfte.

4 THE IMPACT OF DIFFERENT FLOW DIMENSIONS

If we assume that the number of joints per unit area of cross section (N) is the same for the three cases of flow geomtries we get the following relationship between the joint openings:

$$b_{1D} = \sqrt[3]{\frac{r_L - r_w}{r_w \ln\left(\frac{r_L}{r_w}\right)}} \, b_{2D} = \sqrt[3]{\frac{r_L}{r_w}} \, b_{3D}$$

We also now that grouts often has a shear strength (yield stress) and that the maximum penetration of the grout into the rock can be expressed as follows (Wallner 1976, Hässler 1991):

$$I = \frac{\Delta P \, b}{2 \tau_o}$$

where ΔP is the pressure difference acting over the grout and τ_o is the shear strength (yield value) of the grout.

If we want to get the same penetration for all three cases we can either vary the shear strength of the grout (Håkansson et al 1992) or vary the grouting pressure. There might be a considerable difference. Normal values for r_L/r_w might be around 100. For instance this implies that the 3D case has to be grouted with 3 times higher pressure than the 1D case to get the same penetration if the same grout is used. Of course it is possible to adjust the grout properties instead of raising the pressure.

If we look at needed volumes of grout to reach a certain penetration we get the following relations:

$$V_{3D} \approx \frac{2}{3} \frac{r_G}{r_w} \sqrt[3]{\frac{1}{\ln\left(\frac{r_L}{r_w}\right)}} \quad V_{2D} \approx \frac{1}{3} \frac{r_G^2}{r_w^2} \sqrt[3]{\frac{r_w}{r_L}} \quad V_{1D}$$

where r_G is the penetration.

With the same assumptions as before and a value of r_G/r_w of 20 the needed grout volume for the 3D case would be around 500 times the needed volume for the 1D case and around 10 times the needed volume for the 2D case.

The above indicates that it is easier to grout the 1D case than the other cases. This might not be true because the conduits probably are much more difficult to find with the grout holes.

5 CONCLUSIONS

There are several things worth commenting on the subject of the use of Lugeon tests. Obviously it is difficult to directly use the test results as a base for the choice of grout and grouting method.

Over the last 10 years the flow dimensionality has been increasingly discussed in the fields of reservoir engineering and nuclear waste deposits in rock. There are now several methods developed that can evaluate the dimensionality from transient tests (for example Barker 1988) . These tests are executed very much in the same way as the lugeon tests. The evaluation of the tests are also possible to automatize for quick answers at the site. There are also possibilities to estimate the flow dimensions from the actual progress of a grouting session (Hässler 1991).

It is the authors hope that this article will encourage the use of transient hydraulic tests and recordings of grouting progresses to better optimize grouting works. We also hope that the article will stimulate the direction of future research in the grouting field.

6 REFERENCES

Barker, J. A. 1988. A generalized radial-flow model for pumping tests in fractured rock., Water Resources Research, Vol. 24, pp. 1796-1804

Håkansson U., Hässler L., Stille H. 1992. Rheological properties of microfine cement with additives. Proc. of ASCE Speciality Conference on Grouting, Soil Improvement and Geosynthetics, New Orleans, 1992. New York:ASCE.

Hässler L. 1991. Grouting of Rock - Simulation and Classification. Ph.D. Thesis, Dept. of Soil and Rock Mechanics, Royal Institute of Technology, Stockholm, Sweden.

Wallner M. 1976. Propagation of sedimentation stable cement pastes in jointed rock. Rock Mechanics and Waterways Construction, University of Achen, BRD.

Grouting in Rock and Concrete, Widmann (ed.) © 1993 Balkema, Rotterdam, ISBN 90 5410 350 7

Grouting of jointed rock – A case study
Injizieren von geklüftetem Fels – Eine Fallstudie

T. Janson, H. Stille & U. Håkansson
Department of Soil and Rock Mechanics, Royal Institute of Technology, Stockholm, Sweden

ABSTRACT : The development of the grouting's fundamental mechanics have been going on at the Department of Soil and Rock Mechanics, Royal Institute of Technology Stockholm, for some time. The aim of the research is to control and also bound the grout volume and still fulfill the requirement for tightness. A significant parameter for the tightness is the penetration length, which is dependent on the pressure, grout mix and the joint aperture. The grout volume can be estimated from the penetration length and rock porosity, estimated with respect to the grout. With respect to these connections three theoretical models have been developed. All models are based on that the joint system can be described as discs of different types. The grouted part of such a disc can be described by the penetration length and the joint apertures. The three models are based on the different assumptions to describe the joint geometry. The models have been used to calculate the grout volume, joint opening and penetration length. Control of the models was carried out by comparing the calculated results with the measured volume and other observations from grouting in Äspö Hard Rock Laboratory at Oskarshamn in Sweden. The grout mixes have been tested with respect to their rheological properties and used in the model.

ZUSAMMENFASSUNG : Grundlegende Forschung über das Injizieren von Fels wird am Institut für Boden- und Felsmechanik seit längerer Zeit betrieben. Dessen Ziel ist es, die Injiziermenge zu kontrollieren und zu begrenzen aber trotzdem die geforderte Dichte zu gewährleisten. Ein wichtiger Parameter für die Dichte ist die Eindringticfte, die vom Druck, der Mörtelzusammensetzung und der Klüftung abhängt. Das Injektionsvolumen kann abhängig von der Eindringtiefe und der Felsporosität geschätzt werden. Diesbezüglich wurden drei theoretische Modelle entwickelt. Alle Modelle bauen darauf auf, daß die Klüftungen als Schreiben verschiedener Art beschrieben werden können. Der Mörtelanteileiner solchen Schriebe kann durch die Eindringtiefe und die Kluftöffnung beschrieben werden. Die verschiedenen Modelle unterschreiben sich in der Art wie die Klüftungsgeometrie beschrieben wird. Die Modelle wurden verwendet um das Injektionsvolumen, die Klüftung und die Einringtiefe zu berechnen. Die Überprüfung der Modelle erfolgte durch Vergleich der berechneten Werte mit den gemessenen Injektionsvolumen, die Klüftung und die Eindringtiefe zu berechen. Die Überprüfung der Modelle erfolgte durchVergleich der berechneten Werte mit den gemessenen Injektionsvolumen und anderen Parametern bei Versuchen im Äspö Hart-Fels Laboratorium in Oskarshamn, Schweden. Die rheologischen Eigenschaften der in den Modellen verwendeten Mörtemischungen wurden untersucht.

1. INTRODUCTION

The purpose of grouting is to create a watertight zone around a tunnel. The grouted zone around the tunnel should fulfill the tightness requirement corresponding to achieve a certain penetration length and a filling of the aperture of the joints. From theoretical point of view the penetration length is depending on the pressure, grout mix and the joint aperture (Hässler 1991). The grout volume is depending on the penetration length and the rock porosity for used mix. With consideration of these assumptions three different simplification models have been development. These models can describe the grout take, joint aperture and grout penetration.

2. THEORY FOR THE MODELS

General

If the models should be valid, following conditions must be fulfilled.
- The flow of grout has to come to a standstill.
- The flow is laminar.
- The pressure difference, between grouting pressure and pore pressure, ΔP as well as the grout yield value, τ_0, is known and have a constant value during the grouting.

2.1 Model I

Assumption

The joint can be described like a disc with a constant aperture b, see figure 1. The grout is injected into the disc. The joint plane intersects the drill hole and the grout mix will fill the disc symmetrically. The flow of grout has come to a standstill when the penetration reach the maximum length I_n. The number of injected discs along the drill hole, N, is supposed to be the same as the number of conductive joints. The aperture b for the disc can be calculated from measured value of the transmissivity (T), of the borehole.

Connections

The penetration length for the grout is expressed as :

$$I_n = \frac{\Delta P \cdot b}{2 \cdot \tau_0}$$

The grout volume for all the discs along the borehole can be calculated according to :

$$V = I_n^2 \cdot \pi \cdot b \cdot N$$

From the water loss measurements the transmissivity can be expressed as (Gustafson 1986):

$$T = \frac{N b^3 \cdot \rho_w \cdot g}{12 \cdot \mu_w}$$

Based on the relationship between the penetration length (I_n) and the number of discs (N) along the drill hole and transmissivity, the predicted grout volume can be estimated as :

$$V_{hole,cal} = \left(\frac{\Delta P}{2 \cdot \tau_0} \right)^2 \cdot \frac{12 \cdot T \cdot \mu_w}{\rho_w \cdot g} \cdot \pi$$

From the estimated number of conducted joints the aperture b can also be determined as well as the penetration for the grouting.

2.2 Model II

Assumption

This model is an analytic simplification of Hässlers (1991) geometrical model. The principle is that the grout flows from the drill hole out in a systems of channels in the rock. The channel system are simplified and projected down to a disc. The area of the injected channels adds up and is described as a section of the disc. The disc section is defined by an angel α and is assumed to have a constant opening b_g, see figure 2 . The disc openings b_g is in principle calculated from the invert average

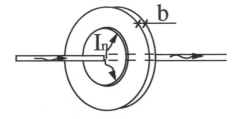

Figure 1. Model I The penetration takes place in a symmetric disc.
Figur 1. Modell I Das Eindringen findet in einer symmetrischen Schribe statt.

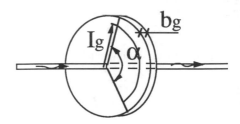

Figure 2. Model II The penetration takes place in a part of a disc.
Figur 2. Modell II Das Eindringen findet in einem Teil der Schreibe statt.

sum from the channels aperture. The section angel is called the spreading angel and is dependent of the geology. Further it is assumed that water flows in the same channels as the grout.

Connections
The penetration length for the grout in the section can be expressed as :

$$I_g = \frac{\Delta P \cdot b_g}{2 \cdot \tau_0}$$

The grout volume for all the discs along the borehole can be calculated according to :

$$V = I_g^2 \cdot b_g \cdot \frac{\alpha}{2} \cdot N$$

The water loss measurement in a drill hole can be expressed as (Hässler 1991) :

$$Q_w = \frac{\Delta P_w \cdot N b_g^3}{12 \cdot \mu_w} \cdot \frac{\alpha}{\ln(1 + \alpha \cdot L/W)}$$

Where ΔP_w is the water over pressure calculated from a certain distance L from the drill hole. W is the length of joint opening crossing the drill hole wall, normally assumed to be the circumference of the drill hole.
From the water loss measurement and the estimation of the spreading angel a prediction of the grout volume can be carried out as :

$$V_{hole,cal} = \left(\frac{\Delta P}{2 \cdot \tau_0}\right)^2 \cdot \frac{12 \cdot Q_w \cdot \mu_w}{2 \, \Delta P_w} \cdot \ln\left(1 + \frac{\alpha \cdot L}{W}\right)$$

From the estimated number of conducted joints the aperture b_g can also be determined and based on that the penetration length for the grouting.

2.3 Model III

Assumption
Model III is a development of model I. In model III the rock is describe more in detailed than the other models. Grouting in model III is based on the assumption that the grout flows out from the drill hole and into a system of discs, see figure 3. Primary discs crossing the borehole and the

secondary discs crossing the primary. The geometry of discs and penetration of grout are described by a number of geological parameters. The parameters takes into consideration the difference of penetration between grout and water in different joint openings and also the effect of secondary discs. The penetration is not a straight way in rock either. The volume is calculated from $\bar{b}$, β and N_g. The discs have an average opening $\bar{b}$, where $\bar{b}$ can be described as the product of a parameter θ and the estimated hydraulic openings b from the water loss measurement:

$$\bar{b} = \theta \cdot b$$

The parameter β is quotient between total disc area and area from the primary disc. N_g is the number of discs which are grouted and intersects the drill hole and can be determined as the product of a reductions parameter k_1 and the number of water conductive discs :

$$N_g = k_1 \cdot N$$

The penetration length is calculated from b_g and k_3. Where k_3 is a parameter for the curve way in the disc, accordingly the quotient between the curve way and the radial way for the penetration. The parameter k is described as the quotient between the channel openings b_g and the hydraulic openings from water loss measurement :

$$k = b_g / b$$

The disc openings b_g is in principle calculated from the invert average sum from the channels aperture.

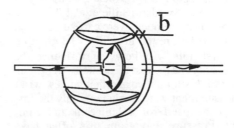

Figure 3. Model III The penetration takes place in a system of discs.
Figur 3. Modell III Das Eindringen findet in einem System von Schrieben statt.

Connections

The penetration length for the grout in the disc system can be expressed as:

$$I_k = \frac{\Delta P \cdot b_g}{2 \cdot \tau_0 \cdot k_3}$$

The grout volume for the disc system along the borehole can be determined as:

$$V_{hole,\,cal} = I_k^2 \cdot \bar{b} \cdot \pi \cdot \beta \cdot N_g$$

From the water loss measurements the transmissivity can be expressed as:

$$T = \frac{Nb^3 \cdot \rho_w \cdot g}{12 \cdot \mu_w}$$

From the transmissivity and determination of the five different geology parameters (k_1, k, k_3, θ, β) the predicted grout volume can be estimated as:

$$V_{hole,\,cal} = \left(\frac{\Delta P}{2 \cdot \tau_0}\right)^2 \cdot \frac{12 \cdot T \cdot \mu_w}{\rho_w \cdot g} \cdot \frac{k_1 \cdot k^2 \cdot \theta \cdot \beta \cdot \pi}{k_3^2}$$

From the estimated number of conducted joints and the parameter k the aperture b_g can also be determined and based on that the penetration length for the grouting.

3. APPLICATION OF THE MODELS

3.1 General

The basic material for the calculation and comparison of the three models is based from grouting in Äspö Hard Rock Laboratory at Oskarshamn in Sweden (Stille et al 1993). The water loss measurements were carried out in all the borehole before the grouting. The grouting of the first hole will however influence the transmissivity for the rest of the holes in the fan. Because of this influence and that no new water loss measurement were carried out only the first hole can be used for calculation and comparison. Pressure and grout mix often have normally been varied during the grouting of a hole. The pressure and yield value used in the calculation have been determined as the mean value for the hole.

Calculation

The accuracy of the models have been checked in three different aspects. First the grout take has been calculated based on the measured value of the transmissivity and an estimation of the geological parameters in model II and III. For model II the value of the spreading angle, α, has been supposed to vary with the rock quality and gives a value between 0,2 - 0,8. For model III the values given in table 1 has been used.

Table 1. Parameters used in model III

Tabelle 1. Im Modell III verwendete Parameter.

Consideration	Parameter	Value
Number of joints	k_1	0,05 - 0,3 : Depending on rock quality and type of joints
Grouted opening	k	4,0
Grouted path	k_3	1,5
Average opening	θ	1,0
Secondary joint plane	β	Number of joint sets divided by two

The joint opening and grout penetration is the second and the third aspects. The joint opening for the grout take and grout penetration have been calculated both based on the measured transmissivity as well as the actual grout take and an estimation of the number of joints based on the geological mapping (Stanfors et al 1992). For model III only four of five parameters must be estimated where the parameters k_1, k_3, θ and β were evaluated according to the table 1 while parameter k was assumed not to have a constant value.

3.2 Grout properties

Several different mix designs have been investigated in order to find suitable combinations. The rheological properties and the effect of different manipulations were evaluated by using a Brookfield Rheoset, rotational viscometer (Håkansson et al 1991). Two types of grouts have been tested (Stille et al 1993).

Firstly "Stabilo grout" was used which has been used for many grouting works in Sweden and with good experiences. Later a highly accelerated grout with calcium chloride was developed to better stand the higher water pressure encountered in the tunnel.

"Stabilo Grout"
The mix design with grouting cement, bentonite, plasticizer and silicate, Stabilodur F1, gives possibilities to suitable flow properties for achieving a limited penetration. It was decided to use the "Stabilo grout concept" based on a finer grinded cement, Mikrodur P by Dyckerhoff in order to improve the penetration in finer joints. The following mixes were tested see table 2.

Table 2. The tested mixes for "Stabilo grout".
Tabelle 2. Die untersuchten Mischungen von "Stabilo grout".

Name	Std	Styv	Acc	Micro
Cement	Degerhamn grout			Microdur
W/C	1,3	1,0	1,0	2,0
Bentonite [%]	2,0	3,0	2,0	2,0
Plasticizer [%]	1,0	2,0	1,0	1,0
Silicate [%] (% solution)	3,0 (40)	4,0 (40)	3,0 (40)	2,0 (40)

The mixing order is (W+C+Pl+Bt+Si) except for "accelerated" which had the following order (W+Bt+C+Pl+Si). The results of the tests with are shown in figure 4.

Calcium chloride
In order to meet the condition in the tunnel, new grouts must be designed and tested. The new mix were with calcium chloride to achieve a more accelerated grout with the following compositions, Degerhamn grouting cement, W/C 1,0, bentonit 2,0 % and 15,0 % chloride. The test results are presented in figure 5. It is clear from the study that by adding calcium chloride, the setting time can be altered as desired within a relatively wide range. However, the grout will be very sensitive to

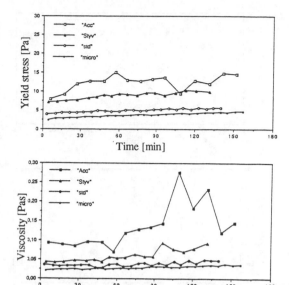

Figure 4. Tests on Stabilo Grout of different mix design.
Figur 4. Versuche mit Stabilo Mörtel mit verschiedenen Mischungsverhältnissen.

disturbance during the first part of the hardening process. If the grout is disturbed during the initial phase the hardening will be delayed for several hours. As long as the mixing is going on the hardening process will be delayed and the grout will have flow properties similar to ordinary cement grout but with a higher yield stress.

3.3 Result

Grout volume
The calculated volumes have been compered to the measured values for the three different models. The quotient between the predicted and measured grout volume has been calculated and shown in figure 6.

At this stage it was assumed that an acceptable models shall not give a predicted grout take which differ more than a factor 2 from the measured value. The reliability of the different models expressed as the ratio between accepted calculated results to all investigated fans is given in table 3.

425

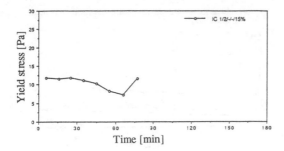

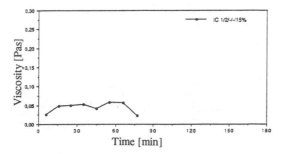

Figure 5. Tests on cement grout with calcium chloride.
Figur 5. Versuche mit Zementmörtel mit Kalziumchlorid.

Joint opening
The calculated joint openings are given in figure 7 for the different models. The joint opening for the grout penetration must be higher than 3 - 5 times the grain size of the grouting material corresponding to in our case around 0,2 mm. Lower calculated values has therefor been determined to be unrealistic and therefor not acceptable.

Table 3. Comparison between the models average, median and reliability with consideration to the quotient V_{cal}/V_{grout}.
Tabelle 3. Vergleich zwichen den Mittelwerten sowie Zuverlässigkeit der Modelle in Bezug auf das Verhältnis V_{cal}/V_{grout}.

Model	I	II	III
Average	2,38	3,52	1,70
Median	1,12	2,05	0,99
Reliability ($0,5<V_{cal}/V_{gro}<2$)	43%	33%	57%

Table 4. Comparison between the models average, median and reliability with consideration to the joint opening.
Tabelle 4. Vergleich zwisvhen den Mittelwerten und Medianwerten sowie Zuverlässigkeit der Modelle in Bezug auf die Kluftöffnung.

Model	I	II	III
Average [$m*10^{-6}$]	72	115	337
Median [$m*10^{-6}$]	70	119	265
Reliability ($>200*10^{-6}$ m)	0 %	24 %	67 %

Even in this case the reliability of the models has been defined as the ration between accepted calculated result to all investigated bore holes, see table 4.

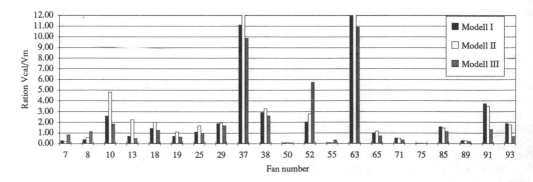

Figure 6. The result of the quotient between calculated volume and grout take.

Figur 6. Resultat der Verhältnisse zwischen berechnetem und tatsächlichem Mörtelverbrauch.

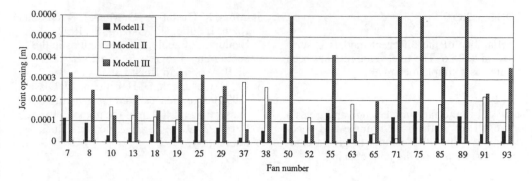

Figure 7. The result of the joint opening. Figur 7. Resulat der Kluftöffnung.

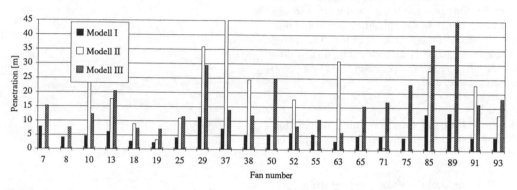

Figure 8. The result of the penetration length. Figur 8. Resultat der Eindringtiefe.

Table 5. Comparison between the models average, median and reliability with consideration to the penetration length.
Tabelle 5. Vergleich zwischen den Mittelwerten und Medianwerten sowie Zuverlässigkeit der Modelle in Bezug auf die Eindringtiefe.

Model	I	II	III
Average [m]	5,95	16,17	17,39
Median [m	4,92	11,05	15,31
Reliability interval 5 - 15m	52 %	19 %	48 %

Grout penetration
The calculated grout penetrations are given in figure 8 for the different models. For the grout penetration the acceptable value is something between 5 to 15 m which was the actual span measured in the three water bearing zones at

Table 6. Comparison between the models reliability with consideration to the volume, joint opening, penetration length and the average.
Tabelle 6. Vergleich der Zuverlässigkeit zwischen den Modelle in Bezug auf die Volumen, Klüftung, Eindringtiefe sowie deren Mittelwerte.

Model	I	II	III
Volume [%]	43	33	57
Joint opening [%]	0	24	67
Length [%]	52	19	48
Average [%]	32	25	57

Äspö Hard Rock Laboratory. Like in the two earlier cases the reliability has been defined in the same way, see table 5.

427

4. CONCLUSION

The reliability of the different models is summarised in table 6. The total reliability has been calculated as a simple mean value.

Model I
The model is the far simplest one and gives a good result for both grout volume and penetration but on the other hand will give a completely wrong value on the joint opening.

Model II
The model is very sensitive and give in seven of the cases completely wrong results. Minor errors in measured grout take and water loss can give large errors in the spreading angle and thus in the penetration and joint opening. The model on the other hand is also developed to follow the complete grouting operation from first pump strike to the refusal.

Model III
The model is the most complex model and needs a good knowledge of the geology to be used

For all the models the shear stress of the grout is a very important parameter and a small error in the measurement and estimations of this value can give large error in the estimated grout take and penetration. The models give, however, such a promising result that further research work has to be carried out to study the different parameters by evaluating more case studies as well as further theoretical analysis.

REFERENCES

Gustafson, G. (1986), Geohydrologiska förundersökningar i berg. BeFo 84:1/86, Stiftelsen Bergteknisk Forskning. Stockholm, 1986

Håkansson, U. Hässler, L. Stille, H. (1991), Mätmetodik för injekteringsmedels reologiska egenskaper. BeFo 241:1/91, Stiftelsen Bergteknisk Forskning. Stockholm, 1991

Hässler, L. (1991), Grouting of Rock Simulation and Classification. Department of Soil and Rock Mechanics, Royal Institute of Technology, Stockholm, 1991

Stanfors, R. Gustafson, G. Munier, R. Olsson,P. Rhen, I. Stille, H. Wikberg, P. (1992), Evalution of geological predictions in the acces ramp. Progress Report 25-92-02, SKB-Äspö Hard Rock Laboratory, 1992

Stille, H. Gustafson, G. Håkansson, U. Olsson,P (1993), Experiences from the grouting of the section 1-1400 m of the tunnel. Progress Report 25-92-19, SKB-Äspö Hard Rock Laboratory, 1993

Grouting in Rock and Concrete, Widmann (ed.) © 1993 Balkema, Rotterdam, ISBN 90 5410 350 7

Water tests to identify the hydraulic characteristics of gneiss in the foundation of a rockfill dam in Greece

Wassertests zur Ermittlung der hydraulischen Eigenschaften des Gneises im Untergrund eines Steinschuettdammes in Griechenland

E.C. Kalkani & D.Ch. Lambropoulos

Department of Civil Engineering, National Technical University of Athens, Greece

ABSTRACT: The water tests at the foundation rock of the Thissavros rockfill dam in northern Greece, were performed to identify the hydraulic characteristics of gneiss in the foundation of the dam. In this work the relation of joint width and spacing, as well as the water-take of the Lugeon tests to the rock depth is defined, and conclusions are drawn regarding the type of flow in the rock mass and the existence of non permeable zones.

Two definite zones of low permeability at depths 40 - 60 m and 100 - 120 m were identified, as well as the flow characteristics of water in the rock of the foundation.

ZUSAMMENFASSUNG: Die Wassertests am Felsuntergrunds des Thissavros Steinschuettdammes in Nordgriechenland wurden durchgefuehrt, um die hydraulischen Eigenschaften des Gneis des Dammuntergrunds zu ermitteln. In der vorliegenden Arbeit wird das Verhaeltnis von Kluftweite und -abstand sowie der Wasseraufnahme der Lugeon-Tests zur Felstiefe bestimmt und Rueckschluesse auf die Art der Stroemung in der Felsmasse und das Vorhandensein von undurchlaessigen Bereichen gezogen.

Zwei abgegrenzte Bereiche geringer Durchlaessigkeit in Tiefen von 40 - 60 m und 100 - 120 m sowie die Stroemungs-charakteristik des Wassers im Felsuntergrund wurden bestimmt.

Die Charakteristik des Kluftsystems, das hydraulische Verhalten der Felsmasse im Untergrund unter dem Steinschuett damm am Nestos Fluss in Nordgriechenland bestimmt, wird dargestellt. Bild 1 zeigt ein Bild des Thissavros Dammes, der zur Zeit im Bau ist. Der Felsuntergrund besteht aus Gneis, der zu dem metamorphen Felsgestein des Rhodopemassivs in Nordgriechenland gehoert.

Das Datenmaterial stammt aus vier Probebohrungen aus dem Kerngebiet des Dammes. Zwei der Bohrloecher sind auf der Hoehe von 210 m, wo die Dammsohle am Flussbett liegt, und die zwei anderen auf der Hoehe von 255 m an den Widerlagern des Dammes. Die Hoehe der Dammkrone liegt bei 390 m, was eine maximale Dammhoehe von 180 m ergibt.

Die vorliegende Arbeit ist Teil der Forschungsarbeiten zur Festlegung der Wasseraufnahme und der Moertelmischung fuer den Injektionsschirm des Steinschuettdammes unterhalb des zentralen Tonkernes. Der Moertelschirm wird ueblicherweise vor dem Bau des Dammes durch hydraulische Tests, Waessern und/oder Messung des elektrischen Widerstandes geprueft. Die Probebohrloecher sind in der Mitte des Schirms gelegen, und in bestimmten Abstaenden, wobei ein moeglicher negativer Befund nicht notwendigerweise eine unzureichende Funktion des gesamten Moertelschirms bedeutet (Verfel, 1989).

In dieser Arbeit wird die Beziehung zwischen Kluftweite und -abstand sowie der Wasseraufnahme der Lugeon-Tests im Verhaeltnis zur Felstiefe bestimmt, und Rueckschluesse auf die Art der Stroemung in der Felsmasse gezogen.

Fig. 1. The Thissavros rockfill dam under construction (shown from downstream).
Bild 1. Der Thissavros-Steinschuettdamm im Bauzustand (vom Unterlauf gesehen).

1 INTRODUCTION

The characteristics of the joint system which affect the hydraulic behaviour of the rock mass in the foundation of the Thissavros rockfill dam on the Nestos river, in northern Greece is presented. A picture of the Thissavros dam presently under construction is shown in Fig. 1. The foundation rock is gneiss, which belongs to the metamorphic rocks of the Rhodope massif in northern Greece.

The data refers to four exploratory boreholes at the core trench of the dam. Two of the boreholes are at el. 210 m, where the foundation of the dam is at the riverbed, and the two other are at el. 255 m on the abutments of the dam. The elevation of the crest of the dam is at el. 390 m, which gives the maximum height of the dam equal to 180 m.

The present work is part of the exploratory tests to define the water-take and the grout mix for the grout curtain of the rockfill dam below the central clay core. The grout curtain, prior to the construction of the dam is usually checked by means of hydraulic tests, logging, and/or measurements of electric resistance.

The exploratory boreholes are located in the middle of the curtain, and at certain distances, where a possibly negative finding does not necessarily mean an unsatisfactory function of the entire grout curtain (Verfel, 1989).

In this work the relation of joint width and spacing, as well as the water-take of the Lugeon tests to the rock depth is defined, and conclusions are drawn on the type of flow in the rock mass.

2 METHOD OF ANALYSIS

The hydraulic characteristics of a given rock mass depend on the presence of joints. The main factors that affect the type of water flow in the rock mass are the orientation and spacing of joints, the persistence, roughness and wall strength, the aperture and the filling of joints (ISRM, 1981).

In the case of the Thissavros rockfill dam the hydraulic behaviour of the foundation rock and the relevant conclusions were based on the examination of four exploratory boreholes with core recovery and on the water tests performed

430

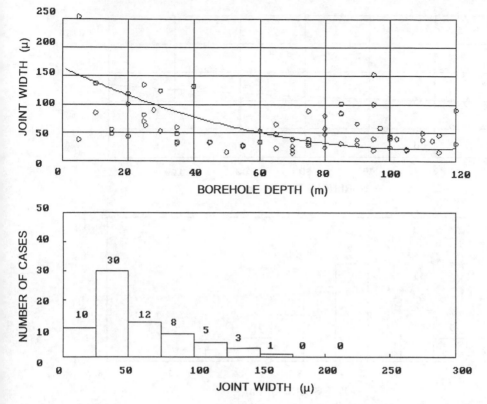

Fig. 2. Dispersion diagram of joint width versus depth and histogram of number of
         cases versus joint width.
Bild 2. Verteilung der Kluftbreite ueber der Tiefe und Histogramm der Anzahl der Werte
         ueber der Kluftbreite.

at the same boreholes. The total length of
the four boreholes is 453 m, and the water
tests of the boreholes consisted of 255
stages of Lugeon tests.

   The following correlations were examined:
- variation of joint width to depth,
- variation of joint spacing to depth,
- variation of water-take to depth,
- type of hydraulic flow according to the
  water tests.

To define the joint width the total number
of joints was measured at each 5-meter-long
stage of the Lugeon test. After
calculating the coefficient of permeability
k from the water-take of the Lugeon tests,
the width of the joints was calculated
using the equation (Snow, 1968):

$$\delta = \frac{15\ n\ k}{(p\ g\ \lambda)exp(1/3)} \quad \ldots \ldots \quad (1)$$

where:
δ = width of joints, μ,

n = water viscosity, $10^{\wedge}(-3)$ N s/ m^2 at 20
    degrees centigrade,
k = coefficient of permeability, m/s,
p = water density, 1000 kg/m^3,
g = acceleration of gravity, 9.806 m/s^2,
λ = number of joints per meter length.

To calculate the correlation coefficients
and the least squares regression
curves of the variables joint width,
spacing and water-take, the statistical
theory was used (Cox, 1987). The
presentation of results in the form of
dispersion curves and histograms was done
by using the appropriate software (Elliot,
1988). The interpretation of the Lugeon
water-take tests is performed using routine
methods (Houlsby, 1976).

3 DATA PROCESSING

The data processing was performed on a
personal computer. The data of the four

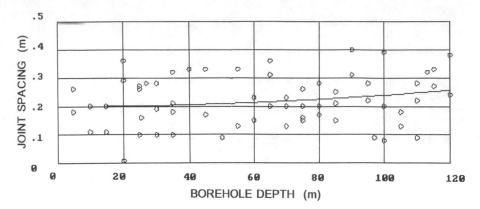

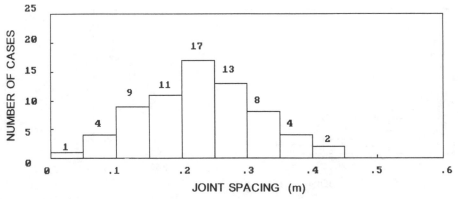

Fig. 3. Dispersion diagram of joint spacing versus depth and histogram of number of cases versus joint spacing.

Bild 3. Verteilung des Kluftabstandes ueber der Tiefe und Histogramm der Anzahl der Werte ueber dem Kluftabstand.

boreholes were processed together in order to emphasize the findings, while data processing separately for each borehole gave similar results.

The dispersion diagrams, the histograms, the regression curves and the calculations of the correlation coefficients were done using the appropriate software (Cox, 1987, Elliot, 1988).

3.1. Variation of joint width to depth.

The joint width was calculated using equation (1) for each 5-meter-long stage of the water test by measuring the number of joints at that stage from the core of the boreholes.

The variation of joint width versus depth is shown in the dispersion diagram of Fig. 2. A total of 69 cases are presented. The joint width varies from 12 to 250 µ and a decrease of the width is present with the increase in depth. The second order

regression curve is the best estimate of joint width y(x), as a function of depth x:

$$y(x) = 163.04 - 2.57\,x + 0.01\,x^2, \quad . \quad . \quad (2)$$

which indicates a definite decrease of joint width with increase in depth.

Also, it is evident that at a depth of 40-60 m the joint width is less than 40 µ, below 40 m the width is up to 150 µ and in one case 250 µ, while below 60 m the joint width is up to 100 µ and below 100 m the joint width is less than 50 µ. Hence, there are two zones of tight rock at depths of 40-60 m and 100-120 m.

The histogram of Fig. 2 indicates that 58% of the cases have joint widths less than 50 µ, and that a concentration of cases, almost 43%, have joint widths of 25-50 µ. Smaller percentages of cases have joint widths greater than 50 µ but less than 150 µ.

The correlation coefficient of joint width versus depth has an estimate equal to

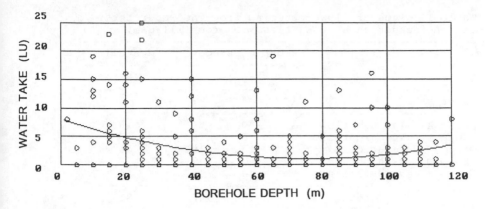

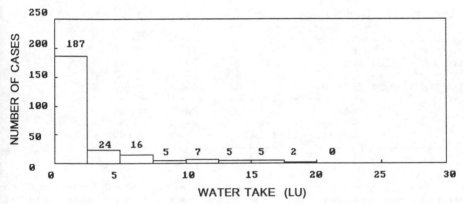

Fig. 4. Dispersion diagram of joint permeability versus depth and histogram of number
         of cases versus joint permeability.

Bild 4. Verteilung der Kluft-Durchlaessigkeit ueber der Tiefe und Histogramm der
         Anzahl der Werte ueber der Kluft-Durchlaessigkeit.

R = -0.41, which indicates a bad
correlation of joint width to depth.

3.2. Variation of joint spacing to depth

The joint spacing was measured directly
from the cores of the boreholes.
    The variation of joint spacing versus
depth is shown in the dispersion diagram of
Fig. 3. A total of 69 cases are presented.
The joint spacing varies from 0.05 to 0.40
m and the increase in spacing is related to
the increase in depth. The first order
regression curve is the best estimate of
joint spacing y(x), as a function of depth
x:

$$y(x) = 0.20 + 0.000035 \ x, \quad \ldots \ldots \ldots (3)$$

which indicates a definite increase of
joint spacing with increase of depth.
    Also, it is evident that at a depth of

40-60 m the joint spacing is 0.10-0.33 m,
at less than 40 m the spacing is up to 0.40
m, while below 60 m the joint spacing is up
to 0.40 m. Hence, the zones of tight rock
at 40-60 m depth and at 100-120 m depth
have joint spacing equal to 0.10-0.33 m.
    The histogram of Fig. 3 indicates that
61% of the cases have joint spacing less
than 0.25 m, and a concentration of cases,
almost 25%, have joint spacing 0.20-0.25 m.
Smaller percentages of cases have joint
spacings greater than 0.25 m and less than
0.45 m.
    The correlation coefficient of joint
spacing versus depth has an estimate equal
to R = 0.23, which indicates a bad
correlation of joint spacing to depth.

3.3. Variation of joint water-take to
      depth

The joint water-take was defined by using

433

Table 1. Type of flow resulting from the water tests at the foundation of the Thissavros rockfill dam.

	Number of stages		% overall
tight	136		53
water-take: 1 - 4 LU	#	%	
Group A - laminar flow	40	51	16
Group B - turbulent flow	8	10	3
Group C - dilation	15	19	6
Group D - wash out	5	7	2
Group E - void filling	10	13	4
total	78	100	31
water-take: 5 LU or more	#	%	
Group A - laminar flow	4	10	1
Group B - turbulent flow	15	36	6
Group C - dilation	12	29	5
Group D - wash out	4	10	2
Group E - void filling	6	15	2
total	41	100	16
overall total	255		100

the Lugeon water tests. A total of 255 stages of water tests were performed. It should be remembered that 1 LU (Lugeon Unit) is the absorbed water flow per unit length of the borehole with diameter of 76 mm at water pressure of 1 MPa, expressed in liters/min, 1 LU = 1 liter/min/m at 1 MPa pressure.

The variation of joint water-take versus depth is shown in the dispersion diagram of Fig. 4. A total of 251 cases are presented. The joint water-take varies from 0 to 25 LU and a decrease of the water-take is present with depth. The second order regression curve is the best estimate of joint water-take $y(x)$, as a function of depth x:

$$y(x) = 8.11 - 0.185 x + 0.001 x^2, \quad . . (4)$$

which indicates a decrease of water-take with depths up to 90 m and then an increase due to the large value of 8 LU at 120 m depth that affects the regression curve.

Also, it is evident that at a depth of 40-60 m the joint water-take is less than 5 LU, above 40 m the water-take is up to 25 LU, while below 60 m the joint water-take is up to 19 LU, and below 100 m the joint water-take is less than 5 LU. Hence, the two zones of tight rock are also identified at 40-60 m and at 100-120 m depth.

The histogram of Fig. 4 indicates that 84% of the cases have joint water-take of less than 5 LU, and the majority of cases, almost 75%, have joint water-take of less than 2.5 LU. Smaller percentages of cases almost 16% have joint water-take greater than 5 LU and less than 20 LU.

The correlation coefficient of joint water-take versus depth has an estimate equal to R = -0.23, which indicates a bad correlation of joint water-take to depth.

3.4. Characteristics of hydraulic flow

The interpretation of the water-take is made according to Houlsby (1976). The results are shown in Table 1, where the number of stages for each type of flow is indicated.

From the 255 stages of water tests shown in Table 1, 53% of the stages indicate water-take of less than 1 LU, and hence the corresponding portions of the boreholes are characterized as impermeable. Where the water-take is more than 1 LU, the flow may be characterized as laminar or turbulent, or the water-take may be due to joint dilation, wash out, or void filling.

Low permeabilities of 1 - 4 LU were present at 31% of the stages, while higher permeabilities greater than 5 LU were

434

present at 16% of the stages.

Laminar flow is present in more stages at low permeabilities less than 4 LU in 16% of the stages, while it is almost absent at permeabilities higher than 5 LU at 1% of the stages. Although wash out is present only at 2% in both low and high permeabilities, void filling is present at 2-4% of the stages. Dilation is present at a high percentage of 5-6% in both low and high permeabilities and turbulent flow is present at a high percentage of 6% only at high permeabilities, while it is as low as 3% at low permeabilities.

4 DISCUSSION

The widths of the joints calculated from the water-take of the rock mass range from 12 μ to 250 μ, values characterized as very small. It has been shown that the widths of the joints do not vary linearly with depth, but are subject to an underground zoning, characterized by alternating zones of open and closed joints.

Two main zones can be identified with closed joints between 40-60 m and 100-120 m, with widths ranging from 12 μ to 50 μ.

According to some researchers, only joints with widths greater than 35 μ can operate as water passages. Hence, it is certain that there are two impermeable zones with joint widths mostly below 35 μ at depths as indicated previously. These zones are also verified by the water tests.

The measured spacing on the joints of the exploratory boreholes ranges between 0.05 - 0.74 m, which according to the ISRM guidelines (ISRM, 1981) is characterized as from close to wide spacing.

It is useful to compare the actual observed distance of the joints of the exploratory boreholes to the theoretical which is calculated by using statistical methods. According to other researchers (Mood and Graybill, 1963) the number of joints crossed by equal lengths of randomly oriented boreholes, is different from part to part, but this number follows a Poisson's distribution. Also, the frequency of presence of zero water-takes depends on the mean number of joints present at each test stage.

In the water tests presented, 17% of the stages had zero water-take. Hence, from the 90 five-meter long stages (453 m / 5 m) the 75 stages included 127 joints. The mean theoretical number of joints per stage is equal to 127/75 or 1.7 joints/stage, a number which coincides to a mean spacing of joints equal to 3 m (5 m / 1.7).

It is evident in the case of the examined rock mass, that the spacing of joints measured follows a Poisson's distribution as shown in the histogram of Fig. 3. The theory giving a 3 m mean value of joint spacing may lead to overestimation of the mean spacing of joints, which actually ranges from 0.05 m to 0.40 m, with a mean value of approximately 0.25 m. Also, as indicated by the regression curve of Fig. 3, although the spacing of joints increases with depth, no strong correlation is actually present.

In the examined rock mass, the water-take ranges from 0 - 25 LU, while there is no evident correlation of its variation with depth. However, the existence of two relatively impermeable zones at 40-60 m depth and at 100-120 m depth is obvious. The two zones are shown not only in the dispersion diagram of Fig. 4, but also in the dispersion diagram of Fig. 2, indicating very small joint widths. While the joint width were not measured directly, they were calculated from the water-takes of the hydraulic tests and the number of joints per stage. Both dispersion diagrams indicate the two zones of reduced joint width and water-take.

The relation of the type of hydraulic flow to the water-take was confirmed for the Thissavros foundation rock. For water-takes of 1-4 LU, the most usual type of flow is the laminar at a rate of 16% of the total number of stages, while for water-takes larger than 5 LU the turbulent flow is prominent at a rate of 6% of the total number of stages.

5 CONCLUSIONS

The water tests performed at the foundation rock of the Thissavros rockfill dam in northern Greece, indicated two definite zones of low permeability at depths 40-60 m and 100-120 m. The variation of joint width, joint spacing and water-take versus borehole depth indicated not only the two low permeability zones, but also the hydraulic characteristics of gneiss in the foundation of the dam.

REFERENCES

Cox, S. T. 1987. Program CURVEFIT. Version 2.10-0.
Elliot, A. C. 1988. Program KWIKSTAT for data analysis. Version 1.3.
Houlsby, A. 1976. Routine interpretation of the Lugeon water test. Q. J. of Engn. Geology, 9: 303-313.
ISRM. 1981. Rock Characterization Testing and Monitoring. ISRM Suggested methods. Ed. E. T. Brown.

Mood, A. M. and Graybill, F. A. 1963.
  Introduction to the theory of statistics.
  2nd Edition. Mcgraw Hill Book Co. N.Y.
Snow, D. T. 1968. Rock fracture spacings,
  openings, and porosities. J. Soil
  Mechanics and Foundation Division. ASCE.
  94: 73-91.
Verfel, J. 1989. Rock grouting and
  diaphragm wall construction. Elsevier,
  Amsterdam. 532.

Grouting in Rock and Concrete, Widmann (ed.) © 1993 Balkema, Rotterdam, ISBN 90 5410 350 7

An experimental study of the grouting effect on deformability of rock masses

Versuche zur Untersuchung der Wirkung von Injektionen auf die Verformbarkeit von Felskörpern

Kohkichi Kikuchi, Yoshitada Mito, Naoto Yoshino & Takuya Naruse
Kyoto University, Japan

Hideya Suzuki
Chubu Electric Power Co., Inc., Nagoya, Japan

ABSTRACT: The aim of consolidation grouting of dam foundation is to improve mechanical property and permeabilityof foundation. Generaly, the check for improvemant on permeability is usually carried out using the check hole, however, the check for improvement on mechanical propety is not. Therefore, the improvement on deformability could have not have not been applied to the mechanical design of foundation.

The authors has made an in situ experiment in order to examine the grouting effects on the deformability of rock masses. Borehole expansion tests using grout hole were carried out before and after grouting.

As the results, the following were obtained:

1. The deformation of rock masses after grouting shows an increase of improvement by 1.3 -4.0 times of initial value.
2. More deformable before grouting, higher improvement on deformability is expected.

1 INTRODUCTION

Recently, with the lack of foundations that have high quality for constructions, more or less unstable rock masses would be selected as dam foundations. Therefore, the improvement of rock masses (foundation treatment) is inavoidable in order to secure the stability of foundations.

Consolidation grouting is mainly carried out at the foundation of concrete dam such as arch dam and gravitation dam. The aim of the consolidation grouting is to improve the deformation and intensity of rock masses by filling up crack with grout milk. It is also effective for controlling seepage flow at the part where the hydroulic gradient is the highest.

The effect of consolidation grouting is usually comfirmed by refering the results of permeability tests. In the case that the discontinuities in rock masses are filled with grout milk, the permeability would be low. Such a comfirmation method is not enough to grasp the mechanical properties of rock masses quantitatively. Therefore, the mechanical effects of grouting could have not been applied to the design of the foundation. If the degree of mechanical improvement of rock masses can be obtained quantita-

tively, the effective design could be carried out. In this study, the authors has made an in situ experiment in order to examine the grouting effects on the deformability of rock masses.

2 METHOD OF IN-SITU TESTING

In order to examine the grouting effect on the deformability of rock masses, we have made an in-situ experiment as follows.

2.1 Outline of testing yard

A square field which length of one side is 4.3 meters is secured. The rock mass of the testing yard is composed of Cretaceous rhyolites. The unconfined compressive strength of intact rock is about 90 to 120 MPa. Joints which surface are weathered are distribut ed in dense.

In order to prevent the cement milk from flowing out of the testing yard, the boreholes are dug along the side of the square at 20 centi-meters intervals as shown in Fig. 1 and the high-viscous cement milk is injected into the rock mass. The locations of grouting holes are shown in Fig. 2.

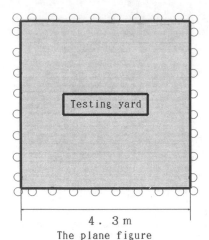

4. 3 m

The plane figure

The cross sectional view

4. 0 m

Fig. 1 Outline of testing yard

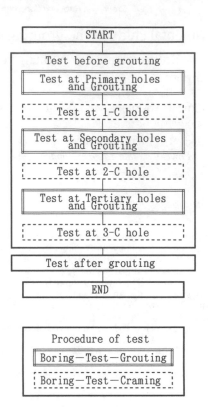

START

Test before grouting

Test at Primary holes and Grouting

Test at 1-C hole

Test at Secondary holes and Grouting

Test at 2-C hole

Test at Tertiary holes and Grouting

Test at 3-C hole

Test after grouting

END

Procedure of test

Boring—Test—Grouting

Boring—Test—Craming

Fig. 3 Flow-chart of in situ testings

2.2 Investigation Method

2.2.1 Surface survey

In order to grasp the condition of the rock mass, rock mass classification and joint survey have been carried out on the surface of the testing yard.

2.2.2 In-situ testings using the borehole

In-situ testings using the boreholes are carried out as shown in Fig. 3.

At first the borehole No 0-c has been dug with the length of 4 meters and the Borehole Hammer Testing, the Borehole Expansion Testing (Lateral Loading Testing), and the Lugeon water pressure testings have been carried out. After that, the boreholes No. 1-1, 1-2, 1-3, 1-4 (primary holes) have been dug and the same testings and grouting have been carriedout. In the same way, only the testings have been carried out at the boreholes No. 1-c (check hole for primary holes), 2-c (check hole for secondary holes), and 3-c (check hole for tertiary holes), and the testings and grouting have been carried out at the bo

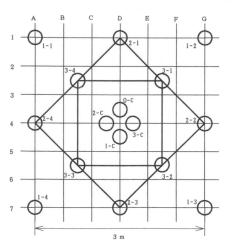

3 m

Fig. 2 Locations of grouting holes

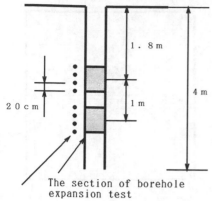

The section of borehole
expansion test

The point of borehole hammer test

Fig. 4   The schematics of in situ testings
at a borehole

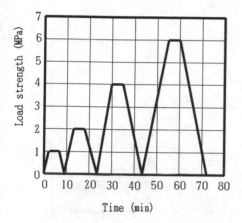

Fig. 5 The pattern of expansion
(the borehole expansion testings)

reholes No. 2-1, 2-2, 2-3, 2-4 (secondary ho
les), 3-1, 3-2, 3-3, and 3-4 (Tertiary holes).
After making these testings, we re-dug all
the boreholes and made the same testings at
the same position of the boreholes in order
to grasp the grouting effect on the deformability and the permiability of rock masses.

(1) Borehole Hammer Testing

A test system using "Rock Test Hammer" is
often used to evaluate rock mass properties,
simply. The Strike Response Value is measured by using this machine, and the modulus of
elasticity is expected. So, in order to
apply this principle to the borehole, the
authors develop a new machine "Borehole
Hammer".

At first, the hammer has been fixed to the
borehole wall at a depth of 3.3 meters by
using the fixation unit. After making the 5
strikes, the velocity waveforms has been
recorded, the acceleration waveforms and the
elastic waves through the rocks which is
obtained by 5 strikes. Then, pulling up the
sonde at an intervalof 20 centimeters to a
depth of 1.3 meters, the same testings have
been made at each point.

(2) Borehole Expansion Testing

After making the Borehole Hammer Testings,
the borehole expansion testings have been
carried out at 2 portions per borehole. The
length of expansion part is 60 centimeters.
The depths of the centre of expansion part
are established 1.8 meters and 2.8 meters as
one expansion part overlaps 4 strike points
shown in Fig. 4. The expansion apparatus is
a lateral loading testing machine. The velocity of expansion is 0.5 MPa/min as shown in
Fig. 5.

(3) Method of Lugeon Water Presure Testings

After making the Borehole Expansion Testings,
the Lugeon water presure testing have been
made. The water presure is established at
0.1 MPa and 0.2 MPa.

3 EXAMINATION

3.1 Grouting effect on deformation.

Fig. 6, Fig. 7 and Fig. 8 show the relationship between the moduli of deformation, the
moduli of elasticity, and strike response
value before and after grouting, respectively.
In each figures, there are fitted lines
with the least upper limit and the lowest
limit. We can find a tendensy that the range
between the least upper limit and the lowes
limit becomes narrower with rising of mechanical propeties. The lower the mechanical
propeties before grouting, the higher the
mechanical propeties after grouting. The
higher the mechanical propeties before grouting, the lower the mechanical propeties
after grouting. this result shows that;
"In the case that the rock masses are weak,
the strength rises largely. In the opposit
case, the strength scarecely rises."
Such tendency supposed by ranges between
the least upper limit and the lowest limit
shows that mechanical propeties after grouting can increase from 1.4 to 4 times larger
than that before grouting. Especially, the
modulous of elastisity and strike responce
value of the holes No. 2-1, 2-2, 2-3 and 2-4
or the holes No. 3-1, 3-2, 3-3 and 3-4 rise
largely.
In order to investigate such tendency in
detail, Fig. 9-11 shows the ratio of proper-

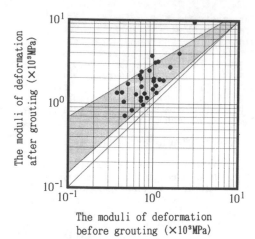

The moduli of deformation after grouting (×10³MPa)

The moduli of deformation before grouting (×10³MPa)

Fig. 6 Relationship between the moduli of deformation before and after grouting

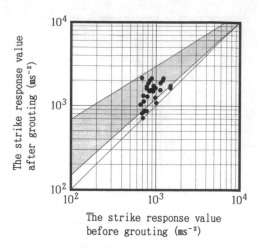

The strike response value after grouting (ms⁻²)

The strike response value before grouting (ms⁻²)

Fig. 8 Relationship between the strike response value before and after grouting

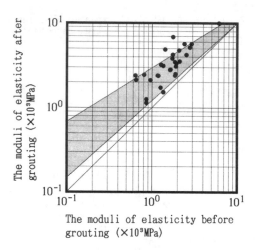

The moduli of elasticity after grouting (×10³MPa)

The moduli of elasticity before grouting (×10³MPa)

Fig. 7 Relationship between the moduli of elasticity before and after grouting

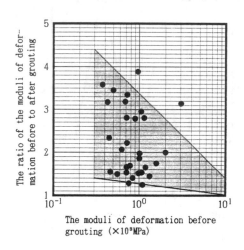

The ratio of the moduli of deformation before to after grouting

The moduli of deformation before grouting (×10³MPa)

Fig. 9 The ratio of the moduli of deformation before to after grouting

ty before to after grouting.

In these figures the fitted lines with the least upper limit are shown. In the case that the mechanical propety before grouting is low, ratio of it is high. In the opposit case, ratio of it is low. To put it into concretely, in the case that the value of propety before grouting is about 500 MPa, the value of propety after grouting can be 4 times larger than that before grouting. Similarly, 1000 MPa ;3times, 2000 MPa ;2.5 times.

Fig.6-11 prove following result;

"In the case that the rock masses are weak, the strength rises largely. In the opposit case, the strength scarcely rises."

And these figures are very important when we think about the improvement of the deformation.

4 CONCLUSION

Recentry, in the construction of civil structures, there are few hard rock foundat-

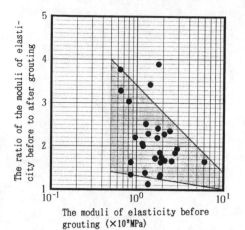

The ratio of the moduli of elasticity before to after grouting

The moduli of elasticity before grouting (×10⁸MPa)

Fig. 10 The ratio of the moduli of
elasticity before to after grouting

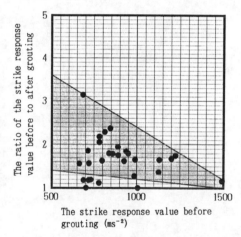

The ratio of the strike response value before to after grouting

The strike response value before grouting (ms⁻²)

Fig. 11 The ratio of the strike response
value before to after grouting

(1) The deformation of rock masses can be im proved by the grouting.

(2) In the case that the rock masses are wea k, the ratio of the moduli of deformation be fore to after grouting is high. In the oppos it case, low. In other word, the weak rock m assescan be improved more effectively than g ood rock masses

It is necessary that many data about the deformation on rock masses before and after grouting are accumulated.

ions. We must construct civilstructures on s oft rock foundations. That is why we must gr out the rock foundations and improve the mec hanical property and water permeabilities of rock masses.

As designing the foundation, the improve ment of the permeability of rock masses are usually taken into account. However, the imp rovement of the deformation of rock masses a re seldom taken into account.

In this study, the basic conclusion abou t the improvement of the deformation on rock masses are obtained as follows.

Grouting in Rock and Concrete, Widmann (ed.) © 1993 Balkema, Rotterdam, ISBN 90 5410 350 7

Bruchmechanische Aspekte bei Verpreßvorgängen

Some fracture mechanics-based assessments of grouting

H. N. Linsbauer
Technische Universität Wien, Österreich

KURZFASSUNG: Der Prozeß von strömungs- und strukturmechanischen Vorgängen beim Verpressen von Rissen bzw. Klüften in Beton und Fels ist gekennzeichnet einerseits durch den gekoppelten Mechanismus von Strömung im Spalt und Deformation der umgebenden Matrix und andererseits durch die Auswirkung dieses Vorgangs auf die Stabilität des Gesamtsystems.

Für die hydromechanische Beschreibung des Strömungsvorganges in Rissen (Spalten) sind speziell im Fall von "zähen" Verpreßfluiden die Navier-Stokeschen Bewegungsgleichungen anzuwenden. Die kritische Beanspruchung des Systems durch (den "Lastfall") Verpressen kann mit den Methoden der Bruchmechanik beurteilt werden.

Die Modellierung des Problems erfolgt anhand eines physikalischen Modells, die Lösung mit numerischen Methoden (Differenzenverfahren). Die Rückkopplung des Verformungseinflusses des Festkörpers (Spaltafweitung) auf den Fließvorgang erfolgt iterativ.

Als maßgebendes Kriterium für eine Abschätzung der Injektionsauswirkungen auf das Tragverhalten ist die Erfassung des zeitlichen Druckaufbaus im Spalt anzusehen. Parallel dazu wird der Verlauf des Spannungsintensitätsfaktors (kennzeichnender Wert für das Stabilitätsverhalten des Risses) ermittelt.

Im vorliegenden Bericht werden Parameteruntersuchungen betreffend Rißgeometrie und Materialverhalten (Länge, Ausgangsspaltweite, Querschnittsverlauf, Makrorauhigkeit, Elastizitätsmodul) einerseits und Eigenschaften der Verpreßflüssigkeit (Dichte, Viskosität, Kohäsion) andererseits vorgestellt.

Die theoretischen Untersuchungen wurden durch Rorströmungsversuche in Hinblick auf eine Kalibrierung ergänzt.

ABSTRACT: The coupled process between fluid mechanics and structural mechanics in case of grouting in concrete or rock may be characterized by flow mechanisms in gaps accompanied with surface deformations as a consequence on one side and interaction with the structure (grouting-system behavior) on the other side.

The postulated one-dimensional flow within a gap generally may be described by the Navier-Stokes equation. The critical effects of grouting consequences to the fractured (fissured) system may by assessed by means of fracture mechanics methods.

The physical modelling and mathematical description by differential equations results in a solution of the problem using finite difference methods. The back-coupling of the surface deformation (crack opening) on flow behavior is done in iterative form.

The essential criterion for stability considerations during grouting is the understanding of the time-dependant crack-surface pressure distribution. Parallel the knowledge of the stress-intensity factor development is required.

Within the present paper results of parameter investigations concerning crack-geometry and material behavior (length, starter-width, characteristic of cross-section, surface roughness, modulus of elasticity) and fluid attributes (density, viscosity, Bingham-character) respectively, are presented.

The theoretical investigations are completed by hydraulic experiments (viscous flow in pipes) for calibration purpose.

1 EINLEITUNG

Die komplexen physikalischen Vorgänge beim Injizieren von Klüften (Rissen) in Fels (Beton) sind generell drei mechanischen Disziplinen zuzuordnen:

Strömungsmechanik	Strömung in Spalten (porösen Medien)
Festkörpermechanik	Flüssigkeitsdruckbedingte Rißaufweitung
Bruchmechanik	Grenzdruckprofil (Stabilität des Systems)

Die hydromechanische Beschreibung des Strömungsvorganges in Spalten (Rissen) erfolgt mittels Kontinuitäts- und Bewegungsgleichungen, deren Ansätze den jeweiligen Gegebenheiten des Problems entsprechen müssen, wobei stets der instationäre Charakter des Verpreßvorganges zu beachten ist.

Aus der für die Aufrechterhaltung des Verpreßgutflusses erforderlichen Druckbeaufschlagung resultiert eine von der Geometrie, der Steifigkeit sowie vom Umgebungsdruck des betrachteten Bereichs abhängige Spaltaufweitung. Dieser gekoppelte Vorgang stellt einen interaktiven Prozeß zwischen der Strömungs- und der Strukturmechanik dar.

Aus technischer Sicht gesehen ist der Injizierprozeß derart auszulegen, daß das Verpressen ohne die Gefährdung einer weiteren Rißausbreitung bzw. ohne Induzierung von bauwerksgefährdenden Zwängsspannungen erfolgt. Das wiederum erfordert Kenntnisse über das Deformationsverhalten im Rißbereich und führt als weitere Konsequenz zu einer bruchmechanischen Abschätzung der Rißstabilität hinsichtlich einer verpreßbedingten Rißausbreitung.

Eine detaillierte Planung eines Verpreßvorganges basierend auf den oben angeführten Methoden ist bestenfalls nur unter klassischen Versuchsbedingungen möglich. Eine erfolgreiche Verpressung in der Praxis ist weitestgehend von der Erfahrung der Planenden und Ausführenden abhängig, wobei Analysen der Meßdaten (Einpreßdruckprofil, Einpreßrate, Reichweite, etc.) weitere wertvolle Erkenntnisse liefern.

Die vorliegende Untersuchung stellt einen Beitrag für die Abschätzung des Einflusses einzelner Parameter auf den Verpreßvorgang dar:

Geometrie	Spaltbreite (Rohrdurchmesser)
	Spaltkonfiguration (parallel, linear, parabolisch)
	Spaltlänge
	Rauhigkeit
Festes Medium	Elastizitätsmodul
Verpreßgut	Dichte
	Viskosität
	Kohäsion
	Kompressibilität
Druck	Initialdruckcharakteristik
	Maximaldruck

Das maßgebende Kriterium dieser Studie stellt die zeitliche Entwicklung der Spannungsintensität im Bereich des Spalt- bzw. Rißendes dar. Dadurch werden Aussagen über kritische Phasen während des Injiziervorganges ermöglicht.

Parallel zur Entwicklung des Rechenprogrammes wurde in Hinblick auf eine "Kalibrierung" der Fließvorgang mit assoziierter Druckverteilung im Rahmen von Rohrversuchen simuliert.

Die Spaltsrömungsroutine konnte mit Versuchen anhand eines Stahlplattenmodells (Feder 1992) verglichen werden.

2 GRUNDLEGENDE BEZIEHUNGEN

Die hydromechanische Beschreibung des Strömungsvorganges erfolgt generell über die Navier-Stokes'schen Bewegungsgleichungen deren Lösung für das Spaltproblem zum bekannten "kubischen" Fließgesetz führt. Die dazu parallel laufende Betrachtung des Stabilitätsverhaltens des gesamten Systems (Rißausbreitungsgefahr) während des Injiziervorganges hat mit Methoden der Bruchmechanik zu erfolgen. Die Verknüpfung (Koppelung) der strömungs- und strukturmechanischen Vorgänge erfolgt über die Rißaufweitung "w" und führt z.B. für den einfachen Fall eines elliptischen Rißbandes (Perkins und Kern 1961) zu einer partiellen Differentialgleichung 2. Ordnung mit entsprechenden Anfangs-und Randbedingungen:

$$\frac{\partial^2 w^4}{\partial x^2} \frac{G}{64 \; \nu \, \rho \, (1-\nu)} - \frac{\partial w}{\partial t} = 0 \qquad (1)$$

mit w=Rißweite, G=Schubmodul, ν=kinematische Viskosität, ρ=Dichte und ν=Querdehnzahl.

Als bruchmechanisches Stabilitätskriterium ist entweder das für die "Linear Elastische Bruchmechanik" (LEBM) geltende K-Konzept

$$K = \frac{2}{\sqrt{\pi\,a}} \int_0^a dp(x)\, f(\tfrac{x}{a})dx \qquad (2)$$

(K=Spannungsintensitätsfaktor, a=Rißlänge)
oder die dem Barenblatt'schen Kohäsionsmodul
(Bruchprozeßzone) entsprechende Randbedingung

$$\left[\frac{\partial w}{\partial x} \right]_{(x=a)} = 0 \qquad (3)$$

zu definieren.

Außer dem Perkins-Kern Modell wurden weitere
Rißkonfigurationen unter Einbeziehung zusätzlicher
Parameter (variabler Fließexponent, Flüssigkeits-
übertritt in das umgebende Medium, Porewasser-
druckverhältnisse, etc.) in mathematischer Form behandelt.

Die Entwicklung dieser Verfahren wurde durch die
Anforderungen der Mineralölindustrie vehement be-
schleunigt und ist unter der Bezeichnung "Hydraulic
Fracturing" zu einem hohen Standard ausgereift.

Der Injiziervorgang als "Sanierungsprozeß" ist bis
zum Erreichen des kritischen Zustandes
(Rißausbreitung) identisch mit den Vorgängen beim
"Hydraulic Fracturing".

3 KONZEPT

Das in der vorliegenden Studie entwickelt Konzept
basiert auf der Beschreibung der Strömungsvorgänge im Spalt als Filterströmung mit dem
Darcy'schen Ansatz. Die hier verwendete grundlegende Gleichung für den eindimensionalen Fall
(t,x) lautet

$$\frac{\partial p}{\partial t} = \nabla(\,\alpha \,\nabla p) \qquad (4)$$

wobei p den Druckverlauf darstellt und der Beiwert
α Terme für die Durchlässigkeit, den spezifischen
Speicherkoeffizient, das Widerstandsgesetz nach
Louis (1967), die Aufbereitung des Bingham'schen
Verhaltens nach Wallner (1976) etc. enthält.

Die flüssigkeitsdruckinduzierte Ausgangsrißöffnung
wird für jeden Zeitschritt aus den Beziehungen für
den halbunendlichen Riß ermittelt und die
Rißgeometrie linear über die Rißlänge angesetzt.
Gleichzeitig erfolgt die zeitbezogene Ermittlung des
Spannungsintensitätsfaktors. Die Koppelung ergibt
sich automatisch im darauffolgenden Zeitschritt. Die
Differentialgleichung wurde mittels eines impliziten

Differenzenverfahrens für verschiedene   Anfangs-
und Randbedingungen gelöst.

Für die Abschätzung einer Gefährdung des Trag-
systems kann ein bruchmechanisches Stabilitätskriterium herangezogen werden. Dies erfordert zusätzlich zum theoretisch ermittelten Spannungs-
intensitätsfaktor die Kenntnis des entsprechenden
Materialkennwertes (Linsbauer 1987). Im Zuge der
Ausarbeitung dieser Untersuchung wurden Bohrkerne aus einem Staumauerbeton mit bruchmechanischen Testmethoden untersucht (Linsbauer
1990).

4 PARAMETERSTUDIE - ERGEBNISSE

Eine gezielte theoretische Untersuchung erfordert
die Überprüfbarkeit der entwickelten Routine
anhand eines Experiments. Für diesen Zweck bot
sich   das   einfach   zu   verwirklichende
Rohrströmungsmodell an. Die Untersuchung erfolgte
für Wasser und Öle verschiedener Viskositäten
anhand einer 2 bis 4 m langen Teststrecke in einem
Plexiglasrohr mit einem Durchmesser von 3.0 mm.
Aufgezeichnet wurden die Drücke an 4 Meßstellen
sowie der Fortschritt der Flüssigkeitsfront an
zwischenliegenden Meßpunkten. Die Kalibrierung
des Rechenmodells basierte auf der Ankunftszeit der
Flüssigkeit am Ende der Teststrecke und erfolgte in
Form einer verschmierten Rauhigkeit.

Der Druckverlauf für das Öl-320 (dynamische
Viskosität bei 20° C   1160 mPas)   in den
entsprechenden Meßpunkten ist   aus der Abb.1
ersichtlich.

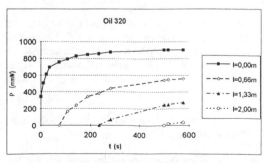

Abb.1   Rohrmodell - Druckverlauf in den Meßpunkten MP1
bis MP4

Fig.1   Pipe model - pressure versus length in MP1 to MP4

445

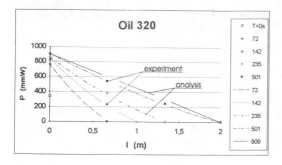

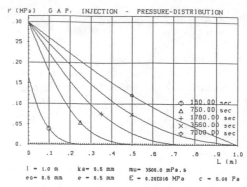

Abb.2 Rohrmodell - Druckverlauf entlang der Meßstrecke.
Vergleich Experiment und Rechnung

Fig.2 Pipe model - pressure versus time over the length.
Comparison of experiment and analysis

Der Vergleich mit der Rechnung basierend auf der erwähnten Kalibrierung ist in der Abb.2 dargestellt.

Ähnliche Ergebnisse resultierten aus den anderen Versuchen mit Ölen unterschiedlicher Viskositäten.

Von den Parameterstudien werden drei typische Fälle vorgestellt, wobei jeweils eine Flüssigkeit mit einer Viskosität von $\mu=3500$ mPas postuliert wurde.

In der Abb.3 ist der zeitliche Druckverlauf (Maximaldruck 0.30 MPa) für einen am Ende offenen Spalt mit einer Länge von 1 m und einer Weite von 0.5 mm ersichtlich. Nach Erreichen des Spaltendes (ungef. 3500s) erfolgt der weitere Druckaufbau bis zum stationären Zustand.

Die Abb.4 zeigt analoge Verhältnisse für eine Spaltweite von 0.2 mm und einem maximalen Einpreßdruck von 14 MPa. Dieser Versuch könnte als Vergleich zu den Versuchen von Feder (1992) dienen.

Eine typische Rißuntersuchung (Geometrie linear auf die Rißspitze verlaufend) mit einer Ursprungsrißöffnung von 0.5 mm ist in den Abb.5a und Abb.5b dargestellt. Die druckinduzierte Rißöffnung ergibt sich bei einem Elastizitätsmodul von 20000 MPa zum Gesamtwert von 1.4 mm.

Der assoziierte Verlauf des Spannungsintensitätsfaktors K ist aus der Abb.5b ersichtlich.

Aus den beiden Diagrammen könnte unmittelbar auf die zulässige Intensitätsdauer des aufgebrachten Maximaldruckes unter Zugrundelegung des Rißzähigkeitswertes für das entsprechende Material (Beton, Fels) geschlossen werden.

Abb.3 Zeitlicher Druckverlauf im offenen Spalt
(Spaltweite e_0=0.5 mm, Rauhigkeit ka=0.5 mm,
Kohäsion c=5 Pa)

Fig.3 Pressure versus length in open end gap
(width e_0=0.5 mm, roughn. ka=0.5 mm,
cohesion c=5 Pa)

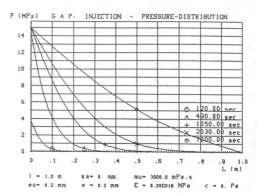

Abb.4 Zeitlicher Druckverlauf im offenen Spalt
(Spaltweite e_0=0.2 mm, Rauhigkeit ka=0.0mm,
Kohäsion c=0 Pa)

Fig.4 Pressure versus length in open end gap
(width e_0=0.2 mm, roughn. ka=0.0 mm,
cohesion c=0 Pa)

5 SCHLUSSBEMERKUNG

Die generelle Bedeutung derartiger Parameterstudien für das komplexe Gebiet der Verpressung von Rissen und Spalten ist in erster Linie in der Möglichkeit einer Abschätzung des Einflusses einzelner Parameter absolut und imVergleich zu sehen. Basierend auf den aufgezeigten analytischen Möglichkeiten wären weitere experimentelle Ergebnisse entsprechend den Versuchen von Feder (1992) bzw.

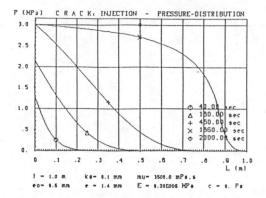

P (MPa)   C R A C K :   INJECTION - PRESSURE-DISTRIBUTION

I = 1.0 m   ka= 0.1 mm   mu= 1500.0 mPa.s
eo= 0.5 mm   e = 1.4 mm   E = 0.20E005 MPa   c = 0. Pa

Abb.5a   Druckverlauf für einen einseitig geschlossenen Riß
          mit linear auf Null verlaufender Rißgeometrie
          (e_0=0.5 mm, Endöffnung e=1.4 mm)

Fig.5a   Pressure versus length for an edge crack with linear
          geometry (e_0=0.5 mm, final width e=1.4 mm)

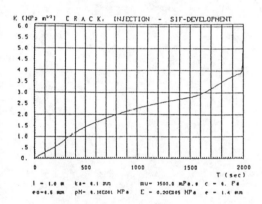

K (MPa m^{1/2})   C R A C K :   INJECTION - SIF-DEVELOPMENT

I = 1.0 m   ka= 0.1 mm   mu= 1500.0 mPa.s   c = 0. Pa
eo=0.5 mm   pH= 0.10E001 MPa   E = 0.20E005 MPa   e = 1.4 mm

Abb.5b   Zeitlicher Verlauf des Spannungsintensitätsfaktors
          entsprechend den Druckverläufen von Abb.5a

Fig.5b   Stress intensity factor (SIF) versus time due to the
          pressure development of Fig.5a

Meßprotokolle aus der Praxis für eine Auswertung
wünschenswert.
Eine Erweiterung auf rotationssymmetrische Probleme mit entsprechenden Experimenten ist vorgesehen.

Der Autor bedankt sich beim Arbeitskreis
"Injektionen im Beton" für die gute Zusammenarbeit,
sowie bei der Hochschuljubiläumsstiftung der Stadt
Wien (H-90/92) für finanzielle Unterstützung und
bei seinen Mitarbeitern.

REFERENCES

Feder, G. 1992. Spaltströmungsversuche
(unveröffentlicht)

Perkins, T.K. und Kern, L.R. 1961. Widths of hydraulic Fractures. J. Petr. Techn.13: 937-948

Louis, C. 1967. Strömungsvorgänge in klüftigen
Medien und ihre Wirkung auf die Standsicherheit
von Bauwerken und Böschungen im Fels. Dissertation, Technische Hochschule Karlsruhe,.

Wallner, M. 1976. Ausbreitung von sedimentationsstabilen Zementpasten in klüftigen Fels.
Veröffentlichungen des Institutes für Grundbau,Bodenmechanik, Felsmechanik und Ver-
kehrswasserbau der RWTH Aachen (Hrsg.
W.Wittke), 2: 47-163

Linsbauer, H.N. 1987. Das Tragverhalten von Be-
tonbauwerken des konstruktiven Wasserbaus -
Einfluß von Rißbildungen. Inst.f.konstr. Wasserbau, TU-Wien, Nr.21:1-173

Linsbauer, H.N. 1990. Sperre Kölnbrein - Ermittlung
bruchmechanischer Materialkennwerte. Arbeitskreis - Injektion von Rissen in Beton.Report:1-17

Wasserkraftanlage Cleuson-Dixence (VS, Schweiz): Versuchsinjektionen

Hydroelectric power project Cleuson-Dixence (VS, Switzerland): Grouting tests

M. Mercier & H. Détraz
Bonnard & Gardel, Ingénieurs-conseils SA, Lausanne + Sion, Schweiz

P. Egger
Institut für Boden- und Felsmechanik, Ecole Polytechnique Fédérale, Lausanne, Schweiz

ZUSAMMENFASSUNG: Im Rahmen der Vorarbeiten für die unterirdische Wasserkraftanlage Cleuson-Dixence (VS; Schweiz) wurden zu Versuchszwecken Konsolidierungs- und Abdichtungsinjektionen in den metamorphisierten und tektonisierten Schichten der Großen-St.-Bernhard-Decke durchgeführt. Diese Erkundungsarbeiten erlaubten, die für die vorliegenden Gebirgsarten bestgeeigneten Injektionsmethoden zu ermitteln. Dank Vergleichsversuchen im unbehandelten und hierauf im verpreßten Gebirge konnten die erhaltenen Ergebnisse quantifiziert werden. Die Standfestigkeit des injizierten Gebirges wurde während der Ausbruchsarbeiten nach der Injektion der beiden Versuchsblöcke in wahrer Größe beurteilt. Hinweise auf die zur Durchörterung von Störungszonen in einigen ähnlich gelagerten Fällen angewandten Maßnahmen runden das Bild ab.

ABSTRACT: Within the studies for the planned underground hydroelectric power plant of Cleuson-Dixence (Wallis, Switzerland), grouting tests were carried out in the metamorphized and tectonized rocks of the Grand-St-Bernard series in order to check their ability to being consolidated and impermeabilized. These preliminary works permitted the definition of the best suited grouting methods for the rock types encountered. The obtained results were quantified with the help of comparative field tests carried out respectively before and after grouting. The stability of the grouted rock mass was evaluated during the excavation of the two test plots after grouting. References to some other projects where fault zones had to be crossed close the paper.

RESUME: Dans le cadre du projet du futur aménagement hydroélectrique souterrain de Cleuson-Dixence (Valais, Suisse), des injections de consolidation et d'étanchement ont été réalisées, à titre d'essai, au sein des séries métamorphisées et tectonisées de la nappe du Grand-St-Bernard. Ces travaux de reconnaissance ont permis de définir les méthodologies de traitement les plus appropriées aux terrains en présence. Les résultats obtenus ont pu être quantifiés grâce aux essais comparatifs effectués in-situ en terrain vierge puis en terrain injecté. La stabilité des terrains injectés a été appréciée, en vraie grandeur, lors de l'excavation qui a suivi l'injection des deux plots d'essais. Un bref rappel des méthodes de construction adoptées sur quelques chantiers présentant des conditions géotechniques semblables, termine la communication.

1 EINLEITUNG

1.1 *Beschreibung des Bauvorhabens*

Das Projekt Cleuson-Dixence stellt eine Ergänzung der bestehenden Einrichtungen der Wasserkraftanlage Grande Dixence dar. Mit einer Ausbauleistung von 1180 MW wird die neue Anlage das Potential des größten Stausees der Schweiz optimal zu nutzen gestatten und dem Eigentümer (L'Energie de l'Ouest-Suisse SA - Grande Dixence SA) eine zur Deckung des winterlichen Energiebedarfs ausreichende Spitzenproduktion zur Verfügung stellen.

Diese im wesentlichen unterirdische Anlage

umfaßt die folgenden wichtigsten Bauteile :

1. eine neue Wasserfassung, hergestellt mittels einer Tunnelbohrmaschine durch die Staumauer Grande Dixence;

2. einen gebohrten Triebwasserstollen (Ausbruch-ϕ 5,60 m; Innen-ϕ 4,80 m) zwischen der Staumauer und der Dent de Nendaz (15 km);

3. einen 300 m hohen Wasserschloßschacht, herkömmlich aufgefahren in den Schichten der Dent de Nendaz;

4. einen großteils gebohrten gepanzerten Schrägschacht (Ausbruch-ϕ 4,80 m; Innen-ϕ 3,20 m; Länge 4,3 km; Höhenunterschied 1500 m);

5. ein unterirdisches Krafthaus (Volumen 150'000 m^3), in Schräm- und Sprengvortrieb aufgefahren und mit drei 400-MW-Maschinengruppen bestückt.

Die geologischen und technischen Unsicherheiten bezüglich der Anordnung dieser verschiedenen Bauteile innerhalb der metamorphisierten und tektonisierten (Überschiebungen, Verschuppungen) Serien des Pennins veranlaßten den Bauherrn, entsprechende Aufschlußarbeiten (Bohrungen, Erkundungsstollen, Versuchsblöcke) durchzuführen. Diese Vorarbeiten wurden vor allem im Bereich des künftigen Schrägschachts verdichtet, dessen Bau ohne Zweifel die größte Aufmerksamkeit erfordern wird.

1.2 *Ziel der Injektionsversuche*

Die Injektion der Versuchsblöcke Tracouet und Drotché verfolgte - in fallender Reihenfolge der Wichtigkeit - folgende Zwecke:

1. verschiedene Injektionstechniken versuchsmäßig anzuwenden, die während des Vortriebs in lokal stark verwitterten und tektonisierten (mylonitisierten) Gesteinen zur Anwendung gelangen könnten;

2. die Größenordnung der Verbesserung der mechanischen Gebirgseigenschaften im injizierten Bereich und den Einfluß der Konsolidierung eines Gebirgsrings auf das Verhalten des Felshohlraums abzuschätzen;

3. die Interpretationsmöglichkeiten der Bohrparameteraufzeichnungen (BPA) zu testen, um festzustellen, ob diese sinnvoll als systematisches Erkundungsverfahren beim Vortrieb angewandt werden könnten;

4. die geometrischen Zwänge zu definieren,

die sich durch die Durchführung von Konsolidierungs- und Abdichtungsinjektionen während des Vortriebs, in etwa 20-Meter-Abschnitten, ergeben, um die künftige Tunnelbohrmaschine nach Möglichkeit an diese Art von Behandlung anpassen zu können;

5. eine Reihung der in den vorliegenden Gebirgsarten bestgeeigneten Behandlungsmethoden vorzunehmen, in Abhängigkeit von den erzielten Ergebnissen hinsichtlich einer Verbesserung der mechanischen Eigenschaften und der Abdichtung des Gebirges;

6. eine Kosten- und Zeitschätzung für die Injektionen vorzunehmen;

7. die von den Unternehmen eingesetzten Geräte und Methoden zu testen.

2 BESCHREIBUNG DER INJEKTIONSARBEITEN

2.1 *Geologische Verhältnisse*

Die Geologie der Versuchsblöcke wurde vom Technischen Büro Norbert auf Grund von Kern- und Vollkronenbohrungen ermittelt.

Der Versuchsblock Tracouet liegt in tektonisierten Serizit- und Chloritschiefern mit untergeordneten Myloniten des Phyllit-Verrucano Fig. 1a und 2).

Der Versuchsblock Drotché liegt zum Teil in arenisierten Quarziten, zum Teil in tektonisierten und verwitterten Serizit- und Chloritschiefern des Rand-Trias (Fig. 1b).

Die beiden Versuchsblöcke unterscheiden sich auch hinsichtlich der Überlagerung, die in

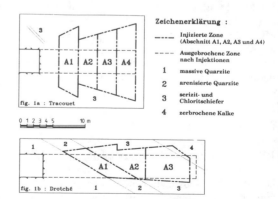

Fig. 1a und 1b: Längsprofile der Versuchsblöcke
Longitudinal sections of the test blocks

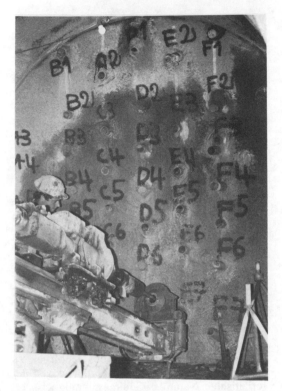

Fig. 2: Tracouet - Bohransatzpunkte / Drilling
pattern

Tracouet maximal 80 m beträgt, während sie in
Drotché 350 m übersteigt.

Die mechanischen Kennwerte der während
der Arbeiten angetroffenen Felstypen sind in
der Fig. 3 zusammengestellt.

2.2 Geometrie der Versuchsblöcke

Die beiden Versuchsblöcke wurden quer zum
Streichen der Schichten angeordnet. In Anbetracht der lokalen geometrischen Verhältnisse
wurde ihre Länge auf 15 m bzw. 23 m beschränkt. In Tracouet betrug die Dicke der Puf-
ferzone (nicht injizierte Sicherheitszone zwischen Ortsbrust und dem ersten Verpreß-
abschnitt) 3 m, in Drotché im Mittel 7 m. Der
theoretische Durchmesser des Verpreßkörpers
betrug am Ende beider Versuchsblöcke 8 m .

3 ABLAUF DER ARBEITEN

3.1 Allgemeines

Das Projekt für die Injektionsarbeiten (Erstellen
des Bohrplans, Wahl der Behandlungsmethode)
wurde von den beiden mit den Arbeiten
betrauten Unternehmen (SIF-Groutbor/Bachy/
Solétanche und Stump-Rodio) ausgearbeitet
und vorgeschlagen . Es sei darauf hingewiesen,
daß diese nach Augenschein der lokalen geologischen Verhältnisse ihre ursprüngliche Absicht, die Injektionen mittels Manschettenrohren
durchzuführen, fallen ließen und im Hinblick auf
einen gesicherten Arbeitsablauf für ein Verpressen mit offenen Bohrungen (Packer am Bohrlochmund) optierten.

3.2 Versuchsblock Tracouet

Der Versuchsblock Tracouet wurde in vier je 3
m lange Abschnitte unterteilt (theoretisch be-
handeltes Gebirgsvolumen 880 m^3) und der
Bohrplan auf eine Behandlung in drei zeitlich
getrennten Phasen (Bohrung plus Wiederaufbohren für jeden Abschnitt und jede Behandlungsphase) ausgelegt. Die Injektion wurde mit Bentonit-Zement-Mischungen in den
Phasen 1 und 3 und mit einem Verpreßgut mit
verbesserter Eindringung in der dazwischen-
liegenden Phase durchgeführt. Die Verpreßdrücke betrugen 1 bis 3 MPa in den beiden ersten Phasen, die das Verfüllen und Abdichten
der Klüfte zum Ziel hatten, und wurden in der
dritten Phase zur Aufsprengung des Gebirges
("claquage") bis auf 10 MPa gesteigert. Die
Charakteristika der verschiedenen Injektionsphasen sind in der Fig. 4 angeführt.

3.3 Versuchsblock Drotché

Auf Grund der Gebirgsinhomogenitäten umfaßte der Versuchsblock Drotché nur drei Abschnitte variabler Länge (theoretisch behandel-
tes Volumen 470 m^3), und der Bohrplan wurde
auf eine Behandlung in zwei getrennten
(Abschnitt 1) bzw. zeitlich gruppierten Phasen
(Abschnitte 2 und 3; ohne Wiederaufbohren
zwischen beiden Phasen) ausgerichtet. Bei der
Verpressung des Versuchsblocks Drotché
kamen drei Arten von Verpreßgut zur
Anwendung: eine Bentonit-Zement-Mischung,

eine solche mit verbesserter Eindringung und eine Mikrofeinzement-Suspension. Die Charakteristika der verschiedenen Injektionsphasen sind in der Fig. 5 zusammengestellt.

Unter Berücksichtigung des oben Angeführten erforderte der Bohrplan für den Block Tracouet 1450 lfm Bohrungen (einschließlich Wiederaufbohren), jener für den Block Drotché 1024 lfm.

3.4 Kontrolle der Injektion

An beiden Versuchsorten erfolgte die Kontrolle der Injektionen mittels der von den jeweiligen Unternehmen entwickelten Datenerfassungsanlagen. In Abhängigkeit von den eingegebenen Sollwerten üben diese Anlagen eine automatische und kontinuierliche Überwachung des Injektionsvorgangs aus und zeichnen gleichzeitig seinen Ablauf auf.

Die zahlreichen Möglichkeiten, die diese Anlagen sowohl zur Kontrolle der Injektionsarbeiten (Veranschaulichung des laufenden Vorgangs, Steuerung und automatisches Abschalten der Pumpen bei Überschreitung der Vorgabewerte) als auch für die Baustellendokumentation (numerische und graphische Synthese der Ergebnisse) bieten, gewährleisten allen Beteiligten Sicherheit und Qualität.

4 ERGEBNISSE DER INJEKTIONSVERSUCHE

Die Ergebnisse der beiden Injektionskampagnen wurden unabhängig voneinander auf Grund der vor und nach der Behandlung durchgeführten Messungen (Pressiometer- und Lugeonversuche, Aufzeichnung der Bohrparameter) geprüft. Die beim Ausbruch der beiden Versuchsblöcke gemachten Beobachtungen erlaubten, das Verhalten des injizierten Gebirges in natürlicher Größe zu beurteilen.

4.1 Die Injektionen in Zahlen

In Tracouet betrug der mittlere Verfüllungsgrad für alle Verpreßgutarten 9 %; unter Berücksichtigung der Zusammensetzung der verschiedenen Mischungen entspricht dies einer Aufnahme von 174 kg Zement pro lfm Bohrloch.

In Drotché belaufen sich die entsprechenden Werte auf 8 % und 29 kg/lfm in den arenisierten Quarziten bzw. auf 7 % und 30 kg/lfm in den Serizit- und Chloritschiefern.

Anmerkung: Obwohl die Zahlen für den Verfüllungsgrad in beiden Blöcken durchaus vergleichbar sind, ist die pro Laufmeter Bohrloch in Drotché aufgenommene Zementmenge wegen des geringen Zementgehalts der verwendeten Mischungen erheblich niedriger.

4.2 Bewertung der Ergebnisse

Die zahlreichen Beobachtungen beim Ausbruch des Versuchsblocks Tracouet zeigten ein sehr heterogenes und und durch zentimeterdicke Mylonithorizonte wechselnder Orientierung abgekapseltes Gebirge. Die verschiedenen Injektionsmischungen drangen vornehmlich in die Klüfte und in geringerem Ausmaß in die Schieferung ein, keinesfalls jedoch in die mylonitisierten Horizonte, und dies unabhängig vom Verpreßdruck.

Außerdem gelang es nicht, durch die Verpressung die Zugfestigkeit des Gebirges zu erhöhen (Verkitten der Kluftkörper wegen der serizitischen Natur des Gebirges unmöglich).

Trotz dieser Schwierigkeiten brachten die Injektionsarbeiten folgende Verbesserungen des Gebirges:

1. Verfüllung der verschiedenen Kluftscharen, Bildung eines Netzes verästelter Mikro-Aufsprengungen normal zur Schieferung und Mikro-Aufsprengungen parallel dazu;

2. starke Verringerung der Gebirgsdurchlässigkeit von durchschnittlich 100 auf 2 Lugeon-Einheiten;

3. Vergrößerung und Homogenisierung des Bohrwiderstands, Bestätigung der Gebirgsabdichtung (auf Grund des Vergleichs der Bohrparameter);

4. bedeutende Erhöhung des Pressiometer-Moduls um den Faktor 4 (mittlerer Wert nach Injektion 175 MPa);

5. Verringerung der Konvergenzen um einen Faktor 4 bis 5 (2 bis 5 mm Konvergenz nach Injektion).

Die am Block Tracouet angewandte Behandlungsmethode erwies sich also als erfolgreich. Sie erlaubte, die angestrebten Ziele ohne besondere Schwierigkeiten - mit Ausnahme der Langsamkeit und folglich der Kosten - zu erreichen. Leider gelang es nicht, durch die Verpressung die Baustellensicherheit entscheidend zu erhö-

Fig. 3

Versuchsblöcke Tracouet und Drotché : mittlere Felskennwerte

	Verrucano			Rand-Trias			
	tektonisiert	stark tektonisiert	mylonitisiert	zerbrochene Quarzite	arenisierte Quarzite	tektonisierte und verwitterte Schiefer	zerbrochene Kalke
Einachsige Druckfestigkeit [MPa]	10.4	5.0	0.2	20.0	ne	ne	10.0
Cohäsion [MPa]	3.0	1.6	0.1	5.2	0.5	ne	2.6
Innerer Reibungswinkel [°]	30	30	25	35	35	ne	35
Elastizitätsmodul [MPa]	3'000	3'000	100	5'000	1'500	ne	4'000
Poisson-Koeffizient (angenommen)	0.4	0.4	0.4	0.3	0.3	ne	0.3

ne : nicht ermittelt

Fig. 4

Injektionsphasen

1 : Grobverfüllung des Gebirges mittels Primärbohrungen
2 : Verpressen der feinen Klüfte mittels Sekundärbohrungen
3 : Aufsprengen des Gebirges mittels Tertiärbohrungen

Abschnitte des Versuchsblocks	Tiefe	Injektionsvorgaben	1	2	3
PUFFERZONE	0 m – 3 m	nicht injiziert			
ABSCHNITT 1 : Serizit- und Chloritschiefer	3 m – 6 m	Vorgang Nr.	1	5	9
		Art des Verpressguts	B - Z	mvE	B - Z
		Z / W =	1	1.5	1
		σ c (28 T.)	10 bis 12 MPa	15 bis 20 MPa	10 bis 12 MPa
		Max. Inj.-Druck :	0,8 MPa	2 MPa	5 MPa
		Max. Fördermenge :	1200 l/h	1200 l/h	2500 l/h
		Max. Volumen :	666 l/lfm	333 l/lfm	333 l/lfm
		Tatsächlich inj. Vol. :	160 l/lfm	60 l/lfm	100 l/lfm
ABSCHNITT 2 : Serizit- und Chloritschiefer	6 m – 9 m	Vorgang Nr.	2	6	10
		Art des Verpressguts	B - Z	mvE	B - Z
		Z / W =	1	1.5	1
		σ c (28 T.)	10 bis 12 MPa	15 bis 20 MPa	10 bis 12 MPa
		Max. Inj.-Druck :	2 MPa	3 MPa	10 MPa
		Max. Fördermenge :	1200 l/h	1200 l/h	2500 l/h
		Max. Volumen :	666 l/lfm	333 l/lfm	333 l/lfm
		Tatsächlich inj. Vol. :	270 l/lfm	180 l/lfm	200 l/lfm
ABSCHNITT 3 : Serizit- und Chloritschiefer	9 m – 12 m	Vorgang Nr.	3	7	11
		Art des Verpressguts	B - Z	mvE	B - Z
		Z / W =	1	1.5	1
		σ c (28 T.)	10 bis 12 MPa	15 bis 20 MPa	10 bis 12 MPa
		Max. Inj.-Druck :	2 MPa	3 MPa	10 MPa
		Max. Fördermenge :	1200 l/h	1200 l/h	2500 l/h
		Max. Volumen :	666 l/lfm	333 l/lfm	333 l/lfm
		Tatsächlich inj. Vol. :	290 l/lfm	245 l/lfm	105 l/lfm
ABSCHNITT 4 : Serizit- und Chloritschiefer	12 m – 15 m	Vorgang Nr.	4	8	12
		Art des Verpressguts	B - Z	mvE	B - Z
		Z / W =	1	1.5	1
		σ c (28 T.)	10 bis 12 MPa	15 bis 20 MPa	10 bis 12 MPa
		Max. Inj.-Druck :	2 MPa	3 MPa	10 MPa
		Max. Fördermenge :	1200 l/h	1200 l/h	2500 l/h
		Max. Volumen :	666 l/lfm	333 l/lfm	333 l/lfm
		Tatsächlich inj. Vol. :	240 l/lfm	285 l/lfm	35 l/lfm

B - Z : herkömmliche Bentonit-Zement-Mischung   mvE : Verpressgut mit verbesserter Eindringung

Fig. 4 : Versuchsblock / Test block Tracouet : Injektionsplan / layout of grouting

---

Fig. 5

Injektionsphasen

1a : Grobverfüllung des Gebirges mittels Primärbohrungen
1b : Aufsprengen des Gebirges mittels Primärbohrungen
Kontinuierliche Injektion zwischen beiden Phasen
2 : Verpressen der feinen Klüfte mittels Sekundärbohrungen

Abschnitte des Versuchsblocks	Tiefe	Injektionsvorgaben	1	2
PUFFERZONE	0 m – 7 m	nicht injiziert		
ABSCHNITT 1 : arenisierte Quarzite	7 m – 10 m	Vorgang Nr.	1	2
		Art des Verpressguts	mvE	MFZ
		Z / W =	0.3	0.3
		σ c (28 T.)	0,2 bis 0,4 MPa	1,5 bis 2 MPa
		Max. Inj.-Druck :	1 MPa	0,5 MPa
		Max. Fördermenge :	400 l/h	350 l/h
		Max. Volumen :	130 l/lfm	195 l/lfm
		Tatsächlich inj. Vol. :	85 l/lfm	125 l/lfm
ABSCHNITT 2 : Serizit- und Chloritschiefer	10 m – 16 m	Vorgang Nr.	3a	3b
		Art des Verpressguts	mvE	B - Z
		Z / W =	0.3	1
		σ c (28 T.)	0,2 bis 0,4 MPa	10 bis 15 MPa
		Max. Inj.-Druck :	3 MPa	3,5 MPa
		Max. Fördermenge :	440 l/h	400 l/h
		Max. Volumen :	63 l/lfm	190 l/lfm
		Tatsächlich inj. Vol. :	90 l/lfm	9,0 l/lfm
ABSCHNITT 3 : Serizit- und Chloritschiefer	16 m – 23 m	Vorgang Nr.	4a	4b
		Art des Verpressguts	mvE	B - Z
		Z / W =	0.3	1
		σ c (28 T.)	0,2 bis 0,4 MPa	10 bis 15 MPa
		Max. Inj.-Druck :	3 MPa	3,5 MPa
		Max. Fördermenge :	440 l/h	400 l/h
		Max. Volumen :	85 l/lfm	255 l/lfm
		Tatsächlich inj. Vol. :	105 l/lfm	0,5 l/lfm

MFZ : Mikrofeinzement-Verpressgut

Fig 5 : Versuchsblock / Test block Drotché : Injektionsplan / layout of grouting

453

hen, so daß der Vortrieb nicht ohne Sicherungsmaßnahmen geschehen konnte (Nachfall von Blöcken aus der Kalotte).

Während des Ausbruchs des Versuchsblocks Drotché zeigte sich, daß das Eindringen der Injektionsmischungen in die arenisierten Quarzite durch das Vorhandensein dünner undurchlässiger schiefriger Horizonte begrenzt wurde. Außerdem erwies sich der sandig-schluffige Anteil der arenisierten Quarzite nicht imprägnierbar. Die Standfestigkeit dieser Feinmaterialien hätte jedoch voraussichtlich durch Aufsprengungsinjektionen verbessert werden können.

Auf Grund dieser ungleichmäßigen Behandlung blieben die arenisierten Quarzite insgesamt, trotz trockener Gebirgsverhältnisse, verhältnismäßig nachbrüchig.

Die tektonisierten und verwitterten Serizit- und Chloritschiefer des Rand-Trias stellten ein lithologisch sehr heterogenes Gebirge mit geringer Durchlässigkeit quer zur Schieferung und hoher Durchlässigkeit parallel dazu dar (Vorkommen von kiesigen arenisierten Quarzithorizonten). Diese stark tektonisierten Gesteine hatten auch eine hydrothermale Verwitterung erlitten, wodurch die seltenen, an der Ortsbrust verschiedentlich sichtbaren Kluftnetze verstopft wurden. Gleiches gilt für die millimetergroßen Hohlräume, die in den Serizit-Chloritschiefern angetroffen wurden; sie wiesen keine Verbindung untereinander mehr auf.

Mit Ausnahme der kiesigen arenisierten Quarzithorizonte erwiesen sich diese tektonisierten und verwitterten Serizit-Chloritschiefer für jedwede Imprägnierung als praktisch undurchlässig.

Die in Drotché erzielten Ergebnisse können wie folgt zusammengefaßt werden:

Arenisierte Quarzite:
1. Nur die kiesigen Horizonte konnten imprägniert werden;
2. die hohe Durchlässigkeit dieses Gebirges konnte nicht maßgeblich verringert werden;
3. deutliche Erhöhung und Homogenisierung des Bohrwiderstandes;
4. der mittlere Pressiometermodul wurde um einen Faktor 2,3 vergrößert (nach Injektion 100 MPa).

Tektonisierte und verwitterte Serizit-Chloritschiefer:
1. Die zahlreichen Vorkommen von kiesigen arenisierten Quarziten innerhalb der Serizit-

Chloritschiefer konnten teilweise imprägniert werden;
2. nur die seltenen nicht durch hydrothermale Verwitterungsprodukte verstopften klüftigen Schieferhorizonte konnten imprägniert werden;
3. die schieferungsparallele Gebirgsdurchlässigkeit blieb weiterhin sehr hoch;
4. deutliche Erhöhung und Homogenisierung des Bohrwiderstandes;
5. der mittlere Pressiometermodul wurde um einen Faktor 2,8 vergrößert (nach Injektion 190 MPa).

Der Ausbruch des Versuchsblocks Drotché erforderte umfangreiche Gebirgsstützungsmaßnahmen während des Vortriebs, gefolgt von einer Betonierung der Sohle (in Drotché nicht injiierte Zone) und der Ulmen (Nachbrüche in den arenisierten Quarziten und Verformungen im Zentimeterbereich).

Die in Drotché durchgeführten Injektionen erlaubten folglich nicht, alle angestrebten Ziele zu erreichen. Die gewählte Behandlungsmethode erwies sich für die angetroffenen Gebirgsverhältnisse unangemessen, da zu sehr auf ihr Absorptionspotential ausgerichtet. Aufsprenginjektionen bleiben das beste Mittel, um die geotechnischen Kennwerte dieses Gebirges zu verbessern. Doch hätten dazu die Verpreßdrücke wesentlich erhöht werden müssen.

Schließlich zeigte sich auch, daß die verwendeten Imprägnierungsmischungen einen zu geringen Zementgehalt aufwiesen (Druckfestigkeit von 0,2 bis 0,4 MPa), um die Konsolidierung des behandelten Gebirges zu gewährleisten.

Dieser teilweise Mißerfolg muß jedoch wegen der zahlreichen angetroffenen technischen Schwierigkeiten (sehr starke Heterogenität in Bezug auf Lithologie und Korngrößen, starke Abkapselung des Gebirgskörpers, geringe Durchlässigkeit, verstopfte Klüfte) relativiert werden; mangels ausreichender Kenntnisse waren diese bei der Ausarbeitung des Injektionsprojekts unterschätzt worden.

4.3 Verwendung von Manschettenrohren

Neben den im offenen Bohrloch "von oben nach unten" durchgeführten Hauptarbeiten wurden in Tracouet und Drotché Injektionsversuche mit Manschettenrohren vorgenommen.

In Tracouet öffneten sich 21 % der Manschetten im massiven und harten bzw. 85 % im stark geklüfteten bis tektonisierten Phyllit-Verrucano mit untergeordneten Myloniten.

Das in Drotché erzielte Ergebnis ist noch ermutigender, da alle Manschetten geöffnet und injiziert werden konnten.

Es sei noch erwähnt, daß das zerbrochene bis tektonisierte Gebirge bei diesen Versuchen vergleichbare Mengen von Verpreßgut aufnahm wie bei den nahegelegenen Hauptarbeiten.

Auf Grund dieser Erfahrungen darf erwartet werden, daß Manschettenrohre mit guten Erfolgsaussichten bei der Behandlung zerbrochenen und tektonisierten Gebirges zur Anwendung gelangen können.

5 FÜR DAS AUFFAHREN DES PANZERSCHACHTS VORGESEHENE BEHANDLUNGSMETHODE

Auf Grund der in den geologischen Verhältnissen des Projekts gewonnenen Erfahrungen können folgende Grundprinzipien für eine qualitativ hochstehende, in vernünftiger Zeit durchführbare Gebirgsbehandlung bei einem Bohrvortrieb angegeben werden:

1. Der Bohrplan umfaßt mindestens zwei konzentrische Kreise (ringförmige Behandlung), wobei der Bohrlochabstand am Blockende 1,5 bis 2 m möglichst nicht übersteigt. Ein dritter Bohrlochkreis muß vorgesehen werden, wenn eine der vorgesehenen Behandlungsphasen eine Abschirminjektion benötigt.

2. Im Normalfall sind zwei Injektionsphasen notwendig und hinreichend, um eine gegebene Gebirgsformation ausreichend zu konsolidieren und abzudichten.

- *Phase 1:* Verfüllung der Klüfte und/oder Imprägnierung des Gebirges
- *Phase 2:* Verdichten des Gebirges durch Aufsprenginjektion. Diese zweite Behandlungsphase kommt erst nach Abbinden des zuvor injizierten Verpreßguts zur Anwendung.

Dieses grundsätzliche Schema muß natürlich den lokalen geologischen Verhältnissen angepaßt werden. So könnte bei undurchlässigem trockenem Gebirge eine einzige Behandlungsphase ausreichen. Wenn hingegen die Konsolidierung des Gebirges durch eine besonders wirksame Abdichtung ergänzt werden muß, wird wahrscheinlich die Zahl der Injektionsphasen erhöht werden müssen.

3. Die erste Behandlungsphase erfolgt mit einer genügend festen (Mindestdruckfestigkeit 5 MPa) Mischung mit verbesserter Eindringung. Eine primäre Verfüllinjektion mit einer klassischen Bentonit-Zement-Suspension dringt nämlich nicht ausreichend gut in die feinen Klüfte ein und erfordert eine zusätzliche Behandlungsphase mit einer Mischung mit verbesserter Eindringung. Aus Kosten- und Zeitgründen ist es günstiger, diese beiden Phasen der Verfüllung und/oder Imprägnierung in eine einzige zusammenzuziehen und die Behandlung mit dem bestgeeigneten Verpreßgut durchzuführen. Dieses wird mit einem Druck von etwa 3 MPa injiziert, wobei die Länge der Behandlungsabschnitte auf etwa 5 m beschränkt wird.

Die zweite Phase (Aufsprengen) erfolgt mit einer klassischen Bentonit-Zement-Mischung unter sehr hohem Druck (> 8 MPa). Die Länge der Abschnitte wird auf etwa 2 m verringert.

Für den Fall, daß eine zusätzliche Behandlung erforderlich wird, kommt ein besonders dünnflüssiges Verpreßgut zur Anwendung. Diese Injektion zur Verbesserung der Gebirgsabdichtung erfolgt vor der Aufsprenginjektion, und die Verpreßdrücke werden auf 1 bis 2 MPa beschränkt.

4. Zur Begrenzung von Zeit und Kosten der Injektionsarbeiten wird einer Behandlung mittels Manschettenrohren der Vorzug gegenüber anderen Injektionsmethoden gegeben, sofern die bei diesem Verfahren vorhandenen Risiken vernünftig eingegrenzt werden können. Im gegenteiligen Fall wird eine Injektion mit offenem Bohrloch "von unten nach oben" bevorzugt gegenüber einer gemischten Methode (offenes Bohrloch/Manschettenrohre mit Packersäcken), sofern nur das Bohrloch ausreichend standfest und die Gefahr einer Packerumläufigkeit gering ist. In den meisten Fällen sind diese beiden Forderungen nach der ersten Behandlungsphase erfüllt.

Wenn schließlich die geringe Standfestigkeit der Bohrung keine andere Lösung zuläßt, wird die Verpressung mit offenem Bohrloch "von oben nach unten" durchgeführt.

6 FALLBEISPIELE FÜR DIE KONSOLIDIERUNG MYLONITISIERTEN GEBIRGES

Zur Abrundung der obigen Ausführungen und zum Vergleich seien im Folgenden einige Beispiele für die Bewältigung von Störungs-

oder ähnlichen Lockergesteinszonen im Tunnel- und Stollenbau kurz vorgestellt:

Beim Bau des Pumpspeicherwerks *Lago d'Avio-Edolo* in Oberitalien mußte 1978/79 die weiträumig durchziehende Tonale-Störung in den beiden Schrägschächten durchquert werden. Das Gebirge präsentierte sich als stark drückender, tektonisierter bis mylonitisierter Glimmerschiefer und erforderte umfangreiche Konsolidierungsinjektionen, die im wesentlichen mit Bentonit-Zement-Mischungen erfolgreich durchgeführt wurden.

Die größten Schwierigkeiten beim Auffahren von Störungszonen rühren meist von Wasser unter hohem Druck her, das die wenig oder überhaupt nicht verkitteten Mylonite oder Kataklasite in einen kaum beherrschbaren Brei verwandelt und des öfteren verheerende Schlammeinbrüche verursacht hat. Typische Beispiele dafür sind der Unterwasserstollen *Guavio* in Kolumbien (Egger 1988) oder die *Valle-Fredda-Störung* des Gran-Sasso-Tunnels (Boutitie & Lunardi 1975); ein ähnliches geomechanisches Verhalten zeigten auch die lockeren vulkanischen Tuffe im *Nakayama-Tunnel* (Egger et al. 1982). In allen Fällen bestanden die zur Durchörterung dieser Zonen ergriffenen Maßnahmen in einer Kombination von Dränagen und Injektionen Bentonit-Zement- und Silikatmischungen).

Im Triebwasserstollen der Anlage *Hongrin-Léman* wurde in einer ausgedehnten Störung praktisch undurchlässige zermalmte Rauhwacke unter 11 MPa Wasserdruck angetroffen (Barbedette 1968), die im Gegensatz zu den obigen Beispielen im Schutz einer Baugrundvereisung aufgefahren wurde.

7 SCHLUSSFOLGERUNGEN

Im Rahmen des Bauvorhabens der Wasserkraftanlage Cleuson-Dixence wurden in zwei Blökken Injektionsversuche unternommen. Diese erlaubten, eine Behandlungsmethode für eine lokale Konsolidierung der Mylonitbereiche während des Vortriebs - insbesondere eines Bohrvortriebs - zu definieren.

Diese Injektionsversuche zeigten, daß die geotechnischen Eigenschaften der stark tektonisierten Serizit- und Chloritschiefer durch eine Kombination von Verfüllungs- und Aufsprenginjektionen verbessert werden können. Durch die Verpressung erhöhte sich der Pressio-

metermodul um einen Faktor 3 bis 4, während die Konvergenzen des Stollens im selben Maß zurückgingen. Hingegen verbesserte sich die lokale Standsicherheit beim Ausbruch praktisch nicht (Ablösen von Blöcken in der Kalotte), da die Zugfestigkeit des Gebirges nicht erhöht werden konnte.

Die Beobachtungen beim Ausbruch im verpreßten Gebirge zeigten, daß die injizierten Mischungen zwar in die Klüfte und die Schieferung eingedrungen waren, wogegen die mylonitisierten, selbst dünnen Horizonte unüberwindliche Hindernisse für die Verpreßmischungen darstellten, unabhängig vom Injektionsdruck. Die Dichte der Injektionsbohrungen muß folglich ausreichend groß sein, um dieser Abkapselung des Gebirges entgegenzuwirken.

Am Abschluß dieser Versuche sei dem Bauherrn (Energie de l'Ouest Suisse SA) für das Vertrauen, das er uns während der Durchführung dieser Erkundungsarbeiten erwiesen hat, und für die Erlaubnis der vorliegenden Veröffentlichung aufrichtig gedankt.

SCHRIFTTUM

Barbedette R. 1968. La congélation des sols. *Schweiz. Bauzeit.* 86: 3-7.

Bonnard & Gardel ing.-cons. SA 1992. Galerie de reconnaissance de Tracouet: rapport sur les injections du plot d'essais TR 186: 1-43.

Bonnard & Gardel ing.-cons. SA 1992. Fenêtre 3a du Drotché (Aménagement Grande Dixence): rapport sur les injections du plot d'essais DR 426: 1-45.

Bonnard & Gardel ing.-cons. SA 1992. Injection des plots d'essais TR 186 (Tracouet) et DR 426 (Drotché): rapport de synthèse: 1-9.

Boutitie, J. & P. Lunardi 1975. Tunnel autoroutier du Gran Sasso. Traversée de la faille de la Valle Fredda. *Travaux*: 50-58.

Egger, P. 1988. Ground improvement measures for crossing a large, heavily water-bearing fault zone. *Proc. Int. Conf. ITA*: 985-990. Madrid.

Egger, P., T. Ohnuki & Y. Kanoh 1982. Bau des Nakayama-Tunnels. Kampf gegen Bergwasser und vulkanisches Lockergestein. *Rock Mech. Suppl. 12*: 275-293.

Grouting in Rock and Concrete, Widmann (ed.) © 1993 Balkema, Rotterdam, ISBN 90 5410 350 7

Transient pressure analysis of 'RODUR' epoxy grouting in concrete and rock at Kölnbrein Dam, Austria

Untersuchungen zur Kluftinjektion und ihren Beziehungen zu physikalischen Modellen der Erdöllagerstättentechnik

Gert Stadler
Insond Gesellschaft mbH, Neumarkt, Österreich

ZUSAMMENFASSUNG: DIE RECHNERISCHEN GRUNDLAGEN, WELCHE GEEIGNET WÄREN DAS FLIEßEN VON INJEKTIONSGUT IN KLÜFTEN ZU BESCHREIBEN, VERFEHLEN DERZEIT NOCH DIE ANWENDUNGSREIFE IM SINNE EINER KONKRETEN NÜTZLICHKEIT FÜR PRAKTISCHE FESTLEGUNGEN VOR ORT. AUSGEHEND VON EINEM NEUARTIGEN INJEKTIONSKRITERIUM VON LOMBARDI, WURDE AM PROJEKT KÖLNBREIN DER ÖSTERR. DRAUKRAFTWERKE VERSUCHT, DIE ANWENDBARKEIT VON PRAKTIKEN AUS DER ERDÖLTECHNOLOGIE ZU PRÜFEN. BESONDERS BEIM INJIZIEREN VON KUNSTHARZEN WURDE DABEI GEFUNDEN, DAß EINE VERBESSERUNG DES WIRKUNGSGRADES, DER REICHWEITEN UND AUSNUTZUNG DES VORGEWÄHLTEN BOHRRASTERS MÖGLICH WIRD.
DIE BEOBACHTUNG INSTATIONÄRER DRUCKZUSTÄNDE WÄHREND DES INJEKTIONS-VORGANGES FÜHRT ZU EINEM "DRUCKFÜHLIGEN" INJIZIEREN. ÜBER DAS ERKENNEN "WIRKSAMER DRÜCKE" IST EINE REALISTISCHE FESTLEGUNG DER NOTWENDIGEN, UND GLEICHZEITIG, UNSCHÄDLICHEN MAXIMALDRÜCKE MÖGLICH GEWORDEN; UND DAMIT DIE EFFEKTIVITÄT DER BEHANDLUNG WESENTLICH VERBESSERBAR GEWORDEN.

SUMMARY: THE FOLLOWING PRESENTATION REFERS TO STUDIES UNDERTAKEN DURING EXTENSIVE GROUTING WORKS AT THE KÖLNBREIN DAM REPAIR PROJECT OF ÖSTERREICHISCHE DRAUKRAFTWERKE UNDER THE PROJECT-AUTHORSHIP OF LOMBARDI. THE INNOVATIVE GROUTING CRITERIA INTRODUCED BY HIM, LED TO FURTHER INVESTIGATIONS BY THE AUTHOR IN APPLYING PETROLEUM ENGINEERING RESERVOIR TECHNIQUES ONTO THE FIELD OF SPECIAL EPOXY GROUTING.
THESE INVESTIGATIONS RESULTED IN THE SUCCESSFUL INTRODUCTION OF OIL-WELL TESTING PROCEDURES FOR THE INTERPRETATION OF GROUT FLOW WHILE CARRYING OUT FISSURE EPOXY GROUTING. THE NEW TECHNIQUES DO HELP IN OBTAINING AN INTIMATE KNOWLWDGE OF EFFECTIVE PRESSURES, REMANENT FORCES IN FISSURES, AND A NEW "DIAGNOSTICS-APPROACH" INTO DISTINGUISHING BETWEEN NECESSARY AND DANGEROUS GROUT PRESSURES, AS WELL AS IN DISCERNING BETWEEN USEFULLY ENGINEERED AND UNACCEPTABLE FISSURE MOVEMENTS.

Seit fast 200 Jahren ist das Injizieren von Boden und Fels Bestandteil der Konstruktions- und Bautätigkeit im Bergbau und Bauwesen.

Durch das Einpressen von sich verfestigenden Suspensionen, Lösungen und Kunststoffen werden Poren und Klufträume gefüllt, Gas- und Wasserwege abgedichtet, Korn- und Gesteinsverband kohäsiv verbunden, und dadurch mit Scherfestigkeit versehen.

Etwa seit 1980 werden umweltbelastende Injektionsstoffe abgelehnt. Damit scheiden viele Möglichkeiten der Behandlung feinkörniger Sedimente und Felsklüfte geringer Öffnungsweite mit bislang herkömmlichen Lösungen und Kunststoffen aus.

Im Falle von Lockergestein übernimmt deren Behandlung vermehrt die, seit etwa 1970 etablierte, Hochdruckboden-vermörtelung (Jet-Grouting).

Im Fels weicht man auf Verwendung von umweltfreundlichen Kunststoffen (z.B. lösungmittelfreie Harze) und stabilen, kohäsiven Feinstbindemittelsuspensionen als Injektionsmittel aus.

Großes Augenmerk richtet sich dabei auf die Verbesserung der Injektionstechniken; und es werden umfangreiche Versuche unternommen, die zum Ziel haben, die sich über Jahrzehnte herausgebildeten, individuellen Feldpraktiken theoretisch zu belegen, und einem allgemeinen Verständnis zuzuführen.

Dabei werden diesen Studien Annahmen zugrundegelegt über:

- Rißweiten, Rißgeometrien
- Anzahl und Verteilung der Risse je Längen- oder Volumeinheit
- Durchtrennungsgrad (Verhältnis offener Rissflächen zur Gesamtfläche)
- Rauhigkeit und Welligkeit
- hydraulische Verhältnisse außerhalb der Injektionsfront
- Verformbarkeit der Kluftränder
- Rheologie der Injektionsmischungen u.v.m.;

Über Laborversuche werden gleichzeitig Bestätigungen dieser Zusammenhänge nach Gesetzen der Ströhmungslehre gesucht.

Die Feldarbeit selbst geht inzwischen, scheinbar unbeeinflußt, ihren traditionellen, eher handwerklichen Weg weiter;

und dies nur, weil einerseits alle erwähnten Parameter, soweit sie im Feld überhaupt ermittelbar sind, sich räumlich mit der Entfernung von der Injektionsstelle ändern, und damit im rechnerischen Sinne unbekannt bleiben;

und andererseits, weil damit allen theoretischen Überlegungen die querverbindende Bestätigung zur Injektionspraxis fehlt.

Die nachfolgende Ausarbeitung soll Bestandsaufnahme über die Möglichkeiten bieten, wie man doch zu quantitativen Aussagen bei Injektionsarbeiten kommen könnte;

andererseits soll sie aber auch neue Wege aufzeigen, wie z.B. physikalische Modelle aus der Erdöllagerstättentechnik auf Injektionsvorgänge im Bauwesen Anwendung finden können, und damit zu einer besseren Kenntnis von Injektionsabläufen geführt wird.

Damit soll die Steuerung des Injizierens schon während der Durchführung, und die frühe Beurteilung eines Erfolges erleichtert werden.

Hier wird also u.a. auf Fragen eingegangen, die sowohl dem Planer wie dem Injekteur im Zuge der Ausführung immer wieder begegnen:

- ob Kanalfluß, ebene oder sphärische Ausbreitung von der Injektionspasse aus in das Kluftgebäude stattfinden,
- ob aus dem Bohrloch über enge Klüfte große Durchlässigkeiten angespeist werden oder umgekehrt,
- ob das Injektionsgut dabei ist, eine sich allseits verjüngende Kluft zu füllen, und was unternommen werden muß, damit unnützes Rißaufweiten verhindert wird.

- Welches dabei diese "gefährliche" Druckschwelle eigentlich ist;
- wo und wie solche kritischen Injektionsdrücke erkannt und gemessen werden können.

- Ob das Injektionsgut eben im Begriff ist, sich durch einen unbeobachteten Austritt an die Oberfläche zu ergießen,
- wann bei "endlosen" Injektionsgutaufnahmen, deren Ausbreitung sich durch keinerlei Verbindungen zu benachbarten Injektionswegen zeigen, abgebrochen, oder auf anderes (und welches) Material umzustellen ist;
- oder, in Summe, wie lange im "Normalfall" der Injektionsvorgang fortzusetzen ist, um eine maximale Verfüllung von der aktuellen Passe und Packerstellung aus zu erreichen.

Herkömmliche Interpretationstechniken von Injektionsdaten oder WAP-Versuchsergebnissen machen sich unter anderem die Entwicklung der Injektionsrate bei konstantem Injektionsdruck, oder den Verlauf des Druckes bei konstanten Raten, gegen die Zeit zunutze.

Z.B. ist eine der Möglichkeiten, das Fließregime auf die sich ausbildende Ausbreitungsform zu interpretieren, aus nachstehender Graphik zu entnehmen: und zwar: einmal zur Form der Ausbreitung,

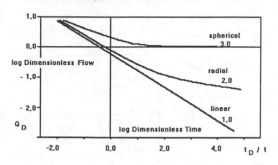

Constant Head Type-Curve

Typenkurven zur Ausbreitungsform
Typecurves on dimension of grout flow

das andere Mal: zur Darstellung von
Randbedingungen des Strömungsbildes,
den sogenannten "boundaries",

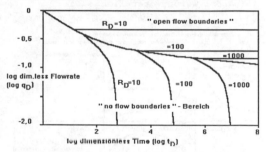

Transiente Fließratendarstellung für offene und geschlossene Ränder

Beispiel für die Darstellung von "flow-
boundaries"
Example for Typecurves relating to
"flow- boundaries

Diese Methoden sind unflexibel insofern, als sie längere Phasen des Injizierens benötigen. Sie sind z.B. besser
für die Erkundung der Randbedingungen
des Strömungsfeldes geeignet, als für
das Management der gerade laufenden
Passe.

Dagegen ist ein rasches, direktes
Reagieren auf die unmittelbaren Verhältnisse eines laufenden Injektionsvorganges eher mit der Analyse von
transienten, also instationären Druck-
entwicklungen möglich, der

"TPA, oder: transient pressure
analysis".

Diese Methode der Beobachtung
instationärer Druckentwicklung nach

Injektionsende oder nach einem Pumpen-
stop, wie sie hier von uns vorgestellt
wird, bietet hohen, unmittelbaren Informationswert bei geringem zusätzlichen Feldaufwand.

Sie hilft u.a., den Bohraufwand bei
gleichem Füllungsgrad der Hohlräume zu
minimieren, und erhöht z.B. die Aus-
sagekraft des gesamte WAP-Versuchs-
aufwandes am Projekt.

Sie stellt die Zusammenarbeit der am
Projekt beteiligten auf eine gemeinsame, sachliche Basis, auf welcher Entscheidungen für ein einvernehmliches
Vorgehen getroffen werden können.
Sie reduziert das bei Injektionen verbreitete Gefühl der Unsicherheit, und
der Befürchtung, unnötigen Aufwand aus
Unwissenheit in Kauf genommen zu haben.
  Im Gegenteil dazu fördert sie die
Überzeugung, daß bei minimiertem Bohraufwand ein maximaler Füllungsgrad erreicht werden kann.

Wenn diese "TPA- Technik" bereits für
die Durchführung der WAP-Versuche in
der Erkundungsphase angewendet würde,
erweiterte sie den Informationswert der
geotechnischen Aufschlüsse wesentlich,
und hilft von vornherein, unnötige
Maßnahmen im Injektionsprogramm zu
vermeiden.
  Andererseits aber, den maximal und
mit Sicherheit zulässigen Injektionsvorgang so zu planen, daß er mit dauer-
haftem Erfolg verbunden bleibt.

Den Kern der hier vorgestellten Überlegungen bilden demnach:

- die Analyse von "shut in"-
  Situationen, ausgelöst über ein
  Abbrechen des Injektionsvorganges,
- die Beobachtung und Aufzeichnung der
  Druckentwicklung gegen die Zeit,
- die Beurteilung dieser Druckentwicklung, sowie der Höhe eines sich
  einstellenden Ruhedrucks, und
  schließlich
- die rechnerische Auswertung über den
  Weg von Typenkurven, zur Ermittlung
  der wirksamen Transmissivität.

Bei diesen Studien sind wir darauf
gestoßen,

- daß der "skin effect", ein aus der
  Erdöltechnologie übernommener Durch-
  lässigkeits-Korrekturterm, für die
  Bestimmung der Kluftaufweitung verwendet werden kann.

459

- Daß andererseits die Zuverlässigkeit
  quantitativer Auswertung
  - auf verläßliche Daten über die
    rheologischen Eigenschaften des
    Injektionsmittels angewiesen ist,
  - und auf einfache Rißgeometrien
    beschränkt bleibt.

- Daß eine rechnerische Beziehung
  zwischen Druckentwicklung gegen die
  Zeit, und Druckausbreitung gegen den
  Radius besteht,

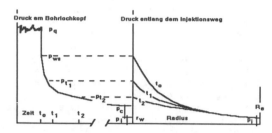

*Typical plot of pressure drop after
shutin versus time and radius*

- auf Grund welcher die Abschätzung von
  Spannungen und Kräften als Folge der
  Injektionstätigkeit möglich wird, und

- die remanenten Spannungen und Verformungen im Gesteinsverband nach
  Injektionsende, sich vor allem als
  eine Funktion der Fließgrenze des
  verwendeten Injektionsmittels
  darstellen.

Diese "Pathologie" von Druckent-
wicklungskurven führt konsequenterweise
zur Möglichkeit einer "Diagnostik", mit
deren Hilfe, vorläufig in erster Linie
empirisch gestützte, qualitative Fest-
stellungen zum Injektionsverlauf getroffen werden können. Sie führen zum
"druckfühligen" Injizieren, oder der

**"pressure sensitive injection, der PSI-
Methode".**

Zusammen also, der "TPA-PSI Methode",
oder dem druckfühligen Injizieren auf
Grund der Analyse transienter Druckzustände.

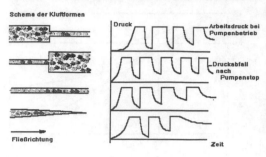

*Beispiele zur Typologie von Druckent-
wicklungskurven in Abhängigkeit von der
Kluftgeometrie.
Typical pressure recovery curves
relating to fissure geometry*

In Ergänzung zu dieser Typologie muß
man sich den Zusammenhang der übrigen
Größen beim Injizieren noch in etwa
folgender Weise vorstellen:

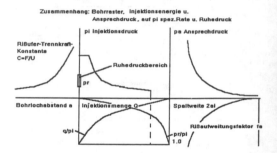

*Genereller, qualitativer Zusammenhang
der meßbaren Größen beim Injizieren von
Klüften.
General diagram on hole spacing,
pressures, rates & fissurewidths*

Ausgegangen sind wir bei unseren Überlegungen von der am weitest verbreite-
ten Sicht auf den regulären, "ordnungs-
gemäßen" Verlauf einer Kluftinjektion
wie folgt (zit. KUTZNER, Esslingen,
1993),

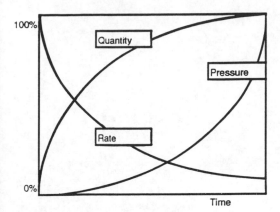

100%

Quantity

Pressure

Rate

0%

Time

Standardverlauf einer Kluftinjektion;
Entwicklung von Menge, Druck und Rate
gegen die Zeit
Standard diagram showing pressure,
rate & quantity versus time

und von einem durch LOMBARDI formu-
lierten neuartigen Abbruchkriterium
(Limite, bei welcher der Injektionsvorgang an der jeweiligen Passe beendet
wird).

Einerseits begrenzt dieser dabei
Druck und Menge, gleichzeitig aber auch
das Produkt aus beiden. Den sich daraus
ergebenden Wert bezeichnet er als Injektionsintensität mit der Dimension:
"Liter.bar je Meter", also einer
spezifischen Energie.

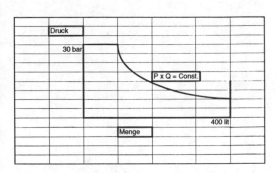

Druck

30 bar

P x Q = Const.

400 lit

Menge

Beispiel für die Festlegung der
Injektionsintensität
Example for defining grouting-
*intensity; p*q=constant.*

Die Erläuterung zu den oben eröffneten
Themen liefert im Vortrag die Darstellung des Phänomens instationärer
Druckentwicklung beim Injizieren,
geht aber auch durch einen großen Teil

an Literatur zu den Fragen von
Injektionsmitteleigenschaften, Kluft-
durchlässigkeit, Strömungsgleichungen und
schließlich, Auswertung von Druck-
entwicklungskurven am Beispiel des
Projektes Kölnbrein.

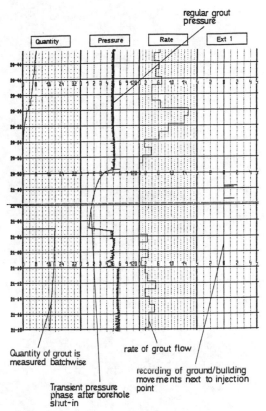

regular grout
pressure

Quantity   Pressure   Rate   Ext 1

Quantity of grout is
measured batchwise

rate of grout flow

recording of ground/building
movements next to injection
point

Transient pressure
phase after borehole
shut-in

Beispiel von AIDEK-Aufzeichnungen
an der Sperre Kölnbrein
Example of autom. grouting data
recording at Kölnbrein dam

Es wird dabei gezeigt, daß für die
Handhabung von Injektionsanweisungen
unterschiedlichster Herkunft und
Auffassung, das Verfahren der Transienten Druckanalyse ein neuartiges,
handliches Werkzeug für den Injekteur werden kann,

und eine willkommene, weil unmittelbare Informationsquelle für den Planer
darstellt, mit deren Hilfe alle Bohr-
lochstrecken einer maximalen Nutzung für
das Erreichen des Injektionsziels zugeführt werden können.

Rißausbreitung ausgehend von Bohrlöchern unter Injektionsdruck

Crack propagation from pressurized boreholes

E. K. Wagner
Tauernkraftwerke AG, Salzburg, Österreich

H. P. Rossmanith
Technische Universität Wien, Österreich

KURZFASSUNG: Hohe Drücke können die Ursache für das Aufreißen von Bohrlöchern sein. Diese Risse beginnen an lokalen Fehlstellen, die immer in Beton und Fels vorhanden sind. Die Rißausbreitung wird durch das schnell abnehmende Feld der aus dem Injektionsdruck resultierenden Ringspannungen gesteuert. Die anfangs beliebig orientierten Rißflächen richten sich bei Rißfortschritt parallel zur Bohrlochachse aus. Ein dreidimensionaler Rißfortschrittsalgorithmus dient zur Berechnung dieser Rißinkremente. Untersucht wird ein Würfel mit einem zentralen Bohrloch. Die Verteilungen der Spannungsintensitätsfaktoren KI, KII und KIII werden entlang der Rißfront ermittelt. Daraus abgeleitete Grenzkurven werden mit insitu-Tests an verschieden Gesteinsarten verglichen.

ABSTRACT: Excessive borehole pressurization may cause damage to the borehole with a large possibility for the initiation of new cracks and/ or the propagation of minute flaws. When these flaws propagate under the control of the borehole pressure-induced symmetrical radially rapidly decaying stress field and crack face pressure they extend along specially curved surfaces trying to align themselves again in a plane the direction of which is governed by the relative magnitude of the pressure in the borehole and the crack. A 3D-crack propagation algorithm serves for the evaluation of the nucleation possibility of borehole cracks as a function of the state of stress. Numerical modelling concentrates on a centrally bored cube with an inclined semi-elliptical surface starter crack subjected to varios boundary conditions. The distributions of the stress intensity factors KI, KII and KIII along the crack periphery as a function of the injection pressure and the state of primary stress have been established. Examples for the behavior of crack initiation and extension due to pressurization are given. The characteristic features of the numerical fracture extension simulation program are the rapid convergence of the sequence of numerically generated intermediate shapes of crack advancements to the final physical shape of the crack. Intermediate simulated crack shapes do not necessarily correspond with physical crack shapes as would be obtained from a slow stable tearing test or from a fatigue test.The results obtained by numerical calculation are compared to known results of field tests.

1 EINLEITUNG

Injektionen werden sowohl zur Bodenverbesserung (Konsolidierungsinjektionen), wie auch zur Reparatur von Rissen eingesetzt. Um eine möglichst gute Injektionsgutaufnahme zu erzielen, werden entsprechend hohe Drücke angewendet. Technisch werden solche flächig wirksam werdenden Injektionen von einem Bohrlochraster ausgeführt. Diese hohen Drücke können nun im Bohr-loch eine Rißinitiation bewirken. Solche Sekundärrisse sind tatsächlich bei Reperaturarbeiten im Talsperrenbereich (Stäuble 1991) aufgetreten. Eine Bewertung dieser Anrißgefährdung (im Zusammenwirken mit einem eventuell vorhandenen Primärspannungszustand), sowie das Aufzeigen über die wahrscheinlichste Lage solcher Risse ist Gegenstand dieser Untersuchung.

Die Methode der linear elastischen Bruch-

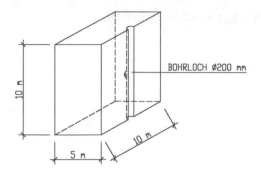

Abbildung 1: Geometrie
Figure 1: Geometry

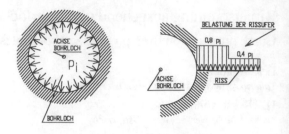

Abbildung 3: Belastung von Bohrloch und Rißufer
Figure 3: Loading of borehole and crack face

mechanik wird verwendet, um das Aufreißen der Bohrlöcher numerisch nachzuvollziehen. Dabei werden die Spannungsintensitätsfaktoren für die drei Rißöffnungsarten berechnet:
Mode-1 ... Öffnen der Rißufer unter Normal-
              spannung
Mode-2 ... Reine ebene Scherbelastung
Mode-3 ... Reine nicht ebene Verscherung
              der Bruchufer normal zur Ebene.
Zur Beschreibung des Problems wird ein würfelförmiger Testkörper (Kantenlänge 10m) mit einer zentralen Bohrung untersucht (Abbildung 1). Zusätzlich wird zur Vereinfachung eine Symmetrie zur Bohrlochachse angenommen. Der Rißfortschritt kann von einer beliebig orientierten Fehlstelle im Fels oder Beton ausgehen. In Abbildung 2 ist der Fall eines achsparallelen Startrisses dargestellt.

Die Bruchflächen sind folgend belastet (Abbildung3):
- Injektionsdruck im Bohrloch
- Injektionsdruck an den Rißufern.

Zusätzlich wird der Einfluß der äußeren Randbedingungen zufolge Primärspannungen auf das Rißwachstum untersucht (Abbildung 4):

- Druckspannungen parallel zur Bohrlochachse
- Druckspannungen normal zur Bohrlochachse
- Druckspannungen parallel zur Symmetrieebene.

Der Spannungszustand verursacht durch den lokalen Einfluß des Injektionsdruckes, sowie durch die Primärspannungen kann nun ein Rißwachstum begünstigen.

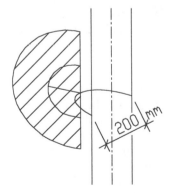

Abbildung 2: Achsparalleler Startriß
Figure 2: Crack parallel to the boreholeaxis

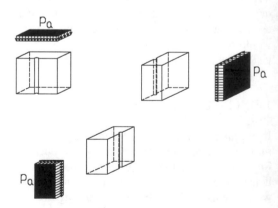

Abbildung 4: Primärspannungszustand
Figure 4: primary stresses

464

2 LAGE DES STARTRISSES

Die Lage des Startrisses kann nun zwischen den beiden Extremlagen:
- Startrißträgerfläche normal zur Bohrlochachse
- Startrißträgerfläche durch die Bohrlochachse

schwanken.

2.1 Startriß in beliebiger Lage

Eine natürliche Fehlstelle im Fels oder Beton wird im allgemeinen Fall räumlich beliebig zur Bohrlochachse orientiert sein. Risse, die von solchen Fehlstellen ausgehen, richten sich bei Rißfortschritt in Ebenen parallel zur Bohrlochachse aus. In Abbildung 5 ist eine Sequenz von Rißinkrementen dargestellt. Deutlich ist das räumliche Verwölben der Rißfläche erkennbar. Dieses Ergebnis ist für den Lastfall Rißuferbelastung und Primärspannung parallel zur Symmetrieebene (entspricht dem Überlagerungsdruck bzw. der Wirkung des Eigengewichts im Beton) ausgewertet. Der Verlauf der Spannungsintensitätsfaktoren für die einzelnen Rißinkremente ist aus Abbildung 6 ersichtlich. Man kann erkennen wie sich das ursprüngliche mixed-mode Problem zu einem reinem Mode-1 Problem wandelt. Die Extremwerte der SIF's KII und KIII jeweils am Rißende sind lediglich auf numerische Ungenauigkeiten zurückzuführen.

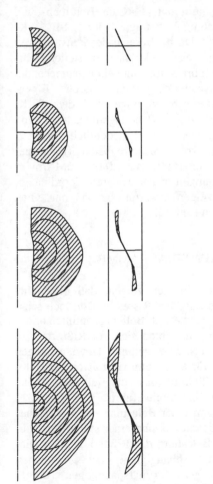

Abbildung 5: Darstellung der Rißinkremente
Figure 5: Propagating crack

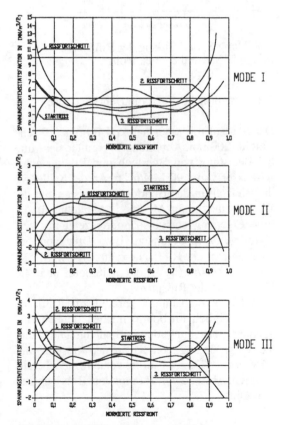

Abbildung 6: SIF entlang der Rißfront
Figure 6: SIF at the crack front

465

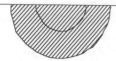

BOHRLOCHACHSE

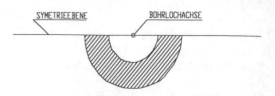

SYMETRIEEBENE   BOHRLOCHACHSE

Abbildung 7: Startrißfläche durch Bohrlochachse
Figure 7: Crackplane in the axis of the borehole

Abbildung 9: Startriß normal zur Bohrlochachse
Figure 9: Crack normal to the axis of the
borehole

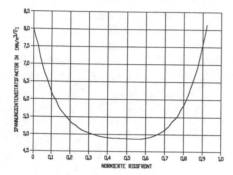

Abbildung 8: SIF entlang der Rißfront
Figure 8: SIF at the crack front

2.2 Startrißfläche durch die Bohrlochachse

Der Startriß mit einer halbkreisförmigen Fläche
(Radius 200mm, Abbildung 7) ergibt in der Mitte
bei Belastung von Bohrloch und Rißufer einen
KI-Faktor von 4.90 MN/m$^{3/2}$. An der Stelle, wo
die Rißfront in das Bohrloch ausbeist, steigt der
Wert auf maximal 8 MN/m$^{3/2}$ (Abbildung 8).
Eine Eigengewichtsvorspannung zeigt nur geringfügige Veränderungen. Eine Druckvorspannung normal zur Rißfläche reduziert den SIF KI
auf 2.5 MN/m$^{3/2}$ in der Rißmitte und 3.6
MN/m$^{3/2}$ an der Rißspitze.
Eine Druckvorspannung parallel zur Rißfläche
bringt lediglich im bohrlochnahen Bereich eine
Spannungserhöhung durch Spaltzug und damit
die Vergrößerung von KI an der Rißspitze auf
8.7MN/m$^{3/2}$. Dieser Einfluß verschwindet bereits
ab einem Abstand von 100mm vom Bohrlochrand
(= Radius).

2.3 Startriß normal zur Bohrlochachse

Hier wird ein kreisringförmiger Riß mit einer
Rißlänge von 200mm (= Bohrlochdurchmesser)
betrachtet (Abbildung 9). Es tritt ein reines
Mode-I Problem auf. Für die Bohrloch- und
Rißuferbelastung beträgt der SIF KI
3.60MN/m$^{3/2}$. Da im Bereich des konstruktiven
Wasserbaues meist Injektionen von steilen Bohrungen ausgeführt werden und ein entsprechender
Primärdruckspannungszustand durch Eigen-
gewicht vorherrscht, verringert sich der SIF KI
auf 1.45 MN/m$^{3/2}$ (Überlagerungsdruck von
6N/mm^2). Dieser Wert liegt deutlich unter der
Rißzähigkeit von Talsperrenbeton (Linsbauer
1991, Brühwiler 1990). von 2.0 bis 4.0MN/m$^{3/2}$.
Primärspannungen in den anderen Koordinaten-
richtungen zeigen erwartungsgemäß nur einen
geringfügigen Einfluß.

3.SICHERHEITSDIAGRAMME

Diese Untersuchungen zeigen, daß die wahrscheinlichste Lage des Risses im Bohrloch achsparallel anzunehmen ist. Beliebig räumlich orien-
tierte Fehlstellen richten sich bei Rißfortschritt
achsparallel aus. Die weiteren Untersuchungen
werden nun für einen Startriß mit einem Halb-
messer von 200mm (das entspricht dem Bohrlochdurchmesser) durchgeführt.
Ausgehend von einer Belastung des Bohrloches
durch Injektionsdruck und einer stufenweise ab-
nehmenden Belastung der Rißufer werden Diagramme in Abhängigkeit der möglichen
Randbedingungen (Primärspannungen) dargestellt.

466

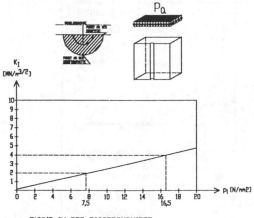

PUNKT IN DER RISSFRONTMITTE

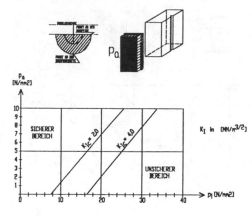

PUNKT IN DER RISSFRONTMITTE

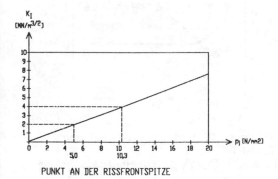

PUNKT AN DER RISSFRONTSPITZE

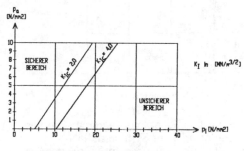

PUNKT AN DER RISSFRONTSPITZE

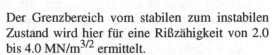

Abbildung 10: Sicherheitsdiagramm - Primärspannung parallel zur Bohrlochachse
Figure 10: Safety diagram - primary stress
parallel to the axis of the borehole

Abbildung 11: Sicherheitsdiagramm - Primärspannung parallel zur Symmetriefläche
Figure 11: Safety diagram - primary stress
parallel to the plane of symmetrie

Der Grenzbereich vom stabilen zum instabilen
Zustand wird hier für eine Rißzähigkeit von 2.0
bis 4.0 MN/m$^{3/2}$ ermittelt.
Vorspannung parallel zur Bohrlochachse:
Diese Belastung bewirkt praktisch keine
Änderung des SIF KI. Daher ist nur der SIF in Ab-
hängigkeit vom Injektionsdruck dargestellt (Ab-
bildung 10).
Vorspannung parallel zur Bohrlochachse:
Hier wirkt der Außendruck rißschließend. Eine
Erhöhung des Außendruckes um 1N/mm^2 erlaubt
die Vergrößerung des Injektionsdruckes um
2N/mm^2 (Abbildung 11).
Vorspannung normal zur Symmetriefläche:
Bei dieser Belastung erhöht der Spaltzug zufolge
Außendruck im bohrlochnahen Bereich den SIF

KI. Bereits bei einem Abstand in der Größe des
Bohrlochdurchmessers (Punkt in Rißfrontmitte)
ist dieser Einfluß nicht mehr erkennbar (Abbildung 12).

4. VERGLEICH RECHNUNG - VERSUCHE

Über die Auswikung des Injektionsdruckes auf
das Rißverhalten von Bohrlöchern wurden an der
Universität von Wisconsin Versuch an verschiedenen Gesteinsarten (Granit und Sandstein)
ausgeführt (Bezalel 1991). In der Abbildung 13
sind die Ergebnisse dieser Versuche dargestell.
Bei sehr kleinem Bohrlochdurchmesser (<20mm)
ist eine ausgeprägte Größenabhängigkeit feststell-

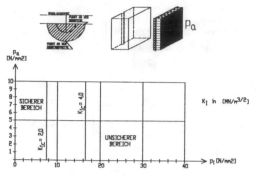

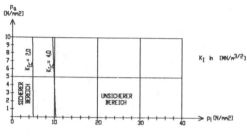

PUNKT IN DER RISSFRONTMITTE

PUNKT AN DER RISSFRONTSPITZE

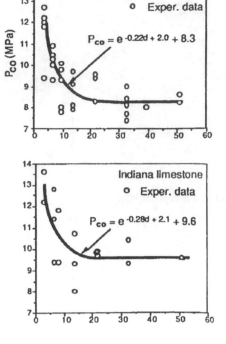

Abbildung 12: Sicherheitsdiagramm - Primärspannung normal zur Symmetriefläche
Figure 12: Safety diagram - primary stress
normal to the plane of symmetrie

Abbildung 13: Versagen von Granit und
Sandstein
Figure 13: Cracking of granite and limestone

bar, für Bohrlochdurchmesser größer als 20mm ergibt sich für beide Gesteinsarten ein konstanter Versagensdruck. Der Wert liegt für Granit bei ca. 8.5 MPa und für Sandstein bei ca. 9.5 MPa. Die höhere Festigkeit von Sandstein ist nach Beschreibung der Versuchsdurchführung auf vorhandene Mikrorisse im Granit zurückzuführen.

Diese Versuche wurden ohne äußere Lasten durchgeführt und lassen sich daher mit dem Sicherheitsdiagramm Abbildung 10 vergleichen. Die aus den Versuchen ermittelten Grenzwerte von ca. 9.0MPa liegen jeweils im dort angegebenen Übergangsbereich.

5. SCHLUSSFOLGERUNG

Schäden die an der Bohrlochwandung durch hohe Injektionsdrücke auftreten können, werden mit der Methode der linear elastischen Bruchmechanik untersucht. Risse richten sich bei Rißfortschritt

unabhängig vom gewählten Primärzustand parallel zur Bohrlochachse aus. Die wesentliche Eigenschaft des verwendeten Rißfortschrittsalgorithmus ist die rasche Konvergenz der numerisch ermittelten Rißinkremente auf die physikalisch erklärbare Rißform. Die numerischen Zwischenschritte stimmen nicht unbedingt mit Rißformen überein, die z.B.aus Bruchversuchen mit langsamer Belastungsgeschwindigkeit ermittelt werden.

DANKSAGUNG

Die Berechnungen wurden mit Unterstützung des Fonds für wissenschaftliche Förderung (Projekt FWF P 6989) durchgeführt. Das an der Cornell-University entwickelte numerische Simulationsprogramm FRANSYS (Ingraffea 1990) wurde eingesetzt.

LITERATUR

Stäuble H.,Wagner E.K. Rossmanith H.P.,Ingraffea A.R. 1991. Cracking of Boreholes in Arch Dams due to High Pressure Grouting. Procc. Int. Conf. on Dam Fracture. Boulder Col.: 585-601

Linsbauer H.N. 1990. Sperre Kölnbrein - Ermittlung bruchmechanischer Materialkennwerte. Bericht für den Arbeitskreis: Injektion von Rissen im Beton.

Brühwiler E. 1990. Fracture of Mass Concrete under Simulated Seismic Actions. Dam Engineering, Vol.I, Issue 3: 153-176

Bezalel C.H. Zhongliang Z. 1991. Effect of borehole size and pressurization rate on hydraulic fracturing breakdown pressure. Ed. Roegiers.-Balkema. Rock Mechanics as a Multidisciplinary Sience: 191-198

Ingraffea A.R. 1990. FRANSYS numerical simulation program. Private communication.

Versuchsschacht Scheibenberg Tunnel – Ein Großversuch im Röttonstein zur Ermittlung optimaler Injektionsparameter

Test shaft Scheibenberg Tunnel – A large test in Röttonstein for determination of optimal grouting parameters

Hartmut Wesemüller
Bundesbahndirektion Hannover, Projektgruppe NBS Hannover der Bahnbauzentrale, Deutschland

Stephan Semprich
Bilfinger + Berger Bauaktiengesellschaft, Tiefbauabteilung Mannheim, Deutschland

ZUSAMMENFASSUNG: Im Zuge der Ausbaustrecke Kassel - Dortmund der Deutschen Bundesbahn ist der Bau des 1 830 m langen Scheibenberg Tunnels und des 409 m langen Röthberg Tunnels nordwestlich von Kassel geplant. Im Bereich der geplanten Tunnel stehen überwiegend Röttonsteine an, die im Zuge der Gipsauslaugung flächig nachgesackt und dabei vielfach völlig zerbrochen sind. Die Gebirgskennwerte mußten daher als so ungünstig eingestuft werden, daß ein bergmännischer Tunnelvortrieb auch bei einer weitgehenden Querschnittsunterteilung nur mit einer vorauslaufenden Gebirgsvergütung zur Anwendung kommen kann. Um dafür die notwendigen Injektionsmaßnahmen optimal planen zu können, wurde im Frühjahr 1993 ein Großversuch ausgeführt: Im Bereich der geplanten Tunneltrasse wurde ein Versuchsfeld mit einer Größe von 50 m² ausgewählt, in dem der Baugrund in einer Tiefe von 8 - 17 m unter Geländeoberfläche injiziert wurde. Den Injektionsarbeiten folgte innerhalb des Versuchsfeldes das Abteufen eines ovalen, 3,50 m x 5,50 m breiten Versuchsschachtes bis zu einer Tiefe von 19,7 m. Parallel zum Schachtaushub wurden im injizierten und nicht injizierten Gebirge geotechnische Versuche durchgeführt. Dazu gehörten vertikale und horizontale Lastplattenversuche sowie die Entnahme von Grobohrkernen mit einem Durchmesser von 60 cm, die anschließend im Großtriaxialgerät auf ihre Verformbarkeit und Festigkeit untersucht wurden. Mit dem Großversuch konnten die zweckmäßigsten Injektionsparameter ermittelt werden. Es wurde nachgewiesen, daß sich die Röttonsteine bei Anwendung spezieller Injektionstechniken soweit vergüten lassen, daß ein Auffahren der für die Eisenbahntunnel großflächigen Querschnitte möglich ist.

ABSTRACT: As part of the work being carried out to upgrade the Kassel-Dortmund line of the German Federal Railway, the construction of two tunnels northwest of Kassel, the 1830m-long Scheibenberg tunnel and the 409m-long Röthberg tunnel, is planned. The area around the planned tunnels is made up predominantly of Röttonstein claystone, large areas of which have collapsed due to the leaching of the gypsum, with the result that many of the layers have broken up. The rock parameters must, therefore, be classified as so unfavourable that driving the tunnels, even with cross-section enlargement carried out in stages, is only possible if ground pretreatment is undertaken beforehand. In order to provide optimum planning of the grouting procedures, a large scale test programme was set up in spring 1993: an area of 50m² was selected for testing purposes close to the site proposed for the tunnels and grouting was carried out at a depth of between 8 and 17m. As penetration of the grout material into the pores of the ground was not anticipated due to the fine-grained nature of the Röttonstein claystone, soil fracturing techniques were employed with relatively high grouting pressures. And at the same time, the spacing of the boreholes, grouting pressure and the grout mix ratios were varied. When the grouting work was complete, there followed the excavation of a 3.50m x 5.50m oval-shaped trial shaft, within the test area, to a depth of 19,7m. While excavation of the shaft was being carried out, in grouted and non-grouted rock, in addition to a thorough geological mapping of the shaft wall, geotechnical field tests were also performed. These included vertical and horizontal plate bearing tests and large scale sampling with 60cm diameter core samples, which were subsequently examined in a large scale triaxial cell for deformability and strength. The entire test programme resulted in optimal grouting parameters being established; they demonstrated that, by using special grouting techniques, pretreatment of the Röttonstein claystone would be sufficient to allow excavation to be carried out with the large cross sections necessary for the railway tunnel.

1 EINLEITUNG

Im Zuge der Ausbaustrecke Dortmund-Kassel der Deutsche Bundesbahn, die für eine Geschwindigkeit von 160 - 200 km/h mit einem Kostenaufwand von rund 1.4 Milliarden DM (Preisstand 01.01.1993) bis zum Jahr 1998 realisiert werden soll, sind zwischen Paderborn und Kassel vier Tunnel mit einer Gesamtlänge von 5.9 km geplant.

Alle Tunnel erhalten einen zweigleisigen Maulquerschnitt mit einer Ausbruchfläche F = 148 m² und werden bergmännisch aufgefahren.

Zwei der Tunnelprojekte, der 1 830 m lange Scheibenberg Tunnel und der 409 m lange Röthberg Tunnel liegen jeweils in einer Linienverbesserung nordwestlich von Kassel. Nach dem derzeitigen Stand der geologischen Vorerkundungen sowie der statischen Voruntersuchungen sind die Kennwerte des überwiegend anstehenden Röttonsteins so ungünstig, daß eine Gebirgsvergütung mit Injektionen vom Planungsteam für erforderlich gehalten wird.

In einem Hearing mit Vertretern der Bauindustrie wurde über geeignete und bewährte Injektionsverfahren diskutiert. Die Arbeitsgemeinschaft Bilfinger + Berger/Ph. Holzmann erhielt von der Deutschen Bundesbahn den Auftrag, einen Großversuch mit Bohr- und Zementinjektionsarbeiten durchzuführen, um optimale Injektionsparameter zu gewinnen. Am Scheibenberg Tunnel wurde dazu ein Versuchsschacht abgeteuft.

2 RÖTHBERG TUNNEL

Nach dem in Bild 1 dargestellten Längsschnitt beträgt die Überlagerungshöhe maximal ca. 18 m über Firste. Es sind 263 m im Tonstein der Rötfolge und 50 m im Buntsandstein der Sollingfolge aufzufahren. Das gesamte Bauwerk befindet sich in der Wasserschutzzone IIIA, im Einflußbereich eines Tiefbrunnens und einer Quellfassung. Eine Beeinträchtigung der Wasserfassungen für den Bau- und Endzustand ist bautechnisch auszuschließen. Deshalb wird der Tunnel als nicht gedraintes System ausgebildet.

2.1 *Geologische Situation*

Die geologischen Verhältnisse (Bild 1) wurden im Rahmen eines Erkundungsprogrammes vom Hessischen Landesamtes für Bodenforschung (1992) beschrieben und bewertet, s.a. Strauß (1993).

Der Tunnel wird danach vier verschiedene Schichtenfolgen durchörtern. Von Nordwest nach Südost betrachtet sind es lehmige Deckschichten, Basalttuff, Sandsteine der Sollingfolge und Röttonstein. Die Tonsteine der Rötfolge sind oberflächig angewittert oder infolge der Salzauslaugung aufgelockert und in zahllose kleine Kluftkörper zerlegt, deren Kantenlängen nur wenige Zentimeter betragen. Sie erscheinen mäßig gut kompaktiert und liegen als halbfeste Tone vor. Die Tonsteine sind wenig kohäsiv und stark brüchig bis bröckelig ausgebildet, abschnittsweise stärker vertont oder plastiziert bzw. von dünnen, weichen Einlagerungen oder auch Dolomitsandadern durchzogen.

Aufgrund der Ergebnisse des Erkundungsprogrammes und ergänzender Laborversuche wurden den Tonsteinen der Rötfolge zunächst folgende Baugrundparameter zugeordnet:

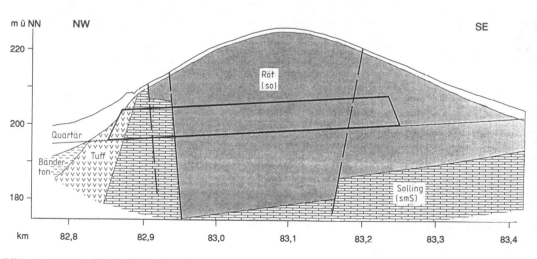

Bild 1 Längsschnitt Röthberg Tunnel

Fig. 1 Longitudinal section Röthberg Tunnel

Seitendruckbeiwert	k_0	= 0,4 - 0,6
Verformungsmodul	E_v	= 15 - 30 MN/m²
Reibungswinkel	φ'	= 30° bzw. 20°
Wichte	γ	= 22 kN/m³
Kohäsion	c'	= 0 bzw. 20 kN/m²

2.2 Vortriebskonzept

Die erwarteten geologischen Verhältnisse erfordern aus Standsicherheitsgründen einen unterteilten Querschnitt. Nach dem Tunnelbautechnischen Gutachten (1992) ist bei einem reinen Kalottenvortrieb aufgrund der geringen Mittragwirkung des Gebirges trotz möglicher Kalottensohle, Kalottenfußverbreiterung und Sicherung der Ortsbrust mit so großen Setzungen zu rechnen, daß eine Entfestigung des Deckgebirges eintreten würde. Die daraus erhöhten Auflagerkräfte würden einen Schubbruch der Kalottensohle im Anschlußbereich der Kalottenfüße verursachen. Deshalb wurde bis dato ein Ulmenstollenvortrieb geplant, der aus mehreren, möglichst kleinen Teilquerschnitten besteht (Bild 2).

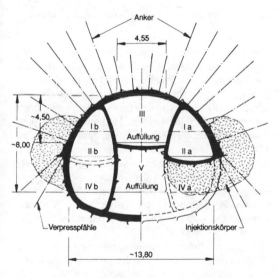

Bild 2   Ausbruchphasen für einen Tunnelvortrieb
im Röttonstein
Fig. 2   Construction stages for tunnelling in
Röttonstein claystone

Bemerkenswert sind hierbei die errechneten Spritzbetonstärken. In der AKL 7A.2 betragen sie bis zu 60 cm für die Außenulme und bis zu 40 cm für die Innenulme. Um die Auflagerkräfte aufzunehmen, werden Verpreßpfähle an jedem Ulmenstollen vorge-

sehen, da ein Plastifizieren des Tons erwartet wird. Ein schneller Ringschluß eines jeden Teilquerschnittes ist erforderlich. Vor der Ausbruchsphase III müssen zudem Injektionen durchgeführt werden, die den Tonstein auf mindestens drei Meter Tiefe zuverlässig vergüten, um die fehlende seitliche Bettung während der nachfolgenden Ausbruchphasen zu gewährleisten.

Von der Wirksamkeit einer Gebirgsverbesserung wird die Entscheidung über weitere Querschnittsunterteilungen in den Phasen IV und V nach Bild 2 und den Umfang der Sicherungsmittel, z.B. der Verpreßpfähle, abhängen.

3 SCHEIBENBERG TUNNEL

Nördlich des Röthbergs, bei Hümme, durchfährt als vorläufige Variante ein 1 830 m langer Eisenbahntunnel den Höhenrücken des Scheibenbergs. Die maximale Überlagerungshöhe beträgt ca. 110 m.

3.1 Geologische Situation

Der Tunnel durchörtet den obersten Teil der Rötfolge, welche im Untersuchungsgebiet eine Mächtigkeit von etwa 55 m hat (Bild 3).

Im Verlauf der Baugrunduntersuchungen für die Ausbaustrecke wurde das Röt sehr detailliert untersucht, mit Aufschlußbohrungen über 1000 m Kernstrecke (s.a. Ingenieurgeol. Gutachten des Hessischen Landesamtes für Bodenforschung, 1992).

Die Gesteine des Röt bestehen hier aus einer Abfolge von unterschiedlich stark gebankten, rotbraunen und z.T. grüngrauen Tonsteinen, in die unregelmäßig dünne Dolomit- bzw. auch Quarzitbänkchen sowie zentimeter bis dezimeter dicke Gipslagen eingeschaltet waren. Die meist kalkhaltigen Tonsteine waren ursprünglich fest bis hart. Infolge der Gipsauslaugungen ist der überwiegende Teil der Röttonschichten großflächig nachgesackt und dabei in sich unterschiedlich stark, vielfach völlig zerbrochen und entfestigt. Laborversuche haben eine deutliche Anisotropie der Druckfestigkeiten ergeben.

Für die Gebirgsausbildung wurden vier Typen beschrieben, wobei auf 80 % der bergmännisch zu erstellenden Strecke die gleichen gebirgsmechanischen Kennwerte (Gebirgstyp IV) wie beim Röthberg Tunnel erwartet werden.

Im Scheibenberg liegen zwei Grundwasserstockwerke vor, und zwar im aufliegenden Muschelkalk (40 - 50 m über Tunnelsohle) und ein unteres im Tunnelniveau. Eine hydraulische Verbindung muß insbesondere bei Störzonen angenommen werden. Der Tunnel wird druckwasserdicht ausgebildet.

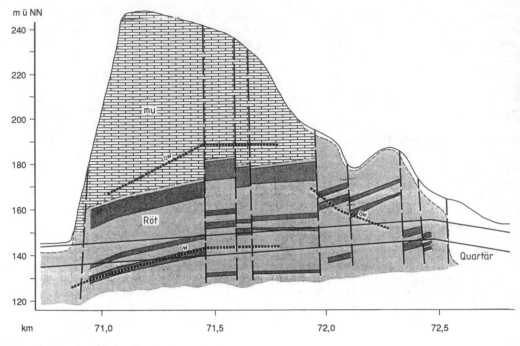

| Bild 3 | Längsschnitt Scheibenberg Tunnel | Fig. 3 | Longitudinal section Scheibenberg Tunnel |

3.2 *Vortriebskonzept*

Bereits vor der abschließenden Bearbeitung eines Tunnelbautechnischen Gutachtens und nach den Erkenntnissen, die bei der Planung des Röthberg Tunnels gewonnen wurden, war klar, daß bei einer vergleichbaren geologischen Situation sich das unter 2.2 geschilderte Vortriebskonzept nur mit einem hochwirksamen Injektionserfolg durchsetzen lassen würde. Eine am Röthberg denkbare offene Bauweise schied hier aus.

Da jedoch Erfahrungen über Injektionen im Röt nur im begrenzten Umfang vorlagen, die Injektionen aber ein entscheidendes Element bei der Herstellung der Tunnel darstellen, wurden für dieses Projekt großmaßstäbliche Versuche im Vorfeld der Bauausführung unumgänglich, s.a. Schetelig et.alt. (1988).

4 VERSUCHSPROGRAMM

Nach den Vorstellungen des Tunnel-Planungsteams des Auftraggebers und aufgrund der Erfahrungen der bauausführenden Arbeitsgemeinschaft wurde geplant, im Bereich des Zugangsstollens Scheibenberg ein Injektionsfeld mit dem Soil-Fracturing-Verfahren anzulegen und einen ovalen Spritzbetonschacht abzuteufen.

Der Schacht sollte einerseits einen neuen, präziseren geologischen Aufschluß liefern und andererseits einen Vergleich von injiziertem und nicht injiziertem Gebirge ermöglichen. Dabei sollten Injektionsraster, -drücke, Rezepturen variiert werden und beim Abteufen des Schachtes in Horizonten, die dem Untergrund der geplanten Tunnel ähnlich sind, an injiziertem und nicht injiziertem Baugrund vergleichende geotechnische Messungen bzw. Versuche vorgenommen werden. Im einzelnen waren es Konvergenzmessungen an der ovalen Schachtwand, horizontale Doppel-Lastplattenversuche, Vertikallastplattenversuche und Großbohrkernentnahmen für Triaxialversuche (Bild 4).

Zur Einrichtung der repräsentativen Horizonte wurde in unmittelbarer Nähe des geplanten Schachtes eine Kernbohrung niedergebracht. Im Zuge der Schachtabteufung ergab sich die im Bild 4 dargestellte detaillierte Schichtenfolge des Röttonsteins.

5 INJEKTIONSVERSUCHE

Aufgrund der Feinkörnigkeit des Röttonsteins schied eine Vergütung des Baugrundes in Form einer Porenrauminjektion aus. Stattdessen bot sich als Injektionsverfahren das Soil-Fracturing Verfahren an. Bei diesem Verfahren dringt das Einpreßgut nicht in die

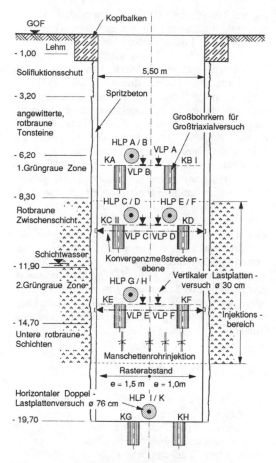

GOF
Kopfbalken
- 1,00 Lehm
Solifluktionsschutt
5,50 m
- 3,20
Spritzbeton
angewitterte,
rotbraune
Tonsteine
Großbohrkern für
Großtriaxialversuch
HLP A / B
VLP A
- 6,20
KA   KB I
1.Grüngraue Zone
VLP B
- 8,30
HLP C / D   HLP E / F
Rotbraune
Zwischenschicht
KC II   KD
VLP C VLP D
Schichtwasser
Konvergenzmeßstrecken-
- 11,90
ebene
2.Grüngraue Zone
HLP G / H
Vertikaler Lastplatten-
versuch ø 30 cm
KE   KF
- 14,70
VLP E VLP F
Untere rotbraune
Schichten
Injektions-
bereich
Manschettenrohrinjektion
Rasterabstand
e = 1,5 m   e = 1,0m
Horizontaler Doppel-
HLP I / K
Lastplattenversuch ø 76 cm
- 19,70
KG   KH

Bild 4 Längsschnitt des Schachtes mit geotech-
         nischem Versuchsprogramm
Fig. 4 Longitudinal section of the shaft with
         geotechnical test programme

einzelnen Poren des Bodens ein, sondern sprengt
aufgrund entsprechend hoch gewählter Verpreßdrücke den Baugrund auf und dringt gleichzeitig in
die dabei entstehenden Risse ein. So bildet sich im
Boden ein Feststoffskelett, das zu einer Erhöhung
des Spannungszustandes im Baugrund und nach
Aushärtung zu einer Erhöhung des Verformungsmoduls und der Scherfestigkeit führt.

5.1 Vorversuch

Den eigentlichen Injektionsarbeiten ging ein Vorversuch in drei dem Schacht benachbarten Bohrungen
voraus. Ihr gegenseitiger Abstand betrug 1 m. Bei
diesem Versuch wurde die Verfahrenstechnik der
geplanten Manschettenrohrinjektion überprüft und in

einem ersten Schritt optimiert. Dazu zählte die Variation des W/Z-Wertes des Verpreßgutes zwischen
0,7 und 0,8. Als Zement wurde ein PZ 35 F verwendet. Für den Abbruch des Verpreßvorganges galten
zwei Kriterien: Ein maximaler Verpreßdruck an der
Pumpe in Höhe von 35 bar oder eine maximale Ver-
preßmenge von 100 l/Verpreßabschnitt. In einer 1.
Injektionsphase wurde zunächst jede 2. Manschette
eines Injektionsrohres verpreßt. Anschließend folgte
in einer 2. Phase die Verpressung der zunächst über-
sprungenen Bereiche. In den drei Bohrungen wurden
insgesamt 2,6 m³ Suspension verpreßt. Bei einer angenommenen Reichweite der Suspension im Fels von
1 m ergibt sich, daß Hohlräume im Röttonstein zu ca.
5 % des Gesamtvolumens mit Injektionsgut verfüllt
wurden. Die Verpreßgutaufnahme war in den einzelnen Austrittshorizonten jedoch sehr unterschiedlich.
In unregelmäßiger Folge wechselten Horizonte, in
denen 100 l Suspension verpreßt wurden mit solchen,
in denen nur eine geringe oder keine Aufnahme erzielt wurde. Eine Abhängigkeit der Aufnahme über
die Tiefe war nicht zu beobachten. Trotz der hohen
Drücke kam es nur vereinzelt zu Umläufigkeiten. Die
Verpreßrate wurde größtenteils zu 10 l/min gewählt.
Eine Erhöhung dieses Wertes führte sofort auch zu
einer Erhöhung des Pumpendruckes. Erwartungsgemäß lagen die einzelnen Verpreßmengen der 1. Injektionsphase deutlich über denen der 2. Phase. Als
Abschluß des Vorversuches wurden örtlich Nach-
verpressungen durchgeführt, die bei durchweg höheren Drücken mit weiteren Verpreßgutaufnahmen
verbunden waren.

Obwohl der Röttonstein nur in einer Tiefe von 8 -
17 m injiziert werden sollte, konnten Hebungen der
Geländeoberfläche nicht ausgeschlossen werden.
Während der Injektionsarbeiten durchgeführte Nivel-
lementmessungen zeigten jedoch keine Bewegungen
der Geländeoberfläche.

5.2 Hauptversuch

Auf der Grundlage der Ergebnisse des Vorversuches
wurde die eigentliche Injektion im Bereich des
nachfolgend herzustellenden Schachtes geplant und
ausgeführt. Dazu wurden insgesamt 42 Bohrungen
hergestellt, die mit Manschettenrohren aus Kunststoff
und einer Mantelmischung ausgebaut wurden. Im
Bereich der einen Hälfte des Versuchsfeldes wurde
der Bohrlochabstand zu 1 m und in der anderen
Hälfte zu 1,5 m gewählt, um Aussagen zum optimalen Abstand machen zu können. Zur Verbesserung
der Filtrationseigenschaften wurde der Suspension
2% Bentonit zugegeben und arbeitstechnisch bedingt
der W/Z Wert auf 0,9 erhöht. Die mit der Bentonitzugabe verbundene Festigkeitsverringerung des er-

475

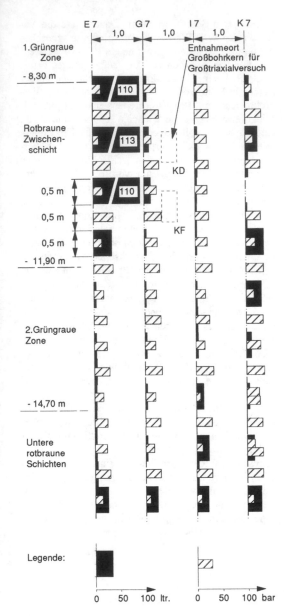

Bild 5 Repräsentative Ergebnisse der Injektion
Fig. 5 Representative results of grouting work

planten Tunnelschale (Ulme) zu vermeiden: Als Abbruchkriterium wurde für die einzelnen Verpreßabschnitte der 1. Injektionsphase maximal 15 bar bzw. 100 l, für die 2. Phase maximal 20 bar bzw. 50 l/Verpreßabschnitt definiert. Um dennoch den Injektionserfolg sicherzustellen, wurde die Einpreßrate auf 8 l/min und weniger reduziert. Zur Unterscheidung des Zementsteins beim Schachtaushub wurde der Suspension der 1. Phase schwarze Pigment-Farbstoffe beigemischt. Die Suspension der 2. Injektionsphase wurde gelb eingefärbt.

Sämtliche Daten der Verpreßarbeiten wurden mit in einem Injektionscontainer installierten Meßgeräten kontinuierlich aufgezeichnet. Insgesamt wurden in den 42 Bohrungen 9,1 m³ Suspension verpreßt. Umgerechnet auf das Gesamtvolumen des von der Injektion erfaßten Felsbereiches ergibt sich damit der verfüllte Hohlraumanteil zu ca. 2 % und damit zu einem wesentlich niedrigeren Wert als er sich beim Vorversuch ergeben hatte. Die mittlere Verpreßgutaufnahme pro Manschette hat ca. 25 l betragen. Allerdings ließ sich auch bei dieser Injektion keine homogene Verteilung des Injektionsgutes im Baugrund erreichen (Bild 5).

6 SCHACHTHERSTELLUNG

Für den Schacht wurde ein elliptischer Grundriß mit den lichten Abmessungen 3,5 m x 5,5 m gewählt, um ggfs. Konvergenzen deutlicher messen zu können, als sie bei einem kreisförmigen Schacht zu erwarten gewesen wären. Zunächst war eine Schachttiefe von 15 m vorgesehen, die jedoch im Zuge der Aushubarbeiten aufgrund der dabei gewonnenen Erkenntnisse auf 19,7 m vergrößert wurde (Bild 4). Die Schachtarbeiten begannen mit der Herstellung eines Kopfbalkens. Anschließend erfolgte der Aushub und die Sicherung der Schachtwandung mit einer 20 bis 30 cm dicken Spritzbetonschale in Meterschritten. Während im oberen Schachtbereich der Aushub mit einem Greifer erfolgen konnte, mußte in größerer Tiefe örtlich auch gemeißelt werden. Parallel zum Aushub wurde die Schachtwandung sorgfältig kartiert und zusätzlich fotografisch dokumentiert. Bei ca. 12 m Tiefe wurde auf einzelnen Trennflächen zirkulierendes Schichtwasser angetroffen (Bild 4). Injektionsgut in schwarzer und gelber Farbe wurde nur in verhältnismäßig geringem Umfang beobachtet. Das Injektionsgut hatte sich ausschließlich in den horizontal liegenden Schichtfugen ausgebreitet. Die maximale Dicke der Zementsteinlagen betrug 1 cm. Vertikale offene Trennflächen des Röttonsteins wurden nicht festgestellt. Auch die Soil-Fracturing Injektion hatte andere vertikale Trennflächen mit den ausgeführten Druckniveaus nicht aufreißen können.

härteten Zementsteins wurde durch die Verwendung eines PZ 45 F kompensiert.

Auch bei dieser Injektion wurde in einer 1. Injektionsphase nur jede 2. Manschette des Injektionsrohres verpreßt, bevor anschließend in einer 2. Phase die dazwischen liegende beaufschlagt wurde. Der maximale Verpreßdruck wurde gegenüber dem Vorversuch reduziert, um eine Überbeanspruchung der ge-

Bild 6   Röttonstein, 12 m unter Geländeoberfläche
Fig. 6   Röttonstein claystone, 12 m below surface

Bild 7   Horizontaler Doppel-Lastplattenversuch
Fig. 7   Horizontal double plate bearing test

Ein für den Baugrund repräsentatives Foto zeigt das Bild 6 im Bereich einer für einen horizontalen Lastplattenversuch freigelegten Wandfläche.

Erwartungsgemäß ergaben parallel zum Schachtaushub durchgeführte Konvergenzmessungen (Bild 4) keine meßbaren Wandbewegungen.

7 LASTPLATTENVERSUCHE

7.1 *Vertikallastplattenversuche*

Das Versuchsprogramm sah insgesamt sechs vertikale Lastplattenversuche (VLP A-F) mit einem Plattendurchmesser von 30 cm vor, die von der FMPA Stuttgart durchgeführt wurden. Je zwei wurden im nicht injizierten Horizont der 1. Grüngrauen Zone sowie in den injizierten Schichten Rotbraune Zwischenschicht und 2. Grüngraue Zone ausgeführt (Bild 4). Die Versuche erfolgten in Anlehnung an die Empfehlung Nr. 6 des AK 19 der DGEG (1985). Eine Auswertung der Meßergebnisse ergibt für den Spannungsbereich zwischen 0,15 und 0,30 MN/m² Verformungsmodul für die Erstbelastung zwischen 30 und 80 MN/m². Der kleinere Wert wurde für die injizierte Rotbraune Zwischenschicht erhalten, während der größte Wert die 2. Grüngraue Zone repräsentiert (Bild 4).

7.2 *Horizontallastplattenversuche*

Es wurden Doppel-Lastplattenversuche mit einem Plattendurchmesser von 76 cm entsprechend der Empfehlung Nr. 6 des AK 19 der DGEG (1985) ausgeführt. Die Anordnung der insgesamt 5 Doppelversuche (HLP A-K) in 4 verschiedenen Horizonten zeigt Bild 4. Bild 7 zeigt eine der beiden Lastplatten

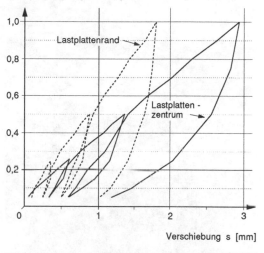

Sohlspannung σ [MN / m²]

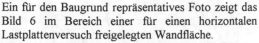

$E_{v\ Erstbel.} = 180\ MN / m^2$ $E_{v\ Ent / Wiederbel.} = 280\ MN / m^2$

Bild 8   Spannungs-Verschiebungslinien des Horizontallastplattenversuchs HLP F
Fig. 8   Stress displacement behaviour of horizontal plate bearing test HLP F

während eines Versuches. In Bild 8 sind beispielhaft die gemessenen Spannungs-Verschiebungslinien des Versuches HLP F dargestellt. Eine Auswertung der Meßergebnisse führt für die beiden oberen Horizonte (Bild 4) zu Verformungsmoduln zwischen 100 und 180 MN/m². Für die beiden unteren Horizonte wurden Werte in der Größenordnung von 300 und 400 MN/m² ermittelt. Ein deutlicher Unterschied hinsichtlich injizierter und nicht injizierter Horizonte konnte nicht festgestellt werden.

477

8 GROßTRIAXIALVERSUCHE AN GROßBOHR-KERNEN

Das Versuchsprogramm sah die Entnahme von acht Großbohrkernen (KA-KH) mit den Abmessungen D = 60 cm und L = 120 cm in vier verschiedenen Horizonten durch den Lehrstuhl für Felsmechanik der Universität Karlsruhe vor (Bild 4). Anschließend sollten diese Proben im Dreiaxialversuch nach der Mehrstufentechnik auf ihre Verformbarkeit und ihre Festigkeit untersucht werden. Bild 9 zeigt einen Kern während der Zerlegung nach dem Dreiaxialversuch. Die Vorgehensweise bei der Kernentnahme und der Durchführung des Großtriaxialversuches ist in der Empfehlung von Natau et. alt. (1989) beschrieben. Das Bild 10 zeigt beispielhaft die für die Probe KB II gemessene Spannungsdehnungslinie und die daraus ermittelten Verformungsmoduln für unterschiedliche Spannungsbereiche.

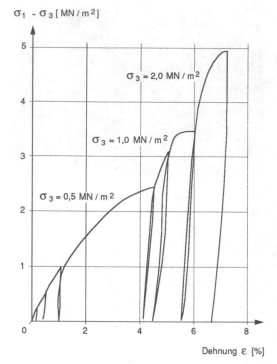

$\sigma_1 - \sigma_3 \,[\,MN\,/\,m^2\,]$

$E_{v \; Erstbel.} = 85 \; MN\,/\,m^2 \; (\sigma_1 - \sigma_3 = 0,0 - 0,8 \; MN\,/\,m^2)$

$E_{v \; Erstbel.} = 60 \; MN\,/\,m^2 \; (\sigma_1 - \sigma_3 = 0,8 - 1,7 \; MN\,/\,m^2)$

Bild 10   Im Großtriaxialversuch ermittelte Spannungsdehnungslinie der Probe KB II

Fig. 10   Stress-strain behaviour of sample KB II from large scale triaxial testing

Bild 9   Bohrkern KB II nach dem Großtriaxialversuch

Fig. 9   Sample KB II after large scale triaxial testing

Die für die beiden Proben KA und KB II ermittelten Scherfestigkeiten sind in Bild 11 dargestellt. Für die beiden nicht injizierten Horizonte haben die Versuche erwartungsgemäß erheblich unterschiedliche Werte ergeben: Der Verformungsmodul der 1. Grüngrauen Zone wurde im Mittel zu ca. 100 MN/m² bestimmt, während er sich für die Unteren rotbraunen Schichten zu ca. 1000 MN/m² ergeben hat. Auch die Scherfestigkeiten lagen für den tiefer liegenden Horizont höher als für die 1. Grüngraue Zone.

Für die beiden injizierten Schichten der Rotbraunen Zwischenschicht und der 2. Grüngrauen Zone ergaben sich die Scherfestigkeiten in gleicher Größenordnung. Unterschiede ergaben sich dagegen hinsichtlich der Verformbarkeit. Für die Rotbraune Zwischenschicht wurden Verformungsmoduln zwischen 100 und 200 MN/m² und für die 2. Grüngraue Zone von 400 und 500 MN/m² gemessen. Damit hat sich auch bei diesen Versuchen der bereits bei den Lastplattenversuchen festgestellte Anstieg des Verformungsmoduls mit zunehmender Tiefe bestätigt.

9 BEWERTUNG DER VERSUCHSERGEBNISSE UND SCHLUßFOLGERUNGEN FÜR DEN ENTWURF DER GEPLANTEN TUNNEL

Die Injektionsversuche haben gezeigt, daß der im Bereich der beiden geplanten Tunnel überwiegend anstehende Röttonstein mit dem Soil-Fracturing

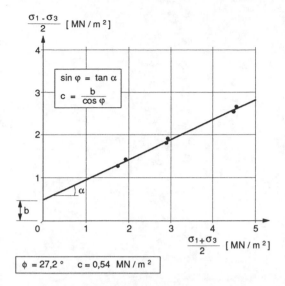

$$\frac{\sigma_1 - \sigma_3}{2} \; [\,MN/m^2\,]$$

$$\sin \varphi = \tan \alpha$$
$$c = \frac{b}{\cos \varphi}$$

$$\frac{\sigma_1 + \sigma_3}{2} \; [\,MN/m^2\,]$$

$\varphi = 27{,}2°$ $c = 0{,}54 \; MN/m^2$

Bild 11   Im Großtriaxialversuch ermittelte Bruch-
spannungen der Proben KA und KB II
Fig. 11   Failure condition of samples KA and KB II
from large scale triaxial testing

Verfahren injizierbar ist. Allerdings ist die Verpreßgutaufnahme mit 2 bis 3 % des gesamten injizierten Baugrundvolumens deutlich niedriger als bei einer Poreninjektion. Da der im Versuchsbereich anstehende Fels keine offenen vertikalen Trennflächen aufwies, breitete sich das Injektionsgut nahezu ausschließlich entlang horizontaler Schichtfugen aus. Die verschieden gewählten Bohrabstände führten zu keinen unterschiedlichen Injektionserfolgen, so daß ein Rasterabstand von 1,5 m als ausreichend angesehen werden kann. Das setzt allerdings voraus, daß die Injektionsbohrungen überwiegend normal zu den das Injektionsgut aufnehmenden Trennflächen verlaufen. Je höher der Verpreßdruck gewählt wird, um so stärker reißt das Gebirge auf und die Verpreßgutmenge steigt an. Der Verpreßdruck ist in seiner Höhe jedoch da begrenzt, wo sich druckempfindliche Bauwerke wie z.B. ein vorauseilender Teilvortrieb eines Tunnels in unmittelbarer Nähe befinden.

Eine Erhöhung der Scherparameter des Röttonsteins aufgrund der erfolgten Injektion konnte mit den durchgeführten Großtriaxialversuchen nicht eindeutig belegt werden. Die für die einzelnen Proben gemessenen Scherparameter streuten dafür zu stark. Außerdem wurde bei der im Anschluß an den Triaxialversuch durchgeführten Zerlegung der Proben nur verhältnismäßig wenig Injektionsgut gefunden. Das wiederum entspricht auch den Erkenntnissen einer detaillierten Auswertung der Verpreßmengen in Bezug auf die Lokalität der entnommenen Großbohrkerne. Dennoch liegen die gemessenen kleinsten Scherparameter mit c = 200 kN/m² sowie φ = 25° deutlich über den bisher aus dem Erkundungsprogramm gewonnenen Kennwerten.

Ebenfalls günstigere Werte als bisher angenommen, ergeben sich für die Verformbarkeit des Röttonsteins. Das gilt sowohl für den injizierten als auch für den nicht injizierten Fels, für den Verformungsmoduln in Abhängigkeit von der Tiefe für die vertikale Richtung zwischen 50 und 1000 MN/m² gemessen wurden. Berücksichtigt man, daß der untere Grenzwert aus den verhältnismäßig kleinmaßstäblichen Vertikallastplattenversuchen resultiert, kann großräumig wahrscheinlich von einem höheren unteren Grenzwert ausgegangen werden. In den untersuchten oberen Horizonten hat sich darüber hinaus für die horizontale Richtung ein gegenüber der vertikalen Belastungsrichtung höherer Verformungsmodul ergeben. Das Anisotropieverhältnis wurde hier zu 2 bis 4 ermittelt und liegt damit in einer Größenordnung, wie es auch von anderen Regionen bekannt ist. Dieses Ergebnis wirkt sich im Tunnelbau z.B. günstig auf die seitliche Bettung aus.

Die Injektion einzelner Horizonte des Röttonsteins führte zu einer Erhöhung des Verformungsmoduls der betreffenden Schichten. Allerdings ließ sich diese Aussage allein aus den Ergebnissen der Lastplatten und Triaxialversuche nicht ableiten, da die vier untersuchten Schichten nennenswert unterschiedliche Eigenschaften aufwiesen, die einen unmittelbaren Vergleich nicht zuließen. Berücksichtigt man jedoch das Ergebnis, daß ein bestimmter Prozentsatz des Baugrundvolumens mit Zementstein in Form horizontaler Lagen verfüllt wurde, so läßt sich daraus rechnerisch eine Erhöhung des Verformungsmoduls bis zum zweifachen Wert für die vertikale Richtung infolge Injektion ermitteln. Der Verformungsmodul für die horizontale Richtung steigt dagegen nur unbedeutend an. Darüber hinaus hat die Injektion zu einer Erhöhung des Primärspannungszustandes im Baugrund und damit gleichzeitig zu einer quasi Homogenisierung der Eigenschaften des Röttonsteins geführt.

Abschließend kann festgestellt werden, daß die in Verbindung mit dem Versuchsschacht und den geotechnischen Versuchen gewonnenen Erkenntnisse den Entwurf der beiden geplanten Tunnels günstig beeinflussen.

LITERATUR

Deutsche Gesellschaft für Erd- und Grundbau e.V.

479

(DGEG), 1985. Empfehlung Nr. 6 des AK 19:
Doppel-Lastplattenversuch im Fels.
Bautechnik 3: 102 - 106.

Hessisches Landesamt für Bodenforschung (HLfB),
1992. Ingenieurgeologisches Gutachen Röthberg
Tunnel, Ingenieurgeologisches Gutachten
Scheibenberg Tunnel. Unveröffentlicht.

Institut für Unterirdisches Bauen (IUB), Universität
Hannover, 1992. Tunnelbautechnisches Gutachten
Röthberg Tunnel. Unveröffentlicht.

Natau, O. und Mutschler, Th. 1989. Suggested
method for large scale sampling and triaxial testing
of jointed rock. Int. J. Rock Mech. Min. Sci.,
Vol 26: 427 - 434.

Schetelig, K. und Semprich, S. 1988. Ingenieurgeologische Aspekte bei der Ermittlung der Gebirgskennwerte im Buntsandstein. Proc. 8. Nat.
Felsmechanik-Symp. Aachen: 171 - 177.

Strauß, R. 1993. Geotechnik des gipsführenden Röts
(Oberer Buntsandstein) in Nordhessen. Proc. 9.
Nat. Tag für Ingenieurgeologie, Garmisch-
Patenkirchen: im Druck.

Grouting in Rock and Concrete, Widmann (ed.) © 1993 Balkema, Rotterdam, ISBN 90 5410 350 7

An analytical solution of ground reaction curves for grouted tunnels

Eine analytische Lösung für einen injektionsverstärkten Tunnel mittels Bodenwiderstandskurven

Yanting Chang
Department of Soil and Rock Mechanics, Royal Institute of Technology, Stockholm, Sweden

Lars Hässler
Golder Geosystem AB, Uppsala, Sweden

ABSTRACT: Ground reaction curve concept is often used in tunnelling engineering for analysing interactions between rock mass and supports. Whereas there are already analytical solutions of ground reaction curves, treating the ground as a homogeneous medium, there is no such a solution for a grouted tunnel, where the ground has two layers of material with different properties. This paper presents an analytical solution, based on the theory of elasticity and theory of plasticity. The rock is considered as a perfectly-plastic medium after it yields, and the dilation of the rock mass is taken into account. This solution can be used not only for analysing grouted tunnels, but also for analysing tunnels with weakened zone caused by blasting. Computer programs of the solution are now available. Comparisons between the theory and numerical calculations are also presented.

ZUSAMMENFASSUNG: Bodenwiderstandskurven werden oft in der Tunnelbautechnik verwendet um die Zusammenwirkung zwischen dem Fels und der Verstärkung zu analysieren. Während es analytische Lösungen für Bodenwiderstandskurven im homogenen Untergrund gibt, fehlen solche Lösungen für den Fall wo der Tunnel von zwei Schichten mit verschiedenen Materialeigenschaften umgeben ist. In dem vorliegenden Aufsatz wird eine analytische Lösung unter Verwendung der Elastizitäts- und Plastizitätstheorie präsentiert. Der Fels wird im Bruchzustand als ideal-plastisches Material behandelt und die Dilatation der Felsmasse berücksichtigt. Die vorgeschlagene Lösung kann auch zur Analyse eines Tunnels mit einer durch Sprengung aufgelockerten Grenzschicht verwendet werden. Berechnungsprogramme zur Lösung des Problems sind vorhanden. Ein Vergleich mit numerischen Lösungen wird auch präsentiert.

1. INTRODUCTION

Grouting is often used in tunnels in order to (i) tight the tunnel, (ii) enhance the strength of the rock mass. In accordance with the New Austrian Tunnelling Method (NATM), the Ground Reaction Curve concept (GRC) is often used to analyse the interaction between the rock mass and supports. Up to date, however, analytical solutions of the ground reaction curves are not available for grouted tunnels. This paper will present such an analytical solution, based on the theory of elasticity and theory of plasticity. The dilation of the rock mass is taken into account and the rock is considered as perfectly-plastic medium when it yields. Moreover, this solution is also applicable for tunnels with weakened zone caused by blasting.

The advantages of such solutions over numerical calculations are
i) it is easy to use,
ii) it is suitable for use *in-situ* because it gives immediate answer,
iii) it is convenient to make sensitivity studies of important parameters.

2. ANALYTICAL MODEL OF GROUTED TUNNEL

The model of a grouted tunnel to be analysed is shown in Figure 1. It is assumed that the tunnel is circular, the initial stress is hydro-static and the rock mass is grouted before tunnelling. At initial state ($p_i = p_o$), both grouted and ungrouted rock mass are elastic. When p_i is decreasing during tunnelling process, plasticity

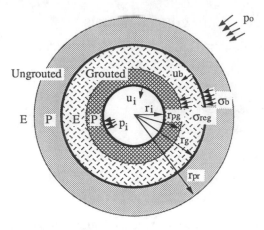

Ungrouted Grouted ub
E P E P Pi ui ri rpg σreg σb
rg
rpr

r_i = radius of the tunnel,
r_g = radius of grouted zone,
r_{pg} = radius of the plastic zone in the grouted zone,
r_{pr} = radius of the possible plastic zone in the
　　　ungrouted rock,
p_i = internal pressure acting on the tunnel wall,
p_o = initial ground stress,
σ_b = radial contact stress at the boundary between
　　　the grouted and ungrouted rock,
σ_{reg} = radial stress at the elastic-plastic boundary in
　　　the grouted rock,
u_i = radius displacement of the tunnel wall,
u_b = radius displacement at the boundary between
　　　the grouted and ungrouted rock.

Fig.1 Analytical model of a grouted tunnel

Analytisches Modell eines injektierten Tunnels

may exist in either grouted zone or ungrouted
zone or both zones. Because the grouted rock
can be fully elastic or elasto-plastic or
completely plastic, and the ungrouted rock can
be fully elastic and elasto-plastic, there will be
six cases that must be taken into account (see
Table 1). In practice all of these cases are not
likely to occur for a particular tunnel project.
But all of them must be considered in the
analytical solution.

In order to obtain the solution of the ground
reaction curve (GRC) (i.e. the relation between
p_i and u_i) of the grouted tunnel, the problem is
treated as a perfect contact problem which can
be described mathematically as

for ungrouted zone $u_{br} = f_1 (\sigma_{br}, p_o)$ (2.1)

Table 1 Six cases in the analytical solution

Case No	Grouted Rock	Ungrouted Rock	Symbol
1	Elastic	Elastic	E/E
2	Elastic	Elasto-Plastic	E/EP
3	Elasto-Plastic	Elastic	EP/E
4	Elasto-Plastic	Elasto-Plastic	EP/EP
5	Fully Plastic	Elastic	P/E
6	Fully Plastic	Elasto-Plastic	P/EP

for grouted zone: $\begin{cases} u_{bg} = f_2 \left(\sigma_{bg}, p_i \right) \\ u_i = f_3 \left(\sigma_{bg}, p_i \right) \end{cases}$ (2.2)

contact conditions: $\begin{array}{l} \sigma_{br} = \sigma_{bg} = \sigma_b \\ u_{br} = u_{bg} = u_b \end{array}$ (2.3)

where
u_{br} = displacement at the grout boundary from
　　　the ungrouted rock side,
u_{bg} = displacement at the grout boundary from
　　　the grouted rock side,
u_i = displacement of the tunnel wall,
s_{br} = radial stress acting on the ungrouted side
　　　at the grouted boundary,
s_{bg} = radial stress acting on the grouted side at
　　　the grouted boundary,
f_1 = a function describing the behaviour of the
　　　ungrouted rock,
f_2 and f_3 = functions describing the behaviour
　　　of the grouted rock.

Function f_1 for elastic and elasto-plastic
conditions has been achieved by different
authors, e.g. Stille (1983, 1989) and Hook &
Brown (1980). While function f_2 and f_3 for
elastic condition can be easily obtained from
the theory of elasticity (e.g. Timoshenko et al,
1970), the solutions for elasto-plastic or fully
plastic conditions have not been published up
to present. When the functions f_1 and f_2 for all
conditions are known, the contact stress σ_b can
be solved by using the contact conditions (2.3),
and then the displacement of the tunnel wall
can be obtained by f_3 .

This paper will present a detailed derivation
of function f_2 and f_3 for all the conditions of the
grouted zone. For the convenience to the
readers and the unity of the paper, the previous
solutions of f_1 will also be given. Then
solutions to all the six cases given in Table 1
will be presented.

3. SOLUTIONS FOR UNGROUTED ROCK

3.1 Elastic condition

By means of the plane strain elastic solution of a circular tunnel, equation

$$u_b = \frac{r_b}{2G}\left[p_o - \sigma_b\right] \tag{3.1}$$

holds for rock mass, where
G = shear modulus of ungrouted rock mass.

3.2 Elasto-plastic condition

For a perfect elasto-plastic material with the Mohr-Coulomb's failure criterion and a non-associated flow rule for the dilatancy after failure, the following solution for a circular tunnel in an infinite medium under hydrostatic initial ground pressure p_o will be achieved (Stille, 1983, 1989).

The deformation of the tunnel surface, u_i, is given as

$$\frac{u_b}{r_b} = \frac{A}{f+1}\left[2\left(\frac{r_p}{r_b}\right)^{f+1} + (f-1)\right] \tag{3.2}$$

$$\frac{r_p}{r_b} = \left[\frac{\sigma_{re}+a}{\sigma_b+a}\right]^{\frac{1}{k-1}} \tag{3.3}$$

where

r_p = radius of plastic zone in the ungrouted rock,

$$\sigma_{re} = \frac{2}{1+k}(p_o + a) - a ,$$

$$A = \frac{1+\mu}{E}(p_o - \sigma_{re}),$$

$$a = \frac{c}{\tan\phi} ,$$

$$k = \tan^2\left(45 + \frac{\phi}{2}\right)$$

c = cohesion of ungrouted rock mass,
ϕ = friction angle of ungrouted rock mass.
E = Young's modulus of ungrouted rock mass,
μ = Poisson's ratio of ungrouted rock mass,
f = the volume expansion after failure given by

$$f = \frac{\tan\left(45 + \frac{\phi}{2}\right)}{\tan\left(45 + \frac{\phi}{2} - \Psi\right)} ,$$

Ψ = dilatancy angle,

4. SOLUTIONS FOR GROUTED ROCK

4.1 Elastic condition

The grouted rock can be treated as a thick tube problem. Under plane strain condition, stresses in an elastic thick tube are given (Timoshenko, 1970)

$$\left\{\begin{array}{c}\sigma_r\\\sigma_t\end{array}\right\} = \left[\begin{array}{cc}\dfrac{r_b^2 - r^2}{r_b^2 - r_i^2}\dfrac{r_i^2}{r^2} & \dfrac{r^2 - r_i^2}{r_b^2 - r_i^2}\dfrac{r_b^2}{r^2}\\[2mm] -\dfrac{r_b^2 + r^2}{r^2 - r_i^2}\dfrac{r_i^2}{r^2} & \dfrac{r^2 + r_i^2}{r^2 - r_i^2}\dfrac{r_b^2}{r^2}\end{array}\right]\left\{\begin{array}{c}p_i\\\sigma_b\end{array}\right\} \tag{4.1}$$

From Hook's law for the plane strain condition, the total strains are given as

$$\left\{\begin{array}{c}\varepsilon_r\\\varepsilon_t\end{array}\right\} = \frac{1-\mu_g^2}{E_g}\left[\begin{array}{cc}1 & -\dfrac{\mu_g}{1-\mu_g}\\[2mm] -\dfrac{\mu_g}{1-\mu_g} & 1\end{array}\right]\left\{\begin{array}{c}\sigma_r\\\sigma_t\end{array}\right\} \tag{4.2}$$

where

E_g = elastic modulus of the grouted rock,
μ_g = Poisson's ratio of the grouted rock.

Because we are interested in the deformation caused by the excavation, the initial strain ε_o existing in the rock mass before the excavation should be subtracted from the total strains. The initial strain is determined by

$$\varepsilon_o = \frac{(1+\mu_g)(1-2\mu_g)}{E_g}p_o = -Lp_o \tag{4.3}$$

$$L = -\frac{(1+\mu_g)(1-2\mu_g)}{E_g}$$

Therefore, from the compatibility conditions, the displacement caused by the excavation is expressed by

$$\frac{u}{r} = \varepsilon_t - \varepsilon_o \tag{4.4}$$

Substituting equations (4.1), (4.2) and (4.3) into (4.4) will give

$$\frac{u}{r} = M(r)\,p_i + N(r)\,\sigma_b + L\,p_o \qquad (4.5)$$

where

$$M(r) = \frac{1-\mu_g^2}{E_g}\left(-\frac{r_b^2 + r^2}{r_b^2 - r_i^2}\frac{r_i^2}{r^2} - \frac{\mu_g}{1-\mu_g}\frac{r_b^2 - r^2}{r_b^2 - r_i^2}\frac{r_i^2}{r^2}\right)$$

$$N(r) = \frac{1-\mu_g^2}{E_g}\left(\frac{r^2 + r_i^2}{r_b^2 - r_i^2}\frac{r_b^2}{r^2} - \frac{\mu_g}{1-\mu_g}\frac{r^2 - r_i^2}{r_b^2 - r_i^2}\frac{r_b^2}{r^2}\right)$$

When r is set to r_i, we will obtain the displacement of the tunnel wall

$$\frac{u_i}{r_i} = M(r_i)\,p_i + N(r_i)\,\sigma_b + L\,p_o \qquad (4.6)$$

and when $r = r_b$ the displacement at the boundary between the grouted and ungrouted rock is given by

$$\frac{u_b}{r_b} = M(r_b)\,p_i + N(r_b)\,\sigma_b + L\,p_o \qquad (4.7)$$

4.2 Elasto-plastic condition

When the grouted rock become plasticized, the rock is divided into two zones, the plastic zone and elastic zone. The radius of the plastic zone is noted as r_{pg} (refer to Figure 1).

From the compatibility conditions, the deformation in the plastic zone caused by excavation can be written as

$$\frac{u}{r} = \varepsilon_t\,, \qquad \frac{du}{dr} = \varepsilon_r \qquad (4.8)$$

where
ε_r , ε_t = relative radial and tangential strain in the plastic zone caused by excavation. These strains consist of a plastic part and an elastic part, i.e.

$$\varepsilon_t = \varepsilon_t^p + \varepsilon_t^e\,, \qquad \varepsilon_r = \varepsilon_r^p + \varepsilon_r^e \qquad (4.8a)$$

The non-associated flow law for the plastic strains is written as

$$\varepsilon_r^p = -f_g\,\varepsilon_t^p \qquad (4.9)$$

where f_g is the volume expansion factor after

failure of the grouted rock. Combining Equation (4.8), (4.8a) and (4.9) will give following differential equation

$$\frac{du}{dr} + f_g\,\frac{u}{r} = \varepsilon_r^e + f_g\,\varepsilon_t^e \qquad (4.10)$$

If we assume that the elastic strains are constant in the plastic zone and equal to the value at the e-p boundary ($r = r_{pg}$), the solution of equation (4.10) can be easily obtained. Then the deformation of the tunnel wall is expressed as

$$\frac{u_i}{r_i} = \frac{1}{f_g + 1}\left(\left(\varepsilon_t^e - \varepsilon_r^e\right)\left[\frac{r_{pg}}{r_i}\right]^{f_g + 1} + \varepsilon_r^e + f_g\,\varepsilon_t^e\right) \qquad (4.11)$$

where the elastic strains can be determined by

$$\varepsilon_t^e = \frac{1-\mu_g^2}{E_g}\left(\sigma_{teg} - \frac{1-\mu_g}{\mu_g}\sigma_{reg}\right) + L\,p_o$$

$$\varepsilon_r^e = \frac{1-\mu_g^2}{E_g}\left(\sigma_{reg} - \frac{1-\mu_g}{\mu_g}\sigma_{teg}\right) + L\,p_o$$

and

$\sigma_{reg}, \sigma_{teg}$ = radial and tangential stress respectively at the e-p boundary, which will be determined later.

Assuming a perfect elasto-plastic material, the radius of the plastic zone r_{pg} in the grouted rock is given by

$$\frac{r_{pg}}{r_i} = \left(\frac{\sigma_{reg} + a_g}{p_i + a_g}\right)^{\frac{1}{k_g - 1}} \qquad (4.12)$$

where

$$a_g = \frac{c_g}{\tan\phi_g}$$

$$k_g = \tan^2\left(45 + \frac{\phi_g}{2}\right)$$

c_g = cohesion of the grouted rock mass,
ϕ_g = friction angle of the grouted rock mass.

The displacement at the outer boundary of the grouted zone ($r = r_b$) is obtained by the compatibility condition and the fact that the outer part of the rock is elastic, e.g.

$$\frac{u_b}{r_b} = \varepsilon_{tb} + L\,p_o$$

$$= \frac{1 - \mu_g^2}{E_g}\left(\sigma_{tb} - \frac{\mu_g}{1 - \mu_g}\sigma_b\right) + L\,p_o \qquad (4.13)$$

where

ε_{tb} = elastic tangential strain at $r = r_b$,
σ_{tb} = tangential stress at $r = r_b$,

Following is to find expressions for the stresses involved in calculating the displacements. The stresses in the elastic zone can be obtained by the same procedure as given the previous section.

$$\left\{\begin{matrix}\sigma_r\\\sigma_t\end{matrix}\right\} = \begin{bmatrix}\dfrac{r_b^2 - r^2}{r_b^2 - r_{pg}^2}\dfrac{r_{pg}^2}{r^2} & \dfrac{r^2 - r_{pg}^2}{r_b^2 - r_{pg}^2}\dfrac{r_b^2}{r^2}\\[3mm] -\dfrac{r_b^2 + r^2}{r_b^2 - r_{pg}^2}\dfrac{r_{pg}^2}{r^2} & \dfrac{r^2 + r_{pg}^2}{r_b^2 - r_{pg}^2}\dfrac{r_b^2}{r^2}\end{bmatrix}\left\{\begin{matrix}\sigma_{reg}\\\sigma_b\end{matrix}\right\} \qquad (4.14)$$

From equation (4.14), the tangential stress σ_{tb} at $r = r_b$ is given

$$\sigma_{tb} = -\frac{2\,r_{pg}^2}{r_b^2 - r_{pg}^2}\sigma_{reg} + \frac{r_b^2 + r_{pg}^2}{r_b^2 - r_{pg}^2}\sigma_b \qquad (4.15)$$

The tangential stress σ_{teg} at $r = r_{pg}$ can be written as

$$\sigma_{teg} = -\frac{r_b^2 + r_{pg}^2}{r_b^2 - r_{pg}^2}\sigma_{reg} + \frac{2\,r_b^2}{r_b^2 - r_{pg}^2}\sigma_b \qquad (4.16)$$

It should be pointed out that σ_{reg} is a dependent variable of σ_b. In order to calculate σ_{tb} and σ_{teg} from equation (4.15) and (4.16), the relation between σ_{reg} and σ_b should be derived. Note that on the e-p boundary, the radial stress σ_{reg} and tangential stress σ_{teg} must meet the Mohr-Coulomb's criterion, i.e.

$$\frac{\sigma_{teg} - a_g}{\sigma_{reg} - a_g} = k_g \qquad (4.17)$$

Eliminating σ_{teg} from equation (4.16) and (4.17) will give the relation

$$\sigma_{reg} = -\frac{(k_g - 1)\left(r_b^2 - r_e^2\right)a_g - 2r_b^2\sigma_b}{\left(r_b^2 - r_e^2\right)k_g + \left(r_b^2 + r_e^2\right)} \qquad (4.18)$$

If the contact stress σ_b is known, all other stresses can be obtained from equation (4.18), (4.15) and (4.17). Then the elastic strains and the radius of the plastic zone can also calculated. Consequently the displacements u_i and u_b can be determined.

4.3 Completely plastic condition

The grouted zone will have no resistance to deformation when it becomes completely plastic, since the grouted rock is assumed to be a perfectly plastic material. In other words, the deformation of a completely plastic rock is arbitrary. However, the grouted rock in our case is bonded to the outer rock mass whose displacements u_b is determined by equation (3.1) or (3.2), i.e. the grouted rock is subjected to a defined displacement u_b. By using the compatibility conditions, the relation between u_b and u_i can be derived

$$u_i = u_b\left(\frac{r_b}{r_i}\right)^{f_g} + \frac{\varepsilon_r^e - f_g\,\varepsilon_t^e}{1 + f_g}\left[r_i - \left(\frac{r_b}{r_i}\right)^{f_g}r_b\right] \qquad (4.19)$$

where ε_r^e and ε_t^e are the elastic strains in the plastic grouted rock, which are assumed to be constant and equal to the value at $r = rb$, i.e.

$$\varepsilon_r^e = \frac{1 - \mu^2}{E}\left(\sigma_b - \frac{\mu}{1 - \mu}\sigma_{tb}\right) \qquad (4.20a)$$

$$\varepsilon_t^e = \frac{1 - \mu^2}{E}\left(\sigma_{tb} - \frac{\mu}{1 - \mu}\sigma_b\right) \qquad (4.20b)$$

where σ_b and σ_{tb} are determined by following equations

$$\sigma_b = \left(p_i + a_g\right)\left(\frac{r_b}{r_i}\right)^{k_g - 1} - a_g \qquad (4.21a)$$

$$\sigma_{tb} = \left(p_i + a_g\right)k_g\left(\frac{r_b}{r_i}\right)^{k_g - 1} - a_g \qquad (4.21b)$$

4.4 Solutions for the six cases of grouted tunnel

Based on the solutions given in the previous sections for grouted and ungrouted rock, we can obtain solutions for each particular case given in Table 1 (see Table 2). To determine which case the tunnel belongs to, the "tree

Table 2 Solutions for the cases

Case	Contact condition	u_i
E/E	Eq.(4.7) & (3.1)	(4.6)
E/EP	Eq.(4.7) & (3.2)	(4.6)
EP/E	Eq.(4.13) & (3.1)	(4.11)
EP/EP	Eq.(4.13) & (3.2)	(4.11)
P/E	u_b = (3.1)	(4.19)
P/EP	u_b = (3.2)	(4.19)

Table 3 Parameters used in the calculations

Initial ground stress	1 MPa
Radius of tunnel	4.7m
Radius of grouted zone	7.7m
Radius of weakened zone	7.7m

Table 4 Properties of rock masses

	Tunnel with grouted zone		Tunnel with weakened zone	
	Origin rock	Grouted rock	Origin rock	Weak rock
E (GPa)	0.1	0.5	0.5	0.1
μ	0.2	0.2	0.2	0.2
ϕ	10°	15°	15°	10°
c (MPa)	0.1	0.2	0.2	0.1
ψ	10°	10°	10°	10°

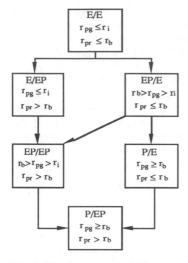

Fig. 2 The "tree approach" to determine the calculation flow

Die Verzweigungs-Methode zur Bestimmung des Berechnungsganges

approach" shown in Figure 2 is used in the computer program.

5. COMPARISON WITH NUMERICAL CALCULATIONS

In order to verity the formulation and the computer programs, ground reaction curves obtained from the solution are compared with that from numerical calculations by FLAC, version 3.22, for both tunnel with grouted zone and tunnel with weakened zone. Parameters used for the calculations and the properties of rock masses are given in Table 3 and Table 4 respectively. The mesh for the numerical model is shown in Figure 3. Results are shown in Figure 4 and Figure 5. It can be seen from the results that a good agreement is obtained

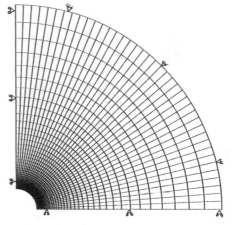

Fig.3 Mesh for numerical calculations

Netz für die numerische Berechnung

between the analytical solution and numerical calculations.

6. CONCLUSIONS

The analytical solution presented in this paper provides a useful tool for obtaining ground reaction curves for grouted tunnels. The advantage of such analytical solutions is that immediate answer can be obtained once the input data are available. Therefore it is very

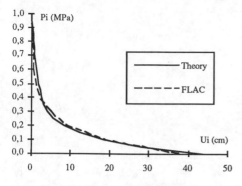

Fig. 4 Comparison between the theory and numerical results for tunnel with grouted zone

Vergleich zwischen der Theorie und dem numerischen Resultat für einen Tunnel mit einer injektierten Zone

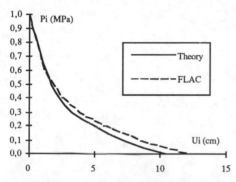

Fig. 5 Comparison between the theory and numerical results for tunnel with weakened zone

Vergleich zwichen der Theorie und dem numerischen Resultat für einen Tunnel mit einer geschwächten Zone

advantageous for the site engineers to use on a tunnel construction site.

The comparison between the results from the analytical solution and numerical calculations shows that a good agreement is obtained. This indicates that the formulation and computer program are reliable.

7. REFERENCES

CHANG, Y., 1992. "Influence of Early-Age Properties of Shotcrete on Tunnel Construction Sequences", Licentiate Thesis, Dept. of Soil and Rock Mechanics, the Royal Institute of Technology, Stockholm, Sweden.

HOEK, E., & BROWN, E. T., 1982. "Underground Excavation in Rocks", The Institute of Mining and Metallurgy.

STILLE, H., HOLMBERG, M. & NORD, G., 1989. "Support of Weak Rock with Grouted Bolts and Shotcrete", Int. J. Rock Mech. Min. Sci. & Geomech. Abstr.,Vol.26, No.1, pp.99-113.

STILLE, H., 1983. "Theoretical Aspects on the Difference between prestressed Anchor Bolt and Grouted Bolt in Squeezing Rock", Proc. of the Int. Symp. on Rock Bolting, Abisko, 28 Aug.-2 Sept.

TIMOSHENKO, S. P ., GOODIER, J. N., 1970. "Theory of Elasticity", Mc Graw-Hill Book Company.

5 Supplement
Anhang

Rheological properties of cement-based grouts – Measuring techniques and factors of influence

Rheologische Eigenschaften von zementbasierenden Injektionsgut – Messtechnik und Einflußfaktoren

Ulf Håkansson
ABV Rock Group KB, Saudi Arabia

Lars Hässler
Golder Associates AB, Sweden

Håkan Stille
Royal Institute of Technology, Sweden

ABSTRACT: Cement is today the most commonly used material for grouting. The suspended cement particles have a significant influence on the penetration performance and rheology of the grout. The present paper concerns the measurement of rheological properties of cement-based grouts and how the properties are influenced and modified by different factors and additives. Important factors are experimental procedure, water/cement ratio, specific surface, temperature and hydration. Manipulations comprises the use of additives such as superplasticizers, bentonite, silica fume, sodium silicate and calcium chloride. In the paper, different measuring methods are shown and their applicability for the assessment of the rheological properties is discussed.

ZUSAMMENFASSUNG: Zement ist heu er das weit verbreitetste Material für Injektionen. Die Eigenschaft der Zementteilchen hat einen entscheidenden Einfluss auf das Eindringverhalten und die Rheologie des Injektionsgutes. Die vorliegende Schrift beschäftigt sich mit der Messung der rheologischen Eigenschaften von auf zement-basierendem Injektionsgut, und wie diese Eigenschaften von verschiedenen Faktoren und Zusätzen beeinflusst und verändert werden. Wichtige Faktoren sind die experimentellen Vorgangsweisen, der Wasser-Zement-Wert, die spezifische Oberfläche, Temperatur und Hydratation. Die Behandlung umfasst die Verwendung von Zusatzmittel, wie Superplasticizers, Bentonit, Silikastaub, Natriumsilikat und Kalziumchloride. In der Schrift werden verschiedene Messmethoden und deren Anwendbarkeit für die Einschätzung der rheologischen Eigenschaften diskutiert.

1 INTRODUCTION

Grouting is generally used as a means of changing or improving the physical properties of a material, e g permeability, deformability or strength. A commonly used method is permeation grouting, whereby a grout material by pressure is forced into voids and fractures of a geological formation or man-made structure. Depending on the prevailing in-situ conditions, various types of grouts are used. The grouts can be divided into two main groups with, in many cases, significantly different characteristics:

*   suspensions

*   solutions

the former incorporates cement-based grouts and the latter most of the available chemical grouts. The two

types differ in characteristics like penetrability/groutability, rheology (i e flow behaviour), strength, durability and intrinsic permeability. Many of the above mentioned characteristics are unknown for both types of grouts (see Table 1) and development in measuring techniques and standards are yet to be undertaken.

Due to ease of preparation and use, wide availability and a relatively low cost cement-based grout is today the most commonly used material for permeation grouting. The strength, durability and intrinsic permeability must be regarded as well documented, and the latter two properties have received considerable attention in nuclear waste management research the last few years (Onofrei et al., 1991). Nevertheless, when it comes to penetrability and rheology of cement suspensions, a lot of research work remains to be done.

It is the suspended cement particles that have a

Table 1   Present knowledge concerning grout characteristics
Tab.1   Der gegen wärtige wisserstand über die Eigenschaften von injektionsgut

Characteristic	Suspensions	Solutions
Penetrability	limited, depending on particle characteristics (size, shape etc)	similar to water
Rheology	non-Newtonian, thixo-tropic, time dependent	Newtonian, time dependent
Stability	separation (sedimentation, filtration)	syneresis (shrinkage)
Strength	well known from cement science (high)	not well known (very low to low)
Durability	well known from cement science (durable)	not well known
Permeability	known from recent research	unknown

significant influence on the penetrability and rheological behaviour of the grout. The particles will affect the flow properties and set a limit to void size that can be penetrated. Moreover, if the particles are separated from the fluid phase due to pressure (filtration) or gravity (sedimentation) the consequence can be an insufficient and unpredictable sealing of the void.

Apart from the penetration performance due to particle size and stability, understanding and controlling the rheology of the grout is essential for theoretical analysis (Hässler, 1991) and for a successful grouting operation. The knowledge will, under the prevailing conditions, facilitate the choice of cement type, additives, grouting pressure, hole spacing and the estimate of grout take (volume). However, it must be clearly stated that the geometry of the conduits that are to be grouted, will always be a very difficult parameter to define.

The present paper reviews the current knowledge concerning the measurement of the rheological properties of cement-based grouts, with emphasis on methods and equipment.

2 RHEOLOGY OF CEMENT GROUTS

The rheological behaviour of cement-based grouts must be considered as very complex. Due to interactions between the two phases and between the particles themselves, the grout can be non-Newtonian, thixotropic and possess a yield stress. The behaviour is also influenced by the hydration of the cement, inevitably causing changes in the properties with time.

Rheology plays an important role in grouting since it determines the relationship between pressure and flowrate. Moreover, a property such as the yield

stress will limit the penetration length, i e the distance the grout will reach at a certain pressure.

The true rheological behaviour is often approximated by a rheological model, incorporating various rheological parameters. Any model used for dense cement grouts must take into account the yield stress that is typical for these type of fluids. The simplest model including a yield stress is the Bingham model

$$\tau = \tau_o + \mu_B \gamma \tag{2.1}$$

another one is the Casson model

$$\tau^{1/2} = \tau_o^{1/2} + \eta_\infty^{1/2} \gamma^{1/2} \tag{2.2}$$

and a third the Herschel-Bulkley model

$$\tau = \tau_o + k \gamma^m \tag{2.3}$$

Since the geometry of the conduits in the media that is to be grouted is so difficult to define (Hässler, 1991), it does not make sense to search for over-sophisticated rheological models in grouting practice. This implies that in many situations the Bingham model is a reasonable approximation of the true rheological behaviour, see Fig.1.

The Bingham model is a two-parameter model, consisting of a plastic viscosity, m_B, and a yield stress, t_o. Both parameters affects the relationship between pressure and flowrate, but as stated earlier the yield stress will also influence the maximum penetration length (Lombardi, 1985), (Hässler, 1991).

2.1 Apparent viscosity

Regrettably, the only property measured and reported by many workers is the so called 'apparent' viscosity. The 'apparent' viscosity denotes a combined effect of the true rheological properties and can, thus, only be used for qualitative comparisons. As an example, the 'apparent' viscosity for a Bingham fluid can be written (see Fig.2)

$$\eta_A = \mu_B + \frac{\tau_o}{\dot{\gamma}} \qquad (2.4)$$

It is obvious that if one Bingham fluid exhibits a higher 'apparent' viscosity than another at a certain shear rate, it can be due to either a higher plastic viscosity or a higher yield stress (or both of them being higher). The 'apparent' viscosity cannot be used for theoretical considerations unless one of the properties and the shear rate are known.

2.2 Yield stress

Inter particle forces between the solids in a suspension result in a yield stress that must be exceeded in order to initiate flow. Below the yield stress the suspension behaves like a (weak) solid. Thus, the yield stress can be regarded as the material property that represents the transition between solid-like and liquid-like behaviour (Keating & Hannant, 1989). Many (dense) suspensions exhibit a yield stress, e g cement paste, cement grouts, bentonite/water mixtures and residue from the mining industry.
The yield stress is defined in British Standards Institution (BS 5168) as: "the stress below which the material is an elastic solid and above which it is a liquid with plastic viscosity".
Many suspensions may not only have one unique yield stress value, but rather a static yield stress (gel strength) at zero shear rate and a dynamic yield stress, lower than the former, under shearing as the internal structure of the suspension breaks down (Keating & Hannant, 1989). The dynamic yield stress reflects the combined result of mechanical breakdown and the build-up due to chemical or physical effects (Haimoni & Hannant, 1988).
The two different yield stress levels are attributed to the thixotropy (see below) of the grout.

2.3 Thixotropy

By definition thixotropy is defined as: "a gradual decrease of the viscosity under shear stress followed by a gradual recovery of structure when the stress is

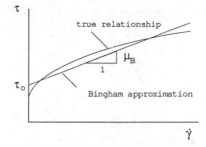

Figure 1.   True rheological relationship together with the Bingham approximation

Die gemessenen rheologische Eigenschaften zusammen mit der Bingham Näherung

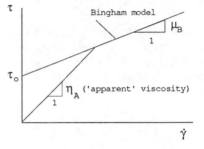

Figure 2.   The Bingham model together with the 'apparent' viscosity

Das Bingham Modell zusammen mit der 'scheinbaren' Viskosität

removed" (Barnes et al. 1989). Thixotropic behaviour is frequently observed in many fine particle suspensions. It is a generally accepted view, that the behaviour is ascribed to isothermal, time-dependent and reversible breakdown of the particulate structure under shear, followed by structural reformation at rest. Structural breakdown involves two opposing processes (Nguyen & Boger, 1985). On the one hand, the applied shear force acts to disrupt structural bonds interlinking the particles or aggregates. Simultaneously, however, shear induced collisions between the elements tend to reform part of the broken bond.
Thixotropy is an important phenomenon in practical grouting since it is the magnitude of the yield stress after a short 'rest' that will determine when the grout has enough strength to withstand the prevailing water pressure. An 'ideal' grout should have a negligible yield stress during penetration combined

with an instant increase when the grout has reached its final position in the void.

3 MEASURING TECHNIQUES

3.1 Capillary viscometer

In a capillary viscometer the fluid flows in circular pipes with a known radius and the pressure difference, ΔP, and flow rate, Q, is registered over a distance, L, of the pipe. The rheological behaviour is assessed by plotting wall shear stress versus wall shear rate. The shear stress at the wall is directly given by

$$\tau_w = \frac{\Delta P\, R}{2L} \qquad (3.1)$$

where ΔP is the pressure difference over the distance, L, and R is the radius of the pipe. The shear rate is more difficult to achieve and incorporates a somewhat tedious operation. The shear rate at the wall is given by

$$\dot{\gamma}_w = \left(\frac{3 + \beta}{4}\right)\left(\frac{4Q}{\pi R^3}\right) \qquad (3.2)$$

where

$$\beta = \frac{d\, log\left(\frac{4Q}{\pi R^3}\right)}{d\, log\, \tau_w} \qquad (3.3)$$

the first factor on the right hand side of eq.(3.2) is known as the Rabinowitsch correction term and the second factor is simply the shear rate of a Newtonian fluid. As seen in eq.(3.3), the factor β can be found by the slope of a log-log plot of Newtonian wall shear rate $(4Q/\pi R^3)$ versus wall shear stress $(\Delta PR/2L)$.

It is important to note that Eq.(3.2) is general and applies to all fluids independent of rheological behaviour and whether or not they possess a yield stress.

If a rheological model is known and applicable the procedure is greatly simplified. The flow rate (and mean velocity) is then found by inserting the rheological model into the following expression,

$$Q = \frac{\pi R^3}{\tau_w^3} \int_{\tau_o}^{\tau_w} \tau^2\, \dot{\gamma}\, d\tau \qquad (3.4)$$

Not much work is reported on the rheology of cement suspensions using capillary viscometers, but a thorough investigation has been conducted on concentrated bauxite residue (Nguyen & Boger, 1983). Although the residue is not a cementitious material, many similarities exists and the measuring problems are the same.

A major problem when using capillary viscometers is slip at the fluid-solid interface, i e at the pipe wall. The problem has been observed by many workers and corrections are made in order to eliminate/minimise the slip effect (Mannerheimer, 1990).

3.2 Rotational viscometer

A commonly used device for rheological measurements is a concentric-cylinder rotational viscometer. The viscometer measures the torque that is needed to maintain a chosen angular speed of a cylinder (bob) that is immersed into the fluid, see Fig.3. From the torque the shear stress can be calculated and by varying the speed, data is produced from which the rheological behaviour can be evaluated. The technique used for the assessment of the rheological behaviour of suspensions is in detail described elsewhere (Nguyen & Boger, 1987), (Håkansson et al, 1991).

The shear stress at the wall of the rotating cylinder is directly calculated from the recorded torque, M, according to

$$\tau_{r\theta} = \frac{M}{2\pi r^2 h_b} \qquad (3.5)$$

where r and h_b is the radius and height of the cylinder, respectively.

An important feature that appears when measuring yield stress fluids with this device is the development of different flow regimes. Since the shear stress in the fluid decreases from the bob to the cup, under the condition where the shear stress at the cup is less than the yield stress, only part of the fluid next to the bob will be sheared while the rest of the material will remain in a solid state (Nguyen & Boger, 1987), (Håkansson et al., 1991). When this occurs, conventional methods are no longer applicable since they do not take into account the presence of a variable effective gap which is smaller than the physical spacing between the bob and the container (cup).

If the yield stress is known a-priori, and thereby the nature of the flow regime the shear rate for the partially sheared case is exactly given by

$$\dot{\gamma}_b = 2\tau_b \frac{d\Omega}{d\tau_b} \qquad (3.6)$$

which is identical to the solution for a single cylinder rotating in an infinite fluid medium. By using a relatively large measuring cup and knowing that the fluid has a yield stress, eq.(3.6) can be used to determine the rheological behaviour by utilising the partially sheared data (Håkansson, 1993).

For the fully sheared case, however, only approximate solutions are available but the error involved in using the approximations is often negligible (Nguyen & Boger, 1987).

The yield stress is often, with this device, assessed indirectly by curve fitting to a rheological model and by extrapolation down to zero shear rate. The accuracy of the procedure is much dependent on the accuracy of the data itself and how the measurement is made. Moreover, in order to achieve an acceptable result it is important to generate data in the low-shear rate range (i e close to zero).

Slip at the fluid-solid interface is also encountered in this geometry and many workers have reduced the problem by roughening the cylinders.

A great advantage with this type of instrument is that the measurement can continue in the same sample for a long time, which implies that the time dependency can be evaluated.

Stress-relaxation method

Although the conventional approach to assess the

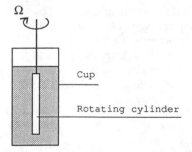

Figure 3.   In a concentric-cylinder viscometer the torque, needed to rotate a submerged cylinder, is measured

In einem Rotationsviskometer wird das Drehmoment gemessen, das für die Bewegung eines eingetauchten Zylinders notwendig ist

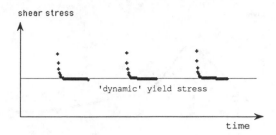

Figure 4.   Typical data from the stress-relaxation method. The minimum value denotes the dynamic yield stress

Typische Daten der Stress-Relaxation Methode. Der Minimal- wert beschreibt die dynamische Fließgrenze

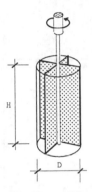

Figure 5.   In the shear-vane method, a four-bladed vane is immersed into the sample and rotated at a constant velocity

In der Shear-Vane Methode wird ein vershaufeliges Blatt in der Probe versenkt und mit eine konstanten Geschwindigkeit rotiert

yield stress involves an indirect extrapolation, the yield stress can also be directly estimated with the rotational viscometer, by a procedure called the stress-relaxation method. The ordinary measuring cylinder is used and rotated at an arbitrarily chosen constant speed for a time that is long enough to achieve a broken down state (i e equilibrium conditions). The rotation is then suddenly stopped, and if the fluid has a yield stress a remaining shear stress will be acting on the cylinder and preventing it from returning to its zero stress position (Nguyen & Boger, 1983). The torque is continuously registered as a function of time and a typical output from four consecutive measurements is shown in Fig.4. The

495

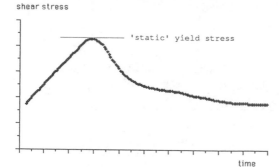

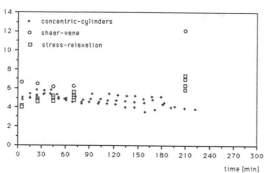

shear stress

'static' yield stress

time

Figure 6.   Typical data from the shear-vane method. The peak value denotes the static yield stress

Typische Daten der Shear-Vane Methode. Der Maximal wert beschreibt die statische Fließgrenze

yield stress [Pa]

+ concentric-cylinders
o shear-vane
□ stress-relaxation

time [min]

Figure 7.   A comparison of the yield stress evaluated with different methods

Ein Vergleich der auf verschieden Messmethoden basierenden Fließgrenze

minimum remaining shear stress is taken as the dynamic yield stress.

Shear-vane

The shear-vane is a direct method to assess the static yield stress. The measurement is analogous to the method used in soil mechanics to evaluate the undrained shear strength of clays in-situ. The above mentioned rotational viscometer is used but a four-bladed vane is utilised as a measuring device instead of the conventional cylinders, see Fig.5.

The vane is slowly rotated with a constant speed (0.1 RPM) and the torque is continuously registered.

The static yield stress is calculated from the maximum ('peak') torque, see Fig.6, by the relationship (Nguyen & Boger, 1985)

$$\tau_o = \frac{2M}{\pi D^3 \left(\dfrac{H}{D} + \dfrac{1}{3} \right)} \tag{3.7}$$

where M is the torque. The relationship is valid assuming a uniform shear stress acting on the cylindrical failure surface.

The vane has been successfully used the last few years, in determination of the yield stress for various fluids including cement slurries, e g Nguyen & Boger (1983), Nguyen & Boger (1985), Haimoni & Hannant, (1988), Barnes & Carnali (1990). A great advantage is that the vane does not disturb the sample when introduced into the grout. Also, slip is prevented since the shearing takes place in the grout itself.

A comparison has been made of the yield stress evaluated with the rotational viscometer using the conventional cylinders assuming the Bingham model, the shear vane and finally the stress-relaxation method, see Fig.7. The cement in the example is a Portland, Micro-cement (Mikrodur P-F, by Dyckerhoff) using w/c = 1 and no additives. As expected the shear vane gives the higher values and the stress-relaxation the lower. The concentric-cylinders should give yield stress values close to the ones obtained by the stress-relaxation method, but as seen in Fig.1 the assumption of the Bingham model will always overestimate the yield stress if the true rheological behaviour is yield pseudo-plastic as was the case in these measurements (Håkansson 1993).

The dynamic yield stress will influence the flow behaviour as long as the grout is in a broken down state, i e during penetration. On the other hand, it is the static yield stress that will determine how high water pressure the grout can withstand when the packer is removed. Consequently both values are of importance in grouting practice.

3.3   Marsh-cone

The Marsh-cone is a simple instrument used in the oil industry to measure the flow performance of drill muds in the field, see Fig.8.

By measuring the time it takes for a certain volume to flow out of the cone the flow properties can be estimated.

However, the only property that can be measured with the Marsh Cone is the 'apparent' viscosity, i e a combination between the true rheological properties. The efflux time can generally be expressed (Håkansson, 1993)

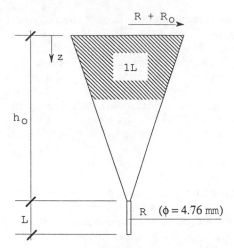

$$T = \int_0^{z_1} \frac{\left(R + R_o \left[1 - \dfrac{z}{h_o} \right] \right)^2}{\bar{v}_z R^2} \, dz$$

(3.8)

where v_z is the mean velocity through the outlet pipe. The used dimensions are shown in Fig.8. Equation (3.8) can only be solved analytically for a Newtonian fluid and for other fluids a numerical approach must be used (Håkansson, 1993).
The density of the grout must be known and it can be estimated by the used of a 'mud-balance', commonly used in the oil industry.

Figure 8.   By measuring the time it takes for a certain volume to flow out of a Marsh-cone, the flow properties can be estimated

Über die Messung der Zeit, die ein bestimmtes Volumen benötigt, um aus einem Marsh-Trichter auszufließen, können die reologische Eigenschaften bestimmt werden

3.4   The Raise-Pipe

A simple way of assessing the yield stress is to utilise the fact that the grout stops when the shear stress at the wall of a pipe equals to the yield stress. The principle has been used in the development of an in-situ measuring device called the 'Raise Pipe' (Håkansson et al, 1992). The grout is poured into a large pipe (R = 30 mm) situated in the middle of the device, see Fig.9, from where it flows up into smaller pipes (R = 2-4 mm) situated at the periphery. If the grout has a yield stress it will not reach the same level in the smaller pipes as the level is in the large pipe. Knowing the density of the grout the yield stress can be estimated by

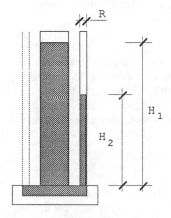

$$\tau_o = \frac{\gamma R}{2} \left(\frac{h_1}{h_2} - 1 \right)$$

(3.9)

where γ is the specific weight of the grout, R the radii of the smaller pipes, h_1 the level in the pipes at the periphery and h_2 the level in the large pipe in the middle.
A similar principle has been used by Lombardi (1985) by dipping a plate with known weight into the grout. If the grout possesses a yield stress a grout layer will remain on the plate when lifted and from the weight difference the yield stress can be estimated.

Figure 9.   The 'Raise-Pipe' is a direct method to estimate the yield tress in-situ

Das 'Steigrohr' ermöglich eine direkte Bestimmung der In-situ-Fließgrenze

4 RHEOLOGICAL RESULTS

In the study a Brookfield Rheoset concentric-cylinder rotational viscometer has been used (Håkansson, 1993). The rheology of the cement-based grouts is assumed to be characterised by the Bingham model.

The measurements are made in such a way that the grout is in a broken down state and hence, it is the dynamic yield stress and plastic viscosity that is assessed. The measurements are made in the same sample for an extended period of time, and thereby the time dependency during hydration can be evaluated. Different factors and additives are investigated in order to see their effect on the rheology of the grout. The results are summarised in Table 2.

Water/cement ratio

The water/cement ratio, i e the concentration, is the single most important factor influencing the rheology. A decrease in the water/cement ratio will result in an exponential increase in both the yield stress and the viscosity.

Cement characteristics

The most significant characteristic is the fineness or specific surface (surface area/weight) of the cement. The specific surface is commonly measured with the Blaine method. An increase in specific surface, i e going towards micro-cement, will inevitably cause an increase in the yield stress and the viscosity.

Temperature

In general a decrease in temperature leads to an increase in viscosity and vice versa. This is also true for cement grouts and seems to be the fact even for the yield stress. However, more work is needed in order to verify the effect of temperature on the yield stress.

Bentonite

Bentonite is regarded as inert, i e it does not participate in the chemical reaction of the cement during hydration. Before the use of micro-cements, bentonite was added to ordinary cement in order to achieve a stable grout. However, by the introduction of micro-cement the role of bentonite is becoming less important. It should also be emphasised that the bentonite particles are often larger than the micro-cement particle and thus should not be used together. Bentonite will influence the rheology of the grout by increasing both the yield stress and the viscosity. Also, by adding bentonite the grout becomes more thixotropic.

Silica fume

Silica fume is a pozzolanic material which like bentonite will enhance the stability of the grout. However, the silica fume particles are much smaller (0.1×10^{-6}m average diameter) than the bentonite particles and can therefore be used together with micro-cement. The use of silica fume will also enhance the durability and decrease the permeability of the grout (Onofrei et al., 1991). The addition of silica fume will increase both the yield stress and the viscosity and to some extent decrease the setting time.

Superplasticizer

Superplasticizers have a significant influence on the rheology of the grout and will decrease both the yield stress and the viscosity. The plasticizer will disperse the cement agglomerates into primary particles and give them a negatively charged surface leading to a repulsion between the particles. However, it must be clearly stated that the time that the plasticizer acts is limited (ranging from 30-60 min) and that the effect decreases continuously from the moment it is added.
Plasticizers also act as weak retarders, i e giving longer setting times.

Sodium Silicate

The addition of Sodium Silicate has a drastic effect on the yield stress of a grout. The Silicate will react with the Calcium that is present in the cement grout as Calcium hydroxide, $Ca(OH)_2$ produced in the cement-water reaction. The Silicate-Calcium reaction is instant and depending on the added amount the yield stress can be varied in a large range. The viscosity is also influenced but to a much lesser extent. The addition of Sodium Silicate will also decrease the setting time somewhat, but it cannot be regarded as an accelerator.

Calcium Chloride

Calcium Chloride is mainly added as an accelerator in order to decrease the setting time of the grout. Depending on the added amount a great variation in setting time, ranging from a few minutes (15-20% by cement weight) to hours. It should also be stated that a small amount (< 1%) can act as a retarder!

| Table 2 | The effect on the rheology by different factors and manipulations |
| Tab.2 | Der Einfluß auf die Rheologie durch verschiedene Faktoren und Zusätsen |

Factor/additive	Effect on yield stress*)	Effect on viscosity*)	Effect on setting time*)
decreased water/cement ratio	+++	+++	-
increased specific surface	++	++	-
decreased temperature	++	+	++
addition of bentonite	++	++	+-
addition of silica fume	+++	++	+-
addition of super-plasticizer	---	---	++
addition of sodium silicate	+++	+	-
addition of calcium chloride	+	+	---

*)	+++ large increase	- - - large decrease
	++ moderate increase	- - moderate decrease
	+ small increase	- small decrease
	+- indifferent	

The rheology is only slightly influenced causing a minor increase in the yield stress and viscosity.

5 CONCLUSIONS

The following general rheological behaviour of cement-based grouts is found:
A yield stress is present in all dense cement grouts. Due to history dependency, two different yield stresses should be defined. Firstly, a static yield stress that must be overcome to initiate flow when the grout has been at rest for a period of time. Secondly, a dynamic yield stress lower than the former and present when the grout, due to shearing, is in a fully broken down state. The phenomenon is attributed to the thixotropy of the grout which is also found i many other dense suspensions.

A non linear relationship between shear stress/shear rate is found in the majority of the tested grouts. This implies that a shear thinning relationship like the Herschel-Bulkley model should be sought if a high degree of accuracy is needed and the geometry of the conduits is known. However, in most grouting applications the geometry is very little known which implies that a simple model like the Bingham is sufficient as it contains the fundamental properties, viscosity and yield stress.

The rheological properties of cement-based grouts are influenced by the applied mixing and measuring techniques, therefore it is important that standards are developed in order to be able to compare results. Due to the complex nature of cement-based grouts it is advisable to use different measuring techniques and to compare the results whenever the properties are estimated. Although the magnitude of the rheological properties are identified today, much more work is needed before a full understanding of the rheology is achieved. Important questions that need answers are for instance:

Which yield stress determines the penetration length, the dynamic or the static?

How does the slip phenomenon, that is frequently observed in the measurement of suspensions, influence the flow and maximum penetration length in practice?

How long time is needed for the maximum static yield stress to develop (i e the thixotropic recovery), when the grout has stopped?

It has been found that the rheological properties can be altered in relatively wide range, the viscosity between 10-250 mPas and the yield stress between 0.5-25 Pa. By knowing how different factors and additives influence the rheological properties a grout can be 'taylor-made' for a certain application and prevailing conditions.

From a practical point of view it is desirable to perform 'on-line' measurements, whereby the significant properties are continuously measured during the grouting application. Research work should be guided in this direction.

ACKNOWLEDGEMENT

The presented research work was financed by the Swedish Rock Engineering Research Foundation (BeFo) and their support is greatly appreciated.

REFERENCES

Barnes, H.A. Hutton, J.F. Walters, K. (1989). An introduction to rheology. Elsevier.

Barnes, H.A. Carnali, J.O. (1990). The vane-in-cup as a novel rheometer geometry for shear thinning and thixo-tropic materials. *Journal of Rheology*, 34(6), 841-866.

Haimoni, A. Hannant, D.J. (1988). Developments in the shear vane test to measure the gel strength of oilwell cement slurry. *Advances in Cement Research*, Vol.1, No.4.

Håkansson, U. Hässler, L. Stille, H. (1991). A Technique for Measuring the Rheological Properties of Injected Grout. Swedish Rock Engineering Research Foundation (BeFo), 241:1/91, Stockholm. (In swedish).

Håkansson, U. Hässler, L. Stille, H (1992). Rheological Properties of Microfine Cement Grouts with Additives. ASCE Specialty Conference on Grouting, Soil Improvement and Geosynthetics, New Orleans 25-28 Feb.

Håkansson, U. (1993). Rheology of fresh cement-based grouts. Ph D thesis, Dept. of Soil and Rock Mechanics, Royal Institute of Technology, Stockholm. (In press).

Hässler, L. (1991). Grouting of Rock-Simulation and Classification. Ph D thesis, Dept. of Soil and Rock Mechanics, Royal Institute of Technology, Stockholm.

Keating, J. Hannant, D.J. (1989). The effect of rotation rate on gel strength and dynamic yield strength of thixotropic oil well cements measured using a shear vane. Journal *of Rheology*, 33(7), 1011-1020.

Lombardi, G. (1985). The role of cohesion in cement grouting of rock. 15:e Congres des Grandes Barrage. Lausanne (ICOLD).

Nguyen, Q.D. Boger, D.V (1983). Yield stress measurements for concentrated suspensions. *Journal of Rheology*, 27(4), 321-349.

Nguyen, Q.D. Boger, D.V (1985:a). Direct yield stress measurement with the vane method. *Journal of Rheology*, 29(3), 335-347.

Nguyen, Q.D. Boger, D.V (1985:b). Thixotropic behaviour of concentrated bauxite residue suspensions. *Rheologica Acta*, 24:427-437.

Nguyen, Q.D. Boger, D.V (1987). Characterization of yield stress fluids with concentric cylinder viscometers. *Rheologica Acta*, 26:508-515.

Onofrei, M. Gray, M. Roe, L. (1991). Cement-based grouts - Longevity studies: Leaching behaviour. Stripa Project, Technical report 91-33, SKB, Stockholm, Sweden.

Author index
Autorenverzeichnis